OPTICAL SCANNING

OPTICAL ENGINEERING

Series Editor

Brian J. Thompson
Provost
University of Rochester
Rochester, New York

---

1. Electron and Ion Microscopy and Microanalysis: Principles and Applications, *by Lawrence E. Murr*
2. Acousto-Optic Signal Processing: Theory and Implementation, *edited by Norman J. Berg and John N. Lee*
3. Electro-Optic and Acousto-Optic Scanning and Deflection, *by Milton Gottlieb, Clive L. M. Ireland, and John Martin Ley*
4. Single-Mode Fiber Optics: Principles and Applications, *by Luc B. Jeunhomme*
5. Pulse Code Formats for Fiber Optical Data Communication: Basic Principles and Applications, *by David J. Morris*
6. Optical Materials: An Introduction to Selection and Application, *by Solomon Musikant*
7. Infrared Methods for Gaseous Measurements: Theory and Practice, *edited by Joda Wormhoudt*
8. Laser Beam Scanning: Opto-Mechanical Devices, Systems, and Data Storage Optics, *edited by Gerald F. Marshall*
9. Opto-Mechanical Systems Design, *by Paul R. Yoder, Jr.*
10. Optical Fiber Splices and Connectors: Theory and Methods, *by Calvin M. Miller with Stephen C. Mettler and Ian A. White*
11. Laser Spectroscopy and Its Applications, *edited by Leon J. Radziemski, Richard W. Solarz, and Jeffrey A. Paisner*
12. Infrared Optoelectronics: Devices and Applications, *by William Nunley and J. Scott Bechtel*
13. Integrated Optical Circuits and Components: Design and Applications, *edited by Lynn D. Hutcheson*
14. Handbook of Molecular Lasers, *edited by Peter K. Cheo*
15. Handbook of Optical Fibers and Cables, *by Hiroshi Murata*
16. Acousto-Optics, *by Adrian Korpel*
17. Procedures in Applied Optics, *by John Strong*

18. Handbook of Solid-State Lasers, *edited by Peter K. Cheo*
19. Optical Computing: Digital and Symbolic, *edited by Raymond Arrathoon*
20. Laser Applications in Physical Chemistry, *edited by D. K. Evans*
21. Laser-Induced Plasmas: Physical, Chemical, and Biological Applications, *edited by Leon J. Radziemski and David A. Cremers*
22. Infrared Technology Fundamentals, *by Irving J. Spiro and Monroe Schlessinger*
23. Single-Mode Fiber Optics: Principles and Applications, Second Edition, Revised and Expanded, *by Luc B. Jeunhomme*
24. Image Analysis Applications, *edited by Rangachar Kasturi and Mohan M. Trivedi*
25. Photoconductivity: Art, Science, and Technology, *by N. V. Joshi*
26. Principles of Optical Circuit Engineering, *by Mark A. Mentzer*
27. Lens Design, *by Milton Laikin*
28. Optical Components, Systems, and Measurement Techniques, *by Rajpal S. Sirohi and M. P. Kothiyal*
29. Electron and Ion Microscopy and Microanalysis: Principles and Applications, Second Edition, Revised and Expanded, *by Lawrence E. Murr*
30. Handbook of Infrared Optical Materials, *edited by Paul Klocek*
31. Optical Scanning, *edited by Gerald F. Marshall*

Other Volumes in Preparation

OPTICAL SCANNING

EDITED BY
GERALD F. MARSHALL

Consultant
Morgan Hill, California

Marcel Dekker, Inc.   New York • Basel • Hong Kong

ISBN 0-8247-8473-1

This book is printed on acid-free paper.

Marcel Dekker, Inc.
270 Madison Avenue, New York, New York 10016

Current printing (last digit):
10 9 8 7 6 5 4 3 2 1

PRINTED IN THE UNITED STATES OF AMERICA

To

Irene,

Clare Margaret and Mark Peter;

Guy Nicholas and Maria Elizabeth,

with love

About the Series

The series came of age with the publication of our twenty-first volume in 1989. The twenty-first volume was entitled *Laser-Induced Plasmas and Applications* and was a multi-authored work involving some twenty contributors and two editors: as such it represents one end of the spectrum of books that range from single-authored texts to multi-authored volumes. However, the philosophy of the series has remained the same: to discuss topics in optical engineering at the level that will be useful to those working in the field or attempting to design subsystems that are based on optical techniques or that have significant optical subsystems. The concept is not to provide detailed monographs on narrow subject areas but to deal with the material at a level that makes it immediately useful to the practicing scientist and engineer. These are not research monographs, although we expect that workers in optical research will find them extremely valuable.

There is no doubt that optical engineering is now established as an important discipline in its own right. The range of topics that can and should be included continues to grow. In the "About the Series" that I wrote for earlier volumes, I noted that the series covers "the topics that have been part of the rapid expansion of optical engineering." I then followed this with a list of such topics which we have already outgrown. I will not repeat that mistake this time! Since the series now exists, the topics that are appropriate are best exemplified by the titles of the volumes listed in the front of this book. More topics and volumes are forthcoming.

Brian J. Thompson
University of Rochester
Rochester, New York

Foreword

This book is about information scanning—the technology of reading or writing information. It fits right into our information age, for scanning is as fundamental to information transfer as vision is to perception. While most information is communicated through images or printed words, most data channels transfer only a series of data elements. Thus, to manipulate the information, a two-dimensional image space must be transformed *to* or *from* a single function of time. That is the job of the scanner. In one form, as a data input device, it samples the information space in a predetermined pattern, deriving a stream of elements of various intensities that represent the original information, making it available for electronic manipulation. That describes only one of two physical tasks. A complementary task, represented by the output device, requires the assembling of a given series of elements in a predetermined pattern, while weighting the intensity of each element appropriately. That is the job of the recorder. Thus, information scanning or recording can employ very similar scanning mechanisms—and when one speaks of a scanner, it may serve reciprocally, to input or to output information.

Over the past 25 years, with the refinement of the laser, much has been done to benefit from its high purity and brightness. These unique features allow the simultaneous achievement of both high resolution and high speed. High purity allows precise focusing to approach diffraction-limited spot sizes, and high brightness allows rapid succession of extremely short exposures and wide bandwidth.

These basic characteristics have nurtured a veritable explosion in the application of the laser for information scanning and recording. This book addresses many of the most prominent options for achieving these high-performance goals.

Thirteen chapters form this cohesive resource. The first three relate, in turn, to gaussian beam characteristics, system lens design, and image quality, forming a framework that clarifies and serves the scanning process. The subsequent chapters cover, in turn, the physical scanning methods—holographic, polygonal, galvanometric, resonant, acoustooptic, electrooptic, and optical disk techniques. Interwoven are two chapters providing expanded coverage of the important factors of windage and bearings of the rotating systems.

This collection of chapters that clarify and express the operational technologies allows a spectrum of objectives to be addressed, from low resolution at low speed to high resolution at high speed. Some scanning methods serve similar needs, while others fit specific tasks, such as bar code reading or optical disk reading. When the system requirements and component capabilities are expressed in sufficient detail, the direction for optimal utilization becomes more apparent. The object of this volume is to offer broad access to the options and to illustrate their effective application.

To further this important selection and utilization process, this volume closes with an Afterword, a summary and overview, where the content of these 13 chapters is interpreted briefly, invoking some key factors in system selection and design and offering some indication of advancing trends.

The mounting challenges to higher performance mean increasing demands on ingenuity. That noble discipline is enhanced with the aid of resources like this.

Leo Beiser

Preface

The aim of this volume is to provide application-oriented engineers and technologists, scientists, and students, with a guideline and a reference to the fundamentals of input and output optical scanning technology and engineering. It brings together the knowledge and experience of 16 international specialists from England, Japan, Scotland, and the United States. Brief biographies of the contributors are included. The Foreword and Afterword by Leo Beiser unify the selected topics of the 13 chapters, and give an overview evaluation of the technologies within the field of optical scanning engineering. Optical scanning technology is a comprehensive subject that encompasses not only the mechanics of controlling the deflection of a light beam but also all aspects that affect the imaging fidelity of the output data that may be displayed on a screen or recorded on paper.

A scanning system may be an input scanner, an output scanner, or a scanner that combines both of these functional attributes. A system's imaging fidelity depends on, and begins with, the reading of the input information and ends with the writing of the output data. Optical scanning intimately involves a number of disciplines: optics, material science, magnetics, acoustics, mechanics, electronics, and image analysis, with a host of considerations.

The book covers gaussian laser beam diameters and divergence, optical and lens design for scanning systems, and scanned image quality. It deals with rotary scanning devices and systems, namely, holographic scanners for bar code readers and graphic arts, polygonal scanners, windage (that is, the aerodynamic aspects), bearings, motors, and control systems associated with high-performance polygonal scanners. *Optical Scanning* treats oscillatory devices and systems; specifically, galvanometric and resonant low-inertia scanners, acoustooptical, and electrooptical scanners, and modulators. It closes with optical disk scanning technology.

The dream is to produce a definitive book on optical scanning, but this is an impossible task to accomplish in this ever more rapidly changing era of technological developments. The best has been done by all the authors, each of whom could have written a volume on his own special subject. The book is complete as an introduction to the field. With the common thread of the subject title, the disparate chapters are brought into perspective in the Afterword.

To assist the reader, measured quantities are expressed in dual units wherever possible and appropriate; the secondary units are in parentheses. The metric system takes precedence over other systems of units, except where it just does not make good sense.

A strong effort has been made for a measure of uniformity in the book with respect to terminology, nomenclature, and symbology. However, with the variety of individual styles of the 16 contributing authors who are scattered across the Northern Hemisphere, I have placed greater importance on the unique contributions of the authors rather than on form.

I extend my thanks to the following persons: Brian J. Thompson, Provost of the University of Rochester, for his patient confidence in inviting me to produce this additional volume on the subject of scanning in this series; my 16 contributing coauthors for their splendid material; and the reviewers of the manuscripts and typescripts, namely,

Robert Basanese	Rofin-Sinar, Inc.
Leo Beiser	Leo Beiser, Inc.
John H. Carosella	Speedring Systems, Inc.
Duane Grant	IBM Corporation
Michael J. Hayford	Optical Research Associates
Ron Hooper	Hooper Engineering Company
Charles S. Ih	University of Delaware
David B. Kay	Eastman Kodak Company
Kathryn A. McCarthy	Tufts University
Robert J. Schiesser	Charles Stark Draper Lab, Inc.

David Strand	Energy Conversion Devices, Inc.
William Taylor	Kollmorgen Corporation
Stanley W. Thomas	Lawrence Livermore Laboratory
Daniel Vukobratavich	University of Arizona
David L. Wright	Spectra-Physics Lasers, Inc.
Francis Yu	The Pennsylvania State University
Ross Zelesnick	RCA, Inc.

Each gave his or her time to critique a script, made helpful comments, and provided excellent suggestions. I thank John H. Carosella of Speedring Systems, Inc., for his indirect support, which I much appreciate. I am also grateful for the generous help and time given, especially in proofreading and organizing the index, by my wife, Irene.

I am pleased to be the coordinator of these works and value the privilege of being the one to share this treatise with my colleagues in the field.

Read, scan, study, and enjoy.

Gerald F. Marshall

Contents

About the Series *v*
Foreword Leo Beiser *vii*
Preface *ix*
Contributors *xix*

1. Gaussian Laser Beam Diameters and Divergence **1**
Gerald F. Marshall

1. Introduction
2. Beam Diameter Definitions
3. Beam Divergence Definition
4. Angles
5. Radiometric Terminology
6. Gaussian Beam
References

2. Optical Systems for Laser Scanners **27**
Robert E. Hopkins and *David Stephenson*

1. Introduction
2. Laser Scanner Configurations

3. Special Requirements for Flat-Field Laser Scanning
4. First-Order Lens Design Considerations
5. Special Optical Design Requirements for Laser Scanners
6. Lens Design Talk
7. A Selection of Laser Scan Lenses
8. Scan-Lens Manufacturing, Quality Control, and Final Testing
9. Closing Comments
References

3. Scanned Image Quality **83**
Donald R. Lehmbeck and *John C. Urbach*

1. Introduction
2. Basic Concepts and Effects
3. Practical Considerations
4. Characterization of Input Scanners
5. Summary Measures of Imaging Performance
References

4. Holographic Scanners for Bar Code Readers **159**
LeRoy D. Dickson and *Glenn T. Sincerbox*

1. Introduction
2. Nonholographic UPC Scanners
3. Holographic Bar Code Readers
4. Other Features of Holographic Scanning
5. Holographic Deflector Media for Holographic Bar Code Scanners
6. Fabrication of Holographic Deflectors
7. Holographic UPC Bar Code Scanners
8. Future Trends in Bar Code Scanning
References

5. Holographic Deflector for Graphic Arts Systems **213**
Charles J. Kramer

1. Introduction
2. Deflector System Performance Criteria
3. Deflector System Parameters
4. Non-Plane-Grating Hologons
5. Basic Plane Grating Hologon Disk Deflector
6. Improved Plane Grating Hologon Disk Deflectors
7. Nondisk Plane Grating Hologon Deflectors
8. Hologon Deflector System Alignment Procedures

9. Design Examples for Hologon Deflector Systems
10. Conclusions
Acknowledgments
References

6. **Polygonal Scanners: Applications, Performance, and Design** 351
Randy J. Sherman

1. Introduction
2. Types of Polygonal Scan Mirrors
3. Applications to Scanning
4. Considerations of Quality
5. Methods of Rotating Polygons
6. Testing Polygonal Scanners for Performance
References

7. **High-Performance Polygonal Scanners, Motors, and Control Systems** 409
Gerald A. Rynkowski

1. Introduction
2. Basic Considerations
3. Manufacturing and Engineering Design Considerations
4. Scanner Subsystem Design
5. Typical Designs
References

8. **Windage of Rotating Polygons** 451
Joseph Shepherd

1. Introduction
2. Windage
3. Direct Measurement of Windage
4. Methods of Predicting Windage
5. Model Testing
6. Results of Model Tests Performed on Rotating Polygons
7. Discussion
8. Conclusion
9. Appendix
References

9. **Bearings for Rotary Scanners** 477
Joseph Shepherd

1. Introduction
2. Gas Bearings

3. Ball Bearings
4. Other Bearing Types
5. Procurement of Rotary Scanners/Bearings
6. Bearing-Related Optomechanical Errors
7. Performance Comparison
References

10. Galvanometric and Resonant Low-Inertia Scanners **525**
Jean I. Montagu

1. Introduction
2. Applications of Low-Inertia Scanners
3. Critical Design Issues
4. Scanner Elements, Selection, and Design
5. Performances of Low-Inertia Scanners
References

11. Acoustooptic Scanners and Modulators **615**
Milton Gottlieb

1. Introduction
2. Acoustooptic Interactions
3. Materials for Acoustooptic Scanning
4. Design and Fabrication of Transducers
5. Applications of Acoustooptic Scanners
References

12. Electrooptical Scanners **687**
Clive L.M. Ireland and *John Martin Ley*

1. Introduction
2. Theory of the Electrooptic Effect
3. Properties and Selection of Electrooptic Materials
4. Principles of Electrooptic Deflectors
5. Electrooptic Deflector Designs
References

13. Optical Disk Scanning Technology **779**
Tetsuo Saimi

1. Introduction
2. Applications of Optical Disk Systems
3. Basic Design of Optical Disk Systems
4. Semiconductor Laser

5. Focusing and Tracking Techniques
6. Radial Access and Drive Technique
7. Appendix
References

Afterword Leo Beiser *831*

Index *835*

Biographies *859*

Contributors

Leo Beiser Leo Beiser, Inc., Flushing, New York

LeRoy D. Dickson Storage Systems Products Division, IBM, San Jose, California

Milton Gottlieb Science and Technology Center, Westinghouse Electric Corporation, Pittsburgh, Pennsylvania

Robert E. Hopkins Optizon Corporation, Rochester, New York

Clive L.M. Ireland Lumonics Ltd., Rugby, Warwickshire, England

Charles J. Kramer Holotek Ltd., Rochester, New York

Donald R. Lehmbeck Webster Research Center, Xerox Corporation, Webster, New York

John Martin Ley Leysop, Basildon, Essex, England

Gerald F. Marshall Consultant, Morgan Hill, California

Jean I. Montagu General Scanning Inc., Watertown, Massachusetts

Gerald A. Rynkowski Speedring Systems Inc., Rochester Hills, Michigan

Tetsuo Saimi Matsushita Electric Industrial Co., Ltd., Kodoma, Osaka, Japan

*Joseph Shepherd** Ferranti International, Silverknowes, Edinburgh, Scotland

Randy J. Sherman Lincoln Laser Company, Phoenix, Arizona

Glenn T. Sincerbox Almaden Research Center, IBM, San Jose, California

David Stephenson Optics Division, Melles Griot, Rochester, New York

John C. Urbach Strata Systems Inc., Portola Valley, California

* *Current affiliation:* Consultant, Edinburgh, Scotland

1

Gaussian Laser Beam Diameters and Divergence

Gerald F. Marshall

Consultant,
Morgan Hill, California

1.1 INTRODUCTION

Lasers are notably intense sources of electromagnetic radiation. A laser radiates at one or more wavelengths in the spectrum according to its design, but is essentially considered monochromatic. It emits a very narrow cone of radiation that is referred to as a laser beam (Figure 1.1). A typical helium-neon (HeNe) laser emits red light at a wavelength of 632.8 nm [25 μin], and has a nominal divergence of 1 mrad [3.44 min of arc] and a nominal diameter of 1 mm [39.4 thousandths of an inch, or mils]. Measuring the diameter and divergence of a laser beam may be likened to determining the width and spread of a smoke plume or the dimensions of a fog patch. Where do the boundaries begin and end? In this chapter the meaning and definitions of the diameter and divergence of a laser beam are addressed.

1.1.1 Laser Modes

Lasers operate, and emit beams, in different modes. The mode that shall be addressed is the simplest, and the one in which the beam has a gaussian wavefront [1, pp. 156–158], namely, the lowest-order transverse electromagnetic (TEM_{oo}) mode. Typically in this mode the characteristics of the beam are radially symmetric about its axis, which is the principal direction of propagation.

1.1.2 Gaussian Characteristics

At any perpendicular cross section of the beam the *irradiance* (power density) E distribution is gaussian. It is at a maximum E_0 on the axis and decreases with distance from the axis (Figure 1.1). The term *power density* is customarily used by laser scientists and engineers, instead of irradiance (Section 1.6.1). The word *density* connotes three dimensions, but in this context it is *areal* density, which denotes two dimensions. To assist the reader who is more familiar with the term *power density* rather than *irradiance*, the expression *power density* is included in parentheses.

1.1.3 Ambiguous Terms

There are several terms in vogue used to express and define the diameter of a laser beam. Some of these terms are misnomers, and some are ambiguous; to the uninitiated both are as misleading as the directions given by some road signs, unless one is thoroughly familiar with the territory. Interrelated with this confusion

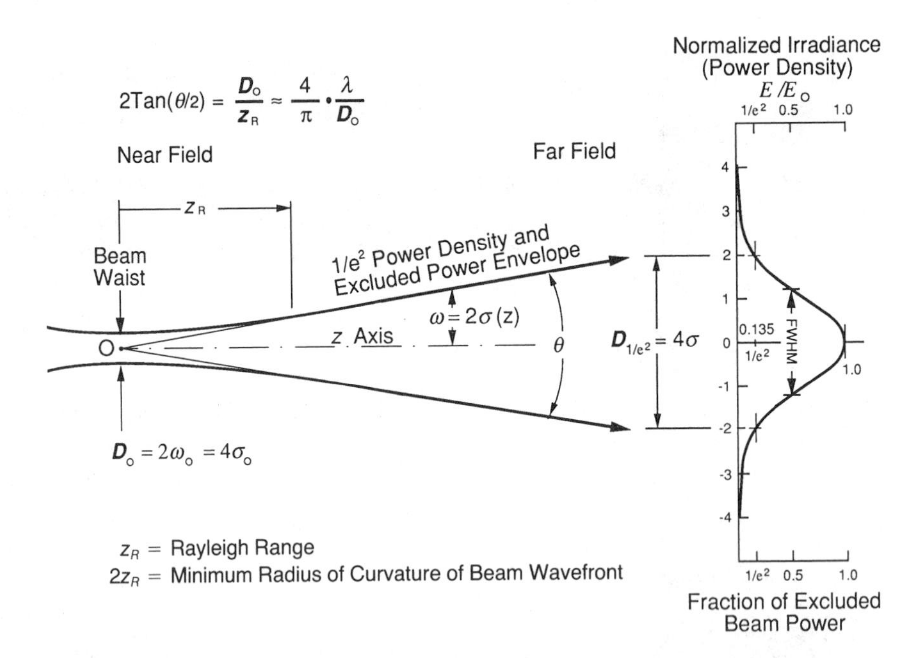

Figure 1.1 Gaussian laser beam divergence angle θ is the far-field full-cone angle of the $1/e^2$ (0.135) irradiance (power density) envelope. The apex of the conical envelope is at the beam waist. At any beam cross section the irradiance (power density) has a radially symmetric gaussian distribution profile, as depicted on the right.

in defining the diameter of a laser beam is the seeming misuse of the more accepted radiometric terminology [2, 3]. The terminology itself is seemingly always in a ''state of flux'' because it is frequently under revision [4, pp. 289–301]. The origin of these misnomers may be attributed to the cross fertilization of terminology from scientists and engineers with backgrounds of experience in other disciplines, wherein the terms are valid. To avoid misunderstandings, awareness and tolerance of these apparent inconsistencies are important.

1.1.4 Section Outlines

The definitions of beam diameters and divergence are deliberately presented in the first few sections so that readers will not have to wade through mathematics before reaching what they want to know at a glance.

1. The three most useful laser beam diameters and their definitions and an outline of the procedures for measuring beam diameters are presented in Section 1.2. See Table 1.1.
2. The beam divergence definition and an outline of the procedure for measuring beam divergence are given in Section 1.3.
3. The three types of angles used in beam divergence and radiometry are defined in Section 1.4.
4. The relevant CIE (Commission International de l'Eclairage) radiometric terms that are most widely accepted internationally, and which are used to develop the beam definitions, are reviewed and clarified in Section 1.5 and summarized in Table 1.3.
5. The mathematics and calculations that are used in the earlier sections to define the beam diameters and divergence are covered in Section 1.6.

1.2 BEAM DIAMETER DEFINITIONS

From Table 1.1 and Figure 1.2 several different radii may be chosen to define the diameter of a gaussian laser beam. The three most widely used definitions are given.

1.2.1 Full-Power Beam Diameter $D_{99.97\%}$

The *full-power beam diameter* may be effectively approximated to that diameter of the laser beam core that contains all but $1/e^8$ ($<0.034\%$) of the total beam power (Figure 1.2). See Section 1.2.3. It also corresponds identically to the distance between diametrically opposite points of an irradiance (power density) distribution profile in a section normal to the beam axis, at which the irradiance (power density) is $1/e^8$ ($<0.034\%$) of the axial irradiance (power density); see Figure 1.1.

Table 1.1 Irradiance, Excluded Power, Encircled Power, and Beam Diameter

Beam diameter in standard deviations σ_z $2k$	Exponential function $\exp[-k^2/2]$ =	Normalized irradiance (power density) (%) E/E_0 = $g(k)$ =	Excluded beam power fraction (%) $1 - h(k)$	Encircled beam power fraction (%) $h(k)$	Symbology for the beam diameter $D_{h(k)\%}$	$D_{1/e^{(k^2/2)}}$
0	$1/e^0$	100	100	0	Axis	
1.3490	$1/e^{0.228}$	79.65	79.65	20.35	$D_{20\%}$	
2	$1/e^{0.5}$	60.65	60.65	39.35	$D_{39\%}$	
$2\sqrt{2 \ln 2}$	$1/e^{0.693}$	50	50	50	$D_{50\%}$	D_{FWHM}
2.8102	$1/e^{0.987}$	37.26	37.26	62.74	$D_{63\%}$	
$2\sqrt{2}$	$1/e^1$	36.79	36.79	63.21	$D_{63\%}$	$D_{1/e}$
4	$\mathbf{1/e^2}$	**13.53**	**13.53**	**86.47**	$\mathbf{D_{86.5\%}}$	$\boldsymbol{D_{1/e^2}}$
6	$1/e^{4.5}$	1.11	1.11	98.89	$D_{99\%}$	
8	$1/e^8$	0.034	0.034	99.966	$D_{99.97\%}$	D_{1/e^8}
∞	$1/e^\infty$	0	0	100	$D_{100\%}$	

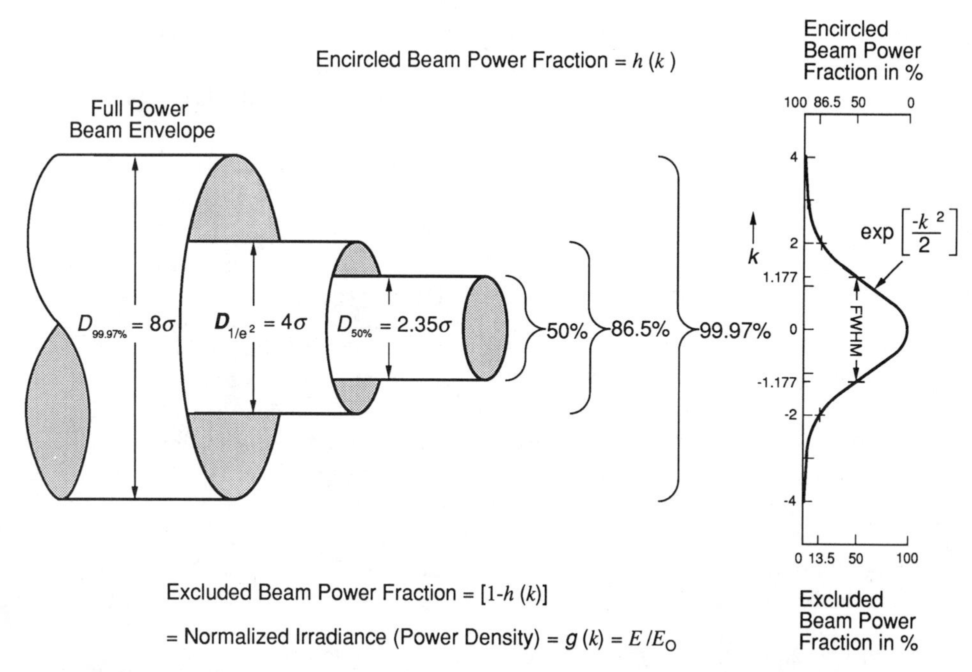

Figure 1.2 The normalized irradiance (power density) $g(k)$ value at a gaussian beam envelope identically equals the fraction of excluded beam power. For example, the $1/e^2$ (0.135) normalized beam irradiance (power density) envelope excludes the $1/e^2$ (0.135) fraction of the total beam power ϕ_∞. Therefore, the envelope must encircle 86.5% of the total beam power. Likewise, the 0.5 (50%) normalized beam irradiance (power density) envelope excludes and encircles 50% of the total beam power.

Let σ represent the standard deviation. If k is a multiplying factor that expresses the beam radius in units of the standard deviation σ, then for $k = 4$ (see Table 1.1) the beam diameter is expressed by

$$2k\sigma = 8\sigma = D_{1/e^8} = D_{\phi=99.97\%} \approx D_{\phi=100\%} \tag{1}$$

1.2.2 Half-Power Beam Diameter $D_{50\%}$

The *half-power beam diameter* is that diameter of the laser beam core that contains, and therefore also excludes, 50% of the total radiant beam power (Figure 1.2). It also corresponds identically to the distance between diametrically opposite points of an irradiance (power density) distribution profile in a section normal to the beam axis, and at which the irradiance (power density) is half (50%) of the axial irradiance (power density). Thus, this beam diameter is frequently called the *full-width at half maximum* (FWHM) diameter (Figure 1.1).

Let σ represent the standard deviation. If k is a multiplying factor that expresses the beam radius in units of the standard deviation σ, then for $k = 1.1774$ (see Table 1.1) the beam diameter is expressed by

$$2k\sigma = 2.355\sigma = D_{1/e^{0.693}} = D_{\phi=50\%} \tag{2}$$

1.2.3 1/e² Excluded Power Beam Diameter *D*

The *1/e² excluded power beam diameter* is that diameter of the laser beam core that contains all but $1/e^2$ (13.5%) of the total radiant beam power (Figure 1.2). It also corresponds identically to the distance between diametrically opposite points of an irradiance (power density) distribution profile in a section normal to the beam axis, at which the irradiance (power density) is $1/e^2$ (13.5%) of the axial irradiance (power density) (Figure 1.1).

Let σ represent the standard deviation. If k is a multiplying factor that expresses the beam radius in units of the standard deviation σ, then for $k = 2$ (see Table 1.1) the beam diameter is expressed by

$$2k\sigma = 4\sigma = D_{1/e^2} = D_{\phi=86.5\%} = \boldsymbol{D} \tag{3}$$

This $1/e^2$ excluded power beam diameter is one-half the effective full-power beam diameter D_{1/e^8}. Hence,

$$\boldsymbol{D} = \frac{D_{1/e^8}}{2} = \frac{D_{\phi=99.97\%}}{2} \tag{4}$$

Unless otherwise stated, the $1/e^2$ excluded power beam diameter $\boldsymbol{D}$ is *the* diameter of a laser beam most generally understood when referring to the diameter of a laser beam. However, the expression is customarily contracted to and referred to as the *1/e² diameter*.

$\boldsymbol{D}_0$ represents the $1/e^2$ excluded power beam diameter at the beam waist, $z = 0$.

1.2.4 Measurement of Beam Diameters

The beam diameter may be determined in several different ways using suitably calibrated radiometric detectors (Figure 1.3). Alignment is important for all methods to obtain accurate results. There are a number of commercially available instruments that are designed to measure beam diameters and display beam irradiance (power density) profiles [5].

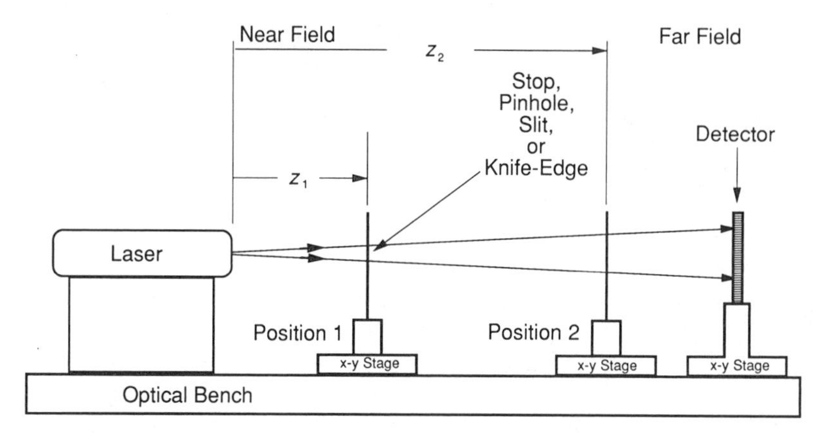

Figure 1.3 Test arrangement for measuring laser beam diameters and divergence by stop, pinhole, knife-edge, and slit methods.

Stops

Measure and plot the truncated beam power ϕ as a percentage of the total measured beam power against the diameter D_ϕ of a set of apertures as the beam power is stopped down to less than 50% (Figure 1.4). A calibrated iris may be used instead of aperture stops. Then

$$\boldsymbol{D} = D_{\phi=86.5\%} \approx 1.70 D_{\phi=50\%} \tag{5}$$

and

$$D_{50\%} = D_{\phi=50\%} = D_{\text{FWHM}} \tag{6}$$

Pinhole

Measure and plot the transmitted beam power ϕ distribution as a percentage of the peak power against micrometer position displacement x of a pinhole aperture as it moves along a beam diameter (Figure 1.4). Then

$$\boldsymbol{D} = [(x_2)_{\phi=13.5\%} - (x_1)_{\phi=13.5\%}] \tag{7}$$

and

$$D_{50\%} = [(x_2)_{\phi=50\%} - (x_1)_{\phi=50\%}] = D_{\text{FWHM}} \tag{8}$$

The pinhole diameter should be no more than about 10% of the expected beam diameter $\boldsymbol{D}$ being measured; see Figure 1.4.

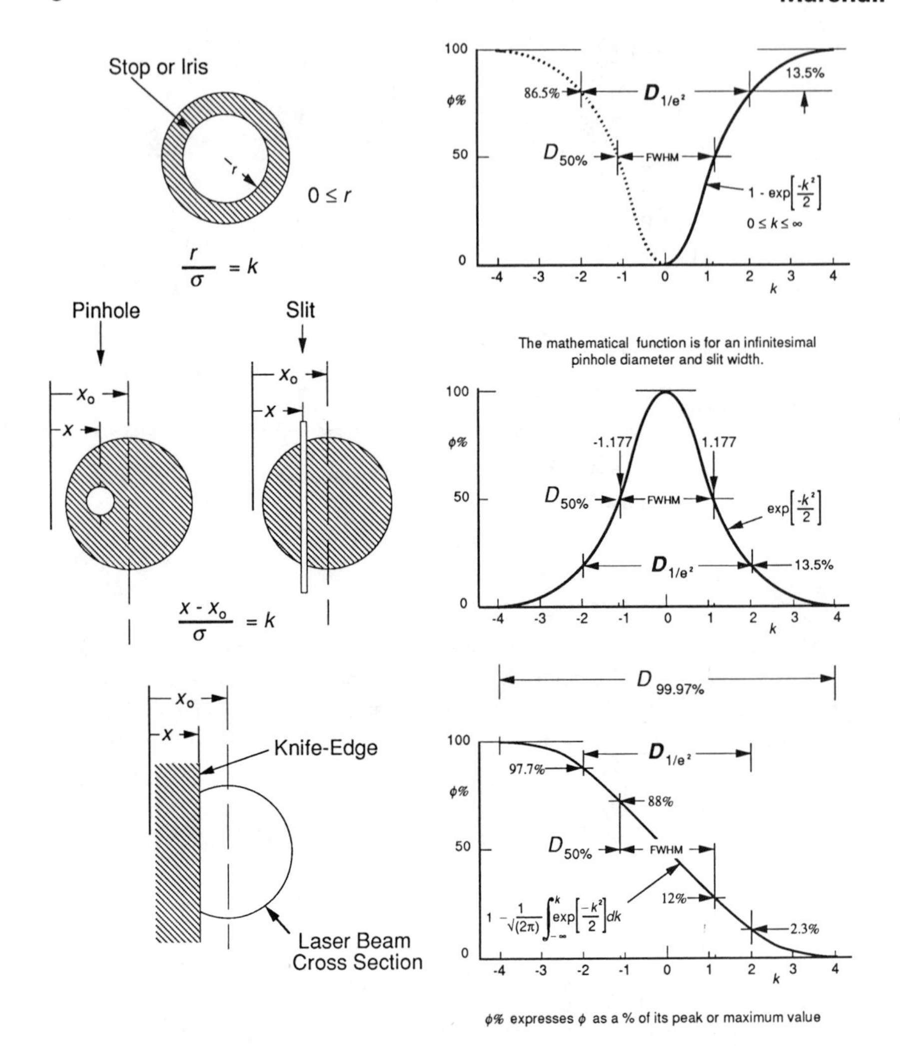

Figure 1.4 Stop, pinhole, knife-edge, and slit methods for measuring the beam diameters produce different radiant flux distribution characteristics.

Knife-Edge

Measure and plot the residual clipped beam power ϕ as a percentage of the total measured beam power against micrometer position displacement x of a knife-edge as it eclipses the beam. See Figure 1.4 and Table 1.2. Then

$$\boldsymbol{D} = [(x_2)_{\phi=2.3\%} - (x_1)_{\phi=97.7\%}] = D_{1/e^2} \tag{9}$$

and

$$D = 2[(x_2)_{\phi = 15.9\%} - (x_1)_{\phi = 84.1\%}] = 2D_{1/e^{0.5}} \tag{10}$$

and

$$D_{50\%} = [(x_2)_{\phi = 12\%} - (x_1)_{\phi = 88\%}] = D_{\mathrm{FWHM}} \tag{11}$$

Table 1.2 Encircled Power Beam Diameter Determined from Beam Power Eclipsed by a Knife-edge

Radius	Diameter	Beam power fraction (%)			Symbology	
in standard deviations σ		Uneclipsed position of knife-edge		Encircled diameter	for the beam diameter	
k	$2k$	x_1	x_2	$(x_2 - x_1)$	$D_{h(k)\%}$	$D_{1/e^{(k2/2)}}$
0	0	50	50	0		Axis
0.6745	1.3490	75	25	20.35	$D_{20\%}$	
1	2	84.134	15.866	39.35	$D_{39\%}$	
$\sqrt{2 \ln 2}$	2.3548	88.048	11.952	50	$D_{50\%}$	D_{FWHM}
1.4051	2.8102	92	8	62.74	$D_{63\%}$	
$\sqrt{2}$	$2\sqrt{2}$	92.135	7.865	63.21	$D_{63\%}$	$D_{1/e}$
2	**4**	**97.725**	**2.275**	**86.47**	$\boldsymbol{D_{86.5\%}}$	$\boldsymbol{D_{1/e^2}}$
3	6	99.865	0.135	98.89	$D_{99\%}$	
4	8	99.997	0.003	99.966	$D_{99.97\%}$	D_{1/e^8}
∞	∞	100	0	100	$D_{100\%}$	

Slit

The slit is equivalent to two rigidly coupled knife-edges. Measure and plot the transmitted beam power ϕ distribution against micrometer displacement x of a narrow parallel slit as it scans through the beam. Then

$$D = [(x_2)_{\phi = 13.5\%} - (x_1)_{\phi = 13.5\%}] = D_{1/e^2} \tag{12}$$

and

$$D_{50\%} = [(x_2)_{\phi = 50\%} - (x_1)_{\phi = 50\%}] = D_{\mathrm{FWHM}} \tag{13}$$

The width of the slit should be less than 20% of the expected beam diameter $\boldsymbol{D}$ being measured in order to keep the convolution error below 2% [6], and a length of at least twice $\boldsymbol{D}$.

Linear Array

Measure and display the beam power density distribution along a beam diameter incident on a suitably calibrated detector array. The pixel elements of the array need to be small compared with the diameter being measured, and the array should have a length no less than twice the expected beam diameter ***D*** being measured.

Two-Dimensional Array

Measure and display the beam power density distributions of a beam cross section incident on a suitably calibrated two-dimensional detector array, say 32×32. The pixel elements of the array need to be small compared with the cross-sectional area of the beam being measured, and the array should have an area containing no less than twice the expected beam diameter ***D*** being measured.

1.2.5 Measurement Sensitivity

From each of these methods data are obtained that have to be evaluated and analyzed to determine the desired beam diameter. Although the $1/e^2$ beam diameter ***D*** is desirable, in practice it is not easy to accurately determine the associated 13.5% of excluded beam power or irradiance (power density). This is because of ''noise'' and because of the slow rate of change in irradiance (power density) with distance along the tails of the gaussian distribution characteristic (Figures 1.2 and 1.4).

Half-Power Beam Diameter

By far the simplest is the determination of the half-power beam diameter $D_{50\%}$, FWHM beam diameter. This is for three good reasons, namely:

1. The 50% encircled power diameter or the FWHM beam diameter can be read almost directly by the stops, pinhole, and slit methods given in preceding subsections as can, in principle, the $1/e^2$ beam diameter ***D***.
2. At the half-power positions in the gaussian characteristic profile the rate of change in irradiance (power density) along a diameter is close to its maximum rate of change, and thereby provides a high sensitivity for measurement (Figures 1.2 and 1.4).
3. For diameters $D_{50\%}$ and ***D*** there is, respectively, 50% and 13.5% of the excluded beam power, or irradiance, respectively, in the two tails of the generated gaussian characteristic that is being clipped (Figure 1.2). The residual 13.5% of the power in the tails, which has to be detected and clipped, is closer to being comparable to the percentage level of experimental error of measurement.

The method in the subsection on knife-edges requires careful analysis to determine the half-power beam diameter $D_{50\%}$ (D_{FWHM}).

$1/e^2$ Excluded Power Beam Diameter

All methods require careful alignment, measurement, calibration, and rigorous analysis to determine the $1/e^2$ beam diameter $\boldsymbol{D}$ because of the relative measurement insensitivity mentioned in the preceding subsection.

Knife-edge Method. In the knife-edge method, which progressively truncates segments from the entire beam, note that the percentage power clip levels, 2.3% and 97.5% for $\boldsymbol{D}$, are markedly different to the 13.5% irradiance (power density) level for $\boldsymbol{D}$ when using the stop and pinhole methods. See Figure 1.4, Tables 1.1 and 1.2. Likewise, note that the percentage power clip levels for a beam diameter $D_{50\%}$ are 12% and 88%.

Similarly, the percentage clip levels for a beam diameter $D_{1/e^{0.5}}$, which corresponds to a beamwidth of 2σ, are approximately 16% and 84%. See equation (10), Figure 1.4, and Table 1.2.

$$\boldsymbol{D} = 2D_{1/e^{0.5}} \tag{14}$$

Slit Method. In the slit method, which progressively apertures long narrow samples of flux across the entire beam, the resultant measured distribution is also gaussian. This is because of the unique mathematical properties of the gaussian function. Thus the percentage clip level is also at 13.5% of the peak power level, as in the case of the pinhole method.

1.3 BEAM DIVERGENCE DEFINITION

The divergence θ of a gaussian laser beam is the plane angle (Section 1.4.2.), expressed in milliradians (mrad), that the $1/e^2$ beam diameter $\boldsymbol{D}$ subtends at the laser. This definition presumes that the beam is continuously diverging and that the beam waist is located at or near the laser output port. Furthermore, it assumes that the beam is uninterrupted by optical elements and that the diameter is measured at a distance far away from the laser, in the far-field of the beam $z > 3z_R$, where z_R represents the Raleigh range.

1.3.1 Rayleigh Range z_R

The *Rayleigh range* z_R is the distance to that position on the beam axis at which the beam wavefront has a minimum radius of curvature $R_{\min}$ on either side of the beam waist. $R_{\min}$ equals twice the Rayleigh range z_R; also $D_R = \sqrt{2}\boldsymbol{D_0}$. Also,

$$z_R \approx \frac{\pi \boldsymbol{D_0}^2}{4\lambda} \tag{15}$$

1.3.2 Calculation of Divergence θ

The divergence θ (in mrad) is given by

$$2 \operatorname{Tan}\left(\frac{\theta}{2}\right) = \frac{\boldsymbol{D}_0}{z_R} \approx \frac{4\lambda}{\pi \boldsymbol{D}_0} \tag{16}$$

For small angles, say $\theta < 10°$, the divergence (in mrad) is $4/\pi$ times the laser beam output wavelength λ (in nm) divided by the nominal beam diameter $\boldsymbol{D}_0$ (in mm), at the laser output port where the beam waist is presumed to be. Thus

$$\theta \approx \frac{\boldsymbol{D}_0}{z_R} \approx \frac{4\lambda}{\pi \boldsymbol{D}_0} \tag{17}$$

1.3.3 Measurement of Beam Divergence θ

The beam divergence may be determined by measuring the beam diameter $\boldsymbol{D}$ at positions z_1 and z_2 in the uninterrupted beam, and at positions at which the distance z is at least three times greater than the Rayleigh distance z_R. The distance $z_1 - z_2$ between the positions in the far field of the beam should be chosen such that one diameter is at least twice the other diameter. The divergence is then the arcsine of the difference in diameters divided by the distance between the positions (Figure 1.5):

$$\theta = \operatorname{Arcsin}\left[\frac{\boldsymbol{D}_2 - \boldsymbol{D}_1}{z_2 - z_1}\right] \times 1000 \text{ (mrad)} \tag{18}$$

1.3.4 Beam Quality

As stated in Section 1.1, this chapter deals only with a laser beam that has a symmetrical spherical gaussian wavefront of the lowest-order TEM_{oo} mode. In practice, beams are not likely to be pure gaussian. Real beams are most likely to contain modes of higher orders that reduce the quality of the beam with respect to characteristics of a pure gaussian beam. One such dimensionless measure of beam quality is the quantity M^2 [7].

The M^2 value of a beam compares the real beam characteristics to that of a theoretically pure gaussian beam of the type that is being discussed in this chapter. The pure gaussian beam is chosen to have the same waist diameter at the same position. An M^2 value of 1.0 indicates perfect correlation with a pure gaussian beam. Larger values of M^2 suggest the presence of higher-order modes and provide a measure of the increase in the angle of divergence over that of a pure gaussian beam with the same waist diameter [8].

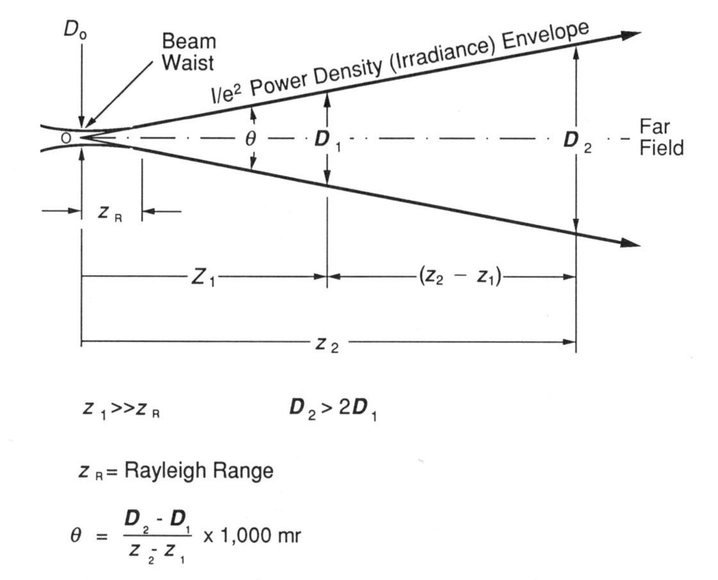

Figure 1.5 The beam divergence may be determined by measuring the beam diameter D in the far field at distances z_2 and z_1 from the beam waist in an uninterrupted beam. The ratio of z_2 and z_1, and therefore the beam diameters D_2 and D_1, should be at least 2.

1.4 ANGLES

Several angles are referred to in the course of defining divergence and radiometric terms. These are (1) the angle of view, (2) plane angle, and (3) solid angle.

1.4.1 Angle of View α

The *angle of view* is the angle between a specified direction and the normal to the emitting surface.

1.4.2 Plane Angle β

Consider the plane angle that a circular arc subtends at its center of curvature. The plane angle in circular measure is the ratio of the length of the arc to its radius of curvature (Figure 1.6).

$$\beta_{rad} = \frac{\text{Length of the arc}}{\text{Radius of curvature of the arc}} \tag{19}$$

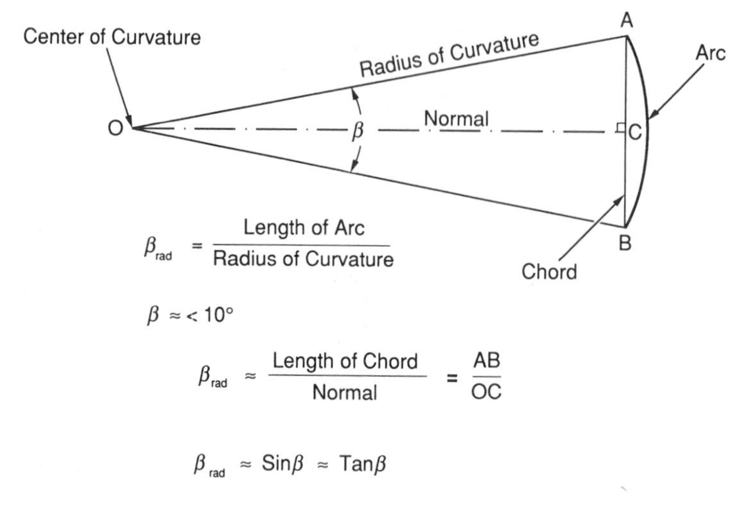

Figure 1.6 Plane angle β defined as a ratio expressed in radians.

The unit in circular measure of a plane angle is the radian, and is dimensionless. A plane angle β_{deg} expressed in degrees is converted to a plane angle β_{rad} expressed in radians by the following relationship:

$$\beta_{rad} = \frac{\pi}{180}\beta_{deg} \tag{20}$$

For small angles, say to $\beta \approx 10°$ (175 mrad),

$$\beta_{rad} \approx \text{Sin } \beta \approx \text{Tan } \beta \tag{21}$$

When $\beta = 10°$ (175 mrad), the approximation $\beta_{rad} \approx \text{Sin } \beta$ in equation (21) leads to a lower value of β_{rad} that is in error by about 0.5%.

Correspondingly, when $\beta = 10°$ (175 mrad), the approximation $\beta_{rad} \approx \text{Tan } \beta$ in equation (21) leads to a higher value of β_{rad} that is in error by about 1.0%. Likewise, when β is small, the length of the arc approaches the length of the chord of the arc, that is, the projected length of the arc. Thus, for small plane angles,

$$\beta_{rad} = \frac{\text{Length of the chord of an arc}}{\text{Distance to the chord}} \tag{22}$$

As a reference, the diameters of the Moon and the Sun both subtend at the Earth's surface an angle of approximately 1/2°, which corresponds to 8.7 mrad. Thus a laser beam with a divergence of 1 mrad (3′ 26″) and the beam directed at such a distant object as the Moon would irradiate (illuminate) an area on the Moon's surface that has a diameter one-ninth that of the Moon.

1.4.3 Solid Angle, Ω

Inherent in the definitions of radiometric terms is the concept of solid angle Ω. A solid angle is the three-dimensional counterpart of the two-dimensional plane angle. Instead of the ratio of lengths in a plane, it is the ratio of areas in a space that is determined (Figure 1.7).

Consider the solid angle that a spherical cap subtends at its center of curvature. The solid angle in circular measure is the ratio of the surface area of the cap to its radius of curvature squared (Figure 1.7):

$$\Omega = \frac{\text{Surface area of the cap}}{\text{Radius of curvature squared}} \tag{23}$$

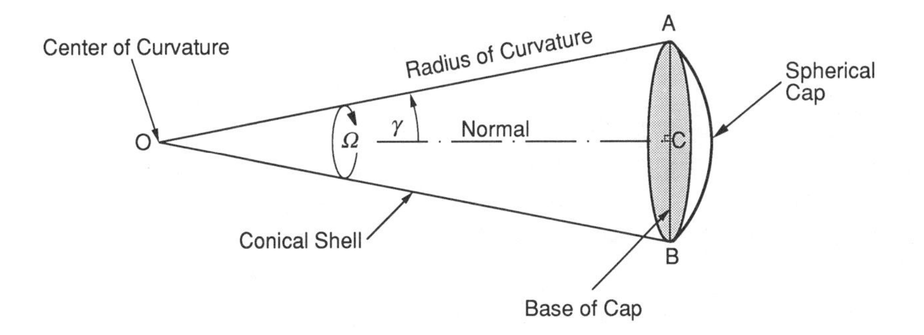

$$\Omega = \frac{\text{Surface Area of Cap}}{(\text{Radius of Curvature})^2} = 2\pi\,(1 - \text{Cos}\,\gamma) = 4\pi\,\text{Sin}^2(\gamma/2)$$

$$\gamma \approx< 10°, \qquad \Omega \approx \frac{\text{Area of Base}}{(\text{Normal})^2} = \frac{\pi}{4}\left[\frac{\text{AB}}{\text{OC}}\right]^2$$

Figure 1.7 Solid angle Ω defined as a ratio expressed in steradians.

The unit in circular measure of a solid angle is the steradian, and is dimensionless, as is the radian for the circular measure of a plane angle.

A spherical cap also subtends a regular cone at its center of curvature. If the full cone angle is 2γ, then the solid angle subtended at the apex, the center of curvature of the cap, is given by

$$\Omega = 2\pi(1 - \text{Cos}\,\gamma) = 4\pi\,\text{Sin}^2\left(\frac{\gamma}{2}\right) \tag{24}$$

For small angles, say to $\gamma = 10°$, $\Omega = 95.5$ mstr.

$$\Omega \approx \pi\,\text{Sin}^2\,\gamma \approx \pi\,\text{Tan}^2\,\gamma \tag{25}$$

When $\gamma = 10°$ (175 mrad), the approximation $\Omega \approx \pi\,\text{Sin}^2\,\gamma$ in equation (25) leads to a lower value of Ω that is in error by about 0.77%. Correspondingly, when $\gamma = 10°$ (175 mrad), the approximation $\Omega \approx \pi\,\text{Tan}^2\,\gamma$ in equation (25) leads to a higher value of Ω that is in error by about 2.25%. Likewise, when γ is small, the surface area of the cap approximates the area of the base of the cap, which is the projected area of the cap or the apparent area as seen by the eye at the viewing distance at the center of curvature. Thus, for small solid angles,

$$\Omega = \frac{\text{Projected area of an object}}{\text{Distance to the object squared}} \tag{26}$$

Equation (26) corresponds to $\Omega \approx \pi\,\text{Tan}^2\,\gamma$, given in equation (25).

As a guideline reference, the Moon's disk has a diameter that subtends approximately 1/2° (8.7 mrad) at the Earth's surface. Substituting $\gamma = 0.25°$ (4.36 mrad) into equation (23) leads to the Moon's disk subtending a solid angle of 59.8 μstr at the Earth's surface. The whole visible sky, a celestial hemisphere, subtends 2π radians at the Earth's surface.

1.5 RADIOMETRIC TERMINOLOGY

Radiometry is the technology surrounding the measurement of electromagnetic radiant power (watts) and energy (joules) across the entire spectrum of wavelengths. Photometry, which involves the sensitivity of the eye and has its own units, is that part of radiometry that deals only with the visible portion of the electromagnetic spectrum ranging approximately from 400 to 720 nm [15.8 to 28.4 μin] [9, p. 189]. In this wavelength band, because of the spectral response characteristic of the eye, the conversion factor between photometric and radiometric units has a nonlinear relationship with wavelength [1, pp. 45–46]. Since radiometry covers the whole spectrum, the subject shall be discussed in the radiometric terms, and for reference the photometric terms are included in parentheses, where appropriate. A summary of the radiometric terms is presented in Table 1.3.

Table 1.3 Summary of Relevant Radiometric Terms

Term	Symbol	Unit
Radiant flux Radiant power	ϕ	Watts
Radiance Radiant power emitted per unit solid angle in a specified direction θ, per unit projected area in that specified direction, of a source, a receiver, or an intermediate intersecting reference surface.	L_α	Watts/meter^2/steradian
Radiant intensity Radiant power emitted per unit solid angle in a specified direction α from a source.	I_α	Watts/steradian
Irradiance Radiant power incident per unit area at a surface or at an intermediate intersecting reference surface.	E	Watts/meter^2

1.5.1 Radiometric Terms

The important radiometric properties to consider are those associated with

1. Radiation emanating from a source
2. Radiation traversing a medium
3. Radiation incident at a surface

The CIE terms to review are given in the following.

1.5.2 Radiant (Luminous) Flux ϕ

The term *radiant flux*, symbolized by ϕ, represents the radiant power in a beam expressed in watts, and is often simply referred to as flux (Figure 1.8). The term

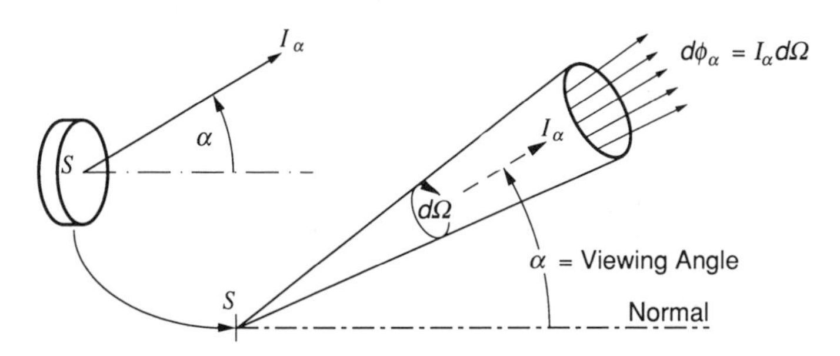

Figure 1.8 Viewing angle α and radiant flux ϕ.

luminous flux represents the luminious power in a beam expressed in lumens, the photometric unit. Provided that there is no ambiguity with radiant flux, these terms are simply abbreviated to flux.

There is no simple relationship between photometric and radiometric units, but luminance flux ϕ_v and spectral radiant flux $\phi_{e\lambda}$ may be related as follows:

$$\phi_v = \int_{380\text{ nm}}^{780\text{ nm}} \phi_{e\lambda} V(\lambda)\, d\lambda \tag{27}$$

where $V(\lambda)$ is the spectral luminous efficiency function expressed in lumens per watts [10, 11]. Beyond each side of the visible waveband, into the invisible ultraviolet and infrared regions, $V(\lambda) = 0$ and, therefore, no lumens are contributed to ϕ_v in equation (27).

1.5.3 Radiance (Luminance) L_α

Radiance, symbolized by $\boldsymbol{L_\alpha}$, corresponds nonlinearly to the photometric term *luminance* and the subjective term *brightness* that is perceived with the eye in the visible region of the spectrum. Radiance is a function of the angle of view α and is therefore a vector quantity (Figure 1.9). Radiance (luminance) $\boldsymbol{L_\alpha}$ is also an invariant parameter for the conservation of flux ϕ within a bundle of rays that emanate from a small source object of finite area as it traverses an optical system of nonabsorbing media [9, pp. 104, 105].

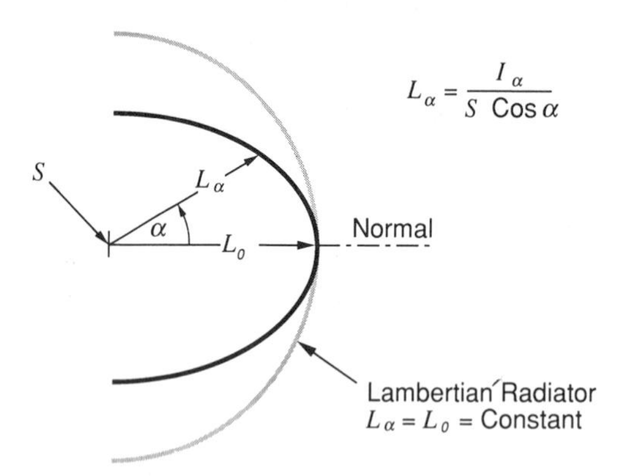

Figure 1.9 Radiance L_α. A lambertian radiator has the same radiance L_o in all directions and therefore is independent of the viewing angle α.

1.5.4 Radiant (Luminous) Intensity I_α

The term *radiant intensity*, symbolized by $\boldsymbol{I_\alpha}$, and the term *radiance* ($\boldsymbol{L_\alpha}$) both relate to the radiant properties of a source and, therefore, to one another by

$$(\text{Radiant intensity})_\alpha = \iint_S (\text{Radiance}_\alpha \text{ Cos } \alpha)\, dS \tag{28}$$

$$I_\alpha = \iint_S (L_\alpha \text{ Cos } \alpha)\, dS \tag{29}$$

where dS is an elementary area of a finite source of area S and α represents the angular direction from the normal (Figure 1.10).

1.5.5 Lambertian Radiator

The surface of an isotropically diffuse radiator in the form of a uniformly radiating luminous disk will look equally bright from any angle of view α from the normal to the disk. That means $\boldsymbol{L}_\alpha$ is a constant and, therefore, equals $\boldsymbol{L}_0$. Such a uniform radiator is referred to as a *Lambertian radiator* or *emitter* (Figures 1.8 and 19). Then because $\boldsymbol{L}_\alpha$ = constant = $\boldsymbol{L}_0$ for a Lambertian radiator, from equation (29),

$$\boldsymbol{I}_\alpha = \boldsymbol{L}_0 S \text{ Cos } \alpha \tag{30}$$

$$\boldsymbol{I}_\alpha = \boldsymbol{L}_0(\text{projected area of the disk}) \tag{31}$$

The projected area (S Cos α) of the disk is also the apparent area of the disk that is seen by the eye from a viewing direction α.

When $\alpha = 0$,

$$\boldsymbol{I}_0 = \boldsymbol{L}_0 S \tag{32}$$

Substituting into equation (31) leads to

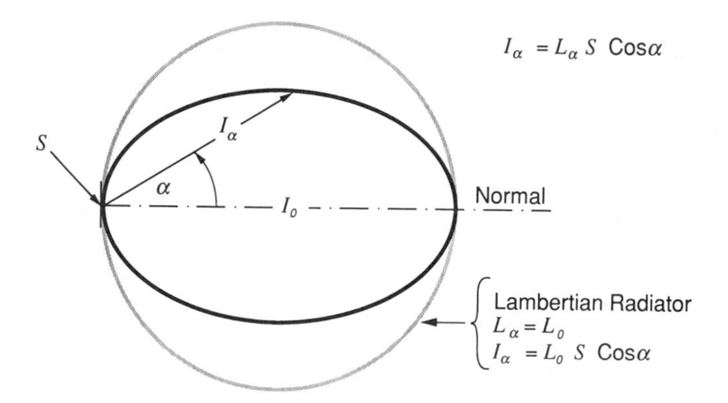

Figure 1.10 Radiant intensity I_α. a lambertian radiator has a radiant intensity that is a cosine function (I_o Cos α) of the viewing angle α.

$$I_\alpha = I_0 \operatorname{Cos} \alpha \tag{33}$$

Thus, from equation (33), the radiant intensity I_α of a Lambertian radiator decreases according to the cosine of the angle of view α. Hence, radiators and reflective surfaces are said to be Lambertian when their radiant intensity distributions with respect to the angle of view closely approximate a cosine function (Figure 1.10).

1.5.6 Irradiance *E*

Irradiance, symbolized by E, is the radiant *power density*—that is, the radiant power per unit area associated with a bundle of light rays as it traverses a plane or is incident on a plane (Figure 1.11).

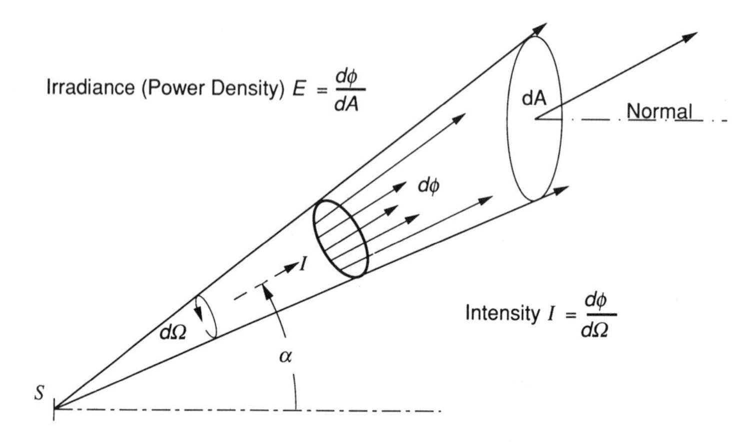

Figure 1.11 Irradiance E is the radiant power per unit area associated with a bundle of light rays as it traverses a plane or is incident on a plane.

1.6 GAUSSIAN BEAM

With respect to cylindrical coordinates, let the direction of propagation of a laser beam be collinear with the z-axis with the origin O at the beam waist (Figure 1.12). for a radially symmetric beam, all points P with radial coordinate r are expressed by $P(r,z)$.

Consider any perpendicular cross section of the beam through the axial point $Q(0,z)$. The radial distribution of radiant flux density traversing this plane is gaussian with a standard deviation σ. Because the beam is either diverging or converging, σ is a function of z; therefore

$$\sigma = \sigma(z) = \sigma_z \tag{34}$$

This is expressed by

$$\sigma(z) = \sigma_0 \left[1 + \left(\frac{\lambda z}{4\,\sigma_0^2} \right)^2 \right]^{1/2} \tag{35}$$

where λ = wavelength of the radiation

σ_0 = beam waist standard deviation for $z = 0$.

σ_z = beam standard deviation at a distance z.

Figure 1.12 shows a laser beam envelope for which $r(k,z) = k\sigma(z)$ and in which k is a chosen constant. Since r is always a function of z, it shall henceforth be dispensed with as a subscript. Thus $r_k = k\sigma_z$.

Hence, the beam envelope diameter $D = 2r_k = 2k\sigma_z$ at any beam section may be defined by a multiple ($k > 1$) or a fraction ($k < 1$) of the standard deviation σ_z of the gaussian beam power density distribution. In other words, the laser beam diameter may be normalized and expressed in terms of the standard deviation σ_z at any cross section by choosing a value for k. It is now the selected value of k, $k = 1, 2, \ldots$, that defines the threshold criterion for the beam boundaries. Hence, the diameter of a laser beam at any cross section is in units of standard deviation σ_z.

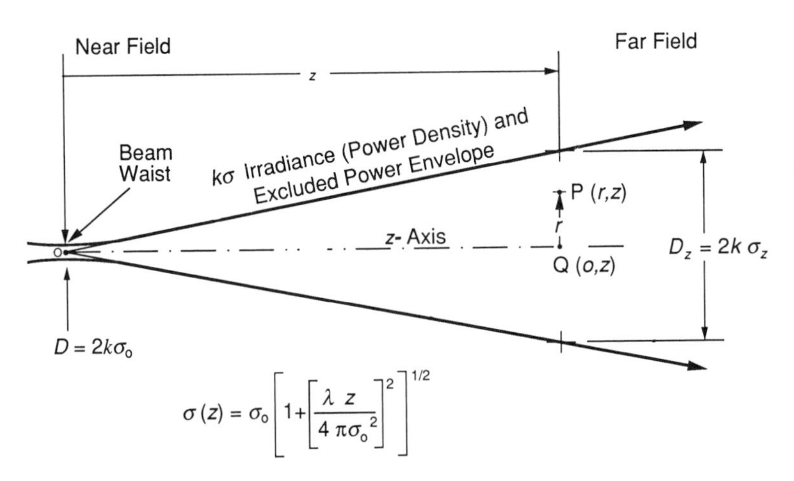

Figure 1.12 Beam diameter D is expressed in multiples $2k$ of the standard deviation $\sigma(z)$ of the irradiance (power density) E distribution profile.

1.6.1 Beam Irradiance

The radiant flux $\Delta\phi$ in a beam can be intercepted and explored with a radiometric detector that has a minute active area ΔA compared with the cross-sectional area of the laser beam, for example, at the point P(r,z) (Figure 1.12). By the definition already given in Section 1.5.6 and Table 1.3, the quotient

$$\frac{\Delta\phi}{\Delta A} = \frac{d\phi}{dA} \tag{36}$$

represents the irradiance $E(r,z)$ incident on the detector surface and, therefore, the irradiance of the beam at all points P(r,z). The three-dimensional radially symmetric gaussian distribution of $E(r,z)$ in the perpendicular plane through the point Q($0,z$) resembles a Mexican sombrero with a flattened brim [12, p. 295, Figure 2]. Mathematically it is expressed by

$$E(r,z) = E(0,z)\exp\left[\frac{-(r/\sigma_z)^2}{2}\right] \tag{37}$$

in which $E(0,z)$ represents the axial and peak beam irradiance. Dividing equation (37) by $E(0,z)$ leads to equation (38), which expresses the irradiance $g(k)$ that is normalized to the peak irradiance and shown in Figure 1.1.

$$g\left(\frac{r}{\sigma_z}\right) = \frac{E(r,z)}{E(0,z)} = \exp\left[\frac{-(r/\sigma_z)^2}{2}\right] \tag{38}$$

Let $r = k\sigma_z = r_k$ and substitute into equation (38), to give

$$g(k) = \frac{E(r_k,z)}{E(0,z)} = \exp\left(\frac{-k^2}{2}\right) \tag{39}$$

Understandably, because of the interdisciplinary influences, other terms and symbology than the acknowledged and preferred radiometric term for irradiance (power density) and its appropriate symbol E are sometimes heard and written, such as intensity with its symbol I [13]. See Section 1.1.3.

1.6.2 Beam Diameter Criteria

For a diameter $D = 2r_k$ of a laser beam, any value of k may be chosen. As the beam diverges from the beam waist or converges toward a beam waist, the beam diameter increases or decreases, respectively, but the criteria for defining the beam diameter remain the same. The important criteria for selecting the value of k for the beam envelope diameter are likely to be governed by

1. Encircled beam power, that is, the radiant beam flux ϕ_k contained within the beam envelope of relative diameter k.
2. Total beam power, that is, the total radiant beam flux ϕ_∞, $k = \infty$.
3. Fraction $h(k)$ of the encircled radiant flux ϕ_k.
4. Fraction $1 - h(k)$ of the excluded radiant flux $\phi_\infty - \phi_k$.
5. Irradiance (power density) variation $\Delta E(r,z)$ across a chosen beam diameter.

1.6.3 Beam Power ϕ_k

The total beam power is determined by substituting $k = \infty$. Then the total beam power (total radiant flux) is expressed by ϕ_∞. For any chosen beam envelope that is determined by a k value, the sum of the encircled beam power ϕ_k and the excluded beam power $\phi_\infty - \phi_k$ must equal the total beam power ϕ_∞. Both the encircled and the excluded beam powers are fractions of the total beam power.

1.6.4 Encircled Beam Power Fraction *h*(*k*)

The *encircled beam power fraction* $h(k)$ of the encircled radiant flux ϕ_k with respect to the total radiant flux ϕ_∞ is given by

$$h(k) = \frac{\phi_k}{\phi_\infty} \equiv \frac{\text{Encircled beam power}}{\text{Total beam power}} \tag{40}$$

See Figure 1.2.

1.6.5 Excluded Beam Power Fraction 1 − *h*(*k*)

The complementary *excluded beam power fraction* $1 - h(k)$ of the excluded radiant flux $\phi_\infty - \phi_k$ with respect to the total radiant flux ϕ_∞ is given by

$$1 - h(k) = \frac{\phi_\infty - \phi_k}{\phi_\infty} \equiv \frac{\text{Excluded beam power}}{\text{Total beam power}} \tag{41}$$

It will now be shown that, for a gaussian beam, the normalized irradiance $g(k)$ given in equation (39) also exactly corresponds to the excluded fraction $1 - h(k)$ of the beam power, such that the righthand sides of equations (39) and (53) are identical; thus

$$g(k) \equiv 1 - h(k) \tag{42}$$

1.6.6 Encircled Beam Power Derivation

Consider the radiant flux $d[\phi(r)]$ encircled between two envelopes of radii r and $r + dr$ at any plane section such as that through the axial point Q(0,z):

$$d[\phi(r)] = 2\pi r\, dr\, E(r,z) \tag{43}$$

Since

$$2\pi r\, dr \equiv \pi d(r^2) \tag{44}$$

equation (43) then becomes

$$d[\phi(r)] = \pi E(r,z)\, d(r^2) \tag{45}$$

Substituting for $E(r,z)$ from equation (37) gives

$$d[\phi(r)] = \pi E(0,z)\left\{\exp\left[\frac{-(r/\sigma_z)^2}{2}\right]\right\} d(r^2) \tag{46}$$

Integrating from $r = 0$ to $r = r_k = k\sigma_z$ gives

$$\int_0^{\phi_k} d[\phi(r)] = \pi E(0,z) \int_0^{r_k} \left\{\exp\left[\frac{-(r/\sigma_z)^2}{2}\right]\right\} d(r^2) \tag{47}$$

$$[\phi(k\sigma_z)]_0^{\phi_k} = \pi E(0,z)\left[\exp\left[\frac{-(r/\sigma_z)^2}{2}\right]\right]_0^{r_k} \tag{48}$$

Substituting $r_k = k\sigma_z$ gives

$$\phi_k = \pi(\sqrt{2}\sigma_z)^2\, E(0,z)\left[1 - \exp\left(\frac{-k^2}{2}\right)\right] \tag{49}$$

in which ϕ_k represents the encircled radiant flux for a beam envelope of radius r_k.

The total radiant flux ϕ_∞, that is, the total beam power, is obtained by substituting $k = \infty$ into equation (49). This leads to

$$\phi_\infty = \pi(\sqrt{2}\sigma_z)^2\, E(0,z) \tag{50}$$

Dividing equation (49) by equation (50) gives the fraction of encircled beam power $h(k)$. This leads to

$$h(k) = \frac{\phi_k}{\phi_\infty} = 1 - \exp\left(\frac{-k^2}{2}\right) \tag{51}$$

Therefore, for a fixed value of k, $h(k)$ is constant, independent of z, and satisfies all sections of the beam envelope.

1.6.7 Determination of Axial Irradiance from Total Beam Power

The coefficient $\pi(\sqrt{2}\sigma_z)^2$ of the axial irradiance (power density) $E(0,z)$ in equation (50) represents a beam cross-sectional area of radius $\sqrt{2}\sigma_z$, which is also equal to $\boldsymbol{D}/(2\sqrt{2})$. Thus the axial irradiance (power density) may be derived from a determination of the standard deviation σ_z and the total power ϕ_∞ of a gaussian beam. Rearranging equation (50) leads to

$$\text{Axial irradiance} = \frac{\text{Total beam power}}{\text{Beam cross-sectional area}} = \frac{\phi_\infty}{\pi(\sqrt{2}\sigma_z)^2} \tag{52}$$

1.6.8 Excluded Beam Power Derivation

As stated in Section 1.6.5, it follows that if $h(k)$ is the encircled beam power then the excluded beam power fraction must be $1 - h(k)$. By substituting for $h(k)$ from equations (51) into equation (41), the fraction $1 - h(k)$ of excluded radiant flux from the beam envelope is given by

$$1 - h(k) = \frac{\phi_\infty - \phi_k}{\phi_\infty} = \exp\left(\frac{-k^2}{2}\right) \tag{53}$$

1.6.9 Equating 1 − *h*(*k*) to *g*(*k*)

Now equation (39) is

$$g(k) = \frac{E(r_k,z)}{E(0,z)} = \exp\left(\frac{-k^2}{2}\right) \tag{54}$$

Inspection and comparison of equations (53) and (54) show that

$$1 - h(k) \equiv g(k) \equiv \exp\left(\frac{-k^2}{2}\right) \tag{55}$$

Thus the fraction of excluded beam power exactly corresponds to the relative power density, normalized to the axial peak power density (Figure 1.2).

The function $\exp(-k^2/2)$ in equation (55) is a gaussian distribution independent of z. Hence, the fraction of the excluded beam power and the normalized irradiance (power density) decrease in an identical manner as increasing values for the relative beam radius k are chosen, where k is a measure of the number of standard deviations σ. This convenient identity is entirely due to the unique mathematical properties of the gaussian function.

Table 1.1 gives $1 - h(k)$ and $g(k)$ for selected values of k. Inspection of Table 1.1 shows that for all practical purposes the effective beam envelope of the total beam power may be limited to a diameter of eight standard deviations, that is, 8σ, $k = 4$, at which the beam envelope encircles all but 0.034% of the total beam power. Further study of Table 1.1 shows that the beam envelope which contains 86.5% of the total beam power corresponds to a diameter of four standard deviations, that is, 4σ, $k = 2$, at which the beam envelope encircles all but 13.5% ($1/e^2$) of the total beam power. Similarly, the full width at the 13.5% ($1/e^2$) level of the normalized irradiance distribution profile corresponds to a width of four standard deviations, that is, 4σ, $k = 2$.

1.6.10 Conclusion

The accepted diameter ***D*** of a gaussian laser beam may be defined in two different ways that give identical results provided that the laser operates in the TEM_{oo} mode.

1. **The beam diameter *D*** is the diameter of the beam envelope that encircles 86.5% of, and excludes $1/e^2$ (13.5%) of, the total beam power.
2. **The beam diameter *D*** is the full width of the normalized irradiance (power density) distribution profile at the $1/e^2$ (13.5%) level of the peak value.

REFERENCES

1. M. Young, *Optics and Lasers*, Springer-Verlag, Berlin and New York, 1977.
2. F. E. Nicodemus, ed., *Self-Study Manual on Optical Radiation Measurements*. Part 1: *Concepts*, Chaps. 4 and 5, National Bureau of Standards, Washington, D.C., p. 3, 1978.
3. F. E. Nicodemus, J. C. Richmond, and J. J. Hsia et al., *Geometrical Considerations and Nomenclature for Reflectance*, National Bureau of Standards, Washington, D.C., p. 35, 1977.
4. G. F. Marshall, Gaussian laser beam diameters. In *Laser Beam Scanning* (G. F. Marshall, ed.), Marcel Dekker, New York, 1985.
5. T. C. Laurin, Beam profilers, laser. In *Photonics Buyers' Guide® to Products & Manufacturers*, Book 2, 36th ed. Laurin, Pittsfield, MA, p. 63, 1990.
6. J. M. Fleischer and J. M. Darchuk, Standardizing the measurement of spatial characteristics of optical beams. In *Laser Beam Radiometry*, SPIE, Vol. 888, p. 63, 1988.
7. A. E. Siegman, New developments in laser resonators. In *Optical Resonators* (D. Holmes, ed.), Proc. SPIE, Vol. 1224, pp. 2–14, 1990.
8. W. Woodward, A new standard for beam quality analysis. In *Photonics*, Laurin, Pittsfield, MA, pp. 139–144, 1990.
9. F. A. Jenkins and H. E. White, *Fundamentals of Optics*, 2nd ed., McGraw-Hill, New York, 1951.
10. F. L. and L. S. Pedrotti, *Introduction to Optics*, Prentice Hall, Englewood Cliffs, NJ, p. 15, 1987.
11. J. F. Snell, Radiometry and photometry. In *Handbook of Optics* (W. G. Driscoll and W. Vaughan, eds.), McGraw-Hill, New York, pp. 1.4, 1.5, 1978.
12. G. F. Marshall, Gaussian laser beam diameters. In *Laser Beam Scanning* (G. F. Marshall, ed.), Marcel Dekker, New York, 1985.
13. G. F. Marshall, Scanning devices and systems. In *Applied Optics and Optical Engineering*, Vol. 6 (R. Kingslake and B. J. Thompson, eds.), Academic Press, New York, p. 218, 1980.

2

Optical Systems for Laser Scanners

Robert E. Hopkins

Optizon Corporation, Rochester, New York

David Stephenson

Optics Division, Melles Griot, Rochester, New York

2.1 INTRODUCTION

The many applications of laser scanning usually depend on optical elements to direct and focus the laser beam. This chapter discusses the interaction between laser scanning requirements and the conditions imposed on the optical systems. The optical components to be discussed include lenses, mirrors, and prisms.

A brief review of first-order and third-order lens design theory is offered, for it helps to describe the optical systems which can meet the scanning requirements.

Several representative optical systems with their characteristics are listed along with drawings showing the lenses and ray trajectories. Some of the optical systems used for scanning require special methods for testing and quality control. These methods are discussed.

2.2 LASER SCANNER CONFIGURATIONS

Figure 2.1 shows three ways to use lenses to focus a scanning laser beam. They are called *postobjective*, *preobjective*, and *double-pass* scanning [1–4].

2.2.1 Postobjective Scanning

Postobjective scanning has an advantage over preobjective scanning because the lens works on axis. For many applications the lens can be a simple plano-convex

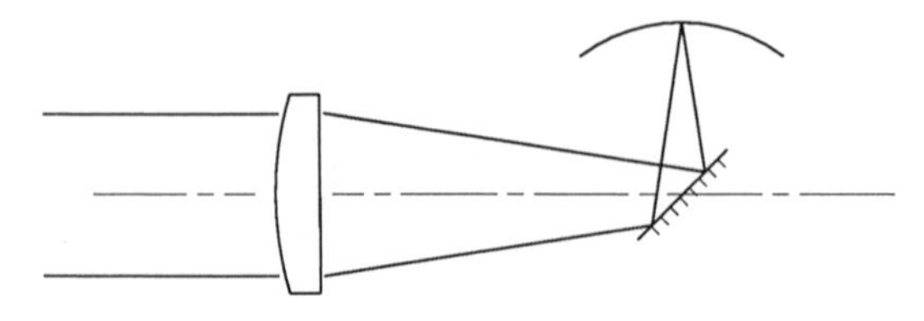

Post-Objective Scanning

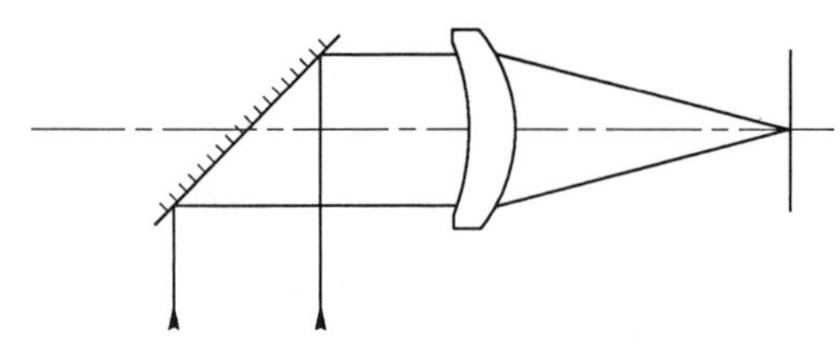

Pre-Objective Scanning

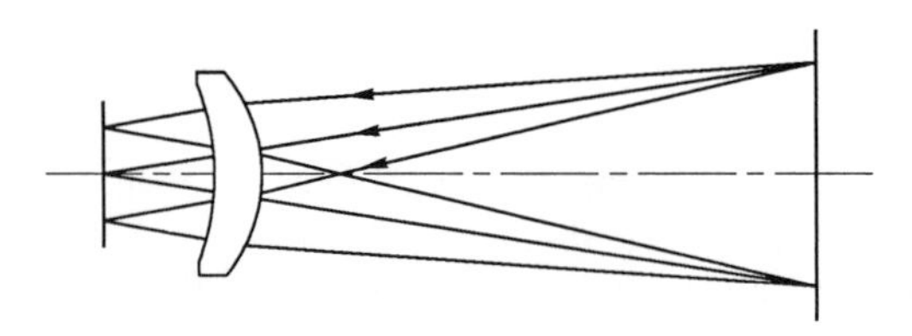

Double-Pass Scan

Figure 2.1 Three common types of optical scanning systems.

lens. High-resolution systems, which require a large numerical aperture, can be fully corrected with a doublet. The disadvantage of postobjective scanning is that the material to be scanned has to be on a curved surface.

2.2.2 Preobjective Scanning

Preobjective scanning requires flat-field lenses. Since the scanning element is out in front of the lens, these objectives are different from photographic lenses, which have their aperture located inside the lens. For a scan lens, the scanning element is the entrance pupil, and there must be clearance space between it and the mount holding the first element.

2.2.3 Double-Pass Scanning

Retrofocus scanning systems can be compact since they do not need a separate collimating lens. The feed beam is introduced from beyond one end of the scan line, and sometimes above or below the scan line. It is collimated as it passes through the lens, and then is redirected back through the lens at the angle determined by the mirror. This system is not quite as symmetrical as it appears, because the feed beam enters from the edge of the lens and operating field. Any aberrations introduced into this beam are added to the aberrations at all the field angles in the scan. Therefore, it cannot be assumed that the double pass will correct all the coma and distortion errors.

2.2.4 Internal Drum Scanners

From an optical point of view, internal drum scanning is recommended for systems requiring high-quality results. When the scanning element and the lens are in a fixed relation to each other and are rotated together about the optical axis of the incident laser beam, simple lenses can be used. The scanning spot traces a complete circle on the inside of a tube. The lens may be placed in front of or behind the turning mirror. A complete two-dimensional raster may be generated by translating either the lens-mirror unit or the drum. This is an ideal system for inspecting the inside surface of a tube by translation of the tube for reading and writing. Documents can be inserted on the inside of the tube.

2.2.5 Most Commonly Used Systems

Presently, it appears that the most commonly used systems, and those being developed, involve preobjective scanning; these systems require flat-field lenses. The special conditions described in the next section must be considered during the design of the lenses.

2.3 SPECIAL REQUIREMENTS FOR FLAT-FIELD LASER SCANNING

Preobjective laser scanners introduce optical design factors that are not present in most photographic and photolithography applications. This section describes the terminology and unique requirements of laser scan-lens design.

2.3.1 The F-θ Condition

In order to maintain uniform exposure on the material being scanned, the image spot must move at uniform velocity. As the deflector turns by the angle $\theta/2$, the beam is deflected by θ, where the angle θ is measured from the optical axis of the lens. Since polygon and holographic scanners rotate at a constant velocity, the reflected beam will rotate at a constant angular velocity. The scanning spot will move along the scan line at a uniform velocity if the displacement of the spot is linearly proportional to the angle θ. The displacement Y of the spot from the optical axis should follow the equation

$$Y = F \cdot \theta \tag{1}$$

The constant F is approximately the focal length of the lens. Figure 2.2 is a plot of the image height plotted against the scan angle for a normal lens which is corrected for distortion. The curve departs from a straight line which is required by an F-θ lens for a constant velocity of scan. As the field angle increases, the distortion-free lens images points too far out on the scan line, causing the spot to move too fast near the end of the scan line. Fortunately scan lenses have negative (barrel) third-order distortion; the image height curve lies below the F–tan θ curve. The distortion can be adjusted to match the F-θ image height at

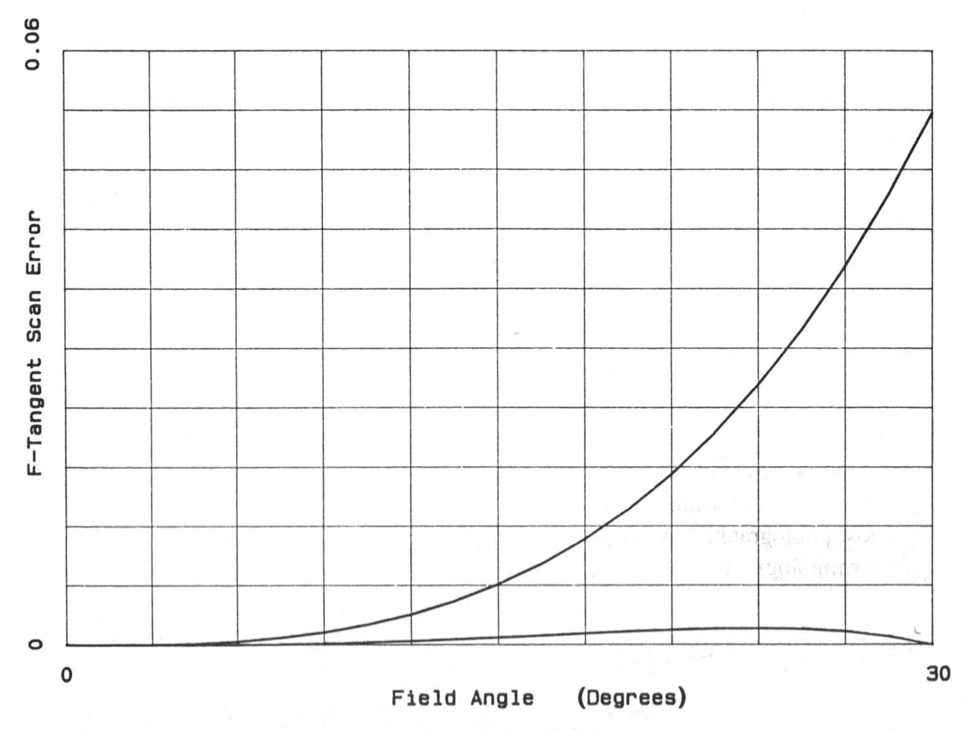

Figure 2.2 A plot showing the difference between F-tan θ and F-θ distortion correction.

the edge of the field. As Figure 2.2 shows, the departures from linearity are reduced. The error may be reduced further by choosing a straight line to equalize the plus and minus departures. This means selecting a value of F that differs from the paraxial focal length. The focal length that minimizes the departures from linearity is called the *calibrated focal length.*

When the field angle is as large as $\pi/6$ radians (30°), the residual departures from linearity may still be too large for many applications. The departures can be further reduced by balancing negative third-order distortion against positive fifth-order distortion. Lens designers will recognize this technique as similar to the method of reducing zonal spherical aberration by using strongly collective and dispersive surfaces, properly spaced. In this case the zonal spherical aberration of the chief ray must be reduced. Figure 2.3 shows an example of this correction.

This high-order correction should not be carried too far, since the velocity of the spot changes rapidly near the end of the scan and may result in unacceptable changes in exposure. It may also turn circles into ellipses. The error is called *local distortion*. This distortion results in a change in the spatial frequencies near the end of the scan line. The standard observer can resolve frequencies of 10 line pairs/mm, [254 lines/inch], but it is even more sensitive to variations of frequency in a repetitive pattern. Variations of frequency as small as 10% may be detected by critical viewing. The linearity specification is often expressed as a *percent error* (the spot position error divided by the required image height). For example, the specification often reads that the F-θ error must be less than 0.1%. This means that the deviations must be smaller and smaller near the center of the scan line. It is not reasonable to specify such a small error for prints near the center of the scan. The proper specification should state rate of change of

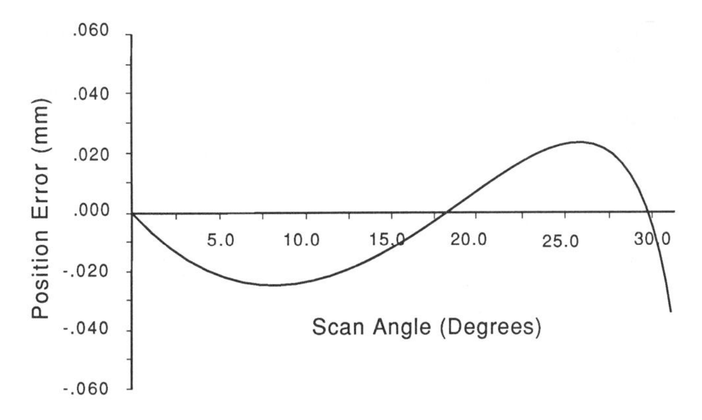

Figure 2.3 A plot showing how the calibrated focal length, third order and fifth order distortion can be adjusted to improve the F-θ correction in a scan lens.

the scan velocity and the allowable deviation from the baseline of Figure 2.2. More detail on this subject may be found in Ref. 5.

2.3.2 Irradiance of The Image Spot

There are subtleties to consider in calculating the image irradiance produced by a scan lens, as opposed to that of a normal camera lens. The aperture stop of scan lenses should be located on the device that deflects the incoming beam. In galvonometric and polygonal laser beam scanners, the turning mirrors do not alter the circular diameter of the incoming beam as the deflection changes. This is different from a camera lens, which has a fixed aperture stop perpendicular to the lens optical axis. The oblique beam in a camera is foreshortened by the cosine of the angle of obliquity on the aperture stop. The off-axis beam passing through the scan lens is, therefore, slightly larger than the one from a fixed aperture.

Most lens design programs do not automatically take this into account, so it is necessary to use the proper tilts in the design program to maintain the beam diameter at each field angle to be optimized. This can be done in the multiconfiguration setup available in most of the commercial design programs. The design program then optimizes several versions of the design simultaneously. Section 2.6.1 further discusses how multiconfiguration design procedures can be used in scan-lens design.

2.3.3 Gaussian Beam Talk

Chapter 1 contains detailed discussions on the optimal properties of gaussian beams. The following section covers similar topics expressed from the point of view of the lens designer.

Before getting into this subject it is helpful to define the terms *F*/NO. and *numerical aperture*. The *F*/NO. of a lens is equal to the lens focal length divided by the design aperture diameter of the input beam. There is a problem with the use of this widely used term. When a lens works at infinite conjugate, as scan lenses usually do, *F*/NO. describes the cone angle of the light focused on the image plane. The sine of half of this cone angle is called the *numerical aperture* (NA) when the image is formed in air. The relation between the numerical aperture and *F*/NO. is

$$F/\text{NO.} = \frac{0.5}{NA} = \frac{F}{D_L} \tag{2}$$

When a lens works at real finite conjugates, there is a numerical aperture and an *F*/NO. for the object side and also for the image side. The *F*/NO. of the lens

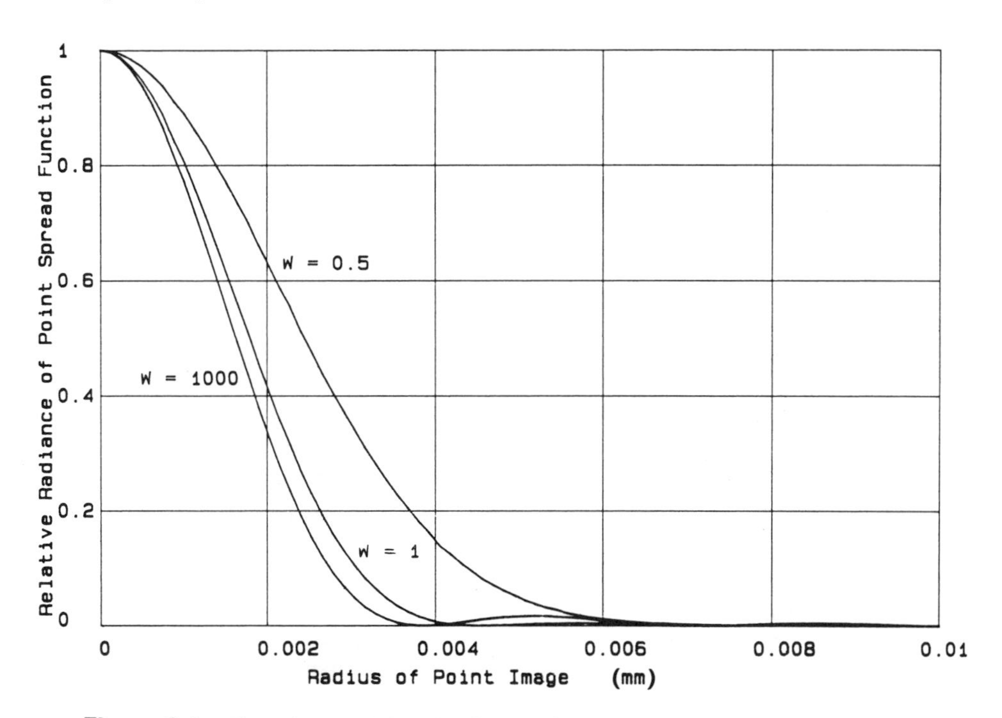

Figure 2.4 The point spread function for a perfect wave front with various truncations $W = D_B/D_L$.

is determined by adding or subtracting the absolute values of the numerical apertures on both sides of the lens and using equation (2) to convert it to the *F*/NO. The following discussion uses *F*/NO. and numerical aperture frequently. Since the discussion applies to scan lenses, both terms refer to the image-side cone angle.

Laser scanning systems usually use collimated gaussian feed beams with some truncation. *Truncation* means that the diameter of the gaussian beam is restricted by a hard aperture. In laser beam scanners the hard aperture is usually located in the input collimator. Figure 2.4 shows how a diffraction-limited wavefront is affected with different truncation ratios (W). W is defined as the ratio of the diameter of the gaussian beam D_B to the diameter of the lens aperture D_L. The beam diameter is often defined at the $1/e^2$ irradiance level. It is important to remember that a scan lens does not have a fixed aperture stop—the lens diameters are much larger than the design aperture because the lenses have to pass the oblique ray bundles from the exterior design aperture. The entering collimated beam, often called the feed beam, usually determines the aperture. The diameter of the beam should be no larger than the diameter of the largest beam for which the lens can provide the required image quality, which is usually a diffraction-limited wavefront. This is called the *design aperture* and is the value to use for the D_L when calculating W. For scan lenses, D_L does not refer to the actual physical diameter of the scan lens. This oddity can be confusing, so it is well to consistently refer to D_L as the *design aperture*.

The image spot diameter formed by a gaussian beam is given by

$$d = k\lambda(F/\mathrm{NO.}) \tag{3}$$

Figure 2.5 [2] shows how the diameter of the image of a point source is affected with different amounts of the truncation ratio W. The value of k depends on the truncation ratio W and the level of irradiance in the image spot used to measure the diameter of the image. The figure shows two criteria for the image diameter d. One is for the $1/e^2$ irradiance level, and the other is for the 50% irradiance level. Mathematical equations for these two cases may be found in Ref. 5.

$$k_{FWHM} = 1.021 + \frac{0.7125}{(W - 0.2161)^{2.179}} - \frac{0.6445}{(W - 0.2161)^{2.221}} \tag{4}$$

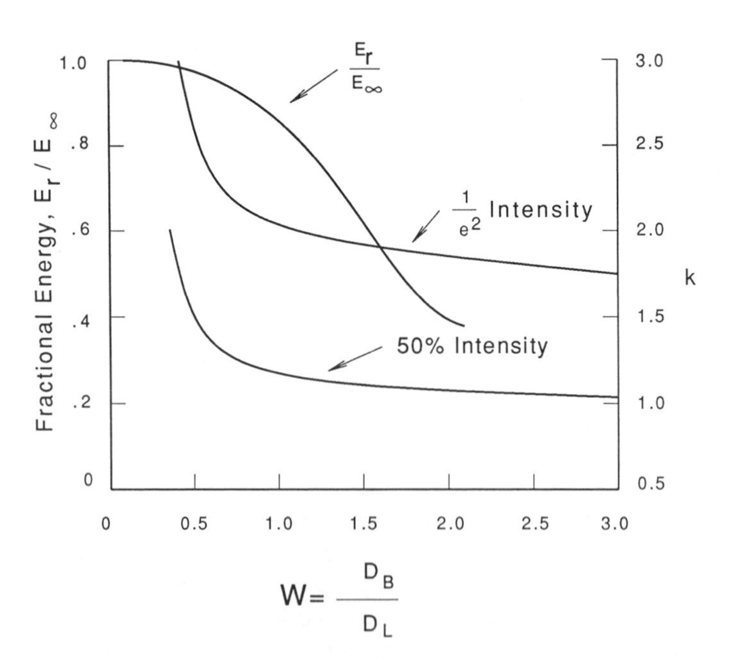

Figure 2.5 Curves showing how the spot diameters are related to the truncation ratio.

$$k_{1/e^2} = 1.6449 + \frac{0.6460}{(W - 0.2816)^{1.821}} - \frac{0.5320}{(W - 0.2816)^{1.891}} \quad (5)$$

Figure 2.5 also shows the loss of energy resulting from different levels of truncation. A truncation ratio of 1 generally provides a reasonable trade-off between spot diameter and conservation of total energy. With $W = 1$, the following equations can be used to estimate a spot diameter:

$$d_{1/e^2} = 1.83\lambda(F/\text{NO.}) \quad (6)$$

and

$$d_{(50\%)} = 1.13\lambda(F/\text{NO.}) \quad (7)$$

There are several points to consider when deciding what truncation ratio to use. The *Airy disk* is the name given to the image formed at the center of a perfect spherical wavefront of uniform irradiance. The Airy disk formula for the spot diameter has 2.44 for the constant k, so it appears that the gaussian beam image is smaller than the Airy disk. The gaussian beam coefficient 1.83, however, refers to the 13.5% irradiance level in the image, while the Airy disk formula relates to the diameter of the first zero in irradiance. The irradiance distributions for the two cases are shown in Figure 2.4. The curves show that the Airy disk pattern is narrower than the truncated gaussian beam, all the way to zero intensity. On the other hand, the Airy disk image has more energy out in the wings of the image than does the gaussian beam. This is shown better in the encircled energy curves in Figure 4 of Chapter 7 in the first edition of *Laser Beam Scanning*.

It is clear that heavy truncation ($W \gg 1$) gives the smallest spot size, provided the exposure is adjusted so that the receiving material does not record the flare light in the diffraction rings. Many designers believe that a low value of W is a better compromise, for there is less light loss and less danger of image spread from the rings formed by a heavily truncated beam.

Engineers developing scanning systems often use the concept of spot diameter. The specifications typically call for a spot diameter measured at a specified intensity level. The $1/e^2$ and the 50% intensity levels are commonly used. The maximum allowable growth of spot size across the length of the scan line is also included in the specification. This custom probably has come into use because the theoretical spot sizes at various intensity levels are well known for diffraction-limited lenses. There are also commercially available instruments which measure the *line spread function*. It is well to note, however that the line spread function is different from the point spread function of the point image which is given by equations (3) to (7). The line spread function is calculated by itegrating the irradiance as a slit passes over the point image. The line spread function of the Airy disk does not have zeros in the irradiance distribution and is a more appropriate criterion to use when the spot is constantly moving during the exposure.

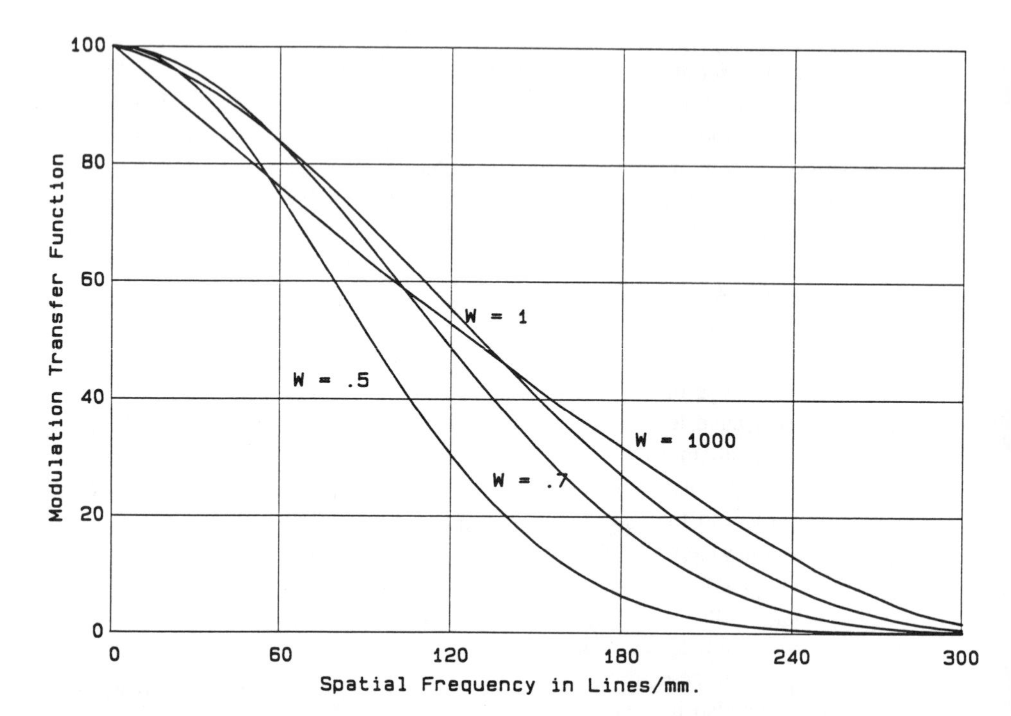

Figure 2.6 MTF curve for a perfect image with truncation ratios ranging from 1000 to 0.5.

Addressability is an important term widely used in laser scanning. It refers to the least resolvable separation between two independent points on a scan line. When the concept of spot diameter is used to describe optical performance, it is difficult to know how close the two spots can be to recognize them as separate points. Electrical engineers tend to think in terms of Fourier analysis, which suggests the concept of the *modulation transfer function* (MTF) [6].

We believe that the MTF specification offers advantages in describing the optical performance of laser scanners. Figure 2.6 shows a plot of the MTF for perfect images formed with truncations W of 1000 (the Airy disk), 1, 0.7, and 0.5. It is clear that the lower values of truncation have higher MTF for the low frequencies. The best value for W is close to 1. At this truncation the MTF is highest out to the 50% level for a frequency which is 0.43 of the theoretical cutoff frequency for the design aperture. This suggests that the principle of design to follow is to use W close to 1 and design to as small an F/NO. as possible, consistent with the performance and cost considerations.

The curves illustrated in Figure 2.6 were calculated using the Oslo design and evaluation program (*Sinclair Optics*). The calculations were made on the central image from an *F*/5 parabola. The cutoff frequency for this aperture is 316 line pairs/mm. With *W* set at 1 the MTF values are above 50% out to 132 line pairs/mm. [6700 dots/inch]. Not many scanning applications require this large number of dots/in. A common specification, however, is 1000 dots/in, which would require approximately an *F*/30 lens system.

The foregoing rule is based on a perfect image. In attempting to increase the MTF at frequencies below 0.43 of the cutoff frequency, problems with aberration eventually occur in the large design apertures required. Fortunately, the small values of *W* mean that the intensity of the rays near the edge of the aperture are reduced, so the acceptable tolerance on the wave abberations can be relaxed. It is not as easy to give a rule-of-thumb tolerance on the wavefront errors because it depends on the type of aberration. The high-order aberrations near the edges of the pupil will have less effect than will low-order aberrations, such as out-of-focus or astigmatism errors.

There is usually considerable confusion between engineers who design laser scanning systems and the lens designers who are asked to design the system. The difficulty arises from whether to use the concept of the point spread function, the line spread function, or MTF.

If the designer thinks in terms of writing the image with independent image points, then the point spread function is a convenient concept. The spot sizes then would appear to be determined by the exposure. As the exposure increases, the spot diameter also increases. With a light exposure the spots can be made small, and the observable spot size is determined by the irradiance level in the point spread function. It then should be necessary to evaluate the point spread functions of a lens at several fields of view to see the effect of exposure. When a lens has aberration, the image will be spread out beyond the Airy disk. This reduces the irradiance in the center of the point spread function. Most of the programs that calculate the point spread function normalize the spread functions to unit irradiance at the center of the image. However, when evaluating the images for a lens at several field angles, the point spread functions should be plotted without normalizing to unity at the center of the point spread function. In this way one can see how the spot diameters change with constant beam irradiance across the field. The problem with the above concept is that the laser beam, as it writes, is constantly moving. This means that the image will be smeared along the scan line. As the spot moves, the beam irradiance level is constantly being modulated in order to write the information required. This lends credence to the idea that the MTF concept is more appropriate. The problem is that the MTF assumes that the recording medium records the irradiance level linearly over the complete range of exposures. This may or may not be true.

It is important for the lens designer to discuss these differences with the system designer and make sure that all concerned understand the slippery slopes we stand on. Prior to gaining experience with complete systems which have been built and thoroughly tested, it is important for designers and users to have good communications. It usually pays to overdesign by at least 10% on initial ventures into laser scanning system development. The time to be most critical of a new design is in the first prototype design and in the testing of the first complete system. Optical lens systems do not deteriorate, nor do they improve with age.

2.3.4 Depth-of-Focus Considerations

Another important consideration in laser scanning systems is the *depth of focus* (DOF). The classical DOF for a perfectly spherical wavefront is given by

$$\mathrm{DOF} = \pm 2\lambda (F/\mathrm{NO.})^2 \tag{8}$$

This widely used criterion is based on one-quarter of a wave departure from a perfect spherical wavefront. Many specifications state that the spot size diameter must be constant within 10% across the entire scan line. One would expect that a corresponding tolerance on MTF would be 10% of the limiting frequency specified. The manufacturers of scanning systems also impose a lower limit to the tolerable depth of focus. There is no simple formula to relate depth of focus with this requirement, but MTF calculations for several focal plane positions provide the pertinent data.

Figure 2.7 shows the MTF curves of an $F/5$ parabola under the following conditions.

A. The perfect image with uniform irradiance, at a wavelength of 632.8 nm [24.9 microinches], across the entire design aperture ($W = 1000$).
B. The perfect image with $W = 0.85$.
C. The same image as A, but with a focal shift of 0.063 mm [0.0394 inches]. This corresponds to a wavefront error of half of a wave at the maximum design aperture.
D. The same image and truncation W as B, but with focal shift of 0.063 mm.
E. The image from a parabola with aspheric deformation added to introduce a half-wave of fourth-order wavefront error at the edge of the design aperture. $W = 1000$, no focus shift.
F. The same as E with $W = 0.85$, no focus shift.

The truncation value of $W = 0.85$ was used for this example instead of 1.0 in order to help reduce the influence of the aberrated rays near the edge of the design aperture. These curves show that the depth of focus is slightly improved by truncating at this value. They also show that half a wave of spherical aberration does not have as serious an effect on the depth of focus as does an equivalent amount of focus error. Therefore it is most important to reduce the Petzval curvature and astigmatism in a scan lens, because these aberrations cause focal shift errors.

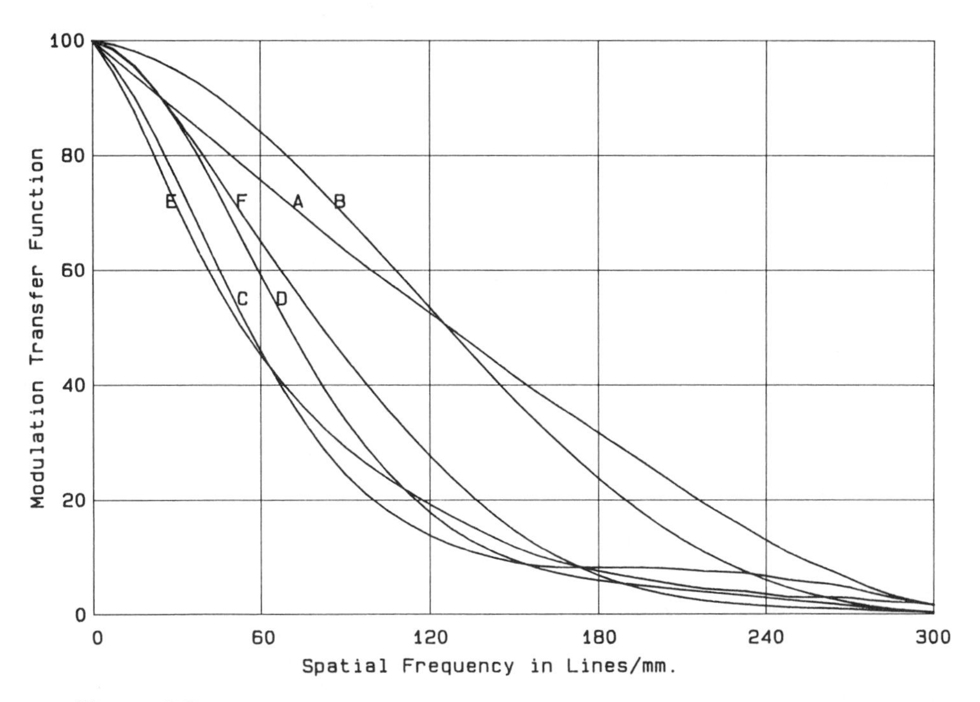

Figure 2.7 MTF curves showing the effect of focus shift, spherical aberration, and truncation conditions.

2.3.5 Beam Diameter, Scan Angle, Scan Length, and Number of Pixels

The total number of pixels along a scan line is a measure of the optical achievement, given by

$$n = \frac{2D\theta}{k\lambda} \tag{9}$$

where n = L/d (usually in micrometers [μm])
d = spot diameter (mm)
L = length of scan (often given in inches)
D = diameter of lens design aperture (mm)
θ = the half scan angle (degrees)

It is important to define the criterion one uses for defining d. Optics people, when making claims for their lenses, tend to use the Airy disk diameter, and laser people tend to use either the $1/e^2$ criterion or the 50% irradiance level.

2.4 FIRST-ORDER LENS DESIGN CONSIDERATIONS

The optical system in a scanner should have a well-considered first-order layout. This means that the focal lengths and positioning of the lenses should be determined before any aberration correction is attempted. Most of the optical systems to be discussed in this section will first be described as groups of thin lenses. The convention used for thin lenses is described in most elementary books on optics [7,8].

The thin-lens solutions can be discussed by using the graphical method shown in Figure 2.8. The diagram shows an axial ray which is parallel to the optical axis. This represents a collimated beam entering the lens. The negative lens refracts the axial beam upward to the positive lens. The positive lens then refracts the ray to the focal plane, which is the writing plane for the laser beam.

The second focal point of the negative lens is at $F_{2,a}$. This point is located by extending the refracted axial ray backward from the negative lens until it meets the optical axis. The second focal point of the positive lens may be determined by drawing a construction line through the center of the positive lens

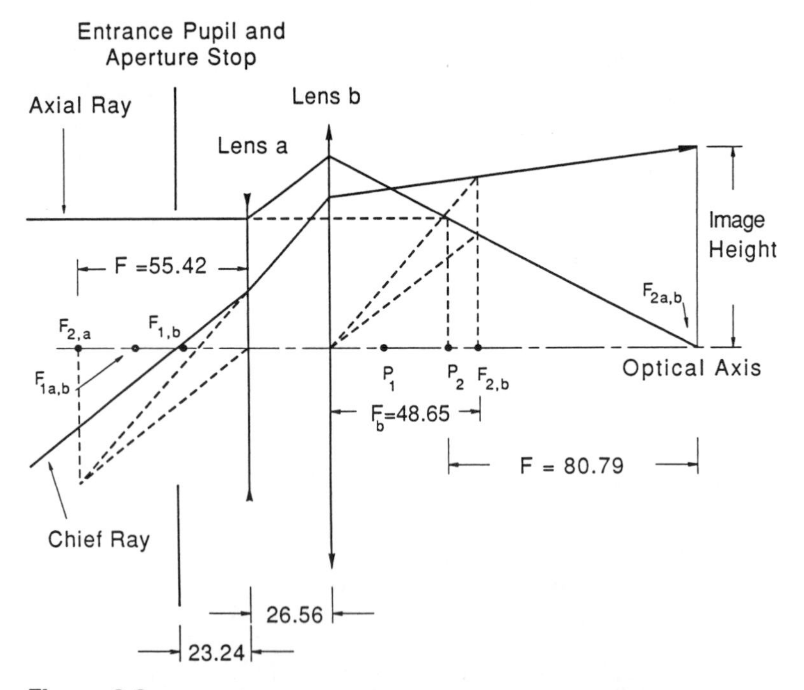

Figure 2.8 A graphical solution to a system of thin lenses. The focal lengths of the individual lenses and the total system can be easily determined.

parallel to the axial ray as it passes between the positive and negative lenses. Since the two lines are parallel, they must come to focus in the focal plane of the positive lens. The focal lengths of the two lenses are now determined. The front and back focal lengths of each lens are equal because the lenses are in air. The focal points $F_{1,a}$, $F_{2,a}$ and $F_{1,b}$, $F_{2,b}$ are now located.

The diagram also shows the construction for finding the second principal point P_2. The distance P_2 to F_2 is the focal length of the negative-positive lens combination. The drawing was made on the Versacad drafting program. A person skilled at using this program can make optical calculations of systems with high accuracy.

The chief ray is next traced through the two-element system. This is done using the concept that two rays which are parallel on one side of a lens must diverge or converge to the second focal plane of the lens. The chief ray enters the lens system after it passes through the aperture stop of the system. In this case the aperture stop is the entrance pupil of the system. In the case of scan lenses, this aperture is usually located at the scanning element. The aperture stop is out in front of the lens, which is in contrast to a photographic lens, where the aperture stop is usually located between the lenses. This is the basic reason why a photographic lens should not be used for scan lenses. It is also one of the reasons why scan lenses are limited in the field angles they can cover.

The completed diagram labels the lens focal lengths. The system focal length is 80.78, $F_a = -55.42$, and $F_b = 48.63$. The *Petzval radius* is given by the formula

$$P = \frac{1}{1/n_aF_a + 1/n_bF_b} \tag{10}$$

The Petzval radius is 3.3 times the focal length and it is curved towards the lens. This is not flat enough for an $F/20$ system when the lens has to cover a long scan line. Equation (14) (described in Section 2.6.5) provides a formula for estimating the required Petzval radius for a given system. When the Petzval radius is too short, the field has to be flattened by introducing positive astigmatism, which will cause an elliptically shaped writing spot. The Petzval radius is a fundamental consideration in laser scan lenses and becomes a major factor that must be reckoned with in systems requiring small spot sizes. Small spots require a large numerical aperture or a small F/NO [2]. Observations to be made from this layout include:

The distance from the entrance pupil to the lens is 23.24 or 28.77% of the focal length of the scan lens.

If the entrance pupil is moved out to F_1, the chief ray will emerge parallel to the optical axis and the system will be telecentric. This condition has several advantages, but the b lens must be larger than the scan length and the large amount of refraction in the b lens will introduce negative distortion, making it difficult to also meet the F-θ condition.

Reducing the power of the negative lens or decreasing the spacing between the lenses will allow for a longer distance between the lens and the entrance pupil; but this will introduce more inward-curving Petzval curvature.

This brief discussion illustrates some of the considerations involved in establishing an initial layout of lenses for a scanner. One must decide, on the basis of the required spot diameter and the length of scan, what the Petzval radius has to be in order to achieve a uniform spot size across the scan length. When field flattening is required, it is necessary to introduce more negative power in the system. The most effective way to do this is to insert a negative lens at the first, second, or both focal points of a positive-focal-length scan lens (Figure 2.9a). In these positions they do not detract from the focal length of the positive lens, so the Petzval curvature can be made to be near zero when the negative lens has approximately the same power as the positive lens. The negative lens at the second focal point, however, must have a diameter equal to the scan length, and it will introduce positive distortion if it is displaced from the focal plane. This distortion will make if difficult to meet the F-θ condition. A negative lens located at the first focal point of the lens is impractical, since there would be no distance between the lens and the position of the scanning element.

The next best thing to do is to place a single negative lens between the positive lens and the image plane. When the negative lens and the positive lens have equal but opposite focal lengths and the spaces between the lenses are half the focal length of the original single lens, then the focal lengths of two lenses are $+0.707F$ and $-0.707F$. This condition is shown in Figure 2.9b,c. The first and second focal points of the two systems are indicated in the drawings. The system with the positive lens in front is a telephoto lens, and the one with a negative lens in front is an inverted telephoto. The telephoto lens has a long working distance from the first focal point to the lens, while the inverted telephoto has a long distance from the rear lens to the image plane. The question now is, ''which is the better form to use for a scan lens?''

It is well known that a telephoto lens has positive distortion, while the inverted telephoto lens has negative distortion. Scan lenses which have to be designed to follow the F-θ law must have negative distortion. This suggests that the preferred solution is the type shown in Figure 2.9c, even though it makes a much longer system from the last lens to the focal plane and the entrance pupil distance is considerably shorter. Most of the lenses in use do employ a negative lens in front and are basically inverted telephoto lenses.

Often the clearance required for the scanning element causes aberration correction problems. A telecentric design provides more clearance. Strict telecentricity may introduce too much negative distortion because the positive lens has

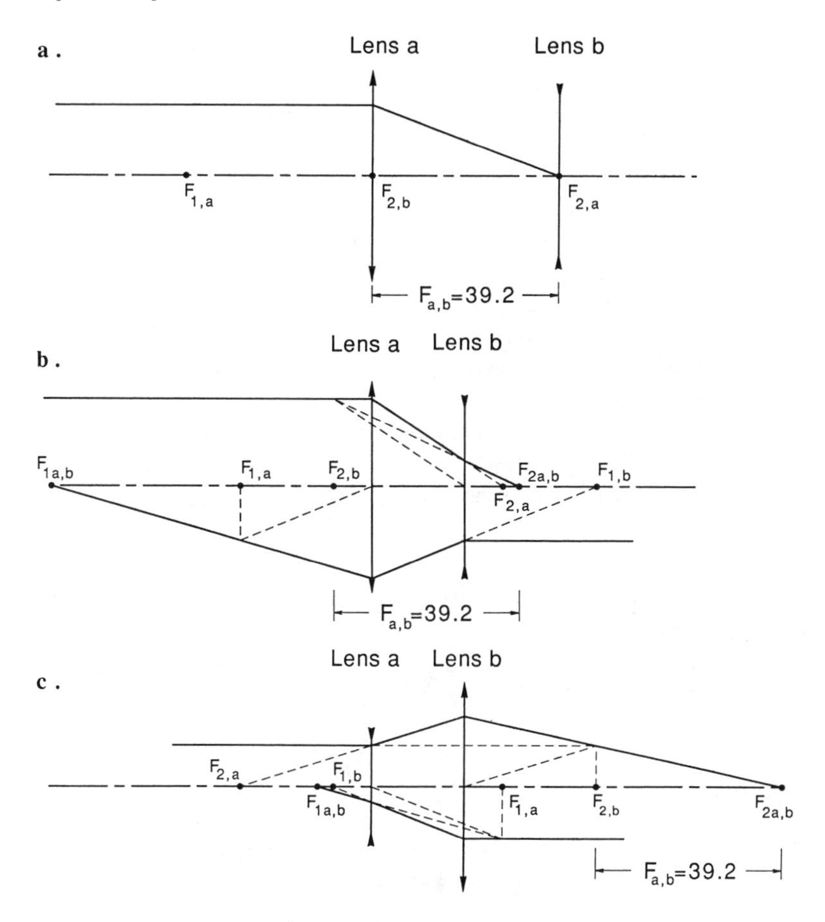

Figure 2.9 Three two-lens solutions for zero field curvature.

to bend the chief ray through a large angle. When there is a tight tolerance on the F-θ condition it is better to move the aperture stop closer to the first lens. It has been our experience that it is difficult to achieve a ratio of the overall length of the system, from the scan element to the image plane, to the lens focal length less than 1.6. The characteristics of several scan lenses are described in Ref. 5; few have a smaller ratio. In cases where the distance from scanning element to the first lens surface has to be longer than the focal length, it is advantageous to use the telephoto configuration. However, it will be difficult to make the lens meet the F-θ condition. Systems like this have been used for galvanometer scanning. It is particularly useful for XY scanning systems where more space is needed between the aperture stop and the lens.

The lenses used in the above example are extreme lenses to illustrate the two cases. In most designs the Petzval radius is not set to infinity. A Petzval radius of 10 to 50 times the focal length is usually all that is needed. The two lenses are also usually made of different glass in order to follow the Petzval rule: to increase the Petzval radius, the negative lenses should work at low aperture and have a low index of refraction and the positive lenses should work at high aperture and have a high index of refraction. It has been pointed out [9] that if the incoming beam is slightly diverging, instead of being collimated, it increases the radius of the Petzval surface. The diverging beam in effect adds positive field curvature. The idea has occasionally been used in systems, but the focus of the collimator lens has to be set at the correct divergence. It is not as convenient to set as a strictly collimated beam.

Some of the modern lenses which are required to image small spot diameters (2 to 4 μm) [7.9 to 15.8 microinches] use negative lenses in front of and behind the positive lens in order to correct the Petzval curvature. Examples are shown in Section 2.7.

2.5 SPECIAL OPTICAL DESIGN REQUIREMENTS FOR LASER SCANNERS

This section discusses specific design requirements for different types of laser scanners.

2.5.1 Galvanometer Scanners

Galvanometer scanners are used extensively in laser scanning systems (Chapter 10). From an optical point of view they have advantages:

1. The scanning mirror can rotate about an axis in the plane of the mirror. The mirror can then be located at the entrance pupil of the lens system and its position does not move as the mirror rotates.
2. The F-θ condition is often not required, for the shaft angular velocity of the mirror can be controlled electronically to provide uniform spot velocity.
3. The galvanometer systems are suitable for XY scanning.

The principal disadvantage is that they are limited in writing velocity.

The galvanometer scanners provide the easiest way to design an XY scanner. The two mirrors, however, have to be separated from each other, and this means the optical system has to work with two separated entrance pupils, with considerable distance between them. This in effect requires that the lens system be aberration corrected for a much larger aperture than the laser beam diameter. A system demanding large aperture and field then will have different degrees of distortion correction for the two directions of scan. In principle the distortions

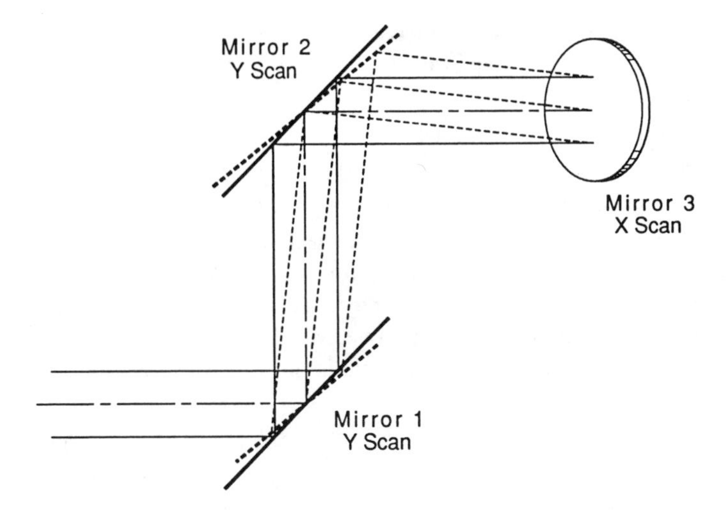

Figure 2.10 Three galvanometer mirrors for XY scanning.

can be corrected electronically, but this adds considerable complexity to the equipment.

It is often suggested that a telescope relay system may be placed between the two mirrors (Chapter 10). This, however, adds more field curvature to the scanning system, which is the basic aberration limiting optical scanners. Any system requiring a large number of image points should avoid extra relay lenses, which add Petzval field curvature.

It is possible, with the use of three galvanometers, to do XY scanning and have a single entrance pupil. One system of mirrors is illustrated in Figure 2.10 [10]. This system adds complexity, since the two mirrors used to make the Y scan have to turn by different angles and have to be synchronized to run together. Mirror 2 also has to be larger than the other two mirrors. This system has probably never been put into practice in a commercial scanner.

For precision scanning, the galvanometer mirror may have some wobble, which can be corrected with cylindrical optics, as described in the section on polygon scanning.

2.5.2 Polygon Scanning

Some precision scanners require severe uniformity of scanning velocity, sometimes as low as 0.1%, with addressability of a few microns. These requirements and high-speed scanning velocities force systems into high-speed rotating elements which scan at high uniform velocity. Polygon and holographic scanners have been most commonly used in these applications. A schematic of a polygon scanning system is illustrated in Chapter 6.

Special Design Requirements for Polygons

The optical effects which must be considered in the design of lens systems for polygons are bow, beam displacement, and cross scan errors.

Bow. The incoming and exiting beams must be located in a single plane which is perpendicular to the polygon rotation axis. Error in achieving this condition will displace the spot in the cross-scan direction by an amount which varies with the field angle. This results in a curved scan line, which is said to have *Bow*. The spot displacement as a function of field angle is given by the equation

$$E = F \sin \alpha \left[\frac{1}{\cos \theta} - 1 \right] \tag{11}$$

where F is the focal length of the lens, θ is the field angle, and α is the angle between the incoming beam and the plane which is perpendicular to the rotation axis.

The optical axis of the focusing lens should be coincident with the center of the input laser beam, hereafter referred to as the *feed beam*. Any error will introduce bow. The bow introduced by the input beam not being in the plane perpendicular to the rotation can, however, be compensated for, to some extent, by tilting the lens axis.

Some system designers have suggested using an array of laser diodes to simultaneously print multiple rasters. Only one of the diodes can be exactly on the central axis, so all the other diode beams will enter and exit the scanner out of the plane normal to the rotation axis, so bow will be introduced. The amount will increase for the diodes farther away from the central beam.

Beam Displacement. A second peculiarity of the polygon is that the facet rotation occurs around the polygon center rather than the facet face. This causes a facet displacement and a displacement of the collimated beam as the polygon rotates. (See Figure 2.11). This displacement of the incoming beam means that the lens must be well corrected over a larger aperture than the laser beam diameter.

Cross-Scan Errors. The polygons usually have pyramidal errors in the facets as well as some axis wobble. These cause cross-scan errors in the scan line. These errors must be small, for the eye is extremely sensitive in spotting them. Some printing systems are now seeking 1000-dots/in (39.4-dots/mm) performance. This means that the maximum cross-scan error in this case should be much less than 25 μm [100 microinches]. A system with no cross-scan error correction using a 700mm [27.6 inches] focal length lens, would require pyramid errors no larger than 2 arc seconds [0.167 mrad].

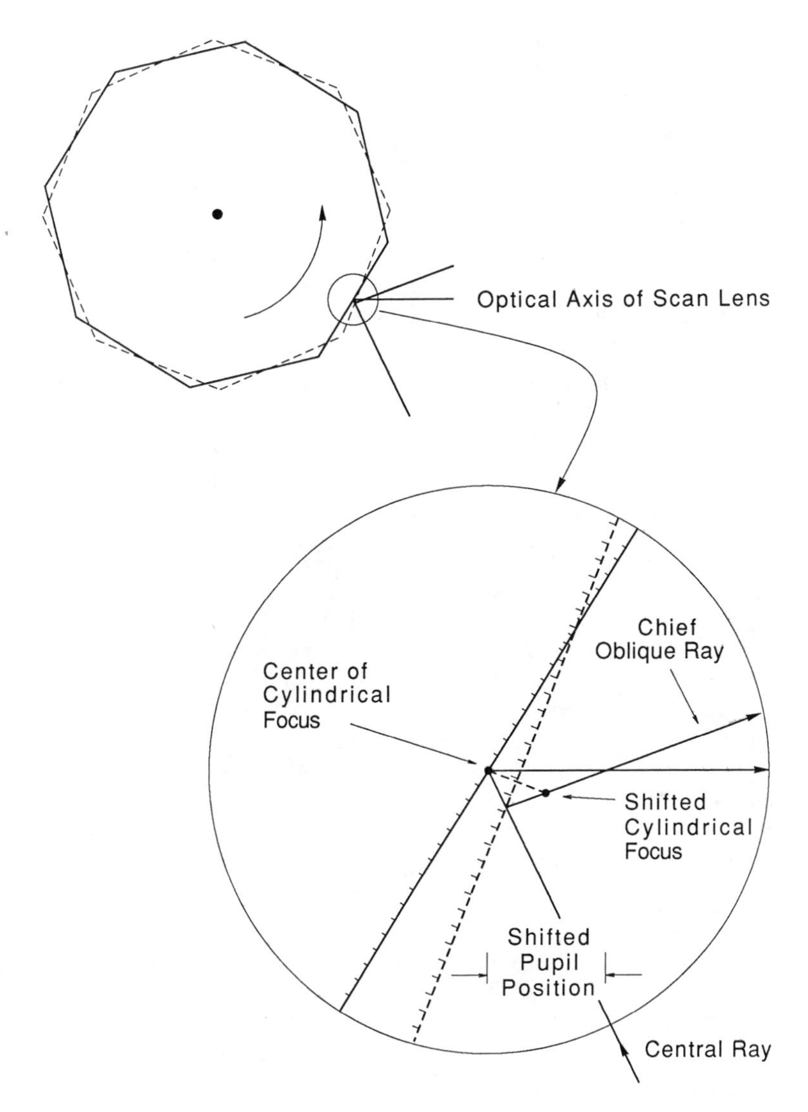

Figure 2.11 Facet rotation around the polygon center causes a translation of the facet, resulting in a beam displacement.

The Correction of Cross-Scan Errors

Three ways to correct for the cross-scan error in polygons are

1. The use of cylindrical lenses
2. Anamorphic beam on the facet
3. The retroreflective prism

Cylindrical Lenses for Wobble Correction. The diagram in Figure 2.12 shows how the cylindrical lenses reduce the wobble in the facet of a polygon. The top figure is the YZ plane, which shows the length of the scan line.

The lower section shows the XZ plane. In this plane the laser beam is focused on the facet of the mirror by a cylindrical lens in the collimator beam. It then diverges as it enters the focusing lens. The focusing lens cannot focus the beam to the image plane because it is a lens with rotational symmetry. In order to form a round image in the scan plane, the beam in the XZ plane must focus with the same numerical aperture as in the YZ plane. To do this a cylindrical lens must be placed as shown in the lower diagram. The placement of the cylindrical lens has to be in the position shown so that after the refraction the ray can reach the

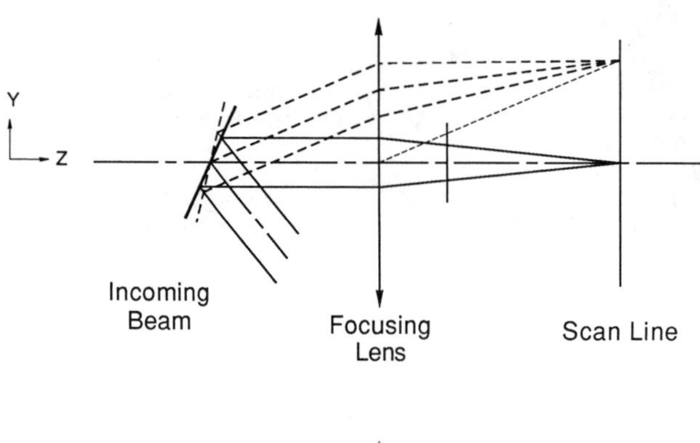

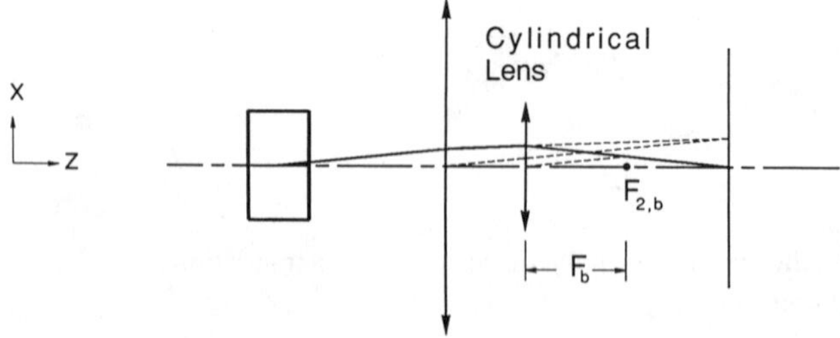

Figure 2.12 The use of a cylindrical lens to focus a line on the facet can reduce the cross scan error caused by facet wobble.

axis, at the image plane, at the same angle as the focusing lens in the top picture. The position of the cylindrical lens depends on the distance from the facet to the all-spherical focusing lens, and the numerical aperture of the cylindrical lens that focuses a line image on the facet. From the diagram it is clear that as the numerical aperture of the prefacet focusing lens is reduced, the cylindrical lens after the focusing lens moves towards the image plane.

If the numerical aperture of the feed beam, in the XZ plane, is made larger than shown in Figure 2.12, the diverging cone of light, after the reflection from the facet, may expand to the full diameter of the beam in the YZ plane before it reaches the all-spherical focusing lens. In this case the cylindrical lens can be placed in front of the all-spherical focus lens. Its focal length may be selected to collimate the XZ fan and restore a round beam. The all-spherical focusing lens can then focus the beam to a round spot on the scan plane.

When the facet rotates to direct the light to the edge of the scan, the distance to the spherical lens increases. In the XZ plane the optical system is focusing the beam from a finite object distance. When the facet rotates to direct the light to the edge of the scan, the object distance increases so there is a conjugate change. As the scan spot moves from the center of scan, the object distance in the XZ plane increases The image conjugate distance is therefore shortened. The consequence is that astigmatism is introduced in the final image. The sagittal focal surface is made inward curving. To compensate for this the all-spherical focusing lens must be able to introduce enough positive astigmatism to eliminate the total astigmatism.

It has been shown [11] that this induced astigmatism may be reduced by placing a toroidal lens between the facet and the all-spherical lens. In the YZ plane, the toroidal surface should have its center located on the facet. In the XZ plane the curve should be adjusted to collimate the light. This solution, however, imposes severe procurement and alignment problems. It is necessary to have a sufficiently large production order to pay for the cost of special tooling.

An Anamorphic Collimated Beam for Reduction of Facet Wobble. Figure 2.13 shows a system for wobble correction. In the YZ plane the diameter of the collimated feed beam is D_1, while in the XZ plane the beam diameter is D_2. As in the previous case using cylindrical lenses, it is necessary to have a round beam converging on the image plane. This can be done by using a negative cylindrical lens close to the spherical focusing lens and then a second positive cylinder to restore the circular beam. The diagrams show the two focal lengths of the system as F_{yz} and F_{xz}. They are in the ratio of the diameters of the incoming collimated anamorphic beam. Since the cross scan error is proportional to the focal length, the effect of the wobble is reduced in proportion to the ratio D_2/D_1. The system does introduce some congugate shift astigmatism, but it eliminates the bowtie error, since collimated light is incident on the facet. The bow tie effect is caused

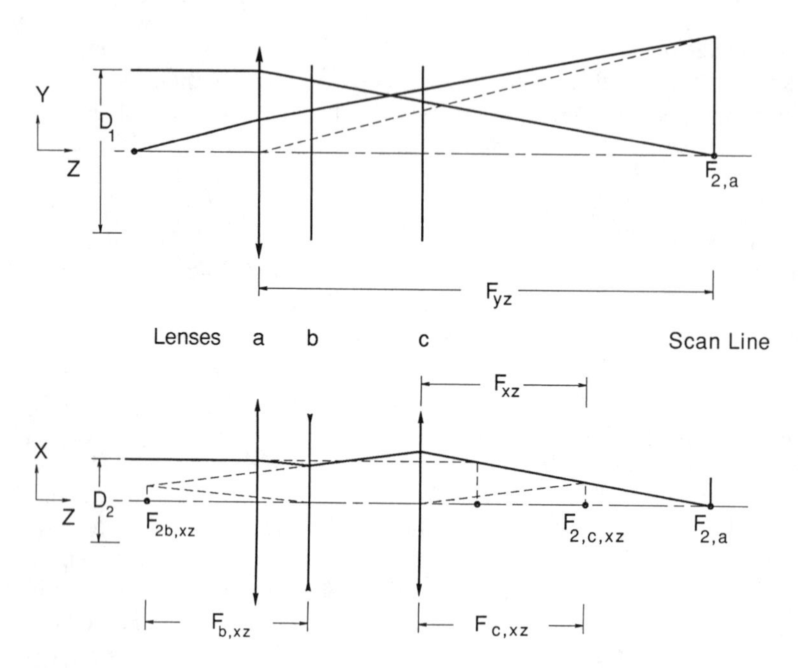

Figure 2.13 An anamorphic beam incident on the facet will also reduce cross-scan error.

when the beam is focused on the facet face. The beam however is not incident perpendicular to the facet face so one end of the line focuses in back of the facet face and the other end focuses in front. The result is the two ends of the line image are out of focus while the center is in focus. This makes it appear similar to a bowtie.

The cylindrical lens powers are reduced by placing the negative cylinder close to the all-spherical focusing lens and placing the positive cylinder as close to the image plane as practical. The position of the positive cylindrical lens, however, must consider such things as bubbles or defects on the surfaces of the lens. The beam size is extremely small when the lens is placed close to the focal plane and the entire beam can be blocked with a dust particle.

The Use of a Retroreflective Prism to Correct for Facet Wobble. Chapter 8 includes a description of a system using a retroreflective prism which reflects the scanning beam back on to the facet face before it passes to the focusing lens. Figure 8.14 shows a schematic of the system. The drawing in Figure 2.16 shows a plan view of a facet face in two positions of rotation. It also includes an elevation view, drawn to show that the retroreflective prism has a 90° [$\pi/2$ radians] roof

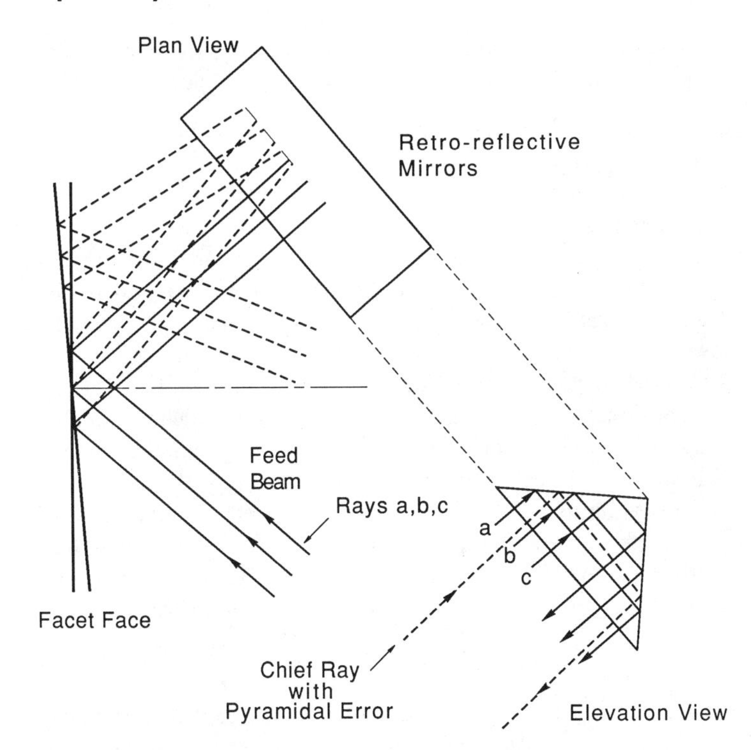

Figure 2.14 A retroreflective prism can be used to reduce cross-scan error due to facet wobble.

edge. The drawing was made using two mirrors instead of a prism. This was done to simplify the tracing of a chief ray when there is pyramidal error in the polygon. The incoming beam comes in on a top level in the plan view. It is reflected by the facet face, enters the prism, and reflects from the top reflecting surface to the bottom reflecting surface of the roof edge and back to the facet face exactly parallel to the feed beam. It finally reflects off the facet face and returns parallel, but below the top beam, before it hits the facet face. This return beam then passes through the scan lens. The elevation view of the roof edge also shows a ray at a small angle with respect to the center of the incoming beam. This represents a ray which would be deviated by pyramidal error in the polygon. This ray passes through the prism, and the beam returns parallel to the ray deviated by the facet face. It will then reflect from the facet and return parallel to the feed beam. The cross-scan error from facet pyramidal error or axis wobble is canceled out. The diagram also shows a beam passing through the prism when the facet has turned through an angle. It shows that the facet face has to be more than twice the aperture required to reflect the beam in a single reflection system. The facet has to be large in order to keep the retroreflective beam on the facet.

Summary. Axis wobble and pyramidal error cause serious problems by introducing cross-scan errors. There are ways to reduce the cross-scan errors, but many other problems are introduced. The use of cylinders results in procurement and alignment problems. The conjugate shift problem is difficult to visualize because the entire line image on the facet is not in focus and the analysis becomes complex. The only way to determine accurately the combination of all the effects—the pyramidal error in the facets, the translation of the facet during its rotation, the bowtie effect, and the conjugate shift—is to ray-trace the system and simulate the precise locations of the facet as it turns through the scanning positions. This can be done using the *multiconfiguration mode*, which is available in most of the large optical design programs. The multiconfiguration technique of design is discussed in more detail in Section 2.6.1.

There are, however, limits to what can be accomplished. For the highest-quality requirements the polygon pyramidal errors and axis wobble should be reduced sufficiently to eliminate the need for the use of cylindrical elements; they are difficult to make, mount, and align, and they are expensive.

Polygon Efficiency of Scan

Figure 2.15 shows one facet of a polygon for scanning. The facet face AB subtends an angle of 2θ at the center C of the polygon. The collimated feed beam is at the angle I_0 with respect to the facet normal. When the polygon rotates through an angle of $\theta/2$, the reflected beam will be swept through the angle θ, which is half the total scan angle.* The dotted facet shows this case. The incoming beam diameter is then restricted to D if there is to be no vignetting over the entire scan. From the geometry it can be shown that the beam diameter D is related to the inscribed radius of the polygon R_P and the feed angle I_0 by the equation

$$D = \frac{2R_P}{\cos\theta}\left[\cos I_0 \cdot \sin\left(\frac{\theta}{2}\right) - \sin I_0 \cdot \left(\cos\left(\frac{\theta}{2}\right) - \cos(\theta)\right)\right] \tag{12}$$

An approximate expression is

$$D = \frac{2R_P}{\cos\theta} \cdot \cos I_0 \cdot \sin\left(\frac{\theta}{2}\right) \tag{13}$$

---

* It is necessary to be careful about using the phrase "scan angle." Most optics people use "half scan angle," which is the angle subtended from the center of scan to one edge of scan. Optics people do this because most systems have rotational symmetry about the optical axis and they are concerned with correcting the aberrations on one side of the scan, knowing that the other side will also be corrected.

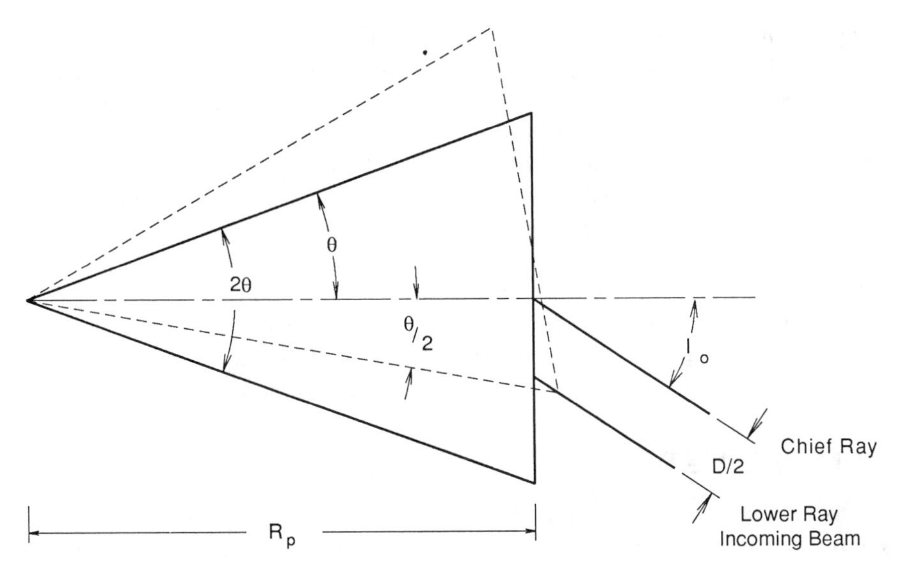

Figure 2.15 A diagram of one facet of a polygon in its central position and in a rotated position showing the maximum beam diameter that can be reflected with no vignetting.

A sample polygon and lens design might have the specifications:

2000 dots/in [12.8 micrometers] for the $1/e^2$ spot diameter.

The *F*/NO. of the lens for this spot diameter is 11.05, given by equation (4).

Scan length 18 in [457.8 mm.]

Assume I_0 to be 30°.

In order to use a small beam diameter the half scan angle should be as large as possible; see equation (9). Twenty-five degrees is about the practical limit for this *F*/NO. A seven-sided polygon requires a facet angle of 51.4° [.9 radians] so the half scan angle (θ) is 25.7° [0.45 radians]. The system then has the properties: $F = 510.4$ mm, $D = 46.2$ mm, and $R_P = 148.4$ mm. These requirements place severe demands on the optical system and the polygon. It is doubtful that a normal compact lens can be designed. It appears that the only feasible way is to use a long-field flattening lens close to the focal plane. The problem will be discussed in the last section on limiting optical designs.

The facet face shown in Figure 2.15 has to be 2.9 times the diameter of the incoming beam. This shows that a small polygon requires a small feed beam diameter. To scan a given number of image points on a single scan line, there is a trade-off between the scan angle θ and the diameter of the feed beam (see equation (9)). The search for system compactness drives the field angle to larger and larger values. Scan angles above 20° [0.349 radians] increase the difficulty in correcting the *F*/NO. The conflict can be somewhat resolved by using a smaller angle of incidence I_0, but then there may be interference between the incident beam and the lens mount. A wise compromise between these variables requires close cooperation between the optical and mechanical engineering effort.

The polygon works at 50% efficiency. This is because the beam is deflected by twice the rotation angle of the facet. The rotating polygon is scanning the recording material only half of the time. In order to transfer information faster, it is necessary to rotate the polygon faster. This increases the velocity of the scan and halves the exposure on the recording material. To make up for this, the irradiance of the laser beam must be doubled.

A pyramidal polygon rotates and reflects the scan beam at the same angular rotation. However, they are much more expensive to make, so they are used primarily in high-precision scanners which can absorb the high cost.

2.5.3 Double-Pass Scan Lens

In using double-pass optical systems with polygon scanners, the feed beam comes in from an off-axis point just beyond the end of the scan line. It passes through the scan lens, is collimated, reflects from the facet face, and then is focused by the scan lens. Their main advantages are that they reduce the number of lens elements, they can be fit into a compact package, and the lenses are less sensitive to thickness and index of refraction errors. They have no real advantage from an optical design point of view. The lens must be well corrected beyond the scan field so that the beam will be well collimated at the facet face. If it is not, facet translation will introduce astigmatism.

It is necessary to consider the reflected light from the incoming beam passing through the lens. Figure 2.16 shows an example of a serious reflected light problem. Some of the lens surfaces may also reflect the light directly back to the scan plane line. One way to reduce the chance of ghost images returning to the scan line is to have the incoming beam out of the plane of the polygon normal. However, this may cause intolerable bow in the scan line, unless compensated for with a prism.

2.5.4 Holographic Scanning Systems

Holographic scanning systems are discussed in Chapters 4 and 5. From an optical designer's point of view, they have an advantage in that the need for wobble correction can be reduced without resorting to cylindrical components. On the other hand, they usually require some bow correction. This can be done by using a prism component between the holographic element and the lens. The prism introduces bow to balance out the bow in the same way that a spectrographic prism adds curvature to the spectral lines. The lens can be tilted and decentered as an alternative method for reducing the bow. Except for this difference, the

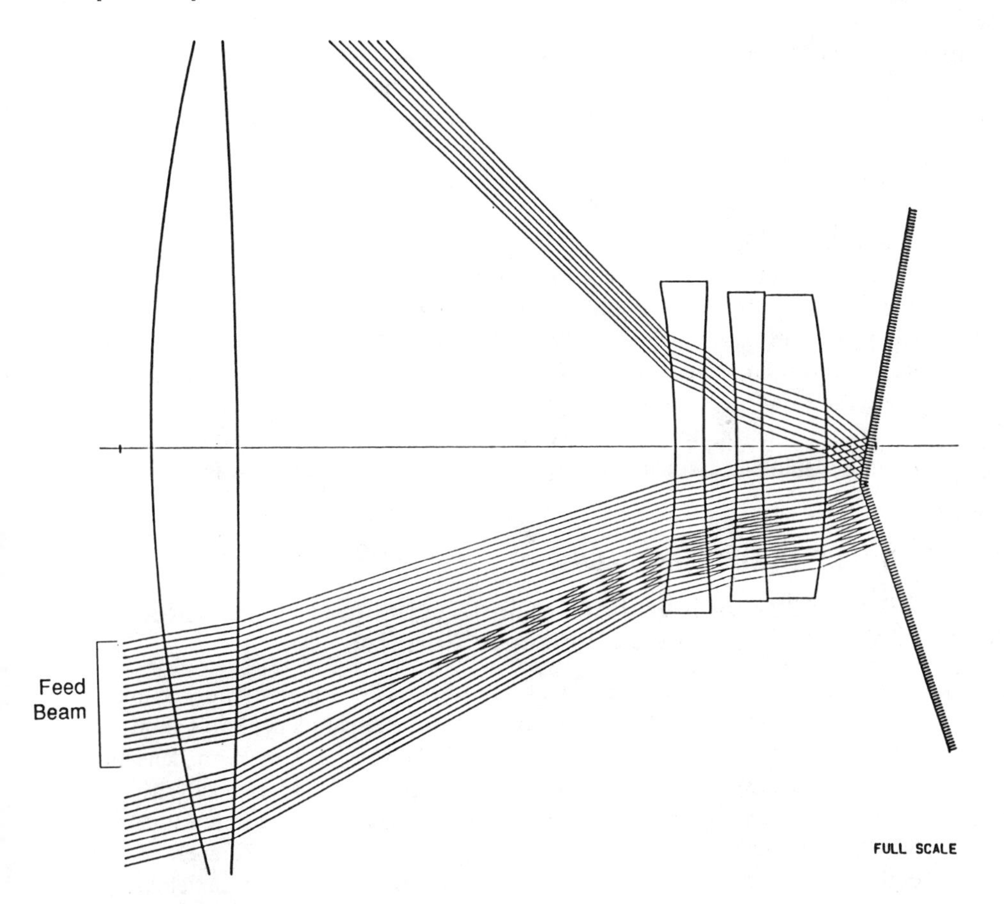

Figure 2.16 A multi-configuration setup of double-pass scan lens showing how the feed beam is split into unwanted beams.

design of the lenses for some holographic scanners follows the same principles as for polygon or galvanometer systems. Like the pyramidal polygon, some holographic scanners rotate at the same speed as the scan beam resulting in a high angular duty cycle.

2.5.5 Internal Drum Systems

As stated before, the internal drum systems are least demanding on the optical systems, for the lenses do not have to cover a wide field. Most of the burden is shifted to the accurate mechanical alignment of the turning mirror.

The concept of the internal drum scanner can be applied to a flat-bed scanner by using a flat-field lens. This then becomes the equivalent of a pyramidal scanner with only one facet. All of these systems have common alignment requirements.

The nominal position of the turning mirror is usually set at 45° [0.785 radians] with respect to the axis of rotation. It does not have to be exactly 45° [0.785 radians] as long as the collimated feed optical beam enters parallel to the axis of the rotation. The latter condition is needed to eliminate bow in the scan line. In a perfectly aligned system the ray that passes through the nodal points of the lens must meet the deflecting mirror on the axis of rotation. When the lens is placed in front of the turning mirror, the second nodal point of the lens must be on the rotation axis of the mirror. When the lens is placed after the turning mirror, its first nodal point must be on the ray that intersects the mirror on the rotation axis. There is some advantage in placing the lens between the mirror and the recording plane. In this position the lens has a shorter focal length, and the bow resulting from any error in the nodal point placement of the lens is reduced.

2.6 LENS DESIGN TALK

2.6.1 Multiconfiguration Design Techniques Using Tilted and Decentered Reflective Surfaces

Early in the design of a scan lens it must be decided how rigorously the geometry of the moving deflector needs to be modeled. A simple model is often all that is needed, "simple" meaning that the actual method by which the beam deflection is introduced need not be included in the lens design model. This typically involves optimizing the lens to perform at several field angles, which is a different approach than taken for a standard photographic objective only in that the aperture stop of the system is defined external to the lens and the chief ray for each field passes through the center of this stop. The examples in Sections 2.7.1 through 2.7.7 demonstrate simple models. In a simple model, the beam deflection method is assumed to introduce only angular motion. It neglects any beam displacement that may be occurring due to the deflection method.

In practice, the reason that all parallel bundles appear to pivot about the center of this external aperture-stop surface is that there is some type of beam deflector rotating nearby that a fixed beam is striking. If the mechanical rotation axis of this deflector intersects the plane of the mirror facet and the optical axis of the scan lens, then the simple model accurately represents what is really occurring. This is nearly true for galvanometer-based systems: the mechanical rotation axis is so close to the plane of the mirror facet that in most situations the difference is negligible.

Multifaceted holographic deflectors fortunately do not emulate all aspects of a polygon; the hologon does not cause beam displacement, even though its mechanical rotation axis is some distance from its facets. Polygons do exhibit this behavior. The simple model suffices for the design of a lens for a hologon-based scan system, though complex truncation and multifacet illumination effects are ignored.

Tilting to Scan

There is often no substitute for the introduction of a mirror surface in the lens design model. This single optical element is tilted from one configuration (or "zoom position") to another to generate the angular scanning of the beam prior to the lens. Modeled in this way, there is no field angle in the usual sense. The beam prior to the rotating mirror is stationary. This avoids a modeling inaccuracy that is largely due to the photographic emphasis historically exhibited by commercial lens design software.

When a beam having a circular cross section reflects off a plane mirror, the reflected beam has the same cross section. This is not true with the simplistic model involving field angle and a fixed, external aperture stop. It is this aperture that is constant in size in the simplistic model, and, if circular, bundles from off-axis fields will be elliptical and compressed in the scan direction by the cosine of the field angle. As the field angle approaches 30° this is especially significant. At some stages of the design this foreshortening can be eliminated by tricking the software into enlarging the bundle in the scan direction to compensate. It usually happens, however, that not all analyses may be run using this work-around. Diffraction calculations relying on the tracing of a grid of rays prior to a Fourier transform are especially notorious for being unforgiving of these types of simplified models.

A constant beam cross section at all scan angles may also be maintained by simply bending the optical axis at the entrance pupil. This is often a good compromise in model complexity that does not require the use of reflective surfaces. Some holographic deflectors do not faithfully emulate a tilted mirror: beam cross section does change as a function of scan angle. Circular input gives rise to elliptical output, and the ellipse changes its orientation with scan angle.

Including a Polygon in the Model

A polygon in a lens design model is simply a mirror at the end of a relatively long arm. All rotation is performed about the end of the arm opposite to the mirror. The arm length, the location of the arm's rotation axis relative to the scan lens, and the amount of rotation about the pivot may all be optimized during the design. The facet shape may be specified by putting appropriate clear aperture specifications on the mirror surface at the end of the arm.

Because the mechanical rotation axis rarely intersects the optical axis of the scan lens, and the mirror facet is so far from the rotation axis, polygons behave in such a geometrically complex fashion that it is often best to just model them rigorously so that all their complicated mirror surface tilts, shifts, and aperture effects are automatically accounted for. Specific pupil shifts and aperture effects could be computed and specified for each configuration, but this does not exploit the full potential of the multiconfiguration setup. When set up in a general fashion, the actual constructional parameters of the polygon or parameters governing its interface with the feed beam and scan lens may be optimized simultaneously as the scan lens is being designed. This has often led to better system solutions than simply combining devices in some preconceived way.

If the location of the entrance pupil is defined as where the chief ray intersects the optical axis of the scan lens, then the pupil location changes as the polygon rotates. Equivalently, this makes it necessary for the lens to be well corrected over a larger aperture than the feed beam diameter. The effect is greater for polygons having few facets than for those having many. High-aperture (large NA) lenses are especially susceptible to this effect.

Determining where the real entrance pupil is for each configuration, or how much larger of a beam diameter the lens should really be designed to accommodate, is difficult. It is especially difficult if the polygon is to be designed at the same time as the scan lens! When this pupil shift is included in the lens design model by rigorously modeling the polygon geometry, then the effect on lens performance may be accurately assessed. More importantly, the lens being designed may be desensitized to expected pupil shifts and, if the polygon is being designed, the pupil shifts may be minimized. As maximum scan is approached, the facet size may be insufficient to reflect the entire beam. Asymmetrical truncation or vignetting occurs, which may modify the shape of spot at the image plane. Accurate aperture modeling is especially important for accurate diffraction-based spot profile calculations. In a rigorous polygon model, by putting aperture specifications on the surface that represents the reflective facet surface, all vignetting by the facet as a function of polygon rotation will be automatically accounted for.

By having a rigorous polygon model implemented, it is also possible to further evaluate what really happens to the section of the beam which misses the facet. Classifying the rays that miss as vignetted is really an oversimplification. In reality these rays will likely reflect off an adjacent facet. This stray light beam may enter the lens and find its way to the image surface. Stray light problems can ruin a system. Double-pass systems are especially susceptible to this design flaw. The multiconfiguration setup can be used to evaluate stray light problems and suggest baffle designs. Figure 2.16 is a multiconfiguration setup for a double-pass scan lens. The feed beam is the wide beam coming in through the lower part of the lens. Some of the beam is reflected from the top facet toward the top of the lens. The rest of the beam is reflected from the lower part facet and enters the lower part of the lens. This beam should be blocked from entering the lens. Both reflected beams should be blocked from exiting the lens.

Dual-Axis Scanning

When more than one galvanometer is used to generate a two-dimensional scan at the image plane it is usually necessary to use a multiconfiguration setup. Unless a third galvanometer is used to make it appear that both the X-scan and the Y-scan originate at the same location on the optical axis, as shown in Figure 2.10, then tilts must be used in conjunction with a multiconfiguration setup to accurately model the skew bundles through the lens. The optical axis may simply be tilted, or reflective surfaces representing the galvanometer mirrors may be tilted. The latter approach is usually worth the effort in order to visualize the problem and avoid mechanical interferences (Fig. 2.17).

Double-Pass Scanning

This class of problems tends to be the most difficult to model correctly. A multiconfiguration setup is essential regardless of the type of deflector that is to be used. A diverging beam must be fed into the scan lens from a position that does not interfere with the output beam. The feed beam is stationary, and the output beam is in motion due to the rotating deflector. The final determination of where the feed beam should be aimed, and from where, is best evaluated by the lens designer using a multiconfiguration setup.

The simplest double-pass schemes to model are those where the feed and output beams are all in the same plane: the feed beam is simply introduced so that it enters from beyond maximum scan. Keeping the feed beam spot-forming lenses and mirrors out of the way of the output beam is relatively straightforward. This is not, however, the optimum solution optically. The feed beam must typically pass through the scan lens at a high field angle. In doing so, coma and astigmatism are introduced. This aberrated beam is then presented to the scan lens for the return pass. The symmetry principle does not apply; it should not be expected that aberrations due to the return pass will compensate for aberrations caused by the first pass.

A better solution, but more difficult to model, involves feeding the system close to the scan-lens axis but not in the same plane as the output beam. The off-axis aberrations imparted upon the feed beam are thereby minimized. Feed beam components are more difficult to keep out of the path of the output beam, and the reflection angles off the deflector are now compound, which would be difficult to handle in the absence of a rigorous deflector model, especially since the feed geometry may change from one iteration to the next as the system is optimized.

Scan line bow will result at the image plane as the consequence of the compound reflection. This may be corrected by using a pyramidal deflector, in which case the feed and output will be on the same side of the scan-lens axis, or by using a prism between the scan lens and deflector through which the beam passes twice. Using the prism, the feed and output will be on opposite sides of the scan-lens

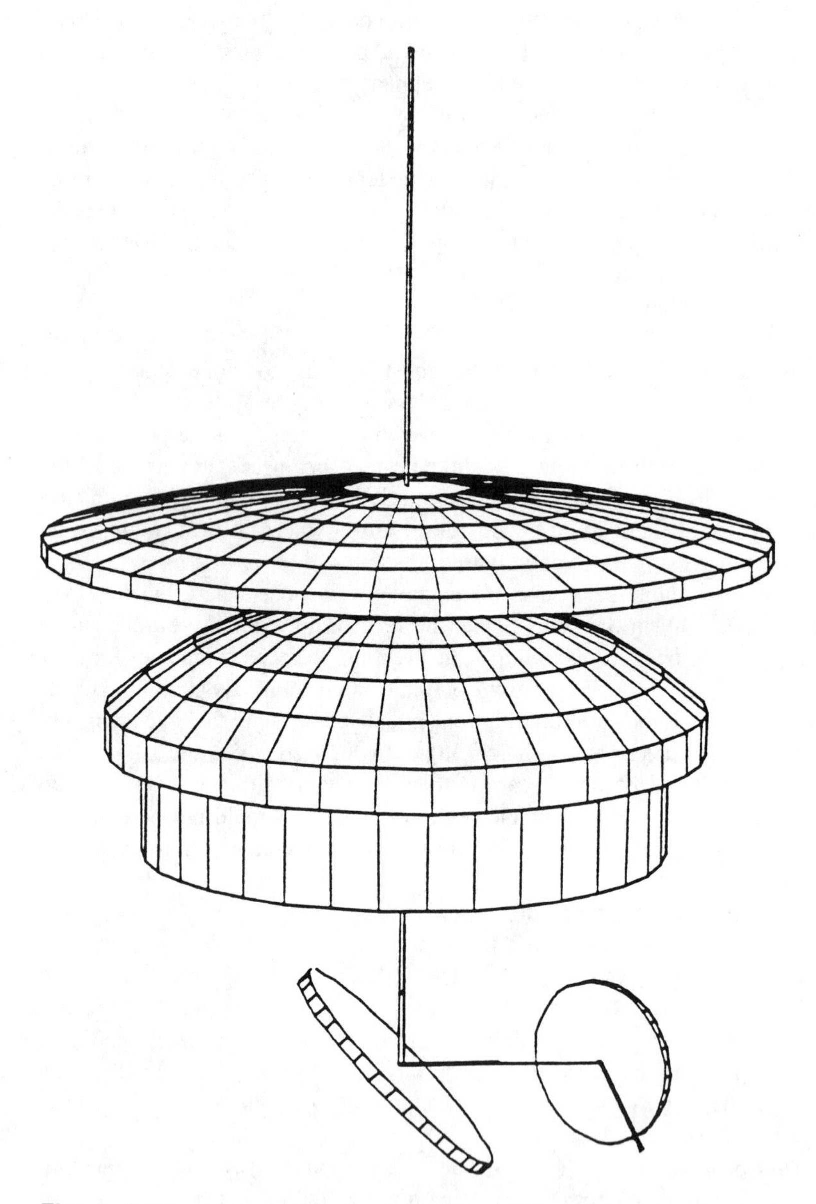

Figure 2.17 A three dimensional view of an XY scanning system.

axis. Both the deflector pyramid angle and the prism angle will be small, on the order of 10° [0.1745 radians] or less. The prism method has been the most cost-effective. Simply tilting the nonpyramidal deflector or the lens will not remove the bow.

Example Single-Pass Polygon Setup

The key to the full exploitation of the multiconfiguration design method is to include the rotation axis when modeling a polygon. Begin either with a raw laser beam or with an expanded laser beam, reflect off a fold mirror, define the polygon and reflect off it, and then continue through the scan lens.

The following CODE V* lens file for the lens illustrated in Figure 2.18, shows how to set up a polygon and lens in a commercial lens design program. Standard catalog components were chosen: the six-sided polygon is Lincoln Laser's PO-06-16-037, and the two-element sectioned scan lens is Melles Griot's LLS-090. The combination will create 300 dpi output using a laser diode source.

A. Start with a collimated, expanded laser beam of the required diameter and fold its path (with a −31.5° [0.550 radians] mirror tilt, here) to obtain the desired feed angle with respect to the planned optical axis of the scan lens (63° [1.10 radians], here). The aperture stop should be defined here, prior to the polygon, on surface 1. Any truncation of the gaussian input beam should be done at the aperture stop. Do not flag the polygon surface as the aperture stop, since some software will automatically ray-aim each bundle to pass through the center of the stop surface.

B. Define a reference point that will be on the optical axis of the scan lens. It is convenient to have its location where the facet would intersect the axis when the polygon is rotated for on-axis evaluation. The surface should be tilted (63° [1.10 radians], here) so that any subsequent thickness would be along the optical axis.

C. Go to the polygon rotation center and tilt so that any subsequent thickness would be radial from the polygon center toward the facet surface. To get to the polygon center from the reference point, use a combination of surface thickness and decenters; travel in right angles for simplicity. It is convenient to choose a tilt that will cause the polygon to be rotated into position for on-axis evaluation. Here, we must translate −16.9 mm [0.665 inches] away from the lens along its axis and then decenter up, along Y, with YDE = 10.4 mm [0.409 inches]. Tilting about X in the YZ-plane with ADE = −31.5° (63/2) [0.550 radians] points to the facet.

D. Now, before going to the polygon facet, any additional tilt about X (ADE)

* CODE V is a trademark of Optical Research Associates, Pasadena, CA.

```
RDM;LEN
TITLE ' LINCOLN PO-6-16-37 MELLES GRIOT LLS-090  90mm F/50 31.5-deg
P-468'
EPD   1.8145
PUX   1.00000
PUY   1.00000
PUI   0.135335
DIM M
WL  780.
CA
YAN 0.

SO      0. 0.1E20                    ! Surface 0      }
S       0. 50.8                      ! Surface 1      }
  STO                                                 }  A
S       0. -25.4 REFL                ! Surface 2      }
  XDE 0. ;   YDE 0. ;  BEN                            }
  ADE -31.5 ;   BDE 0. ;   CDE 0.                     }

S       0. 0.                        ! Surface 3      }
  XDE 0. ;   YDE 0.                                   }  B
  ADE 63. ;   BDE 0. ;   CDE 0.                       }

S       0. -16.892507                ! Surface 4      }
S       0. 0.                        ! Surface 5      }  C
  XDE 0. ;   YDE 10.351742                            }
  ADE -31.5 ;   BDE 0. ;   CDE 0.                     }

S       0. 19.812                    ! Surface 6      }
  XDE 0. ;   YDE 0.                                   }  D
  ADE -15.5 ;   BDE 0. ;   CDE 0.                     }
S       0. -19.812 REFL              ! Surface 7      }

S       0. 0.                        ! Surface 8      }
  XDE 0. ;   YDE 0. ;  REV                            }  E
  ADE -15.5 ;   BDE 0. ;   CDE 0.                     }

S       0. 16.892507                 ! Surface 9      }
  XDE 0. ;   YDE 10.351742 ;  REV                     }  F
  ADE -31.5 ;   BDE 0. ;   CDE 0.                     }

S       0. 7.                        ! Surface 10     }
S       -49.606 4.5 SK16_SCHOTT      ! Surface 11     }
S       0. 6.35                      ! Surface 12     }
S       0. 5.35 SFL6_SCHOTT          ! Surface 13     }  G
S       -38.633 104.3402458543       ! Surface 14     }
  PIM                                                 }
SI      0. -0.62535298                                }

ZOOM 7                                                }
ZOOM ADE S6 -15.5 -11. -7.5 0. 7.5 11. 15.5           }  H
ZOOM ADE S8 -15.5 -11. -7.5 0. 7.5 11. 15.5           }
GO

CIR S2 2.5                                            }
REX S7 4.7625                                         }
REY S7 11.43                                          }
REX S11 5.                                            }
REY S11 5.1                                           }
REX S12 5.                                            }
REY S12 7.6                                           }  I
REX S13 5.                                            }
REY S13 14.9                                          }
REX S14 5.                                            }
REY S14 15.8                                          }
CIR S2 EDG 2.5                                        }
REX S7 EDG 4.7625                                     }
REY S7 EDG 11.43                                      }
GO
```

is specified. This is a multiconfiguration parameter: each configuration will have a different value specified for this additional tilt. Here, ADE = − 15.5° [0.271 radians] is specified to cause the polygon to rotate into position for maximum scan on the negative side. This is really the polygon shaft rotation angle; for nonpyramidal polygons the reflection angle (scan angle) changes at twice the rate of the shaft angle. Once the image surface is defined, the system will be defined to have seven configurations, and a different ADE value for this surface will be specified for each (step ''H'').

Translate to the facet using the polygon inscribed radius for thickness, 19.8 mm here. Now reflect. This reflection occurs at the first real surface that the beam encounters since the fold mirror. All other surfaces have been ''dummy'' surfaces where no reflection or refraction takes place. It is on this reflective surface that aperture restrictions may be imposed that describe the shape of the facet. Use a thickness specification on this surface to go back to the polygon center following the reflection.

E. Some commercial software programs have a ''return'' surface that one could now use to get back to the reference point defined in step B, prior to defining the scan lens. Here, a more conservative approach is taken that can be used with any software package.

   Undo the additional polygon shaft rotation that was done in step D. The REV flag in CODE V internally negates the angle.

F. Continuing to move back to the reference point defined in step B, undo the polygon shaft tilt that sets it for on-axis evaluation (ADE = 31.5° [0.550 radians]), decenter down, along Y, back to the scan lens axis (YDE = − 10.4 mm [0.409 inches]), and translate toward the lens along its axis to the reference. This places the mechanical axis back to the same location that it was in step B, before the reflection off the polygon. Here, the REV flag changes the signs of the tilt and decenter specifications and performs the tilt before the decenter.

G. Define the scan lens. The reference surface defined in step B, and returned to in steps E and F, is approximately the location of the entrance pupil. The thickness at the image surface is the focus shift from the paraxial image plane.

H. Having now specified a valid single-configuration system to the software, the system is redefined to have seven configurations (ZOOM 7) and the parameters that change from one configuration to another are listed. Due to the way that the polygon was modeled, rotating it to a different position is simply a matter of changing the parameter that represents the shaft rotation angle. Here, this is ADE on surfaces 6 and 8.

I. Since these are catalog components, the clear apertures are available and are specified here. The rectangular aperture specifications for the polygon facet are given for surface 7.

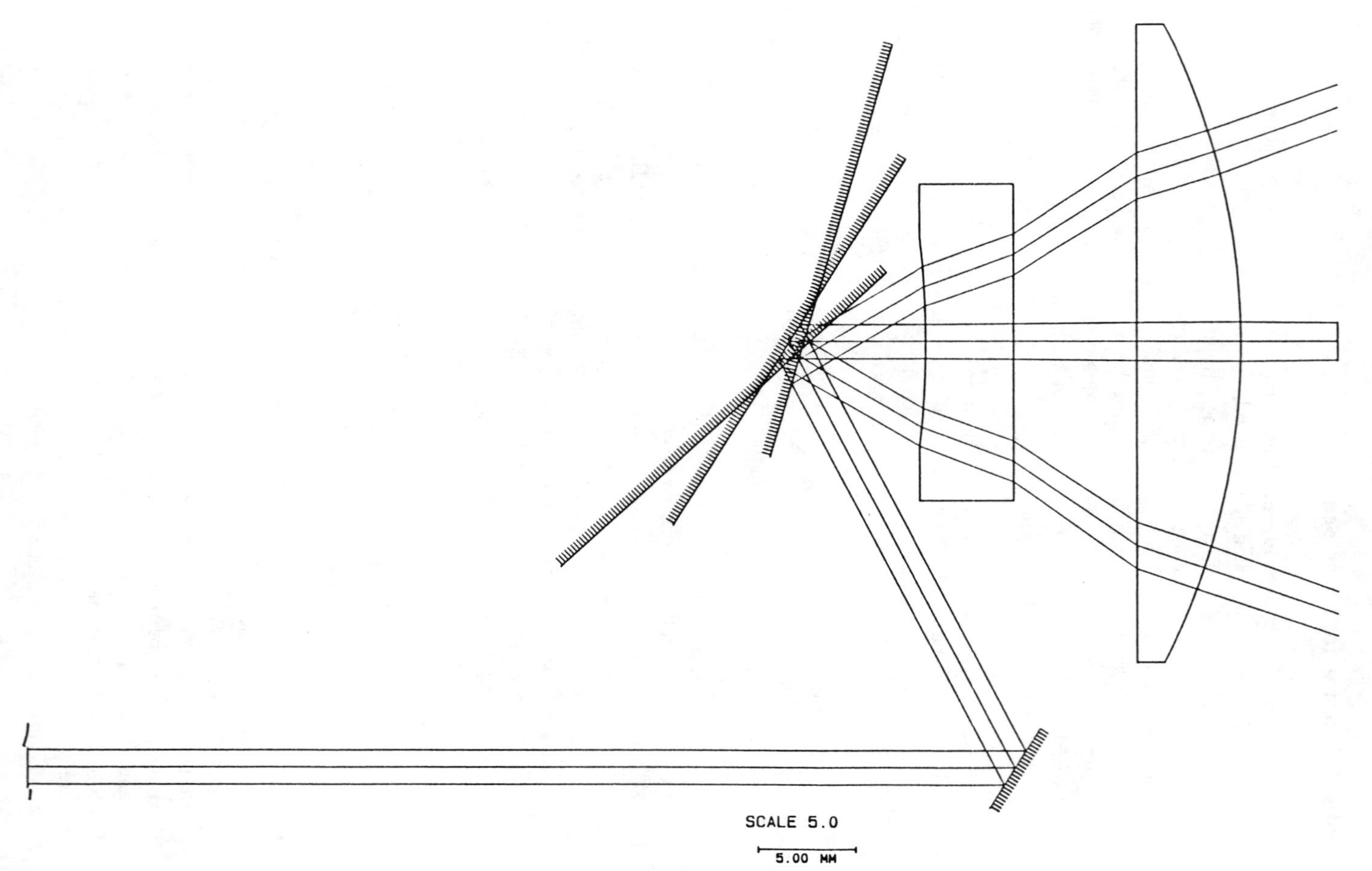

Figure 2.18 A multi-configuration, single-pass polygon system.

2.6.2 The Correction of First-Order Chromatic Aberrations

First-order aberrations not mentioned previously are the chromatic aberrations of axial and lateral color. The correction of these aberrations is usually difficult for scan lenses. Occasionally, two types of chromatic correction are called for. Some specifications call for simultaneous scanning with two wavelengths; no change of focus is permitted. These lenses have to be color-corrected for the two wavelengths; they have to be achromatic. This is difficult to do in a scan lens, again because the aperture stop is out in front of the lens.

The basic way to correct axial and lateral color is to make each element into an achromatic cemented doublet. To make a positive lens achromatic it is necessary to have a positive and negative lens with glass of different dispersions. The positive lens should have low-dispersion glass and the negative lens should have high-dispersion glass. The negative focal length lens reduces the positive power, so the positive lens power must be approximately double what it would be if not achromatized.

This halves the radii so the thickness must be increased in order to maintain the lens diameter. In scan lenses, the lens diameters are determined by the height of the chief ray, so the lenses are much larger in diameter than indicated by the axial beam. As the thicknesses are increased to reach the diameters needed, the angles of incidence on the cemented surfaces are increased, resulting in high-order chromatic aberrations. When the angles of incidence in an achromatic doublet become too large, the doublet has to be split up and made into two achromatic doublets. It is safe to say that asking for simultaneous chromatic correction will more than double the number of lenses.

Some specifications ask for good correction for a small band of wavelengths where small differences due to color can be corrected by refocus or by moving the elements. These systems do not need full color correction, and they can be designed to meet more demanding requirements.

The most demanding scan lenses are achievable because they are used with strictly monochromatic laser beams.

2.6.3 Third-Order Aberration

The understanding of the source of aberrations and their elimination comes from third-order theory. An adequate description of the theory is beyond the scope of this chapter. The following references should be consulted by anyone interested in the actual design of optical systems [7, 12–14]. The following discussion of scanning systems will involve mention of these aberrations, so in this section some of the rules of thumb concerning them will be offered in order to provide some guidelines for the design of scan lenses.

Third-order theory describes the lowest-order monochromatic aberrations in an optical system. Any real system will usually have higher-order aberrations, but third-order aberrations are important for there are thin lens and surface-by-surface equations which tell the designer the source of aberrations. The third-order aberrations are easy to calculate, and in most designs the third-order aberrations have to be close to zero. A brief description of these aberrations follows:

Spherical aberration
Coma
Astigmatism
Distortion

Spherical Aberration. This is an aberration that occurs in the center of the field of the lens; it occurs on the optical axis. It causes a rotationally symmetrical blurred image of a point object on the optical axis. It is the only aberration that occurs on the optical axis, but, if present, it will also appear at every object point in the field. But the other aberrations will be added to it, so the images will no longer be strictly rotationally symmetric.

Coma. This is the first asymmetrical aberration that appears for points close to the optical axis. Coma gets its name from the shape of the image of a point source: the image blur is in the form of a comet. The coma aberration blur varies linearly with the field angle and with the square of the aperture diameter.

Astigmatism. When this aberration is present, the meridional fan of rays (the rays shown in a cross-sectional view of the lens) focus at the *tangential focus* as a line which is perpendicular to the meridional plane. The sagittal rays (rays in a plane perpendicular to the meridional plane) come to a different line focus which is perpendicular to the tangential line image. This focus position is called the *sagittal focus*. Midway between the two focal positions, the image is a round *blur* circle with a diameter proportional to the numerical aperture of the lens and the distance between the focal lines. The third-order theory shows that the tangential focus position is three times as far from the Petzval surface as the sagittal focus. This is what makes the Petzval field curvature so important. If there is Petzval curvature, the image plane cannot be flat without some astigmatism. One can therefore predict approximately what the field curvature should be to meet the requirements. The astigmatism and the Petzval field sag both increase with the square of the field. They increase faster than the coma and become the most troublesome aberrations as the field (length of scan) is increased.

Distortion. In third-order theory distortion is a displacement of the chief ray from the paraxial image height predicted from the first-order calculations through the equation $\bar{Y} = F \tan(\theta)$. In an aberration-free design, the center of the energy concentration is on the chief ray. The third-order displacement of the chief ray from the paraxial image height varies with the cube of the image height. The percent distortion varies as the square of the image height.

Section 2.3.1 noted that the distortion has to be negative in order to meet the F-θ condition. Third-order distortion refers to the displacement of the chief ray. If the image has any order of coma, it is not rotationally symmetric. The position of the chief ray may not represent the best concentration of energy in the image; there may be a displacement. Here the specification for linearity of scan becomes sticky. If there is lack of symmetry in the image, then how does one define the error? If MTF is used as a criterion, this error is a phase shift in the tangential MTF. If one uses an encircled energy criterion, then what level of energy should be used? When a design curve of the departure from the F-θ condition is provided, it usually refers to the distortion of the chief ray. The designer must therefore attempt to reduce the coma to a level which is consistent with the specification of the F-θ condition.

2.6.4 Third-Order Rules of Thumb for Understanding Lens Design

Collective surfaces [8, p. 106] almost always introduce negative spherical aberration. A collective surface bends a ray above the optical axis in a clockwise direction (see Figure 2.19). There is a region where a collective surface introduces positive spherical aberration. This occurs when the axial ray is converging to a position between the center of curvature of the surface and its aplanatic point. When a converging ray is directed at the aplanatic point the angles of the incident and refracted rays, with respect to the optical axis, satisfy the equation $N \sin U = N' \sin U'$, and no spherical or coma aberrations are introduced. Unfortunately this condition is usually not accessible in a scan lens. Surfaces with positive spherical aberration are important because they are the only sources of positive astigmatism.

Dispersive surfaces always introduce positive spherical aberration. A dispersive surface bends rays above the axis in a counterclockwise direction. In order to correct spherical aberration it is necessary to have dispersive surfaces which can cancel out the undercorrection from the collective surfaces.

The coma can be either positive or negative, depending on the angle of incidence of the chief ray. This makes it appear that the coma should be relatively easy to correct, but in the case of scan lenses it is difficult to correct the coma to zero. The basic reason is that the aperture stop of the lens is located in front of the lens. This makes it more difficult to find surfaces that balance the positive and negative coma contributions.

As the field increases, the astigmatism dominates the correction problem. The astigmatism introduced by a surface always has the same sign as the spherical aberration. When the lenses are all on one side of the aperture stop, this makes it difficult to control astigmatism and coma. A lens with a positive focal length

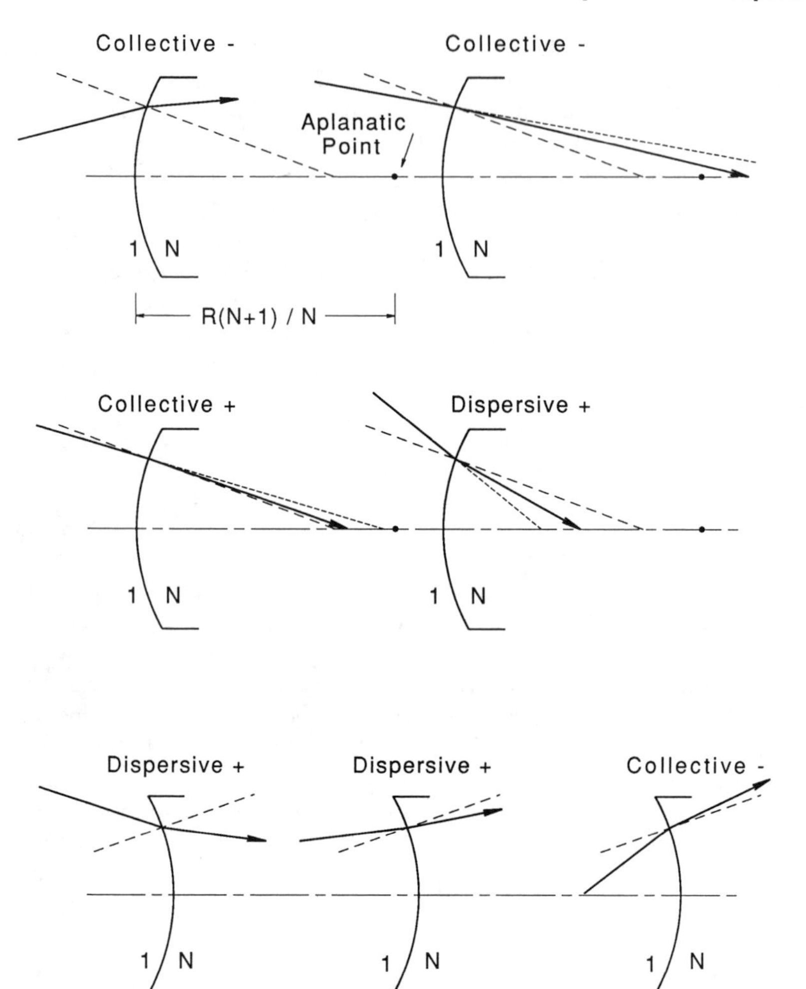

Figure 2.19 This diagram shows several surfaces that are collective and dispersive. The collective surfaces, with one exception, introduce negative spherical aberration. All the dispersive surfaces add positive spherical aberration.

usually has an inward-curving field so the astigmatism has to be positive. This is the reason that a designer must have surfaces that introduce positive spherical aberration. Since distortion is an aberration of the chief ray, surfaces which are collective to the chief ray will add negative distortion and dispersive surfaces will add positive distortion.

2.6.5 The Importance of the Petzval Radius

Even though the Petzval curvature is a first-order aberration, it is closely related to the third-order because of the 3:1 relation with the tangential and sagittal astigmatism. It is not possible to eliminate the relationship by merely setting up the lens powers so that the Petzval sum is zero. By doing this the lens curves become so strong that higher-order aberrations are introduced, causing further correction problems. For this reason it is important to set up the initial design configuration with a reasonable Petzval field radius, and the designer should continually note the ratio of the Petzval radius to the focal length. Based on third-order astigmatism, an estimate of the minimum Petzval radius for a flatbed scan lens can be derived. The following formula may be used as a guideline:

$$\frac{\text{Petzval radius}}{F} = \frac{L^2}{32}\lambda(F/\text{NO.})^2 F \tag{14}$$

where L = total length of scan line and F = the lens focal length.

The next section, which describes some typical scan lenses, lists this ratio as a guideline for each application. The ratio is only an approximation. Lenses operating at large field angles or small F/NO. values will have high-order aberrations not accounted for in the equation. Depending on the type of correction, the final designed ratio may be higher or lower than given by equation (14).

A bit more should be said about the Petzval rule, which was covered in Section 2.4. The rule is: The Petzval field curvature is reduced by the negative lenses working at low aperture and the positive lenses working at large aperture. The negative lenses should have a low index of refraction, and the positive lenses should have a high index of refraction.

In most monochromatic scan lenses, the negative lenses will have a lower index of refraction than the positive lenses. The positive lenses usually have an index of refraction above 1.7. The negative lenses usually have values around 1.5. Most of the modern lens design programs can now vary the index of refraction during the optimization. Occasionally in a three or more element lens, the optimized design violates this rule and one of the positive lenses turns out to have a lower index of refraction than the others. It is not known why this is, but it may mean the design has more than enough Petzval correction, so it lowers the index of one of the elements in order to correct some other aberration. It may mean, however, that one of the positive lenses is not needed.

2.7 A SELECTION OF LASER SCAN LENSES

Scan lenses with laser beams have been considered almost from the time of the development of the first laser. Laboratory models for printing data transmitted from satellites was underway in the late 1960s. Commercial applications began coming out in the early 1970s, and laser printers became popular in the early 1980s. The range of application is now steadily increasing. As far as the lens designs are concerned, it is a fairly new field and there is room for new design concepts.

The lenses presented in this section were selected to show what appears to be the trend in development. The spot diameters are getting smaller, the scan lengths longer, and the speed of scanning higher. Some of the lenses near the bottom of the list are beginning to exhaust our present design and manufacturing capabilities. The newest requirements are reaching practical limits on the size of the optics and the cost of the fabrication and mounting of the optics.

It appears that the future designs will have to incorporate mirrors and lenses with large diameters, and new methods for manufacturing segmented elements will be needed.

The lenses shown in this section start with some modest designs for the early scanners and progress to some of the latest designs. Two of the designs were obtained from patents. This does not mean that they are fully engineered designs. The rest of the designs are similar paper designs. This means that the designer has the problem "boxed in"—all the aberrations are in tolerance and under control. This, however, is only the beginning of the engineering task of preparing the lenses for manufacture. This phase is a lengthy process of making sure that all the clear apertures will pass the rays, that the thicknesses of the lenses are not too thick or too thin. The glass type selected has to be checked with the availability of the glass, and the experience of the shop with it. The design has to be reviewed to consider how the lenses are to be mounted. Some of the lenses may require precision bevels on the glass. A redesign may be able to avoid this costly step. This section also includes a few comments about the designs as regards practicality.

Table 2.1 contains a summary of their attributes. The key for the entries is

- F — Focal length of the lens (mm.)
- F/NO. — The ratio F/D
- L — Total length of scan line (mm)
- ROAL — Overall length from the entrance pupil to the image plane relative to the focal length (mm.)
- RFWD — Front working distance relative to the focal length F
- d — Diameter of the image of a point at the $1/e^2$ irradiance level (micrometers)
- L/d — Number of spots of a scan line
- RBcr — Paraxial chief ray bending relative to the input half angle
- RPR — The ratio of the Petzval radius to the lens focal length
- REPR — The estimated Petzval radius relative to the lens focal length from equation (14)

Table 2.1

Section	F	F/NO.	L	ROAL	RFWD	d	L/d	RBcr	RPR	REPR	NS/I	NO.el
7.1	300	60	328	1.4	0.13	70	4700	.66	−11	−5	370	2
7.2	100	24	118	1.4	0.06	28	4300	.34	−12	−12	920	3
7.3	400	20	310	1.6	0.17	23	13000	.49	−26	−36	1100	3
7.4	748	17	470	1.4	0.06	20	23000	.29	−15	−51	1200	3
7.5	55	5	29	2.1	0.44	5.8	5000	.84	−32	−30	4300	3
7.6a	48	3	16	2.9	0.67	2.4	6700	1.1	−14	−30	11000	6
7.6b	48	3	16	2.9	0.67	2.4	6700	1.1	−14	−30	11000	5
7.7	52	2	20	4.2	0.39	4	5000	1.0	−56	−16	6350	14

NS/I	The number of spots per inch
NO.el	The number of lens elements

In the following lens descriptions all the spot diameters refer to the diameter of the spot at the $1/e^2$ intensity level. The number of spots on a line were calculated assuming that the spots are packed adjacent to each other with no overlap or space between them.

2.7.1 A 300-Spot/Inch Office Printer Lens, λ = 633 nm [24.9 microinches]

See Figure 2.20. This patent (U.S. Patent 4,179,183, Tateoka, Minoura; Dec. 18, 1979) was assigned to Canon Kabushiki Kaisha. The patent contains a lengthy description of the design concepts used in developing a whole series of lenses. Fifteen designs are offered with the design data along with plots of spherical aberration, field curves, and the linearity of scan. The design shown in Figure 2.20 is example 6 of the patent. The design data were set up and evaluated, and

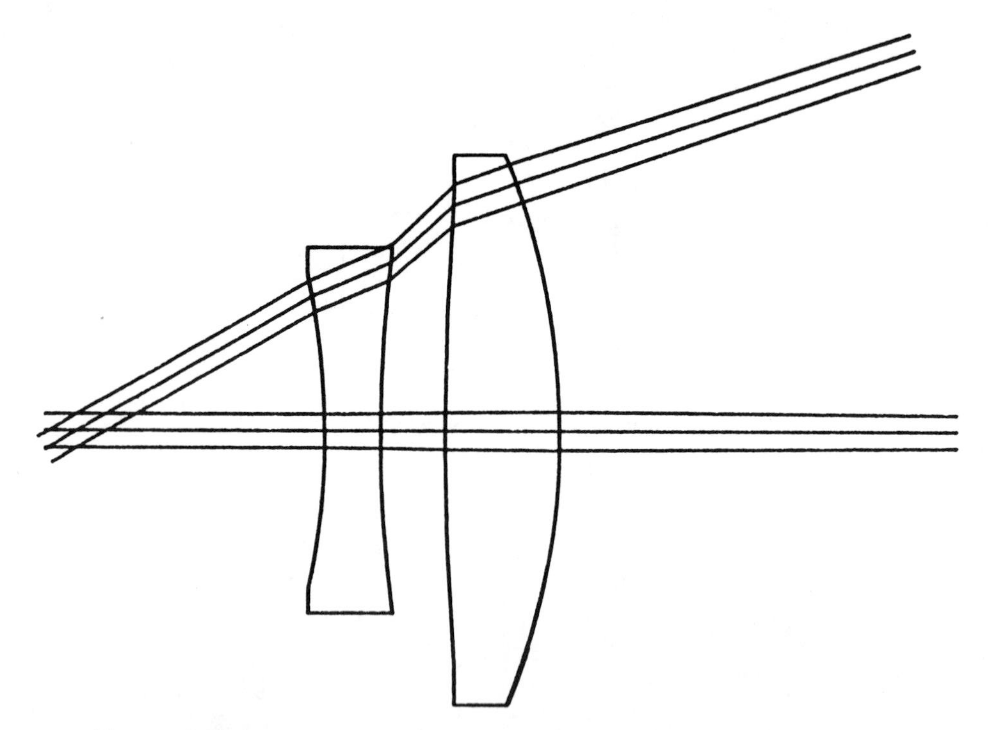

Figure 2.20 Lens 1. U.S. Patent 4,179,193 Tateoka, Minoura: F = 300 mm., $F/60$, L = 328 mm.

the results agree well with the patent. The focal length is given as 300 mm. The aberration curves appear to be given for the paraxial focal plane, and the linearity is shown to be within 0.6% over the scan. However, if one selects a calibrated focal length of 301.8 mm [11.8 inches] and shifts the focus 2 mm [0.079 inches] in back of the paraxial focus, it appears that the lens is well corrected to within $\lambda/4$ OPD (Optical Path Difference) and the linearity to within 0.2%. This lens may be similar to the lens used in the now famous Canon laser printer engine. This has an exceedingly wide angle for a scan lens. It has a great advantage in the design of a compact scanner. This printer meets the needs of 300 dots/inch, which is quite satisfactory for high-quality typewriter printing. The secret of the good performance of this lens is the airspace between the positive and negative lenses. There are strong refractions on the two inner surfaces of the lens, which will mean that the airspace has to be held accurately and the lenses must be well centered. We would be surprised if this is the actual lens type used in their commercial unit.

2.7.2 A Wide-Angle Scan Lens, λ = 633 nm [24.9 microinches]

See Figure 2.21. This lens has a 32° [0.559 radians] half-field angle. It has a careful balance of third-, fifth-, and seventh-order distortion, so that at the calibrated focal length it is corrected to be F-θ to within 0.2%. To do this the lens uses strong refraction on the fourth lens surface and refraction on the fifth surface to achieve the balance of the distortion curve. The airspace between these two surfaces controls the balance between the third- and fifth-order distortion. This would also be a relatively expensive lens to manufacture. The design may be found in U.S. Patent 4,269,478; it was designed by Haru Maeda and Yuko Kobayashi and was assigned to Olympus Optical Co. Japan.

2.7.3 A Semiwide Angle Scan Lens, λ = 633 nm [24.9 microinches]

See Figure 2.22. This lens shows how lowering the *F*/NO. to 20 and increasing the scan length increases the sizes of the lenses. This design is a Melles Griot product (Designer D.S.) It is capable of writing 1096 spots/in and is linear to less than 25 micrometers. The large front element is 128 mm [5.04 inches] in diameter. It will transmit 2.8 times as many information points as the number 1 lens. This lens requires modest manufacturing techniques, but, as shown in the diagram, the negative lens may be in contact with the positive lens next to

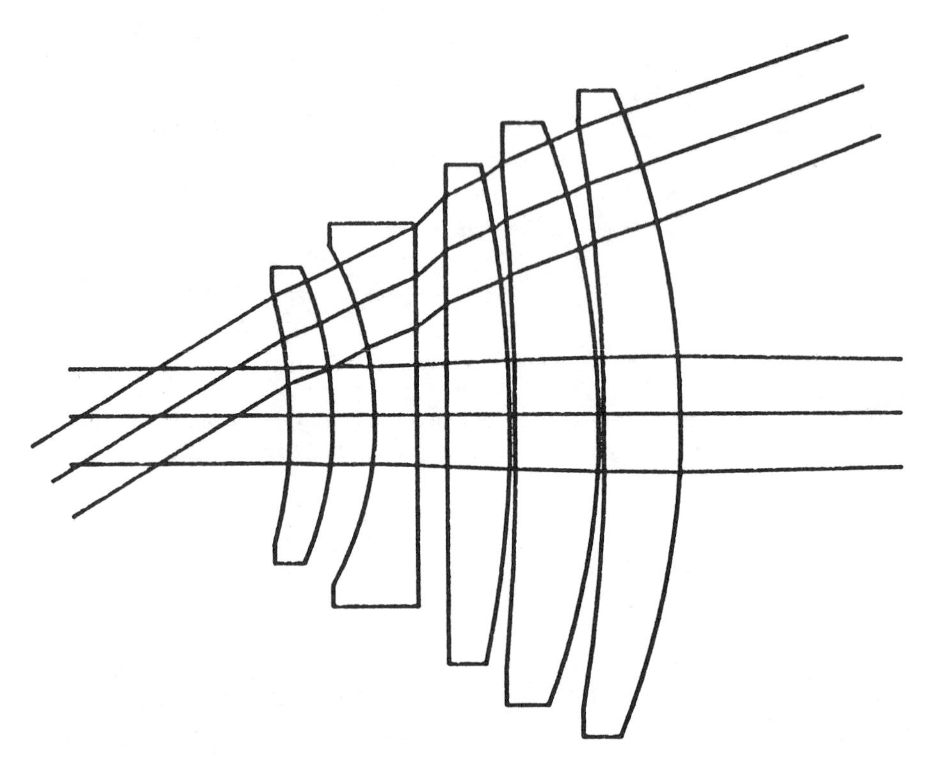

Figure 2.21 Lens 2. U.S. Patent 4,269,478 Maedo-Yuko: $F = 100$ mm., $F/24$, $L = 118$ mm.

it. Most optical shops would not make this with glass-to-glass contact. Either the airspace should be increased, or some careful mounting has to be considered.

2.7.4 A Moderate Field Angle Lens with a Long Scan Line, $\lambda = 633$ nm [24.9 microinches]

See Figure 2.23. One of the authors (R.E.H.) designed this lens. It has a half scan angle of 18° [0.31 radians], covers a 20-in [0.79 mm] scan length, and can write a total of 23,100 image points. This lens was designed for a holographic scanner. It has a short working distance between the holographic scan element and the first surface of the lens. This makes it more difficult to force the F-θ condition to remain within 0.1%. The working distance was kept short in order to keep the lens diameters as small as possible. The largest lens diameter is 110 mm [4.33 inches]. We have not found a way to improve this lens performance without making the elements considerably larger. It does not help any to just

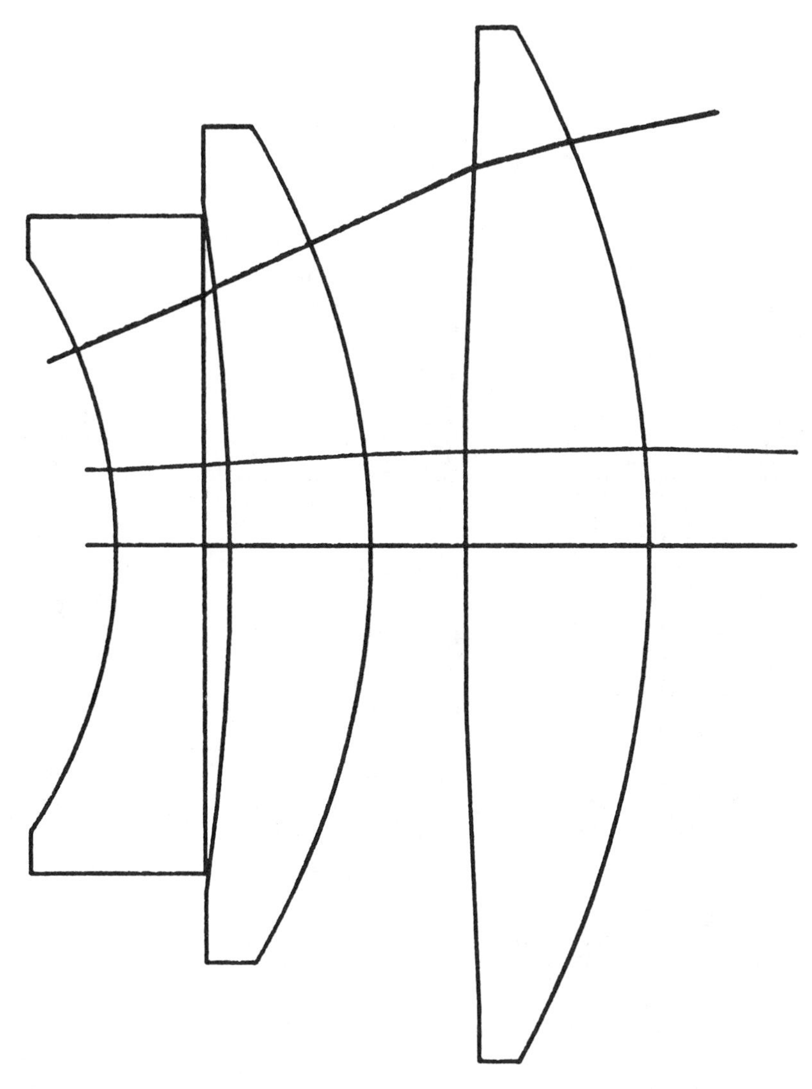

Figure 2.22 Lens 3. Melles Griot, Designer D. Stephenson: $F = 400$ mm., $F/20$, $L = 300$ mm.

add more lenses. The lens performs well in the design phase, but, since it is relatively fast for a scan lens, the small depth of focus makes the lens sensitive to manufacture and mount. The only way to make the lens easier to build is to improve the Petzval field curvature, and this requires more separation between the positive and negative lenses. The result is that the lens becomes longer and

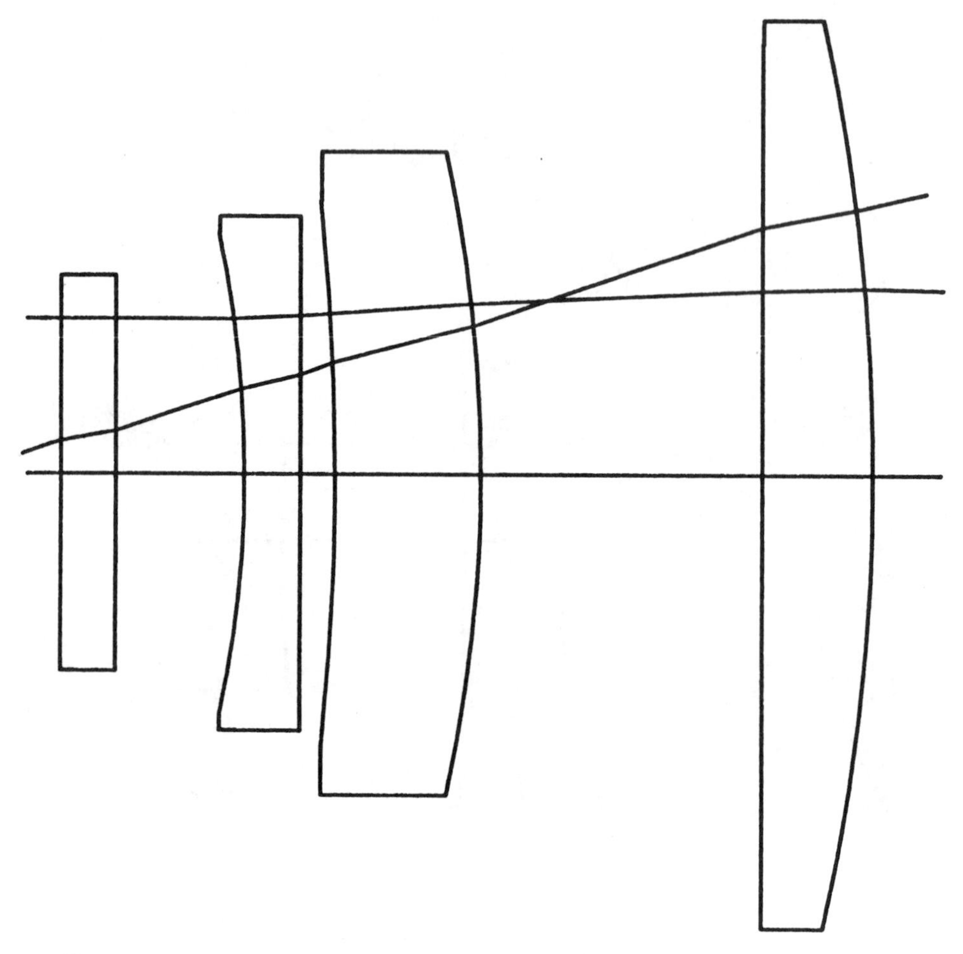

Figure 2.23 Lens 4. Designer R. Hopkins: $F = 748$ mm., $F/17$, $L = 470$ mm.

larger in diameter. Another way is to increase the distance between the scanner and the lens. This will also increase the lens size. We believe that this lens is close to the boundary of what can be done with a purely refracting lens for scanning a large number of image points. To extend the requirements will require larger lenses. It may be possible to combine small lenses with a large mirror close to the focal plane, but the costs would have to be carefully considered. This lens can be built by photographic-quality shops. The lenses do have to be mounted carefully.

2.7.5 A Scan Lens for Light-Emitting Diode, λ = 800 nm [31.5 microinches]

See Figure 2.24. This lens was designed by R.E.H. to perform over the range of wavelengths from 770 to 830 nm [27.5 to 32.6 microinches]. It was designed to meet a telecentricity tolerance of ±2°. The lens could not accommodate the full range of wavelength without slight focal shift. If the focal shift is provided for, the diodes may vary their wavelength from diode to diode over this wavelength region, and the lens will perform satisfactorily. This lens was never fully engineered for manufacture. It would surely be necessary to consider carefully how the negative positive combination would be mounted. Glass-to-glass contact should be avoided.

2.7.6 A High-Resolution Telecentric Lens, λ = 442 nm [17.4 microinches]

See Figure 2.25a, b. This lens did not have a tight F-θ requirement, but it was a troublesome lens to design. The original (R.E.H.) design contained six elements. See Figure 25a. Two of the lenses were meniscus lenses, which are difficult to center, and optical shops complain about them. We discussed this problem with

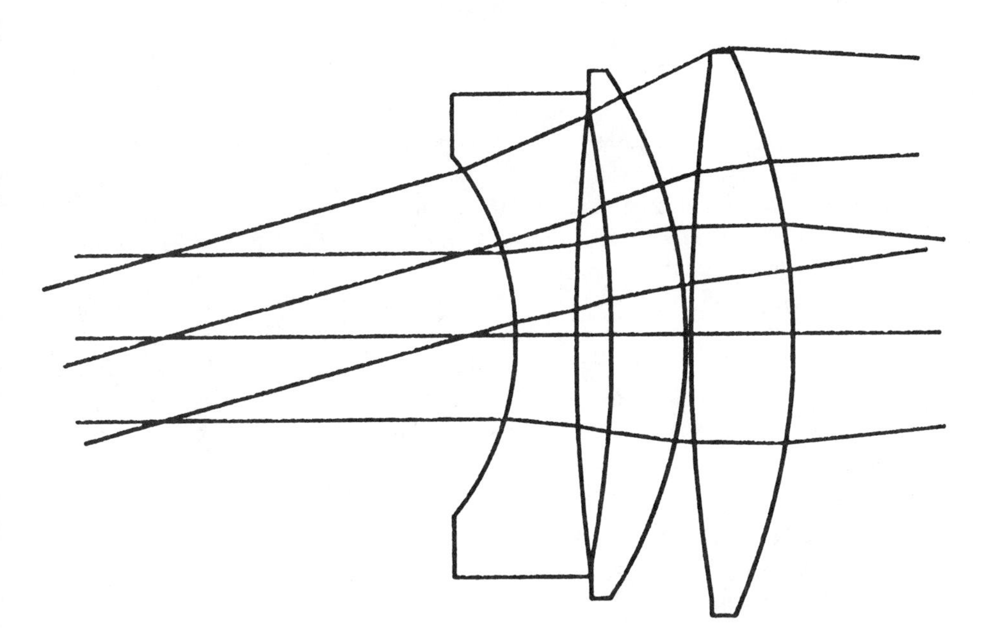

Figure 2.24 Lens 5. Designer R. Hopkins: F = 55 mm., $F/5$, L = 29 mm.

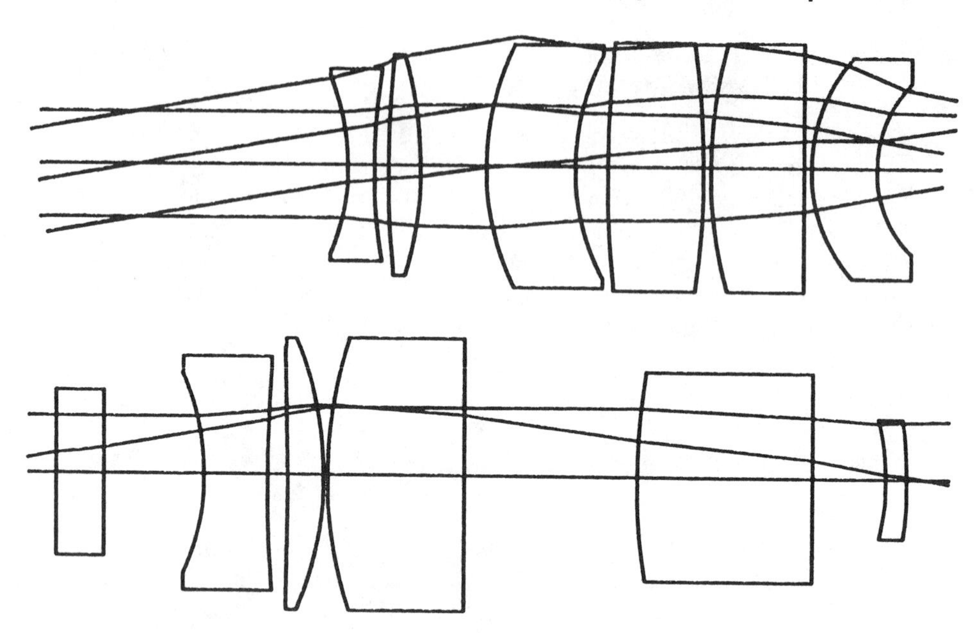

Figure 2.25 Lens 6. Designers R. Hopkins and Berlyn Brixner: $F = 48$ mm., $F/3$, $L = 16$ mm.

Berlyn Brixner and asked him to run it on the Los Alamos computer program. His program was not able to remove the meniscus lenses, but he suggested the removal of one of the other elements. The result was a much more practical lens which actually performed slightly better than the six-element design. See Figure 25b. This illustrates that these scan lenses cannot be improved by merely using extra elements.

This lens images more spots per inch than any of the other lenses, but the scan length is small. This lens represents a precision lens, and it would be necessary to carefully select the shop for its manufacture. The mount design as well as the glass specifications have to be done exceptionally well to avoid having to test and reassemble in order to match the potential performance of the design.

2.7.7 A High-Precision Scan Lens Corrected for Two Wavelengths, λ = 1064 and 950 nm [49.42 and 44.12 microinches]

See Figure 2.26. This is a Melles Griot lens, designed for a galvanometer XY scanner system. It is capable of positioning a 4μm [micrometers] spot anywhere

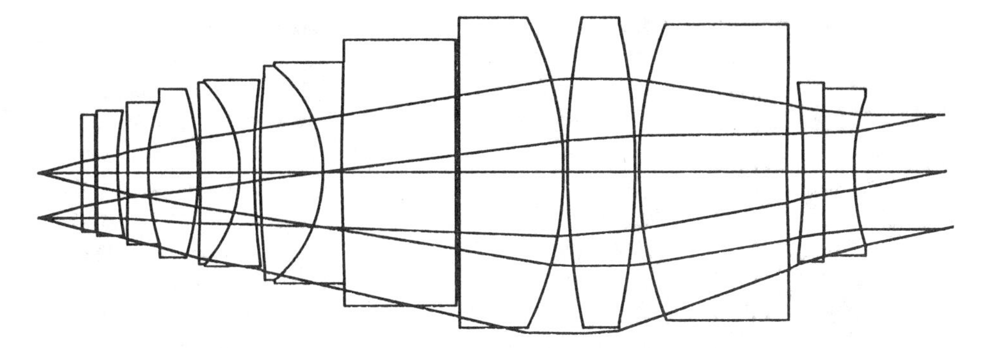

Figure 2.26 Lens 7. Melles Griot, Designer D. Stephenson: $F = 50$ mm., $F/2$, $L = 20$ mm.

within a 20 mm diameter circle. The spot is addressed with 1064nm energy with simultaneous viewing of the object done at 910 to 990 nm. The complexity of the design is primarily due to the need for two wavelength operation, especially the broad band around 946 nm. The thick cemented lenses have to have glass with different dispersions. The Optizon author of this chapter is willing to state that there are probably fewer than five optical shops in this country that have the facilities to build this lens to realize the full design potential.

2.8 SCAN-LENS MANUFACTURING, QUALITY CONTROL, AND FINAL TESTING

The first two designs shown in the previous section require tolerances which are similar to quality photographic lenses. There are a number of shops that can make these lenses. Compared to the lens diameters, the beam sizes are small, and so the surface quality is not difficult to meet.

Designs 2 and 3 do not require the highest quality or precision lenses, but the number of shops that can be depended on to build acceptable lenses of this type is limited. One thing that helps to realize a good yield of acceptable lenses is that the scan line uses only one cross-sectional sweep across the lenses. By rotating the lenses in the cells, some lenses with defects can be accepted by finding the best line of scan. This requires, however, that each lens be mounted correctly in the scanning system. It is important, therefore, to have good communications between the assemblers and the lens builders.

Lenses 5, 6, and 7 have to be built as precision lenses, and there are a limited number of shops that can make these lenses perform up to the design expectations. The surfaces have to be spheres to within quarter-wave quality, and they must be precision-centered and mounted. The shop making them must have precision equipment to maintain quality control through the many steps that it takes to make and mount such a lens. These three lenses require special mounting methods to achieve the required accurate alignment of the lenses in the barrel.

The manufacturer or the buyer of the precision lenses like these (numbers 5, 6, and 7) has to have the equipment for testing the performance of the lenses. This is done by scanning the image with a detector which can measure the spot diameter in the focal plane. The metrology of these tests is not trivial. Attention has to be paid to the straightness of the plane of movement of the detector and its relationship to the lens axis. It is necessary to direct the incoming collimated beam into the lens in exactly the way it will be introduced in the final scanner. If the image beam intersects the image plane at an angle it is necessary to make sure that all the beam passes through the scanner and is detected on the detector. Sometimes this means the scanner has to be rotated so that the beam enters the scanner along its central axis. In this case a correction to the spot size has to be made. If the image is relayed with a microscope objective, it is necessary to be sure that the numerical aperture is large enough to collect all the cone angles. It is well to remember that just because the electronic scanning detector reads a steady value it is not necessarily the correct reading.

Since scan lenses are usually designed to form diffraction-limited images, it is recommended that the lenses be tested using a laser beam with uniform intensity across the design aperture of the lens. The image of a point source should be diffraction-limited and its dimensions are calculable. The image can be viewed visually, and the Airy disk diameters can be measured. Departures from a spherical wavefront as small as 1/10 of a wavelength are easily detected. It is possible to see the effect of excessive scattered light. This test should be done to make sure that the electronic scanning detector is giving reasonable answers. Unfortunately this is not possible to do when the wavelength is 1 μm.

2.9. CLOSING COMMENTS

The trend in laser scanning systems is ever-increasing scan lengths and larger number of spots per inch. We believe the above examples of designs are showing the present-day practical boundaries for scan lenses. If the boundaries are to be expanded, it will mean that some new concepts will be needed. The most obvious ways are to explore the use of combinations of lenses near the scanner and mirrors close to the image plane. Since the mirrors are near the image plane, they will be large, but they only need to be slices of round mirrors. Methods will have to be found for economically making these elements. Another way is to allow the refracting lenses to be larger in diameter and reduce the cost by learning to build segments of lenses. This has always been a "no-no" because of the cost of manufacturing and mounting the segments. Optical shops do not presently have the equipment to do it. These ideas will not be accepted until there is a market of sufficient size to pay for the development of the tooling required. If that comes about, the industry will surely be able to meet the challenge.

Optical systems always seem to be limited by the glass shop. The reason is that special requirements call for special equipment, and there has to be a large enough market to pay for the tooling. One outstanding illustration of underestimating the value of conventional optical systems was in the development of the photolithographic lens used in making large-scale integrated circuits. Millions have been spent on new systems such as electron beam and x-ray lithography; a pittance has been spent on optical systems. Gradually, over 30 years, optical lithography has provided the tools to make mass production integrated circuits. The lens development has been to make do with present equipment and attempt to plow back meager profits into new equipment. Today, optical systems are making most of the large production orders, and some people see no real likelihood of a new substitute for large-scale production. Glass optical systems have the tremendous advantage of being passive; they just sit there passing information with no wear and tear on the parts. It takes great care to make them right, but the information transport in bits per unit time can be enormous.

REFERENCES

1. L. Beiser, Laser scanning systems, in *Laser Applications*, Academic Press, N.Y. Vol. 2, p. 53.
2. M. J. Buzawa, *Laser Focus*, Sept. p. 82.
3. J. H. Rosen, *Rand Report*, #R-925-ARPA, April 1972.
4. J. C. Urbach, T. S. Fisli, and G. K. Starkweather, *Proc. IEEE*, Vol 70: No. 6: 597.
5. Melles Griot, *Laser Scan Lens Guide*, Rochester, NY, 1987.
6. W. B. Wetherell, The calculation of image quality. In *Applied Optics and Optical Engineering*, Academic Press, New York, 1980, Vol. 7.
7. R. E. Hopkins and R. Hanau, *MIL-HDBK-141*, U.S. Government Printing Office. Now out of print. A photocopy version is available from Sinclair Optics, 6780 Pittsford-Palmyra Rd., Fairport, NY 14450.
8. R. Kingslake, *Optical System Design*, Academic Press, New York, 1983.
9. R. E. Hopkins and M. J. Buzawa, *SPIE*, Vol. 15, No.2, 1976.
10. A. Dewey, *IBM Tech. Discl. Bull.*, *17*, 2743, 1976.
11. Fleischer, Latta, Rabedeau, *IBM Jrnl. of Res. and Des.*, *Vol. 21 #5*, 479, 1977.
12. R. Kingslake, *Lens Design Fundamentals*, Academic Press, New York, 1978.
13. W. J. Smith, *Modern Optical Engineering*, McGraw-Hill, New York, 1966.
14. W. T. Welford, *Aberrations of Symmetrical Optical Systems* Academic Press, London, 1974.

3

Scanned Image Quality

Donald R. Lehmbeck

Webster Research Center, Xerox Corporation, Webster, New York

John C. Urbach

Strata Systems Inc., Portola Valley, California

3.1 INTRODUCTION

3.1.1 Imaging Science for Scanned Imaging Systems

The purpose of this chapter is to present some of the basic concepts of image quality and their application to scanned imaging systems. We, like so many others, follow in the path pioneered over a half century ago by the classic 1934 paper of Mertz and Gray [1]. Without going into the full mathematical detail of that paper and many of its successors, we shall attempt to bring to bear some of the modern approaches that have been developed both in image quality assessment and in scanned image characterization.

Numerous and diverse technologies are used in scanned imaging systems. They provide both the designer and the user of such systems with an enormous array of choices and trade-offs. Building on a foundation of imaging science, we shall attempt to provide a framework in which to sort out the many image-quality issues that depend on these choices.

It is our intent not to show that one scanner or technique is better than another, but to describe the *methods* by which each scanning system can be evaluated to compare to other systems and to assess the technologies used in them. This chapter therefore deals primarily with such matters as the sharpness or graininess of an image and not with such hardware issues as the surface finish of a polygon, the speed of a modulator, or the efficiency of charge transfer in a charge-coupled device (CCD) imager. The classic imaging science approach has been modified somewhat to handle electronic imaging. The methods used are reviewed here in enough detail to enable the reader to obtain a working technical knowledge of the subject.

To keep the scope of the discussion within reasonable bounds, we have limited the technical depth and detail as well as the historical development of a number of the well-established measurements and methods described.

References will be given for the benefit of readers wishing to obtain more details.

3.1.2 Scope

Scanning is considered here in the general context of electronic imaging. An electronic imaging system usually consists of an input scanner which converts an optical image into an electrical signal (often represented in digital form). This is followed by electronic hardware and software for processing or manipulation of the signal and for its storage and/or transmission to an output scanner. The latter converts the final version of the signal back into an optical (visible) image, typically for transient (soft) or permanent (hard copy) display to a human observer.

The emphasis in this chapter will be on the input scanner. Output scanners and diverse systems topics will be dealt with mainly by inference, since many input scanner considerations and metrics are directly applicable to the rest of a complete electronic imaging system. It is planned that a future edition of this chapter will include an explicit treatment of image quality for output scanners. A few direct comments on output scanners are included here as required in the present context. After this initial introduction, we shall describe the basic concepts and phenomena of image scanning in a tutorial manner (Section 3.2). This can be omitted by readers already familiar with these fundamentals. Following a brief treatment of certain practical issues (Section 3.3), we proceed to a more detailed discussion of the performance of input scanners that produce multilevel (gray) signals (Section 3.4). We conclude this chapter with an examination (Section 3.5) of various summary measures of imaging performance, most readily applied to gray-level scanners, but often of some value for other scanners, and indeed, within certain limits, for complete systems incorporating both input and output scanners as well.

An important branch of electronic imaging which will not be included directly is the area of computer-generated imaging. In this field, input scanning is omitted, and an original image is generated by the computer.

Input scanners that generate binary signals constitute another important special class and must be evaluated by unique measurement methods [2] beyond the scope of this chapter.

While the focus here is on imaging modules and imaging systems, scanners may of course be used for purposes other than imaging, such as digital data recording. We believe that the imaging science principles used here are sufficiently general to enable the reader with a different application of a scanning system to infer appropriate knowledge and techniques for these other applications.

Types of Scanners

All input scanners convert one- or (usually) two-dimensional image irradiance patterns into time-varying electrical signals. There are two general types of technology involved in building an input scanner:

1. Flying spot scanners such as a laser or CRT.
2. Image integrating and sampling devices such as those found in many forms of electronic cameras and electronic copying devices. The latter in turn may be traditional vacuum tube imagers with continuous (or discrete) photosensitive surfaces scanned by an electron beam, e.g., vidicon, plumbicon, etc., or the more modern solid state type, such as a CCD array.

Other types of scanners can be described by selecting characteristics from each category in Table 3.1.

The temporal signals produced by these scanners can be in one of two general forms, either (1) binary output (a string of on and off pulses), or (2) gray-scale output (a series of electrical signals whose amplitude varies). Likewise, the spatial characteristics of these scanners may be either analog or digital or some com-

Table 3.1 Types of Major Options Characterizing Various Input Scanners

Types of Sampling Mechanism	
Flying spot	Continuous fast scan; discrete slow scan
Sampling arrays	Discrete fast scan and slow scan
	Discrete fast scan; continuous slow scan
Single detector with apertures	Discrete or continuous, either or both
Types of Input Transports	
None (camera)	Used with 2-dimensional sampling arrays
Flat bed	1- or 2-dimensional
Rotating drum	2-dimensional by translation of drum or head
Scanning Motion	
Discrete	
Continuous	
Types of Output Signals	
Analog gray	
Digital (discrete) gray	
Binary	

bination of the two. The term *digital* here refers to a system in which each picture element (pixel) must occupy a discrete spatial location; an analog system is one in which a signal level varies continuously with time, without distinguishable boundaries between individual picture elements. A two-dimensional analog system is usually only analog in the more rapid direction of scanning and is discrete or digital in the slower direction, which is made up of individual raster lines. Television typically works in this fashion. In one form of solid state scanner the array of sensors is actually two-dimensional with no moving parts. Each individual detector is read out in a time sequence, progressing one raster line at a time within the two-dimensional matrix of sensors. In other systems a solid state device, arranged as a single row of photosites or sensors, is used to detect information one raster line at a time. In these systems either the original image is moved past the stationary sensor array, or the sensor array is scanned across the image to obtain information in the slow scan direction.

Another configuration for an input scanner involves a single detector (usually a photomultiplier tube) collecting light from an adjustable aperture which has the input image projected on it. If the original object is a paper document, this document is moved past the optical system, being attached either to a rotating drum or to a flat bed stage that moves in the x and y directions. These scanners are usually much slower than the previously described types, but can have extremely high precision and resolution. They are widely used in the graphic arts, in electronic circuit patterning, and in precision measurement applications. In the case of the rotating drum type of graphic arts scanner, the signal may be fed directly to an output scanner which writes to film attached to the other end of the same rotating drum. Table 3.1 summarizes these types of input scanners, showing the options in several major categories of design that have image-quality implications.

In this chapter we will concentrate on the totally digital solid state scanners using sampling arrays, because, in our judgment, they are the most commonly encountered forms of input scanners. The reader should be able to infer many things about the other forms of scanners from these examples.

Types of Characterizations

We shall, as noted, concentrate on applications that involve capturing and writing images. For other applications, the relative importance of various measurement criteria would change. The basic principles of imaging science used throughout this discussion can be found in a variety of textbooks on imaging science, optics and image processing, a few of which are listed in the bibliography. Several important technical journals in this area, as well as proceedings of selected conferences, are listed in the bibliography.

The general approach here is to describe the relevant imaging characteristics with appropriate functions in sufficient detail so that one can cascade physical performance attributes from one system element to another. It is important to recognize that the straightforward cascading of such descriptors as MTF and granularity (covered later) can only be done rigorously for linear, stationary (uniform in time and space) systems. Generally, in complex imaging systems there are many nonlinearities. In addition, few input scanners, even if they are linear, are truly stationary. It may sometimes be necessary to develop specialized metrics, building on the information presented here, to handle the nonlinear and nonstationary cases. For conceptual simplicity we generally assume that straightforward cascading works adequately.

In evaluating the quality of any imaging system, it is essential not only to have a mathematical and physical description of the imaging process but also to include some understanding of the perceptual (human vision) aspects of the images created by that process. This in turn can be broken down into several levels of understanding:

1. Human perception of what can and cannot be seen
2. The psychophysical relationships, which describe the connection between a visual response and the magnitude of the imaging characteristic being varied
3. The purely subjective or qualitative evaluation of the image, which is highly dependent upon its ultimate use

3.1.3 The Context for Scanned Image Quality Evaluation

The basic building blocks of an electronic imaging system are sketched in Figure 3.1. They are the image input terminal (IIT), the image processing subsystem (including storage and transmission functions), and the output scanner or complete image output terminal (IOT). An additional important building block (not further considered here) is a character generator or other form of computer image synthesizer that acts as an additional source of information for the IOT.

It should be noted that since the IIT does not itself generate a display of any kind, it cannot be assessed by a human observer without some transducer for producing a visible image. This approach is valid for visual evaluations of an input scanner only to the extent that the display can be calibrated and its effects removed from any observations. The same is true for examination of fonts by these methods since that source of data is not directly visible either. It therefore often becomes necessary to evaluate the input scanner or a font set through the "eyes" of the IOT in an overall systems context. In this chapter we shall address many of these topics to some extent, but concentrate on measurements and characterization.

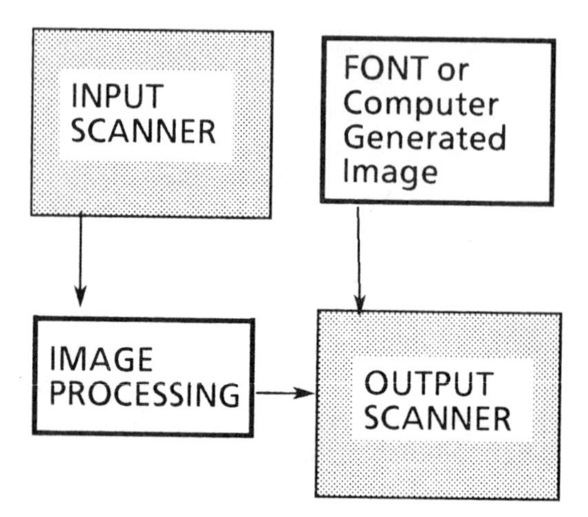

Figure 3.1 Major components of electronic imaging system.

3.2 BASIC CONCEPTS AND EFFECTS

3.2.1 Sampling and Reconstruction

The basic electronic imaging system performs a series of image transformations sketched in Figure 3.2. In this figure we see that an object such as a photograph or a document with lines and text is converted from its analog nature to a digital form by a raster input scanner (RIS) and then quantized and processed with various digital techniques. This digital image is further transformed into information that can be displayed or transmitted, edited or merged with other information in the electronic and software subsystem (ESS). Subsequently a raster output scanner (ROS) converts the digital image into an analog form, typically light falling on some type of photosensitive device. This converts the analog optical image into an analog reflectance pattern on paper, or into some other permanent display medium, i.e., into the final output image. What follows assumes optical output conversion, but nonoptical direct-marking processes (e.g., ink jet, thermal transfer, etc.) can be treated similarly. Thus, the input scanner acts somewhat like an analog-to-digital (A to D) converter and the output printer like a digital-to-analog (D to A) converter. There are many opportunities to manipulate the digital image in between. Therefore, while one often thinks of electronic imaging or scanned imaging as a digital imaging process, we are really concerned in this chapter with the imaging equivalent of A to D and D to A processes.

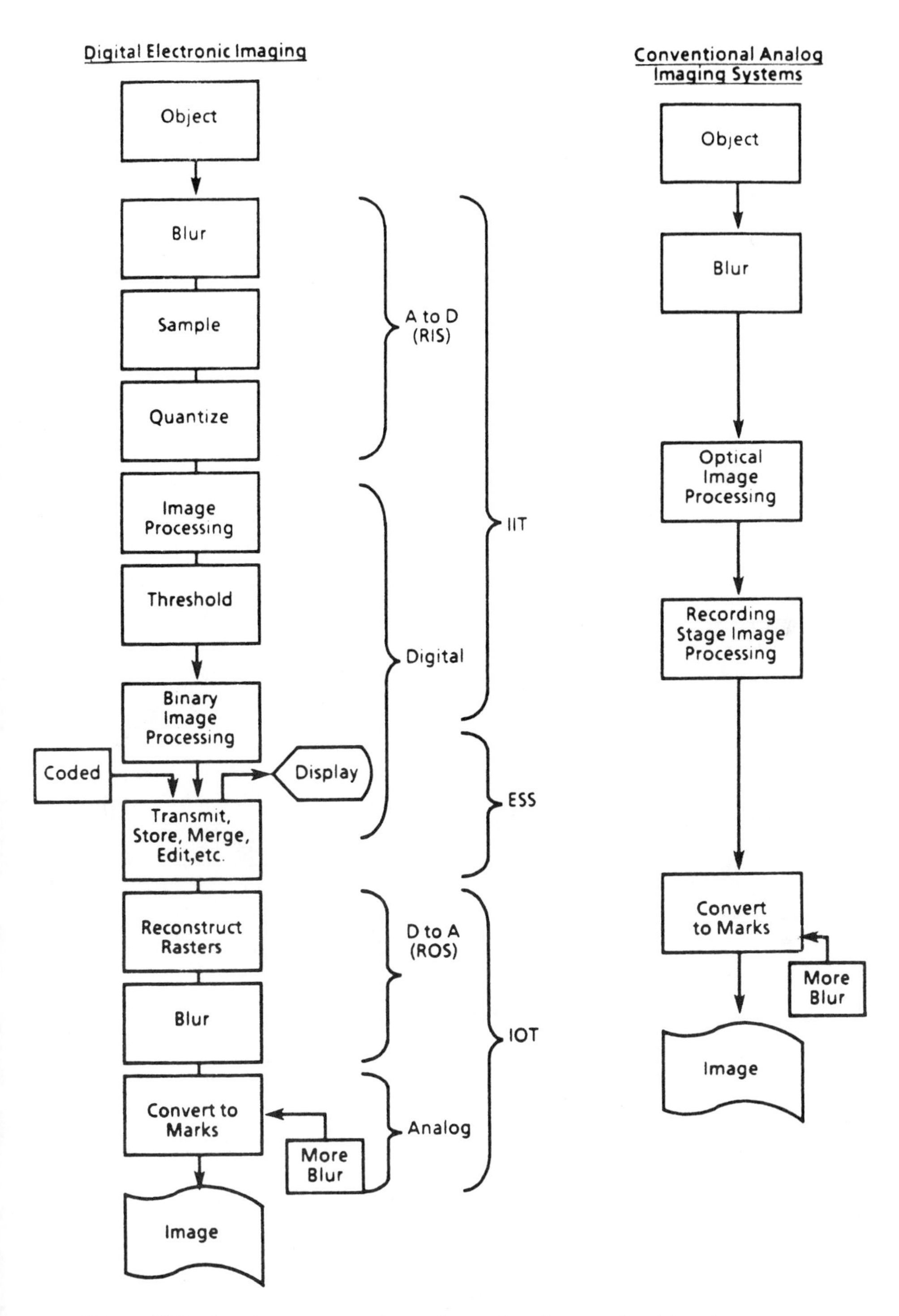

Figure 3.2 Imaging transformations: electronic and conventional imaging systems.

Structure of Elementary Digital Images

Let us turn our attention to the microscopic structure of this process and pay particular attention to the analog-to-digital domain of the input scanner, following the general approach of Mertz and Gray [1]. To understand this, let us examine Figure 3.3. For illustration purposes, Figure 3.3 shows four different aspects of the input scanning image transformations. In the first we have shown a representation of typical microstructures that can occur in an input object: there is a sharp edge, a "fuzzy" edge (ramp), and a narrow line. Part b shows the optical image, which is a blurred version of the input object. Note the relative heights of the two pulses and the sloping edges that were previously straight. Part c

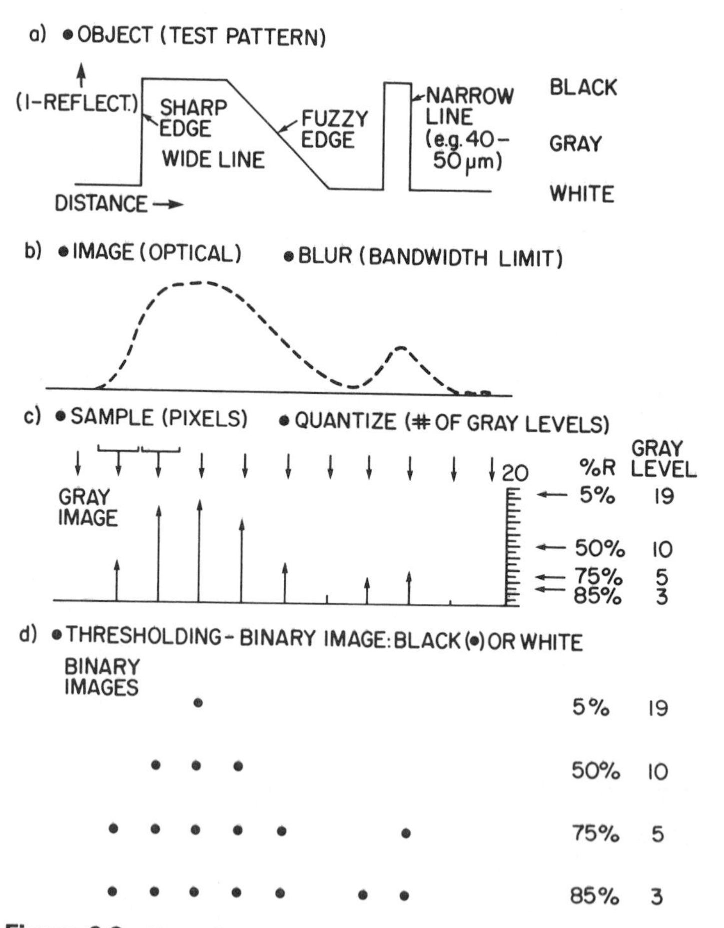

Figure 3.3 Formation of a binary image from a continuous-tone object.

attempts to represent the blurred image with a series of discrete signals, each being centered at the position of the arrows. This process is referred to as sampling. Each sample has some particular height or gray value associated with it. If there is a finite number of gray levels such as 10, 32, 128, or even 1000, then the signal is said to be quantized. If these individual samples can have any level whatsoever, then the system is said to be analog (when the finite number is very large, the quantized signal resembles the analog case). Typically the output of a charge-coupled device would look something like this on the face of an oscilloscope. It would be an analog representation. If passed through an analog-to-digital converter, this signal would become a quantized or digital signal. The fact that it is sampled, that is, the different portions of the signal are represented by different signal pulses, of different heights, indicates sampling and not quantization in the classic sense. Each of these individual samples starts out as a picture element, often referred to as a pixel or pel. A sampled and multilevel quantized image is often referred to as a gray image. When the quantization is limited to two levels, it is termed a binary image.

One of the most common forms of image processing is the conversion from a gray to a binary image. In this process a threshold is set at some particular gray level, and any pixel above that level is converted to one type of signal, for example, black. Any pixel whose gray value is below that level is converted to the other signal, for example, white. Four threshold levels are shown in part c on the gray-level scale at the right and are indicated in part d as four different binary images, one for each of the four different thresholds. In part d each black pixel is represented by a dot, and each white pixel is represented by the lack of a dot. For example, if the threshold level is 5%, only one black pixel is present, that being associated with the center of the larger of the two blurred peaks. As we move down to 50% threshold, three pixels are turned black, and only at 75% threshold do we get one black pixel representing the narrow line. Notice at the 85% threshold, the narrow line is now represented by two pixels (i.e., it has grown), but the wider and darker pulse has not changed in its representation. It is still 5 pixels wide. Notice that the narrow pulse grew in an asymmetric fashion and that the wider pulse, which was asymmetric to begin with, grew in a symmetric fashion. The resulting dot patterns show one line of a sampled binary image. These patterns are merely associated with the location of the sampling arrows, shown in part c, relative to the shape of the blur and the location of the features of the original document. However, they are quite characteristic of the problems encountered in digitizing an analog document into a finite number of pixels and gray levels. The last transformation, that of moving to a thresholded binary image, is a highly nonlinear one.

Figure 3.4 represents the same type of process using real images. In part a we see some gray text being displayed on a monitor. The black line drawn through the letters "Illu" represents a single raster line whose gray profile is plotted in

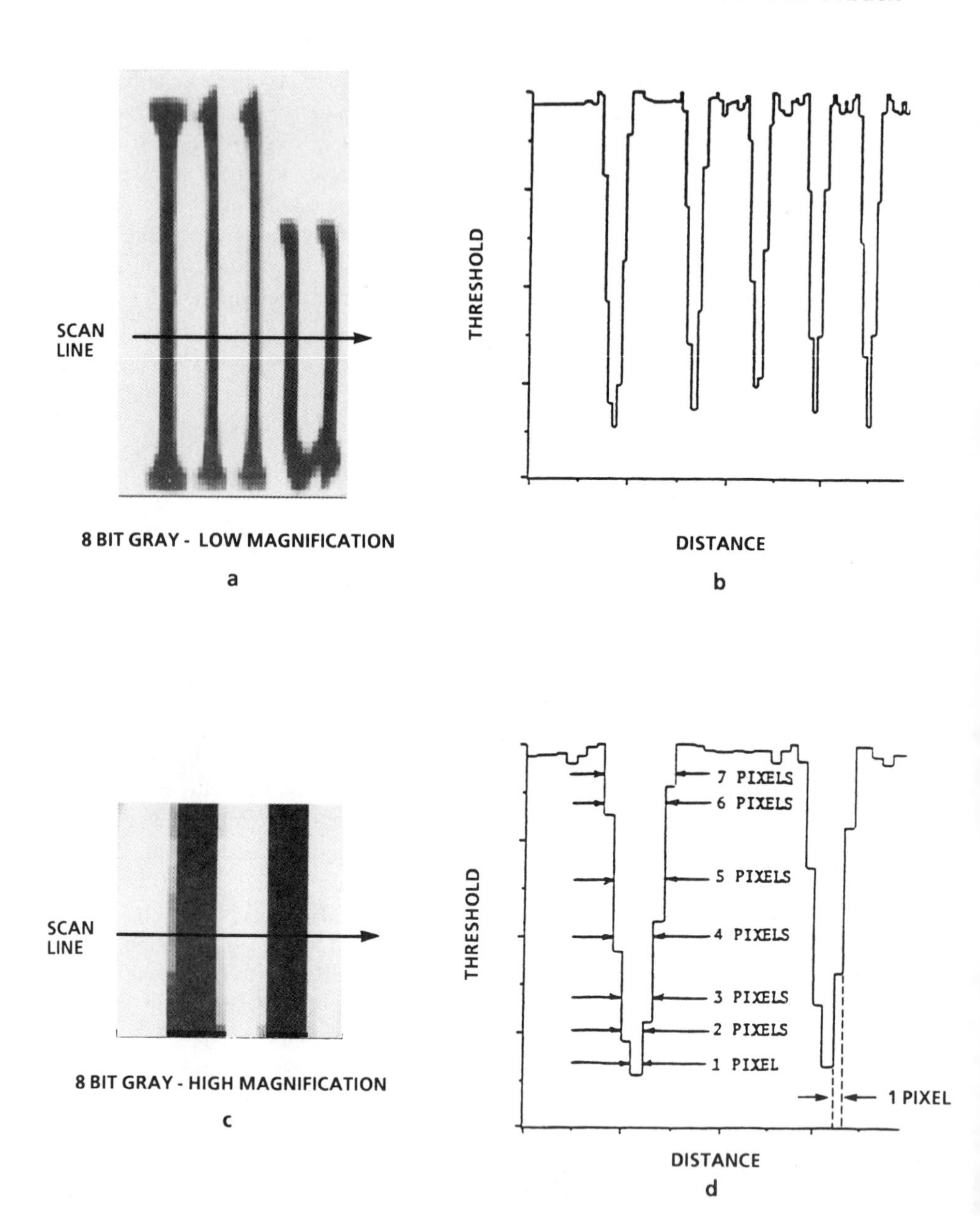

Figure 3.4 Effects of sampling, quantization, and thresholding on line-copy image representations.

part b. The same type of plot is shown at higher magnification in parts c and d. In part d, the width of the letter I is denoted at various threshold levels. The reader can see that the width of the binary image can vary anywhere from 1 to 7 pixels, depending on the selection of threshold.

Figure 3.5 returns to the same information shown in Figure 3.3, except that here we have doubled the frequency with which we sampled the original blurred optical image. There are now twice as many pixels, and their variation in height is more gradual. The gray file looks a little bit more like the original profile. There is also a smoother, more ''natural'' transition in the binary images as we move through the threshold sequence. Note also that two of these higher-frequency samples happen to fall at the minimum between the two pulses, creating two near-zero-level pixels (and thus enhance our ability to see some of the fine detail). It also just happened that one of these higher-frequency samples fell closer to the peak of the narrower pulse. Therefore, in this particular instance, increased resolution is responsible for the binary image detecting that pulse at a lower threshold. This illustration shows the general results that one would expect from increasing the density at which one samples the image. That is, one sees somewhat higher detail in both the gray and the binary images with higher sampling frequency.

This is, however, not always the case when examining every portion of the microstructure. Let us look more closely at the narrower of the two pulses in Figure 3.6. Here we see the sampling occurring at two locations, shifted slightly with respect to each other. These are said to be at different sampling phases. In phase A the pulse has been sampled in such a way that the separate pixels near

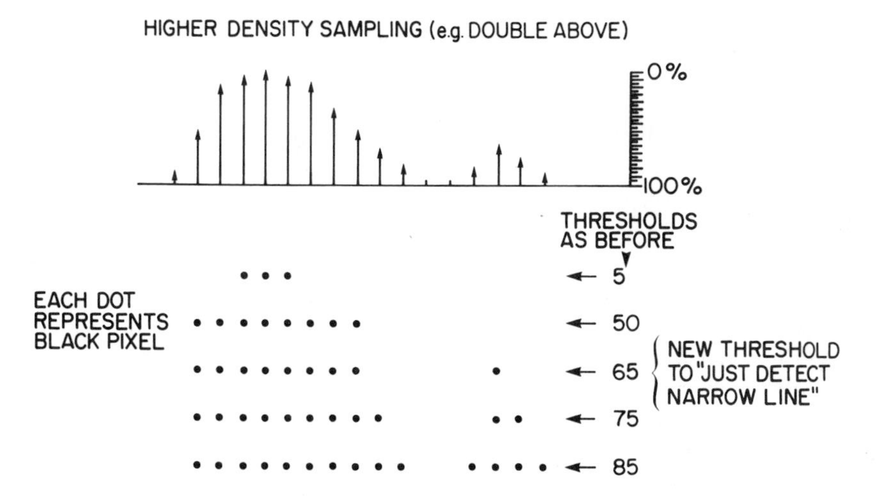

Figure 3.5 Effect of increased sampling frequency on binary image representations.

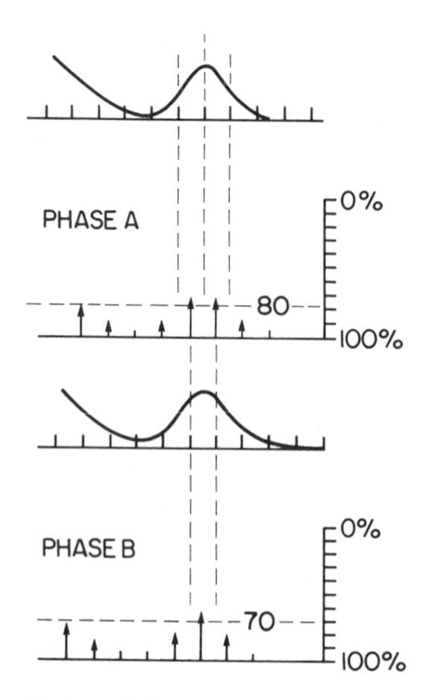

Figure 3.6 Effect of sampling phase on binary image representations.

the peak are identical to each other in their intensity, and in phase B one of the pixels is shown centered on the peak. When looking at the threshold required to detect the information in phase A and phase B, different results are obtained for a binary representation of these images. Phase B would show the detection of the pulse at a lower threshold (closer to ideal) and phase A, when it detects the pulse, would show it as wider, namely as 2 pixels in width.

Consider an effect of this type in the case of an input document scanner, such as that used for facsimile or electronic copying. While the sampling array in many input scanners is constant with respect to the document platen, the location of the document on the platen is random. Also the locations of the details of any particular document within the format of the sheet of paper are random. Thus the phase of sampling with respect to detail is random and the type of effects illustrated in Figure 3.6 would occur randomly over a page. There is no possibility that a document covered with some form of uniform detail can look absolutely uniform in a sampled image. If the imaging system produces binary results, it will consistently exhibit errors on the order of 1 pixel and occasionally 2 pixels. The same is true of a typically quantized gray image, except now the errors are primarily in magnitude and may, at higher sampling densities, be less objectionable. In fact, an analog gray imaging process, sampling at a sufficiently high

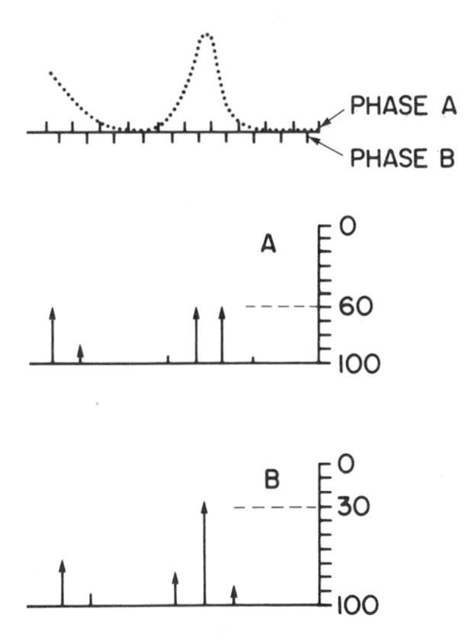

Figure 3.7 Effect of sampling phase on the binary representation of a sharper image (less blur).

frequency, would render an image with no visible error (see the next subsection).

Continuing with the same basic illustration, let us consider the effect of blur. In Figure 3.7 we have sketched a less blurred image in the region of the narrower pulse and now show two sampling phases A and B, as before, separated by half a pixel width. Two things should be noted. First, with higher sharpness (i.e., less blur), the threshold at which detection occurs is higher. Second, the effect of sampling phase is much larger with the sharper image.

The Sampling Theorem

By means of these illustrations we have shown the effects of sampling frequency, sampling phase, and blur at a very elementary level. We will now turn our attention to the more formal description of these effects in what is known as the sampling theorem. For these purposes we will assume that the reader has some understanding of the concepts of Fourier analysis or at least the frequency-domain way of describing time or space, such as in the frequency analysis of audio equipment. In this approach, distance in millimeters is transformed to frequency in cycles per millimeter (cycles/mm). A pattern of bars spaced 1 mm apart would result in 1 cycle/mm as the fundamental frequency of the pattern. If the bars were represented by a square wave, the Fourier series showing the pattern's various harmonics would constitute the frequency-domain equivalent. Figure 3.8 (after Ref. 3, p. 70) has been constructed from such a point of view. In Figure 3.8a we see an analog input document represented by the function $f(x)$. This is a signal extending in principle to $\pm\infty$ and contains upon analysis many different frequencies. It could be thought of as a very long microdensitometer trace across an original document. Its spectral components, that is, the relative amplitudes of sine waves which fit this distribution of intensities, are shown plotted in Figure 3.8b. Note that there is a maximum frequency, in this plot of amplitude versus frequency, at w. It is equal to the reciprocal of λ (the wavelength of the finest detail) shown in Figure 3.8a. This is the highest frequency that was measured in the input document. The frequency w is known as the bandwidth limit of the input document. Therefore the input document is said to be band-limited. This limit is often imposed by the width of a scanning aperture which is performing the sampling in a real system.

We now wish to take this analog signal and convert it into a sampled image. We multiply it by $s(x)$, a series of narrow impulses separated by Δx as shown in Figure 3.8c. The product of $s(x)$ and $f(x)$ is the sampled image, and that is shown in Figure 3.8e. To examine this process in frequency space, we need to find the frequency composition of the series of impulses that we used for sampling. The resulting spectrum is shown in Figure 3.8d. It is, itself, a series of impulses whose frequency locations are spaced at $1/\Delta x$ apart. For the optical scientist this may be thought of as a spectrum, with each impulse representing a different order; thus the spike at $1/\Delta x$ represents the first-order spectrum, and the spike at zero represents the zero-order spectrum. Since we multiplied in distance space in order to come up with this sampled image, in frequency space, according to the convolution theorem, we must convolve the spectrum of the input document with the spectrum of the sampling function to arrive at the spectrum of the sampled image. The result of this convolution is shown in Figure 3.8f. We can now see the relationship between the spectral content of the input document and the spacing of the sampling required in order to record that document. Because the spectrum of the document was convolved with the sampling spectrum, the negative side of the spectrum folds back from the first-order over the positive side of the zero-order spectrum. Where these two cross is exactly halfway between the zero- and first-order peaks. It is a frequency ($1/2\Delta x$) known as the Nyquist frequency. If we look at the region in Figure 3.8g between zero and the Nyquist frequency, the region we would reserve for the zero order information, we see that there is "leakage" from the negative side of the first order down to the frequency. $[(1/\Delta x) - w]$, where w is the band limit of the signal. Any frequency above that point contains information from both the zero and the first order and is therefore corrupted or mixed, often referred to as *aliased*.

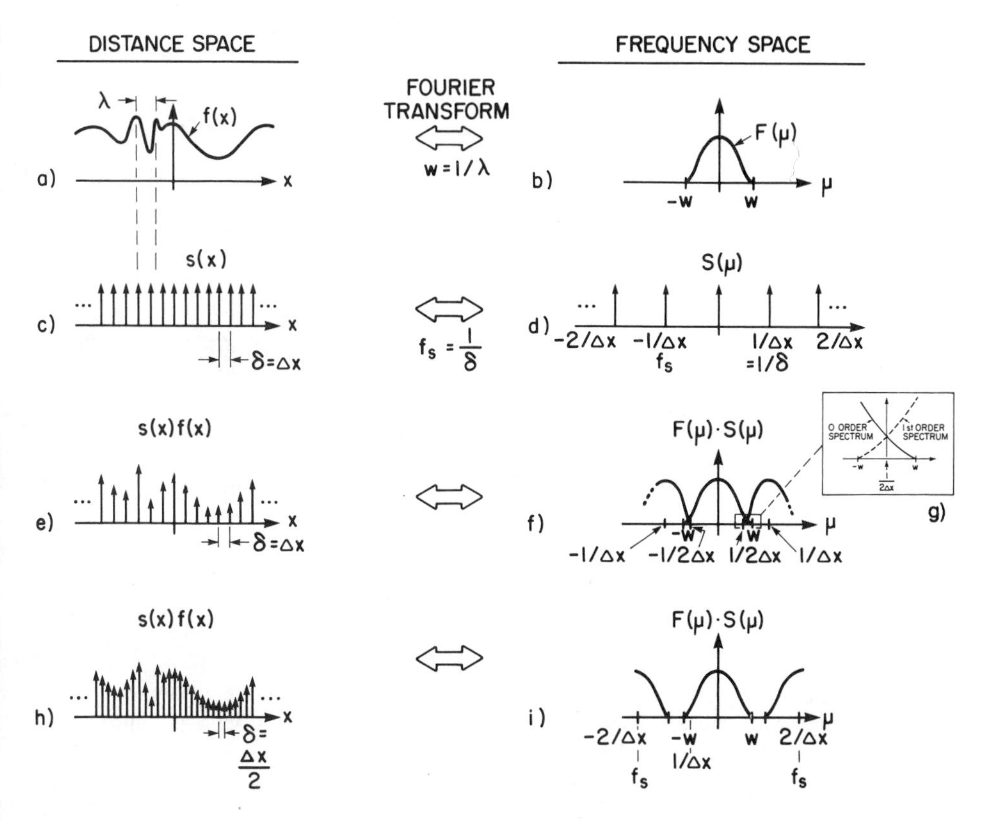

Figure 3.8 The Fourier transformation of images and the effects of sampling frequency: the origin and prevention of aliasing: (a) original object; (b) spectrum of object; (c) sampling function; (d) spectrum of sampling function; (e) sampled object; (f) spectrum of sampled object; (g) detail of sampled object spectrum; (h) object sampled at double frequency; (i) spectrum of object sampled at double frequency.

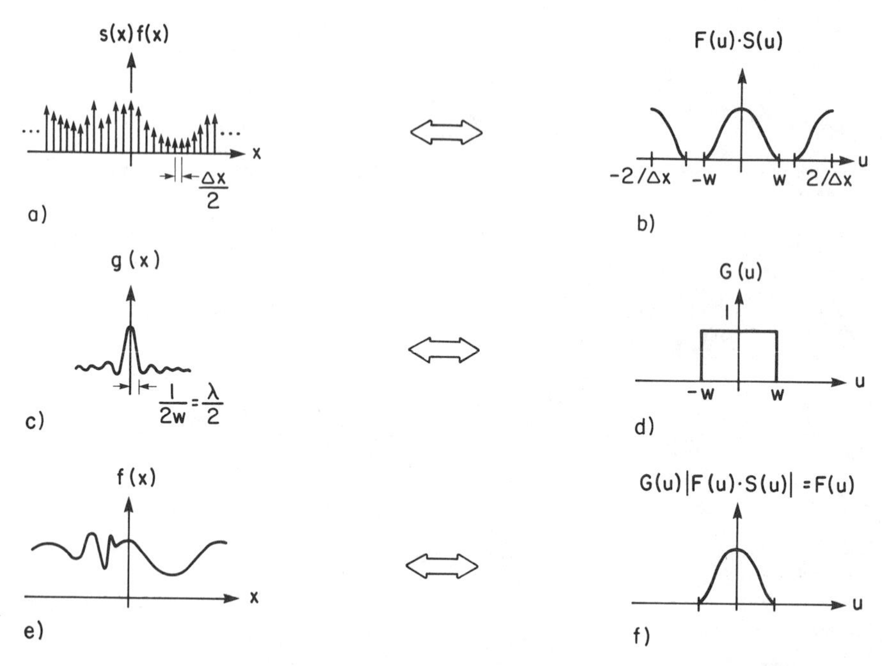

Figure 3.9 Recovery of original object from properly sampled imaging process: (a) object sampled at double frequency (from Figure 8(h)); (b) spectrum of "a" (from Figure 8(i)); (c) spread function for rectangular frequency filter function; (d) rectangular frequency function; (e) recovered object function; (f) recovered object spectrum.

Should we desire to avoid the problem of aliasing, one must sample at a finer sampling interval, as shown in Figure 3.8h. Here the spacing is one half that of the earlier sketches, and therefore the sampling frequency is twice as high. This also doubles the Nyquist frequency. This merely separates the spectra more distinctly by spreading them out by a factor of 2. Since there is no overlap of zero and first orders in this example one can recover the original signal quite easily by simply filtering out the higher frequencies representing the orders other than zero. This is illustrated in Figure 3.9, where a rectangular function of width $\pm w$ and amplitude 1 is multiplied by the sampled image spectra, resulting in recovery of the original signal spectra. When inversely Fourier transformed, this would give the original signal back. See Figures 3.9e and 3.8a.

We can now restate Shannon's [4] formal sampling theorem in terms that apply to sampled imaging: if a function $f(x)$ representing either an original object or the optical/aerial image being digitized contains no frequencies higher than w cycles/mm (this means that the signal is band-limited at w) it is completely determined by giving its values at a series of points 1/2w mm apart. It is formally required that there be no quantization or other noise and that this series be infinitely long; otherwise windowing effects at the boundaries of smaller images may cause some additional problems (e.g., digital perturbations from the presence of sharp edges at the ends of the image). In practice, it needs to be long enough to render such windowing effects negligible.

It is clear from this that any process such as imaging by a lens between the document and the actual sampling, say by a CCD sensor, can band-limit the information and ensure accurate effects of sampling with respect to aliasing. However, if the process of band-limiting the signal in order to prevent aliasing causes the document to lose information that was important visually, then the system is producing restrictions that would be interpreted as excessive blur in the optical image. Another way to improve on this situation is, of course, to increase the sampling frequency, i.e., decrease the distance between samples.

We have shown in Figure 3.9 that the process of recovering the original spectrum is accomplished by a filter having a rectangular shape in frequency space (Figure 3.9d). This filter is known as the reconstruction filter and represents an idealized reconstruction process. The rectangular function has a $(\sin x)/x$ inverse transform in distance space (see Figure 3.9c) whose zero crossings are at $\pm N\ \Delta x$ from the origin where $N = 1, 2, \ldots$ This and other filters with flat MTFs are difficult to realize in incoherent systems because of the need for negative light in the sidelobes in distance space. This filter need not be purely rectangular in order to work. It should be relatively flat and at a value near 1.0 over the bandwidth of the signal being reconstructed (also difficult and often impossible to achieve). It must not transmit any energy from the two first-order spectra. If the sampling resolution is very high and the bandwidth of the signal is relatively low, then the freedom to design the edge of this reconstruction filter is relatively great and therefore this edge does not need to be as square. From a practical point of view the filter often is the MTF of the output scanner, typically a laser beam scanner, and is not usually a rectangular function but more of a gaussian shape. A non-rectangular filter, such as that provided by a gaussian laser beam scanner, alters the shape of the spectrum that it is trying to recover. Since the spectrum is multiplied by the reconstructing MTF, this causes some additional attenuation in the high frequencies, and a trade-off normally is required in real designs.

Quantization, Noise, and the Eye

Now that we have seen how the distance dimension of an input document may be digitized into discrete pixels, it is important to recognize that the image requires quantization into a finite number of discrete gray levels. From a practical standpoint this quantization is usually accomplished by some form of analog-to-digital converter, which usually will quantize the signal into a number of gray levels

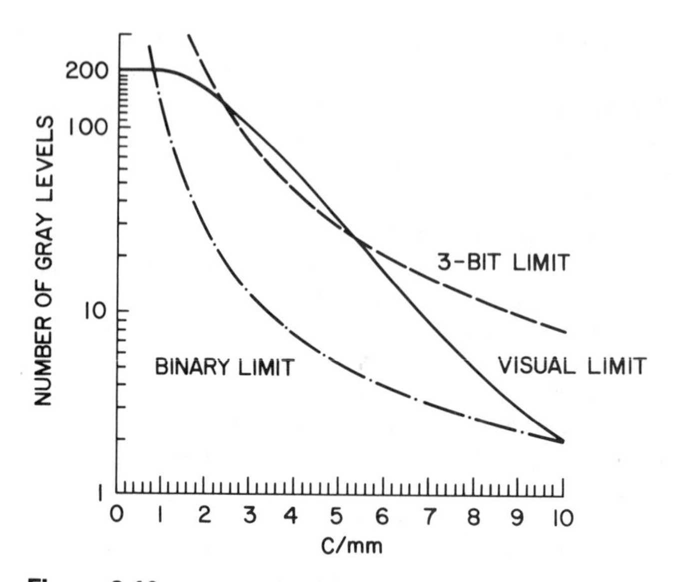

Figure 3.10 Visually distinguishable number of gray levels versus spatial frequency, with corresponding 2- and 3-bit/pixel limits.

which is some power of 2. From the imaging science point of view, one can set some limits on the useful number of quantization levels based upon noise in the system or upon the ultimate goal of distinguishable gray levels seen by the human eye. Both approaches have been explored in the literature and involve complex calculations and experimental measurements of the eye's ability to detect changes in gray. Results shown in Figure 3.10 (after Ref. 5) plot the number of visually distinguishable gray levels against the spatial frequency at which they can be seen.

This curve was derived from a very conservative estimate of the visual system frequency response and may be thought of as an upper limit on the number of gray levels required by the eye. Plotted on the same curve are performance characteristics for 20 pixels/mm (500 pixels/in.) digital imaging systems that produce 3-bits/pixel and 1-bit/pixel (binary) images. These were obtained by use of a generalized algorithm to create halftone patterns (see ''Halftone System Response and Detail Rendition,'' page 107) at different spatial frequencies. The binary limit curve shows the number of effective gray levels for each frequency whose period is 2 halftone cells wide. The 3-bit limit assumes each halftone cell contributes 2^3 gray values, including black. Roetling also integrated the visual response curve to find the average of 2.8 bits per pixel as a good upper bound for the eye itself. Note that his general halftoning aproach using 3 bits/pixel and 20 pixels/mm (500 pixels/in.) also approximates the visual limit. Other specialized halftoning techniques using multiple gray levels per pixel have been proposed [6], which produce more gray levels per pixel at the lower frequencies. They reduce high-frequency performance while using the same number of bits/pixel but quantized in very nonuniformly spaced gray steps. Use of the eye response with various halftoning methods represents an approach to defining practical quantization limits for scanned imaging.

Another approach to setting quantization limits is to examine the noise in input photographic images, assuming in so doing that the quantization is input bound and not output bound by the visual process as in the foregoing approach. The basic principle for describing the useful number of gray levels in a photograph involves quantizing the density scale of the photographic process into steps whose size is based on the granularity of the imaging process. To a first order, granularity is the standard deviation of the density fluctuations, which is directly proportional to the square root of the aperture area in the measuring instrument or scanner.

Table 3.2 shows the number of distinguishable gray levels reported in the literature by various authors, and their conversion to the number of distinguishable gray levels per pixel based on some assumptions made here. In the table we have converted the number of distinguishable levels per square millimeter reported in the literature for the film to the number of distinguishable levels in perfect enlarged

Table 3.2 Number of Distinguishable Gray Levels Reported for Several Silver Halide Films and the Corresponding Number of Bits per Pixel under Various Conditions

		Case →	A	B	C	D	E
		Enlargement →	10x	10x	3.3x	10x	3.3x
		effect apert →	100 μm	100 μm	100 μm	50 μm	50 μm
		K →	10	3	3	3	3
		apert at film* →	10 μm	10 μm	30 μm	5 μm	16 μm
Film Type	Spread Diam (d) μm	# of Levels for d @ K = 3	Bits per Pixel				
Kodak 5454 microfilm[1]	12.5	20	1.60	3.3	5.0	2.3	4.0
Kodak Pan X[1]	15	10	0.70	2.4	4.1	1.4	3.1
EG&G XR (color)[2]	35	67	0.50	2.2	3.9	1.2	2.9
Kodak Plus X[1]	---	10	0.23	2.0	3.7	1.0	2.7
Kodak Royal X[1]	27	6.6	0.14	1.8	3.5	0.8	2.5

Approximate Input Quantization Limits in bits/pixel for Photographic Documents in a 20 ℓ/mm scanner.

1 Altman & Zweig ('63)

2 Lehmbeck ('67)

* Assumes perfect enlargement of grains such that effective square aperture of this width on enlargement sees same granularity as the indicated aperture on film given by effective aperture ÷ enlargement

** Unlike 100 μm unaliased cases, 50 μm square aperture cases are for a highly aliased 20 ℓ/mm scanner.

photographic prints of these films. These are then scanned. A scanner aperture equivalent to that in the Roetling visual calculations was used, i.e., an unaliased 20 samples/mm (500 samples/in.) scanning system with an aperture of 100 × 100 μm (4 × 4 mils). Values for an unaliased 40 samples/mm (1000 samples/in.) system or an aliased 20 samples/mm (500 samples/in.) system using a 50 × 50 μm (2 × 2 mils) aperture are also given. Estimated values for 10X enlargements and 3.3X enlargements made from the associated films are indicated in the table. (This estimate is an oversimplification of the enlarging process ignoring significant nonlinearities and MTF effects in that process.) A practical range of quantization levels is, therefore, anywhere from about 0.1 (column A, Royal X) to 5 (column C, 5454 film) bits per pixel. Of the pictorial films listed (now mostly obsolete), Kodak Panatomic X places the greatest demands on input scanner performance because of its low granularity and high resolution. Hence we select it as representative of the limits for quantization and select the 3.3X enlargement as being the most demanding case. Using the three-standard-deviation criterion, this suggests a limit of 4.1 bits/pixel in an unaliased system or 3.1 bits/pixel in a heavily aliased system. This compares with the rate of 2.8 bits/pixel found by Roetling for a visual limit. Hence, a reasonably high quality photographic product, Panatomic X, used in a modest-quality scanner treated as an input bounded system is compatible with a similar scanner bounded by visual limits on output.

If we assume that the rest of the electrooptical imaging system, including the input and output scanners, is going to add some noise to the imaging process, then it would be desirable up front to have more information or more gray levels in order to allow for further degradation while still obtaining an output that has at least 2.8 bits/pixel.

To help the reader make his own calculation of the information capacity and therefore the ultimate information content of the input to the scanner, a brief description of image information is given below. A more detailed examination of this topic will be given in Section 3.5.7. Equation (1) defines image information, H, as

$$H = \frac{1}{a}\,\mathrm{Log}_2\left[\frac{\text{probability of density message being correct}}{\text{probability of a specific density as input}}\right] \tag{1}$$

where a is area of the smallest resolvable unit in the image (i.e., 2 × 2 pixels, based on the sampling theorem) and the Log factor is the classic definition of information in any message [7]. To convert this into more useful terms, the numerator is set equal to p, the reliability or probability that a detected level within a set of distinguishable gray levels is actually the correct one, and M is the number of distinguishable gray (i.e., quantization) levels. Equation (1) then takes on the form

$$H = \frac{1}{a}\,\mathrm{Log}_2\,\frac{p}{1/M} \underset{p \to 1}{\approx} \frac{\mathrm{Log}_2 M}{a} \tag{2}$$

Assuming a high reliability such that p approaches unity, the simplification on the right results.

To apply this to a photographic document, one can approximate M as

$$M = \frac{L}{K\sigma_a} \tag{3}$$

where L is the density range or practical upper limit of the photographic material, K represents the number of standard deviations of density to be included in each distinguishably different quantization level, and σ_a is the standard deviation of density at a nominal density level. This latter must be measured with a microdensitometer whose aperture area is equal to a. An approximation useful in comparing different photographic materials uses the standard deviation at a mean density of approximately 1 to 1.5 and equation (4) results:

$$H \approx \frac{1}{a}\,\mathrm{Log}_2\left[\frac{L}{6\sigma_a}\right] \tag{4}$$

where K was set to ± 3 $(=6)$, leading to $p = 0.997$ (≈ 1). Since the standard deviation of density is strongly dependent on the mean density level, it is more accurate and also common practice to measure the standard deviation at several average densities and segment the density scale into adjacent nonequal distinguishable, density levels separated by the K standard deviations of density at that level [8–10]. Using this more precise method for counting distinguishable density levels, Table 3.2 was constructed for unaliased and aliased 20 pixels/mm (500 pixels/in.) scanners. The former have a 100 μm (4 mil) square aperture, giving a = 10,000 μm^2 (1.6×10^{-5} in.^{2}), while the latter use a 50μm (2 mil) square aperture, so a = 2500 μm^2 (4×10^{-6} in.^{2}). In both types, $K = 3$, i.e., $\pm 1.5\sigma$.

If the input scanner itself is very noisy, then the σ_a term must be the combined effect of both input noise and scanner noise. This will be covered in Section 3.4.3. Further information on this methodology may be obtained from Refs. 9–11 and Section 3.5.7.

3.2.2 Major System Effects

Blur

As we have seen in the preceding discussion of sampling, blur is a powerful factor in determining the information in an image. In the input scanner, blur is caused by the optical system, the size and properties of the light-sensing element, and by mechanical and timing factors involved in motion. This blur determines whether the system is aliased and the contrast of fine details in the gray video image prior to processing. It can be described conveniently by a series of modulation transfer functions or other metrics that relate generally to the sharpness of optical images. Blur in an output scanner is caused by the size of the writing spot, e.g., the laser beam waist at focus, and by the spreading of the image in any marking process such as xerography or photographic film. It is also affected to some extent by motion of the beam relative to the data rate and the rate of motion of the light-sensitive receptor material. This blur directly affects the appearance of sharpness in the final hard-copy image that is presented to the human visual system. It also affects contrast and the results from resolution targets.

Blur for the total system, both input scanner and output scanner, is not easily cascaded since the intervening processing of the image information is extremely nonlinear. This nonlinearity may give rise to such effects as a blurred input image looking very sharp on the edges for a binary output print because of the small spot size and high MTF of the marking process in the output scanner. In such a case, however, the edges of square corners look rounded and fine detail such as serifs in text or textures in photographs may be lost. This is roughly analogous to what happens when an out-of-focus image is recorded on a high-contrast photographic material such as a lithographic film. Conversely a sharp input scan, e.g., one made with a high MTF lens kept well in focus, but printed with a blurring output system, e.g., one with a large spot, would appear to have fuzzy edges on lines, but the edge noise due to sampling would have been blurred together and would not be as visible as it would in the first case. Moiré,* from aliased images of periodic subjects, however, would still be present in spite of output blur. (Once aliased, no amount of subsequent processing can remove this effect from an image.) Therefore it is apparent that blur can have both positive and negative impacts on the overall image quality.

System Response

There are four ways in which electronic imaging systems display tonal information to the eye or transmit tonal information throughout the system:

A. By producing a signal of varying strength at each pixel, using either amplitude or pulse-width modulation
B. By turning each pixel on or off (a two-level or binary system)
C. By use of a halftoning approach, which is a special case of binary imaging in which the threshold for the white-black decision is varied in some struc-

---

* Superposition of periodic patterns such as a halftoned document (see 2.2.3) and the sampling grid of a scanner results in new and striking periodic patterns in the image commonly called moiré patterns (see Bryngdahl [12]).

tured way over very small regions of the image, simulating continuous response

D. By hybrid halftoning combining the halftone concept in C with the variable gray pixels from A (e.g., see Refs. 5 and 6)

From a hardware point of view, the systems are either designed to carry gray information on a pixel-by-pixel basis or to carry binary (two-level) information on a pixel-by-pixel basis. Since a two-level imaging system is not very satisfactory in many applications, some context is added to the information flow in order to obtain pseudo-gray using the halftoning approach.

The basic way in which all imaging systems are characterized for gray-level performance is some plot of the output response as a function of the input light level. The output may characteristically be volts or digital gray levels for a digital input scanner and perhaps intensity or darkness or density of the final image for an output scanner. The correct choice of units depends upon the application for which the system response is being described. There are often debates as to whether such response curves should be in units of density or optical intensity, brightness, visual darkness, etc. For purposes of illustration see Figure 3.11.

Here we have chosen to use the conventional photographic characterization of output density plotted against input density using normalized densities. Curve A shows the case of a binary imaging system in which the output is white or zero density up to an input density of 0.6, at which point it becomes black or 2.0 output density. Curve B shows what happens when a system responds linearly in a continuous fashion to density. Because the input is equal to the output here, this system would be linear in reflectance, irradiance, or even Munsell value

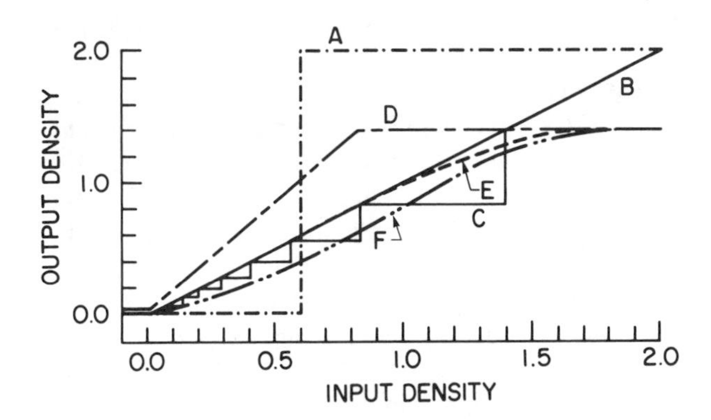

Figure 3.11 Some representative input/output density relationships for (a) binary imaging response; (b) linear imaging response; (c) stepwise linear response; (d) saturation-limited linear response; (e) linear response with gradual roll-off to saturation; (f) idealized response curve for best overall acceptability.

(visual lightness units). Curve C shows a classic abridged gray system attempting to write linearly but with only eight levels of gray. This response becomes a series of small steps but because of the choice of density units, which are logarithmic, the sizes of the steps are very different. Had we plotted output reflectance as a function of input reflectance, the sizes of the steps would have been equal. However, the visual system that usually looks at these tones operates in a more or less logarithmic or power fashion, hence the density plot is more representative of the visual effect for this image. Had we chosen to quantize in 256 gray levels, each step shown would have been broken down into 32 smaller substeps, thereby approximating very closely the continuous curve for B.

When designing the system tone reproduction, there are many choices available for the proper shape of this curve. The binary curve, as in A, is ideal for the case of reproducing high-contrast information because it allows the minimum and maximum densities considerable variation without any change to the overall system response. For reproducing continuous tone pictures, there are many different shapes for the relationship between input and output, some of which are shown in Figure 3.11. If, for example, the input document is relatively low contrast, ranging from a 0 to 0.8 density, and the output process is capable of creating higher densities such as 1.4, then the curve represented by D would represent a satisfactory solution for many applications. However, it would create an increase in contrast represented by the increase in the slope of the curve relative to B which gives one-for-one tone reproductions at all densities. Curve D is clipped at an input density greater than 0.8. This means that any densities greater than that could not be distinguished and would all print at an output density of 1.4.

In many conventional imaging situations the input density range exceeds that of the output density. The system designer is confronted with the problem of dealing with this mismatch of dynamic ranges. One approach is to make the system respond linearly to density up to the output limit, for example, following curve B up to an output density of 1.4 and then following curve D. This generally produces unsatisfactory results in the shadow regions for the reasons given earlier for curve D. One general rule* is to follow the linear response curve in the highlights region and then to roll off gradually to the maximum density in the shadow regions starting perhaps at a 0.8 output density point for the nonlinear portion of the curve as shown by curve E. Curve F represents an idealized case arrived at by Jorgenson [13], who found the curve resembling F to be a psychologically preferred curve among a large number of the curves he tried for lithographic applications. Notice that it is lighter in the highlights and has a midtone region whose slope parallels that of the linear response. It then rolls off

---

* Attributed to J. A. C. Yule by M. Southworth in private communication, 1983.

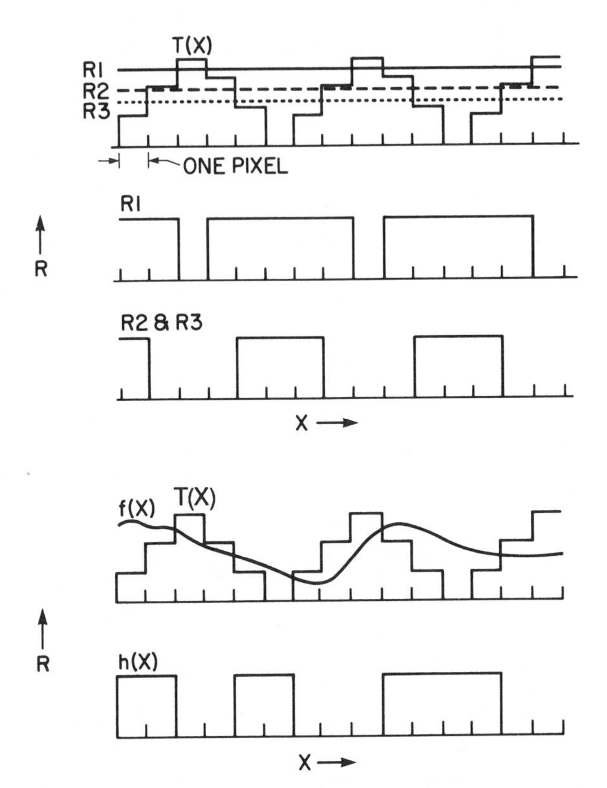

Figure 3.12 Illustration of halftoning process using quantized-threshold-level halftone dots.

much as the previous case toward the maximum output density at a point where the input density reaches its upper limit.

Halftone System Response and Detail Rendition

One of the advantages of digital imaging systems is the ability to completely control the shape of these curves to allow the individual user to find the optimum relationship for his photograph in his application. This can be achieved through the mechanism of digital halftoning as described below. Other studies of tone reproduction, largely for photographic and graphic arts applications, include those of Jones and Nelson [14], Jones [15], Bartleson and Breneman [16], and two excellent review articles, covering many others, by Nelson [17].

The halftoning process can be understood by examination of Figure 3.12. In the top of this illustration two types of functions are shown plotted, against distance x, which has been marked off into increments 1 pixel in width. The first set of functions are three uniform reflectance levels, R1, R2, and R3. The second function $T(x)$ is a plot of threshold versus distance, which looks like a series of up and down staircases. Any pixels whose reflectance is above the threshold will be turned on, and any one that is below the threshold for that pixel will be turned off. Also sketched in Figure 3.12 are the results for the thresholding process for R1 on the second line and then for R2 and R3 on the third line. The last two are indistinguishable for this particular set of thresholding curves. It can be seen from this that the reflectance information is changed into width information and thus that the method of halftoning is a mechanism for creating dot growth or spatial pulse width modulation over an area of several pixels. Typically such patterns are laid out two-dimensionally. An example is shown in Figure 3.13.

Q1 Q2

13	6	7	14	20	27	26	19
12	1	2	8	28	32	31	25
5	4	3	9	21	29	30	24
16	11	10	15	17	22	23	18
20	27	26	19	13	6	7	14
28	323	31	25	12	1	2	8
21	29	30	24	5	4	3	9
17	22	23	18	16	11	10	15

Q3 Q4

8 X 8 Spiral Halftone Matrix

a.

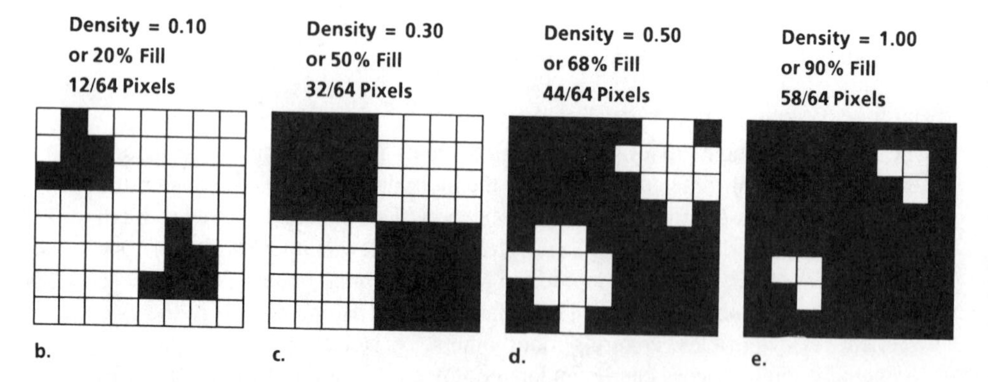

Figure 3.13 Example of two-dimensional quantized halftone pattern, with illustrations of resulting halftone dots at various density levels.

This thresholding scheme emulates the printer's 45° screen angle, which is considered to be favorable from a visual standpoint because the 45° screen is less detectable than the same 90° screen. Other screen angles may also be conveniently generated by a single string of thresholds and a shift factor that varies from raster to raster [18]. The numbers in each cell in the matrix represent the threshold required in a 32-gray-level system to turn the system on or off. At the bottom four thresholded halftone dots (parts b–e) are shown for illustration. There are a total of 64 pixels in the array but only 32 unique levels. This screen can be represented by 32 values in a 4 × 8 pixel array plus a shift factor, or by 64 values in a single 8 × 8 pixel array.

What is perhaps less obvious is that high-spatial-frequency information can be recorded by this type of halftoning process. This can be seen in Figure 3.14. Part b of the figure shows the bit map representation of the original document as it has been captured by an electronic imaging system. The gray levels of 3 and 22 represent background and foreground levels or light and dark pixels. They are the only ones available in this pattern, and they are laid out somewhat in the shape of the letter F. In parts c, d, e, and f we see the resultant thresholded images when the different gray levels given for light and dark pixels are tested against the threshold (halftone) matrix, part a, turning black (B) those that are equal or larger than the halftone matrix element. In part d we again assigned a value of 3 to the light areas but set 29 for the dark parts of the letter. This progresses in parts e and f to contrasts 2/31 and 0/32. Dotted circles indicate errors in correctly reproducing background or foreground pixels as black or white. It is seen that as the contrast of the input increases the errors made by the screen in detecting all of the necessary components go down. One and 2 pixel lines and serifs are readily reproduced by this process. This phenomenon has been described as "partial dotting" by Roetling [19] and others. The illustrations in parts g and h of the same figure show three narrow 1-pixel-wide lines (whose contrast is 30 for the line and 3 for the background field) at different angles and shifts with respect to the halftone threshold matrix. It can be seen, here, that angle and shift information can also be recorded through the process of screening of high-contrast detail.

The halftone matrix described represented 32 specific thresholds. This pattern can also be considered as a sequence in which a series of other threshold values would turn on. Thus the values in the matrix of Figure 3.13a may be thought of as a sequence 1, 2, 3, and 4, etc., which might actually be threshold levels 5, 6, 9, and 14 or some other combination. A gray-level system that had more than 32 levels to use for halftoning would selectively distribute these levels over the sequence defined by the matrix, and thereby determine the mapping of input reflectance to the output area coverage of the black pixels. In such a system, the careful selection of these sequences and thresholds gives good control over the shape of the tone reproduction curve, as will be seen below.

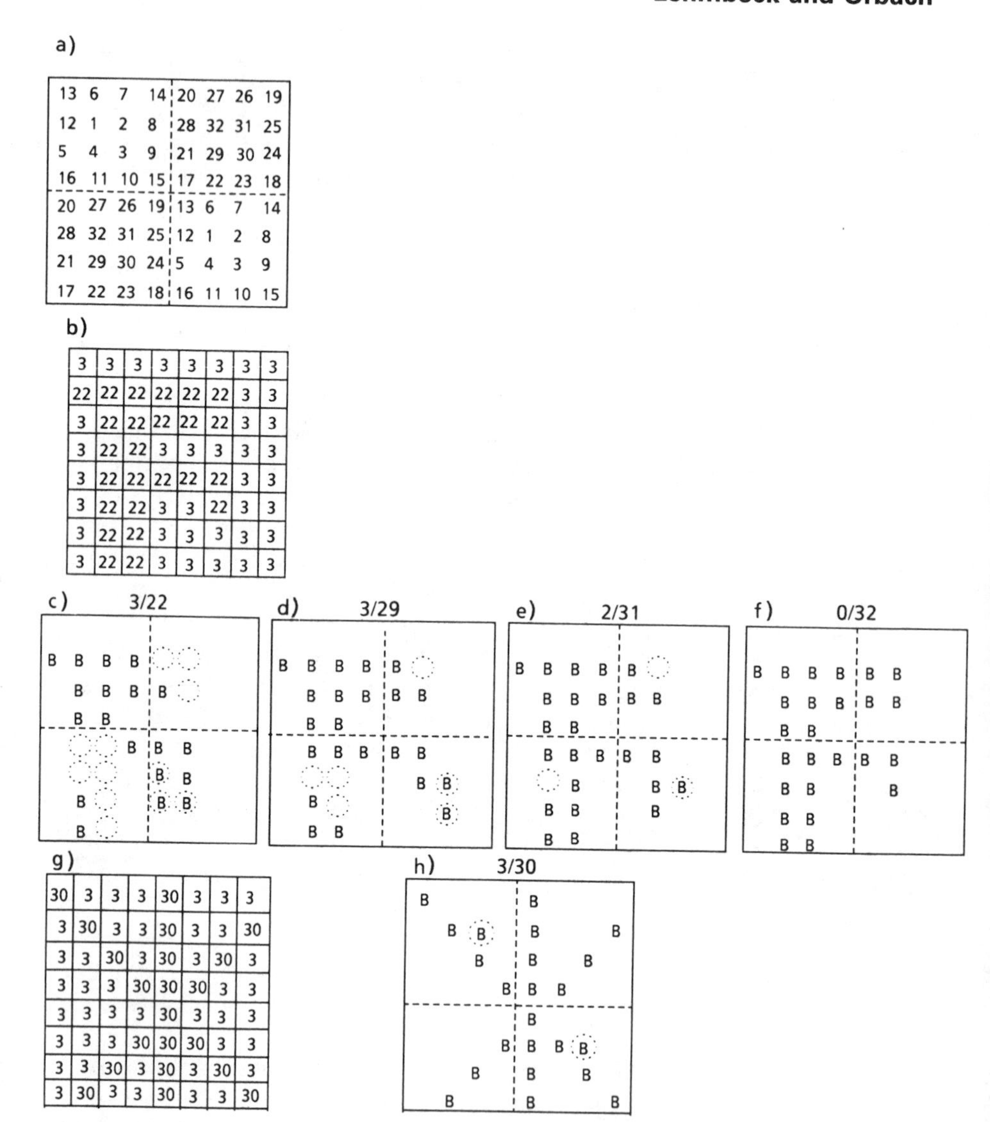

Figure 3.14 Illustration of the preservation of image detail finer than the scale of the halftone pattern (''partial dotting''): (a) the threshold matrix for screen used in (c)–(f) and (h) which are binary screened images; (b) pattern of gray pixels in the image leading to the binary screened image in (c)–(f) showing the explicit combination of gray values, 3 and 22, used for (c). The alternate gray image values for (d)–(f) are shown above each figure; (g) gray values in image leading to (h). If gray value in image b or g is larger than or equal to screen element in a, turn pixel black. An empty dotted circle indicates an error of not writing black. B indicates a black pixel, blank is white in (c)–(f) and (h). A circled B indicates an error of writing black when white is desired. B indicates a black pixel, blank is white in (c)–(f) and (h).

There are many other methods for converting binary images into pseudo-gray images using digital halftoning methods of more complex form ([20, 20a]). These include alternative dot structures, i.e., different patterns of sequences in alternating repeat patterns, random halftoning, and techniques known as error diffusion. In his book *Digital Halftoning*, Ulichney [21] describes five general categories of halftoning techniques:

1. Dithering with white noise (including mezzotint)
2. Clustered dot ordered dither
3. Dispersed dot ordered dither (including "Bayer's dither")
4. Ordered dither on asymmetric grids
5. Dithering with blue noise (actually error diffusion)

He states that "spatial dithering is another name often given to the concept of digital halftoning. It is perfectly equivalent, and refers to any algorithmic process which creates the illusion of continuous tone images from the judicious arrangement of binary picture elements." The process just described in Figures 3.13 and 3.14 falls into the category of a clustered dot ordered dither method (number 2) as a classical rectangular grid on a 45° base.

The reader is cautioned that the densities described in Figure 3.13 only apply to the case of perfect reproduction of the illustrated bit maps on nonscattering material. In reality each pattern of pixels must be individually calibrated for any given marking process. The spatial distribution interacts with various noise and blurring characteristics of output systems to render the mathematics of counting pixels to determine precise density relationships highly erroneous under most conditions. This is even true for the use of halftoning in conventional lithographic processes, due to the scattering of light in white paper and the optical interaction of ink and paper. These relationships have been addressed in the literature, both in a correction factor sense [22, 23] and in a spatial frequency sense [24–26]. All of these methods involve various ways of calculating the effect that lateral light scattering through the paper has on the light reemerging from the paper between the dots.

The effects of blur from the writing and marking processes involved in generating the halftone, many of which may be asymmetric, require individual density calibrations for each of the dot patterns and each of the dithering methods that can be used to generate these halftone patterns. The control afforded through the digital halftoning process by the careful selection of these patterns and methods enables the creation of any desired shape for the tone reproduction curve for a given picture, marking process, or a given application.

Noise

Noise can take on many forms in an electronic imaging system. First of all there is the noise inherent in the digital process. This is generally referred to as either sampling noise associated with the location of the pixels or quantization noise associated with the number of discrete levels. Examples of both have been considered in the earlier discussion. Next there is electronic noise associated with the electronic components from the sensor to the amplification and correction circuits. As we move through the system the digital components are generally thought to be error-free and therefore there is usually no such noise associated with them. Next, in a typical electronic system, we find the output scanner itself, often a laser beam scanner. If the system is writing a binary file, then the noise associated with this subsystem generally has to do with pointing of the beam at the imaging material and is described as jitter, pixel placement errors, or raster distortion of some form (see the next subsection). Under certain circumstances exposure variation will also produce noise, even in a binary process. For systems with gray information, there is also the possibility that the signals driving the modulation of exposure may be in error, so that the ROS can also generate noise similar to that of granularity in photographs or streaks if the error occurs repeatedly in one orientation. Finally we come to the marking process, which converts the laser exposure from the ROS into a visible signal. This marking process noise, which generally occurs as a result of the discrete and random nature of the marking particles, generates granularity.

An electronic imaging system may enhance or attenuate the noise generated earlier in the process. Systems that tend to enhance detail with various types of filters or adaptive schemes will also likely enhance noise. There are, however, some processes that search through the digital image identifying errors and substitute an error-free pattern for the one that shows a mistake [27]. These are sometimes referred to as noise removal filters. Noise may be characterized in many different ways, but in general it is some form of statistical distribution of the errors that occur when an error-free input signal is sent into the system. In the case of imaging systems, an error-free signal is one that is absolutely uniform. Examples would include a sheet of white microscopically uniform paper on the platen of an input scanner, or a uniform series of laser-on pulses to a laser beam scanner, or a uniform raster pattern out of a perfect laser beam scanner writing onto the light-sensitive material in a particular marking device. A typical measure of these systems would be to evaluate the standard deviation of the output signal in whatever units characterize it. A slightly more complete analysis would break this down into a spatial frequency or time-frequency distribution of fluctuations. For example, in a photographic film a uniform exposure would be used to generate images whose granularity was measured as the root mean square fluctuation of density. For a laser beam scanner it would be the root mean square fluctuation in radiance at the pixel level for all "white" raster lines.

In general, the factors that affect the signal aspect of an imaging system positively, affect the noise characteristics of that imaging system negatively. For example, in scanning photographic film the larger the sampled area, as in the case of the microdensitometer aperture, the lower the granularity. At the same time, the image information is more blurred, therefore producing a lower contrast and smaller signal level. In general the signal level increases with aperture area and the noise level (as measured by the standard deviation of that signal level) decreases linearly with the square root of the aperture area or the linear dimension of a square aperture. It is therefore very important when designing a scanning system to understand whether the image information is being noise limited by some fundamentals associated with the input document or test object or by some other component in the overall system itself. An attempt to improve bandwidth, or otherwise refine the signal, by enhancing some parts of the system may, in general, do nothing to improve the overall image information, if it is noise which is limiting and which is being equally amplified. Also, if the noise in the output writing material is limiting, then improvements upstream in the system may reach a point of diminishing returns.

In designing an overall electronic imaging system it should be kept in mind that noises add throughout the system, generally in the sense of an RSS (root of the sum of the squares) calculation. The signal attenuating and amplifying aspects, on the other hand, tend to multiply throughout the system. If the output of one subsystem becomes the input of another subsystem, the noise in the former is treated as if it were signal in the latter. This means that noise in the individual elements must be appropriately mapped from one system to the other, taking into account various amplifications and nonlinearities. In a complex system this may not be easy; however, keeping an accurate accounting of noise can be a great advantage in diagnosing the final overall image quality. We will expand on the quantitative characterization of these various forms of signal and noise in the subsequent parts of this chapter.

3.3 PRACTICAL CONSIDERATIONS

Several overall systems design issues are of some practical concern including the choice of scan frequency and motion errors as well as other nonuniformities. They will be addressed here in fairly general terms.

3.3.1 Scan Frequency Effects

It has generally been thought that the spatial frequency, in rasters or pixels per inch, that is used either to create the output print or to capture the input document is a major determinant of image quality. The most commonly encountered raster spacings are

96, 100, 192, 200, 240, 300, 384, 400, 480, 500, 600, 720, 800 pixels/in.
(3.78, 3.94, 7.56, 7.87, 9.45, 11.81, 15.12, 15.75, 18.90, 19.7, 23.6,
28.35, 31.5 pixels/mm)

In the range 1000 to 2500 pixels/inch (39.4–98.4 pixels/mm) there is a broad class of very high resolution systems that no particular frequencies tend to dominate at this time.

If a scanned imaging system is designed so that the input scanning is not aliased and the output reconstruction faithfully prints all of the information presented to it, then the scan frequency tends to determine the blur, which largely controls the overall image quality in the system. This is frequently not the case, and, as a result, scan frequency is not a unique determinant of image quality. In general, however, real systems have a spread function or blur that is roughly equivalent to the sample spacing, meaning they are somewhat aliased and that blur correlates with spacing. However, it is possible to have a large spot and much smaller spaces (i.e., unaliased), or vice versa (very aliased). The careful optimization of the other factors at a given scan frequency may have a great deal more influence on the information capacity of any electronic imaging system and therefore on the image-quality performance than does scan frequency itself. To a certain extent, gray information can be readily exchanged for scan frequency. We shall subsequently explore this further when dealing with the subject of information content of an imaging system.

Table 3.3 shows some applications for different scan frequencies that are often found in the literature and in commercial practice. Many of these are common usage patterns but do not reflect ultimate optimization. It should also be recognized that the frequencies listed are for a full-size print or display as seen by the observer and may not reflect the particulars of the given optical system, which either reduces or magnifies the image at some intermediate point in the overall process. For example, a scanner may sample 35mm photographic film at 1800 pixels/in (71 pixels/mm) for subsequent printing at 3 to 10X enlargement.

A more general practice is to design aliased systems in order to achieve the least blur for a given scan frequency. Therefore, another major effect of scan frequency concerns the interaction between periodic structures on the input and the scanning frequency of the system that is recording the input information. These two interfere, producing beat patterns at sum and difference frequencies leading to the general subject of moiré phenomena. Hence, small changes in scan frequency can have a large effect on moiré.

Input and output scanning frequencies also affect magnification. A 300 pixels/in (11.8 pixels/mm) electronic image printed at 400 pixels/in (15.7 pixels/mm) is only three-fourths as large as the original, while one printed at 200 pixels/in (7.87 pixels/mm) apears to be enlarged 1.5X. One-dimensional errors in scan frequency cause anamorphic magnification errors.

One of the major considerations in selecting output scan frequency is the number of gray levels required from a given range of halftone screens. Recall the discussion of Figure 3.13. Dot matrices from 4×4 to 12×12 are shown in Table 3.4 at a range of frequencies from 200 to 1200 raster lines per inch

Table 3.3 Typical Applications of Commonly Used Scan Frequencies Reported in the Literature for Binary Imaging Processes

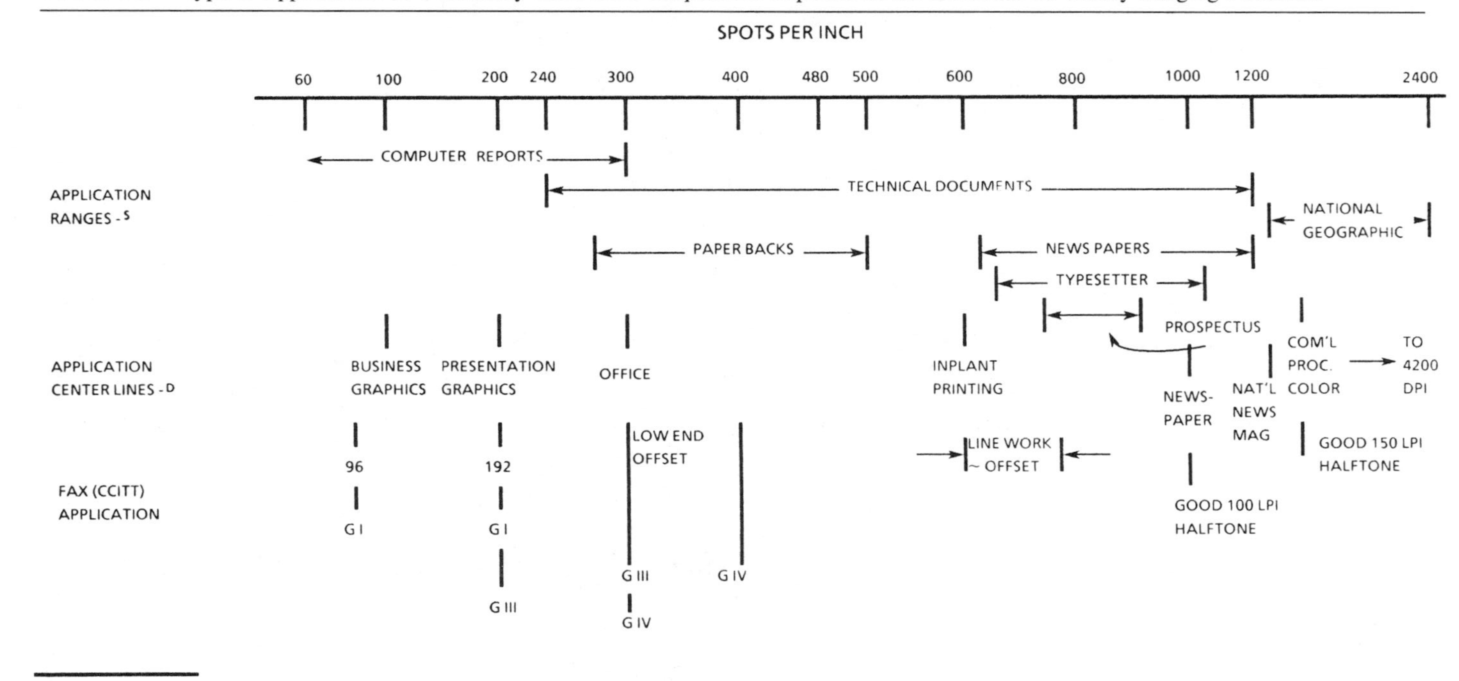

S Seybold Report on Publishing Systems, Vol. 16, #11, P. 1 (Feb 87)

D Dunn Report on EPPS (Electronic Publishing and Prepress Systems), VOL. 5, #1, P. 1 (Jan 87)

Table 3.4 Relationships among Halftone Patterns, Gray Levels Potentially Generated by Each, and Scan Frequencies in Halftone Printing. Units Are Screen Frequencies in Dots/Inch for the Halftone Pattern

Matrix in pixels →	4x4	3x3	4x4	6x6	5x5	8x8	6x6	10x10	12x12	
Angle →	45°	90°	90°	45°	90°	45°	90°	45°	45°	
# gray levels →	9	10	17	18	25	32	36	50	72	
Scan frequency in LPI:										
1200	426	400	300	282	240	212	200	170	141	practical upper lim (<200 LPI)
1000	352	333	250	236	200	176	167	142	118	
800	284	267	200	188	160	142	133	114	94	
600	212	200	150	141	120	106	100	85	71	
500	176	166	125	118	100	88	84	71	59	65 LPI lower visibity limit
400	142	133	100	94	80	71	67	57	47	
300	106	100	75	71	60	53	50	43	36	
200	71	67	50	47	40	35	33	28	24	

(7.87 to 47.2 raster lines per millimeter). For example, a 10 × 10 matrix of thresholds can be used to generate a 50-gray-level, 45° angle screen (2 shifted 5 × 10 submatrices) whose screen frequency is shown in the eighth column in Table 3.4. Also indicated in the table is the approximate useful range for the visual system, represented by a lower limit of a 65-dot/in (2.56 dots per millimeter) halftone screen commonly found in newspapers of the past decade [this has recently begun to move toward 85 dots/in (3.35 dots per millimeter)] and the upper bound representing a materials limit around 175 dots/in (6.89 dots per millimeter), which is a practical limit for many lithographic processes. This table assumes that the scan frequency is not subdivided into higher addressability components.

3.3.2 Placement Errors or Motion Defects

Since the basic mode of operation for most scanning systems is to move or scan rapidly in one direction and slowly in the other, there is always the possibility of an error in motion or other effect that results in locating pixels in places other than those intended. Figure 3.15 shows several examples of periodic raster separation errors, including both a sinusoidal and a sawtooth distribution of the error. These are illustrated at 300 raster lines/in (11.8 per mm) with ±10 through ±40 μm (±0.4 through ±1.6 mils) of spacing error, which refers to the local raster line spacing and not to the error in absolute placement accuracy. Error

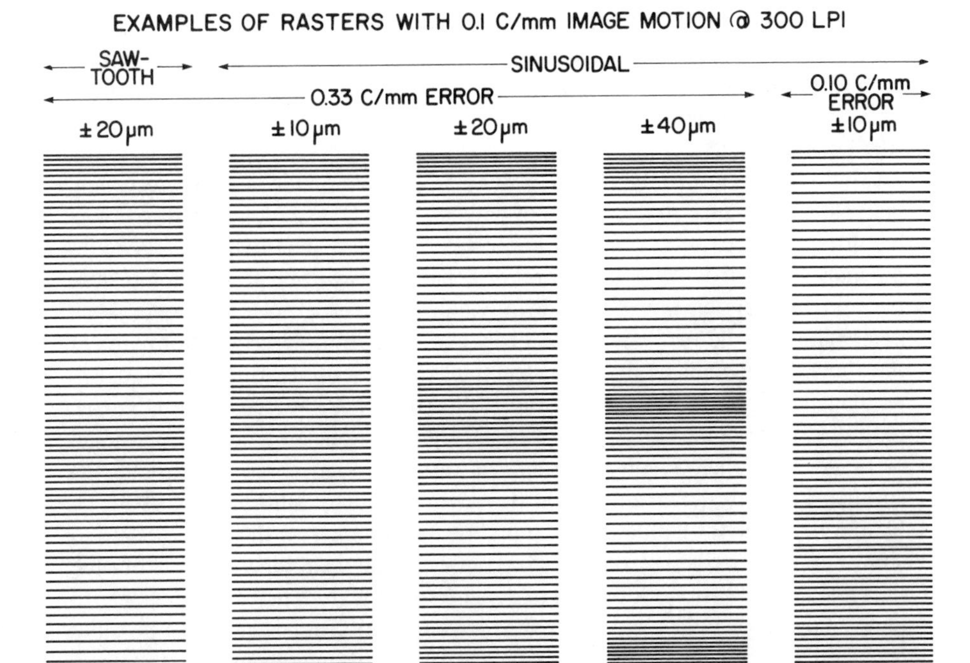

Figure 3.15 Illustrations of sinusoidal and sawtooth raster spacing errors.

frequencies of 0.33 cycles/mm (8.4 cycles per inch) and 0.1 cycles/mm (2.5 cycles per inch) are illustrated.

For input scanners, which convert an analog signal to a digital one, the error takes the form of a change in the sampling of the analog document. Since sampling makes many mistakes, the sampling errors due to motion nonuniformity are most visible in situations where the intrinsic sampling error is made to appear repeatable or uniform, and the motion error, therefore, appears as an irregular change to an otherwise uniform pattern. Long angled lines that are parallel to each other provide such a condition because each line has a regular periodic phase error associated with it, and a motion error would appear as a change to this regular pattern. Halftones that produce moiré are another example, except that the moiré pattern is itself usually objectionable so that a change in it is not often significant. For patterns with random phase errors such as text, the detection of motion errors is more difficult. Effects that are large enough to cause a 2-pixel error would be perceived very easily; however, effects that produce less than 1-pixel error on the average would tend to increase the phase errors and noise in the image generally and would therefore be perceived on a statistical basis. Many identical patterns repeated throughout a document would provide the opportunity to see the smaller errors as being correlated along the length of the given raster line which has been erroneously displaced, and would therefore increase the probability of seeing the small errors.

Motion errors in an output scanner that writes on some form of image-recording material can produce several kinds of defects. In Table 3.5, several attributes of the different types of raster distortion observables are shown. The first row in the matrix describes the general kind of error, i.e., whether it is predominantly a pixel placement error or predominantly a developable exposure effect or some combination of the two. The second row is a brief word description or name of the effect that appears on the print. The third row describes the spatial frequency region in cycles/mm in which this type of error tends to occur. The next row indicates whether the effect is best described and modeled as one-dimensional or two-dimensional. Finally, a graphical representation of an image with the specific defect is shown in the top row, while the same image appears in the bottom row without the defect. The first of the columns on the left is meant to show that if the frequency of the error is low enough then the effect is to change the local magnification. A pattern or some form of texture that should appear to have uniform spacings would appear to have nonuniform spacings and possibly the magnification of one part of the image would be different from that of another. The second column is the same type of effect except the frequency is much higher, being around 1 cycle/mm (25 cycle per inch). This effect can then change the shape of a character, particularly one with angled lines in it, as demonstrated by the letter Y.

Moving to the three right-hand columns, which are labeled as developable exposure effects, we have three distinctly different frequency bands. The nature and severity of these effects depends in part on whether we are using a "write white" or "write black" recording system and on the contrast or gradient of the recording material. The first of these effects is labeled as structured background. When the separation between raster lines increases and decreases, the exposure in the region between the raster lines where the gaussian profile writing beams overlap increases or decreases with the change. This gives an overall increase or decrease in exposure, with an extra large increase or decrease in the overlap region. Since many documents that are being created with a laser beam scanner have relatively uniform areas, this change in exposure in local areas gives rise to nonuniformities in the appearance in the output image. In laser printers, which are used to output computer information, for example, the text is generally presented against a uniform white background. In a positive "write white" electrophotographic process, such as is used in many xerographic printers, this background is ideally composed of a distribution of uniformly spaced raster lines that expose the photoreceptor so that it discharges to a level where it is no longer

Table 3.5 Effects of Motion Irregularities (or Defects or Errors) on the Appearance of Scanned Images

	Pixel Placement Error		Combination		AM/FM Developable Exposure Effects		
Effect On Print	Spacing Non-Uniformity	Character Distortion (&F.S. Jitter)	1/2 Tone Non-Uniformity	Line Darkness Non-Unif.	Structured BAckground	Ragged/ Structured Edges	Unsharp Images
Frequency	< .5	.5-2	.1-6	.005-2	.005-8	1-8	4-24
Type of Effect	1D	2D	2D (1D)	1D (2D)	1D	2D	2D

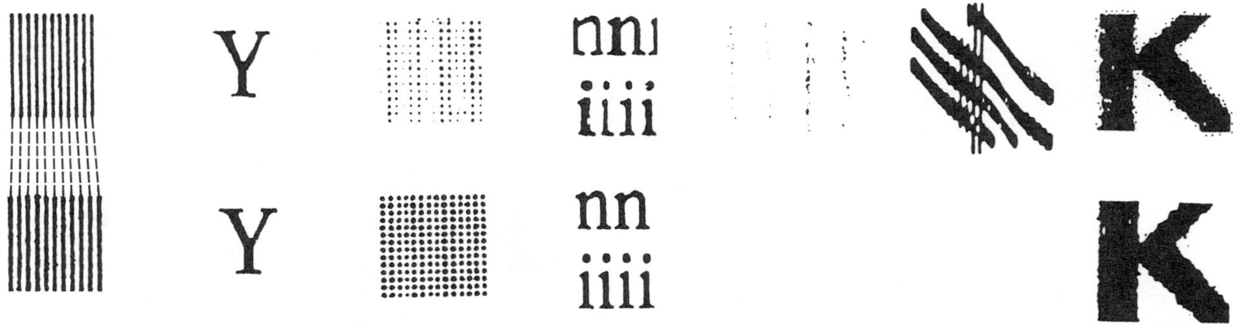

1 2 3 4 5 6 7

XEROX Private Data

developable. As the spacing between the raster lines increases, the exposure between them decreases to a point where it no longer adequately discharges the photoreceptor, thereby enabling some weak development fields to attract toner and produce faint lines on a page of output copy. For this reason among others, some laser printers use a reversal or negative "write black" form of electrophotography in which black (no light) output results in a white image. Therefore, white background does not show any variation due to exposure defects, but solid dark patches often do.

An amplitude allowed for these exposure variations can be derived from minimum visually perceivable modulation values and the gradient of the image-recording process [28, 29a]. In the spatial frequency region near 0.5 cy/mm (13 cy/in), where the eye has its peak response at normal viewing distance, an exposure modulation of 0.004 to 0.001 $\Delta E/E$ is shown to be a reasonable goal for a color photographic system with tonal reproduction density gradients of 1 to 4 [29].

If the frequency of the perturbation is on the order of 1 to 8 cycles/mm (25 to 200 cycles/in), and especially if the edges of the characters are slightly blurred, it is possible for the nonuniform raster pattern to change the exposure in the partially exposed blurred region around the characters. As a result nonuniform development appears on the edge and the raggedness increases as shown by the jagged appearance of the wavy lines in the figure. The effects are noticeable because of the excursions produced by the changes in exposure from the separated raster lines at the edges of even a single isolated character. The effect is all the more noticeable in this case because the darkened raster lines growing from each side of the white space finally merge in a few places. The illustration here, of course, is a highly magnified version of just a few dozen raster lines and the image contained within them. In the last column we see small high-frequency perturbations on the edge which would make the edge appear less sharp. Notice that structured background is largely a one-dimensional problem, just dealing with the separation of the raster lines, while character distortion, ragged or structured edges, and unsharp images are two-dimensional effects showing up dramatically on angled lines and fine detail. In many cases the latter require two dimensions to describe the size of the effect and its visual appearance.

Visually apparent darkness for lines in alphanumeric character printing can be approximately described as the product of the maximum density of the lines in the character times their widths. It is a well-known fact in many high-contrast imaging situations that exposure changes lead to linewidth changes. If the separation between two raster lines is increased, the average exposure in that region decreases and the overall density in a write white system increases. Thus, two main effects operate to change the line darkness. First, the raster information carrying the description of the width of the line separates, writing an actually wider pattern. Second, the exposure level decreases, causing a further growth in the linewidth and to some extent causing greater development, i.e., more density. The inverse is true in regions where the raster lines become closer together. Exposure increases and linewidth decreases. If these effects occur between different strokes within a character or between nearby characters, the overall effect is a change in the local darkness of text. The eye is generally very sensitive to differences of line darkness within a few characters of each other and even within several inches of each other. This means that the spatial frequency range over which this combination of stretching and exposure effect can create visual differences is very large, hence the range of 0.005 to 2 cycles/mm (0.127 to 50 cycles/in). Frequencies listed in the figure cover a wide range of effects, including some variation of viewing distance as well. They are not intended as hard boundaries but rather to indicate approximate ranges.

Halftone nonuniformity follows from the same general description as given for line darkness nonuniformity except we are now dealing with dots. The basic effect, however, must occur in such a way as to affect the overall appearance of darkness of the small region of an otherwise uniform image. A halftone works on the principle of changing a certain fractional area coverage of the halftone cell. If the spatial frequency range of this nonuniformity is sufficiently low, then the cell size changes at the same rate that the width of the dark dot within the cell changes; thus the overall effect is to have no change in the percent area coverage and only a very small change in the spacing between the dots. Hence, the region of a few tenths to several cycles/mm (several to tens of cycles/in) is the domain for this variable. It appears as stripes in the halftone image.

The allowable levels for the effects of pixel placement errors on spacing nonuniformity and character distortion depend to a large extent upon the application. In addition to application sensitivity, the effects that are developable or partially developable are highly dependent upon the shape of the profile of the writing spot and upon amplification or attenuation in the marking system that is responding to the effects. Marking systems also tend to blur out the effects and add noise, masking them to a certain extent.

3.3.3 Other Nonuniformities

There are several other important sources of nonuniformity in a raster scanning system. First of all, there is a pixel-to-pixel or raster-to-raster line nonuniformity of either response in the case of an input scanner or output exposure in the case of an output scanner. These generally appear as streaks in an image when the recording or display medium is sensitive to exposure variations. These, for example, would be light or dark streaks in a printed halftone or darker and lighter streaks in a gray recorded image from an input scanner looking at a uniform area of an input document. A common example of this problem in a rotating polygon output scanner is the effect of facet-to-facet reflectivity variations in the polygonal mirror itself. The same exposure tolerances described for motion errors above apply here as well.

Another form of nonuniformity is sometimes referred to as jitter and occurs when the raster synchronization from one raster line to another tends to fail. In these cases a line drawn parallel to the slow scan direction appears to oscillate or jump in the direction of the fast scan. These effects, if large, are extremely objectionable. They will manifest themselves as raggedness effects or as unusual structural effects in the image, depending upon the document, the application, and the magnitude and spatial frequency of the effect.

3.4 CHARACTERIZATION OF INPUT SCANNERS

In this section we will discuss the theory of performance measurements and various algorithms or metrics to characterize them, the scanner factors that govern each, some practical considerations in the measurements, and visual effects where possible. This can be divided into discussions of scanners that generate a variable (gray) output signal where some general imaging science applies using linear analysis, and those that generate binary output where the signal is either on or off (i.e., is extremely nonlinear) and where specialized methods apply [2]. For purposes of this chapter, only the former will be considered.

3.4.1 Tone Reproduction and DC Systems Response

Unlike many other imaging systems where logarithmic response (e.g., optical density) is commonly used, the tonal rendition characteristics of input scanners are most often described by the relationship between the output signal (gray) level and the input reflectance or brightness. This is because most such systems respond linearly to intensity and therefore to reflectance. A representative such relationship is shown in Figure 3.16. In general this curve can be described by two parameters, the offset, O, against the output gray level axis and the gain of the system Γ, which is defined in the equation in Figure 3.16. Here g is the output gray level, and R is the relative reflectance factor. If there is any offset at all in terms of the output gray response, then the system is not truly linear despite the fact that the relationship between reflectance and gray level may follow a straight-line relationship. This line must go through the origin to make the system linear.

Often the maximum reflectance of a document will be far less than the 1.0 shown here. Furthermore the lowest signal may be significantly higher than 1% or 2% and may frequently get as high as 10% reflectance. In order to have the maximum number of gray levels available for each image, some scanners offer an option of performing a histogram analysis of the reflectances of the input

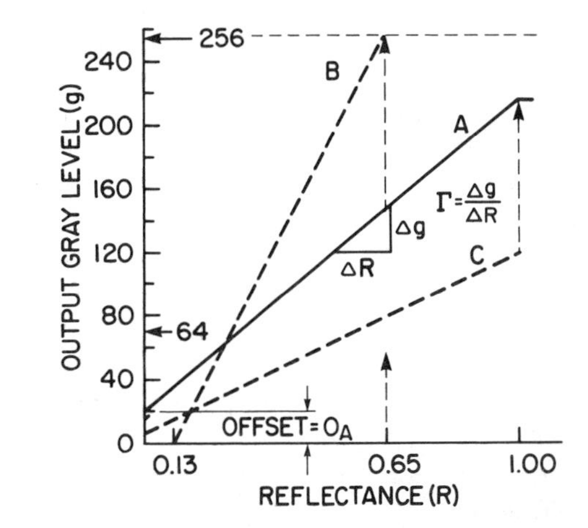

Figure 3.16 Typical input scanner response, illustrating definitions of "gain" and "offset."

document on a pixel-by-pixel or less frequently sampled basis. The distribution is then examined to find its upper and lower limits. Some appropriate safety factor is provided, and new offset and gain factors are computed. These are applied to stretch out the response to cover as many of the total (256 here) output levels as possible with the information contained between the maximum and minimum reflectances of the document. Other scanners may have a full gray-scale capability from 4 to 12 bits (16 to 4096 levels). In the figure, curve A would represent the typical general-purpose gray response for a scanner while curve B represents a curve adjusted to handle a specific input document whose minimum reflectance was 0.13 and whose maximum reflectance was 0.65. Observe that neither of these curves is linear. This becomes very important for the subsequent forms of analysis in which the nonlinear response must be linearized before the other measurement methods can be applied properly. It is also possible to arrange the electronics in the video processing circuit so that equal steps in gray are not equal steps in reflectance, but rather are equal steps in some units that are more significant visually. A logarithmic analog-to-digital converter is sometimes used to create a signal proportional to the logarithm of the reflectance or to the logarithm of the reciprocal reflectance (which is the same as the density). Many scanners for graphic arts applications function in this manner. These systems are highly nonlinear, but may work well with a limited number of gray levels to simulate the response of the eye.

Many input scanners operate with a built-in calibration system that functions on a pixel-by-pixel basis. Using this system, a particular sensor element that has greater responsivity than others may be attenuated by adjusting either the gain or the offset of the system or both. This would ensure that all photosites (individual sensor elements) respond equally to some particular calibration values, usually a light reflectance (e.g., white strip of paint). It is possible in many systems for the sensor to be significantly lower in responsivity in one place than another. A maximum responsivity may perform as shown in curve A while a less sensitive photosite may have the response shown in curve C. If curve C was captured with an analog-to-digital converter, the maximum signal range it contains has only 120 gray levels. This can be operated upon by a digital multiplier to effectively double each gray level, thereby increasing the magnitude of the scale to 220 or 240, depending upon how it handles the offset. Note that if some of the elements of a one-dimensional sensor responded as curve C, others as A, with the rest in between, then this system would exhibit a kind of one-dimensional granularity or nonuniformity, whose pattern depends upon the frequency of occurrence of each sensor type. This introduces a quantization error varying spatially in 1-pixel-wide strips, and ranging from strips with only 120 steps to others with 240 steps covering the same distribution of output tones.

The way to measure tone reproduction is to scan an original whose reflectance varies smoothly and continuously from near 0 to near 100% or at least to the lightest "white" that one expects the system to encounter. The reflectance is evaluated as a function of position, and the gray value from the scanner is measured at every position where it changes. Then the output of the system can be paired with the input reflectance at every location and a map drawn to test each gray response value for its associated input reflectance. A curve like Figure 3.16 can then be drawn for each photosite or for some average across all photosites. The use of a conventional step tablet where there are several discrete density levels provides an approximation to this analysis but does not allow the study of every one of the discrete output gray levels, especially when a system of 32 or more gray levels is under investigation. For a typical step tablet with 0.15 reflection density steps, 50% of the gray values are measured by only two steps, 0.15 and 0.3 density (or 50% reflectance). Thus a smoothly varying density wedge is more appropriate for the evaluation of an electronic input scanner. A major factor in the analysis of input scanner tone reproduction has to do with the concept of averaging. Most scanners operate with sufficiently small detector sites or sensor areas that they respond to input granularity. Thus, a single measurement will not suffice to get a solid area response to a so-called uniform input. Some degree of averaging is required, depending upon the granularity and noise levels of the input test document and the electronic system.

To evaluate the quality of response of a system like this, one merely needs to look for the straightness of the curve shown in Figure 3.16 and the intercept or offset as it is indicated in the figure. The question of what the slope and maximum gray value should be for a given input reflectance has no correct answer, but obviously the region over which the system responds should encompass the expected reflectance range of the input document. For systems that scan transparent documents, the reflectance axis is readily changed to a transmittance axis. In systems that do not adjust automatically to the input target (see Gonzalez and Wintz [3] for discussion of automatic threshold or gray scale adjustment), having the maximum point equal to 100% input reflectance is often a waste of gray levels since there are relatively few, if any, documents whose real reflectance is 100%. A value somewhere between 70 and 90% would be more representative of the upper end of the range of real documents.

An offset in the positive direction can be caused either by an electronic shift or by stray optical energy in the system. If the electronic offset has been set equal to zero with all light blocked from the sensor, such as by closing a shutter in front of it, then any offset measured from an image can be attributed to optical energy. Typical values for flare light, the stray light coming through the lens, would range from just under 1% to 5% or more of full scale [30]. While offset from uniform stray light can be adjusted out electronically, signals from flare light are often document-dependent, showing up as errors in a black region only when it is surrounded by a large field of white on the document. Therefore, correction for this measured effect in the particular case of an analytical measurement with a gray wedge or a step tablet surrounded by a white field may produce a negative offset for black regions of the document that are surrounded by grays or dark colors. If, however, the source of stray light is from the illumination system, the optical cavity, or some other means that does not involve the document, then electronic correction is more appropriate. Methods for measuring the document-dependent contribution of flare have been suggested in the literature and involve procedures that vary the surround field from black to white while measuring targets of different widths [30].

A major point of confusion in the testing of input scanners and many other optical systems that operate with a relatively confined space for the illumination system, document platen, and recording lens. This can be thought of as a type of integrating cavity effect. In this situation, the document itself becomes an integral part of the illumination system, redirecting light back into the lamp, reflectors, and other pieces of the illumination system. The document's contribution to the energy in the illumination depends on its relative reflectance and on optical geometry effects relating to lamp placement, document scattering properties, and lens size and location. In effect the document acts like a position-dependent and nonlinear amplifier affecting the overall response of the system. One is very likely to get different results for Figure 3.16 if the size of the step tablet or gray wedge used to measure it changes or if the surround of the step tablet or gray wedge changes between two different types of measurements. It is best therefore, to make a variety of measurements to find the range of responses for a given system. These effects can be anywhere from a few percent to perhaps as much as 20%, and the extent of the interacting distances on the document can be anywhere from a few millimeters to a few centimeters (fraction of an inch to somewhat over one inch). Relatively little has been published on this effect because it is so design-specific, but it is a recognized practical matter for measurement and performance of input scanners [31].

As was pointed out in the earlier discussion of tone reproduction, the visually unequal steps that result from a linear scaling of reflectance to output gray level may make a system with a limited number of gray levels appear to have unsatisfactory performance. Such a system may perform better if the mapping from reflectance to gray level were constrained into unequal-size reflectance steps. For such a system it would be important to measure the response curve before and after that mapping to ensure, first, linearity for the fundamental sensor portion and, second, the nonlinear mapping required by the system design. Gain described above would appear in the nonlinear mapping as a variable, being different at different signal levels. The types of measurement effects that occur in these cases are largely an exercise in conversions, quantization errors, and, ultimately, mathematics, given the linear responsivity of the fundamental sensors, whether they be photomultiplier tubes, photodiodes or charge-coupled devices (it is noteworthy, however, that a photomultiplier system and some solid state sensors can be set up to have a fundamental, nonlinear response, which is sometimes used to advantage by system designers). The details of designing and evaluating these systems are beyond the scope of this chapter.

3.4.2 MTF and Related Blur Metrics

We will now return to the subject of blur and refer to Figure 3.3. If the narrow line object profile shown at the top (Figure 3.3a) were reduced in width until the only change seen in the resulting image (Figure 3.3b) with further object width reduction is that the height of the smaller peak changes but not the width of its spreading, then we would say that the peak on the right of Figure 3.3b was a profile of the line spread function for the imaging system. To make a measurement of this we would simply record it, making sure there was no other image information present in the nearby area. (To be completely rigorous about this definition of the line spread function, we would actually use a narrow white line rather than a black line.) In the case illustrated, the spread function after quantization would be shown in Figure 3.3c as the corresponding distribution of gray pixels.

There are several observations to be made about this illustration which underscore some of the problems encountered in typical measurements. First, the wide line to the left is too close and interferes with the ability to measure the leftmost portion of the line spread function. This line should be several pixels further to the left in order to avoid such interference. Second, the quantized image is highly asymmetric while the profile of the line shown in Figure 3.3b appears to be more symmetric. This is due to sampling phase and requires that a measurement of the line spread function must be made, adjusting sampling phase in some manner. This is especially important in the practical situation of evaluating a fixed sampling frequency scanner.

The third area to notice is the limited amount of information in any one phase. It can be seen that the smooth curve representing the narrow object in Figure 3.3b is only represented by three points in the sampled and quantized image.

The averaging of several phases as suggested above would improve on this measurement, increasing both the intensity resolution and the spatial resolution of the measurement. One of the easiest ways to do this is to use a long narrow line and tip it slightly relative to the sampling grid so that different portions along its length represent different sampling phases. One then collects a number of uniformly spaced sampling phases, each being on a different scan line, while being sure to cover an integer number of complete cycles of sampling phase. One cycle is equivalent to a shift of one complete pixel. The results are then combined, and a better estimate of the line spread function is obtained. (This is tantamount to increasing the sampling resolution, taking advantage of the one-dimensional nature of the test pattern.) This is done by plotting the recorded intensity for each pixel located at its properly shifted absolute position relative to the location of the line. This can be visualized by considering the two-phase sampling shown in Figures 3.6 and 3.7. There the resulting pixels from phase A could be interleaved with those from phase B to create a composite of twice the spatial resolution. Additional phases would further increase resolution.

Generally speaking, the factors that affect blur for any type of scanner include (see Table 3.6) the optical design of the system, motion of the scanning element during one reading, electronic effects associated with the rise time of the circuit, the effective scanning aperture size, and various electrooptical effects in the detection of the signal. For solid state scanners, the major effects are aberrations and focus for the lens that images the document onto the solid state sensor, the size of the sensor aperture, the circuit, and motion in the slow scan direction. For charge coupled devices, the charge transfer efficiency as the signal is shifted out of the sensor and electronic diffusion effects within the solid state device itself also affect blur. The circuits that handle both the analog and the digital signals, including the A to D converter, may have some restrictive rise times and other frequency response effects that produce a one-dimensional blur. In the absence of nonlinearities, the individual spread functions associated with each of these effects can be convolved with each other to come up with an overall system line spread function.

For engineering analysis, use of convolutions and measurements of spread functions are often found to be difficult and cumbersome. The use of an optical transfer function (OTF) is considered to have many practical advantages from both the testing and theoretical points of view. The optical transfer function is

Table 3.6 Factors Affecting Input Scanner Blur

Factors
Solid State Scanners
Lens aberrations as functions of
wavelength
field position, orientation
focus distance
Sensor
aperture dimensions
charge transfer efficiency (CCD)
charge diffusion
leaks in aperture mask
Motion of sensor during/reading
Electronics rise time (frequency response)
Flying Spot Laser Beam Scanner
Spot shape at document
Lens aberrations (as above)
Polygon aperture or equivalent
Motion during reading
Sensor or detector circuit rise time

the Fourier transform of the line spread function. The value of optical transfer function analysis is that all of the components in a linear system can be described by their optical transfer functions, and these are multiplied together to obtain the overall system response. The method and theory of this type of analysis has been covered in many journal articles and reference books (see, for example, Refs. 11, 32, 33). Obtaining the transform of the line spread function has many of the practical problems associated with measuring the line spread function itself plus the uncertainty of obtaining an accurate digital Fourier transform using a highly quantized input.

There are several other methods for measuring the optical transfer function. These include taking the derivative of the edge profile in the image of a very sharp input edge. This generates the line spread function, and then the Fourier transform is taken. It should also be mentioned at this point that a Fourier transform of the spread function contains two terms that comprise the optical transfer function. One characterizes the modulation (modulus), and the other gives the phase, both as a function of frequency. For most characterizations of imaging systems the modulus, i.e., the modulation transfer function (MTF), is more significant. The phase transfer function, however, may be of interest and can be tracked either by careful analysis of the relative location of target and image in a frequency-by-frequency method, or by direct computation from the line spread function.

There are several approaches involving the use of specially selected input targets whose spatial frequency content is very high. The frequency composition of the input target is characterized in terms of the modulus of the Fourier transform, $M_{in}(f)$, and of its spatial radiance profile the frequency decomposition of the output image is similarly characterized, yielding $M_{in}(f)$ and $M_{out}(f)$. Dividing the output modulation by the input modulation yields the modulation transfer function as shown:

$$MTF(f) = \frac{M_{out}(f)}{M_{in}(f)} \tag{5}$$

The success of this depends upon the ability to characterize both the input and the output accurately.

A straightforward method to perform this input and output analysis involves imaging a target of periodic intensity variations and measuring the modulation on a frequency-by-frequency basis. If the target is a pure sine wave of reflectance or transmittance, i.e., it has very low harmonic content, and the input scanner is linear, then the frequency-by-frequency analysis is straightforward. The modulation is obtained directly, measuring the output gray values g', and the corresponding input reflectance (or transmittance or intensity) values, R, values of equation (6) for each frequency pattern at the appropriate high (max) and low (min) points.

Expanding the numerator and denominator for equation (5) and the case of sinusoidal patterns and linear systems yields

$$MTF(f) = \frac{(g'_{max}(f) - g'_{min}(f))/(g'_{max}(f) + g'_{min}(f))}{(R_{max}(f) - R_{min}(f))/(R_{max}(f) + R_{min}(f))} \tag{6}$$

when the prime is used to denote gray response which has been corrected for any nonlinearity as described below.

Figures 3.17 and 3.18 show an example of this process. In Figure 3.17a, we see the layout of a commercially available sinusoidal test target [34].* The sinusoidal distributions of intensity (reflectance) are located in different blocks in the center of the pattern and uniform reflectance patterns of various levels are placed in the top and bottom rows to enable correction for nonlinearities should there be any. Parts *b* and *c* show enlargements of the pattern in part *a*, selecting a lower and a higher frequency. Parts *f* and *g* are enlargements of a gray monitor display of the electronically captured image of the same parts of the test target. Parts *d*, *e*, *h*, and *i* show profiles of the sinusoidal patterns immediately above them. To calculate a modulation transfer function, the modulation of each pattern is measured using equation (6), finding the average maximum and minimum

* Sine Waves Inc., 236 Henderson Drive, Penfield, NY 14526.

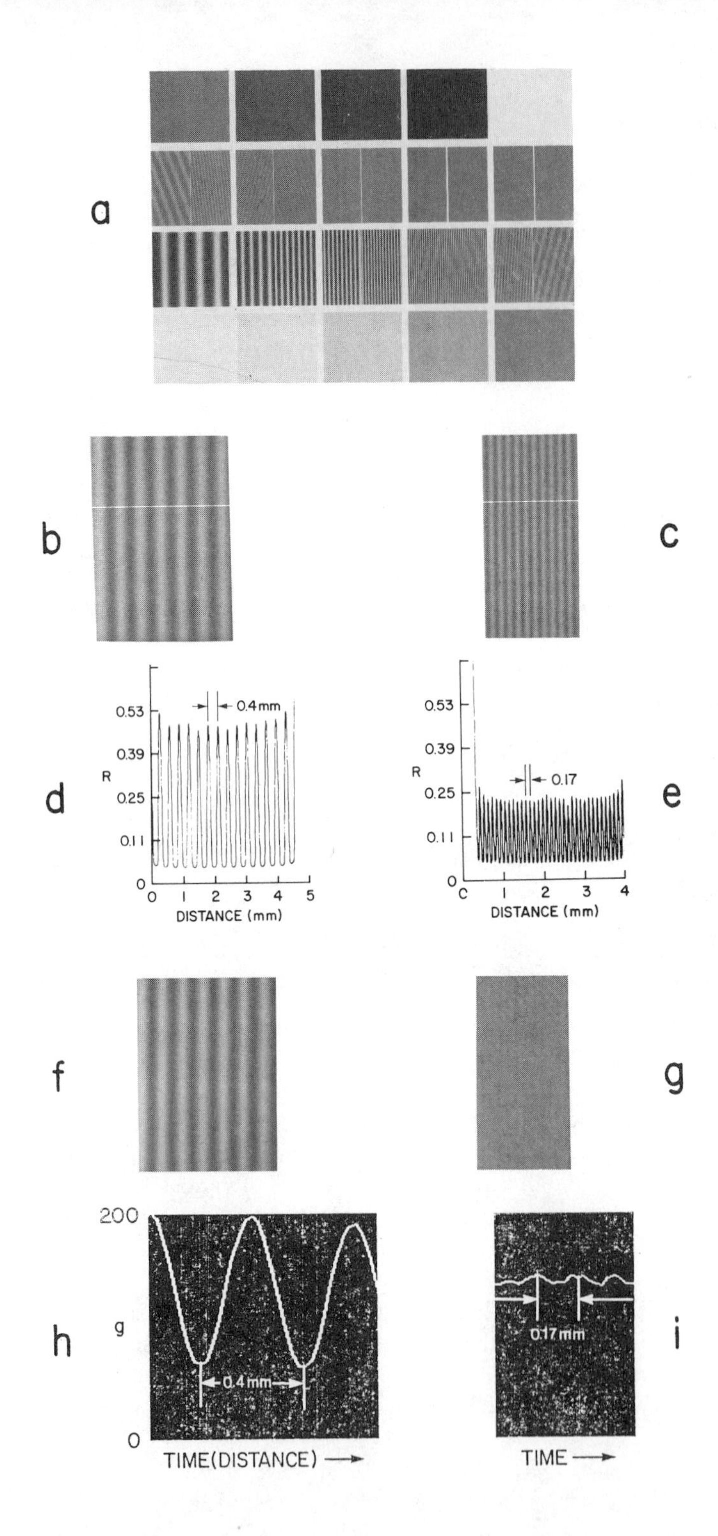
a
b
c
0.4 mm
0.53
0.39
R
0.25
0.11
0
0 1 2 3 4 5
DISTANCE (mm)
d
0.17
C 1 2 3 4
e
f
g
200
g
0.4 mm
0
TIME(DISTANCE) ⟶
h
0.17 mm
TIME ⟶
i

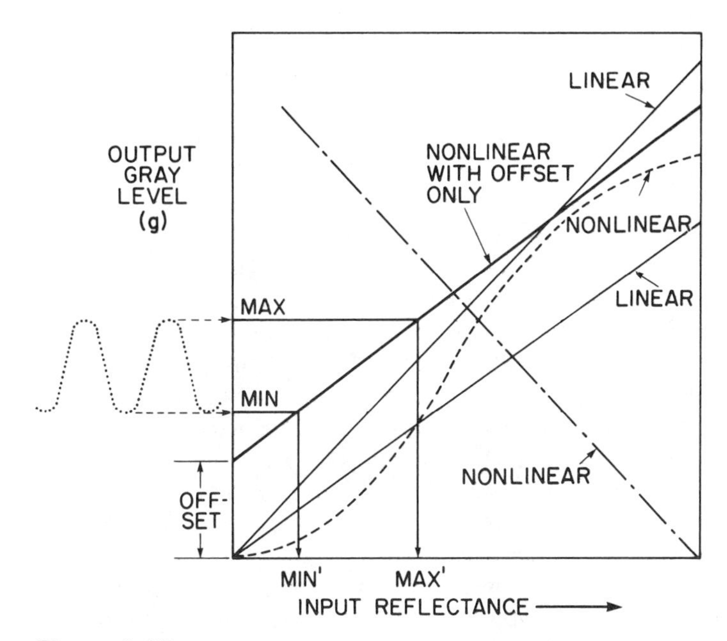

Figure 3.18 Examples of linear and nonlinear large-area response curves with illustration of output modulation correction for offset using effective gray response at max′ and min′.

values for each separate frequency. These modulations, plotted on a frequency-by-frequency basis, described the MTF. From a practical standpoint it is important to tip the sine waves slightly as seen in parts *d* and *e* to cover the phase distributions as described above under the spread function discussion. A new higher-resolution image can be calculated by interdigitating samples from the individual scan lines which each appear phase-shifted with respect to the sine wave.

Figure 3.18 shows several examples of linear and nonlinear response curves and describes the case for the nonlinearity with offset, showing the method for correction. Here the maximum and minimum values for the sine waves are unfolded through the response curve to arrive at minimum and maximum input reflectances. To obtain corrected modulation (i.e., using g' in equation (6)), each output gray level must be modified by such an operation. If the response curve for the system was one of those indicated as linear in the figure, then no correction is required. It is important to remember that while the scanner system response

---

Figure 3.17 Illustrations of sine wave test targets and their degraded images. Part a is the full pattern of patches and sine waves. Enlargements are shown in parts b and c, with their microreflectance profiles shown in d and e. Degraded images of the same target are shown enlarged on a display monitor in f and g, while profiles of a few selected cycles are shown in h and i.

(pp. 122–123) may obey a straight-line relationship between output gray level and the reflectance, transmittance, or intensity of the input pattern, it may be offset due to either optical or electronic biases (e.g., flare light, electronic offset, etc.) This also represents a nonlinearity and must be compensated.

As the frequency of interest begins to approach the sampling frequency in an aliased input scanning system, the presence of sampling moiré becomes a problem. This produces interference effects (p. 104) between the sampling frequency and the frequency of the test pattern. When modulation is computed from sampled image data using maxima and minima in equation (6), errors may arise. An example is the case where the test pattern image frequency is exactly one half the sampling frequency, i.e., the Nyquist frequency. In this case, when the sampling grid lines up exactly with the successive peaks and valleys of the sine wave, we get a strong signal indicating the maximum modulation of the sine wave. When the sampling grid lines at the midpoint between each peak and valley of the same sinusoidal image (phase shifted 90° relative to the first position), each data point will be the same, and no modulation whatever results. There is no right or wrong answer to the question of which phase represents the true sine wave response, but the analog or highest value is often considered as the true modulation transfer function. Each phase may be considered as having its own sine wave response. Reporting the maximum and minimum frequency response or reporting some statistical average are both legitimate approaches, depending upon the intended use of the measurement. It is common practice to represent the average or maximum and the error range for the reported value.

The analog modulation transfer function, on the other hand, is only given as the maximum curve, representing the optical function before sampling. Therefore, the description of upper and lower phase boundaries for sine wave response shows the range of errors in the measurement of the modulation transfer function which one might get for a single measurement. This strongly suggests the need to use several phases to reduce error if the analog MTF is to be measured. Mathematically, phase errors may be thought of as a form of microscopic nonstationarity complicating the meaning of MTF for sampled images at a single phase [7, p. 18; 35, p. 382; 36, p. 139]. The use of information from several phases reduces this complication by enabling one to find the correct analog MTF which obeys the principle of stationarity. See Figure 3.19 for a typical example, showing errors for sine waves whose period is a submultiple of the sampling interval.

In the case of a highly quantized system, meaning one in which there are a relatively small number of gray levels, quantization effect becomes an important consideration in the design and testing of the input scanner. The graph in Figure 3.20 shows the limitation that quantization step size, E_q, imposes on the measurement of the modulation transfer function using sine waves. The number of gray levels used in an MTF calculation can be maximized by increasing the contrast of the sinusoidal signal that is on the input test pattern and by repeated measurements in which some analog shifts in signal level are introduced to cause the quantization levels to appear in steps between the discrete digital levels and

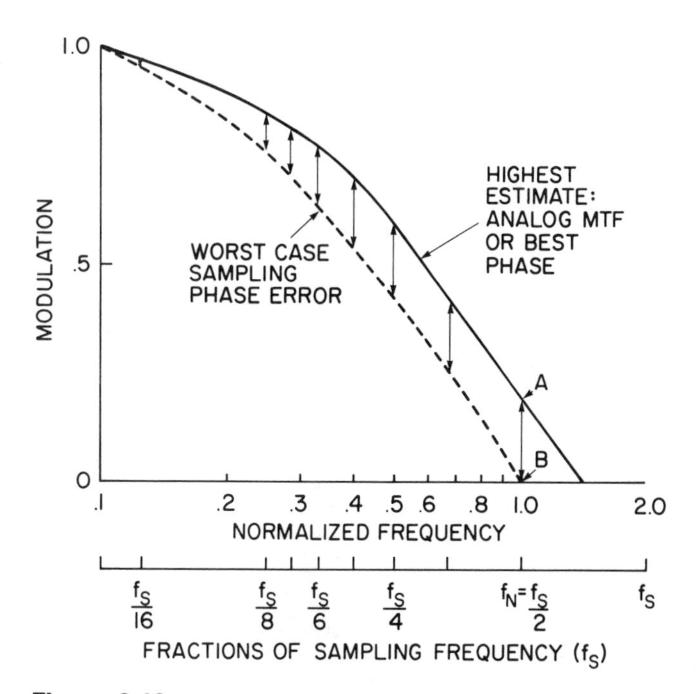

Figure 3.19 Example of the possible range of measured sine wave response values of an input scanner, showing the uncertainty resulting from possible phase variations in sampling.

therefore at different points on the sinusoidal distributions. The latter could be accomplished by changing the light level or electronic gain.

Averaging over several rows or raster lines is also useful in keeping down the noise and in displaying all the phase information, especially if one tips the sinusoid a degree or two to change the sampling phases of successive measurements. It is also important to note that because the actual modulation transfer function can vary over the field of view, a given measurement may only apply to a small local region over which the MTF is constant. (This is sometimes called an isoplanatic patch or stationary region within the image.) To further improve the accuracy of this approach, one can numerically fit sinusoidal distributions to the data points collected from a measurement, using the amplitude of the resultant sine wave to determine the average modulation. Taking the Fourier transform of the data in the video profile may be thought of as performing this fit automatically. The properly normalized amplitude of the Fourier transform at the spatial frequency of interest would, in fact, be the average modulation of the sine wave that fits the video data best.

Another approach to the measurement of the MTF of an input scanner would involve the application of square-wave test patterns (as opposed to sinusoidal

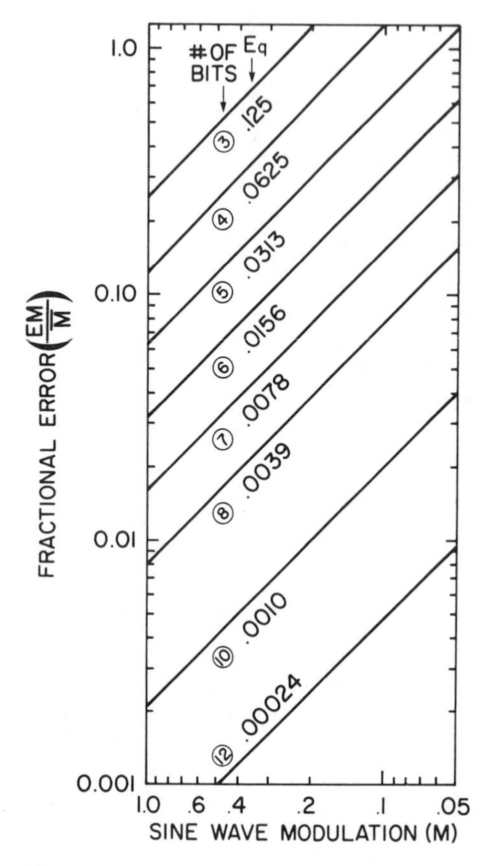

Figure 3.20 Errors in MTF measurements, showing the effects of modulation at various quantization errors. E_M is in zero to peak units.

distributions). Square-wave test patterns are commonly found in resolving power test targets. To apply these correctly, one must take advantage of the knowledge of the square wave nature of the test pattern and unfold for the harmonics introduced by this square wave [37]. It is generally advisable to measure the target's actual harmonic content rather than to assume that it will display the theoretical harmonics of a perfect mathematical square wave. Likewise, other patterns of known spectral content can be calibrated and used.

As a matter of practical interest, several spatial frequency response measurements of the human visual system are shown in Figure 3.21. The work of several authors is shown [38–43]. The curves shown have all been normalized to 100% at their respective peaks to provide a clearer comparison. Except for the various

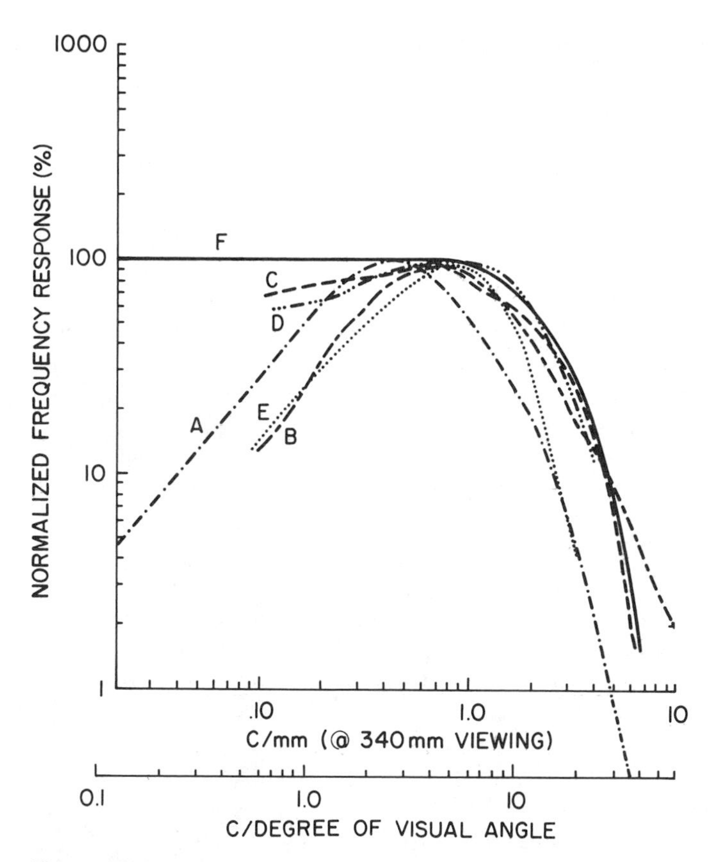

Figure 3.21 Measured spatial frequency response of the human visual system, according to (a) Campbell [38], (b) Patterson (Glenn et al.) [69], (c) Watanabe [41], (d) Hufnagel (after Bryngdahl) [70], (e) Gorog et al. [39], (f) Dooley [43]. All measurements normalized for 100% at peak and for 340mm viewing distance.

normalizing factors, the ordinates are analogous to a modulation transfer factor of the type described by equation (6); however, MTFs are applicable only to linear systems, which the human eye is not. The visual system is in fact thought to be composed of many independent, frequency-selective channels [44, 45] which, under certain circumstances, combine to give an overall response as shown in these curves. It will be noted that the response of the visual system has a peak (i.e., modulation amplification relative to lower frequencies) in the neighborhood of 6 cycles/degree (0.34 cycles/milliradian) or 1 cycle/mm (25 cycles/in.) at a standard viewing distance of 340 mm (13.4 inches). The variations among these curves reflect the experimental difficulties inherent in the measurement task and may also illustrate the fact that a nonlinear system such as human vision cannot be characterized by a unique MTF.

These visual frequency response curves suggest that the performance of an imaging system could be improved if its frequency response could be increased at certain frequencies. It is not possible with most passive imaging systems to create amplification at selected frequencies. The use of electronic enhancement, however, can impart such an amplified response to the output of an electronic scanner. Amplification here is meant to imply a high-frequency response which is greater than the very low frequency response or greater than unity (which is the most common response at the lowest frequencies). This can be done by convolving the digital image with a finite-impulse response (FIR) electronic filter that has negative sidelobes on opposite sides of a strong central peak. The details of FIR filter design are beyond the scope of this chapter, but the effects of two typical FIR filters on the system MTF are shown in Figure 3.22.

3.4.3 Noise Metrics

Noise in an input scanner, whether the scanner is binary or multilevel gray, comes in many forms (see Sect. 3.2.2, pp. 111–113). A brief outline of these can be found in Table 3.7. Various specialized methods are required in order to discriminate and optimize the measurement of each.

In this table we see that there are both fixed and time-varying types of noise. They may occur in either the fast or the slow scan direction and may either be additive noise sources or multiplicative noise sources. They may be either totally random or they may be structured. In terms of the spatial frequency content, the noise may be either flat (white), i.e., constant at all frequencies to a limit, or may contain dominant frequencies, in which case the noise is said to be colored. These noise sources may be either random or deterministic; in the latter case, there may be some structure imparted to the noise.

The sources of the noise can be in many different components of the overall system, depending upon the design of the scanner. Instances of these may include the sensor of the radiation, or the electronics, which amplify and alter the electrical signal, including, for example, the A to D converter. Other noise sources may be motion errors, photon noise in low-light-level scanners, or noise from the illuminating lamp or laser. Sometimes the optical system, as in the case of a laser beam input scanner, may have instabilities that add noise. In many scanners there is a compensation mechanism to attempt to correct for fixed noise. This typically utilizes a uniformly reflecting or transmitting strip parallel to the fast scan direction and located close to the input position for the document. It is scanned, its reflectance is memorized by the system, and it is then used to correct or calibrate (pp. 123–124) either the amplifier gain or its offset or both. Such a

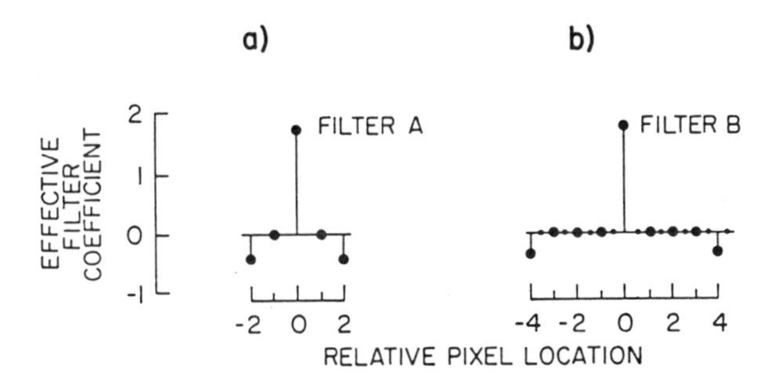

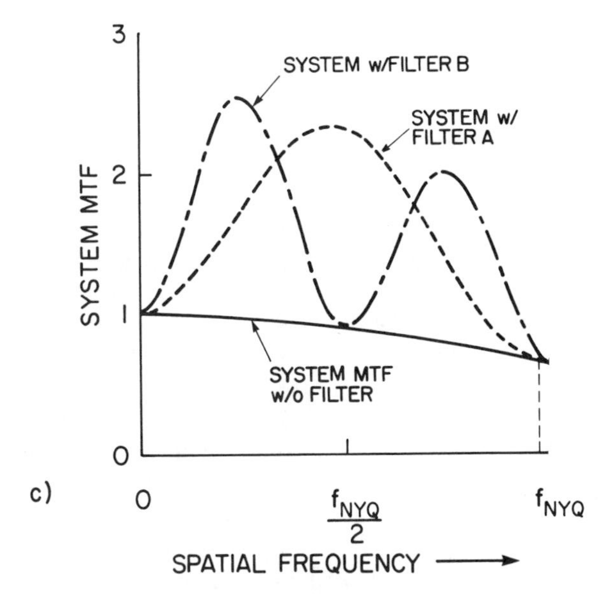

Figure 3.22 Two examples of the enhancement of system MTF using electronic finite-impulse response filters in conjunction with an input scanner: (a) the one-dimensional line spread function for filter A; (b) filter B is shown; (c) shows the effect of these filters on the MTF of a high-quality scanner (aliased).

calibration system is, of course, subject to many forms of instabilities, quantization errors, and other kinds of noise. Since most scanners deal with a digital signal in one form or another, quantization noise must also be considered.

In order to characterize noise in a gray output scanner, one needs to record the signal from a uniform input target. The most challenging task is finding a

Table 3.7 Types and Sources of Noise in Input Scanners

NOISE Category			
DISTRIBUTION	Fixed with platen	Time Varying	
TYPE OF OPERATION	Multiplicative	Additive	
SPATIAL FREQUENCY	Flat (white)	Colored	
STATISTICAL DISTRIBUTION	Random	Structured	Image Dependent
ORIENTATION			
	-fast scan	-slow scan	none (2-dimensional)
SOURCES:	Sensor	Electronic System	Motion Error
	Calibration Error	Photons	Lamp, Laser Controls
	Optics		

the signal from a uniform input target. The most challenging task is finding a uniform target with noise so low that the output signal does not contain a large component due to the input or document noise. In many of these scanners, the system is acting much like a microdensitometer, which reacts to such input noise as the paper fibers or granularity in photographic, lithographic, or other apparently uniform samples.

The basic measurement of noise involves understanding the distribution of the signal variation. This involves collecting several thousand pixels of data and examining the histogram of their variation or the spatial frequency content of that variation. Under the simplifying assumptions that we are dealing with noise sources that are linear, random, additive, and flat (white), a typical noise measurement procedure would be to evaluate the following expression:

$$\sigma_s^2 = \sigma_t^2 - \sigma_o^2 - \sigma_m^2 - \sigma_q^2 \tag{7}$$

where σ_s = the standard (s) deviation of the noise for the total scanner system
σ_t = the total (t) standard deviation recorded during the analysis
σ_o = the standard deviation of the noise in the input object (o) measured with an aperture that is identical to the pixel size
σ_m = the standard deviation of the noise due to measurement (m) error
σ_q = the standard deviation of the noise associated with the quantization (q) error for those systems which digitize the signal

This equation assumes that all of the noise sources are independent. Removing quantization noise is an issue of whether one wants to characterize the scanner with or without the quantization effects, since they may in fact be an important characteristic of a given scanner design. The fundamental quantization error [35] is

$$\sigma_q^{\,2} = \frac{2^{-2b}}{12} \tag{8}$$

where b = the number of bits to which the signal has been quantized.

As it stands, the second and third terms in equation (7) give the performance of the analog portion of the measurement and would include the properties of the sensor amplification circuit and the A to D converter as well as any other component of the system that leads to the noise noted in the table above. The term σ_q characterizes the digital nature of the scanner.

Equation (7) is useful when the noise in the system is relatively flat with respect to spatial frequency or when the shape of the spatial frequency properties of all of the subsystems is similar. If, however, one or more of the subsystems involved in the scanner is contributing noise that is highly colored, that is, has a strong signature with respect to spatial frequency, then the analysis needs to be extended into frequency space. This approach uses Wiener or power spectral analysis [11, 46]. A detailed development of Wiener spectra is beyond the scope of this chapter. However, it is important here to realize its basic form. It is a particular normalization of the spatial frequency distribution of the square of the signal fluctuations. The signal is often in optical density (D), but may be in volts, current, reflectance (R), etc. The normalization involves the area of the detection aperture responsible for recording the fluctuations. Hence units of the Wiener spectrum as used in photographic applications are often $[\mu m\ D]^2$ and can be $[\mu m\ R]^2$ (see Ref. 46), the latter being more appropriate for scanners because they respond linearly to reflectance, or, more generally, to irradiance.

3.5 SUMMARY MEASURES OF IMAGING PERFORMANCE

Attempts have often been made to take the general information described above and reduce it to a single measure of imaging performance. While none of the resulting measures provides a single universal figure of merit for image quality, each brings additional insight to the design and analysis of an imaging system and each has found some level of success in a range of applications. Perceived image quality, however, is a psychological reaction to a complex set of trade-offs and visual stimuli. There is a very subjective, application-oriented aspect to this reaction which does not readily lend itself to analytical description. In some instances where the application is sufficiently well bounded by identifiable visual tasks, an overall measure of image quality can be found. This narrowly applicable metric is a result of psychological and preference research and usually involves extensive measurement of a limited class of imaging variables. We have never found such a figure of merit with broad applicability.

Instead, in an attempt to help the engineer control or design his systems, we shall describe a number of metrics. Each individual metric, in many cases, is suited to optimizing one or two subsystems and is valuable in its own right. For given applications, several of these can be combined to explore a trade-off space and perhaps construct a measure of overall image quality for that application. We shall describe some examples of how these have been constructed so that the reader has a good starting point for his own applications. Since genuine image-quality attributes are really preference features, market research is usually required and will not be covered in this section. The building blocks are offered here to enable the reader to build his own regression equation or other means for connecting physical quantification and description of an image with its subjective value.

The metrics are described here in their general form, as they would apply to analog imaging systems such as cameras and film, and have not been particularized (except in one or two cases) to the digital imaging conditions, in order to simplify their treatment in this summary section. To the extent that the scanning systems in question are unaliased and have a large number of gray levels associated with them, the direct application of analog metrics is valid. In general, it should be remembered that digital imaging systems are not symmetric in slow and fast scan orientations in either noise or spatial frequency response (MTF). Therefore what is given below in one-dimensional units must be applied in both dimensions for successful analysis of a digital input scanner. These concepts can be extended to an entire imaging system with little modification if the subsequent imaging modules, such as a laser beam scanner, provide gray output writing capability and generate no significant sampling or image conversion defects of their own (i.e., they are fairly linear). Since full gray-scale input scanners are usually fairly linear, most of these concepts can be applied fairly directly to them, with the qualification that some display or analysis technique is required to convert the otherwise invisible electronic image to a visual or numerical form.

All of these measures involve the concept of the signal-to-noise ratio. Some deal directly with the terms described above, while others are oriented toward a particular application, and still others use more generalized constructs. Tailoring the metric to a specific application involves finding those signal and noise characteristics that are most relevant to the intended application.

Most of these summary measures can be described by curves like the illustrative ones shown in Figure 3.23. Here we show some common measures for both signal and noise such as intensity, modulation or (modulation)^2, generically called F, plotted as a function of spatial frequency f. A signal $S(f)$ is shown generally decreasing from its value at 0 spatial frequency to the frequency f_{max}. A limiting or noise function $N(f)$ is plotted on the same graph starting at a point below the signal; it too varies in some fashion as spatial frequency increases. The various unifying constructs (metrics) involve very carefully considered approaches to the

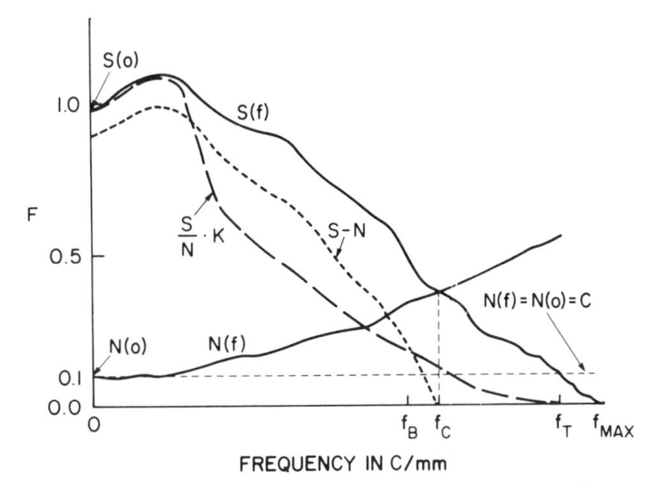

Figure 3.23 Signal, noise, and various measures of the relationships between them, plotted as functions of spatial frequency.

relationship between S and N, to their respective definitions, and to the frequency range over which the relationship is to be considered, along with the frequency weighting of that relationship.

3.5.1 Basic Signal-to-Noise Ratio

The simplest of all signal-to-noise measures is the ratio of the mean signal level $S(0)$ to the standard deviation $N(0)$ of the fluctuations at that mean. If the system is linear and the noise is multiplicative, this is a useful single number metric. If the noise varies with signal level, then this ratio is plotted as a function of the mean signal level to get a clearer picture of performance. Hypothetical elementary examples of this are shown in Figure 3.24 in which are plotted both the multiplicative type of noise at 5% of mean signal level (here represented as 100) and additive noise of 5% with respect to the mean signal level. It can be seen why such a plot is important in evaluating a real system. It should be noted that in some cases, multiplicative or additive noise may vary as a function of signal level for some important design reason. In comparing signals to noise, one must also be careful to ensure that the detector area over which the fluctuations are collected is appropriate for the application to which the signal-to-noise calculation pertains. This could be the size of the input or output pixel, of the halftone cell, or of the projected human visual spread function. The data must also be collected in the orientation of interest. In general, for scanned imaging systems there will

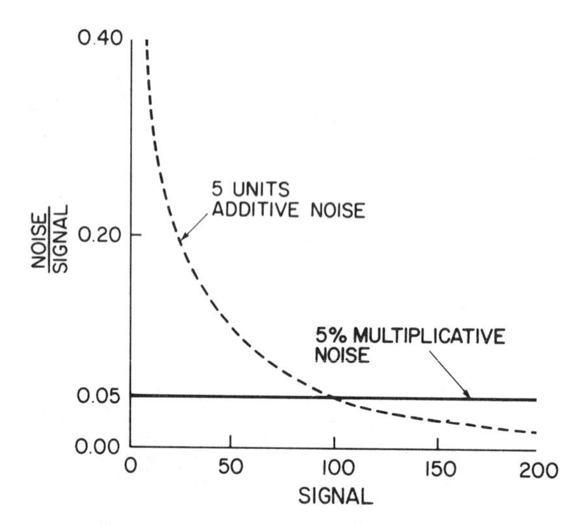

Figure 3.24 Relative noise (noise/signal) as a function of signal level for additive and multiplicative noise.

be a different signal-to-noise ratio in the fast scanning direction than in the slow scanning direction.

3.5.2 Detective Quantum Efficiency and Noise Equivalent Quanta

When low light levels or highly noise limited situations occur, it is desirable to apply the concepts of detective quantum efficiency (DQE) and noise equivalent quanta (NEQ). These fundamental measurements have been extensively discussed in the literature [11]. The expression for DQE and its relationship to NEQ is

$$\mathrm{DQE} = \frac{\mathrm{NEQ}}{q} = \frac{A(\log_{10} e)^2}{W(0)} \frac{\gamma^2}{q} \tag{9}$$

This form of the equation relates to a typical photographic application in which the signal is in density units, where density is the negative log of either transmission or reflectance, and γ is the slope of the relationship between the output density and the input signal (expressed as the $\log_{10}$ of exposure). W is the Wiener spectrum of the noise given here at 0 spatial frequency (where it is equal to $A\sigma_s{}^2$), and q is the number of exposure quanta collected by the detector whose area is A. For a more general case, equation (9) is rewritten as equation (10) where the units have been expressed in terms of a general output unit O, which could represent voltage, current, number of counts, gray level, etc.

$$\text{DQE} = \frac{q(dO/dq)^2}{\Delta O^2} \tag{10}$$

where ΔO^2 is the mean-square output fluctuation [11, p. 155; 47]. The term $W(0)$ contains the factor $1/q^2$ which, through manipulation, leads to q in the numerator of equation (10). The term in parentheses is simply the slope of the characteristic curve of the detection system (see Figure 3.16), indicating the change in the output divided by the change in the input. It is noted that this is related to the gain, Γ, of the system and appears as a squared factor in the equation. If we set this gain to a constant r in arbitrary output units, and assume the distribution of fluctuations obeys normal statistics, then we can rewrite equation (10) as

$$\text{DQE} = \frac{r^2 q}{\sigma_0^2} \tag{11}$$

where σ_0 represents an estimate of the standard deviation of the distribution of the output fluctuations. It can now be seen that this expression differs in a significant way from the simple construct used above, namely the mean divided by the standard deviation. Here the average signal level q is divided by the square of the standard deviation (i.e., the variance) and contains a modifier that is related to the characteristic amplification factors associated with a particular detection system, namely r, which also enters as a square. It should also be pointed out that detective quantum efficiency is an absolute measure of performance since q is an absolute number of exposure events, i.e., number of photons or quanta. Returning to equation (9), we can now see that under these simplifying assumptions the noise equivalent quanta can be represented by

$$\text{NEQ} = \frac{r^2 q^2}{\sigma_0^2} \tag{12}$$

The concepts just described can be extended to the rest of the spatial frequency domain in the form

$$\text{NEQ} = \frac{Ar^2 q^2[\text{MTF}(f_x, f_y)]^2}{W(f_x, f_y)} \tag{13}$$

Provided the response characteristics of the system are linear, equation (13) reduces to equation (14) since $r \times q$ is equal to O, the output in whatever units are required:

$$\text{NEQ} = \frac{AO^2[\text{MTF}(f_x, f_y)]^2}{W(f_x, f_y)} \tag{14}$$

For illustration purposes, Figure 3.23 shows all of the above constructs. For DQE, the generic signal measure $S(0)$ can be thought of as the numerator of equations (10) and (11) and $N(0)$ may be thought of as the denominator. Similarly, for noise equivalent quanta, $S(0)$ represents the numerator and $N(0)$ the denominator of equation (12). For the spatial-frequency-dependent version of noise equivalent quanta, $S(f)$ represents the numerator of equations (13) or (14) and $N(f)$ represents the denominator of the same equations. The dashed curve in Figure 3.23, KS/N, represents the ratio of $S(f)$ to $N(f)$ normalized to $S(0)$. K is an arbitrary constant which, for this illustration is set equal to 0.1. It is seen that this function continues to the cutoff frequency f_{max}. It should also be observed that this relationship between the signal characteristic and the noise characteristic can vary with signal level, as shown in Figure 3.24, and hence a full functional description requires a three-dimensional plot, making S the third axis of an expanded version of Figure 3.23.

3.5.3 Application-Specific Context

The above descriptions are frequently derived from the fundamental physical characteristics of various imaging systems, but the search for the summary measure of image quality usually includes an attempt to arrive at some application-oriented subjective evaluation, correlating subjective with objective descriptions. Applications that have been investigated extensively include two major categories: those involving detection and recognition of specific types of detail and those involved in presenting overall aesthetically pleasing renderings of a wide variety of subject matter. These in turn have centered around a number of imaging constraints, which can usually be grouped into the categories of display technologies and hard-copy generation. Many studies of MTF have been applied to each [48, 49]. All of these studies are of some interest here. Note that modern laser beam scanning tends to focus on the generation of hard copy where the raster density is hundreds of lines per inch and thousands of lines per image compared with the hundreds of lines per image for early CRT technology used in the classical studies of soft display quality.

3.5.4 Modulation Requirement Measures

One general approach characterizes $N(f)$ in Figure 3.23 as a "demand function" of one of several different kinds. Such a function is defined as the amount of modulation or signal required for a given imaging and viewing situation and a given target type. In one class of applications, the curve $N(f)$ is called the threshold detectability curve and is obtained experimentally. Targets of a given format but varying in spatial frequency and modulation are imaged by the system under test. The images are evaluated visually under conditions, and using criteria, required by the application. Results are stated as the input target modulation required (i.e., "demanded") for being "just resolved" or "just detected" at each frequency. It is assumed that the viewing conditions for the experiment are optimum and that the threshold for detection of any target in the image is a function of the target image modulation, the noise in the observer's visual system, and the noise in the imaging system preceding the observer. At low spatial frequencies this curve is limited mostly by the human visual system, while at higher frequencies imaging system noise as well as blur may determine the limit.

One such type of experiment involves measuring the object modulation required to resolve a three-bar resolving power target. For purposes of electronic imaging, it must be recalled that the output video of an input scanner cannot be viewed directly, and therefore any application of this method must be in the systems context, including some form of output writing or display. This would provide additional noise restrictions. If the output noise limitations came from photographic film, for example, as in some laser output scanners, then the results for a variety of different photographic films similar to those shown in Figure 3.25 would apply. These specific examples are offered as an illustration of the broad range of possible results. They would have to be reworked to account for laser wavelength and exposure times appropriate to the scanner and for currently available photographic materials.

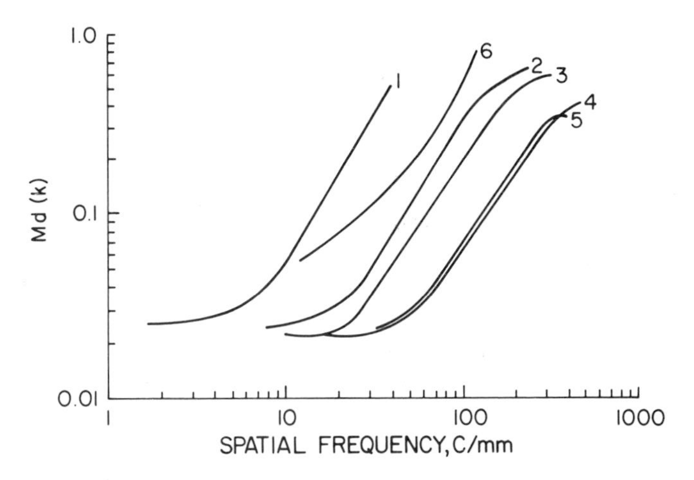

Figure 3.25 Object modulation threshold required to resolve a three-bar target as a function of spatial frequency, illustrated by examples of several silver halide films. Photographic film types used to generate these curves included Eastman Kodak aerial film types 8403 (1), 3400 (2), SO-226 (3), 3404 (4), and SO-243 (5). All were exposed to daylight spectral distribution through a Wratten No. 12 filter at 0.017 s and developed in D19 developer for 8 min at 68°F. Curve 6 is an experimental aerial reversal color film (after [50]).

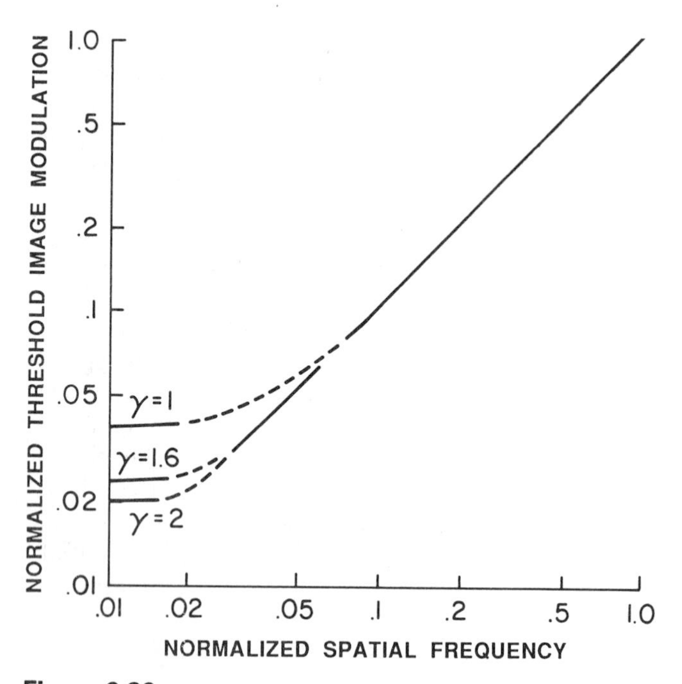

Figure 3.26 Generalized detectability threshold as a function of spatial frequency for various process gains (γ's) (after [50]).

Figure 3.26 shows a generalized photographic detectability threshold curve in which the contrast or gain of the system (photographic gamma, here, where $\gamma = \Delta D/\Delta \text{LogE}$) is shown to have an important low spatial frequency visual effect. The high spatial frequency portion of this curve is an asymptote established by the granularity of the output writing system. It is independent of gamma.

The curves in Figures 3.25 and 3.26 could be applied directly to the analysis of an input scanner, if the major source of degrations in the overall system is photographic film. If the major noise source is not photographic film, then curves of the kind described here would have to be derived for the key system elements, paying particular attention to the noise characteristics and the MTF of both the input scanner and the output display or printer. The output could be a CRT display of some type, such as a video monitor with gray-scale (analog) response. Another likely output would be a laser beam scanner writing on xerography. The details for measuring and using the demand function can be found in Scott [50] for the example of photographic film and in Biberman [48, especially Chapter 3] for application to soft displays.

3.5.5 Area under the MTF Curve (MTFA)

Modulation detectability, while useful for characterizing systems in task-oriented applications, is not always useful in predicting overall image-quality performance for a broad range of imaging tasks and subject matter. It has been extended to a more general form through the concepts of the threshold quality factor [51] and MTFA [52, 53]. These concepts were originally developed for conventional photographic systems used in military photointerpretation tasks [51]. They have been generalized to electrooptical systems applications for various forms of recognition and image-quality evaluation tasks, mostly involving soft displays [48]. The concept is quite simple in terms of Figure 3.23. It is the integrated area between the curves $S(f)$ and $N(f)$ or, equivalently, the area under the curve labeled S-N. In two dimensions, this is

$$\text{MTFA} = \int_0^{f_{c_x}} \int_0^{f_{c_y}} [S(f_z,f_y) - N(f_x,f_y)]\, df_x\, df_y \tag{15}$$

where S is the MTF of the system and N is the modulation detectability or demand function as defined above, and f_{c_x} and f_{c_y} are the two-dimensional "crossover" frequencies equivalent to f_c shown in Figure 3.23.

This metric attempts to include the cumulative effects of various stages of the scanner, films, development, the observation process, the noise introduced into the perceived image by the imaging system, and the limitations imposed by psychological and physiological aspects of the observer by building all these effects into the demand function $N(f)$. Extensive psychophysical evaluation and correlation has confirmed the usefulness of this approach [53] for recognition of military reconnaissance targets, pictorial recognition in general, and for some alphanumeric recognition. We are not aware of any particular work connected with this metric on the overall aesthetic aspects of image quality.

3.5.6 Measures of Subjective Quality

Several authors have explored the broader connection between objective measures of image quality and the overall aesthetic pictorial quality for a variety of pictorial subject matter as encountered in amateur and professional photography. The experiments to support these studies are very difficult to perform, requiring extremely large numbers of observers to obtain good statistical measures of subjective quality. The task of assessing overall quality is less well defined than the task of recognizing a particular pattern correctly, as in most of the studies cited above. It would appear that no single measurement criterion has become universally accepted by individuals or corporations working in this area. Below we shall discuss a few of the key descriptors, but we do not attempt to list them all.

Many of the earlier studies tended to focus on the signal or MTF-related variable only. In one such series of studies [54, 55], S in Figure 3.23 is defined as the modulation of reflectance on the output print (for square waves) divided by the modulation on the input document (approximately 0.6 for these experiments). The quality metric is defined as the spatial frequency at which this ratio falls to 0.5. This is indicated in the figure by the frequency f_B for the curve S as drawn. In these studies, a landscape without foreground was rated good if this characteristic or critical frequency was 4 to 5 cycles/mm (100 to 125 cycles per inch), but for a portrait 2 to 3 cycles/mm (50 to 75 cycles per inch) proved adequate. Viewing distance was not a controlled variable. By using modulation on the print and not simply MTF, the study has included the effects of tone reproduction as well as MTF. Granularity was also shown to have an effect, but was not explicitly taken into account in the determination of critical frequency.

A number of studies have shown that the visual response curve discussed earlier and a measure of $S(f)$ can be connected to arrive at an overall quality factor. See for example system modulation transfer actuance (SMT Acutance) by Crane [71] and an improvement by Gendron [72] known as cascaded area modulation transfer (CMT Acutance). One metric known as the subjective quality factor (SQF) [56] defines an equivalent passband based on the visual MTF having a lower (initial) cutoff frequency at f_i and an upper (limiting) frequency of f_l. Here, f_i is chosen to be just below the peak of the visual MTF, and f_l is chosen to be four times f_i (two octaves above it). For prints that are to be viewed at normal viewing distance [i.e., about 340 mm (13.4 in.)], this range is usually chosen to be approximately 0.5 to 2.0 cycles/mm (13 to 50 cycles per inch).

The MTF of the system is integrated as follows:

$$\text{SQF} = \int_{f_x=.5}^{2} \int_{f_y=.5}^{2} S(f_x, f_y)\, d(\log_{10} f_x)\, d(\log_{10} f_y) \tag{16}$$

Note that the frequencies are then weighted reciprocally by the differential of the logarithm (see Figure 3.27).

This function has been shown to have a high degree of correlation with pictorial image quality over a wide range of picture types and MTFs. It is possible that a demand function similar to that described in the MTF concept could be applied to further improve the performance. The SQF metric is applied to the final print as it is to be viewed and may be scaled to the imaging system, when reduction or enlargement is involved, by applying the appropriate scaling factor to the spatial frequency axis. It should be noted that there is a slight difference between the upper band limit of this metric at 2 cycles/mm (50 cycles per inch) and the critical frequency described above in Biedermann's work for landscapes, which is in the 4 to 5 cycle/mm (100 to 125 cycles per inch) region. But there is good agreement for the portrait conclusions of the earlier work, which cites an upper

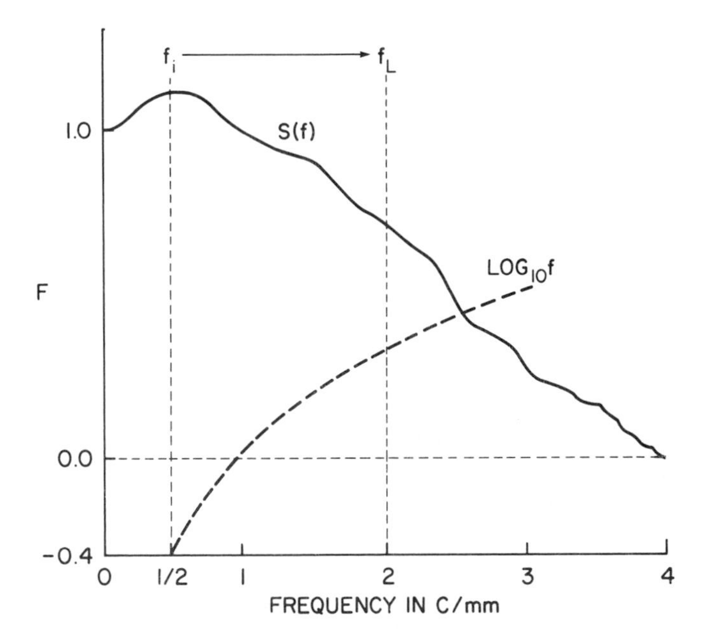

Figure 3.27 The logarithmic weighting function and the passband used to obtain the SQF from the signal function S(f).

critical frequency of 2 to 3 cycles/mm (50 to 75 cycles per inch). The differences between these metrics, then, is that the SQF tends to be measuring the amplitude of given frequencies and the critical frequency method tends to be measuring frequencies at a given amplitude of the system MTF. Both metrics acknowledge the importance of granularity or noise without directly incorporating granularity in their algorithms. Granger [57] discusses some effects of granularity and digital structure in the context of the SQF model, but calls for more extensive study of these topics before incorporating them into the model.

It is clear that when the gray content and resolution of the system are high enough to be indistinguishable from an analog imaging system, then these techniques, which are general in nature, should be applicable. The quantitative levels at which this equivalence occurs vary broadly. Usually 32 to 512 levels of gray suffice, depending on noise (higher noise requires fewer levels), while resolution values typically range from 100 to 1000 pixels/in (4 to 40 pixels per millimeter), depending on noise, subject matter, and viewing distance.

3.5.7 Information Content and Information Capacity

There are numerous articles in the imaging science literature that analyze imaging systems in terms of information capacities and describe their images as having various information contents. Some of these were reviewed earlier in the subsection "Quantization, Noise, and the Eye." This approach uses all of the concepts developed above for the MTF and the Wiener spectrum and can incorporate the human visual system as well. It produces results in bits/area that are directly related to the task of moving electronic image data from an input scanner to an output scanner. Much of the work done in this area is related to the photographic image, and that work will be briefly addressed later in this chapter. The basic equation for information content of an image is given [58] by

$$H_i = \frac{1}{2}\int_{-\infty}^{\infty} \mathrm{Log}_2 \left(1 + \frac{\Phi_S(f)}{\Phi_N(f)}\right) df \tag{17}$$

where H_i is the information content of the image, Φ_S is the Wiener spectrum of the signal, Φ_N is the Wiener spectrum of the system noise, and f is the spatial frequency, usually given in cycles per millimeter. This equation is in one-dimensional form for simplicity, in order to develop the basic concepts. For images these concepts must be extended to two dimensions. Unlike the work dealing with the photographic image, the assumption of uniform isotropic performance cannot be used to simplify the notation to radial units. For scanners the separation of the orthogonal x and y dimensions of the image must be preserved.

An alternative method for calculating information capacity that does not include explicit spatial frequency dependence but does explicitly handle probabilities has also been used [9, 10, 59] and served as the basis for our discussion of quantization. Here equation (2) is rewritten as

$$H_i = N\, \mathrm{Log}_2\, pM \tag{18}$$

where N is the number of independent information storage cells per unit area. It may be set equal to the reciprocal of the effective area of the image spread function. Here, p is the reliability with which one can distinguish the separate messages within an information cell, and M is the number of messages per cell. In this case the number M is determined by the number of statistically significant different gray levels that can be distinguished in the presence of system noise at the reliability p, using noise measurements made with the system spread function.

Generalizing equations (2) and (3) to the "generic" units of Figure 3.23, L is set equal to S_0 and σ_a is set equal to σ_s for the maximum signal and its standard deviation, respectively. We select a spread function for an unaliased system equal to 2 pixels by 2 pixels and translate this to frequency space using the reciprocal of the sampling frequencies f_{s_x} and f_{s_y} in the x and y directions. This gives a generalized, sampling-oriented version of equations (2) and (18) as

$$H_i = \frac{f_{s_x} f_{s_y}}{4}\, \mathrm{Log}_2 \left[\frac{S_0}{K\sigma_S}\right] \tag{19}$$

where S and σ_s are measured in the same units. K can be set to determine the reliability for a given application. Values from 2 [10] to 20 [9] have been proposed for K for different applications; 6 is suggested here, making $p = 0.997$. This assumes that σ_s is a constant (i.e., additive noise) at all signal levels. If not, then the specific functional dependence of σ_s on S must be accounted for in determining the quantity in the brackets, measuring the desired number of standard deviations of the signal at each signal level over the entire range [8, 10]. While this approach predicts text quality and resolving power [59, 71] and deals with the statistical nature of information, it does not (as noted above) permit the strong influence of spatial frequency to be handled explicitly.

Equation (17) may be expanded to illustrate the impact of the MTF on information content, giving

$$H_i = \frac{1}{2}\int_{-\infty}^{\infty} \mathrm{Log}_2 \left(1 + \frac{k^2 \Phi_i(f)|\mathrm{MTF}(f)|^2}{\Phi_N(f)}\right) df \tag{20}$$

where $\Phi_i(f)$ is the Wiener spectrum of the input scene or document and MTF (f) is the MTF of the imaging system (assumed linear) with all its components cascaded. At this point we need to begin making some assumptions in order to carry the argument further. The constant k in the equation is actually the gain of the imaging system. It converts the units of the input spectrum into the same units as the spectral content of the noise in the denominator. For example, a reflectance spectrum for a document may be converted into gray levels by a k factor of 256 when a reflectance of unity (white level) corresponds to the 256th level of the digitized (8-bit) signal from a particular scanner; the noise spectrum is in units of gray levels squared.

Various authors have gained further insight into the use of these general equations. Some of those investigating photographic applications have extended their analysis to allow for the effect of the visual system [60]; others have attempted to apply some rigor to the terms in the equation that are appropriate for digital imaging [61, 62]. Others have worked on image-quality metrics for digitally derived images [63], but some have tended to focus on the relationship to photointerpreter performance [64]. Several of these authors have suggested that properly executed digital imagery does not appear to be greatly different from standard analog imagery in terms of subjective quality or interpretability. One almost always sees these images using some analog reconstruction process to which many analog metrics apply. It therefore seems reasonable to combine some of this work into a single equation for image information and to hypothesize that it has some direct connection with overall image quality when applied to a scanner whose output is viewed or printed by an approximately linear display system. It must also be assumed that the display system noise and MTF are not significant factors or can be incorporated into the MTF and noise spectra by a single cascading process. A generic form of such an equation is given below without the explicit functional dependencies on frequency in order to show and explain the principles that follow (expanding on the analysis in Ref. 60).

$$H_i = \frac{1}{2}\int_{-\infty}^{\infty}\int_{-\infty}^{\infty} \mathrm{Log}_2 \left[1 + \frac{k^2\Phi_i|\mathrm{MTF}|^2 R_1{}^2}{\{1 + 12\,(f_x{}^2 + f_y{}^2)\}\{[\Phi_a + \Phi_n + \Phi_q]\,R_2{}^2 + \Phi_E\}}\, df_x\, df_y\right] \tag{21}$$

Let us begin by examining the numerator. Several authors have attempted to multiply the modulation transfer function for the system by a spatial frequency response function for the human visual system to arrive at an appropriate weighting for the signal part of equation (21). Kriss and his co-workers [60] observed that

> a substantial increase in the enhancement beyond the eye's peak response produced larger improvements in overall picture quality than did equivalent increases in enhancement at the peak of the eye's response. The pictures with large enhancement at the eye's peak response were "sharper," but were also judged to be too harsh. These results indicate that the human visual system does not act as a passive filter and that it may weigh the spatial frequencies beyond the peak in the eye's response function more than those at the peak.

Lacking a good model for the visual system's adaptation to higher frequencies as described above, Kriss et al. proposed the use of the reciprocal of the eye frequency response curve as a weighting function, $R_1(f)$, that could be applied to the numerator.

The conventional eye response $R_2(f)$ should be applied in the denominator to account for the perception of the noise, since the eye is not assumed to enhance noise but merely to filter it. The noise term, Φ_N, in the earlier equations has been replaced by the expression in the square brackets (see below) and multiplied by $R_2^2(f)$. Examples of functions R_1 and R_2 are shown in Figure 3.28 for a 1-ft viewing distance. The reciprocal curve, R_1, is set equal to 0.0 at 8 cycles/mm (200 cycles/inch) in order to limit this function [the figure only shows it to 6 cycles/mm (150 cycles/in)].

Next let us examine the noise effects themselves. A major observation is that noise in the visual system within one octave of the signal's frequency tends to affect that signal. It can be shown that the sum in the first curly brackets in the denominator of equation (21) provides a weighting of noise frequencies appropriate to this one-octave frequency-selective model for the visual system [60]. Frequency-selective models of the visual system have been described by several authors [44, 65]. The present construct for noise perception was first described by Stromeyer and Julesz [66]. A term for the Wiener spectrum of the noise in

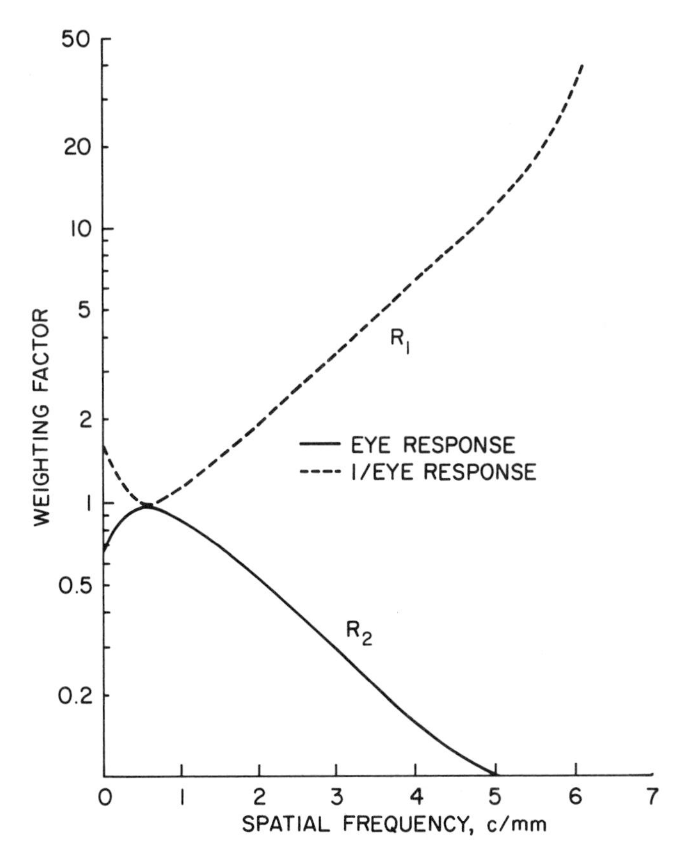

Figure 3.28 Spatial frequency response R2 of human eye, and reciprocal eye response R1 to be used as weighting functions in equation (21).

the visual process, Φ_E, has been added to the second factor in the denominator to account for yet one more source of noise. It is not multiplied by the frequency response of the eye, since it is generated after the frequency-dependent stage of the visual process.

The factor in square brackets in the denominator contains three terms unique to the digital imaging system [62]. These are Φ_a, the Wiener spectrum of the aliased information in the passband of interest; Φ_n, the Wiener spectrum of the noise in the electronic system, nominally considered to be fluctuations in the fast scan direction; and Φ_q, the quantization noise determined by the number of bits used in the scanning process. We have thus combined in equation (21) important information from photographic image-quality studies, which include vision models and psychophysical evaluation, with scanning parameters pertinent to electronic imaging.

The study of information capacity, information content, and related measures as a perceptual correlate to image quality for digital images is an ongoing activity. By necessity it is focused on specific types of imaging applications and observer types. For example, an excellent data base of images and related experiments on quality metrics exists for aerial photography as used by photointerpreters [64].

Experiments correlating subjective quality scores with the logarithm of the basic information capacity (taking the log of H_i as defined in equation (17)) showed correlation of .87 and greater for subjective quality of pictorial images [67]. Specific MTF and quantization errors were studied. The results were normalized by the information content of the original.

By use of various new combinations of the same factors discussed above, it was possible to obtain even higher correlations. Here a digital quality factor was defined as

$$\mathrm{DQF} = \left[\frac{\int_0^{f_n} \mathrm{MTF}_s(f)\, \mathrm{MTF}_v(f)\, d(\log f)}{\int_0^{10} \mathrm{MTF}_v(f)\, d(\log f)} \right] \times \log_2 \left[\frac{L}{L/M + 2\sigma} \right] \tag{22}$$

where one-dimensional frequency description (f) is used for simplicity (per the original authors) and the subscripts s and v refer to the system under test and the visual process, respectively. L is the density range of the output imaging process, f_n is the Nyquist frequency, M is the number of quantization levels, and σ is the RMS granularity of the digital image using a 10×1000 μm microdensitometer slit. The first factor is related to the subjective quality factor (SQF) described above, and the second factor is related to the fundamental definition of image information capacity in equation (18). A correlation coefficient of .971 was obtained for these experiments, using student observers and pictures showing a portrait together with various test patterns. It must be noted, however, that information capacity or any of these information-related metrics cannot be accepted, without psychophysical verification, as a general measure of image quality when different imaging systems are to be compared [68].

In conclusion, this brief overview of specific quality metrics should give the reader some perspectives on which ones may be best suited to his or her needs. The variety of these metrics, and the considerable differences among them are evidence of the inherent diversity of imaging applications and requirements. Given this diversity, together with the large and rapidly expanding range of imaging technologies, it is hardly surprising that no single universal measure of quality has been found.

REFERENCES

1. P. Mertz and F. A. Gray, *Bell System Tech. J.*, *13*:464 (1934).
2. D. R. Lehmbeck and J. C. Urbach, *Scanned Image Quality*, Xerox Internal Report X8800370, Xerox Corporation, Webster, NY, July 1988.
3. R. C. Gonzalez and P. Wintz, *Digital Image Processing*, Addison-Wesley, Reading, MA, 1977.
4. C. E. Shannon, *Bell System Tech. J.*, *27*:379, 623 (1948).
5. P. G. Roetling, *Proc. SPIE*, *74*:195 (1976).
6. W. Lama, S. Feth and R. Loce, *J. Imaging Tech.*, *15*:130 (1989).
7. S. Goldman, *Information Theory*, Prentice Hall, New York, 1953.
8. J. A. Eyer, *Photog. Sci. Eng.*, *6*:71 (1962).
9. J. H. Altman and H. J. Zweig, *Photog. Sci. Eng.*, *7*:173 (1963).
10. D. R. Lehmbeck, *Photog. Sci. Eng.*, *11*:270 (1967).
11. J. C. Dainty and R. Shaw, *Image Science: Principles, Analysis and Evaluation of Photographic-Type Imaging Processes*, Academic Press, New York, 1974.
12. O. Bryngdahl, *J. Opt. Soc. Am.*, *66*:87 (1976).
13. G. W. Jorgensen (1977). In *Advances in Printing Science and Technology*, Pentech Press, London, p. 109.
14. L. A. Jones and C. N. Nelson, *J. Opt. Soc. Am.*, *32*:558 (1942).
15. L. A. Jones, *Photogr. J.*, *89B*:126 (1949).
16. C. J. Bartleson and E. J. Breneman, *J. Opt. Soc. Am.*, *57*:953 (1967).
17. C. N. Nelson. In *The Theory of Photographic Process*, 4th ed., Macmillan, New York, 1977, p. 536; *Neblette's Handbook of Photography and Reprography: Materials, Processes and Systems*, 7th ed., Van Nostrand Reinhold, New York, 1977, p. 234.
18. T. M. Holladay, *Proc. S.I.D.*, *21*:185 (1980).
19. P. G. Roetling, *J. Appl. Photog. Eng.*, *3*:12 (1977).
20. J. C. Stoffel, *Graphical and Binary Image Processing and Applications*, p. 289, Artech House, Norwood, MA, 1981.

20a. J. C. Stoffel and J. F. Moreland, *IEEE Trans. on Communications*, *COM-29*: 1898 (1981).

21. R. Ulichney, *Digital Halftoning*, The MIT Press, Cambridge, MA, 1987.
22. J. A. C. Yule and W. J. Nielson, *The Penetration of Light into Paper and Its Effect on Halftone Reproduction*, Communication No. 416, Kodak Research Laboratories, 1951.
23. F. R. Clapper and J. A. C. Yule, *J. Opt. Soc. Am.*, *43*:600 (1953).
24. D. R. Lehmbeck, *Preprint Book, 28th Annual Conference of SPSE*, Society of Photographic Scientists and Engineers, Washington, D.C., 1975, p. 155.
25. M. Maltz, *J. Appl. Photog. Engr.*, *9*:83 (1983).
26. J. L. Kofender (1987). "The Optical Spread Functions and Noise Characteristics of Selected Paper Substrates Measured in Typical Reflection Optical System Configurations"; M.S. Thesis, Rochester Institute of Technology, Rochester, NY.
27. K. J. Klees and J. Holmes (1985). Abstract in SPSE Conference Paper Summaries, 25th Fall Symposium, Arlington, VA. p. 199.

28. F. Bestenreiner, U. Greis, J. Helmberger, and K. Stadler, *J. Appl. Photog. Engr.*, *2*:86 (1976).
29. R. R. Firth, D. Kessler, E. Muka, K. Naor, and J. C. Owens, *J. Imaging Tech.*, *14*:78 (1988).
29a. H. Sonnenberg, *Appl. Opt.*, *21*:1745 (1982).
30. D. R. Lehmbeck and J. J. Jakubowski, *J. Appl. Photog. Engr.*, *5*:63 (1979).
31. Canon, Inc., *Fundamentals of Digital Copiers*, Rev. 1, Chapter 5. Reprographic Products Technical Documents Department Report #FY8-1339-D10, Canon Inc., Tokyo 160, Japan, 1984. (Document distributed with Canon 9030 product.)
32. W. J. Smith, *Modern Optical Engineering*, McGraw-Hill, New York, 1966, pp. 308–324.
33. F. H. Perrin, *J. SMPTE*, *69*:151, 239 (1960).
34. R. L. Lamberts, *Appl. Optics*, *2*:273 (1963).
35. A. V. Oppenheim and R. Schafer, *Digital Signal Processing*, Prentice-Hall, Englewood Cliffs, NJ, 1975.
36. J. D. Gaskill, *Linear Systems, Fourier Transforms and Optics*, John Wiley, New York, 1978.
37. F. Scott, R. M. Scott, R. V. Shack, *Photog. Sci and Eng.*, *7*:345 (1963).
38. F. W. Campbell, *Proc. Australian Physio. Pharmacol. Soc.*, *10*:1 (1979).
39. I. Gorog, C. R. Carlson, and R. W. Cohen, *Image Analysis and Evaluation*, *SPSE Conf. Proc.* Society of Photographic Scientists and Engineers. Washington, D.C., (1977) p. 382.
40. O. Bryngdahl, *J. Opt. Soc. Am.*, *56*:811 (1966).
41. H. A. Watanabe, T. Mori, S. Nagata, and K. Hiwatoshi, *Vision Research*, *8*:1245 (1968).
42. W. E. Glenn, K. G. Glenn, and C. J. Bastian, *Proc. S.I.D.*, *26*:71 (1985).
43. R. P. Dooley and R. Shaw, In *Image Science Mathematics Symposium*, Western Periodicals, Hollywood, CA, 1977, p. 12.
44. C. Blakemore and F. W. Campbell, *J. Physio.*, *203*:237 (1969).
45. B. E. Rogowitz, *Proc. S.I.D.*, *24*:235 (1983).
46. R. C. Jones, *J. Opt. Soc. Am.*, *45*:799 (1955).
47. R. Shaw. In *Image Science Mathematics Symposium*, Western Periodicals, Hollywood, CA, 1977, p. 1.
48. L. M. Biberman (ed.), *Perception of Displayed Information*, Plenum Press, New York, 1976.
49. M. Kriss. In *The Theory of Photographic Process*, 4th ed., Macmillan, New York, 1977, chap. 21.
50. F. Scott, *Photog. Sci. and Eng.*, *10*:49 (1966).
51. W. N. Charman and A. Olin, *Photog. Sci. and Eng.*, *9*:385 (1965).
52. H. C. Burroughs, R. F. Fallis, T. H. Warnock, and J. H. Brit, ''Quantitative Determination of Image Quality,'' Boeing Corporation Report D2 114058-1, May 1967.
53. H. L. Snyder, ''Display Image Quality and the Eye of the Beholder.'' In *Image Analysis and Evaluation*, Society of Photographic Scientists and Engineers, Wash-

ington, D.C., 1976, p. 341; "Image Quality and Observer Performance," In *Perception of Displayed Information*, Plenum Press, New York, 1970, p. 87.
54. H. Frieser and K. Biedermann, *Photog. Sci. and Eng.*, *7*:28 (1963).
55. K. Biedermann, *Photog. Korresp.*, *103*:25, 103:41 (1967).
56. E. M. Granger and K. N. Cupery, *Photog. Sci. and Eng.*, *16*:221 (1972).
57. E. M. Granger, *Digital Image Processing*, SPIE Vol. 528, 1985, p. 95.
58. P. B. Felgett and E. H. Linfoot, *Philos. Trans. R. Soc., London*, *247*:269 (1955).
59. C. S. McCamy, *Appl. Opt.*, *4*:405 (1965).
60. M. Kriss, J. O'Toole, and J. Kinard, *Proc. SPSE Conference on Image Analysis and Evaluation*, Toronto, SPSE, Washington, D.C., 1976, p. 122.
61. F. O. Huck and S. K. Park, *Appl. Opt.*, *14*:2508 (1975).
62. F. O. Huck, S. K. Park, D. E. Speray, and N. Halyo, *Proc. SPIE*, *310*:36 (1981).
63. Y. Miyake, K. Seidel, F. Tomamichel, *J. Photog. Sci.*, *29*:111 (1981).
64. J. J. Burke and H. L. Snyder. In SPIE, Vol. 310, Bellingham, WA, 1981, p. 16.
65. M. B. Sachs, J. Nachmias, and J. G. Robson, *J. Opt. Soc. Am.*, *61*:1176 (1971).
66. C. F. Stromeyer and B. Julesz, *J. Opt. Soc. Am.*, *62*:1221 (1972).
67. Y. Miyake, S. Inoue, M. Inui, and S. Kubo, *J. Imaging Tech.*, *12*:25 (1986).
68. J. H. Metz, S. Ruchti, and K. Seidel, *J. Photog. Sci.*, *26*:229 (1978).
69. Patterson, *Proc. S.I.D.*, *27*:4 (1986).
70. Hufnagel, In *Perceptions of Displayed Information*, Plenum Press, New York, 1976, p. 48.
71. J. Crane. *Soc. Mot. Pict. Telev. Eng.*, 73:643 (1964).
72. R. G. Gendron, *J. Soc. Mot. Pict. Telev. Eng.*, *82*:1009 (1973).

BIBLIOGRAPHY

R. Bracewell, *The Fourier Transform and Its Application*, McGraw-Hill, New York, 1985.

F. M. Brown, H. J. Hall, and J. Kosar, *Photographic Systems for Engineers*, Society of Photographic Scientists and Engineers, Washington, D.C., p. 167.

K. Castleman, *Digital Image Processing*, Prentice-Hall, Englewood Cliffs, NJ, 1979.

P. S. Cheatham (ed.), *Proc. SPIE*, Vol. 310, San Diego, CA, SPIE, Bellingham, WA, 1981.

D. Dutton (ed.), *Proc. SPIE*, Vol. 46, *Image Assessment and Specification*, SPIE, Bellingham, WA, 1974.

J. W. Goodman, *Introduction to Fourier Optics*, McGraw-Hill, New York, 1968.

E. M. Granger and L. R. Baker (eds.), *Proc. SPIE*, Vol. 549, *Image Quality: An Overview*, Arlington, VA, SPIE, Bellingham, WA, 1985.

T. H. James (ed.), *The Theory of Photographic Process*, 4th ed., Macmillan, New York, 1977.

R. C. Jennison, *Fourier Transforms and Convolutions for the Experimentalist*, Pergamon Press, New York, 1961.

E. O'Neil, *Statistical Optics*, Addison-Wesley, Reading, MA, 1963.

A. Rose, *Vision: Human and Electronic*, Plenum Press, New York, 1973.

R. Shaw (ed.), *Selected Readings in Image Evaluation*, Society of Photographic Scientists and Engineers, Washington, D.C., 1976.

R. Shaw (ed.), *Image Analysis and Evaluation, SPSE Conf. Proc.* Society of Photographic Scientists and Engineers, Washington, D.C., 1977.

P. N. Slater and R. F. Wagner, *Technical Digest of the SPSE Conference on Image Analysis Techniques and Application*, Tucson, AZ, Society of Photographic Scientists and Engineers, Washington, D.C., 1981.

J. M. Sturge, *Neblette's Handbook of Photography and Reprography: Materials, Processes and Systems*, 7th ed., Van Nostrand Reinhold, New York, 1977.

C. O. Wilde and E. Barrett (eds.), *Image Science Mathematics Symposium*, Western Periodicals, Hollywood, CA, 1977.

SPIE/SPSE *Proceedings of the International Symposium and Exposition of Electronic Imaging Devices and Systems*, 1988–1991.

SPSE International Electronic Imaging Conf., 1984, 1985.

4

Holographic Scanners for Bar Code Readers

LeRoy D. Dickson

Storage Systems Products Division, IBM, San Jose, California

Glenn T. Sincerbox

Almaden Research Center, IBM, San Jose, California

4.1 INTRODUCTION

A bar code is a sequence of dark bars on a light background, or the equivalent of this with respect to the light-reflecting properties of the surface. The coding is contained in the relative widths or spacings of the dark bars and light spaces. Perhaps the most familiar bar code is the universal product code (UPC) which appears on nearly all of the grocery items in supermarkets today. Figure 4.1 is an example of a UPC code.

A bar code scanner is an optical device that reads the code by scanning a focused beam of light, generally a laser beam, across the bar code and detecting the variations in reflected light. The scanner converts these light variations into electrical variations that are subsequently digitized and fed into the decoding unit, which is programmed to convert the relative widths of the digitized dark/light spacings into numbers and/or letters.

The concept of bar code scanning for automatic identification purposes was first proposed by N. J. Woodland and B. Silver in a patent application filed in 1949. A patent, titled "Classifying Apparatus and Method," was granted in 1952 as U.S. Patent No. 2,612,994. This patent contained many of the concepts that would later appear in bar code scanning systems designed to read the UPC code.

Figure 4.1 A typical UPC bar code.

4.1.1 The UPC Code

In the early 1970s, the supermarket industry recognized a need for greater efficiency and productivity in their stores. Representatives of the various grocery manufacturers and supermarket chains formed a committee to investigate the possibility of applying a coded symbol to all grocery items to allow automatic identification of the product at the checkout counter. This committee, the Uniform Grocery Product Code Council, Inc., established a symbol standardization subcommittee whose purpose was to solicit and review suggestions from vendors for a standard product code to be applied to all supermarket items.

On April 3, 1973, the Council announced their choice. The code chosen was a linear bar code that was similar to the one proposed by IBM. The characteristics of this bar code, the now familiar UPC code, are described in detail by Savir and Laurer [1].

The UPC code is a fixed-length numeric-only code. It consists of a pair of left guard bars, a pair of right guard bars, and, in the standard version A symbol shown in Figure 4.1, a pair of center guard bars. Each character is represented by two dark bars and two light spaces. The version A symbol contains 12 characters, 6 in the left half and 6 in the right half. Thus, a version A UPC symbol will have 30 dark bars and 29 light spaces, counting the 6 guard bars— left, right, and center. The first character in the left half is always a number system character. For example, grocery items are given the number system 0, which often appears on the left of the symbol. The last character on the right is always a check character. This sometimes appears to the right of the bar code symbol.

The remaining five characters in the left half of the version A UPC symbol identify the manufacturer of the product. For example, the left-half number 20000 represents Green Giant products. This left-half five-digit code is assigned to the various manufacturers by the Uniform Product Code Council.

The remaining five characters in the right half of the version A UPC symbol identify the particular product. This right-half five-digit code is assigned at the product manufacturer's discretion. For example, Green Giant has assigned the right-half number 10473 to their 17-oz can of corn. Therefore, the complete UPC code for the Green Giant 17-oz can of corn, ignoring the number system character and the check character, is 20000–10473.

There are a number of other properties of the UPC code and symbol that are significant, relative to the design and use of equipment for reading the code. First of all, the left and right halves of the version A symbol are independent. That is, each half can be read independently of the other half and then combined with the other half in the logic portion of the reader to yield the full UPC code. Furthermore, as shown in Figure 4.2, each half of the UPC symbol is "over-square." That is, the symbol dimension parallel to the bars is greater than the symbol dimension perpendicular to the bars. This fact is vital in the determination of a minimum scan pattern for reading the code, as we will see later.

Each character of the code is represented by two dark bars and two light spaces. Individual bars and spaces can vary in width from one module wide to four modules wide. (Note: On all UPC codes, the guard bars are always one module wide and separated by a one-module-wide space.) The total number of modules in each character is always seven. The left-half characters are coded inversely from the right-half characters. As shown in Figure 4.3, for example, if we let white = 0, and black = 1, then the code for the number 2 in the left half is 0010011 and the right half is 1101100. The other numerals are similar combinations of the seven-module set.

The fact that each character is always seven modules wide leads to a second

Figure 4.2 Two UPC half-symbols.

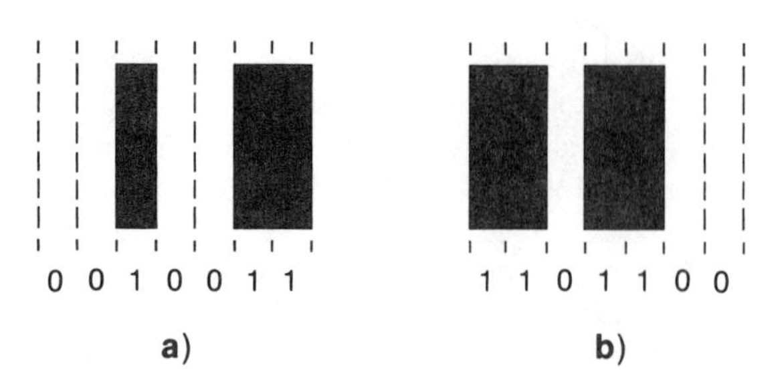

Figure 4.3 Example of the character encoding for the number 2: (a) left-half character; (b) right-half character.

major property of the UPC code: it is self-clocking. Therefore, absolute time measurements are unimportant. What is measured is the time that is required to go from the leading edge of the first black bar to the leading edge of the third black bar (i.e., the first black bar of the next character). This time interval is then divided into seven equal intervals, and the relative widths of the two black bars and the two white spaces are then determined for decoding purposes. Thus, the total width of a character—black-white-black-white—is measured, and the relative widths of the two blacks and the two whites are determined for decoding.

This self-clocking feature is very important in the design of scanners for reading the UPC code. It means that the velocity of a scanning light beam for reading the code does not have to be constant across the full width of the code. The velocity only needs to be reasonably constant across a single character. This means that the UPC code can be read by moderately nonlinear scan patterns, such as sinusoidal or Lissajous patterns. It also means that the code can be read on curved surfaces. Furthermore, the scan lines reading the code do not have to be perpendicular to the bars and spaces in the code. Satisfactory reading can be obtained with scan lines passing through the code at any angle relative to the bars and spaces so long as a single scan line passes completely through a full half-symbol, including the center guard bars and one pair of edge guard bars.

A third property of the UPC symbol of significance to the scanner designer is the size of the symbol, which is allowed to vary from the nominal size [about 1.0 in $\times$ 1.25 in (25.4 mm $\times$ 31.75 mm) for the version A symbol] down to 0.8 $\times$ nominal and up to 2.0 $\times$ nominal. This size variation allows the use of small labels with good print quality on small packages and large labels with poorer print quality on larger packages. From the scanner designer's viewpoint, the small label will establish the minimum bar width to be read, and the large label will set a lower limit on the size of the scan pattern.

The minimum bar width established by the UPC specification, including tolerances, is 0.008 in (0.2 mm). This number establishes the maximum attainable depth of field for the optical reader. In practice, the depth of field of the typical laser scanner designed to read the UPC code will easily meet, and exceed, the 1-in (25.4 mm) depth of field required by the early UPC guidelines. However, this 1-in depth of field did not take into consideration the manner in which the scanners would eventually be used. Depths of field of several inches (100 mm^{+}) are required for today's UPC bar code readers.

Finally, the contrast specification for the UPC symbol requires that the contrast be measured using a photomultiplier detector (PMT) with an S-4 photocathode response curve coupled with a Wratten 26 filter. This combination has a peak response at a wavelength of approximately 610 nm (24 μin), falling to zero at approximately 590 nm (23.2 μin) and 650 nm (25.6 μin). This response includes, not coincidentally, the wavelength of the helium-neon laser, 632.8 nm (24.9 μin). While several of the inks used for printing UPC labels can provide acceptable contrast out to 700 nm (27.56 μin), there are many other inks in use that do not provide acceptable contrast beyond 650 nm (25.6 μin). These inks would preclude general use of longer-wavelength light sources, such as laser diodes.

4.1.2 Other Bar Codes

The UPC code is not widely used in the industrial environment (the manufacturing, warehouse, and distribution applications). Here, the requirements are different from those of the supermarket, so the codes used are different from the UPC code. The preferred codes for the industrial environment are Bar Code 39, Interleaved 2 of 5, and Codabar.

The most common bar code in the industrial environment is the so-called Bar Code 39, or 3 of 9 bar code. This code is fully alphanumeric and self-checking. For a full discussion of Bar Code 39, as well as several other codes, and the definition of such terms as "self-checking," see Allais [2]. The code shown in Figure 4.4 is an example of Bar Code 39.

Bar Code 39 got its name from the fact that it originally encoded 39 characters: the 26 letters of the alphabet, the numbers from 0 through 9, and the symbols

Figure 4.4 An example of Bar Code 39.

-,., and SPACE, plus a unique start/stop character, the asterisk (*). Today it also encodes the four so-called special characters: $, /, +, and %, for a total of 43 characters. However, it is still referred to as Bar Code 39. It is also often called the 3 of 9 code, since the code always consists of nine elements (five dark bars and four light spaces), and three of them are wide with the remaining six narrow. In the primary set of 39 characters, two of the wide elements are dark bars. In the four special characters, the wide elements are all light spaces.

The 2 of 5 code is a subset of the 3 of 9 code. In 2 of 5, only the bars are used for encodation. Two of the five bars are wide, just as in the original 3 of 9 code. The spaces are not used. This code is strictly numeric. The basic 2 of 5 code is not widely used in industry, but a variation of it, called the Interleaved 2 of 5 code, is used extensively for manufacturing and distribution applications. This code uses the bars to encode one character in the standard 2 of 5 code, and then uses the interleaving spaces to encode a second character in 2 of 5 code. This allows more characters to be encoded in a fixed bar-code length than either 3 of 9 code or 2 of 5 code. This code is also only numeric, but, due to the interleaving feature, it can encode nearly 80% more characters per unit length than Bar Code 39, assuming both codes have the same minimum bar width. For this reason, the Interleaved 2 of 5 code is often used where space limitations will not permit the use of Bar Code 39.

A third code that is used extensively in medical institutions, and which was adopted as an early standard by the American Blood Commission for use in identifying blood bags, is Codabar. This code is also frequently seen in some transportation and distribution applications.

4.1.3 Bar Code Properties

From the standpoint of the scanner, the important properties of any bar code are

1. Minimum bar width: generally specified in millimeters or mils (thousandths of an inch) and often referred to as the X dimension.
2. Contrast: a measure of the reflectance of the bars and spaces. Generally expressed in terms of the print contrast signal (PCS) where

$$PCS = \frac{r_s - r_b}{r_s} \qquad (1)$$

where r_s = reflectance of a space

r_b = reflectance of a bar

It should be noted that PCS is usually measured for one particular wavelength of light. In the majority of applications this wavelength is 633 nm (24.9 μin), the wavelength of the helium-neon laser, which is the most common light source for most laser scanners. (Some applications allow PCS to be measured at 900 nm (35.4 μin), the wavelength of some infrared light sources used in some readers.) This is an important point to remember since PCS will vary drastically as a function of wavelength if colored inks or backgrounds are used. In practice, the most important reflectance property of the bar code is the absolute contrast, which is simply the space reflectance minus the bar reflectance (the numerator equation (1)).

3. Code length: the physical length of a bar code is determined by the density of the code (which is determined by the minimum bar width) and the number of characters in the code. The physical length of the code determines how long the scan lines must be and, when combined with the code height, will determine how accurately the scan line must be oriented with respect to the bar code.
4. Code height: the height of the bar code (the dimension parallel to the bars) will determine the angular accuracy required in orienting the scan line relative to the bar code.
5. Bar code quality: this includes both the quality of the printing or etching of the code itself and the quality of the surface on which the code is printed. Obviously, the better the quality of both, the easier it will be for the scanner to successfully scan and decode the bar code.

There is a great deal more that could be said about the bar codes themselves. However, a more detailed analysis of the fundamental properties of the bar codes is beyond the scope of this review and is not really necessary for the purposes of our discussion of bar code scanning.

4.2 NONHOLOGRAPHIC UPC SCANNERS

A block diagram of a typical laser scanner system for reading the UPC code is shown in Figure 4.5. The focused laser beam scans the UPC symbol on a package as the package passes over the read window of the scanner. The laser beam is reflected from the symbol as it passes over the dark bars and light spaces. The diffuse portion of the reflected light is modulated by the reflectivity variations of the symbol (bars and spaces). This light modulation is detected by the photodetector, which converts the light modulation into electric modulation. The electric "signal" is then amplified and digitized and transmitted to the "candidate select" block. This block acts as a filter, allowing only valid UPC half-symbols to pass to the decoder. The decoder converts the signals for each half-symbol into characters, and then combines the characters for the two half-symbols to yield a complete UPC product identification code. The computer then searches its memory for a description and price of the item identified by this UPC code. This information is transmitted back to the checkout terminal, where it appears

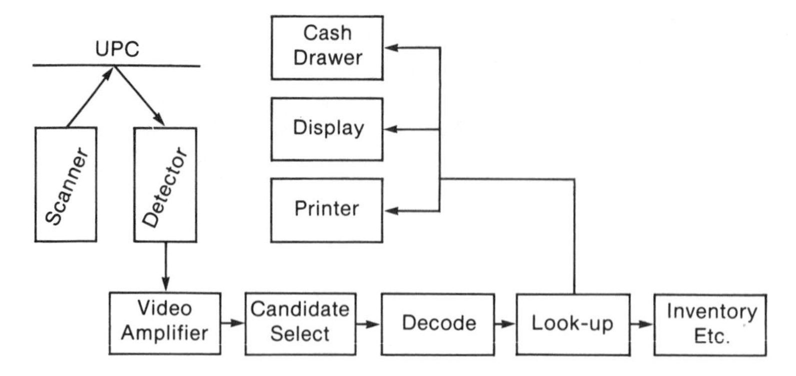

Figure 4.5 Block diagram of a UPC scanner.

on the display and on the customer receipt. Simultaneously, the store inventory is updated to reflect the sale of the identified item. All of this takes place in a few milliseconds.

The focused spot size of the scanning laser beam must be about 0.2 mm in order to be able to read the labels with the smallest bar widths while still yielding adequate depth of field. This requires an optical f-number of 250, which, when combined with scanner geometry, establishes the focusing optics' requirements.

A number of technologies are available for deflecting the focused laser beam in a conventional, nonholographic, UPC scanner. Cost and performance requirements limit the choice to mechanical deflectors—generally either rotating or oscillating mirrors, or a combination of these. The scan pattern created by the laser deflection mechanism must be capable of reading a full-size version A UPC symbol regardless of the orientation of the symbol in the scan window. In other words, the scanner must be omnidirectional. An omnidirectional scanner will allow maximum freedom for the scanner operator when bringing the item across the scan window.

We have already seen that the two halves of the UPC symbol can be read independently of each other. We have also seen that each half-symbol is oversquare. Therefore, the minimum scan pattern that will allow omnidirectional scanning is a pair of perpendicular scan lines in the form of an X (see Figure 4.6a). As the UPC symbol passes over the scan window, at least one of the legs of the X will pass through the entire half-symbol at some point in the window. The figure shows two extreme orientations of the symbol as it is passed over the window. These are the worst-case orientations in that they allow the minimum time for scanning the symbol satisfactorily.

The amount that the half-symbol is oversquare, when combined with the maximum item velocity of 2.54 m/s (100 in/s), determines the minimum pattern

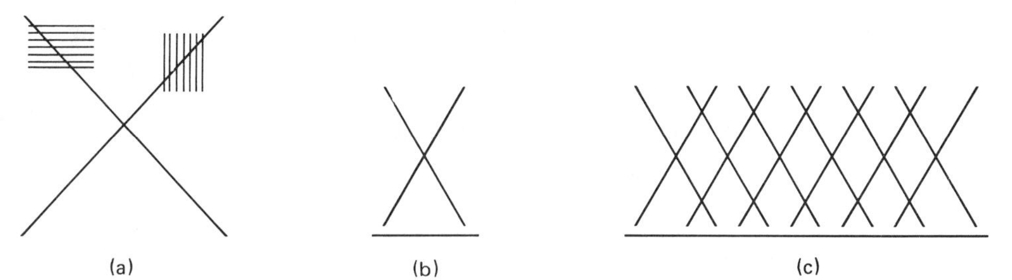

Figure 4.6 Omnidirectional scan patterns in the plane of the window: (a) the minimum scan pattern; (b) basic element of an optimum scan; (c) optimum pattern for a rectangular window.

repetition rate to guarantee at least one good scan through the symbol as it passes across the scan window, regardless of its orientation. The pattern repetition rate, the total scan length, and the width of the smallest UPC module establish the maximum video signal rate seen by the photodetector.

Although the pattern in Figure 4.6a is the minimum scan pattern required to yield an omnidirectional scanner for the UPC symbol, it is not an "optimum" pattern. The pattern repetition rate required to guarantee one scan through a UPC half-symbol moving across this pattern at 2.54 m/s (100 in/s), at the worst-case orientation relative to the scan pattern, is very high. This results in high scanning-spot velocities and subsequently high video signal rates. A "better" scan pattern can guarantee one good scan at lower pattern repetition rates and lower scan velocities. An optimum scan pattern will minimize scan velocity, thereby minimizing video signal rates.

If one could increase the amount of symbol oversquare then one could reduce the pattern repetition rate, and the scan velocity, and still guarantee one good scan through the UPC half-symbol at maximum symbol velocity and worst-case symbol orientation. While the symbol itself cannot be changed, one can effectively increase the symbol oversquare by using a scan pattern where the scan lines are separated by angles less than 90°. Thus, a scan pattern consisting of three scan lines, for example, instead of two could be repeated less often and still be able to read the UPC symbol under the worst-case conditions mentioned above. Increasing the number of scan lines has the effect of increasing the length of the scan pattern, which, by itself, would increase the scan velocity of the scanning spot. However the reduction in the pattern repetition rate is greater than the increase in the scan length. The net result is a better scan pattern, in the sense described above.

Can we continue to improve the scan pattern by adding still more lines? Unfortunately, the answer is no. The reduction in the required pattern repetition rate realized by using four scan lines is nearly offset by the increased scan velocity resulting from the greater length of the scan pattern. The small amount of gain is not enough to justify the increased cost and complexity required to generate the four-line pattern. Beyond four lines, there is no gain. It appears, then, that the optimum scan pattern in the plane of the window would be one based on the three-line pattern in Figure 4.6b. This fundamental three-line pattern, which is the basic element in all of the major UPC scanners marketed today (1991), formed the basis of the first scanner designed to read the UPC code, the IBM 3666 scanner. The linear equivalent of the Lissajous scan pattern used in the IBM 3666 scanner is shown in Figure 4.6c.

4.2.1 Forward-Looking Scanners

Initially, all UPC scanners were conceived as "bottom scanners." That is, the scanning laser beam pointed directly upward to read the UPC symbols on the bottoms of packages as they passed over the scan window. A major problem was encountered in the design of this type of scanner. UPC symbols printed on shiny surfaces were difficult to read because the specular reflection from the shiny surfaces contained no bar-space modulation in the specularly reflected light. In addition, the specular reflection created saturation problems in the photodetector since the specularly reflected light was so much more intense than the diffusely reflected light. In most scanners, the photodetector was located back along the general direction of the outgoing laser beam. Therefore, some solution had to be found which would keep the specularly reflected light from being directed back along the laser beam path.

One solution to this problem is shown in Figure 4.7a. The laser beam is tilted at an angle of approximately 45° relative to the scanner window. In this configuration, the specularly reflected light is reflected away from the photodetector, thereby eliminating the specular reflection problem.

A fringe benefit occurs when this scanner geometry is used. The tilted beam

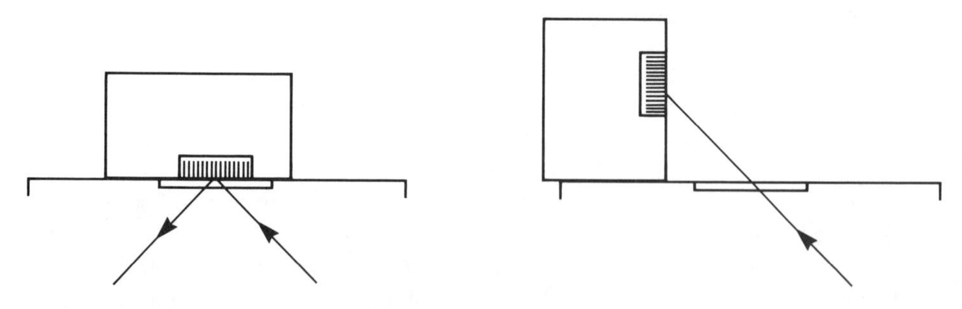

Figure 4.7 Tilted-beam scanning: (a) removing the specular reflection problem; (b) side reading with a forward-looking tilted beam.

can be used to read UPC symbols on the front of packages without tilting the packages forward (see Figure 4.7b). Of course, this increases the depth of field required to read these upright labels, but the laser scanner has the capability to provide a depth of field of several inches (100 mm^{+}), which is usually, but not always, sufficient for side reading with the tilted-beam geometry. Nearly all UPC scanners today employ some form of forward-looking, tilted-beam reading geometry.

4.2.2 Scan Pattern Wraparound

The next development in the evolution of the scan pattern was the introduction of scan pattern wrap-around. Several scan patterns were introduced that took the basic three-line optimum scan element shown above and created it in such a way that the scan lines were directed at the items from points within the scanner that were slightly off to the sides of the package. In these scanners, horizontal lines were projected forward from immediately in front of the item and from directions slightly off to each side. Vertical lines were also projected from slightly off to each side. The pattern projected from the two sides was essentially a cross pattern similar to that shown in Figure 4.8. The scan pattern projected from immediately in front of the item was a horizontal line (not shown in the figure). The overall pattern was created by using a rotating mirror deflector and an array of fixed folding mirrors.

This type of scan pattern was effective in reading the UPC symbol on the scan window because it employed the basic three-line optimum scan element. It was also effective in reading upright items because it projected a pattern of perpendicular horizontal and vertical lines on the front of the items. Such a pattern is effective for upright reading because the bars in the UPC symbol will usually

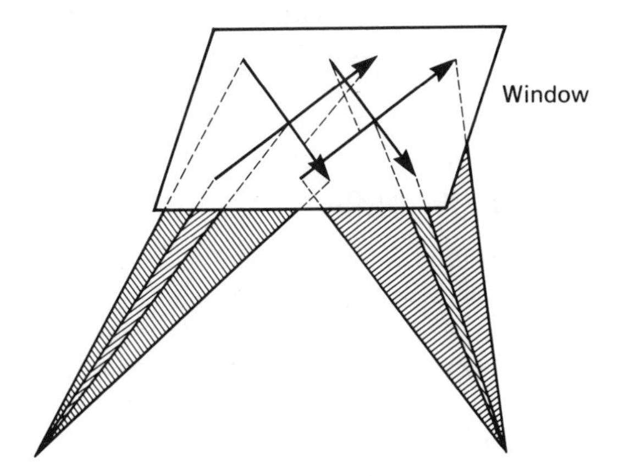

Figure 4.8 Projected wraparound scan pattern.

be either vertical or horizontal when the package is presented to the scanner in this manner.

The major advantage of this type of scan pattern is that it "wraps around" to the sides of the packages to some degree. This means that the operator does not have to align the item as carefully when he or she brings the item across the scan window. The UPC symbol can be on the bottom of the package, on the front of the package, or on the side of the package and still be readable by the scanner. This has a positive effect on operator productivity.

4.2.3 Depth of Field

The forward-looking, multidirectional, three-line scan pattern forms the basis for nearly all present-day UPC scanners, holographic and nonholographic. Unfortunately, the forward-looking feature increases the depth of field requirement considerably. As much as 150 mm (6 in) depth of field may be required to read some bar codes on upright items. Since the codes must, in many cases, be read by a tilted beam, the resultant spot ellipticity on the bar code will increase the effective scanning-spot diameter. This will reduce the depth of field of the scanner.

Providing satisfactory scanning performance over such a large depth of field, with a tilted scanning beam, is a significant challenge to the scanner designer. One means of easing the problem is to design a scanner that can provide more than one focal plane. Such a scanner could focus part of its scan lines close to the scan window and part of its lines further from the scan window, thereby increasing the effective depth of field of the scanner.

Holographic scanning allows the designer to add this additional degree of flexibility to the scanner. The holographic scanning element essentially allows each scan line to be optimally focused to provide increased depth of field and increased flexibility in the placement of beam-folding mirrors for the creation of the scan pattern. This also allows for a more complex, and more effective, scan pattern. The need for greater depth of field and the desire for a more effective scan pattern led to the development of the holographic bar code scanner.

4.3 HOLOGRAPHIC BAR CODE READERS

The concept of holographic scanning has been around for two decades [3], and during this time many different applications have been suggested [4, 5], but few have been demonstrated and an even smaller number have made it into the marketplace. A general review of holographic scanning and various applications can be found in a recent book by Beiser [6] as well as in a previous volume of optical engineering [7]. A discussion of graphic art scanners by Kramer appears in Chapter 5.

Holographic bar code readers first appeared commercially in 1980 with the introduction of holographic UPC scanners by IBM and Fujitsu. Today, IBM, Fujitsu, and NEC manufacture holographic scanners for reading the UPC codes in supermarket and for retail applications. Fujitsu also has marketed a compact version of their holographic scanner that is specifically designed for the low-end retail market (discount stores, drugstores, and convenience stores), while IBM has entered the industrial market with a holographic scanner that can provide either a very large depth of field or a raster scan pattern.

4.3.1 What Is a Holographic Deflector?

Photography is a light-recording process in which a two-dimensional light-intensity distribution incident on a light-sensitive medium is recorded by that medium. In contrast, holography is a light-recording process in which both the amplitude and phase distribution of a complex wavefront incident on the recording medium can be recorded by that medium.

Holography, therefore, differs from photography in that it is able to record all of the information that is needed by the eye, or any other optical system, to interpret the full three-dimensional nature of the original object [8, 9]. This information is accessed when the recording (the ''hologram'') is illuminated by the proper light source—usually, but not always, a laser.

The most common form of hologram creates, when viewed, a three-dimensional image of a complex, three-dimensional object. As will be described below, the reduction of the three-dimensional object to a single point-source produces a special case of particular importance to deflection—a hologram that acts as a lens to focus an incident laser beam. This type of hologram is referred to as a holographic optical element (HOE).

The concept of holographic recording and reconstruction—more importantly, how it deflects light—can best be understood with reference to Figures 4.9 and 4.10. In Figure 4.9, two wavefronts of equal intensity created from a laser are directed to overlap in some region of space where the recording is to be made. If the optical path difference from the point of beam separation to the region of overlap is within the coherence length of the source, the resulting interference pattern will be stationary in both space and time and will have high fringe contrast. The intensity distribution in these fringes can be exposed onto, or more properly into, a suitable photosensitive medium such as a photographic emulsion. After processing, the recording contains a variation in optical density, refractive index, or optical thickness—sometimes a combination of all three—and is the hologram. When this recording is repositioned and illuminated by one of the wavefronts, for example, the diverging wavefront in Figure 4.9, the structure at each point within the hologram diffracts the illuminating light and creates a new wavefront that is identical to the original, second wavefront, in this case a converging and deflected wavefront. This simple HOE is the equivalent of a positive or converging

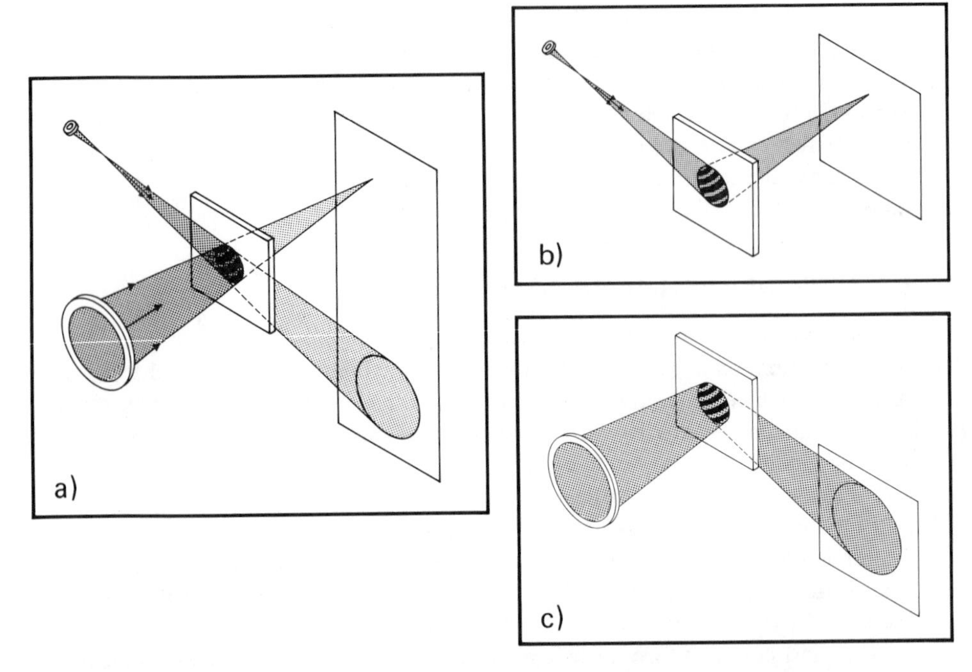

Figure 4.9 Simple holography: (a) recording the hologram; (b) reconstruction of the convergent wavefront (positive lens); (c) reconstruction of the diverging wavefront (negative lens).

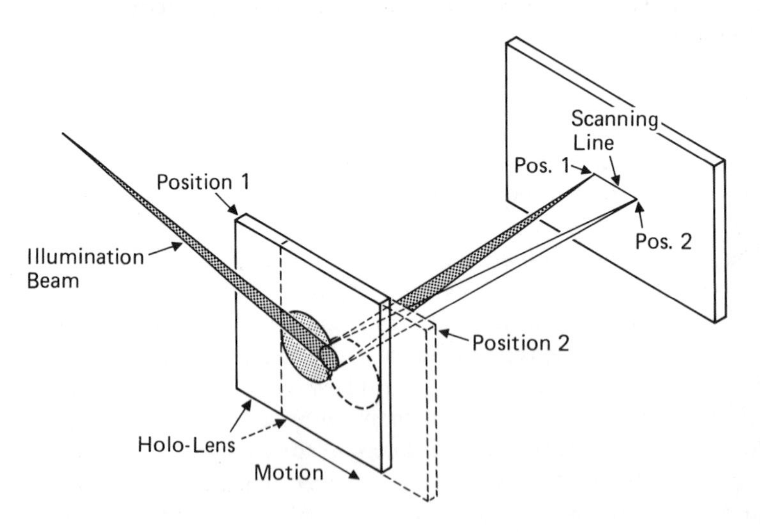

Figure 4.10 Principle of holographic deflection.

lens in combination with a prism converting and deflecting light from a point object to a point image. The efficiency and quality with which this wavefront conversion occurs is directly related to the recording configuration and selection of recording material. Returning to Figure 4.9c, the equivalent of a negative lens is realized by illuminating the same HOE with the converging wavefront, thereby reproducing the original diverging wavefront. Holographic recording of complex multidimensional objects can be treated as the recording of a superposition of individual spherical waves from all the points in the object field.

By using a combination of small area illumination and HOE translation, the reconstruction geometry of Figure 4.9 can be used to create simultaneous light deflection and focusing. This is shown in Figure 4.10,where the HOE is initially located at position 1 and has a small subarea, on the right side, illuminated by a diverging wavefront. The light is focused by this HOE area to a point in the image plane corresponding to the location, with respect the displaced HOE, of the original converging wavefront, labeled "position 1." As the HOE is translated, different subareas of the HOE are passed under the illuminating beam, and the reconstructed image point is caused to translate by the same distance, in this case to position 2. This is completely analogous to the deflection and focusing that would occur if a conventional lens were illuminated off-axis with a collimated beam and the lens displaced normal to its optic axis. A continuous back-and-forth motion in either case produces the same motion in the focused spot.

In practice, however, a continuous rotary motion rather than a reciprocating motion is easier to implement. Higher scan speeds can be realized, and different holograms can be easily accessed. Consequently, most holographic deflectors consist of a number of unique HOEs placed circumferentially as sectors on a glass disk, as shown in Figure 4.11. Other materials can be used, and other geometries besides the disk geometry can be used [7], but, for simplicity, we will restrict our discussion to the glass disk, which is the most common medium and geometry used in holographic scanners today. It should be noted that plane linear gratings, producing prismatic deviation without focal power, must be rotated to generate scanning. Translation of a plane grating in one direction will not produce scanning.

A holographic deflector-disk, when properly illuminated by a laser and rotated about its axis of symmetry, can produce a complex variety of scanning laser beams. The optical and geometrical properties of each of these beams can be distinctly different from all of the others. This is the most important feature of holographic scanning. It is the major feature that distinguishes it from conventional laser scanning technology and, in a bar code scanner, allows the introduction of capabilities that could not be readily achieved with conventional technology [10].

The holographic disk works in the following manner. Each sector, or facet, of the holographic disk is a unique HOE of the type previously described—the

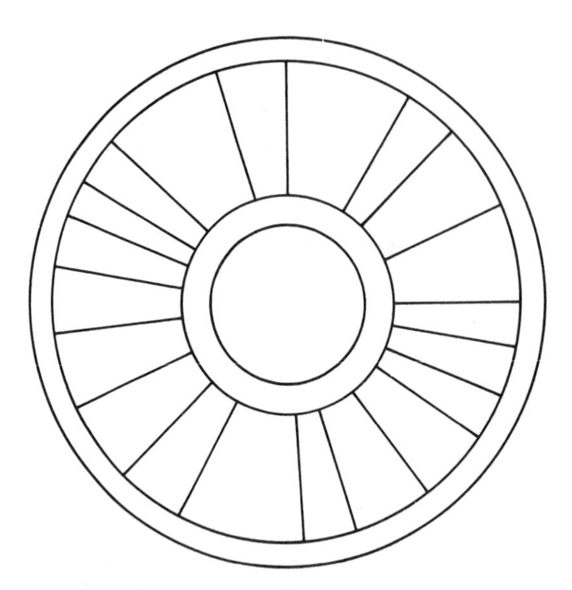

Figure 4.11 A holographic deflector disk.

holographic equivalent of a prism and lens combined. When a facet is illuminated by a laser beam, the beam is diffracted, or bent, by the facet and focused to some point in space (see Figure 4.12). The focal length and bending angle are established during the holographic construction of each facet and may vary from facet to facet.

As the disk rotates, the deflected, focused laser beam scans. When the beam scans across a bar code, some of the diffusely reflected light will return to the facet that generated the scanning beam. The facet now acts as a light-collection lens, combined with a prism, to collect a portion of the reflected light and direct it toward a photodetector [11].

4.3.2 Novel Properties of Holographic Bar Code Scanning

The use of holography in bar code scanning allows the introduction of scanning concepts that are not available to the designer of conventional bar code scanners, at least not in any economically practical design. Such concepts as multiple focal planes, overlapping focal zones, variable light-collection aperture, facet identification, and scan-angle magnification allow holographic scanning to bring to bar code scanning some significant design and performance capabilities.

A conventional bar code scanner contains a lens for focusing the laser beam, a device for deflecting the laser beam, and some optics for collecting a portion

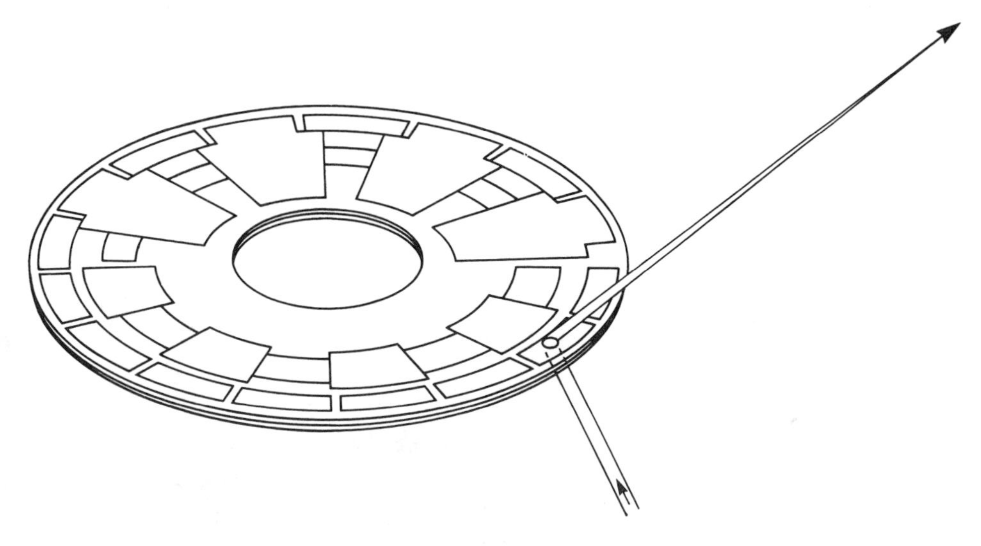

Figure 4.12 Light deflection and focusing by a holographic disk.

of the laser light reflected from the bar code and focusing it onto the photodetector.

In a holographic scanner, all of these properties—focusing, deflecting, and light collection—are contained in the holographic disk. As indicated earlier, these properties may be different in each sector of the holographic disk; thus, a 16-sector holographic disk, for example, would contain 16 unique optical systems. Each of these systems would have its own focal length, scan angle, and light-collection aperture. One revolution of such a holographic disk would produce the equivalent of scanning with 16 different scanners.

Since each facet of the holographic disk may be different, with its own combination of focal length, deflection angle, and facet area, then one complete rotation of the disk will create multiple scan lines with multiple deflection angles, multiple focal lengths, and multiple light-collection systems. This enables the holographic scanner to introduce some novel operational characteristics.

One of the major advantages of using a holographic disk in a bar code scanner is that it can provide a much larger depth of field than would be attainable with a conventional, single-focal-length, bar code scanner. In order to understand this point, we need to review briefly the subject of depth of field.

4.3.3 Depth of Field for a Conventional Optics Bar Code Scanner

In bar code scanning, depth of field is the distance along the laser beam, centered around the focal point of the scanner, over which the bar code can be successfully

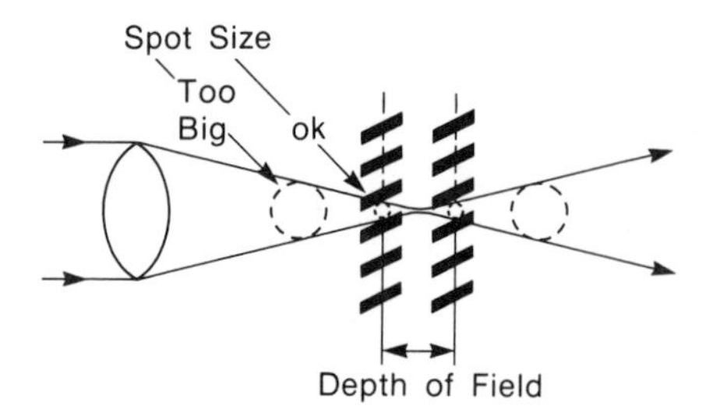

Figure 4.13 Depth of field for a conventional optical system.

scanned. The depth of field of a bar code scanner is established by the beam diameter at the focal point of the scanner, the wavelength of the laser light source, and the size of the minimum bar width in the bar code being read. Once these three parameters are established, there is little that can be done in a conventional bar code scanner to increase the depth of field.

Figure 4.13 illustrates the concept of depth of field. The lens in the scanner focuses the laser beam to a relatively small spot size at the focal point. The diameter of the beam at the focal point is determined by the focal length of the lens, the diameter of the beam at the lens, and the wavelength of the laser being used. When the scanner optical system is properly designed, the spot size at the focal point of the lens will be small enough to be able to scan successfully the bar codes, since the focused spot size will be no larger than the minimum bar width in the bar code.

As one moves the bar code to either side of the focal point, toward or away from the scanner, the spot size increases as the beam becomes out of focus. Eventually, a point is reached where the out-of-focus spot size is larger than the minimum bar width in the bar code. When this occurs, the bar code can no longer be scanned successfully by the beam. The distance between the two points to either side of the focal point where the limit of scanning capability occurs is, by definition, the depth of field.

The major factors determining the depth of field are the spot size at the beam waist, the wavelength of the laser, and the minimum bar width. (For convenience, we will assume throughout this discussion that the minimum space width is the same as the minimum bar width.) Using the notation of Dickson [12] for the variation in beam radius of a propagating Gaussian beam, the $1/e^2$ beam radius r at a distance z from a beam waist of $1/e^2$ radius r_0 is given by

$$r = r_0 \left[1 + \left(\frac{\lambda z}{\pi r_0^2} \right)^2 \right]^{1/2} \tag{2}$$

This can be rearranged to express the depth of field for a given beam radius r and waist radius r_0 as

$$z = \pm \frac{\pi}{\lambda} (r_0^2 r^2 - r_0^4)^{1/2} \tag{3}$$

If we assume that the limit of depth of field occurs where the $1/e^2$ beam diameter is $\sqrt{2}$ times the minimum bar width, $2r = \sqrt{2}\, w_{\min}$, then it is straightforward to show that the depth of field given by equation (3) will be maximized when the $1/e^2$ waist diameter equals the minimum bar width $2r_0 = w_{\min}$. For this maximized condition, the total depth of field is the sum of the "+" and "−" components and becomes

$$\Delta z = \frac{\pi w_{\min}^2}{2\lambda} \tag{4}$$

At the He-Ne wavelength, equation (4) can be approximately expressed as

$$\Delta z = \frac{w_{\min}^2}{16} \tag{5}$$

where Δz is the depth of field in inches when the minimum bar width is in mils or as

$$\Delta z = \frac{w_{\min}^2}{400} \tag{6}$$

where Δz is the depth of field in millimeters when the minimum bar width is in microns.

For example, if the scanner is optimally designed to read bar codes that have a minimum width of 8 mils (200 μm), the depth of field will be 4 in (100 mm). Note that equations (5) and (6) tell us that the depth of field is strongly dependent on the size of the minimum bar width to be read. Therefore, a small minimum bar width is always accompanied by a small depth of field.

Furthermore, if the scanner is not optimally designed for the minimum bar width to be read, the depth of field will be less. This will be true whether the scanner is optimally designed to read either a higher-density bar code or a lower-density bar code. In addition, if the scanner uses a laser with a different wavelength, the depth of field will be multiplied by the ratio of the wavelength of the He-Ne laser to the wavelength of the laser being used. Some infrared semiconductor lasers, for example, have a wavelength of approximately 780 nm (30.7 μin). The depth of field for an optimally designed scanner using such a laser will be about 80% of the depth of field of an optimally designed scanner using a He-Ne laser.

There is very little that can be done in a conventional bar code scanner to increase the depth of field. It is possible, in concept, to use an autofocus scanner, but the reaction time of the autofocus system would have to be very fast to accommodate fast-moving items on a conveyor system. Since all autofocus systems today require mechanical movement of some of the optics of the scanning system, the reaction time may not be fast enough. In addition, such systems will add cost and complexity to the scanner.

One could also add a supplemental optical element to the scanner that could move into position in the laser beam path to change the net focal length of the scanner. This moving element would allow, for example, two different focal lengths to be selected. In practice, this approach would allow only two or three different focal lengths to be selected, giving only a slight increase in the depth of field.

Furthermore, in either an autofocus system or a dual- or triple-focal-length system, only the focal lengths can be easily changed. Ideally, one should also change the aperture of the scanner as the focal length is changed. This would maintain a constant level of light collection over the full range of focus, thereby optimizing the performance of the scanner over the full range of readability. However rapid variation of the light-collection aperture is difficult to accomplish in a conventional bar code scanner.

4.3.4 Depth of Field for a Holographic Bar Code Scanner

The use of holography in a bar code scanner would allow the introduction of a true multi-focal-plane scanner with a variable light-collection aperture. The way this would be accomplished in a holographic scanner is illustrated in Figure 4.14.

Figure 4.14 shows focusing of the laser beam by two consecutive facets on the holographic disk. Each facet will exhibit a conventional depth of field as established by the focal length of the facet, the beam diameter at the disk, and

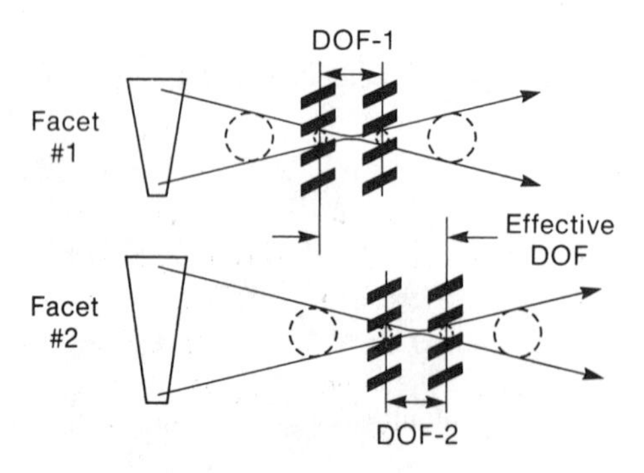

Figure 4.14 Combined depth of field for two holographic facets.

the wavelength of the laser. Notice, however, that the two facets are focused at different distances from the disk. Therefore, while each facet has only a conventional depth of field, the combined depth of field of the two facets is twice as great as for either facet alone, assuming that the focal lengths are chosen so that the end of the depth of field for facet 1 coincides with the beginning of the depth of field for facet 2.

Thus, with only two focal planes, the holographic disk can double the depth of field of a conventional, nonholographic scanner.

If the disk is designed so that all of the facets are focused at different distances, then a much larger overall depth of field can be achieved. For example, if the minimum bar width to be scanned is 0.2 mm (8 mils), a conventional, single-focal-length scanner would have a depth of field of approximately 100 mm (4 in). A properly designed holographic scanner could provide a depth of field of 800 mm (32 in) or more for the same 0.2-mm (8 mils) minimum bar width code.

4.4 OTHER FEATURES OF HOLOGRAPHIC SCANNING

There are other novel features of holographic scanning that are not as obvious as the ability to provide a large depth of field. The major features of holographic scanning are

1. Multiple focal planes
2. Overlapping focal zones
3. Variable light-collection aperture
4. Facet identification
5. Parallel-line raster scanning
6. Scan-angle magnification

We have already discussed the multiple-focal-plane feature and the large depth of field that it provides. Let us now examine the other features to see what they are, how they are produced by the holographic disk, and what capability they provide.

4.4.1 Overlapping Focal Zones

We showed in the previous section how holographic scanning can provide a large depth of field by designing the holographic disk so that the depth-of-field region of each successive facet was contiguous with the depth-of-field regions of the facets immediately preceding it and following it. This may not, in practice, be the best disk design. It may be better to design the holographic disk so that the focal point of one facet coincides with the limit of the depth of field for the preceding and following facets, resulting in an overlapping focal zone design.

The reason why this design may be superior is explained in the following paragraphs.

One of the major contributors to decoding problems in a bar code scanner is the existence, or creation, of noise in the so-called quiet zone, the white, or clear, region immediately preceding and following the bar code. One of the contributing factors to noise in this region, and throughout the bar code, is substrate noise, or paper noise. Paper noise occurs when the size of the focused spot of the scanning laser beam is about the same as the size of the granularity of the substrate material. Paper fibers, for example, can be as large as 0.1 mm (4 mils). For very coarse paper or cardboard, the fiber size can be even greater. For nonpaper substrates, such as for bar codes etched into plastic or metal, the granularity can be greater still.

If a bar code on a noisy substrate is scanned at the focal point of a scanning laser beam, the small in-focus spot will "see" the granularity of the substrate material. This will introduce paper noise on the return light signal that will, in turn, lower the probability of achieving a successful read.

While noise reduction could, in general, be reduced with low-pass electrical filtering, the filter properties would have to be altered for each facet to correct for the differences in spot velocity. That is, a low-pass filter designed to remove noise from the short-focal-length facets would, at the same time, filter out the bar code signals from long focal length facets. Electrical filtering does not appear to be a practical solution for large depth-of-field scanners.

The solution to this problem is to scan the noisy bar code with a slightly out-of-focus spot. This spot will be larger than the in-focus spot but still small enough to read the bar code. This larger spot will, while scanning the bar code, act as a filter to smooth out the surface roughness, effectively lowering the paper noise and increasing the probability of achieving a successful read.

Figure 4.15 shows the analog photodetector signal for a noisy bar code when scanned (a) by an in-focus spot and (b) by a slightly out-of-focus spot. The noise on the signal from the in-focus scanning spot is apparent. The resultant reduction in the noise level due to scanning the same bar code with a slightly out-of-focus scanning spot is equally apparent.

By overlapping the focal zones of the individual holographic facets, as shown in Figure 4.16, we can guarantee that all bar codes will be scanned by both an in-focus scanning spot and one or more slightly-out-of-focus scanning spots. The slightly-out-of-focus spots will be small enough to read the bar codes, but large enough to smooth out the substrate noise.

This in-focus/out-of-focus capability, which would be difficult to implement with conventional scanning technology, is relatively simple to introduce with holographic scanning. One merely selects, during the master holographic disk design phase, the focal length for each of the facets that guarantees the desired amount of focal-zone overlap.

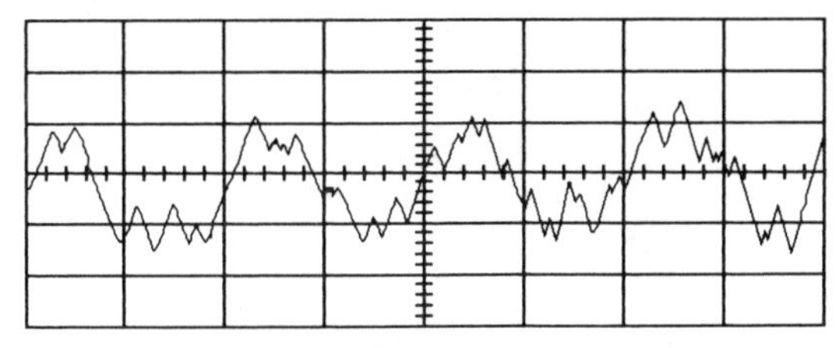

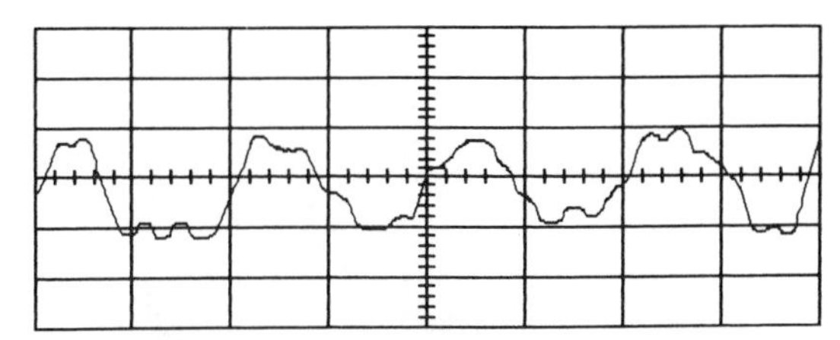

Figure 4.15 Photodetector signals for in-focus and out-of-focus scanning laser beams.

4.4.2 Variable Light-Collection Aperture

There is more to achieving a large-depth-of field scanner than simply providing multiple focal-planes. If, for example, one designs a scanner with a 1-m (40 in) depth of field where the optical throw (closest reading distance) is 200 mm (8

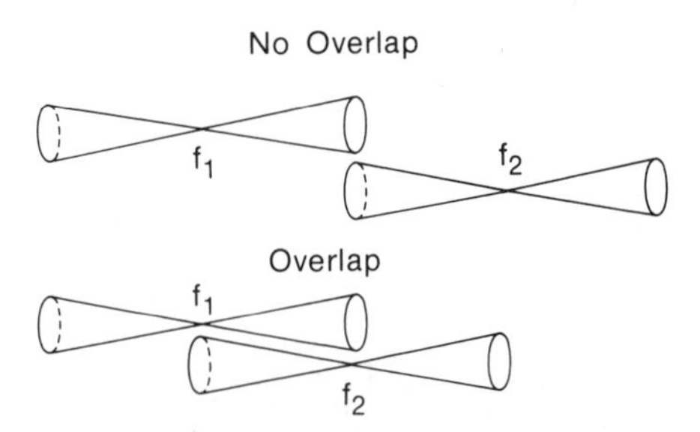

Figure 4.16 Overlapping focal zones of two holographic facets.

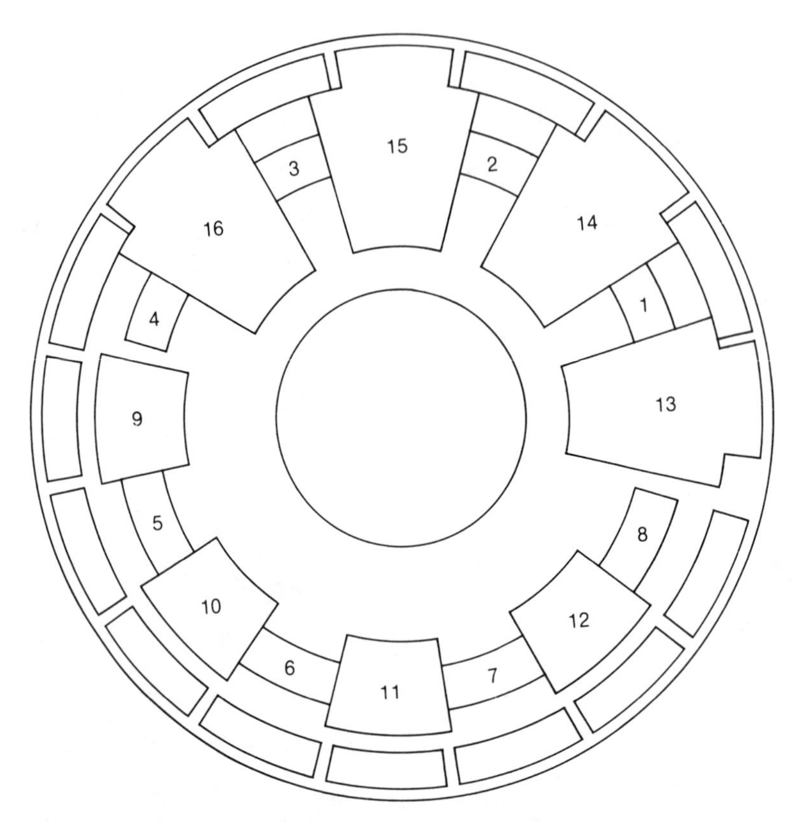

Figure 4.17 Holographic disk showing large variation in facet areas.

in) and the range (farthest reading distance) is 1200 mm (48 in), then the variation in the light level returned to the detector, for bar codes with identical reflection characteristics, will be 36:1, the square of the ratio of the far and near distances. This places a severe dynamic range requirement on the analog electronics in the scanner. The problem is worse in practice since other factors, such as label skew and label reflectivity variations, also affect the amount of light returned.

In order to reduce the variation in the light level of the return light in a multiple-focal-plane scanning system, it would be desirable to vary the light-collection aperture to compensate for changes in the distance to the bar code being scanned. One could then use a relatively small aperture for a near-focus scan line and a relatively larger aperture for far-focus scan line.

Holographic scanning allows one to do exactly that. Figure 4.17 shows a holographic scanning disk designed for focal lengths ranging from $\approx$300 to $\approx$1000 mm ($\approx$12 in to $\approx$40 in).

In this particular disk design, the outer part of each facet is used for the outgoing beam, while only the inner part of each facet is used for collecting the returned light. Notice that the light collection area of each facet is different. The facet with the shortest focal length has the smallest light-collection area while the facet with the longest focal length has the largest light-collection area. The light-collection area of each of the remaining intermediate facets is a direct function of its focal length.

This difference in light-collection area for the near and far facets of the holographic disk allows the light collection to be approximately uniform over the total depth of field of the scanner. This is a major advantage in obtaining decoding accuracy over a large depth of field for a bar code scanning system.

4.4.3 Facet Identification

Note that the disk design in Figure 4.17 incorporates gaps between the outgoing portions of the holographic facets. All of the gaps are the same size except for one, which is larger than the rest. Since the outgoing laser beam is incident on the disk at these outer sections, a detector placed in the proper location above the disk can sense these gaps by measuring the laser power incident on the detector. The larger gap will result in a longer light pulse on the detector, which can be interpreted as a ''home'' pulse. Thus, we can determine which facet we are on. This method of facet identification could be used in several ways to improve the decoding accuracy of the scanner.

If we knew, for example, that we were on a short-focus facet, we could decrease the electrical gain in the analog electronics. If we were on a long-focus facet, we could increase the gain of the analog electronics. This electronic automatic gain control (AGC) would add to the already existing optical AGC, introduced by the variable light-collection aperture, to further improve the decoding accuracy.

We could also vary the internal clock rate from facet to facet to improve resolution. Since the scanner is an angular scanner, the linear velocity of the scanning beam will vary directly with the distance from the scanner. The bit rate seen by the detector while scanning with a long-focus facet would be greater than the bit rate seen when scanning with a short-focus facet, assuming that the code density is the same in both cases. By making the clock rate vary from facet to facet to maintain the optimum clock rate for a given bit rate, we could, once again, improve the decoding accuracy of the scanner.

4.4.4 Parallel-Line Raster Scanning

Raster scanning refers to the capability of a scanner to create an array of parallel, or quasi-parallel, scan lines instead of a single, repeated scan line, thereby producing a two-dimensional scan pattern. Such scan patterns are useful for reading bar codes in which the positional tolerance of the code may be very large. They are also *required* in order to be able to read the multipart AIAG (automotive industry action group) shipping/parts identification label in the picket-fence orientation specified by the AIAG.

Raster scan patterns can be created in two major ways. If an oscillating deflector is used, two such deflectors can be used to create a Lissajous pattern to cover a two-dimensional scan area. If a rotating deflector is used, individual facets of the deflecting element can be tilted at different angles to create a raster of parallel scan lines.

The Lissajous scanner, if properly designed, would have the advantage of random coverage so that, in theory, denser scan coverage could be obtained. The major disadvantage of this type of raster scanner is that, in order to get rapid coverage for fast, horizontally moving items with picket-fence-oriented bar codes, the vertical deflector frequency would have to be high. But this creates a tilt in the scan lines that would cause problems reading bar codes whose height is small and whose length is large. This problem is exacerbated in the AIAG label where the overall multipart label is relatively tall. In this case the vertical deflector amplitude must be large, increasing the tilt of the scan lines (see Figure 4.18).

The rotational raster scanner can be designed to create straight scan lines so that the tilt problem associated with the Lissajous raster scanner is nonexistent (see Figure 4.19). The disadvantage of this type of raster scanner is that the angular placement of the raster scan lines is discrete, so that a dense scan pattern

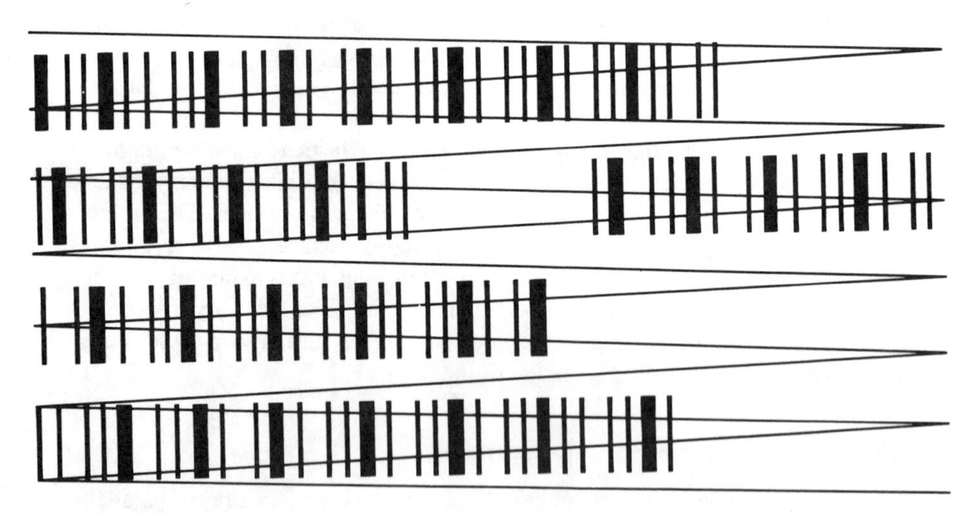

Figure 4.18 Raster scan pattern produced by two oscillating deflectors.

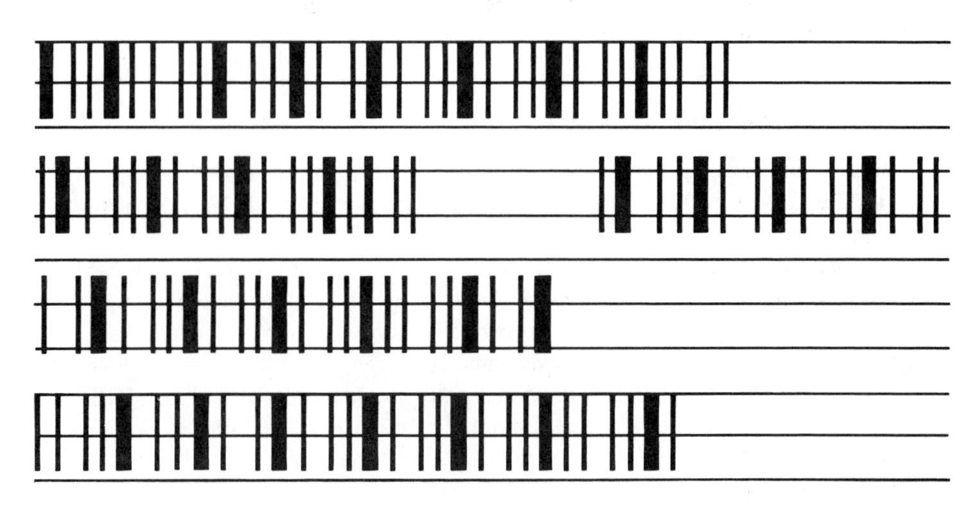

Figure 4.19 Raster scan pattern produced by a rotational deflector.

must be laid down to guarantee that all of the codes will be read at all distances within the depth of field.

The holographic scanner falls into this second category. It can produce discrete, parallel scan lines by introducing a different angle of elevation for the outgoing laser beam for each facet on the holographic disk. For a 16-facet disk, one can create a 16-line, raster scan pattern. The advantage of the holographic disk in this application is that no additional components need to be added to the scanner to create the raster scan pattern. Moreover, there should be relatively little impact on manufacturability for a raster version scanner compared to a multiple focal-plane version. The only difference between the two versions would be in the design of their individual holographic disks. While the holographic scanner actually produces a raster of ''slightly curved'' parallel scan lines, the amount of curvature is negligible in the bar code scanning application. As will be indicated by Kramer in Chapter 5, graphic arts applications generally have a more demanding straightness requirement.

Of course, the raster version scanner would incorporate only a single focal plane since all of the facets would be used to create the raster pattern. One could, however, envision a holographic scanner that would combine some of the features of both of these versions. For example, one could create a 16-facet holographic disk in which an 8-line raster pattern is repeated at two different focal planes to provide a raster scanner with increased depth of field.

Changing the focal point of a holographic raster scanner, or changing from a 16-line, single-focal-plane raster to an 8-line, dual-focal-plane raster, would merely require the substitution of a different holographic disk.

4.4.5 Scan-Angle Multiplication

Holographic scanning disks used in bar code scanners are frequently designed to be illuminated with a collimated beam incident normal to the surface of the holographic disk. This illumination geometry provides a scanning spot that is free from aberrations, particularly at the ends of the scan line. It is a special case of the more general aberration-free illumination geometry for rotationally symmetric systems in which both the recording (reference) and illumination (playback) beams are converging spherical wavefronts, converging toward a common point on the rotational axis of the disk. (A normally incident collimated beam has a point of convergence on the axis at infinity.) Under these conditions, the illuminating wavefront always matches the original reference wavefront throughout the motion of the hologram.

If the holographic disk is designed to be illuminated with a collimated beam inclined at a nonnormal angle, then some amount of aberrations will be introduced in the scanning beam. Each facet of the disk can be designed to still provide zero aberrations at the center of its corresponding scan line, but there will always be aberrations introduced as the disk rotates because of the resulting mismatch between the recording and playback wavefronts. The amount of the aberrations will be dependent on the amount of rotation away from the center of the facet and the amount of tilt of the collimated incident beam.

There is, however, one advantage to tilting the incident beam. It can be shown that a tilted, collimated incident beam will provide a scan-angle multiplication factor relative to an untilted, collimated incident beam geometry. A precise determination of the multiplication factor requires the use of a computer program because of the interdependence of the diffracted beam elevation angle (β in Figure 4.20) and the rotation angle (ϕ_{rot}). A first-order approximation, accurate to a few percent, can be obtained from the following simple relationship. The geometry and terms are defined in Figure 4.20.

$$\phi_{scan} = \phi_{rot}\left(\frac{r}{f}\cos\theta_s + \cos\alpha + \cos\beta\right) \tag{7}$$

The multiplication effect of the tilted collimated incident beam is due to the $\cos\alpha$ term in equation (7); for normal incidence this term goes to zero.

As an example of the effect of the tilted incident beam, consider a holographic facet with f = 350 mm (13.8 in), $\theta_s = 40°$, $\beta = 66°$, and r = 72 mm (2.8 in). For normal incidence, $\phi_{scan} = 0.564\phi_{rot}$, while for an incident beam tilted at an angle of 22° relative to the normal ($\alpha = 68°$ in Figure 4.20), $\phi_{scan} =$

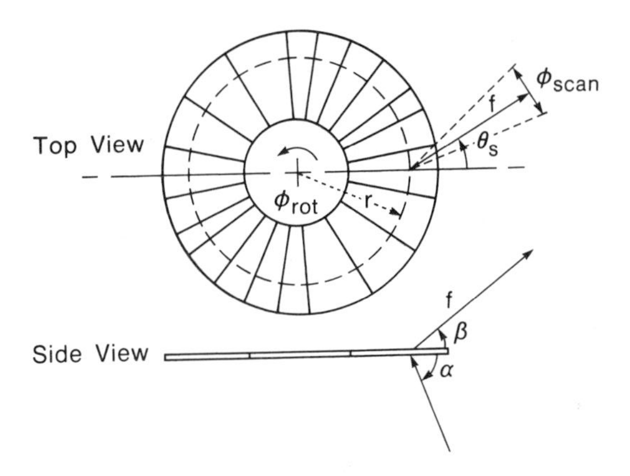

Figure 4.20 Scan-angle magnification parameters.

$0.939\phi_{rot}$. The relative multiplication factor obtained by using the tilted incident beam is 1.66. This is a significant amount of scan-angle multiplication.

Scan-angle multiplication factors of this magnitude provide significant design flexibility. For a given scan length: one could make the disk smaller to produce a more compact unit; keep the disk the same size and add facets to provide more scan lines or to generate a more complex scan pattern; move the disk closer to the window to increase the light collection efficiency of the individual facets; or some combination of all three options. One may, of course, elect to use scan-angle magnification to just generate a longer scan line.

The incident-beam tilt angle, the aberrations, and the scan-angle multiplication factor are interrelated: the smaller the amount of tilt of the incident beam, the smaller the aberrations; conversely, the larger the tilt, the larger the scan-angle multiplication factor. By careful selection of the tilt angle of the incident beam, one can get a significant amount of scan-angle multiplication while maintaining an acceptable amount of aberrations. The amount of aberrations introduced can be determined with a suitable ray tracing program [10]. Acceptable levels of aberrations are established by the individual application.

4.5 HOLOGRAPHIC DEFLECTOR MEDIA FOR HOLOGRAPHIC BAR CODE SCANNERS

All holographic bar code scanners today (1991) use a rotating circular disk as the substrate for the recording medium. Other geometries have been considered, but, for bar code reading applications, the disk geometry offers a number of manufacturing advantages and is generally less expensive.

All of today's holographic bar code readers also operate only in the transmission mode. It would not be impossible to develop a reflective holographic bar code scanner, but the transmission mode provides a simpler design, an easier manufacturing process, and less susceptibility to disk wobble [7].

There are two general types of media suitable for recording holographic optical elements on a disk surface for use in a holographic bar code scanner: surface-relief phase media, such as photoresist, and volume-phase media, such as bleached silver halide and dichromated gelatin. There are advantages and disadvantages associated with both types of media. For a more general review of the wide variety of holographic recording materials, see Bartolini [13] and Smith [14].

The major factors influencing the selection of the type of holographic medium are manufacturing cost, diffraction efficiency and, as will be described, the scan pattern density.

4.5.1 Surface-Relief Phase Media

There are only two significant surface-relief phase media presently being used for holographic scanning disks—photoresist, and copies made either directly from photoresist or from intermediate copies of the photoresist. As will be discussed later, this latter type is probably the least expensive of all to manufacture in a high-volume process and, from a purely cost consideration, would appear to be the best choice for a holographic deflector disk.

The major disadvantage of the low-cost surface-relief material is that its diffraction efficiency is relatively low, on the order of 30%. This is a major drawback in the supermarket scanner application. The low efficiency means that a higher-power laser must be used to get sufficient laser power onto the bar code symbol to get a good reading. The greater cost of the higher-power laser may offset the lower cost of the disk.

It is possible to get high diffraction efficiency using a surface-relief medium [15,16]. It is not possible at the present time, however, to mechanically replicate these high-efficiency surface-relief holograms because of the high aspect ratio of the relief profile. (An example of such a high aspect ratio is shown in Figure 4.21.) This means that original holograms, not inexpensive copies, would have to be used. In some holographic deflector applications, this is acceptable. In general, however, this is not an acceptable alternative for a holographic bar code scanner, due to the higher cost of the disks and their greater susceptibility to physical damage (surface-relief holographic disks cannot be protected by a cover glass in contact with the hologram since an index-matching adhesive would effectively fill and eliminate the surface-relief structure).

The low diffraction efficiency of the mechanically replicated surface-relief material also means that the collected light will be low in a system employing the holographic facets of the disk in a retroreflective mode. This loss in collected light cannot be compensated for by an additional increase in the laser power

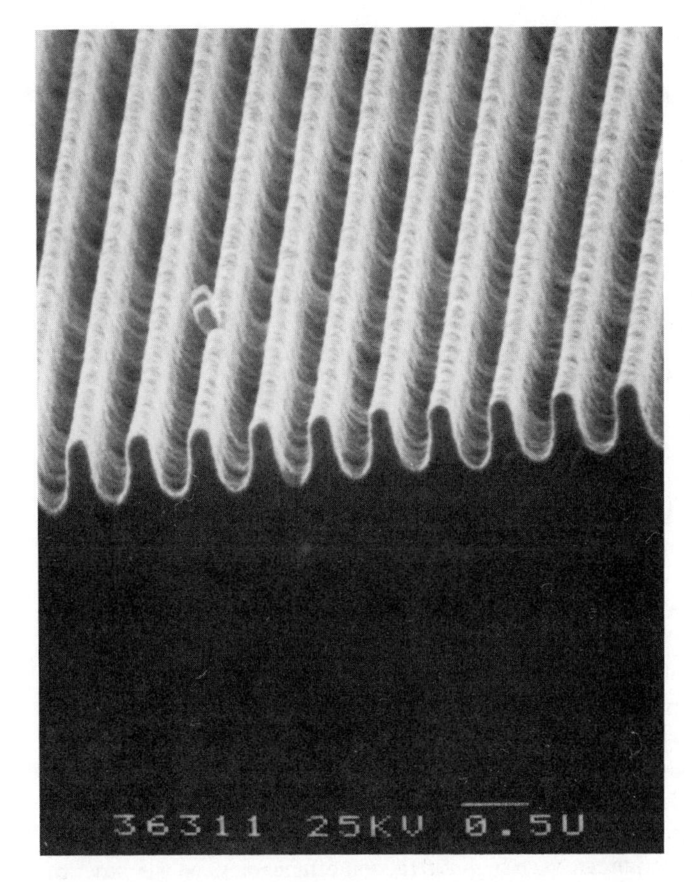

Figure 4.21 Surface-relief hologram in positive photoresist.

because of the limitations established by the federal laser safety standards. The only means left to compensate for the low diffraction efficiency in the collected light is to increase the size of the facets on the holographic disk. But this reduces the total number of facets, hence the total number of independent scan lines, and the subsequent scan pattern density. In some bar code scanning applications, this may be an acceptable trade-off. In other applications, such a trade-off may be unacceptable.

For example, in supermarket/retail bar code scanning applications, the depth-of-field requirement is relatively moderate, so a holographic scanner with as few as two focal planes can provide adequate performance. However, in industrial bar code scanning applications, the required depth of field may be as large as 1 m (40 in), or more, for medium-density bar codes (bar codes with a minimum bar width on the order of 0.3 mm). This kind of depth-of-field requirement can only be met with a scanner that can provide a large number of focal planes. A holographic scanner can be designed to provide this capability, but the number of independent facets on the scanning disk must be as large as possible. Therefore, any reduction in the number of facets imposed by a low-diffraction-efficiency recording medium will reduce the depth of field.

Despite the relatively low diffraction efficiency of the holographic disks produced by mechanical replication, which is a surface relief process, the low cost of such disks makes them very attractive for supermarket and retail bar code scanners in which component cost is a major factor.

4.5.2 Volume-Phase Media

Volume-phase materials are capable of very high diffraction efficiencies, on the order of 80% or more. Such high efficiencies mean that the individual facets on the disk can be relatively small, even when the disk is used in a retroreflective mode. This means that there can be more facets on the disk, which, in turn, means that the scanner can generate more independent scan lines, resulting in a larger depth of field and/or a more complex scan pattern. The higher diffraction efficiency also means that a lower-power laser can be used to generate the scan lines.

There are a number of materials that are suitable for use as volume-phase materials in holographic scanners. The first material that comes to mind is bleached silver halide. In this process, the absorptive structure in a photographic emulsion hologram is chemically converted from metallic silver to a material having a refractive index different than the surrounding gelatin matrix [17–20]. For example, the silver may be rehalogenated by exposure to bromine vapor. Holograms created with this material can have high diffraction efficiencies, on the order of 80% or more [21,22]. Processing is relatively simple, and the holograms are reasonably stable. There are a few bleaches, however, that leave reaction products behind in the emulsion. Some of these products are photosensitive and exhibit printout effects, particularly when subjected to intense ultraviolet irradiation. Nevertheless, moderately efficient holograms can be realized, and the advantages of photographic emulsions, such as extended spectral response and speed, may be exploited. One practical disadvantage associated with this material is that the disks must be coated and sensitized by one of the major companies producing general photographic materials. Such companies are usually reluctant to stock odd substrate shapes (like disks) and coat them to a user's specifications, in quantities that are, for them, relatively small. This creates a very real sourcing problem.

The next most attractive volume-phase material is the photopolymer [23,24]. Cross-linking in these materials is produced when they are exposed to light of relatively short wavelength, blue to ultraviolet. When a photopolymer is exposed to a holographic fringe pattern at the proper wavelength, the periodic variation in light intensity of the fringe pattern produces a corresponding periodic variation in cross-linking in the polymer. When developed, the photopolymer will exhibit a periodic variation in refractive index corresponding to this periodic variation in cross-linking. These materials are relatively stable when exposed to normal levels of ambient light, heat, and humidity.

The main drawback to these materials has been their relatively small change in refractive index, Δn, produced by exposure to light and subsequent processing. This means that, in order to get high diffraction efficiency, the thickness of the photopolymer coating has to be on the order of 50 microns (2 mils). Such a large thickness would make the holographic deflector disk very sensitive to the Bragg angle. That is, very slight deviations in the angle of incidence of the reconstruction beam in the scanner would cause severe reductions in the disk diffraction efficiency. Deviations on the order of $\frac{1}{4}°$ (4.4 mrad) could cut the diffraction efficiency in half [25]. This is generally unacceptable in a product where the total angular manufacturing tolerances could easily be this large. Furthermore, the anticipated mode of operation could cause the effective angle of incidence to vary by $\frac{1}{4}°$ (4.4 mrad) during disk rotation.

A new photopolymer, recently developed by Polaroid [26], has exhibited refractive index changes which are much greater than those of previous photopolymers. Δn's approaching the values obtainable with dichromated gelatin (nearly 10 times as great as earlier photopolymer Δn's) have been obtained. This material has great potential for use as a recording medium for holographic deflectors used in bar code scanners, since high diffraction efficiency should be achievable in relatively thin coatings, on the order of 5 microns (200 μin).

The volume-phase material that has, up to now, been the most successful material for use in holographic deflectors for bar code scanners is dichromated gelatin (DCG) [27–29]. The major advantage of this material, as a medium for holographic deflector disks, is that its diffraction efficiency can be very high (>80%) in a relatively thin (3 to 5 microns) (120 to 200 μin) coating because of its high Δn (0.05 to 0.10). This means that dichromated gelatin can have, simultaneously, high diffraction efficiency and very low Bragg-angle sensitivity. This is a significant advantage from both a manufacturing and an application standpoint.

The major disadvantage of DCG is that it is extremely sensitive to moisture. Holograms made with DCG must be sealed to protect them from environmental moisture.

From the standpoint of the development of a bar code scanner, there is one other disadvantage to DCG. Although it has been around a long time, DCG is the least understood of all the holographic recording media. There are at least three theories that attempt to explain the mechanism of image formation [30–32], and there are as many recipes for processing DCG as there are authors writing on the subject. Many of them start with gelatin which is already coated on photographic plates [33], a procedure which is unacceptable for the same reasons that bleached silver halide is unacceptable: the sourcing problem.

In most large corporations, one will also find considerable resistance to the use of dichromated gelatin. Most chemists feel comfortable with well-understood inorganic materials, such as silicon, and the more traditional organics, such as photoresist, photopolymer, etc. DCG is an organic material whose properties are poorly understood and relatively unpredictable. Gelatin is, after all, made from the skins, bones, and connective tissues of animals. Its properties can vary depending on what the animals ate or where they were raised.

Nevertheless, because of its excellent holographic qualities, DCG is one of the best recording materials for holographic deflectors used in bar code scanners. It is relatively stable when exposed to normal ambient temperatures. However, it is extremely moisture sensitive and must be sealed to protect it from normal ambient humidity.

DCG is generally sensitive only to the short-wavelength portion of the visible spectrum, $\lambda \leq 520$ nm (20.5 μin), and although it is possible to sensitize it to the red end of the spectrum (34–38) only moderate success has been achieved. The primary problem has been removal of the residual sensitizing dye to give a complete phase structure. For bar code scanning applications, where the light source in the scanner is generally a helium-neon laser, the DCG holographic disk must be made as a copy of a master disk that has been created using a helium-neon laser. The DCG copy disk can then be made using any wavelength to which the DCG is sensitive. So long as the wavelength used to create the master disk is identical to the wavelength used in the scanner, there will be no aberrations introduced in the copy process, regardless of the wavelength used in the copy process. We will have more to say about this in the section on disk fabrication.

DCG is processed in a sequence of alcohol/water baths of varying concentrations of alcohol and varying temperatures. Times, temperatures, and concentrations vary, depending on whose process is used.

Diffraction efficiencies obtained with DCG approach the theoretical limits for volume-phase materials. Efficiencies greater than 90% can be readily obtained. Typical diffraction efficiencies for a sealed DCG holographic disk range from 70% to 80%, where the main losses are the Fresnel reflection losses at the air glass interfaces, some Fresnel losses at the internal gelatin/glass interface, and some small amount of scattering in the gelatin.

4.6 FABRICATION OF HOLOGRAPHIC DEFLECTORS

4.6.1 The DCG Holographic Disk

The DCG holographic disk fabrication process is shown schematically in Figure 4.22.

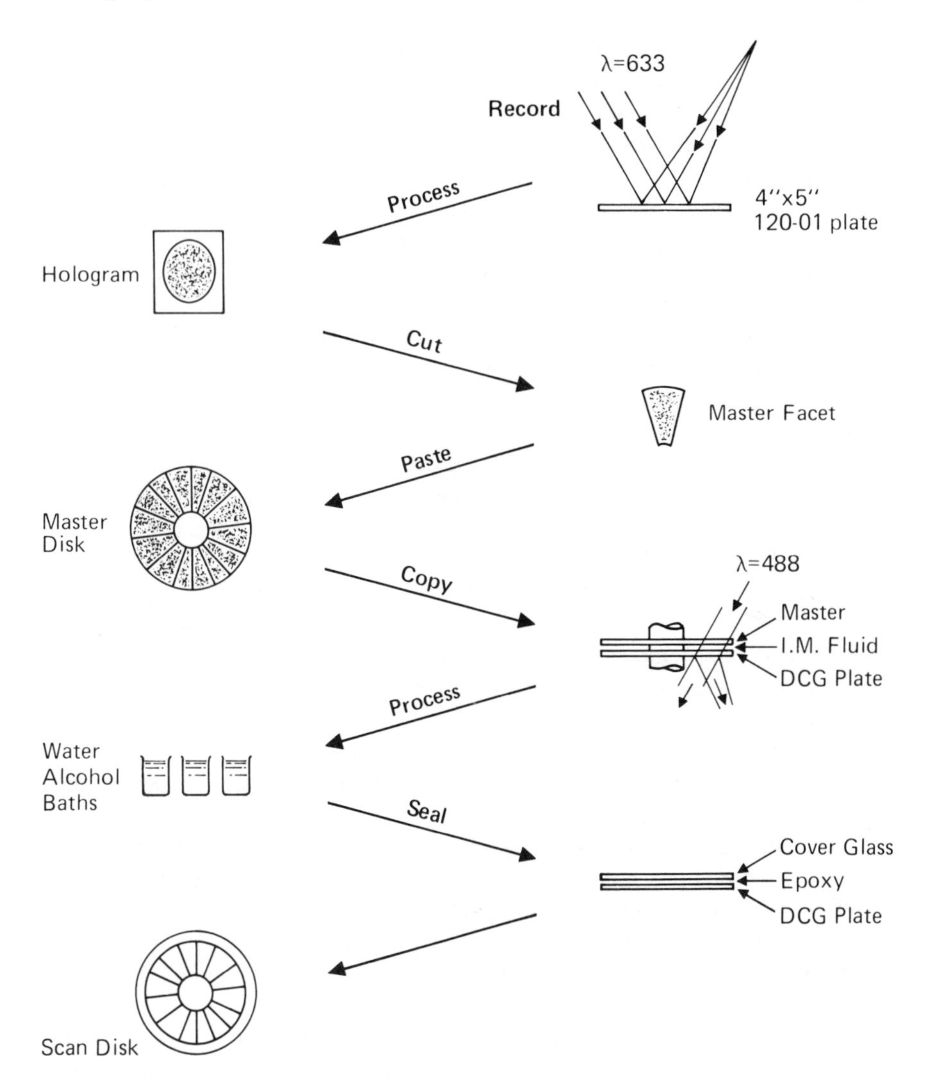

Figure 4.22 Holographic disk fabrication process.

Each facet of the holographic master disk is recorded in a conventional manner using a He-Ne laser and a vibration-isolated table. The facets are recorded on individual 4 × 5 in (100 × 125 mm) silver halide holographic plates. The facets are then cut to size from the 4 × 5 in (100 × 125 mm) plates and mounted emulsion side down, to a glass substrate using an index-matching optical adhesive. When all of the facets are created, cut to size, mounted, and bonded to the disk, we have a master disk.

The DCG disk fabrication process resembles, but is not identical to, a photographic contact copy process. The silver halide master holographic disk is placed in contact with an unexposed DCG-coated copy disk. An index-matching fluid is used between the master disk and the DCG disk. Each facet is then sequentially illuminated in a step-and-repeat exposure process, using an expanded and collimated beam from an argon laser. The angle of illumination of the laser beam is modified for the change in wavelength, 632.8 to 488 nm (24.9 to 19.2 μin), to satisfy the new Bragg condition.

Exposure of each master holographic facet creates an optically identical holographic copy facet in the DCG through the interference of the diffracted beam with the undiffracted zero-order beam. Proper construction of the master facet will yield the required intensity ratio in these two beams for producing high diffraction efficiency in the DCG copy.

Since the master disk was created at a wavelength of 632.8 nm (24.9 μin), and the DCG copy disk is to be used in the scanner at the same wavelength, no aberrations are introduced during the copy process, even though the intermediate copy wavelength is different [39].

When all of the facets of the DCG copy disk are exposed, the disk is processed in a sequence of water and alcohol baths. The details of these baths, and the relative times and temperatures of the liquids, are critical to the process and considered proprietary by those who have invested the time developing a high-yield, stable process.

A major objective in any DCG disk manufacturing operation is to establish a total exposure and development process that provides consistent results and high yield. One of the more difficult problems encountered in attempting to do this is the problem of "gel swell"—the tendency for the exposed and processed gelatin to be thicker than the unexposed, unprocessed gelatin. This residual gelatin swell causes a shifting of the Bragg planes within the thickness of the gelatin so that the angle of the Bragg planes, relative to the surface of the gelatin, is not the same after processing as it was during exposure (see Figure 4.23). This results in a decrease in diffraction efficiency when the reconstruction beam angle matches the reference beam angle. Any attempt to increase the diffraction efficiency by changing the reconstruction beam angle will introduce undesirable aberrations.

Several methods of eliminating the gel swell have been described in the literature [33]. Generally, these involve either some sort of postprocessing chemical treatment or some form of postprocessing baking of the hologram. None of these methods are predictable enough to be suitable in a manufacturing process.

If the gel swell is predictable and consistent, it can be compensated for in the copy process by reducing the calculated angle of incidence of the argon laser copy beam. This increases the Bragg plane tilt angle. After processing, the gel

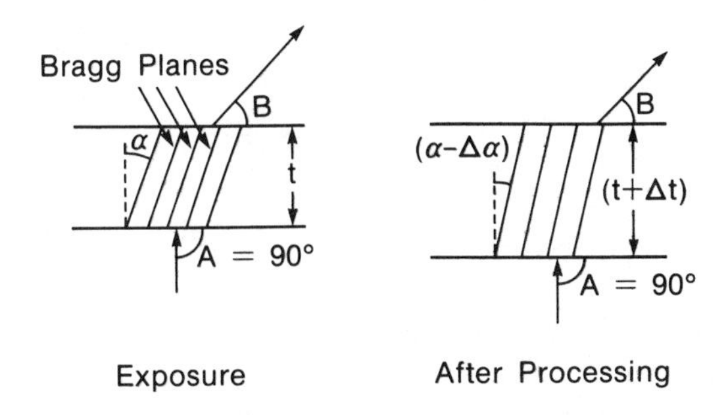

Figure 4.23 Effect of gel swell on the angle of the Bragg planes.

swell will raise the Bragg planes, decreasing the tilt angle until it equals the original value. This process is described in greater detail by Dickson [40].

If the processing methods are well controlled, so that the gel swell is both predictable and consistent, then one can eliminate the gel-swell problem by altering the copy-beam angle. Altering the copy-beam angle, incidentally, has no effect on the optical properties of the holographic copy disk since these are fixed in the surface fringe structure of the master. The copy process will always faithfully reproduce this fringe structure.

After the DCG disk is exposed and processed, several millimeters of the gelatin are stripped from the outer and inner edges to inhibit wicking-in of moisture in the sealed disk. The disk is then sealed with a glass cover disk for protection from moisture, using an index-matching optical adhesive. A metal hub is then bonded to the inner diameter of the disk, and the disk is dynamically balanced.

The optical properties of each and every DCG copy disk will be identical to those of the master holographic disk. The optical characteristics of the holographic scanner are essentially established at the time of the construction of the holographic master. While it is possible to modify these characteristics somewhat through variations in predisk or postdisk optics, this is generally not done in holographic bar code scanners.

4.6.2 The Mechanically Replicated Surface-Relief Holographic Disk

The other primary holographic recording material, photoresist, has a diffracting structure in the form of surface deformation or relief as shown in Figure 4.21 [15,16]. Consideration can therefore be given to using mechanical replication for mass production. This is not to imply that optical copying techniques cannot be used with surface relief, because they certainly can, either as master or copy or both.

Mechanical replication of surface relief is not new, having been developed decades ago for low-cost replication of mechanically ruled gratings. Today, it has become one of the primary manufacturing techniques of the production of high-quality gratings from holographic masters [41].

Although replication of a surface relief can be performed directly from the photoresist master, there is some danger that the photoresist may not stand up to repeated mechanical pressures, elevated temperatures and/or the copy-master release process. Considering the difficulty and potential expense associated with master fabrication, a replication process that permits maximum replication volume is required. This is accomplished by the fabrication of a more durable submaster and usually results in the master itself being sacrificed. A very early technique, borrowed from the audio recording industry uses a metal ''stamper'' to emboss or compression mold the relief into a vinyl thermoplastic [42,43].

In the adaptation of this process [44,45], a very fine-grain layer of nickel or gold is deposited by evaporation or sputtering onto the relief to form a conductive conformal coating, typically a few hundred angstroms thick. Nickel formation is then continued by other methods, such as electrochemical deposition, until a thickness of several hundred microns is achieved. At this point, the outer nickel surface has no significant relief, and it can be attached to a rigid substrate. The sandwich is separated at the nickel/resist interface, and residual photoresist is dissolved away, leaving behind a rigid metal replication of the relief. This structure is a negative of the original and can be used in either the hot-pressing, injection-molding, or epoxy replication processes that will be discussed later.

An alternative method of submaster preparation is to transfer the resist relief downward into its own substrate by radio-frequency (RF) sputter etching or reactive ion etching (RIE) techniques [46,47]. In these methods, the relief surface is removed at a uniform rate by bombardment with accelerated ions or, in the case of RIE, with reactive atoms that react with the substrate molecules to form a volatile gas. The valleys of the resist pattern disappear first and the underlying substrate is exposed and etching occurs. By the time the resist peaks have disappeared, the valley areas are deeply cut into the substrate. Proper choice of photoresist and substrate materials, such as silicon or quartz, and plasma parameters, allows the surface relief to be accurately transferred into the substrate and the cross-sectional shape preserved [48]. These processes result in a submaster having a positive replication of the original master in contrast to the negative shape of the previous nickel submaster.

Once fabricated, these more durable submasters may be used to generate multiple copies. One such method, thermal mechanical embossing or compression molding, is accomplished by pressing the relief into a heated and softened thermoplastic film such as polymethyl methacrylate (PMMA) or polyvinyl chloride

(PVC). Bartolini [44] rolled the submaster together with a vinyl strip between two heated cylinders. Gale [45] used a conventional hot stamping press at 150°C and 3 atm to emboss into PVC sheets. A similar pressing technique was used by Iwata and Tsujiuchi [49] with separation of the copy from the mold performed by sudden cooling and differential contraction.

Replication by pressing tends to introduce considerable strain and other inhomogeneities into the new substrate. These problems can be overcome by using injection molding techniques that have been developed for high-volume, high-quality fabrication of plastic lenses [50]. In this case, the submaster is one surface and an optically polished stainless steel flat is used as the facing, parallel surface. The appropriate polymer is plasticized to a more fluid state than used by compression molding and introduced into the temperature-controlled mold under high pressure [51]. Most of the materials in use are copolymers of PVC and polyvinyl acetate (PVA) and acrylic (PMMA) compounds. The acrylic material has an advantage over vinyl due to the lack of birefringence in the finished substrate and the stability and ease with which it can be machined and polished.

The final alternative is to use a polymer that can be cross-linked by ultraviolet illumination [52]. This technique eliminates the need for high-temperature processing and reduces the possibility of induced stresses and dimensional changes upon cooling/curing. An injection mold apparatus can be adapted for these purposes as long as one plate is sufficiently transparent for the UV illumination. Depending on the use of release agents and relative adhesion, the replica can also be attached directly to a rigid substrate in the same operation.

4.7 HOLOGRAPHIC UPC BAR CODE SCANNERS

4.7.1 The IBM 3687 Point-of-Sale Scanner

A diagram of the optical system of a bar code scanner using a holographic deflector disk is shown in Figures 4.24 and 4.25. This diagram is for the IBM Model 3687 supermarket scanner [10,53,54], introduced in 1980.

Figure 4.24 is a top view of the scanner with the top cover removed showing the laser beam path from the laser to the underside of the holographic disk. Mirrors are designated by the letter M. The lens is designated by the letter L. D is the photodetector, and BX is the beam expander. The box labeled SH is the laser beam shutter. The dotted line in both figures represents the outgoing laser beam path. The solid lines represent the boundaries of the cone of diffusely reflected light collected by the light collection lens L.

Figure 4.25 is a side view of the scanner showing the outgoing beam path, including two of the beam-folding mirrors, and the reflected light cone, above the holographic disk.

The selection and design of the conventional optical components in the 3687

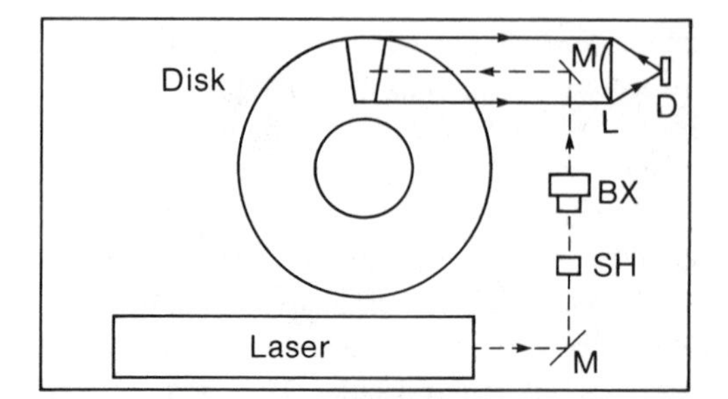

Figure 4.24 3687 scanner top view showing the optical components preceding the holographic disk.

scanner is relatively straightforward. The He-Ne laser has a nominal output beam diameter of 0.75 mm (0.03 in) and a nominal power of 1.45 mW. The laser subassembly includes a beam-folding mirror which deflects the beam 90°.

The beam expander is a Galilean design, consisting of a negative element and a positive element, both made of plastic. Surface curvatures of the positive element are chosen to minimize spherical aberration of the complete beam expander. Spherical aberration is the only aberration of concern in an on-axis optical system using monochromatic light.

The photodetector light-collection lens L is a plastic f/1 aspheric. The plastic

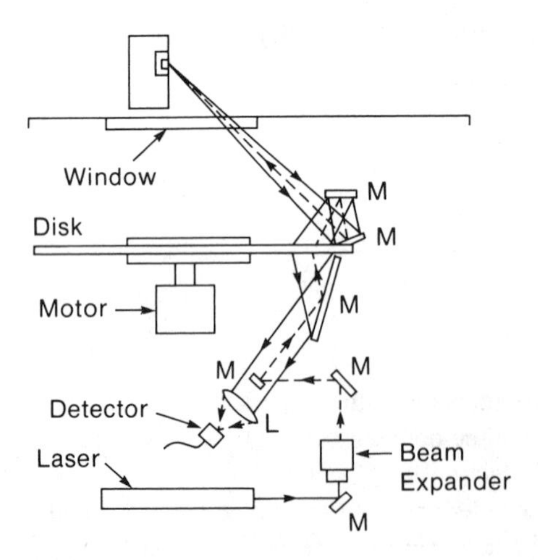

Figure 4.25 3687 scanner side view showing the beam path following the holographic disk.

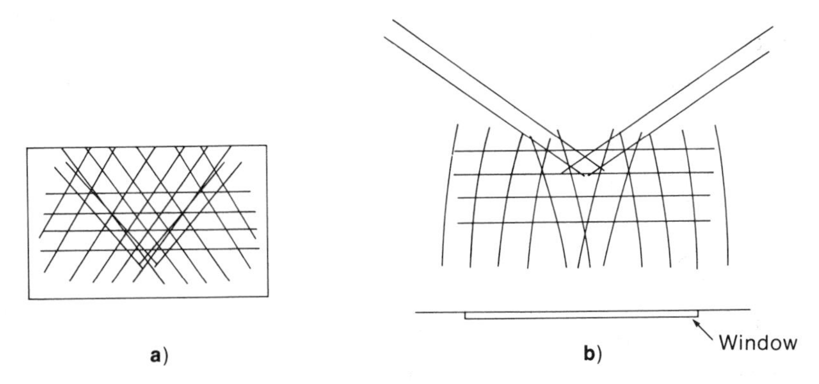

Figure 4.26 The 3687 scan pattern: (a) in the plane of the window; (b) in a plane normal to and in front of the window.

has a red dye introduced to act as a filter for reducing the ambient light reaching the detector. The lens focal length and diameter are both 60 mm (2.36 in).

All of the mirrors, the three predisk beam-folding mirrors and the 11 postdisk pattern-forming mirrors, are first surface mirrors. The resulting scan pattern is shown in Figure 4.26.

Figure 4.27 shows an external view of the 3687 scanner. The high and low rails on either side of the top surface of the scanner contain the item sensors which signal the internal shutter to open when the optical path between the leading edges of the rails is interrupted. (In a later version of this scanner, the item sensor

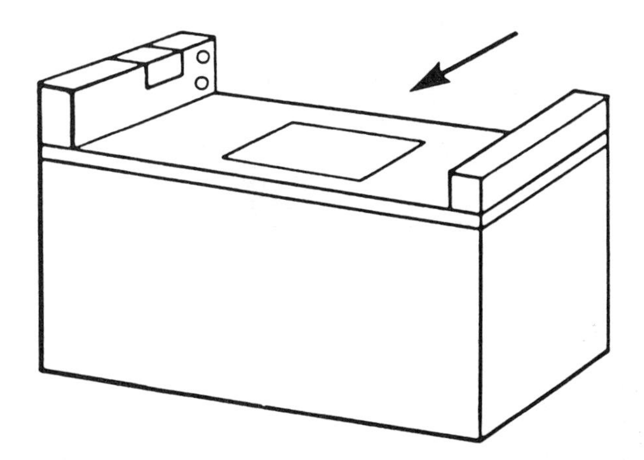

Figure 4.27 The 3687 scanner. The arrow indicates typical motion of items across the scanner surface.

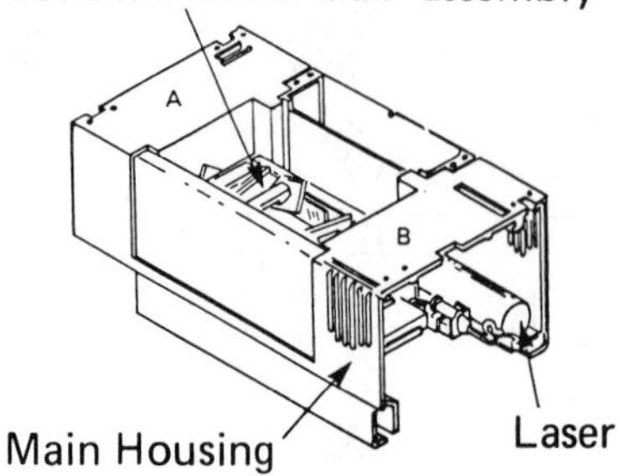

Figure 4.28 Illustration of the 3687 scanner with the top and side covers removed.

rails and the shutter have been removed.). The motion of the items across the scanner surface is in the general direction indicated by the arrow.

Figure 4.28 illustrates the scanner with the top and side covers removed. The postdisk beam-folding mirror subassembly is partially visible in this view. The housing to the left of the optical volume (A in Figure 4.28) contains the power supply, the light-collection lens, the photodetector, and the analog electronics. The logic and the laser power supply are located in the housing to the right of the optical volume (B in Figure 4.28).

Design of the beam-folding mirror subassembly, shown in Figure 4.29, is a

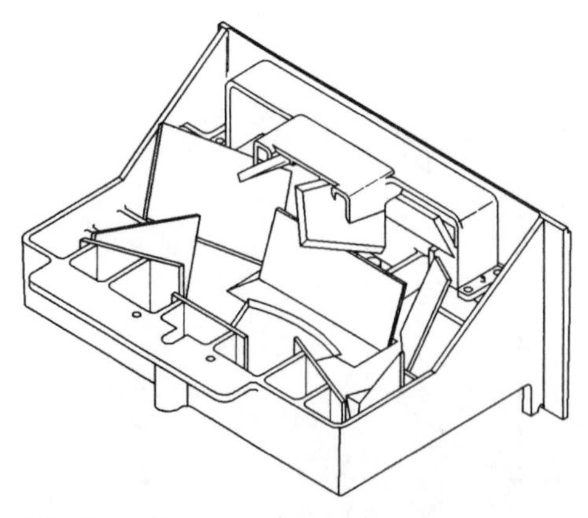

Figure 4.29 The postdisk mirror subassembly of the 3687 scanner.

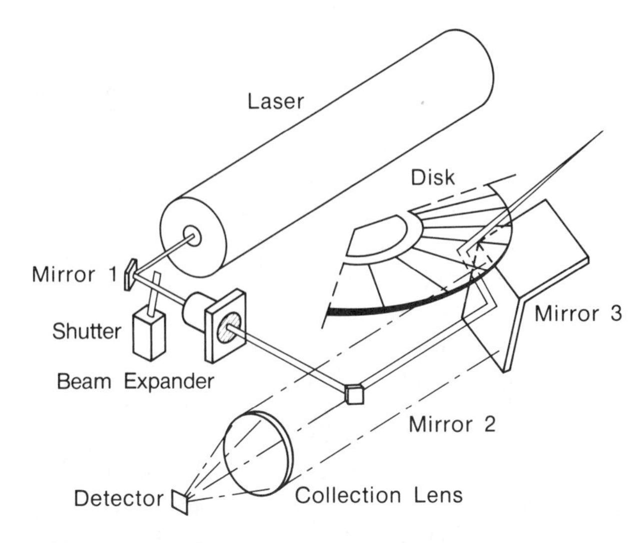

Figure 4.30 Predisk optics of the 3687 scanner.

fairly complex task since the scanner operates in a retroreflective mode. That is, the collected reflected light returns to the photodetector along the same path as the outgoing laser beam. Therefore, the mirror design must take into consideration the light-collection cone of the returning beam as well as the scan length associated with the outgoing beam. Each beam-folding mirror must be big enough to avoid vignetting the return-beam, light-collection cone, but small enough to avoid obstructing the light path to adjacent mirrors for other beams.

A computer program was written to simulate the entire beam-folding mirror subassembly in combination with the holographic disk. It checked the precise location, size, shape, and tilt angle of all postdisk mirrors and calculated the light-collection efficiency back to the holographic disk for a given combination of disk facet sizes, beam deflection angles, and focal lengths. This simulation program was an important tool in the later evolution of the design of the scanner [7].

Figure 4.30 shows the predisk optics for both the outgoing laser beam and the returning reflected light. Mirror 2 is attached to the beam expander and is adjusted so that the outgoing beam strikes mirror 3 at a predetermined position. Mirror 3 is then adjusted so that the beam strikes the holographic disk at a distance of 72 mm (2.83 in) from the disk center and at an angle of incidence of 22° (0.38 rad). The final adjustment is the positioning of the photodetector to center the diffuse reflection image of the focused laser beam in the detector aperture.

4.7.2 The Fujitsu Point-of-Sale Scanner

The Fujitsu holographic point-of-sale scanner, FACOM 3721A, also uses a disk configuration to provide deflection, focusing, retrocollection scanning, and ambient light rejection. Used for scanning the Japanese article number (JAN), development models of these systems were installed in Japanese supermarkets in 1980. As described by Ikeda et al. [55–57], the holographic disk used in this scanner, termed an interferometric zone plate (IZP), is fabricated using a collimated reference with a converging object wavefront. Reconstruction is performed with diverging spherical wave illumination that causes the scanning plane to move away from the original focal plane by the convex lens behavior of the IZP. This introduces a 4× scan length magnification and reduces the required disk size to less than 30 cm (11.8 in) in diameter. Under these conditions, however, the image plane distance varies greatly with deflection angle and reduces the useful operating range to less than ±10° (±0.17 rad) for image distance errors less than 10%. In a technique described by Ikeda et al. [58], an oblique recording and reconstruction geometry is proposed and demonstrated to extend the deflection range to 30° (0.52 rad) with minimum aberration. In the scanner system, a rectangular lattice of 24 × 8 scan lines is generated over a scan field 67.5 mm × 150 mm (2.66 in × 5.91 in) in size with a resolution of 0.26 mm (10 mils). Signal detection of the scattered He-Ne light is performed using lenses embossed into the substrate, adjacent of the hologram ring, that direct the light to a detector/spatial filter assembly. The holographic material and manufacturing process are unreported.

In a more recent publication, Ikeda et al. [59] describe a shallow [16-cm (6.3 in) profile height] scanner designed for reading truncated symbols. The truncated symbols are often used in Japan for source marking of small packages or in-store marking. Several fixed mirrors after the seven-facet, 18-cm (7.1 in) diameter holographic disk create the dense scan pattern emerging from the scanner window at an oblique angle. These mirrors convert the three long and four short scan lines from the hologram to a 13-line scan pattern that can read labels on the bottom or side of the package. A silver halide recording material is used to achieve diffraction efficiencies in excess of 60% (apparently bleached) and emulsion shrinkage compensation is obtained by a 3° (52 mrad) offset of the reference beam during recording.

4.7.3 NEC Point-of-Sale Scanners

The Nippon Electric Company point-of-sale scanner, announced in 1980 as the OBR-80-1, is described by Ono et al. [60] and Ono and Nishida [61] as an embossed surface-relief holographic disk containing seven facets on a 176-mm (6.93 in) diameter substrate. Transmission surface-relief holograms are manufactured by recording a master on Kodak 120-01 plates, and duplication of the amplitude hologram is accomplished by contact printing onto photoresist by the method of Nakano and Nishida [62]. The developed resist is converted to a nickel master by electroforming, and multiple copies are produced by hot pressing into transparent thermoplastic sheets. The resulting diffraction efficiency is approximately 35%. The seven parallel scan lines generated by facet rotation are converted to an 18-line delta pattern by an auxiliary mirror arrangement to produce all-direction symbol reading. In other publications [63,64] the scanner is reported to rotate at 100 rev/s to generate scan angles varying from $\pm 20.9°$ to $\pm 30.4°$ (± 0.365 rad to ± 0.531 rad) with a 70-mm (2.8 in) depth of focus and 0.2-mm (8 mil) resolution over a 180-mm $\times$ 120-mm (7.1 in $\times$ 4.7 in) area. Hologram illumination with a 2.0-mW He-Ne laser provide a backscattered signal with label modulation that is collected and directed to the detector by retrocollection through the scanning facet.

Using this base technology, NEC has developed a new scanner, the OBR-80-5, for omnidirectional scanning of truncated labels [65]. The new scanner features a dense scan pattern consisting of five sets of parallel scan lines created with the aid of postscan mirrors. One scan set is horizontal, and the other four cross at two symmetrical angles on each side. The beam projection angles are approximately 45° and have a depth of field of 100 mm (3.9 in).

In related work, Ono and Nishida [66] describe a multidirectional scanning prototype using a cylindrical drum configuration to create an asterisk-shaped scan pattern. Cross scans, perpendicular to the motion of the hologram, are realized by recording holograms with elliptic and hyperbolic phase distributions and do not require the use of auxiliary postscan mirrors.

4.7.4 Miscellaneous Scanners

A multidirectional scanner has been described by Nishi et al. [67] in which an asterisk-shaped scan pattern is generated by counterrotating the laser beam at an angular speed different than the hologram rotation speed. In this method, the number of scan directions n and the number of scan lines N during one rotation of the laser beam are given by

$$n = \frac{\omega_h}{\omega_l} + 1 \tag{8}$$

$$N = m\left(\frac{\omega_h}{\omega_l} + 1\right) \tag{9}$$

where ω_h and ω_l are the hologram and laser beam rotation speeds and m is the number of holograms. For example, for $\omega_h = 5000$ rpm and $\omega_l = 250$ rpm, 280 scan lines are generated in 20 directions during the laser rotation period of 240 ms. In their prototype, they used a 200-mm (7.9 in) diameter holographic disk replicated from a surface-relief photoresist master. Diffraction efficiencies on the order of 20% are reported. Scanners of this design are intended to read stationary labels.

4.8 FUTURE TRENDS IN BAR CODE SCANNING

It is difficult to predict future trends in this or any other technology today because of the rapid changes that are occurring. In the marketplace, there is a significant trend toward the use of bar code scanning in places other than the supermarket. Retail and discount stores are moving toward bar codes and away from OCR and magnetic stripe. The needs of this customer are somewhat different than those of the supermarket.

The high performance necessary for supermarket applications may not be necessary for retail applications. The average supermarket shopper may have an order consisting of many items, mostly small and easily handled. These items can be smoothly and rapidly handled with a high-performance bar code scanner. Furthermore, the large number of items in each order makes the high-performance scanner a virtual necessity. The average customer in a retail store, on the other hand, will have fewer items, and the items will generally be more cumbersome to handle and more difficult to scan rapidly. This may mean that the trend in scanning in the retail environment will be toward lower-performance, lower-cost scanners.

In some applications, such as at the small area registers that exist at camera and specialty item counters in retail stores, the trend may be toward hand-held scanners, because of the very low item quantities encountered in these applications.

There are some indications that the scanning technology may evolve toward the use of semiconductor lasers. The development of a visible semiconductor laser would have a significant impact on both the cost and the performance of a bar code scanner. Major progress is taking place in the semiconductor laser growth technology, particularly in Japan. The most promising material is InGaAlP, which can emit at wavelengths very near 633 nm (24.9 μin). The most promising material growth technologies are molecular beam epitaxy (MBE) and metal-organic chemical vapor deposition (MOCVD). Either of these technologies is capable of growing good quality InGaAlP devices. Liquid phase epitaxy (LPE), which is the most common growth technology used for GaAs lasers, is unsuitable for growing InGaAlP.

It is vitally important for semiconductor lasers to be able to emit in the visible, red end of the spectrum, if they are to be useful in bar code scanners. The present bar code standards generally require that the printed bar codes have good contrast at, or near, the wavelength of the helium-neon laser. Some of the inks that are presently being used have good contrast at 633 nm (24.9 μin) and very poor contrast at wavelengths above 720 nm (28.3 μin). Therefore, present-day GaAlAs lasers, emitting at wavelengths of about 750 nm (29.5 μin) will not perform satisfactorily for these inks. Unfortunately, since these inks provide very good contrasts at 633 nm (24.9 μin), they are an excellent and frequent choice of the product suppliers.

Since the present UPC ink reflectivity specification has been engrained and widely accepted in the supermarket industry, it is highly unlikely that it will be modified to accommodate the use of infrared semiconductor lasers in UPC bar code scanners.

One must note that the use of a semiconductor laser will introduce some problems for holographic scanners because of their wavelength variability and the dispersive nature of holographic optical elements. These problems should not be severe for holographic bar code scanners, however, because of their relatively moderate resolution requirements.

4.8.1 Nonsupermarket, Nonretail Applications

There is a great deal of activity today in bar code scanning in manufacturing and distribution applications. These applications have significantly different requirements than supermarket and retail applications. In general, the requirements are much more demanding. The environment is much more hostile, and the need for high performance is much greater. The penalty for missing a bar code in the supermarket is a rescan by the operator. The penalty for missing a bar code in an automated manufacturing environment may be the wrong paint color on an automobile.

Requirements for these industrial applications are continuing to develop. Large reading distances, large depth of field, raster scanning capability, the ability to read variable code lengths and variable code types, and the capability to read both high- and low-density bar codes are just a few of the requirements that appear to be important in this application.

The capabilities of holographic scanning would appear to be well suited to the problems encountered in this environment. In industrial applications, the items containing the bar codes are generally on conveyor systems, or mechanical systems that do not employ human intervention to bring the bar code to the scanner. Such systems will generally require depths of field that are much greater than those encountered in the supermarket.

4.8.2 Industrial Applications

The features of holographic scanning that we have described could be used to great advantage in a bar code scanner designed for the manufacturing, warehouse, and distribution environments. The multiple-focal-plane, large-depth-of-field property alone would allow a bar code scanner to be used in ways that heretofore, would have been virtually impossible.

One could, for example, consider using the bar code scanner in an overhead installation for reading medium-density bar codes on a conveyor system where the box heights could vary over a relatively large range (see Figure 4.31). This would allow the scanner to be clear of the area alongside the conveyor system.

An overhead installation would also allow reading the bar codes on a stack of boxes on a pallet after stretch wrap has been applied if the bar code were placed on top instead of on the sides of the stack, as is done today. (Stretch wrap is placed only around the sides of the stack, not on the top. See Figure 4.32.) This cannot be done today since the stack height variation exceeds the depth of field of conventional bar code scanners.

One could also envision an application where the large depth-of-field feature could be used for reading bar codes on hand-held packages or documents. Today this is generally done with a hand-held scanner. The shallow depth of field of conventional fixed-head laser scanners would require the operator to locate the focal zone of the scanner in space whenever he tried to scan a bar code on a hand-held item. This could be tedious and time consuming, particularly if the code is a high-density or medium-density bar code.

If, however, the scanner had a very large depth of field, the operator would only have to position the item so that the red line from the laser scanner, which

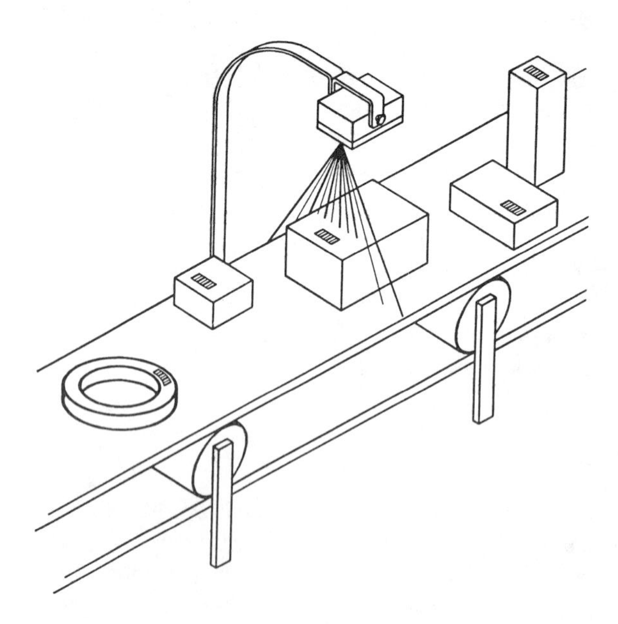

Figure 4.31 Overhead scanning with a large-depth-of-field holographic scanner.

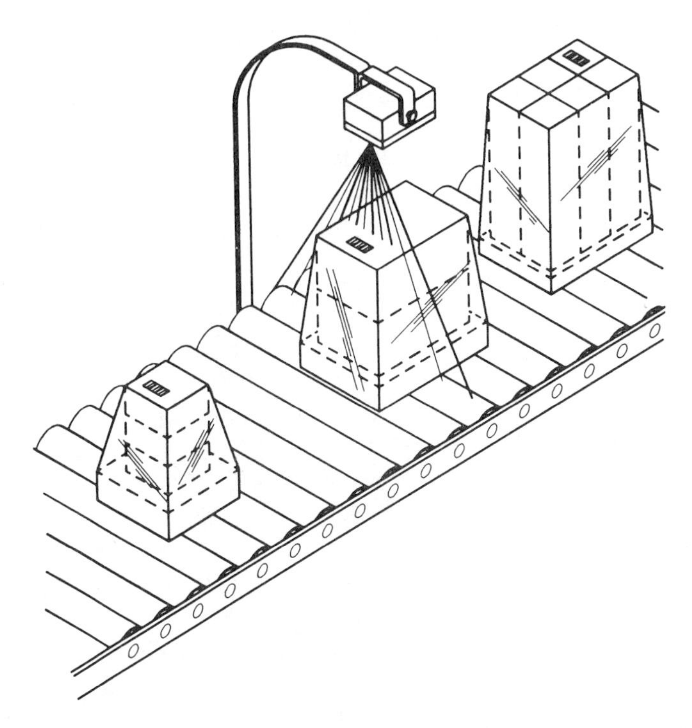

Figure 4.32 Scanning stretch wrap pallets with an overhead, large-depth-of-field, holographic scanner.

is easy to see, intercepts the bar code. The scanner could be positioned overhead in such an application, allowing the operator to have both hands free for moving the packages or documents. This would also allow the work area to be completely free for use in nonscanning operations. Such an installation would have one further obvious advantage. The operator would never handle the scanner.

The multiple-focal-plane, large-depth-of-field feature, combined with the overlapping-focal-zone feature, should provide improved performance in just about any bar code scanning application due to the greater probability of getting a good scan line through the code and the reduced sensitivity to paper noise in the code and substrate. Moreover, these features should be especially advantageous in any application where the bar code and/or substrate material was of relatively poor quality. Since the scanner would scan the bar code with several scan lines, each with a different focal length, the probability of getting a readable scan with one of them would be greatly increased over that of a conventional single-focal-plane scanner.

The multiple-focal-plane feature provides one further advantage which would be more significant at installation than during the actual use of the scanner. Since the holographic scanner can have such a large depth of field, set up of the scanner would be greatly simplified. One would not have to position the scanner at some precise distance from the expected location of the bar codes. Furthermore, if the scanner were moved to a different installation where the distance to the bar codes was different from that in the original location, the scanner optics would not have to be adjusted.

The parallel-line, raster scan feature of holographic scanning would allow the scanner to be used in applications where the positioning of the code may vary significantly. Furthermore, in applications where raster scanning is absolutely essential, as for the picket-fence-oriented AIAG label, the discrete, parallel-line capability of the holographic scanner, achieved by simply changing the holographic disk, could prove to be a very attractive feature.

Overall, the unique capabilities obtainable with holographic scanning—multiple focal planes, overlapping focal zones, variable light-collection aperture, facet tracking, and straight-line raster scanning—would appear to make this technology extremely well suited to industrial bar code scanning applications.

4.8.3 The IBM 7636: An Industrial Holographic Bar Code Scanner

This new holographic bar code scanner is specifically designed for manufacturing, distribution, and warehouse applications. Unlike the supermarket bar code scanners, it does not have omnidirectional and wraparound capability. Instead, the scanner is designed to make use of the capabilities of holographic scanning to solve the different set of problems that exist in this environment as discussed in the preceding section.

Figure 4.33 is a photograph of the 7636 scanner. It is available in two different options. The first option makes use of the multiple-focal-plane, overlapping-focal-zone, and variable-light-collection-aperture features of holographic scanning described in sections 4.3 and 4.4. Sixteen scan lines are created with 16 different focal planes. The focal zones of the 16 facets are overlapped in the manner described in Section 4.1. Each facet has a sector area proportional to the focal length of the facet. These features allow the scanner to provide an extremely large depth of field (75 cm) (30 in) while simultaneously providing uniform response and better immunity to surface noise over the full reading range. These are very attractive features in industrial scanning applications, where bar code quality is relatively poor and environmental conditions are generally harsher than those found in the supermarket. In addition, the large-depth-of-field feature allows scanning without human intervention, as discussed in Section 8.1

The second option, which is obtained by providing a different holographic

Figure 4.33 The IBM 7636 industrial holographic bar code scanner.

disk, is the raster option for reading the AIAG shipping/parts identification label described in Section 4.4. This option generates a 16-scan-line raster scan pattern with a single focal plane for reading the AIAG label in the picket-fence orientation.

Both options make use of facet identification to enhance scanner performance. Both scanners use a 195-mm (7.7 in) diameter rotating holographic disk, using dichromated gelatin as the recording medium.

The capabilities of holographic scanning, as described in previous sections, would appear to make it extremely well suited for industrial applications. So, while the 7636 is the only holographic bar code scanner designed specifically for the industrial environment as of this writing, it seems likely that other industrial holographic bar code scanners will be developed in the near future.

REFERENCES

1. D. Savir and D. J. Laurer, *IBM Systems J.*, *14*:16 (1975).
2. D. C. Allais, *Bar Code Symbology*, Intermec Corporation, 1984.
3. I. Cindrich, *Appl. Opt.*, *6*:1531 (1967).
4. L. Beiser, *Proc. 1975 Electro-Opt. Syst. Des. Conf.*, p. 333, 1975.
5. L. Beiser, E. Darcey, and D. Kleinschmitt, *Proc. 1973 Electro-Opt. Syst. Des. Conf.*, p. 75, 1973.

6. L. Beiser, *Holographic Scanning*, Wiley, New York, 1988.
7. G. T. Sincerbox, *Laser Beam Scanning* (G. Marshall, ed.), Marcel Dekker, New York, p. 1, 1985.
8. D. Gabor, *Nature*, *161*:777 (1948).
9. E. Leith and J. Upatnieks, *J. Opt. Soc. Am.*, *52*:1123 (1962).
10. L. D. Dickson, G. T. Sincerbox, and A. D. Wolfheimer, *IBM J. Res. Dev.*, *26*:228 (1982).
11. R. V. Pole, H. W. Werlich, and R. Krusche, *Appl. Opt.*, *17*:3294 (1978).
12. L. D. Dickson, *Appl. Opt.*, *9*:1854 (1970).
13. R. A. Bartolini, *Proc. SPIE 123*:2 (1977).
14. H. M. Smith, ed., *Holographic Recording Materials*, Springer-Verlag, New York, 1977.
15. H. Werlich, G. Sincerbox, and B. Yung, *Dig. 1983 Conf. Lasers Electro-Opt.*, p. 224, 1983.
16. H. Werlich, G. Sincerbox, and B. Yung, *J. Imaging Tech.*, *10(3)*:105 (1984).
17. G. Rogers, *J. Opt. Soc. Amer.*, *55*:1185 (1965).
18. J. Upatnieks and C. Leonard, *Appl. Opt.*, *8*:85 (1969).
19. K. Pennington and J. Harper, *Appl. Opt.*, *9*:1643 (1970).
20. A. Graube, *Appl. Opt.*, *13*:2942 (1974).
21. N. Phillips and D. Porter, *J. Phys. E.*, *9*:631 (1976).
22. N. Phillips, R. Cullen, A. Ward, and D. Porter, *Photogr. Sci. Eng.*, *24*:120, (1980).
23. B. Booth, *J. Appl. Phot. Eng.*, *3*:24 (1977).
24. E. Chandross, W. Tomlinson, and G. Aumiller, *Appl. Opt.*, *17*:566 (1978).
25. H. Kogelnik, *Bell. Sys. Tech. J.*, *48*:2909, (1969).
26. R. Ingwall, *Proc. SPIE*, *615*:81 (1986).
27. T. Shankoff, *Appl. Opt.*, *7*:2101 (1968).
28. L. Lin, *Appl. Opt.*, *8*:903 (1969).
29. B. J. Chang, *Opt. Eng.*, *19*:642 (1980).
30. D. Meyerhofer, *RCA Rev.*, *33*:111 (1972).
31. D. Samoilovich, A. Zeichner, and A. Freisem, *Photogr. Sci. Eng.*, *24*:161 (1980).
32. S. Sjolinder, *Photogr. Sci. Eng.*, *25*:112 (1981).
33. B. J. Chang and C. D. Leonard, *Appl. Opt.*, *18*:2407 (1979).
34. A. Graube, *Opt. Commun.*, *8*:251 (1973).
35. A. Graube, *Photogr. Sci. Eng.*, *22*:37 (1978).
36. T. Kubota and T. Ose, *Appl. Opt.*, *18*:2538 (1979).
37. M. Akagi, *Photogr. Sci. Eng.*, *18*:248 (1974).
38. T. Kubota, T. Ose, M. Sasaki and M. Honda, *Appl. Opt. 15*:556 (1976).
39. L. Lin and E. Doherty, *Appl. Opt.*, *10*:1314 (1971).
40. L. D. Dickson, U.S. Patent 4,416,505, assigned to IBM, 1983.
41. J. Lerner, J. Flamand, and A. Thevenon, *Proc. SPIE*, *353*:68 (1982).
42. J. C. Ruda, *J. Audio Eng. Soc.*, *25*:702 (1977).
43. H. E. Roys, ed., *Disc Recording and Reproduction*, Dowden, Hutchinson & Ross, Stroudsburg, PA, 1978.
44. R. Bartolini, N. Feldstein, and R. J. Ryan, *J. Electrochem. Soc.*, *120*:1408 (1973).
45. M. T. Gale, J. Kane, and K. Knop, *J. Appl. Phot. Eng.*, *4*:41 (1978).

46. J. J. Hanak and J. P. Russell, *RCA Rev.*, *32*:319 (1971).
47. H. W. Lehman and R. Widner, *J. Vac. Sci. Tech.*, *17*:1177 (1980).
48. S. Matsui, K. Moriwaki, H. Aritome, S. Namba, S. Shin, and S. Suga, *Appl. Opt.*, *21*:2787 (1982).
49. F. Iwata and J. Tsujiuchi, *Appl. Opt.*, *13*:1327 (1974).
50. H. D. Wolpert, *Photonics Spectra*, *17(2–3)*:68 (1983).
51. R. J. Ryan, *RCA Rev.*, *39*:87 (1978).
52. Y. Okino, K. Sano, and T. Kashihara, *Proc. SPIE*, *329*:236 (1982).
53. L. D. Dickson and G. T. Sincerbox, *Proc. SPIE*, *299*:163 (1981).
54. E. A. Moore, *Electronics*, *54*(*14*):139 (1981).
55. H. Ikeda, M. Ando, and T. Inagaki, *Jpn. J. Appl. Phys.* *15*:2467 (1976).
56. H. Ikeda, M. Ando, and T. Inagaki, *Digest of 1978 Conference on Lasers and Electro-Optical Systems*, p. 6, 1978.
57. H. Ikeda, S. Matsumoto, and T. Inagaki, *Fujitsu Sci. Tech. J.*, *15*:59 (1979).
58. H. Ikeda, M. Ando, and T. Inagaki, *Appl. Opt.*, *18*:2166 (1979).
59. H. Ikeda, K. Yamazaki, F. Yamagishi, I. Sebata, and T. Inagaki, *Appl. Opt.*, *24*:1366–1370 (1985).
60. Y. Ono, K. Kosuge, and N. Nishida, *IECE of Japan Tech. Digest*, *2*:217 (1982).
61. Y. Ono and N. Nishida, *Appl. Opt.*, *21*:4542–4548 (1982).
62. M. Nakano, and N. Nishida, *Appl. Opt.*, *18*:3073 (1979).
63. K. Shirakabe, F. Kawamata, H. Miyazaki, T. Yamoda, Y. Yamaguchi, T. Shinoki, Y. Ono, and N. Nishida, *NEC Tech. J.*, *34*:64–66 (1981).
64. K. Shirakabe, Y. Yamaguchi, Y. Tanaka and H. Miyazaki, *NEC Tech. J.*, *35*:11 (1982).
65. H. Miyazaki, K. Shirakabe, K. Yasui, Y. Yamaguchi, Y. Ono, S. Sugama, K. Kosuge, and N. Nishida, *NEC Res. Dev.*, *No. 75*:56 (1984).
66. Y. Ono and N. Nishida, *Appl. Opt.*, *22*:2128–2131 (1983).
67. K. Nishi, K. Kurahashi, and T. Kubo, *Opt. Eng.*, *23*:784 (1984).

5

Holographic Deflector for Graphic Arts Systems

Charles J. Kramer

Holotek Ltd., Rochester, New York

5.1 INTRODUCTION

Laser beam scanners incorporating a rotating holographic optical element (hologon) have been in commercial use since the mid-1970s. Up until about 1987, the most prevalent commercial use of hologons has been in point-of-sale bar code readers (see Chapter 4). Advancements in hologon fabrication techniques and in the performance capabilities of hologon systems have escalated the use of this deflector technology for high-resolution graphic arts imaging applications. By 1990, hologon deflector systems were incorporated in commercial high-resolution imaging equipment used for phototypesetting and photocomposition, computer output microfilm, printed circuit board artwork generation, inspection of printed circuit boards, and xerographic-based prepress proofing and publishing systems.

5.1.1 Electronic Imaging System Requirements

The primary function of any imaging system is the controlled sampling or reconstruction of information. In an electrooptic imaging system, the information is processed either in parallel by a light beam which can simultaneously illuminate many data sites, such as in a CCD input scanner, or sequentially by a light beam which, due to its size, only illuminates one data site at a time. Interest in sequential optical beam scanning has surged since the invention of the laser because the high brightness of the laser enables high-resolution images to be rapidly generated. Advances in image processing software and hardware have created a demand for graphic arts electronic imaging systems that are capable of reading or writing high-quality images at high data rates.

These electronic image writing systems are predominantly required to generate repetitive collinear straight lines with good pixel-to-pixel uniformity and no pixel dropouts. The image writing device of choice for these performance requirements is a flying spot laser scanning system, since the beam can be focused to a fine spot to achieve the high pixel density required for high-resolution imaging applications and the intensity of only a single source need be controlled. Most flying spot laser scanning systems are based on a galvanometer, a rotating polygonal mirror, a hologon, or an acoustooptic or electrooptic deflector. When all scanning system parameters are considered—scan angle, speed, linearity, duty cycle, simplicity, life, etc.—rotating multifaceted deflectors are the preferred laser beam scanning method. The major problem with the multifaceted approach is fabricating the deflector so that all the facets have identical imaging properties.

5.1.2 Deflector Selection Criteria

Because there is often more than one deflector technology solution that can be utilized to satisfy the electronic imaging system design specifications, it is the nontechnical aspects such as system cost, component availability, technology maturity, and designer familiarity that usually determine what deflector approach is used. System cost normally becomes the determining factor in the selection of the deflector technology when there are no significant differences in either performance or availability of technologies.

The cost of a raster scanning system is strongly influenced by what pixel resolution, pixel placement accuracy, and pixel intensity uniformity are required to produce acceptable images of the information to be presented [1]. There is no single rule for determining acceptability, since it is a function of material content (text, graphic, or pictorial), intended use, display media, and system limitations, such as data bandwidth. Table 5.1 on the next page summarizes the performance criteria typically required of raster imaging systems used for pictorial half-tone reproduction and for text and line graphic reproduction applications. Examination of Table 5.1 reveals that primarily it is the cross scan pixel placement accuracy that prevents deflector systems designed for line art reproduction from achieving halftone reproduction requirements.

A common laser beam scanning method for high-speed imaging applications is the polygonal mirror deflector (see Chapters 6 and 7). Facet-to-facet fabrication alignment error and deflector wobble are the prime causes of periodic scan-beam tracking (cross scan) error in polygonal mirror systems. Two basic approaches

Table 5.1 Deflector System Performance Criteria

Performance parameter	Halftone reproduction (%)	Line art reproduction (%)
Scan-line spacing as a fraction of $1/e^2$ spot diameter	50	25–50
Fractional spot size growth	10–15	10–20
Scan linearity error	0.01–0.5	0.01–1.0
Fractional pixel placement error		
Cross scan direction	1–3	10–20
In-scan direction	±50–100	±50–100
Fractional scan intensity variation		
Line to line	±1–3	±4–10
Within a line		
Center-to-edge	±2–5	±4–10
Per inch scan	±0.5–1	±2–5
Residual scan-line bow		
For noncomposite images as a fraction of		
Scan-line length	0.04	0.04
Maximum rate of change per centimeter of scan	0.6	0.6
For composite images as a fraction of		
Scan-line length	0.004	0.004
Maximum rate of change per centimeter of scan	0.04	0.04

have evolved to deal with these deflector errors. The approach used in many of the low-resolution laser page printers is to start with a low-grade polygon/motor assembly having a fairly large (±3 arc minutes) scan tracking error and compensate for this error by means of an anamorphic optical imaging system, as illustrated in Figure 5.1. This approach trades a relatively inexpensive deflector element for a complex optical system that is difficult and expensive to implement at higher resolutions (>600 dots/in) and/or larger scan formats (>355 mm or 14 in).

Polygonal mirror systems used to generate high-resolution, halftone images or large format images usually employ a polygon/motor assembly having very low scan tracking error and an optical imaging system comprised entirely of spherical lens elements. Typical cost is from $5000 to $10000 for commercially available air-bearing-mounted polygonal mirror deflectors that achieve less than

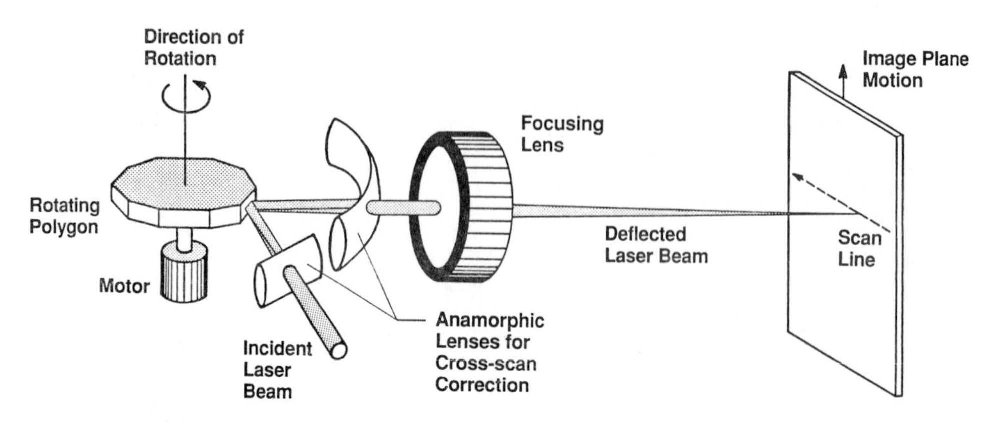

Figure 5.1 Polygonal mirror deflector using an anamorphic optical imaging system.

4 arc seconds of scan-beam tracking error. This cost figure includes the required air support equipment but not the optical imaging system.

Multifaceted hologon deflectors have been developed that solve many of the performance problems encountered with polygonal mirror deflectors. These hologon deflectors are ideally suited for high-speed, high-resolution, halftone imaging applications. The major advantage of these hologon deflectors is that their scan-beam tracking error is virtually insensitive to deflector wobble. When used with a ball-bearing motor, these multifaceted hologon deflectors can achieve scan-beam tracking error performance comparable to the highest-quality air-bearing polygonal mirror systems, but at a fraction of the cost of air-bearing systems. These hologon deflectors achieve this scan tracking performance with optical imaging systems comprised only of spherical lens elements.

Before 1982, there were no graphic arts products on the market that incorporated a hologon deflector and, therefore, no substantial commercial product base by which to judge the merits of this deflector technology. Since 1982, Agfa Compugraphic Division, DatagraphiX Inc., Fela Tec AG, General Optronics Corp., Graphic Enterprises Inc., Holotek Ltd., Indigo Graphic Systems, and Optrotech Ltd. have all introduced products that utilize hologon-based laser beam scanning systems. These companies' products serve the following markets, respectively: phototypesetting and photocomposition, computer output microfilm, printed circuit board artwork generation, xerographic-based paper printer, xerographic-based prepress proofing, OEM imagesetter equipment, xerographic-based CAD printer, and circuit board inspection. Many of these companies' hologon-based products have attained considerable success in the marketplace. It can be concluded that part of this success is due to the performance price advantage provided by the hologon deflector technology.

5.1.3 Chapter Overview

This chapter is divided into eight major sections that review the system performance requirements and system parameter trade-offs that have to be considered when developing specifications, a design, and evaluation techniques for a hologon-based laser beam scan imaging system. Topics covered in these sections include a brief review of image-quality criteria as related to laser beam scanning system performance; deflector system parameters associated with high-resolution imaging; a brief review of non-plane-grating hologon deflector systems; analyses of three different categories of plane grating hologon deflector systems; procedures used to align hologon deflector units to the other optical system components; and an examination of system trade-offs that apply to the design for a hologon-based laser beam scanning system utilized to record computer data on a microfilm format. The author apologizes for the fact that only Holotek hologon deflector components are discussed, but at the time of this writing no other supplier of this type of equipment was known to the author.

5.2 DEFLECTOR SYSTEM PERFORMANCE CRITERIA

Discussion of deflector system performance criteria and related system errors will be limited to the case of a one-dimensional raster scanner that utilizes a flying spot Gaussian laser beam to read from or write images on a moving image surface. The quality of the recorded images depends on the pixel density and the residual errors in the laser beam deflector system and in the image surface transport mechanism. Different image reproduction quality requirements are reviewed in this section in terms of the relationship between pixel density and system pixel placement accuracy, as well as in terms of the scan-to-scan and intrascan variation in scan intensity uniformity.

5.2.1 Scan-Line Spacing

Pixel density is usually stated in dots per inch (dpi), which in the cross scan direction is the scan-line spacing and in the in-scan direction is the scan addressability. Scan addressability refers to the ability to position pixels along the scan line which is determined by the relationship between system electronic bandwidth, scan beam velocity, and image spot size. A large number of factors enter into the determination of the pixel density for an imaging system, many of which are not deflector system performance dependent. Once pixel density is established, image spot size is determined by the trade-off to maximize image resolution while minimizing the visual appearance of imaging system artifacts and errors. This trade-off is very dependent on material content of the information to be reproduced.

In order to minimize the visual appearance of the jaggedness of edges running at an angle other than 0° or 90° to the scan line, the pixel density for text and line graphic reproduction is often increased well beyond what normally would be used when calculated on image spot size alone. Theoretical calculations indicate that for a Gaussian beam the optimum scan-line overlap for most imaging applications occurs when successive scan lines overlap at about the 50% intensity points since this yields a 41% modulation transfer function. When the 50% overlap point is used, the $1/e^2$ intensity diameter S of the scanning image spot is related to pixel cell size d by

$$d = 0.589S \tag{1}$$

Scan resolution is usually specified in terms of $1/d$. Experimental data indicate that due to imaging system positional errors, the optimum scan-line overlap for pictorial halftone reproduction occurs when successive scan lines are spaced by half the $1/e^2$ intensity diameter of the writing spot. For line art reproduction, one can write with a scan overlap that is four times finer than the $1/e^2$ intensity diameter of the writing spot, and thereby reduce the appearance of edge jaggedness and other scan artifacts and errors.

5.2.2 Scan-Beam Tracking Error

Both random and periodic errors in the scan-line spacing and pixel placement within scan lines will degrade image quality because it is the raster structure that imparts image fidelity. Bestenreiner et al. [2] determined that the human visual system can readily detect periodic scanning artifacts caused by fractional line spacing errors as small as 1% for both halftoned binary recording and continuous-tone recording of continuous-tone images. Periodic, high-frequency pixel placement errors beat with the halftone structure of the image to produce lower-harmonic-density fluctuations that appear as streaks or banding in the image. Periodic scan-line spacing errors are caused by either image surface translation velocity errors or by cross scan positional beam errors associated with the deflector system. These deflector-related errors are termed scan-beam tracking errors. Random perturbations in pixel placement position resulting from air turbulence and system vibrations do not contribute to image banding if minimized, and therefore, have essentially negligible influence on both line art and halftone reproduction image quality.

Because the human visual frequency response is only very sensitive to image density fluctuations in the 0.6- to 1.2-cycle/mm range, the spectrum content of periodic system errors is critically important in determining whether they will be observed. Line placement tolerances associated with deflector scan-beam tracking errors are relaxed relative to the scan-line spacing as raster spacing frequencies are increased. This is due to the higher harmonic content of potential periodic scan-line spacing errors and since usually more scan lines comprise a halftone dot. Reducing the number of deflector facets also increases harmonic content of potential periodic scan-line spacing errors. Machine vibrations and air turbulence have a beneficial aspect in that they introduce some randomization into the periodic pattern. The problem is that these random tracking errors can be disastrous if not minimized.

Laser printers used for line art reproduction can have much looser line spacing tolerances. Present commercial low-resolution, line art laser printers have fractional line spacing tolerances of about 10% to 20% [3]. The line spacing tolerance is usually specified as a maximum cross scan positional error for the deflector system as a whole. This means that all scan lines should lie in a band having a width specified by the tolerance. This method of specifying line spacing error gives acceptable results if it is assumed that the errors are distributed essentially uniformly among the deflector facets and that adjoining facets do not have the maximum allowed error with opposite signs.

5.2.3 Scan Jitter

Dynamic pixel placement error within scan lines is termed scan jitter, and is most noticeable as waviness of recorded lines running perpendicular to the scan direction. Jitter results from errors in synchronization between the scan spot position in image space and the electronic pixel clock rate. In rotational deflectors such as a polygon and a hologon, the synchronization errors are primarily caused by scan spot velocity inaccuracies resulting from errors in the deflector rotational speed, facet polar location, and intrafacet differences in scan-beam pointing accuracy.

Scan jitter caused by deflector rotational velocity error can be separated into slow- and fast-varying components. Slow variations in scan jitter are normally caused by deflector motor hunting which is quasicycle with a period of about 2 to 6 Hz. Being a relatively slow varying phenomenon means that deflector motor hunting can be compensated for by altering the electronic pixel clock rate. Variable pixel clock rate is used because the high inertia of rotational deflectors makes it impractical to correct their velocity in a short time period.

Start- and end-of-scan detectors are a commonly used method to measure the time-of-flight of the scan beam across the image plane. Time-of-flight information can be used to correct for pixel placement errors caused by deflector motor hunting. The start-of-scan signal is also used to correct for pixel placement errors caused by facet polar placement error. Many monofacet deflector units incorporate a shaft encoder to measure the variation in deflector motor speed associated with motor hunting. Shaft encoder data can be used in conjunction with a start-of-scan detector signal to calculate the change in the pixel clock rate required to compensate for the deflector motor component of scan jitter.

A pixel grating clock is a less commonly used method to dynamically measure scan-beam position and, thereby, correct for deflector motor hunting and facet polar placement error. Unlike the simpler start- and end-of-scan method, the pixel grating clock system samples the scan-beam flight time and, therefore location, at many points across the scan field. This technique enables multiple point correction within the scan line, which means that localized pixel placement errors—caused by intrafacet differences in scan-beam pointing accuracy and residual scan system linearity error—can be corrected.

Fast variation in scan jitter for ball-bearing-mounted deflectors is caused by irregularities in both the bearings and bearing raceways and by nonuniform distribution and contamination of bearing lubricant. Deflector imbalance contributes to the scan jitter error since it imparts mechanical disturbance to the bearing assembly. Deflector-induced air turbulence is a source of scan jitter for all rotating deflectors. As a result of smooth surfaces, hologon disk deflectors produce less air turbulence than polygonal deflectors, which reduces the scan jitter, windage noise and motor drive power requirements.

Fast variation in scan jitter associated with ball-bearing motors is nonperiodic in nature and has a prime frequency spectrum from 1 to 10 kHz. Being a random high-frequency phenomenon means that it is impractical to compensate for, and therefore it must be minimized. Fast variation in scan jitter should be in the range of one pixel size or less for both line art and halftone reproduction applications.

A rotation rate of 18,000 rpm is a practical upper limit for ball-bearing-mounted deflectors since bearing life deteriorates much faster at high rotation rates. For example, usable bearing life can be expected to be reduced by half if the deflector is operated at 18,000 rpm instead of 9000 rpm. Bearing life for a deflector motor is defined in terms of a scan jitter specification for the deflector system, as opposed to a mechanical catastrophic failure of the bearing assembly. Based on the scan jitter criteria for bearing life, the expected usable life for a ball-bearing deflector motor used for high-resolution imaging is in the range of 5000 to 10,000 h when the deflector unit is operated at 18,000 rpm.

Usable high-speed bearing life can be greatly extended by reducing motor speed whenever high-speed operation (>8000 rpm) is not required and by not turning the motor off but letting it run at a low speed (<4000 rpm) when the rest of the deflector imaging system is either in the standby or off mode. Deflector rotational rate for a given scan beam rate can be reduced by simultaneously deflecting a number of scan beams from the same facet [4].

5.2.4 Scan-Beam Intensity Variations

From the discussion of image density fluctuations associated with scan-beam tracking errors, it can be concluded that the tolerance on periodic scan-to-scan beam intensity variation for halftone reproduction is in the range of 1% and for low-resolution, line art reproduction about 10%. As previously stated, the human visual frequency response is only very sensitive to image density changes in the 0.6- to 1.2-cycles/mm range and, therefore, the spatial frequency content of both the scan-to-scan and intrascan variation in scan intensity uniformity are critically important in determining whether they will be observed. A fairly significant change in scan-beam intensity between the center and edge of scan is usually not visually noticeable as an image density variation when the change occurs at a fairly uniform rate and is essentially the same for all scan lines. Periodic variations in the intrascan intensity uniformity as small as 1% can cause noticeable image density changes when the variations have a spatial frequency content in the 0.6- to 1.2-cycles mm range.

5.2.5 Scan-Line Bow

The departure of the scan line from a perfect straight line is termed scan-line bow. Some level of scan-line bow is present in all deflector systems. Scan-line bow occurs in rotating mirror systems when the scan beam is not perpendicular to the mirror rotating axis and/or when a f-θ scan lens following the deflector has a tilt with regard to the scan beam axial ray. A major effort in the development of hologon deflectors for graphic arts imaging applications was directed at reducing the large scan-line bow that was present in the first hologon deflector systems.

Scan-line bow usually is not noticeable in most imaging applications when each scan has an essentially identical scan-line bow curve, even when the scan-line bow value is many times larger than the scan-line spacing. Lack of visual sensitivity to scan-line bow occurs for this case since the resulting bowed image lines form a self-contained reference frame. Also, printed material is usually not viewed with all wrinkles removed from the printed stock using a well-defined straightedge. The maximum rate of change from straightness of the scan-line bow is more important in most imaging applications than the absolute magnitude of the scan-line bow band.

Minimizing the absolute magnitude of scan-line bow is important in applications where separately generated images are overlapped to form a composite image, such as in color printing or multilayer art work for circuit boards. A common practice is to generate several color separations for an image in parallel across the scan image plane of a wide format recorder, thereby increasing the throughput productivity of the imaging system. Figure 5.2 schematically illustrates what occurs when the four color separations for a color image are generated across the scan field with a scan line having an absolute bow band of 20 μm. For this bow band there is a 15-μm maximum error between corresponding pixels on different separations. This pixel displacement error is comparable to the scan-line spacing for this imaging requirement and, therefore, is considered by many in the industry to be the maximum usable bow band for this application.

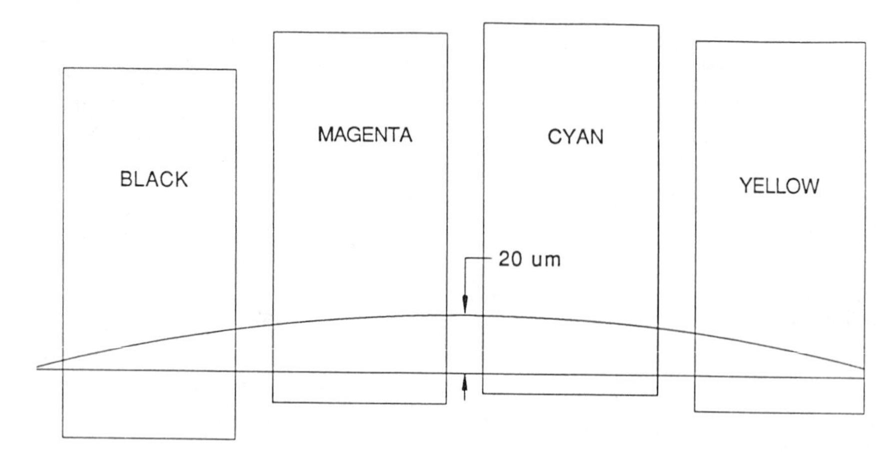

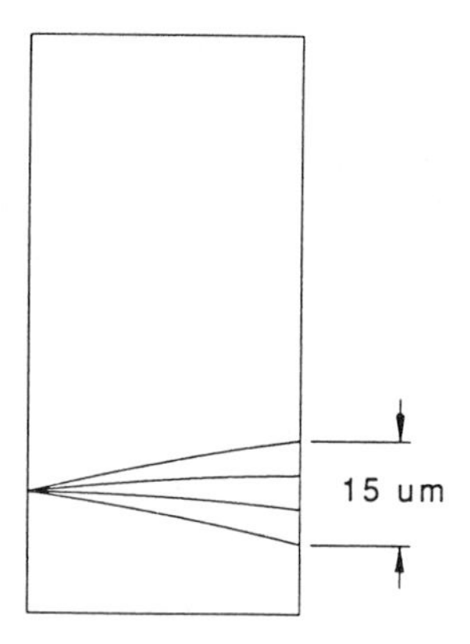

Figure 5.2 Pixel placement error associated with scan-line bow for parallel color image separation application.

The pixel displacement error illustrated in Figure 5.2 could also be caused by a rotation alignment error between separations. Very acceptable looking prints often have registration errors between color separations equal to a halftone dot, which corresponds to 125–200 μm. Based on this observation, the absolute scan-line bow band for the case illustrated in Figure 5.2 could probably be almost 10 times larger if the scan-line bow curve has the curve shape illustrated. If the color separations are generated sequentially in the same image plane region of the scanning system, then a fairly large repeatable scan-line bow can be tolerated because there is no pixel displacement error between separations.

For imaging applications not involving overlapping of separately generated images, the magnitude of residual scan-line bow as a fraction of scan length should be $<0.04\%$ and have a maximum rate of change in slope per centimeter of scan of $<0.6\%$. For imaging applications involving overlapping of separately generated images from noncorrelated scan image fields, the magnitude of residual scan-line bow as a fraction of scan length should be $<0.004\%$ and have a maximum rate of change in slope per centimeter of scan of $<0.04\%$.

5.2.6 Summary of System Performance Requirements

Table 5.1 summarizes the performance criteria typically required of laser beam deflector systems used for halftone and line art reproduction applications. The pixel placement and intensity uniformity specifications in Table 5.1 apply to periodic errors that create image density fluctuations in the 0.6- to 1.2-cycle/mm range and therefore are easily detected by a human observer.

5.3 DEFLECTOR SYSTEM PARAMETERS

Based on the performance criteria listed in Table 5.1, reproduction of halftone images at 1200 dpi requires the scan-beam image spot to have a $1/e^2$ intensity diameter of about 42 μm and have $<15\%$ spot growth for the entire image scan-line length. These scan imaging conditions can be achieved on a flat image plane by a deflector system having the focusing lens following the deflector element. Systems having this optical configuration are termed preobjective deflector systems [5]. For this class of deflector systems the focusing lens can be designed to produce a flat-field, linearized scan. Illustrated in Figure 5.3 is one configuration for a preobjective hologon deflector system that is designed for imaging onto a flat surface. A preobjective hologon deflector system designed for high-resolution imaging onto an internal drum surface was developed by Kramer [6,7].

In this section, the imaging properties of a flat-field, preobjective deflector system are reviewed in terms of the design trade-offs that influence image spot size, image spot size growth, scan linearity, and deflector system duty cycle performance.

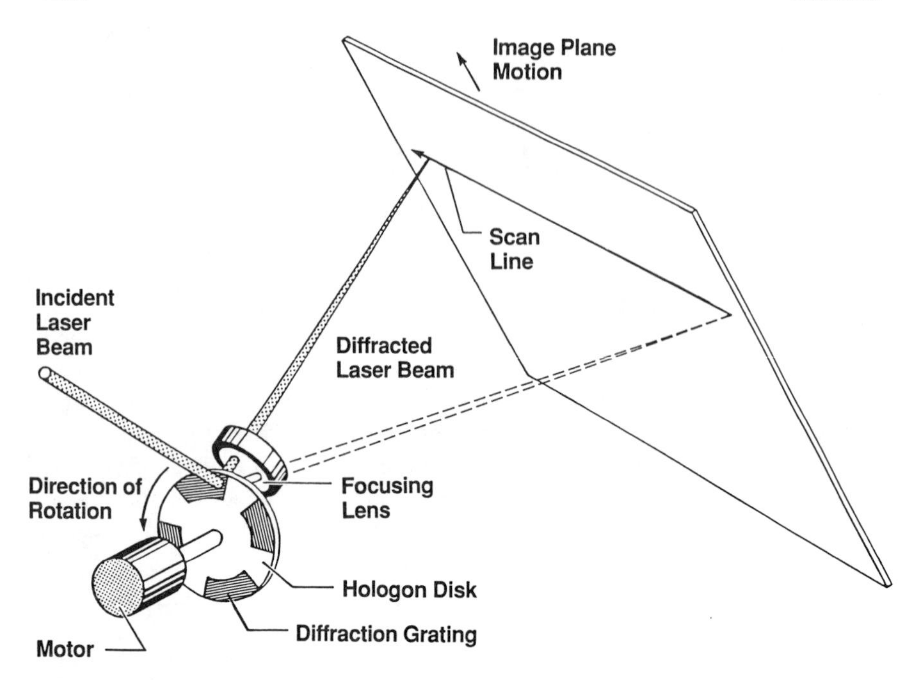

Figure 5.3 Preobjective hologon deflector system designed for flat-field imaging.

5.3.1 Scan-Beam Image Spot Size

An often overlooked design parameter is the effect of Gaussian beam truncation on the imaging properties of a deflector system. The $1/e^2$ intensity diameter S of the scanning spot at the image plane is dependent on both the diameter of the beam incident on the focusing lens and its intensity profile, as follows [8,9]:

$$S = TG\lambda F = \frac{G\lambda f}{A_s} \tag{2}$$

where $T = A_b/A_s$ is the Gaussian beam truncation ratio for the deflector system, A_b and A_s are, respectively, the $1/e^2$ intensity diameter of the Gaussian beam incident on the focusing lens and the diameter of the limiting input aperture to the deflector system, λ is the wavelength of light used, F is the commonly used equivalent f-number of a Gaussian beam given by f/A_b, where f is the focal length of the focusing lens, and G is a parameter dependent on the beam truncation ratio [8,10]:

$$G = 1.6449 + \frac{0.6460}{(T - 0.2816)^{1.821}} - \frac{0.5320}{(T - 0.2816)^{1.891}} \tag{3}$$

The following three examples help illustrate the relationship between S, A_b, and T:

a. Truncation aperture $2 \times 1/e^2$ diameter (untruncated), $G = 2.51$, $T = 0.5$, and

$$S = 1.26\lambda F = \frac{2.51\lambda f}{A_s} \tag{4a}$$

b. Truncation aperture $1.5 \times 1/e^2$ diameter, $G = 2.08$, $T = 0.667$, and

$$S = 1.39\lambda F = \frac{2.08\lambda f}{A_s} \tag{4b}$$

c. Truncation aperture $1.0 \times 1/e^2$ diameter, $G = 1.83$, $T = 1.0$, and

$$S = 1.83\lambda F = \frac{1.83\lambda f}{A_s} \tag{4c}$$

These examples reveal that if system clear aperture is not the prime limiting factor then untruncated operation can provide the smallest S value for a given F value and, thereby, minimize both laser beam energy loss and the magnification factor of the optical beam expander. When system clear aperture A_s is a limiting factor, the smallest value of S is achieved by increasing the value of both A_b and T. Increasing the truncation ratio entails the following disadvantages: loss of beam energy given by $\exp(-2/T^2)$, slightly lower mechanical duty cycle for underfilled multifacet deflector operation, and slight change in spot shape with defocus. Energy loss is only 1.1% and 13.5%, respectively, for truncation at 1.5 and 1.0 $1/e^2$ diameter and, therefore, is usually not a major issue at these T values. Larger T values are usually avoided in deflector systems, since energy loss increases rapidly and the increase in the intensity of the Airy diffraction rings may cause problems. Under the condition of equal spot size S, an untruncated incident beam has the smallest effective beam diameter when calculating underfilled facet mechanical duty cycle. An untruncated beam has approximately a 10% smaller effective beam diameter when calculating mechanical duty cycle than either of the truncation conditions expressed by equations (4b) and (4c).

Equations (4a)–(4c) apply for a truncation aperture that is circularly symmetric about the input beam. Truncation of a Gaussian beam by a knife-edge from one side only, such as a facet edge, produces a more gradual increase in both image spot size growth and spot energy loss. For example, the change in image spot size due to one-sided knife-edge truncation only starts to be observed when the edge is moved to within a distance of less than 1.35 of the $1/e^2$ Gaussian radius of the input beam.

5.3.2 Scan-Beam Depth of Focus

A major source of image spot size degradation in preobjective deflector systems is defocus. Experimental results indicate that for most deflector systems having T in the range of 0.5 to 1.0, the depth of focus, Δf, can be accurately approximated by [8]:

$$\Delta f = \frac{1.8TGS^2\Delta S^{1/2}}{\lambda} \tag{5}$$

where ΔS is the fractional allowable spot size growth by defocus for the $1/e^2$ intensity diameter. The goal for most deflector systems is to keep $\Delta S \leq 0.15$. Equation (5) states that depth of focus is increased for a given spot size by using larger T values to achieve that spot size. This increase in Δf is equal to the change in the value of A_b needed to obtain the spot size for that G value. The increase in Δf between $T = 0.5$ and $T = 1.0$ for equal S values is about 23% less than what would be calculated based on the square of the ratio of these aperture diameters. The smaller increase in Δf can be explained by the reduction in aperture diffraction effects resulting from lower aperture edge illumination as T decreases.

Equation (5) accounts for only the defocus due to Gaussian beam propagation and does not account for residual field curvature of a preobjective flat-field focusing lens. The usable depth of focus of the flat-field deflector system is calculated by subtracting the residual field curvature of the focusing lens from the calculated Δf value. One of the major advantages for the preobjective internal drum scan imaging system illustrated in Figure 5.40 is that its depth of focus does not suffer from scan lens field curvature since its focusing lens is always used on axis. Angular misalignment between the deflector system and the image plane and residual image plane nonflatness reduce the usable depth of focus for both flat-field and internal drum imaging systems.

Inadequate usable depth of focus is often the factor that prevents a deflector system from utilizing a smaller image spot size. Beam truncation ratios of 0.7 to 1.0 should be used in high-resolution halftone imaging applications, since beam energy loss and duty cycle are usually of secondary importance in these applications when compared to the depth of focus requirement.

Laser sources typically have aperture truncation asymmetries that introduce correspondingly asymmetries into the output beam intensity profile. High-speed acoustooptic modulators also introduce an asymmetric truncation in the modulated laser beam intensity profile. Another benefit derived from utilizing a beam truncation ratio of 0.7 to 1.0 is a reduction in the effect of these intensity profile asymmetries on both the image spot shape and size.

5.3.3 Scan Linearity

Compensation for distortion is required in a flat-field flying spot scanning system since the distance of the spot from the center of scan, X, is given by

$$X = f \tan \theta \tag{6}$$

where θ is the scan angle about the center of scan. A rectangular pattern laid down under these conditions would suffer from pincushion distortion in the scan direction because the grid line spacing progressively increases from the center to the end of scan. By introducing the proper amount of barrel distortion into the focusing lens, one obtains an f-θ scan lens that linearizes the scan so that $X = f\theta$. A maximum scan linearity error of about 0.5% is acceptable for most monochrome imaging applications, while less than 0.1% is required for color separation applications.

Residual linearity error data for an f-θ scan lens is presented in terms of the calibrated focal length of the lens. The calibrated focal length of a scan lens is determined by tracing rays through the lens for equal-angle incremental field positions and calculating the focal length that minimizes the f-θ linearity error for the entire design scan field. The calibrated focal length value is usually less than a few percent different from the effective focal length value of the scan lens.

5.3.4 Deflector System Duty Cycle

Up to this point the discussion of deflector parameters has been mainly concerned with system characteristics that contribute to observed image features. Other deflector parameters that are of prime concern when specifying and evaluating a scanning system are deflector radiometric efficiency, deflector rotation rate, and duty cycle (scan efficiency). Even though discussion of these scanning system parameters is based on a multifaceted rotation hologon deflector that functions in the preobjective mode, the derived results apply to all rotation-based deflector devices. Truncation values for $T \leq 1.0$ for the scan beam are obtained by using underfilled facet operation. This operating condition is achieved by using a beam on the facet that is typically 2 to 10 times smaller than the facet width in the scan direction or by facet tracking [11]. Even though facet tracking and pulse imaging can be incorporated into a hologon deflector system, the subject will not be dealt with here.

Mechanical Duty Cycle

System duty cycle is an important, but often miscalculated, deflector parameter. Scan system duty cycle is defined as the percentage of time the laser beam is scanning within the specified image format size. Usually the scan system duty

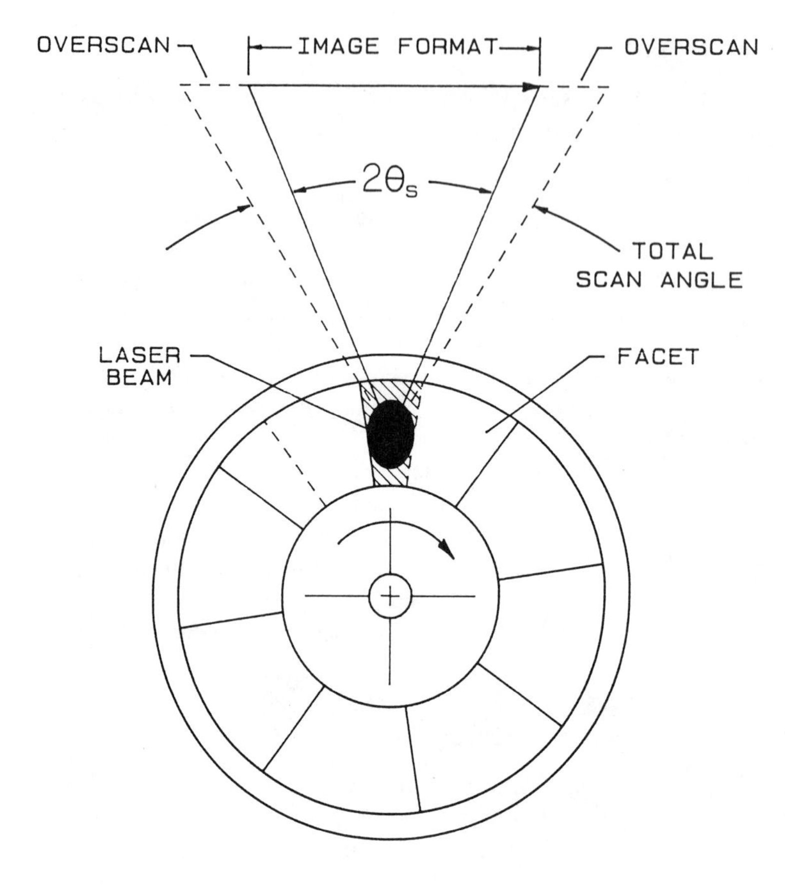

Figure 5.4 Relation between image recording period and total scan period for underfilled facet operation.

cycle for underfilled facet operation is calculated on the basis of the ratio of the input beam diameter to the facet width, which is a measure of the ''optical retrace time'' associated with translation of the facet edge across the laser beam. Scan system duty cycle calculated on this basis is termed mechanical duty cycle, U_m.

The retrace time or deadband is indicated by the shaded area in Figure 5.4. Depicted in Figure 5.4 is a stationary laser beam incident on a rotating hologon disk that has eight grating facets. The reading or writing period of the scan is usually limited to the time that the laser beam resides between the edges of the facet because spot intensity and spot resolution decrease as the facet edge traverses the beam. This recording period is indicated in Figure 5.4 by the pair of dashed facet edges which show the facet position at the beginning of the period, and the solid edges which correspond to the end of the period. Overscan occurring during the retrace time is often used for monitoring purposes such as start- and end-of-scan detectors.

If the recording period is set relative to the $1/e^2$ intensity points of the laser beam on the facet surface, the mechanical duty cycle is

$$U_m = 1 - \frac{k_c A_b}{C} \tag{7}$$

where A_b is the $1/e^2$ intensity laser beam width on the facet in the scan direction, C is the length of the facet in the scan direction as calculated by the mean facet radius r, and k_c is a safety factor to account for truncation effects due to the facet edge. Equation (7) assumes that the facets form a continuous band and that no scan dropout occurs between adjoining facets. Normally the incident beam to a hologon deflector has the same in-scan beam profile dimension on the facet surface that the beam has just prior to intersecting the hologon. This is not the case with a cylindrically shaped (regular) polygon mirror deflector since the incident beam is often coplanar with the deflected beam, as illustrated in Figure 5.1. Consequently, the incident beam must be brought into the facet at an in-scan angle that is larger than the maximum half-scan field angle. The in-scan beam profile width on the facet surface of a hologon deflector having a collimated incident beam can be calculated in terms of scan-beam image spot by equation (2).

As noted earlier, the facet edge can move to within a distance of 1.35 of the $1/e^2$ Gaussian radius for an untruncated input beam before spot size growth is observed. Using this observation as a guide, a value of 1.35 for k_c should be adequate for beam truncation ratios between 0.5 and 0.75, while values of $k_c = 1/T$ should apply to T values ≥ 0.75. The relationship between U_m and deflector system imaging parameters is illustrated by substituting the value of A_b for an untruncated incident Gaussian beam from equation (4a) into equation (7) and setting $k = 1.35$:

$$U_m = 1 - \frac{0.85\lambda f N}{\pi r S} \tag{8}$$

where N is the number of facets on the deflector. It is evident from equation (8) that for fixed values of λ, r, and S, U_m is maximized by minimizing the product of f and N which indicates that the deflector system should have a large scan angle and/or a small number of facets.

Angular Duty Cycle

Another method for calculating the scan system duty cycle of a deflector system is based on the ratio of the angle used to scan the image format, $2\theta_s$, to the total scan angle per facet. This relationship is termed angular duty cycle, U_a, and is also illustrated in Figure 5.4. It is U_a that determines the usable duty cycle for a properly designed deflector system; that is,

$$U_m \geq U_a = \frac{\theta_s N}{\pi M} \tag{9}$$

where $M = M_d M_L$ is the optical angular magnification factor of the deflector system, and M_d and M_L are, respectively, the optical angular magnification factor for the deflector element and of the lens system following the deflector.

The optical angular magnification factor for the deflector system determines how deflector rotation angle, θ_R, is converted into incremental change in scan-line length. For an f-θ scan lens:

$$X = f'\theta_s = M_L f M_d \theta_R \tag{10}$$

where $f' = M_L f$ and is the effective calibrated focal length of the lens system following the deflector element, f is the calibrated focal length of the f-θ scan lens, and $\theta_s = M_d\theta_R$ is the scan angle exiting the deflector. Usually M_L is 1.0 unless a relay lens having magnification properties is incorporated between the deflector and the f-θ scan lens. The value of M_L should not be confused with the ratio of the scan-beam angle exiting the f-θ scan lens to that entering the f-θ scan lens. This ratio becomes zero as the scan lens approaches being telecentric, even though the value of M_L can remain constant. The factor that determines the value of M_d for a given deflector type, mirror or hologon, is the geometric angular relationship between the incident and deflected beam. Typically M_d is either 1 or 2 for polygon deflectors and has a value between 0.7 and 1.42 for hologon deflectors.

Maximizing Scan System Duty Cycle

It is evident from equations (8) and (9) that scan system duty cycle is maximized by increasing θ_s and/or decreasing M and, thereby, reducing the required number of deflector facets. Other benefits derived by minimizing scan lens focal length by means of a large scan angle are reductions in both scanning system package size and in the image plane magnitude of beam displacement errors associated with deflector scan-beam tracking errors and system vibrations. In many scanning applications it is the f-θ scan lens that limits the value of θ_s. A lens which covers a large angular field requires more elements and is more difficult to manufacture and, hence, must be analyzed in terms of the cost trade-offs of the total scanning system. Values of θ_s in the range of $\pm 18°$ to $\pm 27°$ usually provide a good compromise between system performance and cost goals. Maximum allowable scan-line bow is the primary factor limiting θ_s in some hologon deflector configurations, since bow increases directly with scan angle for these configurations.

Scan-line bow usually does not present a problem for these configurations if $\theta_s < \pm 20°$.

System Burst Bandwidth

A prime consideration in any deflector system is to maximize the scan system duty cycle and, thereby, minimize both the electronic burst bandwidth and energy exposure requirements. The burst bandwidth requirement, B, for raster imaging is

$$B = \frac{LR_xW}{U_a} \tag{11}$$

where $L = R_yV$ is the scan-line repetition rate, R_x and R_y are, respectively, the pixel addressabilities in the in-scan and cross scan directions, V is translation rate of the image surface in the cross scan direction, and W is the image format scan length, which for an f-θ lens system is

$$W = 2f\theta_s = \frac{2\pi MU_af}{N} \tag{12}$$

The rotational speed ω of the deflector is

$$\omega = \frac{L}{N} \tag{13}$$

Substituting equations (12) and (13) into equation (11) illustrates that B is just the product of scan-beam image spot velocity, V_s, and pixel density:

$$B = 2\pi Mf\omega R_x = V_sR_x \tag{14}$$

Reducing N increases duty cycle but also increases the deflector rotational rate for a given photoreceptor velocity. In-scan jitter decreases as both ω and the accompanying deflector inertia are increased, due primarily to the reduction in motor hunting. A deflector rotation rate of about 18,000 rpm is a practical upper limit for ball-bearing-mounted deflectors since bearing life deteriorates much faster at high rotation rates.

System Laser Power Requirement

The output power at the laser, P, required to expose a photoreceptor material or read a document is

$$P = \frac{VWI_a}{U_aE} \tag{15}$$

where I_a is the light intensity per unit area needed to expose the photoreceptor or illuminate the document, and E is the optical transmission efficiency of the total scanning system from laser to photoreceptor. A typical value for E is about 40% for a hologon deflector system, assuming that both the acoustooptic modulator and hologon deflector unit each have a 75% radiometric transmission efficiency, and that the other lenses and mirrors in the system have high-efficiency coatings.

5.4 NON-PLANE-GRATING HOLOGONS

Significant research has been devoted to developing practical hologon deflector systems since Cindrich [12] demonstrated a holographic disk deflector in 1967. This activity was spurred by the potential of the technology to solve many of the problems encountered with polygonal mirror deflectors and thereby simplify the deflector system architecture and reduce system costs. Although promising, initial hologon systems were plagued with problems that prevented them from meeting graphic arts imaging requirements. Among the difficulties were bowed scan lines, sensitivity of the scan-beam tracking error to both deflector wobble and centration errors, low radiometric system efficiency, and lack of a flat-field linearized scan. This section briefly reviews three of the many non-plane-grating hologon deflector geometries that were investigated in an effort to develop a device that could meet a variety of imaging requirements. These three hologon geometries illustrate the problems encountered when trying to build a hologon deflector that has focusing power incorporated into its facets. An extensive review of non-plane-grating hologon scanning system configurations is presented by Beiser [13].

5.4.1 Cindrich-Type Hologon Deflector

Figure 5.5 schematically illustrates how a Cindrich-type hologon beam deflector functions. This type of disk geometry is the most desirable hologon configuration from the standpoint of mechanical simplicity, high-speed deflector rotation rate, and economy of fabrication. As depicted, the hologon disk can be sectioned into individual grating facets that function in a manner analogous to the facets of a polygonal mirror deflector. The hologon facets in this figure function in transmission as opposed to reflection and are utilized in the underfilled mode. These grating facets are holographically fabricated using a point source as the object and a normally incident plane wave as the reference beam. When the developed hologon deflector is illuminated with a plane wave which is the conjugate of the reference beam, as illustrated in Figure 5.5, the holographic grating facets function as positive zone lenses. These lens facets diffract part of the incident wave into a wavefront which converges to form an aberration-free image of the point source used to form the facets. Rotating the deflector about an axis perpendicular to the deflector surface causes the reconstructed image spot to scan a circle in space.

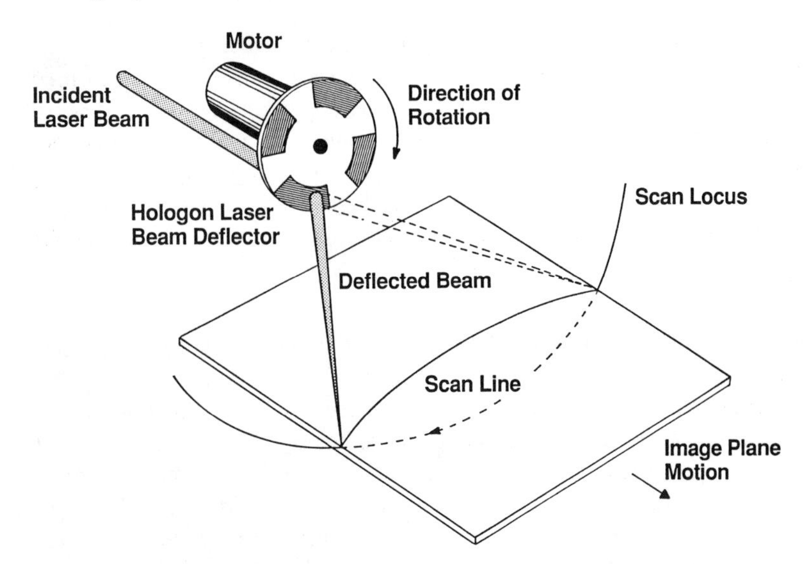

Figure 5.5 Schematic of a Cindrich-type hologon laser beam deflector.

This scanning spot remains aberration-free because the incident reference beam is parallel to the deflector rotation axis, and, therefore, the incident beam has rotational symmetry with respect to the rotating grating facets. For this optically symmetric geometry, scan angle is equal to deflector rotation angle and the deflector angular magnification factor is equal to 1.0.

Significant interest in Cindrich-type hologons for bar code reading applications was created by their inherent dual-function capability of beam imaging and beam deflection. Contributing to this interest was the potential low cost of the hologon and its unique ability to generate a two-dimensional raster scan by means of a single rotational motion [14]. Although suited to bar code reading applications, the hologon geometry in Figure 5.5 is plagued with problems that prevent it from achieving most imaging system performance requirements.

As depicted in Figure 5.5, one problem with this deflector geometry is that a bowed scan line is produced when a flat image surface is utilized. This scan-line bow occurs because the trajectory of the scan beam from the hologon disk deflector in Figure 5.5 is a cone in space. Intersection of this cone scan trajectory by any plane surface produces a conic section, such as the curved scan line depicted in Figure 5.5. The departure ΔZ of this curved scan line from a straight line is given by

$$\Delta Z = \frac{X^2}{2Y'} \cot \theta_d \csc \theta_d \tag{16}$$

where X is the displacement of the scan beam from the center of scan, Y' is the distance from the scanning facet to the center of the scan image plane, measured along the diffracted principal ray, and θ_d is the angle between the diffracted principal ray and the deflector normal. The derivation of equation (16) assumes that both the incident beam and the scan image plane are perpendicular to the hologon disk deflector surface, as depicted in Figure 5.5. Equation (16) gives the value for the sag of a curve which has an effective radius of curvature J given by

$$J = Y' \sin \theta_d \tan \theta_d \tag{17}$$

It is evident that as θ_d approaches 90°, the cone scan-beam trajectory becomes disk-shaped and the scan line for the hologon deflector becomes straight.

For a flat hologon disk deflector, the effective facet aperture dimension in the cross scan direction decreases at a rate equal to cos θ_d. Therefore, the largest practical diffraction angle that can be utilized for this deflector configuration is about 70°. Even with this diffraction angle and $Y' = 545$ mm, the departure from straightness for the Figure 5.5 hologon deflector system is 10.27 mm for $X = 170$ mm. According to Table 5.1, this scan-line bow band is 100 times too large for graphic arts imaging applications. Significant research was directed at analyzing how the Cindrich-type hologon disk deflector could be modified to eliminate the scan-line bow that is inherent in this deflector geometry.

One way to eliminate the scan-line bow problem for the Figure 5.5 hologon deflector system is to use a curved image plane that conforms to the scan locus. This curved image plane geometry offers two additional advantages: the scanned beam is optimally focused for the entire image plane, and the scan displacement in the image plane is linearly related to the deflector rotation angle. However, this curved image plane approach is not feasible in applications where the imaging surface consists of a straight rigid member, such as a xerographic drum.

Other problems encountered with the hologon geometry of Figure 5.5 were sensitivity of the scan-beam tracking error to both deflector wobble and centration errors and lack of a flat-field linearized scan.

5.4.2 Nonflat Hologon Deflectors

If a nonflat hologon element is employed, the diffracted principal ray can be made perpendicular to the deflector rotation axis. Under these conditions, the scan-beam trajectory resides within a plane, so the intersection of this scan trajectory by any plane surface results in a straight scan line. Figure 5.6 shows two scanning geometries that use cylindrically shaped hologon deflector elements to achieve straight scan lines.

By employing reflection grating facets on the circumference of a metallic cylinder, CBS Laboratories [15] was able to build a deflector that operated at very high scan rates with essentially no scan-beam wavefront distortion caused

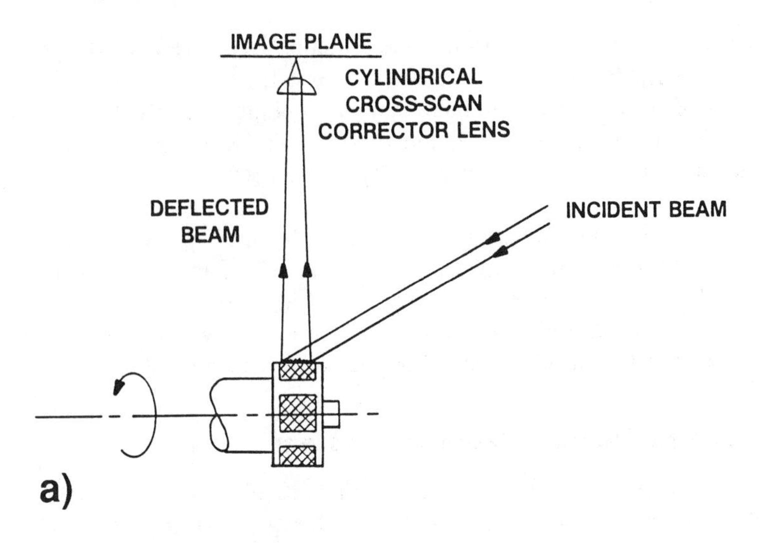

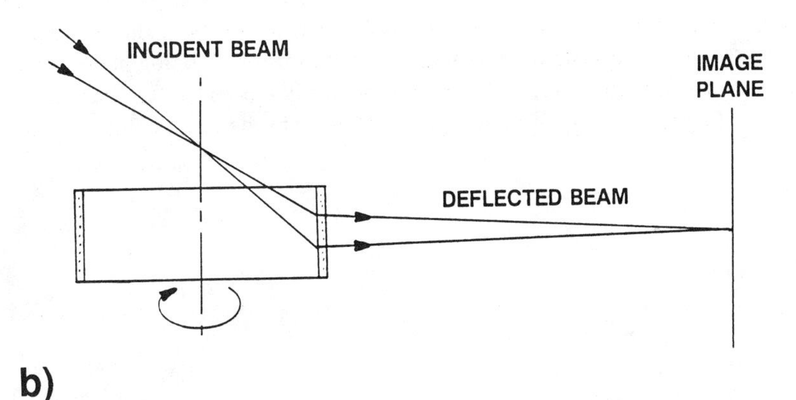

Figure 5.6 Two scanning systems that use cylindrically shaped hologon deflectors to generate a bow-free scan line. System (a) developed by CBS laboratories while system (b) was developed by the IBM corporation.

by centrifugal-force-induced substrate deformation. This deflector was built for a noncommercial reconnaissance image recording application and achieved a resolution of 50,000 pixels/scan at a data rate of 500 Mpixels/s! A cylindrical lens was used to compensate for scan-beam tracking error caused by deflector wobble.

In the IBM device [16], transmission volume grating facets were supported on a hollow glass cylinder. By employing large facets, IBM successfully used the deflector in reading applications to both scan the outgoing beam and to focus scattered light from the target onto a photodetector. IBM did not indicate how to compensate the system for scan-beam tracking error associated with deflector wobble.

Solving the scan-line bow problem via a nonflat hologon element entails abandoning the virtues embodied in the disk geometry—mechanical simplicity and economy of fabrication. Neither of the deflector configurations in Figure 5.6 provide sufficient advantages to offset drawbacks associated with not being a disk geometry and therefore have not been incorporated into commercial products.

5.4.3 Concentric Hologon Deflector System

In a technique developed by C. S. Ih [17], an auxiliary reflector is used to make the principal diffracted ray from a disk hologon deflector perpendicular to the deflector rotation axis, thereby eliminating the scan-line bow problem. Ih showed that theoretically either a convex or concave reflector could be used, and presented experimental data for a system employing a convex mirror. A hologon deflector geometry which initially appeared very promising is the concentric design illustrated in Figure 5.7 [18]. This design deviates from the basic Cindrich-type

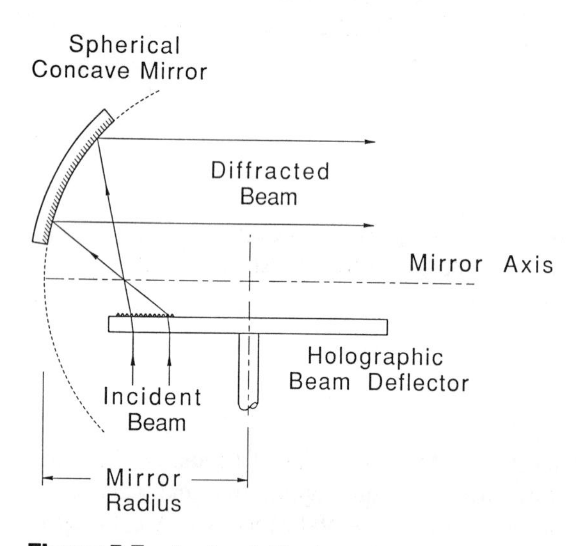

Figure 5.7 Sectional side view of a concentric hologon deflector system that generates a bow-free scan line.

hologon configuration shown in Figure 5.5 by the inclusion of a spherical concave mirror. This mirror is positioned so that its center of curvature resides on the deflector rotation axis. The mirror is angularly oriented so that the collimated scan beam reflected from it is perpendicular to the deflector rotation axis.

The hologon and concave mirror in Figure 5.7 form a symmetric imaging system with regard to the deflector rotation axis. Optical symmetry for the hologon is achieved by having the collimated incident beam be parallel to the deflector rotation axis. The diffracted beam from the hologon grating facet lens is focused to an image point located at a focal point of the concave mirror, thereby being converted into a collimated beam by the mirror. A concave mirror has an infinite number of focal points that reside on a circle concentric with its center of curvature, which for this case is the deflector rotation axis. When the hologon rotates, its scanning image point always resides at a focal point of the concave mirror. Every scan beam exiting the concave mirror is therefore collimated and passes through a point situated on the deflector's rotation axis. This point becomes a virtual entrance pupil for the remainder of the optical imaging system.

Converting the scanning collimated beam into a scanning image spot is readily accomplished by inserting a focusing lens in the scan-beam path. This focusing lens diameter is kept small by positioning the lens at the virtual entrance pupil location. A flat-field linearized scan is achieved by using a flat-field f-θ designed focusing lens. Another advantage of this hologon deflector geometry is that the system scan-beam tracking error can be made insensitive with regard to deflector wobble for hologons operating in either the transmission or reflection mode [19]. Also, because of the newtonian telescope form of the deflector system optical design, the system can be arranged so that the diameter of the scan beam is substantially larger than that of the incident beam, as depicted in Figure 5.7. This allows the deflector system to achieve a very high duty cycle for underfilled facet operation.

Unfortunately, this hologon deflector system design suffers from several significant performance problems. Its major problem is that its performance is extremely sensitive to errors in the centration position of the hologon facets with respect to the deflector rotation axis. These centration errors cause both image spot defocus and scan-beam tracking errors, as illustrated in Figure 5.8. The centration error ΔY in this figure has been exaggerated for illustrative purposes. A ΔY centration error of the hologon facet position (dashed outline) causes a corresponding lateral displacement of the diffracted scan beam (dashed line). Both the undisplaced (solid line) and displaced diffracted scan beams exit the hologon with the same diffraction angle θ_d. The displacement between these parallel beams causes them to intercept different slope regions of the concave mirror. These displaced beams therefore exit the mirror with an angular misalignment between them which converts to a scan-beam displacement error at the image plane of ΔZ, where

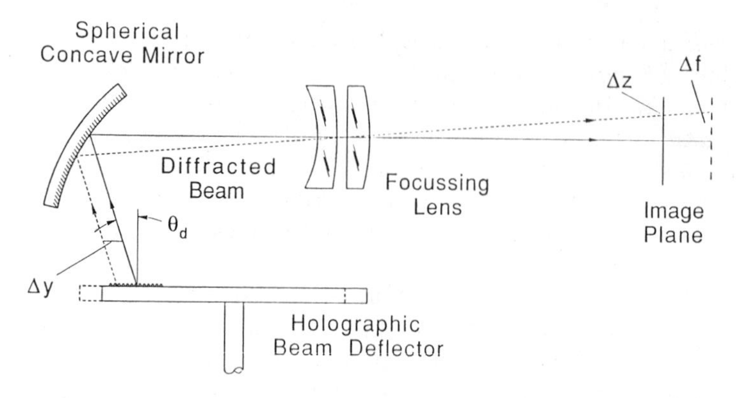

Figure 5.8 Deflector centration errors in the concentric hologon deflector system cause both scan tracking errors and defocus errors.

$$\Delta Z = \frac{f}{f_m} \cos \theta_d \, \Delta Y \tag{18}$$

and f and f_m are the focal lengths, respectively, of the focusing lens and of the spherical concave mirror. For nominal deflector system design parameters, $\Delta Z \approx 7\ \Delta Y$, and therefore, ΔY must be less than about 2.5 μm in order to meet the scan-beam tracking error requirements of Table 5.1 for a laser printer having 300 dpi resolution and used for line art reproduction. This centration tolerance cannot be achieved with a ball-bearing deflector motor.

Also, a centration error causes the scan image point from the hologon facet not to reside at a focal point of the concave mirror. This misalignment between these focal points produces a change in the collimation state of the scan beam incident on the focusing lens which results in an axial shift, Δf, of the scan-beam image spot location. This image shift appears as a defocus error and is

$$\Delta f = \left(\frac{f}{f_m}\right)^2 \sin \theta_d \, \Delta Y \tag{19}$$

For nominal deflector system design parameters, $\Delta f \approx 70\ \Delta Y$, and therefore, ΔY could be as large as about 150 μm and still meet the Table 5.1 spot size growth requirements for a 300-dpi laser printer.

It is evident from these examples that scan-beam tracking error associated with centration errors prevents this hologon deflector geometry from achieving graphic arts imaging requirements. Also, this hologon deflector system is difficult to align for diffraction-limited optical performance. Finally, the focusing power incorporated into the holographic grating facets used in this system and the other non-plane-grating deflector systems, makes it very difficult to make the hologon deflector at one wavelength and use at a different wavelength.

5.5 BASIC PLANE GRATING HOLOGON DISK DEFLECTOR

Analysis of the Cindrich-type hologon deflector systems revealed that many of the imaging performance problems of that deflector geometry were linked to the optical imaging power incorporated in their facets. In his 1987 doctoral thesis [20,21], Herzig showed that when optical imaging power is incorporated into the facets of a hologon disk deflector, the system designer is faced with the trade-off between achieving a smaller diffraction limited image spot size or reducing the scan-line bow band for the deflector system. His results indicated that when a hologon disk deflector has optical imaging power in its facets, it can only achieve a diffraction-limited image spot size of approximately 120 μm for the $1/e^2$ intensity points when the scan-line bow does not exceed 0.03% of the scan-line length for a scan angle of $\pm 20°$. This image spot size performance is suitable only for low- and medium-resolution imaging applications. Also, having optical imaging power in the hologon facets creates a number of manufacturing and system alignment problems.

It has been demonstrated [22–24] that the bow in the scan line generated by a hologon disk deflector can be virtually eliminated if the incident beam is not perpendicular to the deflector surface. For a nonperpendicular incident beam, there no longer exists the condition of optical symmetry with regard to deflector rotation that is present in Cindrich-type hologon deflector systems. Therefore, the type of grating facet selected for this nonsymmetric configuration must possess imaging characteristics that remain constant with relative changes in incident beam orientation. Plane diffraction grating (PDG) facets satisfy this requirement. A PDG is defined as a grating having a flat surface and a constant grating period. It is advantageous to construct a hologon deflector with PDG facets since they make the system scan-beam tracking error insensitive with regard to deflector centration errors, they readily enable the hologon to be made at one wavelength and used at a different wavelength, they make the system very insensitive to small misalignments of the deflector, and they are readily fabricated utilizing either holographic or conventional ruling techniques. Also, high-resolution imaging spot size requirements are achieved with a PDG deflector by the addition of a flat-field, f-θ designed scan lens.

5.5.1 Basic System Parameters

Figure 5.9 reexamines the form of the bow in the scan line for a Cindrich-type deflector when the image plane is rotated 90° from that shown in Figure 5.5. It

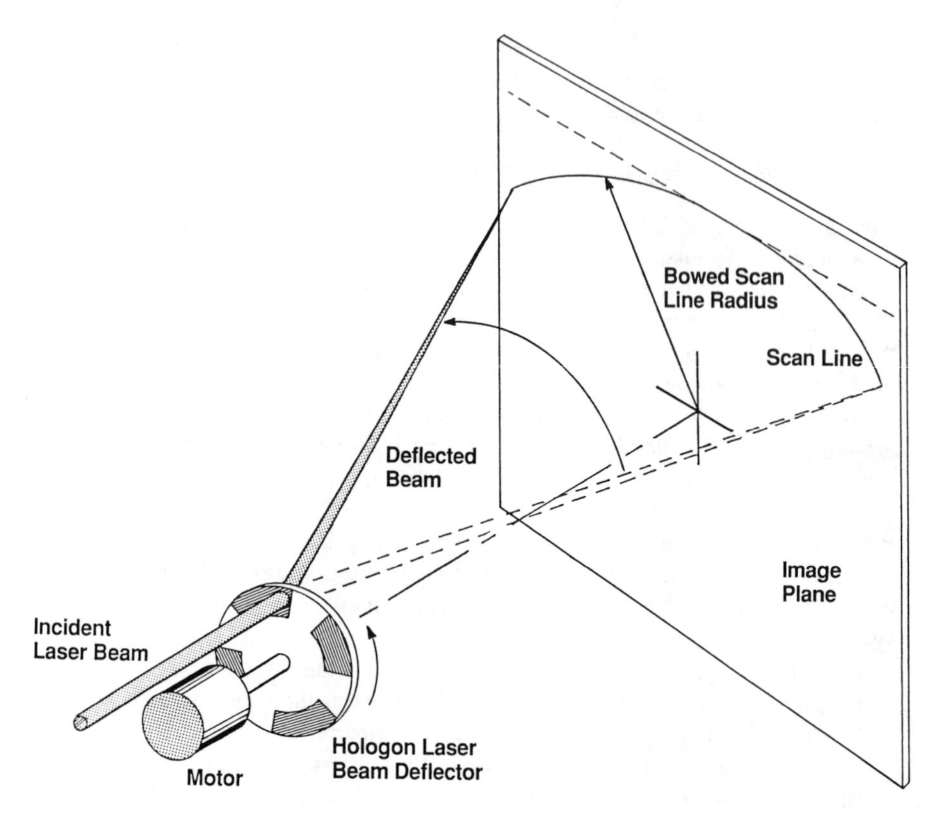

Figure 5.9 Reexamination of the form of the scan bow for a Cindrich-type hologon deflector.

is very apparent from this image plane orientation that the scan-line bow for this deflector geometry is symmetric about the center of scan and increases monotonically with deflector rotation angle. Therefore, the scan line would become straight for this deflector geometry if the diffraction angle of the scan beam increased as a specific function of the absolute value of the deflector rotation angle about the center of scan.

Scan-Beam Diffraction Angle

The relationship between the scan-beam diffraction angle and the incident beam angle for a hologon deflector is provided by the familiar simple form of the grating equation when the incident beam resides in the plane that is perpendicular to the facet grating lines:

$$\sin \theta_i + \sin \theta_{do} = K \tag{20}$$

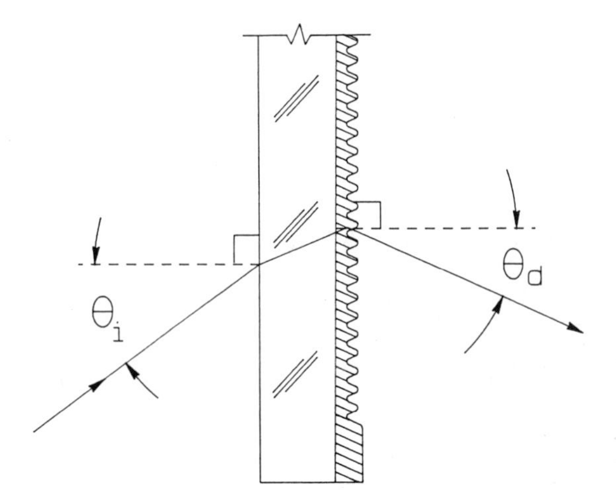

Figure 5.10 The incident and diffracted beam angles illustrate the sign convention for positive angles relative to the grating substrate surface normal. The grating is depicted as being of the sinusoidal surface-relief type.

where θ_i and θ_{do} are, respectively, the angles that the incident and diffracted beams make with respect to the hologon grating surface normal, and $K = \lambda/D$, where λ is the optical wavelength of the beam and D is the facet grating period. The subscript *do* shows that the diffracted beam angle, θ_d, is for the case when the facet grating lines are perpendicular to the plane containing the incident beam angle. Figure 5.10 illustrates the sign convention for positive angles relative to the grating normal. The hologon substrate index of refraction is not included in equation (20) because the incident and diffraction beam angles are measured in the air medium surrounding the hologon element. Higher-order diffraction terms are absent from equation (20) since only the first-order diffracted beam is used for image scanning. Equation (20) gives the diffraction angle for the scan beam of a Cindrich-type hologon deflector by setting $\theta_i = 0$.

It is apparent from the grating equation that in order to increase the scan-beam diffraction angle for the optically symmetric hologon deflector system in Figure 5.9, θ_i must decrease if K is kept constant. It is not practical or elegant to physically alter θ_i to achieve the increase in θ_d required to straighten the scan line for the hologon deflector system in Figure 5.9. Therefore, another solution must be found for increasing θ_d as a function of scan angle. Mathematically it will be shown that when a stationary incident beam is not perpendicular to the hologon surface, it will appear to have a varying incident angle with respect to the coordinate system that rotates with the facet grating lines. This relative change

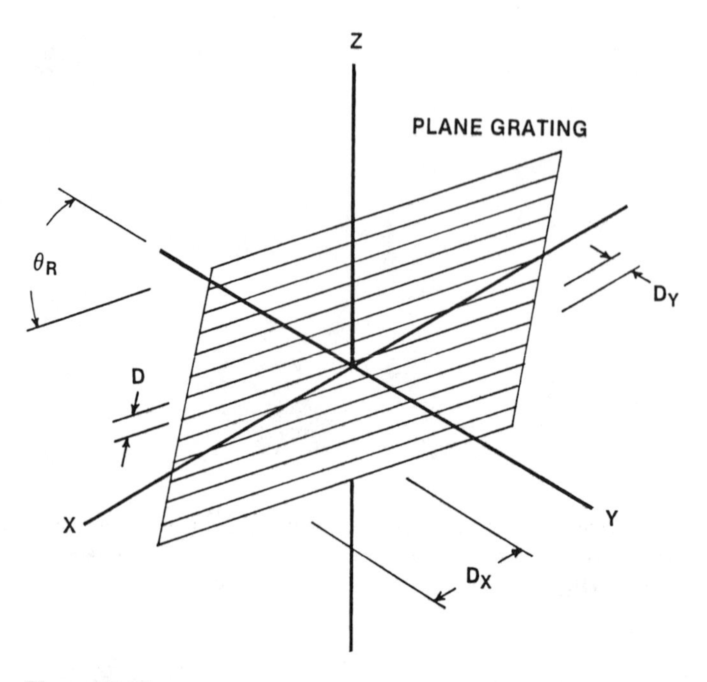

Figure 5.11 The coordinate system used to analyze the diffracted beam from a rotating PDG facet assumes that the PDG facet resides in the *XY* plane and rotates around the *Z* axis. Grating periodicity along the *X* and *Y* axes, D_x and D_y, respectively, vary as a function of grating facet rotation angle, θ_R.

in θ_i can be used to produce the change in θ_d required to straighten the scan line for a hologon disk deflector system.

Inspection of the grating equation reveals that it is a relationship between the optical *k*-vector components of the incident and diffracted beams that reside in the plane of the grating and the grating *k*-vector that is perpendicular to the grating lines. The magnitude of the optical *k*-vector is $k = 2\pi/\lambda$ while that of the grating *k*-vector is $k_g = 2\pi/D$. When deriving equations for the scanning parameters of a disk PDG hologon, it is assumed that the facets lie in the *xy* plane and rotate about the *z* axis, as illustrated in Figure 5.11. It is further assumed that when the deflector rotation angle, θ_R, is zero, the grating lines of the facet are parallel to the *x* axis and k_g is parallel to the *y* axis. Under these conditions, the incident and diffracted beams of the hologon satisfy the following equations:

$$\sin\theta_{ix} + \sin\theta_{dx} = K\sin\theta_R \tag{21}$$

$$\sin\theta_{iy} + \sin\theta_{dy} = K\cos\theta_R \tag{22}$$

where $\sin \theta_{ix}$ and $\sin \theta_{iy}$ are the components of the incident beam k-vector along the x and y axes, respectively, and $\sin \theta_{dx}$ and $\sin \theta_{dy}$ are the components of the diffracted scan beam k-vector along the x and y axes.

The angles that the diffracted scan beam makes with regard to the positive x, y, z axes can be expressed in terms of the directional cosines angles α, β, and γ, respectively. If it is assumed that the incident beam lies in the yz plane, then $\theta_{ix} = 0$ and $\theta_{iy} = \theta_i$, and the directional cosines for the diffracted scan beam are

$$\cos \alpha = k_{dx} = K \sin \theta_R \tag{23}$$

$$\cos \beta = k_{dy} = K \cos \theta_R - \sin \theta_i \tag{24}$$

$$\begin{aligned}\cos \gamma &= k_{dz} = [1 - k_{dx}^2 - k_{dy}^2]^{1/2} \\ &= [1 - K^2 - \sin^2 \theta_i + 2K \sin \theta_i \cos \theta_R]^{1/2}\end{aligned} \tag{25}$$

where k_{dx}, k_{dy}, and k_{dz} are the components of the diffracted scan-beam k-vector along the x, y, and z axes, respectively. Equations (23) and (24) are directly derived from equations (21) and (22), respectively, while equation (25) is derived from equations (23) and (24) by means of the requirement that the sum of the squares of the directional cosines must equal unity. For this case γ is θ_d, so

$$\sin \theta_d = [K^2 \sin^2 \theta_R + (K \cos \theta_R - \sin \theta_i)^2]^{1/2} \tag{26}$$

As expected for $\theta_R = 0$, this equation reduces to equation (20) and $\theta_d = \theta_{do}$. Also as expected, θ_d is constant with regard to θ_R when $\theta_i = 0$ and changes as a function of θ_R when $\theta_i \neq 0$.

Equations (21) and (22) indicate that beam scanning from a rotating PDG can be modeled as diffraction from a two-dimensional grating pattern having periodicities along orthogonal axes that vary periodically in time from D to infinity, as depicted in Figure 5.11. The relative change with θ_R in grating periodicity along the x and y axes of this grating pattern is, respectively,

$$D_x = \frac{D}{\sin \theta_R} \tag{27}$$

$$D_y = \frac{D}{\cos \theta_R} \tag{28}$$

Equations (21)–(26) can be rewritten in terms of these D_x and D_y periodicities. Change in the D_x periodicity is what causes the diffracted beam to scan as the grating facet is rotated.

Scan-Beam Angle

The scan beam of the hologon deflector tracks in essence the angular change in grating k-vector in the same manner that the scan beam of a mirror deflector tracks the angular change in the normal to the mirror surface. In a galvanometer or regular polygonal mirror deflector, the change in scan-beam angle is twice the angular change in the mirror normal, whereas in a hologon deflector, the change in scan-beam angle is typically between 0.8 and 1.42 times the angular change that k_g experiences.

As previously stated, the scan-beam angle θ_s for the Cindrich-type hologon deflector geometry is equal to the deflector rotation angle, and, therefore, the scan beam exactly tracks k_g for this deflector system. It is easy to verify for this deflector geometry that $\theta_s = \theta_R$ for a plane parallel to the disk deflector surface by using the ratio k_{dx}/k_{dy}, since $\theta_i = 0$ and $\theta_d = \theta_{do}$ for all θ_R values. The deflector angular magnification factor M_d is equal to 1.0 for this deflector geometry.

For the general case of a hologon deflector system having an incident beam not parallel to the deflector rotation axis, the scan angle in the plane of the scan beam is equal to $\alpha - 90°$; therefore, equation (23) can be expressed in terms of the scan-beam angle:

$$\sin \theta_s = K \sin \theta_R \tag{29}$$

Except for when $K = 1.0$, M_d varies very slightly as a function of θ_R for this general hologon case. Scan-beam angle increases slightly faster than the product of $K\theta_R$ for values of $K > 1.0$ and is slightly slower than this product for values of $K < 1.0$. This change in M_d with θ_R and K must be accounted for when designing the deflector system f-θ scan lens. Equation (29) can be accurately approximated by $\theta_s \simeq K\theta_R$ for most hologon deflector systems of interest, and, therefore, $M_d \simeq K$.

Changes in rotation rate associated with deflector motor hunting and irregularities in the bearings and bearing raceways of a ball-bearing deflector motor cause a scan jitter error in PDG deflector systems that is equal to K times the variation in rotation rate, so it is desirable to reduce the K value of the deflector. Reducing the deflector K value also improves the scan system duty cycle as indicated by equations (8) and (9).

Scan-Bow Minimization Condition

With the aid of Figure 5.9 it was determined that a bow-free scan line would be achieved if θ_d increased as a specific function of θ_R. Equations (25) and (26) state that θ_d increases with θ_R when $\theta_i \neq 0$. This increase in θ_d with θ_R can be viewed as due to a decrease in the apparent incident beam angle, θ_i', as a function of θ_R. θ_i' is the projection of θ_i onto the plane containing the grating k-vector and the deflector rotation axis, and, therefore, the change in θ_i' is given by

$$\tan \theta_i' = \tan \theta_i \cos \theta_R \tag{30}$$

where θ_i' decreases from θ_i to 0 as θ_R increases from 0° to 90°. Greater scan bow compensation should be achieved by increasing the value of θ_i since this increases the rate of change of θ_i' with θ_R.

Using a differential equation derivation approach [13,23], it can be shown that when the PDG facets of the hologon are fabricated with a K value residing within a specific range, the deflector will generate a scan line that is virtually bow-free for the proper incidence beam angle. Scan-line bow is minimized for this hologon deflector case when the following conditions are achieved at the center of scan:

$$\sin \theta_i = K - \frac{1}{K} \tag{31}$$

$$\sin \theta_{do} = \frac{1}{K} \tag{32}$$

Equation (32) is derived from equation (31) by means of equation (20). Real solutions for equations (31) and (32) exist only for the range, where

$$1 \leq K \leq 1.618 \tag{33}$$

The corresponding maximum and minimum values of θ_i and θ_{do} are

$$\begin{aligned} 0° \leq \theta_i \leq 90° \\ 90° \geq \theta_{do} \geq 38.17° \end{aligned} \tag{34}$$

These solutions are elegant in that they apply to either transmission or reflection gratings and depend only on the relationship between λ, D, and θ_i.

5.5.2 System Configuration

Figure 5.12 presents a sectional side view for the basic system configuration of a PDG transmission deflector that utilizes only the relationship between K and θ_i to produce a virtually bow-free scan line. The folding mirror in this figure is without optical power and is included for system packaging convenience. This mirror does redirect the central scan ray so that it is perpendicular to an image plane that is aligned to the deflector coordinate system; x axis in-scan direction and z axis cross scan direction. An isometric schematic of this deflector system is presented in Figure 5.3 without the folding mirror and the image plane rotated so that the central scan ray is perpendicular to it. Having the image plane essentially perpendicular to the central scan ray is beneficial in that it reduces the magnitude of image plane residual scan-line bow. Normally the image plane is slightly misaligned from being exactly perpendicular to the central scan ray. This is done in order to minimize laser power fluctuations associated with retroreflected feedback light from the image plane.

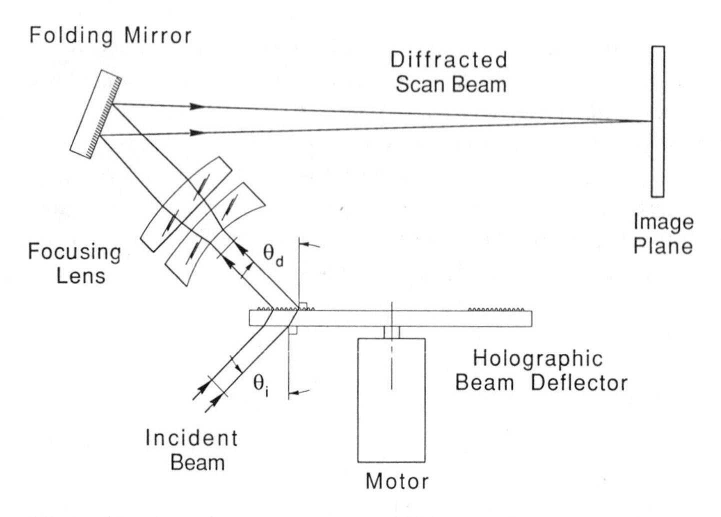

Figure 5.12 Sectional side view of basic plane grating hologon disk deflector system.

Absence of focusing power in the facets of the deflector in Figure 5.12 might be viewed as a disadvantage because a separate focusing lens is required. Having a separate focusing lens ensures scan-to-scan imaging quality uniformity that is difficult to achieve when optical power is incorporated into the facets. Also, it is relatively easy to design and fabricate a focusing lens that provides both flat-field imaging and a linear scan velocity at the image plane (f-θ condition).

The hologon deflector in Figure 5.3 can be thought of as a ''flat polygon,'' in that the plane grating facets function in a manner analogous to the plane mirror facets of the polygon in Figure 5.1. This picture of the hologon as a flat polygon is useful when examining the similarities and differences between these deflectors. A major difference between these deflector technologies is the reflection-versus-transmission nature of the facets. Surface errors in a reflective optical element, whether they are nonflatness or tilt in nature, produce an error in the optical wavefront exiting from the surface that is twice the magnitude of the corresponding surface error. Errors in transmissive optical elements, such as tilt and surface nonflatness, are generally reduced by a factor between 4 and 10000, depending on the optical component and the nature of the error. Wobble in a polygonal mirror deflector causes a scan-beam tracking error that is twice the corresponding deflector wobble angle. Wobble-induced scan tracking error in a transmissive hologon deflector may only be $1/100$ of the deflector wobble angle when $\theta_i \approx \theta_d$ [25]. Solving equations (31) and (32) for this equal angle condition gives $K = 1.4142$, corresponding to $\theta_i = \theta_{do} = 45°$.

5.5.3 Hologon Facet Configuration

As illustrated by Figures 5.3, 5.4, 5.5, and 5.9, the hologon disk can be sectioned into individual grating facets. These facets have identical properties for the PDG hologon deflector when the deflector is used with only a single laser wavelength [26]. Normally the facets form a continuous band, as depicted in Figure 5.4, and thereby maximize scan duty cycle. In all previously presented figures, the facet grating lines are orientated perpendicular to the disk radius that bisects the facet. This facet grating line orientation is termed tangential.

Figure 5.13 presents two hologon disks that only have a single facet on them.

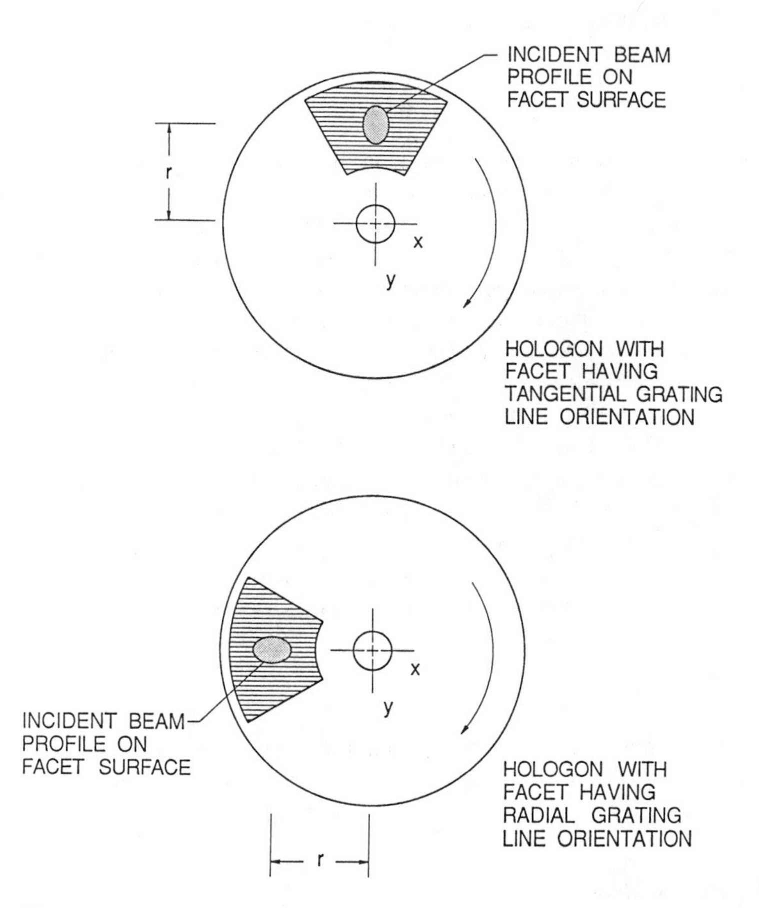

Figure 5.13 Facets on a disk hologon deflector can have either tangentially oriented grating lines or radially oriented grating lines. Beam profile on the radial grating line facet surface has been anamorphically compressed in y direction in order to improve scan duty cycle.

The top hologon in this figure has tangential orientated grating lines while the grating lines in the bottom hologon are rotated 90° from the tangential case and are termed radial. There is no basic difference in the scanning properties for these two facet grating line orientations. For both grating line orientations, the incident beam lies in the yz plane and the beam scans along the x axis. The scan line is symmetric with regard to the disk rotation axis for the tangential case, while the central scan position is offset by the mean facet radius r for the radial case.

With $\theta_i \neq 0$, the incident beam profile on the disk surface is usually an ellipse with the major axis parallel to the y axis, as depicted for the top hologon in Figure 5.13. The diameter of this major axis, A_b', is

$$A_b' = \frac{A_b}{\cos \theta_i} \tag{35}$$

where A_b is the $1/e^2$ intensity beam diameter in the cross scan direction for the incident beam. For the usually elliptical profile case associated with a circular cross section incident beam, the tangential grating line orientation provides better scan duty cycle than the radial orientation. This improved scan duty cycle performance results because A_b' is perpendicular to the facet scan direction for tangential orientation and parallel to the scan direction for radial orientation.

Considerable improvement in scan duty cycle can be achieved for the radial grating orientation case when anamorphic beam compression in the cross scan direction is utilized [13,27,28] since the compression occurs in the direction of the facet rotation. The incident beam profile on the bottom hologon in Figure 5.13 illustrates the magnitude of anamorphic beam compression that can be achieved in the cross scan direction when prism elements are added to the deflector system, as illustrated in Figure 5.32.

The optimum number of facets for a hologon with tangential grating line orientation can be calculated from equations (8) and (9) by setting $U_m = U_a$:

$$N = \frac{\pi r S M}{0.825\ \lambda M f + r S\ \theta_s} \tag{36}$$

It should be kept in mind when solving equation (36) that N must be an integer, and, therefore, other system parameters may have to be modified to achieve an optimized system design.

5.5.4 Scan-line Bow

This subsection analyzes the relationship between scan-line bow, incident beam angle, scan-beam angle, and scan system duty cycle for a PDG hologon disk deflector system.

Scan-Line Length

When a PDG disk deflector system does not have a focusing lens after the deflector, the displacement X of the scan beam from the center of scan for a flat image plane is

$$X = Y' \tan \theta_s = \frac{Y'K \sin \theta_R}{(1 - K^2 \sin^2 \theta_R)^{1/2}} \tag{37}$$

where Y' is the distance from the scanning facet to the center of the scan image plane and equation (29) was used to express θ_s in terms of K and θ_R. If a f-θ scan lens is positioned after this PDG deflector, the scan-line length W for the preobjective deflector system is achieved by setting the lens focal length $f = Y'$, $\tan \theta_s = \theta_s = K\theta_R$, and $W = 2X$ in equation (37):

$$W = 2f\theta_s = 2fK\,\theta_R \tag{38}$$

This equation applies to the case where the f-θ scan lens corrects for the slight change in M_d as a function of θ_R, as calculated by equation (29).

Scan-Line Bow as a Function of Incidence Angle

The deviation from straightness, ΔZ, of the scan-line generated with a PDG disk deflector system is calculated for each point in the scan field by [29].

$$\Delta Z = \frac{Y' \cos^2 \theta_{do}}{\cos \theta_d} [K \cos \theta_R - \sin \theta_i - \cos \theta_d \tan \theta_{do}] \tag{39}$$

where θ_{do} and θ_d are calculated with equation (26). It was assumed in the derivation of this equation that the beam for the center of scan field position is perpendicular to the flat image plane surface. As predicted by equations (16) and (39), there is no scan-line bow for a hologon deflector having $\theta_i = 0$ and $\theta_d = \theta_{do} = 90°$. Calculation of the scan-line bow value for a hologon deflector having $\theta_i = 0$ and $\theta_{do} < 90°$ is less when equation (39) is used as compared with the value achieved with equation (16). This difference in calculated bow values occurs because these equations are based on different image plane orientation conditions with respect to the beam for the center of scan.

Figure 5.14 is a pictorial illustration that shows how the shape for the residual scan-line bow curve changes as the incident angle is varied about the value predicted by equation (31). The curve labeled θ_i in this figure illustrates the scan-line shape when the PDG deflector satisfies equation (31). For the θ_i curve, the scan-line bow is essentially zero for small rotation angles and monotonically increases with rotation angle. For incident angles less than θ_i, the monotonic increase in scan-line bow is faster, as indicated by the θ_1 curve. For incident angles slightly larger than θ_i, the scan line develops three inflection points, as indicated by the curve labeled θ_2. Since the θ_2 curve first goes negative before

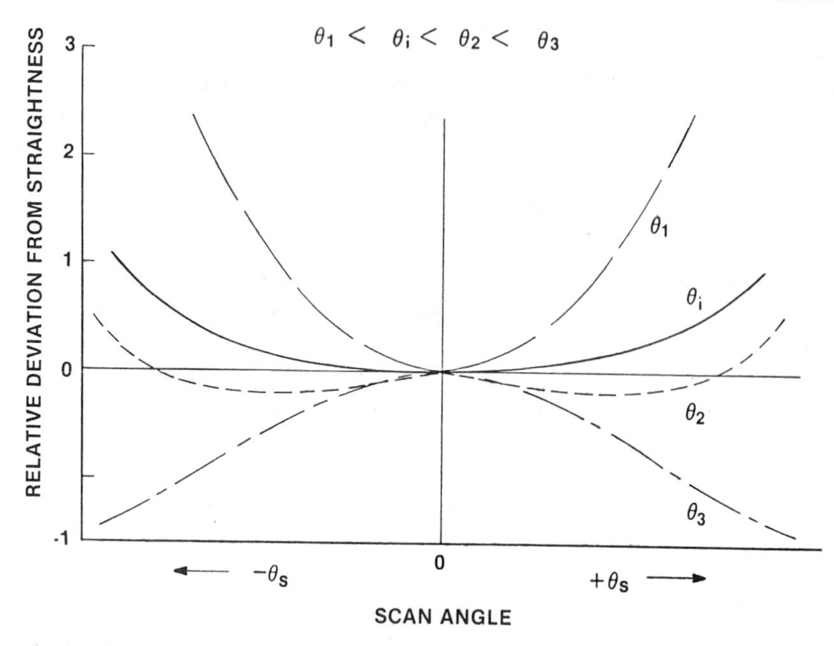

Figure 5.14 Pictorial illustration for the change in shape of scan-line bow as a function of incident beam angle.

turning around and monotonically increasing, its deviation from an ideal straight line can be made less than that for the θ_i curve. For incidence angles much larger than θ_i, the scan line begins to develop a fairly large negative bow, as indicated by curve θ_3.

To achieve the minimum bow in the scan line, the hologon deflector system would be set up to produce the θ_2 curve of Figure 5.14. Since the position of the inflection points of the θ_2 curve determine the maximum bow in the scan line, it is important to be able to predict the location of these points. Equations (31) and (32) can be generalized to account for the condition where the inflection points of the scan line occur at a position other than $\theta_R = 0$. These generalized equations for scan-line bow minimization are [23]

$$\sin\theta_i = \left(K - \frac{1}{K}\right)\sec\theta_R \tag{40}$$

$$\sin\theta_{do} = (1 - \sec\theta_R)\,K + \frac{1}{K} \tag{41}$$

When θ_i and θ_{do} satisfy these equations, the three scan-line bow inflection points occur at the scan field positions corresponding to $\theta_R = 0$, and $\pm\theta_R$.

Normally the hologon deflector system is set up so that the $\theta_R \neq 0$ inflection points for the scan-line bow curve occur at approximately ± 0.71 of the maximum scan-angle positions used for image recording. This condition produces the minimum scan-line bow band corresponding to a scan-line curve that recrosses the reference straight scan line in Figure 5.14 for the end of scan recording position. When it is important to minimize the rate of change in the scan-line bow per unit length of scan, the $\theta_R \neq 0$ inflection points should occur near the end of the recording scan line. This is because the rate of change in the scan-line bow increases rapidly after these inflection points.

Scan-Line Bow as a Function of Scan Angle

Figure 5.15 illustrates the change in magnitude of residual scan-line bow as a function of scan angle for the Figure 5.12 preobjective hologon deflector system having $K = 1.4142$. The three curves in this figure correspond to the calculated scan-line bow for the following f-θ scanning conditions: $\theta_s = \pm 15°$, $\pm 20°$, and $\pm 25°$, for $f = 657$ mm, 493 mm and 394 mm, respectively, and $X = \pm 172$ mm. Only half of the scan-line bow curve is presented in Figure 5.15 because the bow curve is symmetric about the center of scan, as illustrated in Figure 5.14.

The three scan-line bow curves in Figure 5.15 correspond to the minimum scan-line bow band cases for their respective θ_s value. For each curve the $\theta_R \neq 0$ inflection point occurs at about the ± 0.71 scan-field position and the scan-

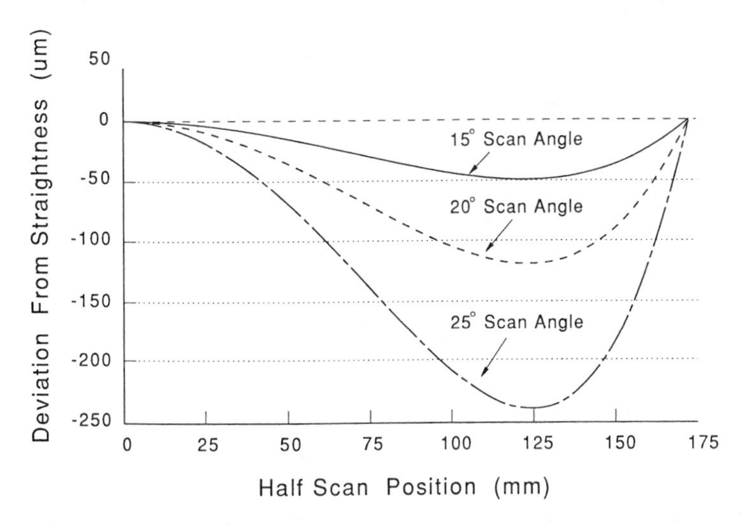

Figure 5.15 Calculated scan-line bow curves for three different scan-beam angles.

line curve recrosses the reference straight scan line (dashed line) at the end of scan. The θ_i values used for calculating these curves are 45.496°, 45.896°, and 46.425°, respectively, for θ_s = 15°, 20°, and 25°. Residual scan-line bow as a fraction of scan-line length for these three θ_s cases are 0.014%, 0.034%, and 0.069%, respectively. The rate of change in the scan-line bow slope for these three θ_s cases is greatest between 0.9 and 1.0 scan-field positions and are, respectively, 0.17%, 0.44%, and 0.9%. It is evident from these three θ_s cases that the scan angle must be kept less than 20° in order to meet the requirements in Table 5.1 for noncomposite images. A scan angle of less than 9° is required to meet the Table 5.1 requirement for composite images.

Minimum Scan-Line Bow versus System Duty Cycle

Figure 5.16 illustrates the calculated scan-line bow curve for a f-θ hologon deflector system having K = 1.4142, θ_i = 45.72°, θ_R = 12.625°, θ_s = 18°, f = 545 mm, and X = 172 mm. These hologon parameters provide the minimum scan-line bow band for this deflector geometry and θ_s value. Total scan-line bow band for this system is 86 μm, which corresponds to 0.024% of the scan-line length and a maximum rate of change of 0.31%. This bow band value is 123 times smaller than what was achieved for the Cindrich-type hologon deflector system when θ_i = 70°, Y' = 545 mm, and X = 170 mm.

It is evident from comparing the Figure 5.16 scan-line bow example with the three scan-line bow examples in Figure 5.15, that the scan-line bow band for

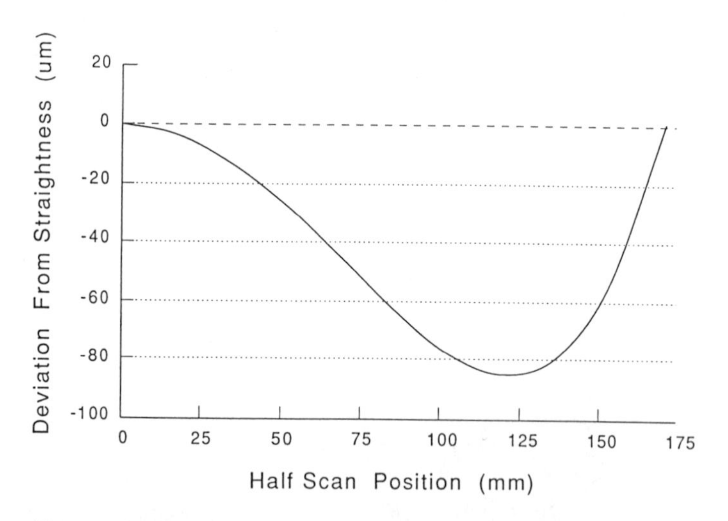

Figure 5.16 Calculated scan-line bow curve for a hologon deflector system having K = 1.4142, θ_i = 45.72°, θ_s = 18°, and f = 545 mm.

the basic PDG hologon disk deflector increases rapidly for scan angles greater than 18°. Equation (9) states that scan system duty cycle is maximized by increasing θ_s. Therefore, $\theta_s \simeq 18°$ would appear to provide the best trade-off between scan-line bow band and scan system duty cycle for the PDG disk deflector configuration.

If the Figure 5.12 hologon deflector system is designed having $M = K = 1.4142$, $r = 40$ mm, $s = 43$ μm, $\lambda = 0.6328$ μm, $f = 545$ mm, and $\theta_s = 18°$ for the end of scan position, then equation (36) states that $N = 8$ provides an optimum system duty cycle solution. These parameters correspond to a system that utilized a 100-mm-diameter hologon disk and was developed by Holotek for phototypesetting and photocomposition imaging applications.

5.5.5 Sources of Scan-Beam Positional Error

This subsection examines the deflector parameters that contribute to cross scan and in-scan positional beam errors for a PDG hologon disk deflector system.

Deflector-Wobble-Induced Scan-Beam Tracking Error

Figure 5.17 presents an exaggerated view of a hologon deflector having a fixed wobble tilt angle $\Delta\phi$ with regard to the deflector rotation axis. As depicted in this figure, fixed deflector wobble produces a periodic change in the magnitude

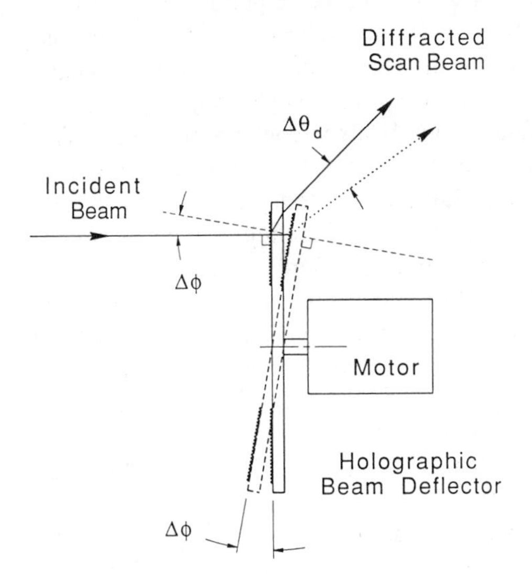

Figure 5.17 Exaggerated view showing a hologon deflector wobbling through an angle $\Delta\phi$ and the resulting change in angle of the diffracted beam.

of the cross scan incident beam angle that is equal to $\pm\Delta\phi$. This change in θ_i produces a corresponding variation in diffracted beam angle with respect to the deflector element. Due to wobble, the deflector is, in turn, changing its angular orientation with respect to the fixed coordinate system associated with the image plane. To account for the change in the deflector coordinates due to wobble, equation (20) can be revised in terms of the fixed image plane coordinates by setting $\theta_i = \theta_i + \phi$ and $\theta_{do} = \theta_d - \phi$, where ϕ is the component of $\Delta\phi$ in the cross scan direction. Differentiating the revised equation (20) with respect to $d\phi$ and setting $d\theta_{z1} = d\theta_d$ give the change in the cross scan beam angle, $d\theta_{z1}$, at the image plane associated with $d\phi$ [25]:

$$d\theta_{z1} = \pm\left[1 \mp \frac{\cos(\theta_i + \phi)}{\cos(\theta_d \mp \phi)}\right] d\phi \tag{42}$$

where the upper signs apply to transmission gratings, the lower signs to reflection gratings, and values for θ_d are calculated using equation (26). It is evident from this equation that the output beam angle of a transmission grating deflector is self-compensated with respect to deflector wobble when $\theta_i = \theta_d$. This self-compensation phenomenon is known as the minimum deviation condition, which is usually associated with light propagating through a prism. Minimum scan-line bow and the minimum deviation condition are simultaneously achieved for the Figure 5.12 deflector geometry when $K = 1.4142$, corresponding to $\theta_i \simeq \theta_d \simeq 45°$. As expected, $d\theta_{z1} = 2d\phi$ when $\theta_i = \theta_d$ for a reflective deflector element.

Theoretical and experimental data for the change in cross scan beam angle, $d\theta_{z1}$, as a function of the change in hologon cross scan tilt angle, $d\phi$, are graphically presented in Figure 5.18 for two transmission hologon configurations [25]. This data is for the center of scan position. Experimental data for a hologon having $\theta_i = \theta_{do} = 45°$ are represented in this figure by the triangles, while data for a Cindrich-type hologon having $\theta_i = 0$ and $\theta_{do} = 60°$ are represented by circles. Theoretical curves for these two hologon geometries were calculated using equation (42) and are respectively labeled in this figure. It is evident from this figure that equation (42) accurately predicts the scan-beam tracking error for small tilt angles of the hologon deflector and that scan-beam tracking error is essentially insensitive with regard to deflector wobble when the hologon is operated at the minimum deviation condition. By comparison, there is a one-to-one relationship between deflector wobble and scan-beam tracking error for the Cindrich-type hologon having $\theta_i = 0$ and $\theta_{do} = 60°$.

Having $\theta_i = \theta_{do} = 45°$ would make the scan-beam tracking error essentially totally insensitive with regard to deflector wobble at the center of scan. However, this angle condition does not produce the minimum scan-line bow band and significantly increases scan-beam tracking error sensitivity to deflector wobble at the end of scan. Scan-beam tracking error sensitivity to deflector wobble is

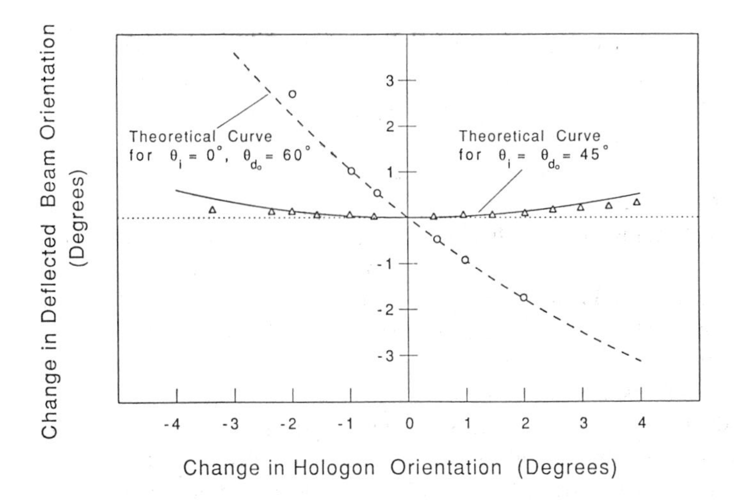

Figure 5.18 Comparison of experimental and theoretical data for the change in the scan-beam angle as a function of hologon wobble angle.

minimized for the entire scan-line length when the absolute difference between θ_i and θ_d is the same for the center and end-of-scan positions. Using equation (26) to calculate the value of θ_d for both the center- and end-of-scan positions for the hologon deflector parameters used to generate the scan-line bow curve in Figure 5.16, $K = 1.4142$ and $\theta_i = 45.72°$, shows that θ_d is 44.298° and 47.11°, respectively, for $\theta_R = 0$ and $\theta_R = 12.625°$. These values correspond to an absolute difference between θ_i and θ_d of approximately 1.41° for both the center- and end-of-scan positions. Therefore, equation (40) can be used to calculate the θ_i value that minimizes scan-beam tracking error sensitivity to deflector wobble for the entire scan-line length. This is because the minimum scan-beam tracking error condition is achieved when the $\theta_R \neq 0$ inflection points for the scan-line bow curve occur at the ± 0.71 scan-field positions. For this minimum scan-line bow condition, the maximum scan tracking error due to deflector wobble occurs at the center- and end-of-scan positions while this error is essentially totally insensitive with regard to deflector wobble at the ± 0.71 scan-field positions.

For the Figure 5.16 hologon deflector system parameters, $d\theta_{z1} = d\phi/41$ for the center- and end-of-scan positions, which means that every minute of deflector wobble translates into 1.46 arc seconds of scan-beam tracking error. The total deflector wobble induced scan-beam tracking error band is twice the fixed deflector tilt angle due to the positive and negative swings of the scan beam caused by the corresponding motion of the deflector element. Deflector wobble is usually specified in terms of a $\pm$ value to account for the swing. The scan-beam tracking error of the hologon deflector for these system parameters is 82 times less sensitive to deflector wobble than a polygonal mirror deflector. It is on the basis of this comparison that the hologon scan-beam tracking error is stated to be insensitive with regard to deflector wobble.

It is evident from preceding analysis that scan-beam tracking error sensitivity to deflector wobble, like the scan-line bow band, increases proportionally with θ_s for the basic PDG hologon disk deflector. For the three sets of hologon deflector system parameters used to calculate the minimum scan-line bow curves in Figure 5.15, $d\theta_{z1} = d\phi/58$, $d\phi/33$, and $d\phi/21$, respectively, for $\theta_s = 15°$, $20°$, and $25°$. It does appear that $\theta_s \simeq 18°$ provides the best system solution for this deflector geometry with regard to scan-line bow, scan system duty cycle, and scan tracking error sensitivity with respect to deflector wobble.

A later subsection describes techniques for reducing the fixed deflector wobble angle to 30 arc seconds or less. A fixed 30-arc-second deflector wobble angle causes a periodic scan-beam tracking error of only 1.46 arc seconds between all the facet scans of a PDG hologon disk deflector having $K = 1.4142$, $\theta_i = 45.72°$, and $\theta_s = \pm 18°$. Deflector wobble causes a maximum change in the scan-beam tracking error performance between adjacent facet scans that is equal to the product of the fixed deflector wobble angle times $\sin(\pi/N)$. Therefore, a fixed deflector wobble angle of 30 arc seconds for these deflector system parameters would cause a maximum periodic scan-beam tracking error between adjacent facets of only 0.37 arc second for a six facet deflector. As expected, there are no scan-to-scan performance errors associated with fixed deflector wobble or other fixed deflector deficiencies for the monofacet case of $N = 1$.

Later, this section describes techniques that can be used in a hologon deflector system to compensate for the periodic scan-beam tracking errors that remain constant with time. Vibrations in the rotor/bearing assembly of a ball-bearing deflector motor are expected to cause less than 3 arc seconds of random change in the fixed deflector wobble tilt angle. This value of random change in the deflector wobble angle causes only a 0.073-arc-second change in the scan-beam tracking error performance of a PDG disk deflector having $K = 1.4142$, $\theta_i = 45.72°$, and $\theta_s = \pm 18°$. This change in performance cannot be detected, and, therefore, it can be assumed that deflector wobble induced scan-beam tracking error for this hologon deflector geometry remains constant with time.

Deflector-Wobble-Induced Scan Jitter

Although not illustrated, the fixed deflector wobble tilt angle in Figure 5.17 causes a periodic change in the in-scan incident beam angle that is equal to $\pm \Delta\phi$. This change in θ_{ix} contributes to the periodic scan jitter error of the hologon disk deflector system. To account for the change in the deflector coordinates due to wobble, equation (21) can be revised in terms of the fixed image plane coordinates by setting $\theta_{ix} = \phi'$ and $\theta_{dx} = \theta_s - \phi'$, where ϕ' is the component of $\Delta\phi$ in the in-scan direction. Differentiating the revised equation (21) with respect to $d\phi'$ gives the change in the scan-beam angle, $d\theta_{s1}$, at the image plane associated with $d\phi'$:

$$\begin{aligned} d\theta_{s1} &= \pm\left[1 \mp \frac{\cos(\phi')}{\cos(\theta_s \mp \phi')}\right] d\phi' \\ &\simeq \pm\left[1 \mp \frac{1}{\cos(K\theta_R)}\right] d\phi' \end{aligned} \tag{43}$$

where the upper signs apply to transmission gratings and the lower to reflection gratings. It is evident from this equation that deflector-wobble-induced scan jitter for a transmission PDG disk deflector is zero for the center scan position and increases monotonically with θ_s. This periodic component of scan jitter is not symmetric about the center of scan since $d\phi'$ changes as a rate equal to $\sin\theta_R$.

Because this component of scan jitter is a function of θ_R, its appearance is not completely eliminated from recorded images generated by a hologon disk deflector system when only a start-of-scan detector signal is utilized to synchronize the start of scan-line pixel data. Consequently, deflector wobble for this pixel synchronization method must be minimized to reduce the appearance of this source of scan jitter. The appearance of this source of scan jitter is eliminated from recorded images when a pixel grating clock system is used to sample the scan-beam position at many points across the scan field.

Equation (43) predicts that a fixed deflector wobble angle of 30 arc seconds causes a total periodic change in θ_s of only 3 arc seconds for the start- and end-of-scan positions of a hologon deflector having $K = 1.4142$, $\theta_i = 45.72°$, and $\theta_s = \pm 18°$. This change in θ_s accounts for the positive and negative swings of the scan beam caused by the corresponding motion of the deflector element. A start-of-scan detector system eliminates the appearance of scan jitter from the start-of-scan image position and reduces the total value for this component of scan jitter to a range of approximately 2 arc seconds and 1 arc second, respectively, for the center- and end-of-scan image positions. This scan-beam angular error corresponds to a ± 2.66-μm in-scan positional beam error for the center-of-scan image position when the deflector system is used with a f-θ scan lens having a focal length of 550 mm. This positional error is acceptable for most high-resolution graphic arts imaging applications.

Techniques Used to Minimize Deflector Wobble

Minimizing deflector wobble is an important goal in any deflector system. Wobble minimization can be achieved in a hologon disk deflector by allowing the rotating disk to align itself perpendicular to the deflector rotor axis by means of the centrifugal force acting on it. This self-alignment was initially accomplished by flexibly coupling the deflector element to the drive rotor by means of an elastomer

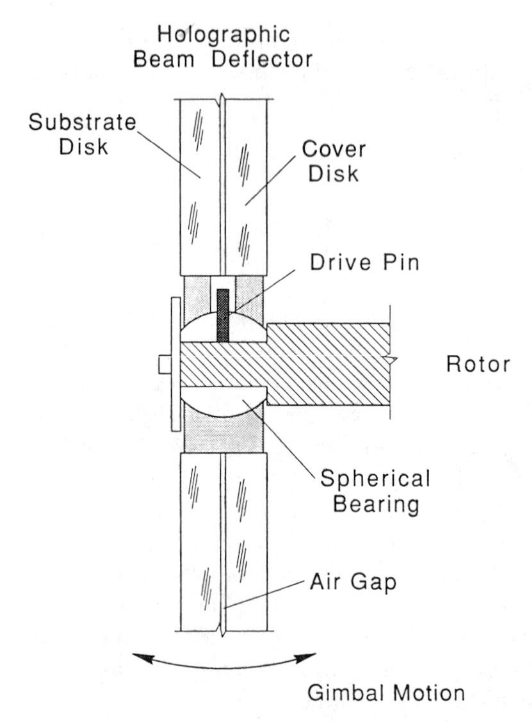

Figure 5.19 Dynamic mounting assembly that allows the hologon disk to align itself perpendicular to the rotor axis by means of the centrifugal force acting on it.

such as rubber [30]. Due to resistance within the elastomer, these flexible couplers only allow the deflector to approach the perpendicular position asymptotically without ever reaching it.

Figure 5.19 illustrates a dynamic coupling assembly that employs a spherical bearing to link the deflector element to the rotor [31]. The spherical curvature of the outer raceway permits the outer part of the spherical bearing to rotate freely about the center of the bearing in any direction. Angular drive is coupled to the deflector element by a drive pin since the internal friction of the bearing is low. Mating the pin to the outer part of the assembly by a slot ensures that the deflector assembly can pivot freely in the direction indicated by the double-headed arrow labeled ''gimbal motion.'' Performance of the bearing assembly mount is determined primarily by the dynamic balance of the deflector element because resistance is usually not a problem. Typical performance for a dynamic mount is 10 to 30 arc seconds of fixed deflector wobble angle.

Fixed deflector wobble angles as low as 10 to 30 arc seconds are routinely achieved for hologon disks that are rigidly hard mounted to ball-bearing-supported rotors. This is accomplished by machining a precision surface on a rigid mounting hub after the hub is incorporated into the rotor ball-bearing assembly. The hologon is tightly held against the precision machined hub surface by a spring washer assembly.

Reflective Hologon Systems That Are Insensitive to Deflector Wobble

Unfortunately, there is no minimum deviation condition for reflection gratings. It is evident from equation (42) that $d\theta_{z1} \geq d\phi$ for a reflection hologon deflector. Schematically illustrated in Figure 5.20 is an optical arrangement that can be utilized to make the scan-beam tracking error of a reflection hologon deflector insensitive with regard to deflector wobble [32]. This arrangement compensates for deflector wobble by means of a passive optical feedback technique.

The incident beam in Figure 5.20 is initially reflected from the deflector surface at an area which is adjacent to the active scanning facet. This adjacent area is part of a flat mirror band extending around the deflector surface. The reflected beam from this mirror band is stationary with respect to deflector rotation except for changes in its angular orientation caused by corresponding change in the orientation of the deflector wobble tilt angle. This reflected beam propagates to a corner cube reflector. This reflector is also known as a triple mirror reflector and has the property that any ray entering the reflector will be retroreflected in a direction that is parallel to the entering ray propagation direction. The beam entering the corner cube is thus inverted, displaced, and retrodirected as an incident beam onto the active scanning facet. After diffraction from the facet, the scan beam is reflected by the fold mirror and focused by the f-θ scan lens to a scanning spot at the image plane.

It is evident from Figure 5.20 that any deviation in the angle of the beam reflected from the deflector mirror band due to deflector wobble will be present

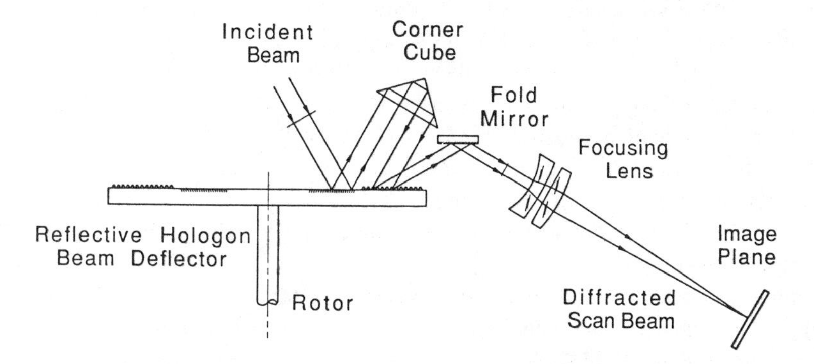

Figure 5.20 Passive optical feedback technique that makes scan-beam tracking error of a reflection hologon insensitive with regard to deflector wobble.

in the retrodirected beam incident on the facet. Incorporating information about the deflector surface tilt angle into the beam incident on the facet enables the final scan beam to be corrected for changes in this tilt angle. When a feedback device such as the one depicted in Figure 5.20 is utilized with a reflective hologon, equation (42) can be rewritten in the following form to include the effect of the feedback on the retrodirected incident beam [32]:

$$d\theta_{z1} = -\left[1 + (1 - 2M_f)\frac{\cos(\theta_i + \phi)}{\cos(\theta_d + \phi)}\right] d\phi \tag{44}$$

where the $2M_f$ term represents the multiplier factor introduced by the feedback process. The value of M_f depends on the feedback device used and is zero when no feedback mechanism is present.

When a corner cube is used as the feedback mechanism, $M_f = 1$, and the scan-beam tracking error for the reflective hologon deflector system is totally insensitive to deflector wobble for only the case where $\theta_i = \theta_d$. In the hologon system in Figure 5.20, θ_i is depicted as being equal to about 30° while θ_d is depicted as being about 50°. When these values are substituted into equation (44) for $M_f = 1$, $d\theta_{z1} = -d\phi/2.88$. This feedback arrangement has reduced the sensitivity of scan-beam tracking error to deflector wobble by 5.76 times, when compared with the same system without feedback. This corner cube feedback arrangement totally compensates for wobble-induced scan jitter for all scan angles.

In order for the feedback arrangement to make the scan-beam tracking error totally insensitive with respect to deflector wobble, the following condition must be met:

$$M_f = 0.5\left[1 + \frac{\cos\theta_d}{\cos\theta_i}\right] \tag{45}$$

Typical M_f would be chosen to satisfy this condition for the θ_d value corresponding to the 0.71 scan-field position, and thereby, minimize the scan-beam tracking error sensitivity to deflector wobble for the entire scan-line length. This condition can be achieved for the general case of $\theta_i \neq \theta_d$ by altering the simple corner cube feedback arrangement illustrated in Figure 5.20 by the attachment of a 30°–60°–90° prism to the entrance area of the corner cube [32]. The attached prism on the corner cube alters the ratio between the change in angle of the beam exiting the corner cube arrangement to the change in angle of the beam entering the corner cube arrangement.

A similar feedback arrangement that is based on a 90° roof mirror reflector assembly has been used to compensate for the scan beam tracking error of a polygonal mirror deflector [33]. Also, a reflective feedback arrangement has been proposed [34] for correcting the scan-line bow from a transmission Cindrich-type hologon deflector.

Scan-Beam Positional Error Due to Grating Periodicity Error

Variation in the facet-to-facet and/or intrafacet grating periodicity contributes to both the scan-beam tracking error and scan jitter error of a PDG deflector system. These variations in facet periodicity occur during the manufacturing of the hologon element and remain constant with time, and, therefore, they cause periodic scan-beam positional errors that repeat at the deflector rotational rate. Facet-to-facet periodicity error is similar to pyramidal alignment error in the mirror facets of a polygon deflector while intrafacet periodicity error is similar to mirror facet nonflatness. If dD is the variation in facet-to-facet and/or intrafacet periodicity, the change in scan-beam tracking angle, $d\theta_{z2}$, due to grating periodicity error is obtained by differentiating equation (20) with respect to dD and setting $d\theta_{do} = d\theta_{z2}$:

$$d\theta_{z2} = \frac{-K^2\, dD}{\lambda \cos \theta_{do}} \tag{46}$$

No significant accuracy is lost by using θ_{do} instead of θ_d in equation (46) since this component of system scan beam tracking error is fairly insensitive to slight changes in θ_d. The change in the scan-beam angle, $d\theta_{s2}$, due to dD is obtained by differentiating $\theta_s \simeq K\, \theta_R$ with regard to dD:

$$d\theta_{s2} = \frac{-K^2\, \theta_R}{\lambda}\, dD \tag{47}$$

To limit the facet grating periodicity induced level of scan-beam tracking error to less than 1 arc second for a hologon deflector having $K = 1.4142$, $\lambda = 0.6328$ μm and $\theta_i \simeq \theta_d \simeq 45°$, $dD \leq 10^{-6}$ μm. This stringent periodicity requirement is achieved by holographically recording the PDG facets in photoresist. Section 5.5.6 describes the diffraction efficiency properties of gratings formed in photoresist. The scan-beam angle error for these hologon deflector parameters and $dD = 10^{-6}$ μm is $d\theta_{s2} = -3 \times 10^{-6}\, \theta_R$, which corresponds to -0.136 arc second for $\theta_R = 12.62°$. This error in scan-beam angle has negligible effect on the performance of the hologon deflector system when used for high-resolution imaging applications.

Substrate Wedge-Induced Scan-Beam Positional Error

Wedge angle in the deflector substrate of a transmission hologon causes the same type of periodic cross scan and in-scan beam positional errors that are associated with fixed deflector wobble angle. The change in scan-beam tracking angle due to substrate wedge angle is [13,23]

$$d\theta_{z3} = n\left[1 - \frac{\cos\theta_1}{\cos\theta_2}\right]\frac{\cos\theta_{do}}{\cos\theta_2}\, d\phi_1 + [n\cos\theta_1 - \cos\theta_i]\,\frac{d\phi_2}{\cos\theta_{do}} \tag{48a}$$

where θ_i and θ_{do} are, respectively, the incident and diffracted beam angles measured in air, θ_1 and θ_2 are the incident and diffracted beam angles measured in the substrate medium of refractive index n, and $d\phi_1$ and $d\phi_2$ are the wedge angles by which the first and second surfaces of the hologon deflector substrate deviate from their ideal parallel position. As was the case for equation (46), no significant accuracy is lost by using only the center of scan values for θ_d and θ_2 in equation (48a) since this component of system scan-beam tracking error is also fairly insensitive to small changes in these diffraction angle values.

The change in the scan-beam angle $d\theta_{s3}$ due to the substrate wedge angle, $d\phi_w = d\phi_1 + d\phi_2$, is obtained by differentiating equation (21) with respect to $d\phi_w$ and setting $d\theta_{ix} = d\phi_w/n$:

$$d\theta_{s3} = \frac{d\phi_w}{n \cos(K\,\theta_R)} \tag{48b}$$

As is the case for the scan jitter error associated with deflector wobble and grating periodicity error, this component of periodic scan jitter is also a function of θ_R. Therefore, variations in scan jitter error associated with substrate wedge cannot be completely corrected for by the start-of-scan detector method.

When $\theta_i \simeq \theta_d$, $d\theta_{z3}$ and $d\theta_{s3}$ are both approximately equal to the substrate wedge angle. When the wedge angle has a uniform prismatic form across the substrate, $d\theta_{z3}$ and $d\theta_{s3}$ have a total error band for all facet scans that is approximately twice the wedge angle due to the positive and negative swings of the scan beam caused by the corresponding change in orientation of the prismatic wedge angle. Normally, the wedge in hologon disks is associated with a difference in the radius of curvature between the substrate surfaces. This curvature difference for hologon substrates used for high-resolution imaging applications is essentially symmetrically centered on the disk axis of rotation and has an effective radius of approximately 2 *km*. There are no scan-beam positional errors caused by the substrate wedge when it has rotational symmetry with regard to the deflector rotation axis.

Scan-beam tracking error and scan jitter error associated with residual asymmetry in the substrate wedge angle in precision manufactured hologon disks is typically < 1 arc second between all facet scans when $\theta_i \simeq \theta_d$. If it is assumed that this residual wedge error is uniformly angularly distributed around the disk, then $d\phi_w$ changes at a rate equal to $\sin\theta_R$. A wedge-induced scan jitter error of 1 arc second between all facet scans can be reduced by means of a start-of-scan detector system to no error for the start-of-scan position and about 0.2 arc second for the end-of-scan position when $\theta_s = \pm 18°$. This level of scan jitter error has negligible effect on the performance of a hologon system used for high-resolution imaging.

Active Technique for Reducing Scan-Beam Tracking Error

Periodic scan-beam tracking error due to hologon facet grating periodicity error and substrate wedge are fixed with respect to time. The maximum expected change in scan-beam tracking error performance due to either deflector vibrations and/or a change in the angle of precession of the deflector rotor is only ± 0.08 arc second for a hologon having $K = 1.4142$, $\theta_i = 45.72°$, and $\theta_s = \pm 18°$. It can therefore be assumed that periodic scan-beam tracking error associated with fixed deflector wobble tilt angle is also fixed with regard to time for this hologon disk deflector geometry. Being constant with time means that the scan-beam tracking error associated with these three hologon deflector deficiencies need only be measured once to generate corresponding correction values for the combined error. This correction data can be stored as digital values in a ROM chip.

The scan-beam tracking error for each scan-field position can be corrected by changing the incident cross scan beam angle into the hologon deflector by an amount that is equal to error. This change in incident angle can be accomplished by means of either an acoustooptic-based minideflector, as shown in Figure 5.50, or a piezo controlled mirror minideflector, as shown in Figure 5.44. An acoustooptic minideflector is preferable to the piezo controlled mirror because the piezo stacks used to change the mirror angular orientation suffer from changes in drive sensitivity with temperature variation and age. The change in beam angle for the acoustooptic minideflector is controlled by the change in frequency of the *RF* acoustic carrier which is altered by means of a voltage-controlled oscillator. This type of acoustooptic minideflector is, in essence, a digital device which is essentially unaffected by temperature variations and aging.

Scan-beam tracking error is usually fairly constant across the scan field for each facet of a PDG hologon deflector operated at the minimum deviation condition. Therefore, adequate scan-beam tracking error correction for all scan-field positions is usually achieved by only correcting the scan tracking error variation between facets for the center-of-scan position. Residual scan-beam tracking error between all facet scans for this single scan-field position correction method is typically <0.75 arc second. This error can be reduced to <0.25 arc second by using multi-scan-field correction points.

Wavelength-Shift-Induced Scan-Beam Tracking Error

The major disadvantage with the hologon laser beam deflector technology is that it must be utilized with a very monochromatic light source. This normally does not present a problem when gas laser sources are used. Amplitude modulating the output intensity of a diode laser causes a temperature variation in the junction region which produces a time fluctuation in the index of refraction of this region. This refractive index change alters the optical path length of the laser cavity

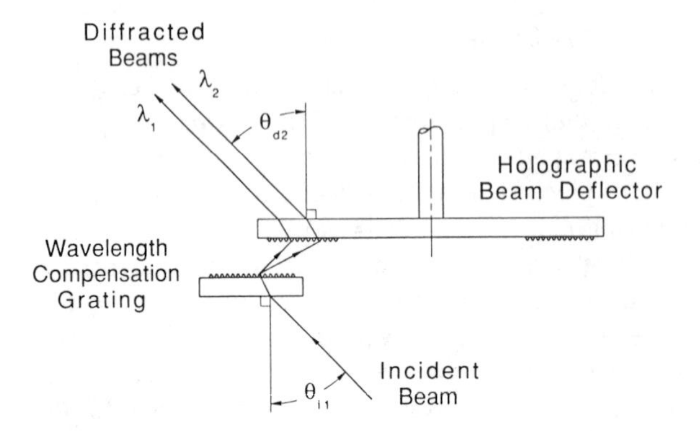

Figure 5.21 Passive optical technique that significantly reduces the scan-beam tracking error of a hologon with respect to wavelength shift.

which causes the wavelength of the lasing mode to change (hop) by approximately 0.2 *nm*. Once initiated, these mode hops take approximately 1 *ns* to complete, and therefore, cause abrupt, unacceptable cross scan and in-scan positional beam errors in hologon deflector systems utilized for graphic arts imaging applications.

Kay [35] showed that wavelength-shift-induced scan-beam tracking errors in hologon deflectors could be essentially compensated for by utilizing a stationary grating having the same periodicity as the hologon, as illustrated in Figure 5.21. For illustrated purposes it will be assumed that the incident beam in Figure 5.21 is composed of two laser wavelengths, λ_1 and λ_2, and that $\lambda_2 > \lambda_1$. Under these conditions, the wavelength compensation grating separates the incident beam into two diffracted beams having different angles with regard to its surface normal. When the wavelength compensation grating is parallel to the hologon facet surface, as depicted in Figure 5.21, the two diffracted beams have incident angles with respect to the hologon surface normal that are equal to the diffraction angles they have with regard to the compensation grating surface normal. It is evident that if the compensation grating and hologon have identical grating periodicity and are aligned parallel, that beams λ_1 and λ_2 in Figure 5.21 leave the hologon with identical diffraction angles no matter what their wavelength difference. These beams are therefore focused to a common image spot by a scan lens following the hologon. This compensation technique is similar in principal to the optical feedback technique used to correct for deflector wobble in a reflective hologon system.

The major technical problem associated with the Figure 5.21 wavelength compensation approach is that only for the center-of-scan position does it completely correct for wavelength-shift-induced scan-beam positional errors. That is, when the hologon facet grating lines are parallel to the grating lines of the wavelength compensation grating. When the hologon rotates from this parallel grating line position, its relative grating periodicity, D_x and D_y, no longer matches that of the wavelength compensation grating, and therefore, wavelength shift introduces both cross scan and in-scan beam positional errors.

In order to calculate what the scan beam tracking error is as a function of wavelength shift, the grating equation is differentiated with respect to wavelength change. For the two grating case in Figure 5.21, the grating equation for each grating will be designated as

Wavelength compensation grating (first),

$$\sin \theta_{i1} + \sin\theta_{d1} = \frac{\lambda}{D_1} \tag{49}$$

Hologon facet grating (second),

$$\sin \theta_{i2} + \sin \theta_{d2} = \frac{\lambda}{D_2} \tag{50}$$

where θ_{in} and θ_{dn} are, respectively, the incident and diffracted beam angles, D_n is the grating periodicity, and the subscript n designates the grating (1 or 2) for which the variables apply. Differentiating equations (49) and (50) with respect to change in wavelength $d\lambda$ and setting $d\theta_{i2} = d\theta_{d1}$ gives the change in θ_{d2} with $d\lambda$ for the center of scan:

$$d\theta_{d2} = \frac{D_1 \cos \theta_{d1} - D_2 \cos \theta_{i2}}{D_1 D_2 \cos \theta_{d1} \cos \theta_{d2}} d\lambda \tag{51}$$

When the compensation grating is not present, D_1 becomes infinite, and equation (51) becomes

$$d\theta_{d2} = \frac{K_2 \, d\lambda}{\lambda \cos \theta_{d2}} \tag{52}$$

where $K_2 = \lambda/D_2$. When the wavelength compensation grating completely corrects for the wavelength shift, $d\theta_{d2} = 0$, and equation (51) becomes

$$D_1 \cos \theta_{d1} = D_2 \cos \theta_{i2} \tag{53}$$

If $D_1 = D_2$, then equation (53) is satisfied when $\theta_{d1} = \theta_{i2}$ and the two grating element surfaces are parallel. When $D_1 \neq D_2$, equation (53) can still be satisfied by having $\theta_{d1} \neq \theta_{i2}$.

When equation (53) is satisfied, the scan-beam tracking error is totally insensitive with regard to wavelength shift at the center of scan, but the error increases with scan angle. An optimally designed system would have the scan-beam tracking error sensitivity with regard to wavelength shift be the same for the center and end-of-scan positions. This is accomplished by minimizing this error for the ± 0.71 scan-field positions. One method for achieving this is to manufacture the wavelength compensation grating with the same absolute grating periodicity as the hologon. The angle of the compensation grating is adjusted in the cross scan direction with respect to the hologon surface so that its wavelength dispersion power matches that of the hologon when θ_R corresponds to the ± 0.71 scan-field positions.

To determine the relative wavelength dispersion of the hologon at the 0.71 scan-field position, the change in hologon diffraction angle as a function of both $d\lambda$ and hologon rotation angle must be calculated. Differentiating equation (26) with respect to $d\lambda$ and setting $d\theta_{i2} = d\theta_{d1}$ for the case of a wavelength compensation grating gives the change in θ_{d2} with both $d\lambda$ and θ_R for this case

$$\frac{d\theta_{d2}}{d\lambda} = \frac{D_1 \cos \theta_{d1}(K_2 - \sin \theta_{i2} \cos \theta_R) - D_2 \cos \theta_{i2}(K_2 \cos \theta_R - \sin \theta_{i2})}{D_1 D_2 \cos \theta_{d1} \sin \theta_{d2} \cos \theta_{d2}} \tag{54}$$

When the wavelength compensation grating completely corrects for the wavelength shift, $d\theta_{d2} = 0$, and

$$D_1 \cos \theta_{d1} = \frac{D_2 \cos \theta_{i2}(K_2 \cos \theta_R - \sin \theta_{i2})}{K_2 - \sin \theta_{i2} \cos \theta_R} \tag{55}$$

For the case where $D_1 = D_2$, equation (55) becomes

$$\cos \theta_{d1} = \frac{\cos \theta_{i2}(K \cos \theta_R - \sin \theta_{i2})}{K - \sin \theta_{i2} \cos \theta_R} \tag{56}$$

Equation (56) can be used to determine the cross scan tilt angle between the hologon and the wavelength compensation grating that will provide a match to the hologon relative periodicity for the ± 0.71 scan-field positions.

The following example illustrates the expected magnitude of the wavelength-shift-induced scan-beam tracking error that is achieved for a hologon deflector system that incorporates a wavelength compensation grating element. This example is based on a 0.2 *nm* wavelength shift associated with a mode hop in a diode laser light source having $\lambda = 670$ *nm*. The scan-beam tracking error due to this mode hop will be calculated for a preobjective hologon disk deflector system having $K = 1.4142$, $\theta_{i2} = 45.72°$, $\theta_R = 12.625°$ for the end-of-scan position, and $f = 545$ mm. Resulting scan-beam tracking error for the case of no compensation grating is 0.325 mm, as calculated from equation (52). When a compensation grating is utilized having $\theta_{d1} = \theta_{i2}$ and $D_1 = D_2$, the residual scan tracking error for the end-of-scan position is 23.7 *um*, as calculated from equation (54). When the compensation grating is tilted to provide a dispersive match to the hologon for the 0.71 scan position, the maximum scan-beam tracking error is reduced to 12.3 μm, as calculated from equation (51) for the center-of-scan position and $\theta_{d1} = 47.74°$.

Even though the residual scan-beam tracking error for the tilted compensation grating case is 26 times less than the uncompensated system, it is too large for most graphic arts imaging applications. Therefore, the diode laser source must either be controlled so that it does not experience mode hopping or the wavelength shift must be controlled from occurring very rapidly. If the wavelength shifts gradually with time, in comparison to the scan rate, both the cross scan and in-scan positional beam errors are distributed among many scan lines and, thereby become unnoticeable in the scan-generated images.

Wavelength-Shift-Induced Scan Jitter

Unfortunately, none of the stationary wavelength compensation grating approaches can correct for the in-scan positional beam error associated with wavelength shift. This is because their dispersion direction is perpendicular to the in-scan beam direction. An accurate estimate of the magnitude of the wavelength-shift-induced scan jitter error is achieved by differentiating $\theta_s \simeq K\theta_R$ with regard to $d\lambda$:

$$d\theta_{s4} = \frac{K\theta_R}{\lambda}\, d\lambda \tag{57}$$

If $\lambda = 670\ nm$, $d\lambda = 0.2\ nm$, $K = 1.4142$, $f = 545$ mm, and the maximum rotation angle is $\theta_R = 12.72°$, the maximum in-scan positional beam error is 51 μm. This error is too large for most imaging applications, and therefore, the diode laser source must either be controlled so that it does not mode hop or the wavelength shifts must be controlled from occurring very rapidly.

It is interesting to note that for a hologon deflector possessing optical symmetry with regard to the deflector rotation angle, such as the Cindrich-type in Figure 5.5, wavelength shift does not produce any scan jitter error, only scan-beam tracking error. Unfortunately, when the compensation grating method is used to reduce the scan-beam tracking error for an optically symmetric hologon deflector, it introduces scan jitter error with wavelength shift. This problem does not occur if either a cylinder lens or a toroidal lens is utilized to correct for the wavelength-shift-induced scan-beam tracking error in an optically symmetric hologon deflector. These anamorphic optical elements introduce their own problems.

5.5.6 Hologon Radiometric Efficiency

Diffraction efficiency of a hologon deflector is dependent upon its K value, the relationship between the incident and diffracted beam angles, and the properties of the holographic recording medium. The primary factor that determines recording material selection for hologon fabrication is the relative stability and homogeneity of the material. The gratings constituting the hologon facets must be fabricated to very high standards of consistency in order to ensure that they do not introduce unwanted levels of scan-beam positional error (equations (46) and (47).

To limit the grating-induced levels of scan-beam tracking error to less than 1 arc second for a hologon disk deflector having $K = 1.4142$, the mean periodicity of the facet gratings must be constant to within approximately 10^{-12} m. Kramer [23,29] concluded that photoresist could best achieve this stringent periodicity requirement, since the latent holographic image remains intact during development, in contrast to most photoprocesses. The unmatched stability and fidelity of photoresist make it an ideal material for fabricating hologons and other diffraction grating components. Holotek has concluded that replication techniques cannot achieve the periodicity requirement for the facet gratings, and therefore, fabricates all hologons as master elements.

Gratings formed in photoresist are of the sinusoidal surface-relief type, as illustrated in Figure 5.22. For illustrative purposes the thickness of the photoresist layer in Figure 5.22 has been greatly exaggerated in comparison to the thickness of the substrate material. Typical thickness for the photoresist layer is about 1.5 to 2.0 μm while the substrate thickness is on the order of 4 mm. The indices of refraction of these two materials are typically fairly close, and are in the range of 1.5 to 1.70. A photomicrographic of a surface-relief grating in photoresist is shown in Figure 4.19 of Chapter 4.

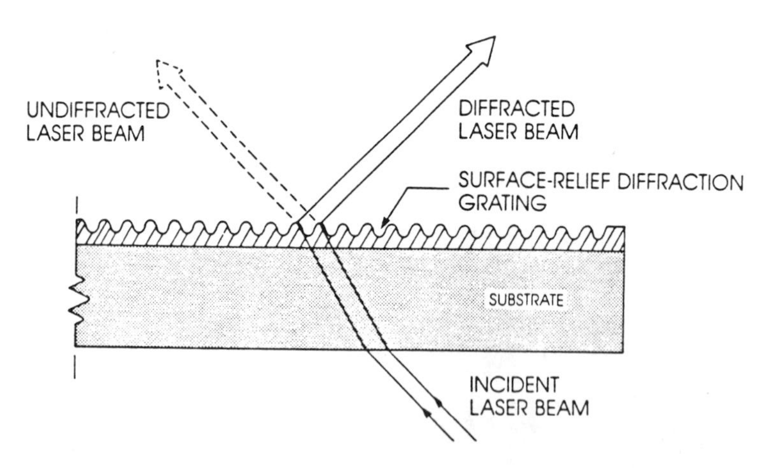

Figure 5.22 Exaggerated sectional side view of a surface-relief diffraction grating formed in photoresist.

Diffraction Efficiency of Surface-Relief Gratings

Having a large λ/D ratio is advantageous when working with surface-relief transmission gratings since diffraction occurs beyond the scalar regime, and therefore, first-order diffraction efficiency can approach 100% [23,36]. Greater than 90% absolute first-order diffraction efficiency is routinely achieved for surface-relief transmission gratings having λ/D ratios between 0.8 and 1.42 [28]. Figure 5.23 presents measured diffraction efficiency data for surface-relief transmission gratings formed in photoresist. This data is presented as a function of λ/D ratio and for the Littrow condition, that is $\theta_i = \theta_d$. The measured data points are shown on the graph as open circles for both polarization states, while the drawn curves represent the best fix to this data. *S*-polarized light has its electric field parallel to the grating lines.

In order to achieve the high diffraction efficiencies indicated in Figure 5.23, the gratings must have a deep groove profile shape. That is, the grating aspect ratio must be greater than about 1.5. This aspect ratio is defined by the grating

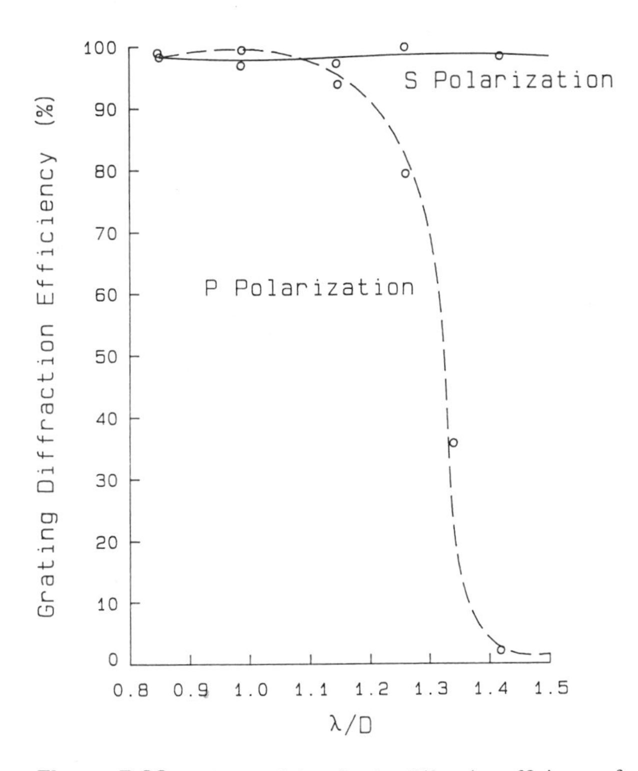

Figure 5.23 Measured data for the diffraction efficiency of surface-relief transmission gratings formed in photoresist.

peak-to-trough groove depth divided by the grating period. A theoretical model for the diffraction efficiency of surface-relief transmission gratings having a deep groove profile shape was presented by Maharam and Gaylord [37]. One problem associated with a 1.5 grating aspect ratio is that no one has been able to faithfully replicate these gratings. Diffraction efficiency curves for surface-relief reflection gratings are similar to those in Figure 5.23 [38]. Surface-relief reflection gratings only require a grating aspect ratio of about 0.35 to achieve 100% diffraction efficiency. Consequently, they can be replicated faithfully utilizing an epoxy casting method.

Radiometric Scan Uniformity

It is evident from the data in Figure 5.23 that when $K = 1.4142$, one can achieve essentially 100% diffraction efficiency for S polarized light while achieving only about 5% diffraction efficiency for P-polarized light. This diffraction efficiency sensitivity to polarization state reduces deflector radiometric efficiency for randomly and circularly polarized light applications. This diffraction efficiency property also produces scan beam intensity variation as a function of scan angle when a linearly polarized light source is used with the hologon deflector. The relationship between diffracted beam intensity, I, and hologon rotation angle is given by

$$I = I_s(E_p \sin^2 \theta_R + E_s \cos^2 \theta_R) + I_p(E_p \cos^2 \theta_R + E_s \sin^2 \theta_R) \quad (58)$$

where E_p and E_s are, respectively, the diffraction efficiencies for P- and S-polarized light, and I_p and I_s are the intensities of the P- and S-polarized light components of the incident beam, respectively.

An S-polarized laser beam is usually used for a $K = 1.4142$ hologon due to the very low diffraction efficiency of P-polarized light for this K value. When the Figure 5.23 data is used in equation (58) for $K = 1.4142$ ($E_s = 1.0$, $E_p = 0.05$), the calculated drop in scan-beam light intensity from the center to the end of scan is 4.6% for S-polarized light and $\theta_s = \pm 18°$. If P-polarized incident light is used, the scan-beam intensity increases from 5% to 9.5% for the center- and end-of-scan positions, respectively, when $\theta_s = \pm 18\%$. If a randomly or circularly polarized laser incident beam is used for this hologon K value case ($I_p = I_s = 0.5$), the scan-beam intensity stays constant over the total range of scan angles and $I = 52.5\%$.

These numbers indicate that for $K = 1.4142$, uniform intensity can be achieved across the scan line at the expense of 47.5% of the incident beam power, while high diffraction efficiency is achieved at the expense of a decrease in scan-line intensity from center to end of scan. It is not unusual to find that the facet diffraction efficiency uniformity varies slightly across the facet aperture. This variation usually results in a gradual decrease (3–5%) in the diffraction efficiency

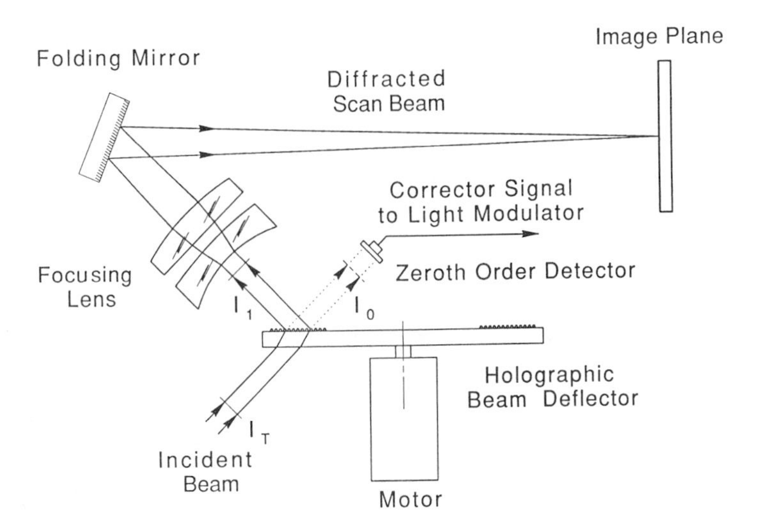

Figure 5.24 Active electrooptic feedback technique that can be used to improve the facet-to-facet and intrafacet radiometric uniformity of a hologon deflector.

from the center to the edge of the facet. If this decrease occurs then the relative end-of-scan efficiency is slightly lower than that calculated by equation (58).

Technique for Improving Radiometric Scan Uniformity

Schematically illustrated in Figure 5.24 is a feedback technique that can be used to improve the facet-to-facet and intrafacet radiometric scan uniformity of a hologon deflector [23]. This technique utilizes a small stationary photodetector to monitor the intensity of the zeroth-order beam which is stationary for all hologon rotation angles. Essentially all the incident beam intensity, I_T, is divided between the first-order and zeroth-order beam intensities, I_1 and I_0, respectively, when only these diffraction orders can exist. The first-order beam intensity for this case is

$$I_1 = I_T - I_0 \tag{59}$$

It is evident from this equation that when I_1 decreases, I_0 increases for a fixed value of I_T. Therefore, one can determine the value of I_1 by monitoring the values of I_T and I_0. This information can be used to alter the value of I_T so that I_1 remains constant between facets and within a facet. In reading systems, this information can be used to adjust the gain of the photodetection system and thereby correct for variations in I_1.

Facet-to-facet and intrafacet diffraction efficiency variations for a hologon deflector are constant with time. They can therefore be corrected for by measuring them once and storing corresponding digital correction values in a ROM chip. This correction technique has advantages when compared with the analog feedback method illustrated in Figure 5.24. Scan intensity variations associated with other system components, such as fold mirrors, scan lens, and the photodetector elements in reading systems, can be accounted for in the stored digital correction values.

Influence of Other Diffraction Orders

When higher-order diffraction beams exist, the analog feedback technique in Figure 5.24 may not function accurately. In addition to taking power from the first-order diffracted beam, these higher-order beams can, by means of multiple reflections within the grating substrate material, generate ghost first-order beams. These ghost beams coherently interfere with the main first-order scan beam and cause scan beam intensity fluctuations. Calculation of the conditions under which higher-order diffraction beams can exist is provided by the generalized form of the grating equation

$$\sin \theta_{md} = \frac{m\lambda - D \sin \theta_i}{nD} \tag{60}$$

where θ_{md} is the diffraction angle measured in the grating substrate medium of refractive index n, and m is the order number of the diffracted beam. Real solutions for equation (60), corresponding to allowable diffraction orders, exist only when

$$\frac{m\lambda - D \sin \theta_i}{n D} \leq 1.0 \tag{61}$$

The condition where the second diffracted order can just exist is calculated by setting equation (61) equal to 1.0 and $m = 2$:

$$K = 0.5(\sin \theta_i + n) \tag{62}$$

If it is assumed that the first diffracted order satisfies the Littrow condition, $\theta_i = \theta_d$, then equation (62) can be rewritten as:

$$K = \frac{2}{3} n \tag{63}$$

When $n = 1.63$, which is the approximate index of refraction of photoresist for $\lambda = 0.6328$ μm, the second diffracted order can not exist for K values greater than 1.087.

Only the first diffracted order can exist for the preferred PDG hologon disk deflector geometry having $K = 1.4142$. Described in Section 5.6 are PDG hologon deflector systems that use K values of less than 1.087, and therefore, generate a second-diffracted-order beam. Fortunately, for these hologon deflector systems, the second-order diffracted beam intensity is negligible when compared to first-order beam intensity for $K > 0.8$ and $\theta_i \simeq \theta_d$.

Hologons having surface-relief gratings and $\theta_i = 0$ are deficient with regard to radiometric efficiency in that both the positive and negative first-order beams exist with equal intensity. A single first-order beam can be achieved for the $\theta_i = 0$ case by using thick volume gratings.

5.5.7 Image Spot Ellipticity

As previously described, the major factors contributing to the restriction on scan angle for the basic PDG hologon disk deflector are scan-line bow and scan-beam tracking error sensitivity with regard to deflector wobble. Another factor that limits scan angle for this hologon geometry is image spot ellipticity. Image spot ellipticity occurs in a hologon disk deflector because the incident beam for underfilled facet operation has an elliptical shape profile on the grating facet (see Figure 5.13) that remains stationary as the scan-beam angle is changed. This elliptically shaped beam profile on the facet surface functions as the entrance pupil for the scan beam. The cross-sectional projection of this elliptically shaped entrance pupil along the scan-beam path is circular for the center-of-scan position when the incident beam to the hologon has a circular cross section. The off-axis projection of the elliptically shaped entrance pupil gives rise to an elliptically shaped scan beam because the stationary entrance pupil has the wrong projected cross-sectional shape to produce a circular beam profile. Scan-beam ellipticity increases with scan angle since the stationary entrance pupil shape becomes progressively misaligned with scan angle from the pupil aperture shape required for a circular beam profile. The scan lens Fourier transforms the elliptically shaped scan beam into an elliptically shaped image spot. The major and minor axes of the resulting elliptical image spot increase and decrease symmetrically in size from the center-of-scan circular cross section value. These major and minor axes are orientated at about 45° to the scan-line direction, with the relative direction of both changing sense (clockwise to counterclockwise) as the scan beam passes through the center of scan.

An approximation for the deviation from roundness, ΔES, for the major and minor axes of the scan beam from a hologon disk deflector having $\theta_i \simeq \theta_d \simeq \theta$ is

$$\Delta ES = \pm(\sec\theta - 1)\tan\theta_s \qquad (64)$$

For $\theta = 45°$ and a scan angle of $\theta_s = 18°$, $\Delta ES = \pm 13.5\%$. This image spot ellipticity either increases or decreases the overlap of successive scan lines orientated at $\pm 45°$ to the scan-line direction. This change in overlap affects the quality of lines and halftone dots orientated in these directions. If it is assumed that a maximum acceptable value for ΔES is $\pm 15\%$, then $20°$ is the largest permitted scan angle for the $\theta = 45°$ case.

In Section 5.6, an auxiliary prism or grating element is added to a PDG disk deflector system to achieve a virtually straight scan-line system having $K = 1.00$ and $\theta_i \simeq \theta_d \simeq 30°$. Under these conditions, ΔES for the hologon element is $\pm 5\%$ for $\theta_s = 18°$. Unfortunately, these prism and grating elements magnify the hologon beam ellipticity problem. The resulting image spot ellipticity for these systems is essentially the same as that achieved for the basic PDG hologon disk deflector geometry.

Technique for Improving Image Spot Ellipticity

One technique that has been shown to be useful to reduce the image spot ellipticity problem is to use an overfilled facet illumination approach instead of the underfilled facet illumination technique illustrated in Figures 5.4 and 5.13. Grating facets in the overfilled illumination approach are fabricated with an elliptical profile aperture shape that is similar to the incident beam profile shape on the tangential grating hologon in Figure 5.13. The Gaussian incident beam for the overfilled case has a $1/e^2$ intensity diameter on the hologon disk that is three or more times larger than the facet aperture dimension in the scan direction. Typically the overfilled incident beam profile on the hologon surface has the same relative size to the facet aperture that the underfilled facet in Figure 5.13 has to the incident beam profile on the facet. Scan-beam light is generated from the entire overfilled facet area as the facet translates across the incident beam profile on the deflector.

The overfilled approach is useful for reducing or eliminating the image spot ellipticity problem because the facet aperture is the entrance pupil for the scan beam. Unlike the stationary beam profile for underfilled facet operation, the overfilled facet aperture changes angular orientation as the scan-beam angle changes. For the case where the grating facet has $K = 1.0$ and $\theta_s = \theta_R$, the overfilled facet aperture exactly tracks the scan beam. Therefore, the scan beam exiting the overfilled facet aperture has the same cross-sectional profile for each scan position. The shape of the facet aperture determines the cross-sectional profile of the scan beam for this case. A circular cross-sectional profile for the scan beam is achieved by using the proper elliptical shape for the overfilled facet aperture.

When $K = 1.4142$, $\theta_s = 1.4142\theta_R$ and the overfilled facet aperture does not exactly track the scan beam. Under these conditions, the scan beam cross-sectional profile changes shape with increasing scan angle, but to a significantly lesser extent than that occurring with underfilled facet operation. For $\theta_s = 20°$ and overfilled facet operation, the facet aperture rotates through an angle of 14.1° when $K = 1.4142$. The angular misalignment between the facet aperture and scan beam for this case is only 5.9°. Equation (64) predicts that ΔES is only 4% if a value of 5.9° is used for θ_s in the equation.

While essentially 100% mechanical duty cycle can be achieved with overfilled facet operation, it is usually not used due to the large light loss associated with the method. Also, scan-line intensity uniformity is usually a problem with this illumination method. This uniformity problem is caused by the facet aperture translating across the Gaussian incident beam intensity profile. When the Gaussian incident beam has a $1/e^2$ intensity diameter that is only three times larger than the facet aperture, the scan-line intensity profile varies from a peak at the center of scan to approximately 25% of that peak at the end-of-scan position. The scan-line intensity uniformity can be improved by expanding the incident beam considerably more and truncating it at a very high intensity point. This approach results in very poor radiometric system efficiency and requires very large numerical aperture collimating optics for the incident beam.

Another disadvantage of the overfilled facet approach is that the facet functions as the entrance pupil for the focusing lens. Due to the translation of this aperture, the focusing lens must be larger in diameter and possibly more complex. In essence, the scan rays appear to come from an apparent entrance pupil located on the opposite side of the hologon disk from which the scan beam exits. For the case where the grating facet has $K = 1.0$ and $\theta_s = \theta_R$, the apparent entrance pupil is stationary and located on the deflector rotation axis.

Many of the problems encountered with the overfilled facet approach are solved by using the centered facet deflector configuration described in Section 5.5.9.

5.5.8 Factor Limiting Hologon Rotation Speed

The required rotational rate ω of the hologon deflector is given by equation (13). As previously stated, a rotational rate of about 18000 rpm is a practical upper limit for ball-bearing mounted deflectors since usable bearing life deteriorates much faster at high rotation rates. Rotational rates as high as 50,000 rpm can be achieved with ball-bearing motors if short operating life is not a problem. Ferrofluid and air bearings can be used to achieve long operating life for rotational rates above 18,000 rpm (see Chapter 9). Air drag becomes a limiting factor for deflector rotation rates greater than about 30,000 rpm since it significantly increases the motor drive power requirements and causes deflector heating. These problems are solved by running the deflector in a partial vacuum.

Rupture strength of the deflector material is of prime importance in hologon systems since transmission deflectors utilizing glass substrates are the preferred embodiment. Glass is a very strong but brittle material that suffers from microcrack propagation which can considerably reduce its rupture strength. A safety rule of thumb is that the calculated rupture strength for a glass disk should be three to five times higher than the maximum stress force that will be induced in the disk by rotation.

Every element of a rotating disk experiences radial stress in the direction of the centrifugal force and a corresponding tangential stress which is significantly larger than the radial stress. For a rotating disk of uniform thickness, the maximum stress, S_t, is the tangential stress along the circumference of the central bore hole in the disk [39]:

$$S_t = 7.1 \times 10^{-6} \rho\omega^2[(\sigma + 3)H^2 + (1 - \sigma)h^2] \tag{65}$$

where S_t is in pounds per square inch (psi), ρ is the density of the disk material in pounds per cubic inch, ω is the rotation rate in revolutions per minute, σ is Poisson's ratio for the disk material, H is the disk outer radius in inches, and h is the central bore hole radius in inches. It is evident from this equation that the dominant factors contributing to S_t are the rotation rate and the outer and inner radii, which should both be minimized to reduce S_t for a given ω value.

Graphically presented in Figure 5.25 is calculated data for the induced stress, S_t, generated in rotating disks of different diameters as a function of rotation speed. Also indicated in this figure is the stress required to rupture four different materials.

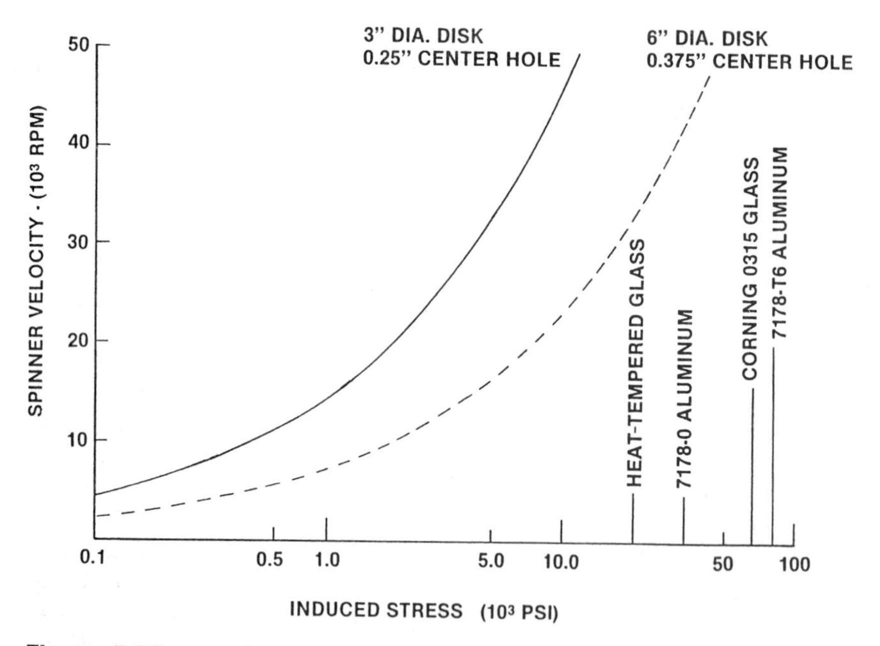

Figure 5.25 Calculated data for the induced stress generated in rotating disks of different diameters as a function of rotation speed. Also included are data for the stress required to rupture different materials.

Applying a four times safety rule to the glass materials listed in this figure indicates that disks fabricated with heat tempered glass are suitable for operating rates up to about 34,000 rpm for the 3-in (76.2mm) diameter and about 18,000 rpm for the 6-in (152.4 mm) diameter. If the Corning 0315 chemically strengthened glass is utilized, then operating rates for these diameter disks can be increased to $>$ 50,000 rpm and about 40,000 rpm, respectively. A 2-to-1 safety factor is considered adequate for aluminum, and therefore, both alloys listed in Figure 5.25 can be used to operate the 6-in-diameter disk at rates up to 30,000 rpm. For safety considerations it is advisable to adequately enclose all rotating deflectors due to the possible catastrophic release of the large energy stored within them.

5.5.9 Centered Facet Hologon Disk Deflector

Ritter [40] proposed a unique hologon disk deflector that employs a single PDG facet centered on the deflector rotation axis. This deflector can be used with the overfilled facet technique described in Section 5.5.7 without incurring the previously stated problems associated with that illumination method. Figure 5.26 illustrates a centered grating facet that is overilluminated by an incident beam that has a circular cross-sectional profile on the hologon disk. Since the facet aperture in Figure 5.26 is centered on the deflector rotation axis, it changes its angular orientation to track the scan beam without translation across the incident beam profile. For this deflector geometry: the scan-beam cross-sectional profile is determined by the facet aperture shape; the scan-beam cross-sectional profile is essentially constant with respect to scan angle; the scan line has very good radiometric uniformity; the deflector can have high radiometric throughput efficiency since the incident beam profile need only be slightly larger than the facet aperture; and the scan-beam entrance pupil is fixed which allows a smaller focusing lens design. Normally, the incident beam profile on the hologon disk would be elliptically shaped as opposed to the circular profile shown in Figure 5.26. When the incident beam profile is elliptically shaped, the radiometric intensity of the scan beam would increase very slightly with scan angle.

The single centered grating facet disk in Figure 5.26 is positioned in the middle of the motor assembly. The motor assembly in Figure 5.26 is depicted as being of the pancake DC torque motor type. This type of motor could also be constructed using circuit board technology, thereby enabling it to be manufactured relatively inexpensively. A pancake motor geometry can also be achieved by using a small gas turbine to actuate the rotation of the disk assembly. The small thickness of the pancake motor enables the focusing lens to be positioned relatively close to the center of scan, thus reducing its in-scan dimension. This centered deflector configuration could also be constructed by placing the grating facet within a large bearing assembly and rotating the inner part of the bearing assembly by a drive belt linked to a motor. Compared with the pancake motor construction, the drive

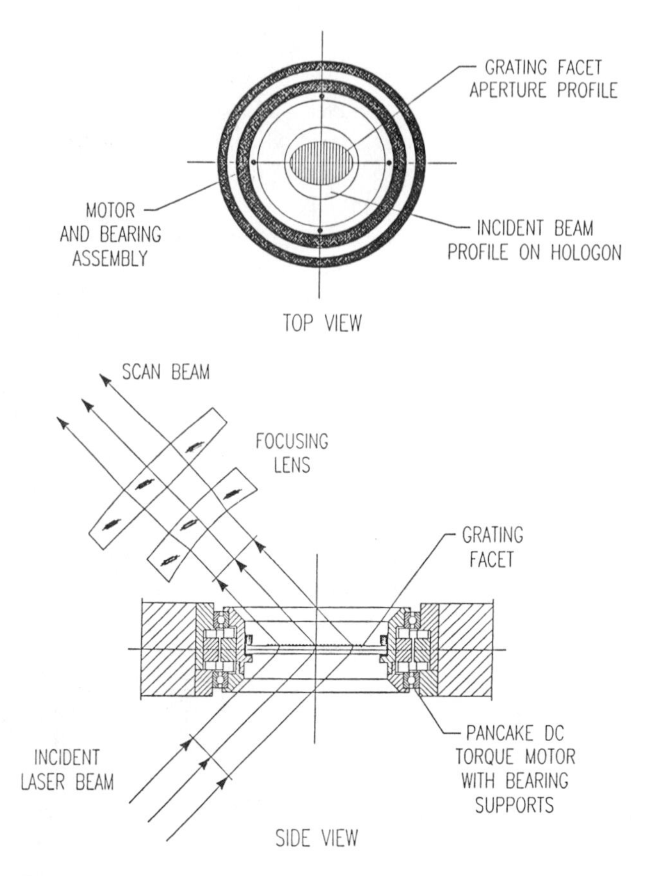

Figure 5.26 Centered PDG facet disk deflector used with overfilled facet illumination.

belt configuration offers greater asymmetry in construction of the deflector, which should enable the focusing lens to be placed nearer to the facet center.

Since the hologon disk for the centered facet case need only be slightly larger than the required facet aperture size, the disk can be relatively small. Therefore, the disk can be rotated to very high speeds without the worry of the disk rupturing due to centrifugal-force-induced stress. The motor assembly in Figure 5.26 could be a gas turbine type incorporating air bearings, thereby enabling the disk to rotate at speeds between 30,000 and 50,000 rpm while achieving good deflector running life. The relatively small disk size of this deflector configuration can present some problems since its rotational inertia is also relatively small. Rotational inertia helps reduce the motor hunting found in all rotating deflector systems. Additional rotational inertia can be added to the deflector assembly and/

or better speed motor controls used with it. One can also use a grating pixel clock to measure rotational speed variations within a scan and, thereby, correct for it.

Imaging Properties of the Centered Facet Disk Deflector

The PDG facet in Figure 5.26 is depicted as having its grating lines orientated perpendicular to the plane of the figure. When the facet is rotated 180° from its depicted position, the grating lines are also perpendicular to the plane of the figure. Under these conditions, the incident beam is diffracted in the same direction as indicated in Figure 5.26. It is evident from this description that two scans are achieved for each complete rotation of the hologon disk in Figure 5.26. This single centered grating facet functions like a sequential dipole radiator. Having $\theta_i \simeq \theta_d$ and $K > 0.8$, ensures that both the scan-beam tracking error is very insensitive with respect to deflector wobble and that the dipole facet only radiates significant energy into the positive first-order diffracted beam. This latter feature of the sequential dipole facet enables high radiometric system throughput efficiency to be achieved.

Having two scans per revolution not only doubles the scan rate but also doubles the scan duty cycle for the system. The scan duty cycle for this deflector configuration is determined by the angular duty cycle, U_a, of the total deflector system. From equation (9), U_a for a dipole grating facet deflector having $\theta_i \approx \theta_d$ is

$$U_a = \frac{2\theta_s}{\pi K} \tag{66}$$

Scan duty cycle for this dipole geometry can only be improved by increasing the scan angle of the system or by reducing the K value of the grating facet. Scan angle for the dipole disc deflector is still limited to $\pm 20°$ due to scan-line bow and scan beam tracking error sensitivity with regard to deflector wobble.

Scan duty cycle of the centered facet disk deflector in Figure 5.26 can be doubled if a second centered PDG facet is superimposed on the first centered PDG facet. This second facet would be orientated to have its grating lines perpendicular to those of the first facet. This centered cross grating deflector produces four scans for each complete rotation of the hologon disk. For $K = 1.4142$ and $\theta_s < 20°$, there is no crosstalk between cross gratings and only a single diffracted beam is generated per recording scan period. Thus high radiometric system throughput efficiency is achieved for this quadrapole grating deflector. This quadrapole deflector can be constructed by closely air-spacing two glass disks, each having a centered PDG facet fabricated on it.

Cross-scan beam tracking error for the centered grating facet deflector is essentially zero if only one scan per disk revolution is used and is determined by the disk wobble angle and wedge angle when two scans per disk revolution are obtained for the dipole configuration. Grating facet periodicity error does not contribute to scan-beam tracking error for the dipole case because the periodicity is the same for both scans. Scan-beam tracking error associated with deflector wobble and disk wedge can be compensated for in the dipole case by mechanically adjusting the static wobble tilt angle of the disk within the deflector assembly. This adjustment is performed by means of set screws located along the disk circumference holding mechanism. A similar screw tilt angle adjustment is used in the pyramidal grating deflector system described in Section 5.7.5. This tilt disk adjustment technique can also be applied to the quadrapole deflector assembly. Grating facet periodicity error does contribute to scan-beam tracking error for the quadrapole deflector configuration.

5.5.10 Summary of System Performance for the Basic PDG Hologon Disk Deflector

The scan imaging performance properties of hologon disk deflectors fabricated with PDG facets were extensively analyzed in this section. This analysis showed that PDG disk deflectors can meet the performance requirements of graphic arts electronic imaging systems used for high-resolution reproduction of line art and halftone images. PDG disk deflectors achieve these performance requirements when used with a ball-bearing motor and an optical imaging system comprising only spherical lens elements.

Having PDG facets makes it easy to fabricate the hologon at one wavelength and use it at another wavelength. Also, it makes the deflector system easy to align and fairly insensitive to small angular misalignment errors. Section 5.8 briefly describes techniques and equipment used to align a hologon deflector unit to the other optical components in the scan imaging system. Typical measured radiometric throughput efficiency for a PDG hologon fabricated in photoresist is >85% when $\lambda = 0.633$ μm, $K = 1.4142$, and S-polarized light is used. Variations in scan-to-scan and intrascan radiometric uniformity for this photoresist hologon are typically both less than $\pm 4\%$.

Most of the analyses and data presented in this section for the dynamic, radiometric, and wavelength performance properties of the basic PDG deflector will be used when examining the performance of the hologon deflector systems described in Sections 5.6 and 5.7.

Preferred Configuration for the Basic PDG Disk Deflector System

The preferred configuration for the basic PDG disk deflector system is a preobjective, transmission hologon deflector having tangential oriented facet grating lines, $K = 1.4142$, $\theta_i = 45.72°$, and $\theta_s = \pm 18°$. This set of parameters maximizes system scan duty cycle while still achieving the scan-line bow requirement, as presented in Table 5.1, for most noncomposite image reproduction applications. Requirements for scan spot size, spot size growth, and scan linearity are mainly controlled in this preobjective deflector system by the design of the f-θ scan lens.

The scan-beam positional error for this preferred hologon deflector geometry is totally insensitive with regard to deflector centration errors and essentially insensitive to deflector wobble, with $d\theta_{z1} = -d\phi/41$ and $d\theta_{s1} = -d\phi'/20$. When 20 arc seconds of deflector wobble is achieved with either a rigid or a dynamic mount, the maximum deflector wobble induced scan tracking error between all facet scans is <1 arc second while the maximum end-of-scan jitter due to deflector wobble is <2 arc seconds.

When the preferred hologon is fabricated with photoresist on a disk having a substrate wedge of <1 arc second, typical measured scan-beam tracking error performance for the hologon mounted on a ball-bearing motor is <2.5 arc seconds for the total facet-to-facet scan band. This result indicates that scan-beam tracking error due to facet-to-facet and/or intrafacet periodicity error is <1 arc second for this hologon, which corresponds to $dD < 10^{-6}$ μm for $\lambda = 0.6328$ μm. When a scan lens having $f = 545$ mm is used with this hologon deflector, the maximum uncorrected scan-beam tracking error between adjacent scan lines is typically <1.7 μm for a six-facet hologon. Periodic scan-line jitter error for this preobjective hologon deflector system is typically $< \pm 2$ μm when a start-of-scan detector signal is utilized to synchronize the start of scan-line pixel data. This periodic cross-scan and in-scan pixel placement error performance is adequate for many halftone imaging applications having a scan resolution of about 1200 dpi. The system scan-beam tracking error can be reduced to <0.25 μm by means of an acoustooptic-based minideflector corrector system, thereby enabling the deflector system to meet the performance requirements for halftone imaging applications having a scan resolution of 2000 dpi.

Disadvantages of the Basic PDG Hologon Disk System

The major disadvantage of the hologon deflector technology is that it must be utilized with a very monochromatic laser source. Wavelength shifts due to mode hopping in a diode laser cause abrupt, unacceptable cross scan and in-scan beam positional errors in a hologon deflector system used for graphic arts imaging applications. While the compensation grating approach of the section "Wavelength-Shift-Induced Scan-Beam Tracking Error" provides fairly good correction for scan beam tracking error due to these mode hops, it does not correct for the associated in-scan error. Therefore, diode laser light sources cannot be used with the hologon for these imaging applications unless the diode laser is stabilized with regard to operating wavelength or is modified so that the wavelength shifts gradually with time in comparison to the system scan rate.

Another disadvantage of the basic PDG hologon disk deflector geometry is that minimum scan-line bow and the minimum deviation condition are only simultaneously achieved when $K = 1.4142$. Having $K = 1.4142$ makes it difficult to fabricate hologon deflectors for the blue end of the wavelength spectrum since the grating periodicity must be approximately 40% finer than the optical wavelength. Hologons fabricated with surface-relief gratings can have essentially 100% diffraction efficiency for S-polarized light while achieving only about 5% diffraction efficiency for P-polarized light when $K = 1.4142$ and $\theta_i \simeq \theta_d$. This diffraction efficiency sensitivity to the polarization state not only reduces system radiometric efficiency for circularly and randomly polarized light applications, but also produces beam intensity variation as a function of scan angle for polarized laser sources. Having K fixed also eliminates the ability to alter the optical angular magnification factor of the deflector, and thereby improve system scan duty cycle. Section 5.6 describes how the K value for a PDG disk deflector can be reduced while still achieving minimum scan-line bow and the minimum deviation condition.

Also, the basic PDG hologon disk deflector system suffers from being limited to scan angles of $< \pm 20°$ for most graphic arts imaging applications. This restriction on scan angle occurs because of the functional dependence on scan angle of the following parameters: scan-line bow, scan-beam positional error sensitivity with regard to deflector wobble, and scan-beam image spot ellipticity. The centered-facet PDG disk deflector described in Section 5.5.9 can be used to minimize scan image spot ellipticity as a function of scan angle.

5.6 IMPROVED PLANE GRATING HOLOGON DISK DEFLECTORS

As previously noted, one major disadvantage of the basic plane grating hologon configuration illustrated in Figure 5.12 is that there is only a single optimum solution corresponding to $K = 1.4142$. Having only a single optimum solution constrains the deflector system design, its performance, and the deflector fabrication process. Reducing the K value for the PDG deflector makes it easier to fabricate, reduces differential polarization diffraction efficiency effects, as illustrated by the data in Figure 5.23, improves system scan duty cycle, as shown by Equations (9) and (29), and reduces scan-beam tracking error sensitivity of the deflector with regard to the following parameters: deflector wobble, as shown by equation (42); facet periodic error, as shown by equation (46); and wavelength shift, as shown by equation (52).

Reducing the deflector K value from 1.4142 to about 1.0 would enable equal diffraction efficiency for S- and P-polarized light to be achieved and would make it easier to fabricate the hologon deflector element. The smaller incident angle ($\theta_i \simeq 30°$) associated with $K = 1.0$, results in a lower Fresnel reflection loss per hologon surface, and therefore, makes it easier to reduce these losses by means of anti-reflection (AR) coatings. Having low reflectance per surface is important when using an air-spaced, sandwiched hologon structure to protect the grating facets, since this considerably reduces scan-beam intensity fluctuations resulting from Fabry-Perot interference effects associated with multiple reflections within the hologon substrate and cover disk. Using P-polarized light is usually advantageous when intensity uniformity is a major priority because this polarization normally provides the lowest reflectance loss per surface. Having $K = 1.0$ would make the scan angle exactly equal to the deflector rotation angle, as shown by Equation (29).

Kramer [41] demonstrated that the K value for the PDG hologon disk geometry in Figure 5.12 could be reduced while maintaining both a virtually straight scan line and the $\theta_i \simeq \theta_d$ relation. This was accomplished by having the optical axis of the f-θ scan lens be tilted with regard to the central scan beam ray direction. Under these conditions, the barrel distortion incorporated into the scan lens to linearize the scan-beam velocity, also introduces bow into the scan line. Bow introduced by the scan lens is used to cancel the bow in the scan beam from the hologon deflector. Essentially a bow-free scan line was achieved with this technique for the following deflector system parameters; scan-lens tilt 9°, $\theta_s = \pm 18°$, and $\theta_i \approx \theta_d \simeq 41.26°$, corresponding to $K = 1.3191$. Scan-lens tilt only permits moderate reduction in the hologon K value that enables both the minimum scan-line bow condition and the minimum deviation condition to be simultaneously achieved. Factors hindering further reduction in the K value include demands on lens aperture size and lens performance. These limitations make 15° a practical upper limit for lens tilt.

It is well known that spectral lines produced by spectrographs and monochrometers that use either a prism or grating element to separate incident light into wavelength components will be curved if the instrument uses a straight entrance slit and will be straight if a curved entrance slit is used [42]. Based on the results achieved using lens tilt, it is logical to assume that either a prism or grating could be used to introduce a bow into a scanning laser beam that is essentially equal and opposite to a bow introduced by the laser beam deflector element, and thereby, reduce the overall system scan-line bow to virtually zero [43]. Kramer [44] demonstrated that scan-line bow from a PDG hologon disk deflector system could essentially be eliminated by incorporating a stationary prism or grating element between the hologon and image plane. Incorporation of either a prism or grating element into a hologon deflector system has enabled the minimum scan-line bow condition and the minimum deviation condition to be simultaneously achieved for hologon K values ranging between 0.7 and 1.4142.

5.6.1 Utilizing Lens Tilt to Reduce Scan-Line Bow

As stated in Section 5.3.3, a f-θ scan lens incorporates barrel distortion to compensate for the pincushion distortion that is present in a flat-field scanning system, and thereby linearize the scan-beam velocity at the image plane with respect to the deflector rotation angle. A potential problem with scan lenses that incorporate barrel distortion is that when the scan beam is incident in the cross scan direction at an angle to the optical axis of the lens, the scan beam is bent in a direction toward the lens optical axis. If the scan beam generates a straight line before entering the scan lens, the bending associated with lens distortion causes the scan line to become bowed. The magnitude of the scan-line bow being proportional to the cross scan angle that the beam makes with regard to lens optical axis. If the scan beam is bowed before entering the lens, then scan bow introduced by lens tilt can be utilized to reduce the scan-line bow for the deflector system.

Figure 5.27 illustrates two of the possible cross scan angular relationships that the scan beam can have with regard to scan lens optical axis (y axis). In one case (solid line), the scan beam makes a negative angle, $-\theta_T$, with the lens axis while in the other case (dotted line), the beam makes an equal but opposite angle, $+\theta_T$, with respect to the lens axis. Figure 5.28 pictorially illustrates the expected shape of the image plane scan lines for these cases when the scan beam entering the scan lens has no scan bow and the scan lens incorporates barrel distortion. Having positive beam tilt angles ($+\theta_T$) gives a scan-line bow curve which is useful for reducing the scan-line bow associated with PDG hologon disks having $K < 1.4142$ and $\theta_i \simeq \theta_d$. Negative beam tilt angles θ_T) give a scan-line bow curve that is useful for reducing the scan-line bow associated with hologons having $K > 1.4142$ and $\theta_i \simeq \theta_d$. Only positive beam tilt angles are considered in the following analysis since the goal is to reduce the K value for the hologon deflector, while maintaining the $\theta_i \simeq \theta_d$ relationship, and thereby improve its performance.

The scan-lens angular orientation in Figure 5.27 is depicted as being fixed

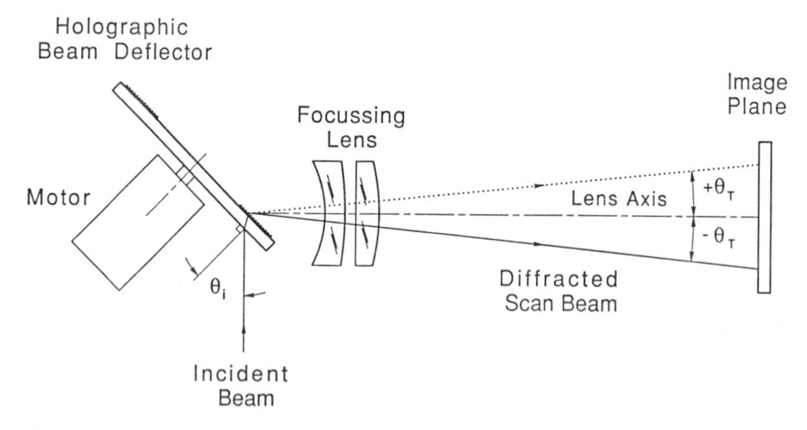

Figure 5.27 Two possible tilt angle relationships that the scan beam can have with regard to the scan-lens optical axis.

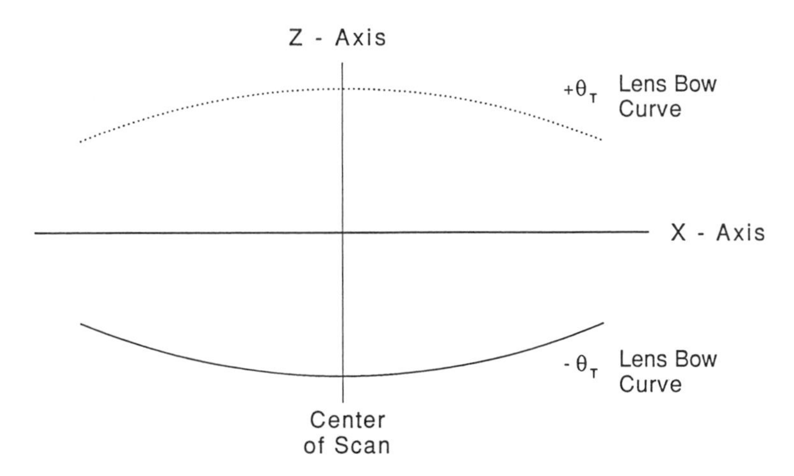

Figure 5.28 Pictorial illustration of scan-line bow caused by barrel distortion of a tilted f-θ scan lens.

while the diffraction angle of the scan beam is varied to generate the positive and negative tilt angles with respect to the lens optical axis. It is obvious that the equivalent scan-beam angular relationships with respect to the lens optical axis are achieved by fixing the cross scan angular orientation of the scan beam and rotating the scan lens about its center in the cross-scan direction. Under these conditions, the lens optical axis is rotated in the counterclockwise direction to generate a positive beam tilt angle and the central scan beam propagates through the middle of the effective lens aperture.

Analysis of Scan-Line Bow with Lens Tilt

A computer ray tracing program was utilized to determine how well lens tilt compensates for the bow in a scan line produced by a PDG hologon disk deflector. The f-θ scan lens used for these ray tracing calculations had three air-spaced lens elements and was designed to have a maximum scan linearity error of 0.1% when used with a hologon deflector having $K = 1.4142$, $\theta_i = 45.72°$, and $\theta_s = \pm 18°$. The lens effective focal length was 545 mm so that the system scan-line bow curve achieved for the hologon plus tilted scan lens could be directly compared with the scan-line bow curve presented in Figure 5.16 for the $K = 1.4142$ case without lens tilt. The computer program was set up to determine what scan-lens tilt angle was required to minimize the scan-line bow for hologons having different K values. The K value of the hologon was stepwise reduced from the 1.4142 value while maintaining both $\theta_s = \pm 18°$ and $\theta_i = \theta_d$ for the ± 0.71 scan-field positions.

These calculations showed that as the value of K decreased from 1.4142, the scan-lens tilt angle had to be proportionally increased to minimize system scan-line bow. These calculations also showed that the magnitude of the minimized scan-line bow band increases as K is decreased from the 1.4142 value. The following example demonstrates the level of scan-line bow correction that can be achieved in a hologon deflector system by means of scan-lens tilt angle. The hologon for this example has $K = 1.3191$, $\theta_i = 41.977°$, and $\theta_R = 13.55°$ for $\theta_s = 18°$. Before the scan lens is tilted, the system scan-line bow for this example increases monotonically with scan angle to a maximum value of -3.67 mm for the end-of-scan position. This scan-line bow band is reduced to 105 μm by negatively rotating the scan lens in the cross scan direction by 8.7°.

Except for having a 23% larger scan-line bow value than that presented in Figure 5.16, the scan-line bow curve for this tilted scan-lens example has the same minimum scan-line bow shape as that shown in Figure 5.16. This shape has the maximum deviation from straightness occurring at about the ± 0.71 scan-field positions and the scan-line bow curve recrossing the reference straight scan line at the end of scan. The scan-line bow band for this example correlates to 0.031% of the scan-line length and has a maximum rate of change of 0.38%. These values meet the requirements in Table 5.1 for noncomposite images, but the rate of change of the scan-line bow is essentially at the upper limit for noncomposite images.

The θ_i value in this example provides for an absolute difference between θ_i and θ_d of approximately 1.42° for both the start- and end-of-scan positions. This absolute angular difference condition only minimizes the scan-beam tracking error sensitivity with regard to deflector wobble for the entire scan-line length and does not correlate, as in the $K = 1.4142$ case, with either the shape or magnitude of the system scan-line bow curve achieved by using scan-lens tilt. For the 1.42° difference in θ_i and θ_d angles, $d\theta_{z1} = d\phi/43$ for this hologon deflector system. This value of $d\theta_{z1}$ is only slightly less than that achieved for the $K = 1.4142$ geometry.

According to the radiometric data in Figure 5.23, essentially 100% diffraction efficiency is achieved for S-polarized light and 35% for P-polarized light when $K = 1.3191$. Substituting these values and $\theta_R = 13.55°$ in equation (60) indicates that the end-of-scan beam intensity is 96.4% of the center-of-scan beam value for S-polarized light. If P-polarized light is used, the beam intensity increases from 35% to 38.6%, respectively, for center- and end-of-scan positions. While there is only a small performance improvement in the intrascan radiometric uniformity for S-polarized light for the $K = 1.3191$ hologon, when compared to the $K = 1.4142$ hologon case, there is significant improvement in the intrascan radiometric uniformity for P-polarized light. Further improvement in both the intrascan radiometric uniformity and scan-beam tracking error performance are achieved by using a prism or grating bow compensation element and reducing K to 1.0.

This example demonstrates the benefits and potential problems associated with modifying the scan-line bow of a PDG disk deflector by means of scan-lens tilt angle. Tilt angle adjustments are usually incorporated into the scan-lens mounting platform assembly of a deflector utilized for composite image applications because the scan-line bow is so sensitive to tilt angle of the f-θ scan lens. Scan lens tilt angle is very useful for making small adjustments to the scan-line bow achieved with hologon deflector systems incorporating either prism or grating bow compensation elements.

5.6.2 Utilizing a Prism Element to Reduce Scan-Line Bow

Figure 5.29 shows the basic spatial relationship between the hologon deflector and the stationary prism element used to compensate for bow in the scan beam exiting the hologon. The reverse principle is being illustrated in Figure 5.29, that is, the prism converting a straight scan line from the hologon into a bowed scan line at the image plane. The straight scan line becomes bowed because the angle between the deflected beam and the normal to the prism face changes as the beam scans across the prism face. This change in input angle causes the scan beam to experience a greater deviation angle as it propagates through the prism. This change in deviation angle produces an image line having its ends bowed in the direction that the beam cross scan angle is deviated by the prism, as illustrated in Figure 5.29.

A better insight into why the scan beam becomes bowed on passing through the prism is obtained by tracing a ray through a prism and calculating how the output angle is changed as a function of a change in input angle. In Figure 5.30, an input ray is traced through a prism having an apex angle A. The relationship between the incident and refracted beam angles is given by Snell's law:

$$n \sin \phi = n' \sin \phi' \tag{67}$$

where ϕ is the angle that the incident ray makes with respect to the normal to the refracting surface in index medium n, and ϕ' is the angle that the refracted ray makes with respect to the surface normal in index medium n'. The relationship between the change in the angle, $d\phi_2$, of the output beam from the prism and the change in the angle, $d\phi_1$, of the input beam to the prism is achieved by differentiating equation (67) with respect to the change in the incident ray angle for each refracting surface of the prism in Figure 5.30, $d\phi_1$ and $d\phi'_2$, and setting $d\phi'_2 = -d\phi'_1$:

$$d\phi_2 = -\frac{\cos \phi_1 \cos \phi'_2}{\cos \phi'_1 \cos \phi_2} d\phi_1 \tag{68}$$

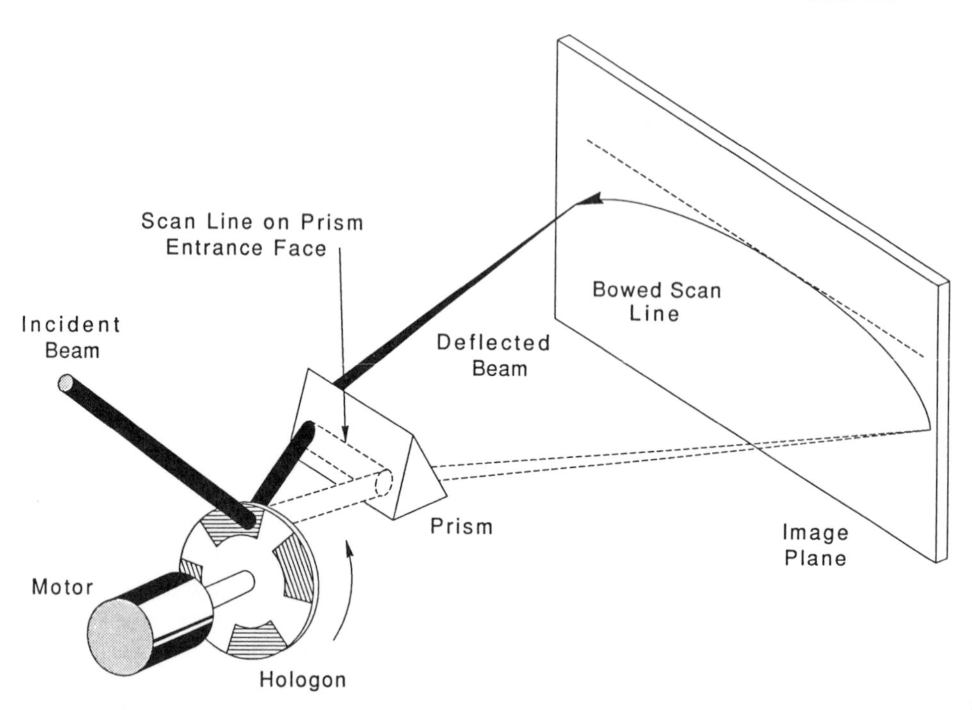

Figure 5.29 Schematic illustration of how a prism converts a straight scan from a hologon deflector into a bowed scan line at the image plane.

It is evident from equation (68) that a parallel plate, $A = 0$, does not introduce a bow into the scan line since $\phi_1 = \phi_2$ and $\phi'_1 = \phi'_2$ and, therefore, $d\phi_2 = d\phi_1$. When $A \neq 0$, change in ϕ_1 resulting from the scan beam changing its angular orientation with regard to the prism input surface normal causes corresponding changes in the angular values in equation (68), thereby causing an alteration in the relationship between $d\phi_2$ and $d\phi_1$ that gives rise to the bow in the scan beam. It is clear that the faster the relationship between $d\phi_2$ and $d\phi_1$ changes, the greater the resulting bow in the scan beam. Parameters that influence the rate of change of this relationship include prism apex angle, prism index, and ϕ_1. When the central scan ray is perpendicular to the prism apex roof line, the scan-line bow is symmetrical about the center-of-scan position.

Analysis of Scan-Line Bow with Compensation Prism

A computer ray tracing program was utilized to determine how well a prism compensates for the bow in a scan line produced by a PDG hologon disk deflector. The computer program was set up to find the prism apex angle as a function of

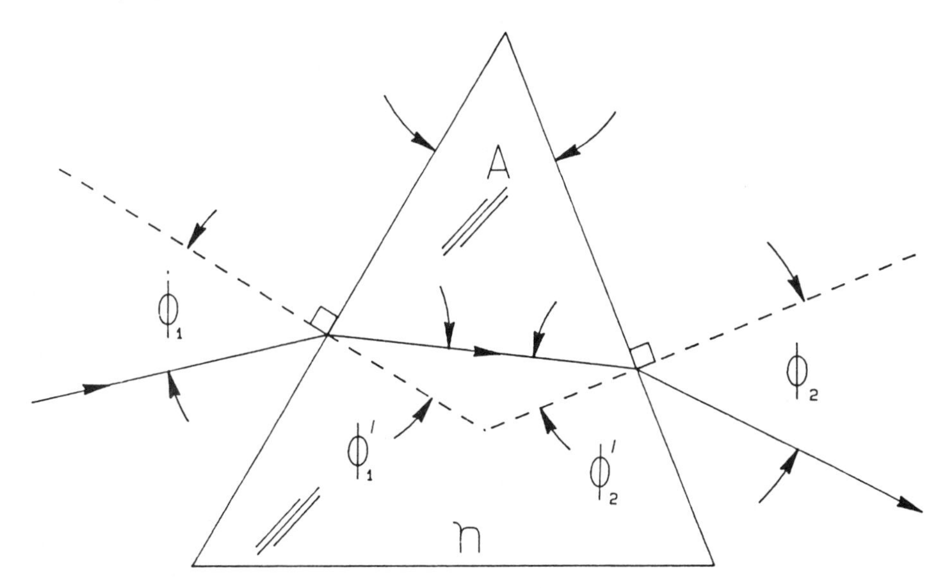

Figure 5.30 Ray trace for propagation of an incident beam through a prism having a refractive index n and an apex angle A.

the ray incident angle to the prism that minimized scan-line bow at the image plane when a scanning beam having a preprogramed scan bow was incident on a prism of refractive index n. The preprogramed scan bow used for these ray tracing calculations was based on the scan beam produced by a hologon having $K = 1.00$, $\theta_i = 30.82°$, and $\theta_s = \theta_R = 18°$.

A f-θ scan lens focal length of $f = 545$ mm was used in the ray tracing calculations so that the system scan-line bow curve achieved for the hologon plus prism element could be directly compared with the scan-line bow curve presented in Figure 5.16 for the $K = 1.4142$ case. These calculations showed that the scan line had a maximum deviation from straightness of 16.33 mm before the prism element is inserted between the hologon and f-θ scan lens. Results of the ray tracing calculations are graphically presented in Figure 5.31 for the case where a prism having an index of refraction of 1.799 is inserted between the hologon and f-θ scan lens. There are three individual curves presented in Figure 5.31 and three corresponding vertical scales. These three curves are labeled apex curve, bow curve, and ratio curve. Their corresponding vertical scales are labeled prism apex angle, residual bow band, and beam ratio factor. All three curves are plotted as a function of the incident beam angle on the prism.

One can determine from the apex curve what prism apex angle is required to minimize the system scan-line bow for a given incident beam angle on the prism.

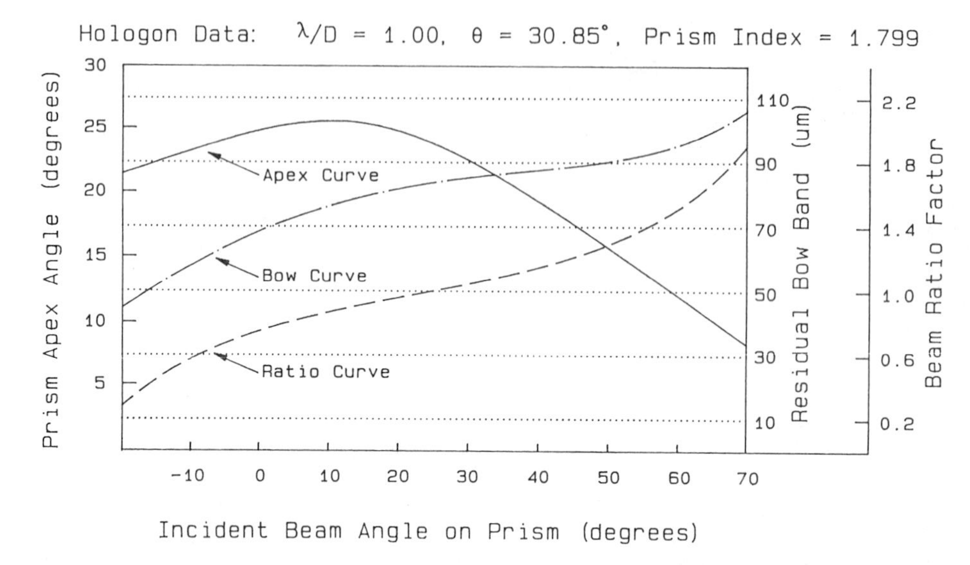

Figure 5.31 Results for ray tracing calculations for hologon plus prism geometries that minimize scan-line bow.

Reducing the index of refraction of the prism material does not change the relative shape of the apex curve, but only increases the prism apex angle required to minimize scan-line bow for each incident beam angle. Designing the system to use incident beam angles in the range of 0° to 20° is advantageous since the prism apex angle is nearly constant over this range and, therefore, small misalignment of the prism with regard to the incident beam angle has negligible effect on the performance of the system.

The residual scan-line bow band achieved for each prism apex angle, and incident beam angle is indicated by the Bow Curve in Figure 5.31. This scan-line bow band goes from a low of about 40 μm to a high of about 105 μm for the incident beam angles plotted. For incident beam angles in the range of 0° to 20°, the scan-line bow band has a mean value of approximately 75 μm, which is approximately 10% less than the bow band present in Figure 5.16 for the $K = 1.4142$ case. Essentially identical minimum scan-line bow band shape to that shown in Figure 5.16 was obtained for the calculated scan-line bow curves corresponding to the hologon plus various prism geometries represented by the data points in Figure 5.31.

For the hologon deflector parameters used in these calculations, there is an absolute difference between θ_i and θ_d of approximately 1.63° for both the start- and end-of-scan positions. This absolute angular difference condition only minimizes the scan-beam tracking error sensitivity with regard to deflector wobble for the entire scan length and does not correlate, as in the $K = 1.4142$ case, with either the shape or magnitude of the system scan-line bow curve achieved with the bow compensating prism element. For the 1.63° difference in θ_i and θ_d angles, $d\theta_{z1} = d\phi/61$ for this hologon deflector. Having $K = 1.00$ reduces the scan-beam tracking error sensitivity with regard to both deflector wobble and facet grating periodicity change by approximately 50% to 60% when compared to that achieved for the $K = 1.4142$ geometry.

According to the radiometric data in Figure 5.23, essentially 100% diffraction efficiency is achieved for both S- and P-polarized light when $K = 1.0$. Consequently, essentially 100% radiometric system throughput efficiency should be achieved when this hologon deflector is used with either randomly or circularly polarized light. Intrascan radiometric uniformity for this hologon deflector should only be dependent on the intrafacet diffraction efficiency uniformity.

Beam Magnification due to Prism and Grating Elements

When prism and grating elements are not used at the minimum deviation condition, the emerging parallel beam will be either larger or smaller than the incident parallel beam. Beam magnification and demagnification in prism and grating elements is anamorphic and occurs only in the direction in which the beam is deviated by the prism or grating element. The ratio R of the output beam diameter to the input beam diameter for a prism is given by

$$R_p = \frac{\cos \phi'_1 \cos \phi_2}{\cos \phi_1 \cos \phi'_2} \tag{69}$$

and for a grating is given by

$$R_g = \frac{\cos \theta_d}{\cos \theta_i} \tag{70}$$

where θ_i and θ_d are, respectively, the incident and diffracted beam angles with regard to the grating normal, as shown in Figure 5.10. The ratio curve in Figure 5.31 gives the R_p value for that particular prism apex angle and incident beam angle combination.

Schematically illustrated in Figure 5.32 is a hologon deflector that incorporates two identical prism elements that are used far from the minimum deviation condition. If it is assumed that the hologon in Figure 5.32 has $K = 1.00$ and $\theta_i = 30.8°$ for $\theta_s = \pm 18°$, and that the prism elements in this figure have an index $n = 1.799$ and an apex angle of approximately 12°, then the data in Figure 5.31 require that the incident beam angle on prism 2 be approximately 60° in order to minimize scan-line bow. Under these conditions, $R_p \simeq 1.6$ for prism 2, and therefore, the diameter of the output beam from prism 2 is magnified 1.6 times in the prism deviation direction when compared to the input beam diameter to

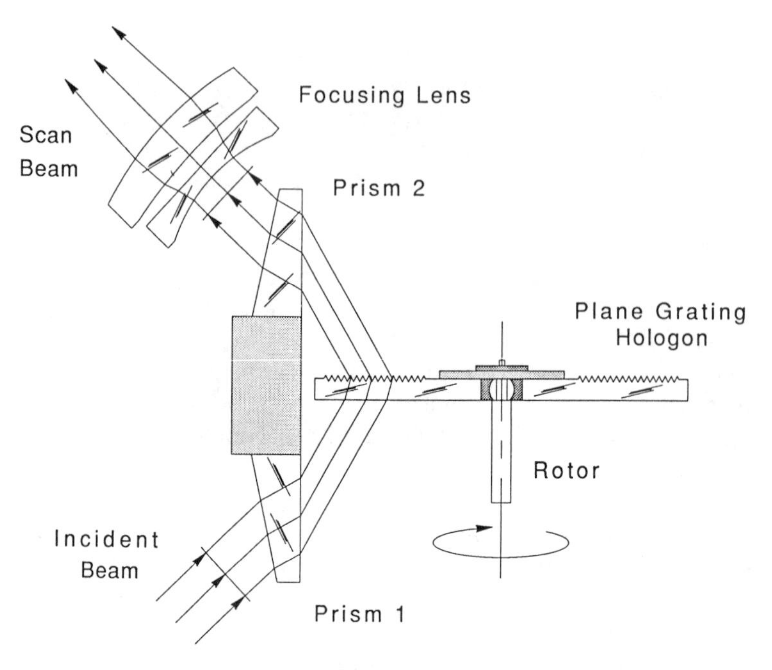

Figure 5.32 Schematic of hologon deflector incorporating two prism elements.

prism 2. Prism 1 in Figure 5.32 is used not only to compensate for the anamorphic beam magnification in prism 2, but being rigidly coupled to prism 2 it also compensates for scan beam tracking error associated with vibrations of prism 2. In essence the two prisms in Figure 5.32 function like a single prism that operates at the minimum deviation condition and has the hologon sandwiched within it.

Having anamorphic beam expansion after the hologon is advantageous in that it can improve both the scan duty cycle and scan beam tracking error of the system. Associated with an increase in beam size is always a decrease in the beam angular magnification factor. The beam angular magnification factor for prism and grating elements is given by 1/R. Having the hologon positioned between the two prism elements in Figure 5.32 reduces its contribution to the system scan beam tracking error by a factor of 1.6, while the remainder of the deflector system components are essentially unchanged in regard to their contribution to system scan beam tracking error. This level of reduction in hologon contribution to system scan beam tracking error can be significant for many deflector system applications.

Having the hologon sandwiched between the prism elements in Figure 5.32 also results in a 1.6 reduction in the cross scan incident beam size on the hologon facet surface. This beam reduction facilitates smaller radial facet size and/or higher hologon scan system duty cycle. Beam compression in the cross scan direction can significantly improve hologon scan duty cycle for the radial grating facet geometry illustrated in Figure 5.13 since the compression occurs in the direction of the facet rotation.

Commercial PDG Hologon Disk Deflector Unit

A sectional side view is presented in Figure 5.33 of a commercial hologon laser beam deflector unit that incorporates a prism to compensate for the bow in the scan beam from the PDG hologon disk deflector element. This deflector unit is designed to operate with a hologon having $K = 1.0$ and $\theta_i = 30.8°$ for a half-scan angle of 18°. It was decided when designing this deflector unit that the prism should operate at the minimum deviation condition, $\phi_1 = \phi_2$, since for this geometry the prism introduces no beam magnification effects for the central scan beam, scan-beam tracking error is very insensitive to prism vibrations, and scan-line bow performance is fairly insensitive to small misalignment errors in the cross scan angular orientation of the prism element. From the data presented in Figure 5.31, the prism apex angle should be approximately 24.7° and $\phi_1 =$ 22.5° if the prism has an index of 1.799 and is used at the minimum deviation condition.

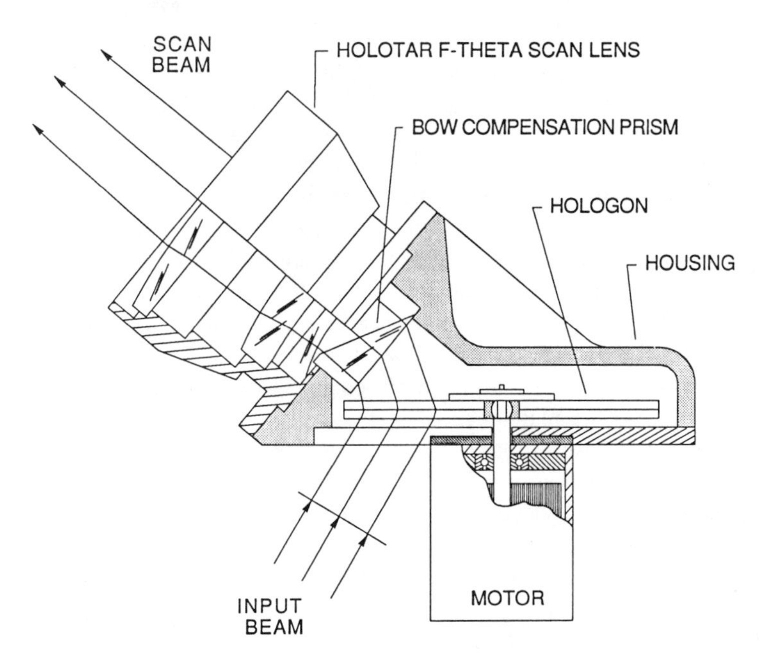

Figure 5.33 Sectional side view of a commercial hologon laser beam deflector unit. (Courtesy of Holotek Ltd., Rochester, New York.)

The hologon in Figure 5.33 is coupled to the ball-bearing, hysteresis synchronous motor by either means of a precision machined rigid hard mount or by a dynamic spherical bearing mount, as depicted in Figure 5.19. Typical scan-beam tracking error performance between all facet scans is usually <2 arc seconds for the Figure 5.33 deflector unit manufactured with an off-the-shelf ball-bearing motor and a f-θ scan lens that contains only spherical lens elements. This level of scan-beam tracking performance is equal to that achieved with the best commercially available air-bearing-mounted polygonal mirror deflector systems. Radiometric throughput efficiency for this hologon deflector is typically $>85\%$ for both S- and P-polarized light when photoresist is used as the holographic recording medium.

Procedures used to align the Figure 5.33 deflector unit to the other deflector system optical system components are briefly described in Section 5.8. Measured performance data for the Figure 5.33 deflector unit equipped with a f-θ scan lens having a focal length of 545 mm is presented in Ref. 8 for radiometric scan uniformity and throughput efficiency, scattered image light, image spot size, scan-line bow, cross scan positional beam error, in-scan positional beam error, and scan-beam velocity linearity error. The measured scan-line bow curve for this deflector unit is, as expected, symmetrical about the center of scan. The scan-line bow curve matches to within experimental measurement error the scan-line bow calculated using the ray tracing program for a scan beam angle of 16.4°, corresponding to the scan angle used in the measurements. The magnitude of the scan-line bow band increases as a strong function of the scan beam angle for this deflector geometry, and therefore, increasing the scan angle from 16.4° to 18°, causes the scan-line bow band to increase from the measured 60-μm value shown in Ref. 8 to the 80-μm value predicted by the data in Figure 5.31.

The Figure 5.33 deflector geometry is unique in that the magnitude of scan-line bow remains fairly constant as the incident beam angle θ_i is varied between 30° and 32°. This insensitivity to small changes in θ_i is important when the deflector is used to generate simultaneous multibeam scans as described in Section 5.7.4 since it reduces differential scan-line bow effects between the simultaneous scanning beams. This insensitivity occurs since changing θ_i alters both the diffracted ray angle θ_d and the magnitude of bow in the scan beam exiting the hologon. The change in θ_d causes the scan beam to propagate at a different angle through both the bow compensation prism and f-θ scan lens. This change in ray angle through these elements produces a corresponding change in the magnitude of the bow compensation introduced by these elements. This change in compensation is essentially equal and opposite to the variation in the magnitude of hologon scan bow caused by changing θ_i. This self-compensation effect indicates that the magnitude of residual scan-line bow for this deflector geometry is only easily altered by varying the angular relationship between hologon, prism and f-θ scan lens. Altering the tilt angle of the f-θ scan lens readily changes the magnitude and shape of the scan-line bow and is a useful way of adjusting the bow curve for the system.

Wavelength-Shift-Induced Scan-Beam Positional Errors

As described in the section "Wavelength-Shift-Induced Scan-Beam Tracking Error," wavelength shifts associated with mode hopping in diode lasers cause both in-scan and cross scan positional beam errors. Reducing the K value of the deflector reduces the scan tracking sensitivity with regard to wavelength shift but does not change the sensitivity of in-scan error to wavelength shift because the deflector must be rotated to a larger angle to achieve the same θ_s value. This larger rotation angle requirement means that the wavelength compensation grating method, described in the aforementioned section is less effective for this case.

The following example demonstrates the expected magnitude of the scan-beam tracking error due to wavelength shift for a deflector system incorporating a PDG disk hologon with $K = 1.0$ and a wavelength compensation grating having $K = 1.0$. It will be assumed that the cross scan angles of the compensation grating for this example is adjusted, according to equation (56), to totally correct the wavelength-shift-induced scan tracking error for the ± 0.71 scan-field positions. If $\theta_i = 30.8°$ and $\theta_R = 18°$ for the end-of-scan field position, the residual wavelength-shift-induced scan-beam tracking error as calculated with equation (54) for the center and end-of-scan positions is $d\theta_{d2} = \pm 0.093 d\lambda/\lambda$, respectively. A maximum scan-beam tracking error of 15 μm is achieved at the image plane if $d\lambda = 0.2$ nm, $\lambda = 670$ nm and $f = 545$ mm. This scan beam tracking error is 22% larger than that achieved for the $K = 1.4142$ hologon system incorporating a wavelength compensation grating.

When no wavelength compensation grating is incorporated into the hologon system, the maximum scan-beam tracking error for the stated system conditions is 186 μm for the $K = 1.00$ hologon case and 321 μm for the $K = 1.4142$ hologon case. The compensated scan-beam tracking error is larger for the $K = 1.0$ case because the deflector must be rotated to a larger angle to achieve the same θ_s value. This larger rotation angle causes a greater dispersive mismatch between the rotating hologon and the stationary wavelength compensation grating, thus reducing the wavelength shift correction capability of the total system.

So far, there has been no accounting for the dispersion properties of the scan bow compensation prism element. The dispersion power of a prism element is small compared to that of a grating element. For the wavelength range of 600 to 800 nm, the prism element in the deflector unit in Figure 5.33 would have approximately 1/400 the dispersive power of the hologon. If $f = 545$ mm, the prism element introduces a cross scan positional beam error of about 0.5 μm for $d\lambda = 0.2$ nm. This error is opposite in direction to the cross scan displacement error introduced by the hologon for this wavelength shift, and therefore, reduces the residual system cross scan displacement error for the end-of-scan position.

5.6.3 Utilizing a Grating Element to Reduce Scan-Line Bow

Replacing the prism element in Figure 5.29 with a grating element does not alter the end effect, in that the straight scan line from the hologon deflector is converted into a bowed scan line at the image plane. A grating element bows the straight scan line for the same reason that the prism does, that is, the incident beam changes its angular orientation with respect to the normal to the grating surface as it scans across the surface. As the angular direction of the scan beam departs from a plane that is perpendicular to the grating lines, the apparent grating line spacing decreases in the same manner that the separation between parallel lines appears to decrease as the lines propagate toward a distant vanishing point. This apparent decrease in grating line spacing causes the scan beam to undergo increasingly more deviation when it passes through the grating. The relationship between the change in grating output beam angle, $d\theta_d$, due to a change in the input beam angle, $d\theta_i$, is obtained by differentiating equation (20) with respect to $d\theta_i$:

$$d\theta_d = -\frac{\cos \theta_i}{\cos \theta_d} d\theta_i \tag{71}$$

Equations (68) and (71) are of the same form, and therefore, one would expect the scan-line bows produced by prism and grating elements to be of the same form. The scan-line bow produced by a grating compensation element is in the direction in which the grating diffracts the beam and is symmetric about the center of scan when the central scan ray is perpendicular to the grating lines. Grating periodic D and incident angle θ_i are the parameters that influence the rate of change between $d\theta_d$ and $d\theta_i$, and therefore, the rate at which bow is introduced into the scan beam.

Figure 5.34 illustrates the basic spatial relationship between the hologon deflector, the stationary auxiliary plane grating element used to compensate for bow in the scan beam exiting the hologon, and the focusing lens used to image the scan beam to a spot at an image plane which is not shown. Only auxiliary plane gratings were used to generate the calculated and measured results reported on here. In theory, the auxiliary grating element could incorporate optical focusing power that could be used to supplement or replace the focusing lens. The hologon and the auxiliary plane grating element in Figure 5.34 are both depicted as functioning at essentially the minimum deviation condition, $\theta_i \simeq \theta_d$.

Analysis of Scan-Line Bow with Compensation Grating

As with the prism element, a computer ray tracing program was utilized to determine how well an auxiliary plane grating element compensates for the bow in a scan line produced by a PDG hologon disk deflector. The ray tracing program was set up to find the auxiliary grating λ/D value as a function of incident angle

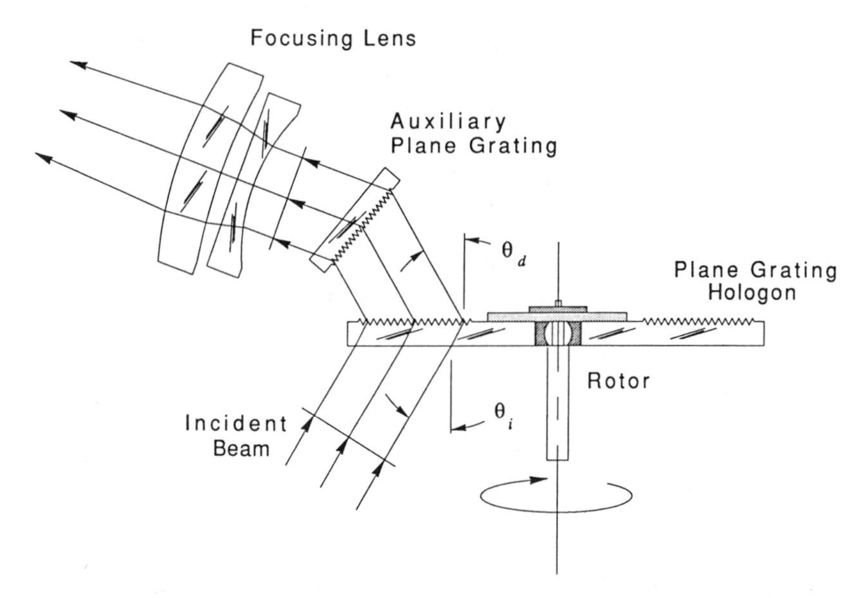

Figure 5.34 Hologon deflector incorporating a grating element for scan-line bow compensation.

on the auxiliary grating that minimized scan-line bow at the image plane when a scanning beam having a preprogramed scan bow was incident on the auxiliary grating. The preprogramed scan bow used for these ray tracing calculations was the same as that utilized in the prism ray tracing calculations, thereby enabling comparison of these two scan-line bow compensation approaches. This preprogramed scan bow was based on the scan beam produced by a hologon having $K = 1.00$, $\theta_i = 30.82°$, and $\theta_s = \theta_R = 18°$.

As noted for the bow-compensating prism case, these hologon deflector parameters minimize the scan-beam tracking error sensitivity with regard to deflector wobble for the entire scan line and $d\theta_{z1} = d\phi/61$ for the hologon deflector. As also noted for the prism case, this absolute angular difference condition only minimizes the scan-beam tracking error sensitivity with regard to deflector wobble for the entire scan length and does not correlate, as in the $K = 1.4142$ case, with either the shape or magnitude of the system scan-line bow curve achieved with the bow-compensating grating element.

A f-θ scan lens focal length of $f = 545$ mm was utilized in the ray tracing calculations so that the computed system scan-line bow band could be directly compared with those achieved with both the prism bow compensation method and that presented in Figure 5.16 for the $K = 1.4142$ case. Results of these ray tracing calculations are graphically presented in Figure 5.35 in a format similar to that used to present the prism bow compensation results in Figure 5.31. Like Figure 5.31, Figure 5.35 contains three individual curves and three corresponding vertical scales. These curves are labeled λ/*D* curve, bow curve, and ratio curve. Their corresponding vertical scales are labeled λ/*D* of auxiliary grating, residual scan-line bow, and beam ratio factor. All three curves are plotted as a function of the incident beam angle on the auxiliary bow compensating grating.

As expected, the λ/*D* curve in Figure 5.35 is similar in shape and in function to the apex curve in Figure 5.31. One can determine from the λ/*D* curve what auxiliary grating λ/*D* value is required to minimize the system scan-line bow for a given incident beam angle on the auxiliary grating. As was true with the prism bow compensation geometry, it is advantageous to design the system to use an incident beam angle on the auxiliary grating in the range of 0° to 15° since the required grating λ/*D* value for bow compensation is fairly constant over this range, and therefore, small misalignment of the grating with regard to the incident beam angle has little effect on the performance of the system. The bow curve in Figure 5.35 gives the residual system scan-line bow band achieved for each auxiliary grating λ/*D* value and incident beam angle combination. The ratio curve in Figure 5.35 gives the R_g value as calculated by equation (70), for each auxiliary grating λ/*D* value and incident beam angle combination.

Comparison of the overall range of residual scan-line bow band values in Figures 5.31 and 5.35 reveals that the auxiliary grating provides slightly better

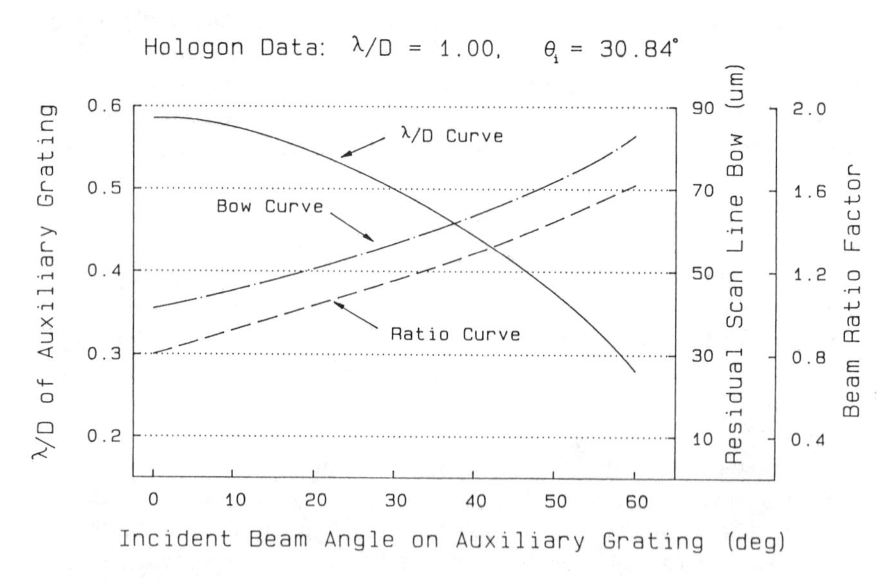

Figure 5.35 Results for ray tracing calculations for hologon plus grating geometries that minimize scan-line bow.

compensation for the scan bow produced by a PDG hologon having $K = 1.00$ and $\theta_i \simeq \theta_d$. A better comparison of the bow correction with auxiliary prism and grating elements is achieved when each bow compensation element is used at the minimum deviation condition, that is, $R_p = R_g = 1.0$. Under these conditions, a residual system scan-line bow band of approximately 80 μm and 50 μm are, respectively, achieved for the auxiliary prism and grating elements. When compared on this basis, the bow-compensating grating element reduces the residual system scan-line bow band for the $K = 1.00$ hologon case by approximately 60% of the value achieved for either the bow-compensating prism method when used with a $K = 1.00$ hologon or for the uncompensated hologon operating at $K = 1.4142$.

Scan-Line Bow as a Function of Hologon *K* Value

Further investigation of the auxiliary grating bow compensation approach as a function of hologon *K* value was prompted by the significant improvement achieved with this approach when compared with that achieved with the prism bow compensation method and the uncompensated hologon operating at $K = 1.4142$. Presented in Figure 5.36 is a graph of the λ/D value required by an auxiliary grating element to minimize the scan-line bow produced by hologons having *K* values of between 0.6 to 1.4142. The data in Figure 5.36 was calculated for the case where the auxiliary grating is used at the minimum deviation condition for the central scan beam and the PDG disk deflector is operated to minimize the scan-beam tracking error sensitivity with regard to deflector wobble for the entire scan-line length when $\theta_s = \pm 18°$. Examination of Figure 5.36 reveals that, as expected, no grating element is needed to minimize the scan-line bow when the hologon *K* value is 1.4142. Figures 5.35 and 5.36 agree in that an auxiliary grating having a λ/D value of approximately 0.55 is required to compensate for the scan-line bow produced by a hologon having $K = 1.00$.

According to the data in Figure 5.36, a hologon having a *K* value of 0.8 is of particular interest in that an auxiliary grating having a λ/D value also of about 0.8 can be used to minimize the scan-line bow produced by the hologon. Having the same grating periodicity for both the hologon and the auxiliary grating element would not only facilitate the fabrication of the deflector system but should also produce a system having essentially no residual scan-line bow since the bow-producing properties of the hologon exactly match those of the auxiliary grating element.

The residual scan-line bow for each of the data points represented by the curve in Figure 5.36 is graphically presented in Figure 5.37 as a function of the hologon *K* value. As postulated, the residual scan-line bow in Figure 5.37 reduces to zero for a hologon *K* value of approximately 0.8. To verify this calculated scan-line bow performance, an experiment was performed to measure the scan-line bow of a deflector system that utilized an auxiliary grating element of $\lambda/D = 0.83$

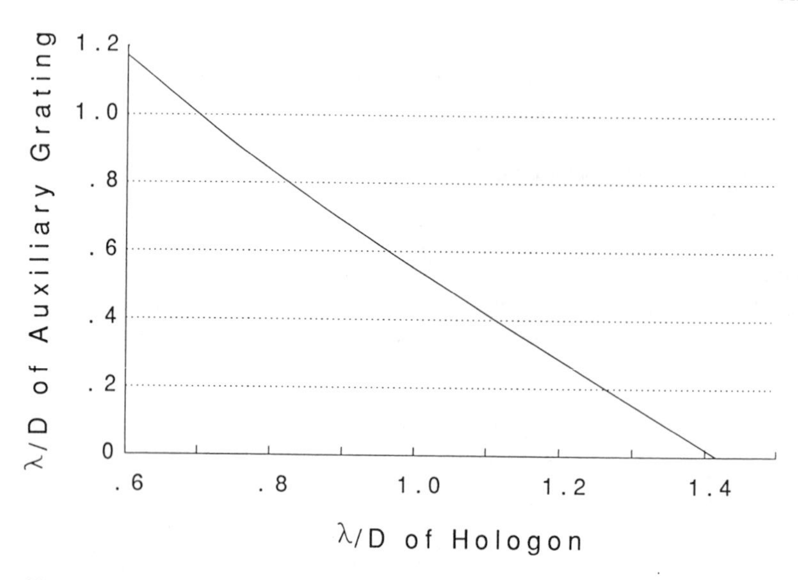

Figure 5.36 Grating versus hologon λ/D value for minimum scan-line bow.

to compensate for the scan-beam bow produced by a hologon having K also equal to 0.83. From the data presented in Figure 5.36, the auxiliary grating has to have a λ/D value of 0.8 if it is to simultaneously operate at the minimum deviation condition for the central scan beam and minimize the scan-line bow for a hologon having $K = 0.83$ and $\theta_s = \pm 18°$. In order to minimize the system scan-line bow for these hologon parameters, the auxiliary grating was orientated to have an incident beam angle of $\theta_i = 17.15°$, which produced a diffraction angle of $\theta_d = 32.35°$. The scan-line bow for the deflector system was measured at the image plane of a f-θ focusing lens having a 545-mm focal length. These measurements were performed using the test setup described in Ref. 8.

Measured data for the deviation from straightness of the scan-line for this deflector system is presented in Figure 5.38 as a function of scan-field position. The maximum measured field position of ± 170 mm corresponds to a scan angle of $\theta_s = \pm 17.9°$. The measured data points in Figure 5.38 are shown as open circles crossed by vertical error bars. All of these error bars have been set equal to ± 10 μm, which is calculated to be the mean error for the test setup measurement in the cross scan direction. The solid curve in Figure 5.38 is the theoretical calculated scan-line bow curve for this deflector system geometry.

Analysis of the data in Figure 5.38 reveals that the majority of the measured data points provide a very close fit to the theoretical scan-line bow curve. Four consecutive measured data points in Figure 5.38, corresponding to field positions -42.5 mm to $+85$ μm, reside on a straight line that is tilted by approximately

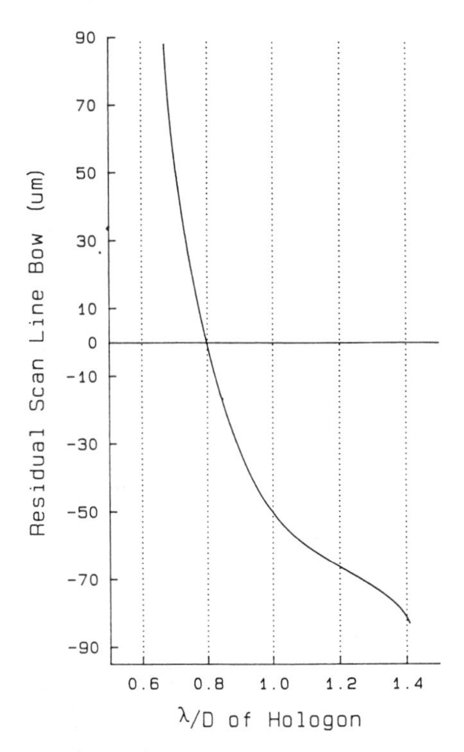

Figure 5.37 Scan-line bow versus hologon λ/D value for corresponding grating data in Figure 5.35.

25 seconds of arc with respect to the overall system scan line. This tilt is probably due to a small skew error in the long translation slide assembly used to make the scan-line bow measurements. From both practical and measurement considerations, the scan line in Figure 5.38 can be assumed to be straight.

Combination of Grating plus Prism Elements

From radiometric considerations, it is desirable to utilize a hologon having $K \geq 1.1$, because the second-order diffracted beam is prevented from existing, as calculated by equation (63). Not only does the second-order diffracted beam take power from the first-order diffracted beam, it can, by means of multiple reflections within the grating substrate material, generate ghost first-order beams. These ghost beams can produce intensity interference effects that translate into scan-beam intensity fluctuations. In a number of deflector system applications it is desirable to use a hologon having $K = 1.0$ since this makes $\theta_s = \theta_R$. One drawback with the auxiliary grating bow compensation technique is that for

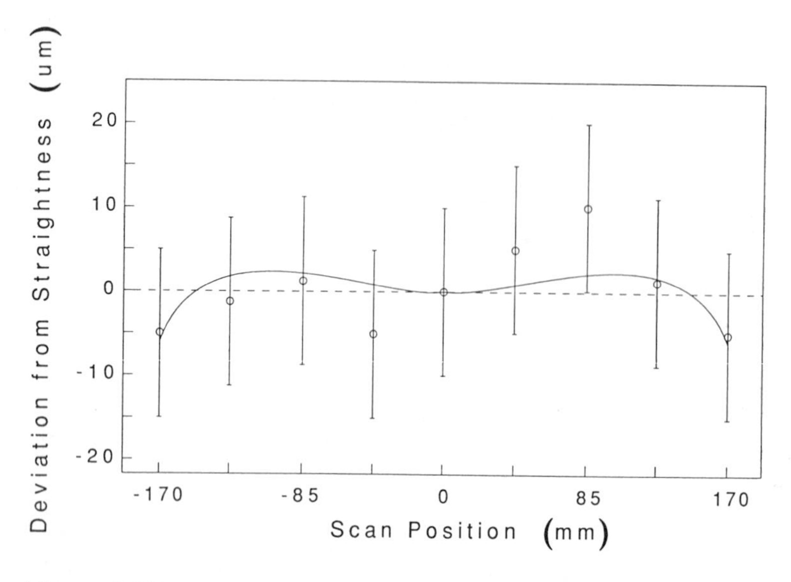

Figure 5.38 Measured data for scan-line bow for hologon having the same λ/D = 0.83 as that used for the bow compensation grating element.

hologons having $K \geq 1.0$, the auxiliary grating has a $\lambda/D \leq 0.6$. Surface-relief transmission gratings cannot be used to achieve high first-order diffraction efficiency for λ/D ratios ≤ 0.6. Also, intensity fluctuations associated with the second-order diffracted beam are a problem.

Figure 5.39 schematically illustrates a scan-line bow compensation technique that first uses a prism element to magnify the scan-beam bow produced by the hologon deflector element, and then an auxiliary plane grating element is utilized to compensate for the magnified scan-beam bow. It should be feasible with this technique to achieve minimum scan-line bow for a hologon having $K = 1.0$ with an auxiliary grating element having $\lambda/D \simeq 1.0$. This combination may produce considerably less scan-line bow band for the $K = 1.0$ hologon case, when compared with the scan-line bow reduction achieved when either an individual prism or auxiliary grating element is utilized by itself. Factors that tend to support the postulation that the Figure 5.39 geometry will provide improved scan-line bow performance include: the extra degree of freedom achieved by using two compensation elements and the bow-producing properties of the hologon would closely match those of the auxiliary grating element, and thereby, provide a better match between some of the higher-order terms involved in the compensation process.

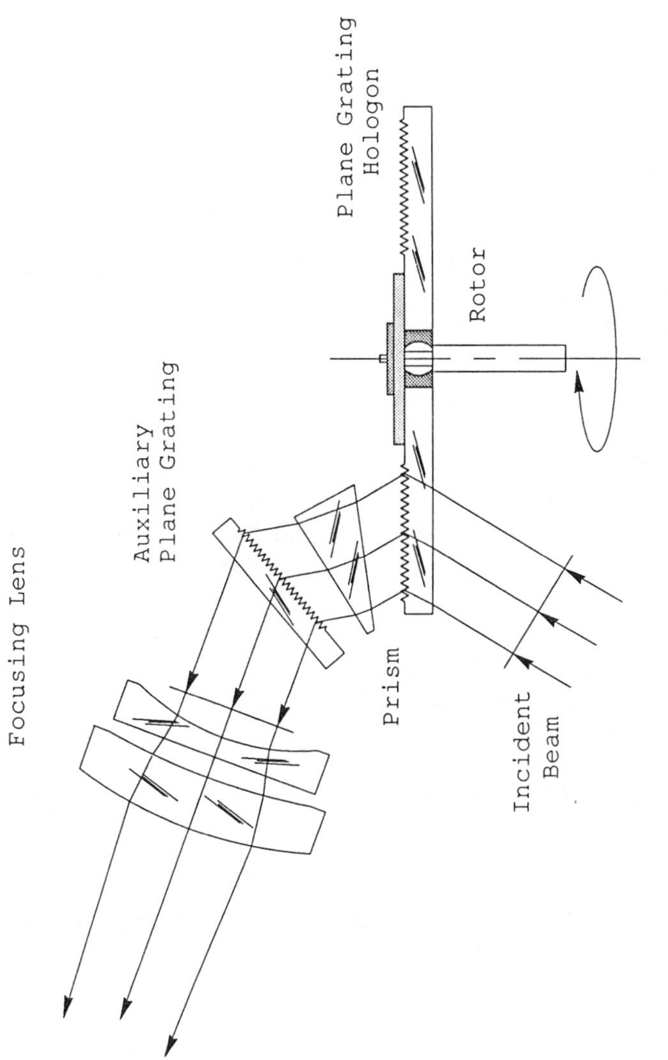

Figure 5.39 Hologon deflector incorporating both a prism and grating bow compensation element.

Wavelength-Shift-Induced Scan-Beam Positional Errors

Hologon in-scan positional error due to wavelength shift is not altered by the presence of the bow compensation grating element because it is stationary and has its grating lines parallel to the scan direction. Deflector system scan-beam tracking error sensitivity with regard to wavelength shift is essentially doubled due to the presence of the bow compensation grating. This is because the dispersive power of the bow compensation grating is of the same magnitude and in the same direction as that of the hologon. Reducing the scan tracking error due to wavelength shift for this combined grating case can be accomplished by utilizing a wavelength compensation grating having dispersive power equal to the dispersive power of the combined two gratings.

To calculate what the scan-beam tracking error is as a function of wavelength change for this three-grating case, we continue the grating equation designation of equations (49) and (50) for the first two grating elements and use the following grating equation notation for the third bow compensation grating:

$$\sin \theta_{i3} + \sin \theta_{d3} = \frac{\lambda}{D_3} \tag{72}$$

Differentiating equation (72) with respect to $d\lambda$ and setting $d\theta_{i3} = -d\theta_{d2}$ give the change in θ_{d3} for the center of scan when the equation (51) value is used for $d\theta_{i2}$:

$$d\theta_{d3} = \frac{D_1 D_2 \cos \theta_{d1} \cos \theta_{d2} + D_3 \cos \theta_{i3}(D_1 \cos \theta_{d1} - D_2 \cos \theta_{i2})}{D_1 D_2 D_3 \cos \theta_{d1} \cos \theta_{d2} \cos \theta_{d3}} \, d\lambda \tag{73}$$

The system is totally corrected for wavelength shifts for the center of scan when $d\theta_{d3} = 0$, corresponding to

$$D_1 \cos \theta_{d1} = \frac{D_2 D_3 \cos \theta_{i2} \cos_{i3}}{D_2 \cos \theta_{d2} + D_3 \cos \theta_{i3}} \tag{74}$$

Once the values for gratings 2 and 3 are determined, equation (74) can be satisfied by a large number of values for D_1 and θ_{d1}. For the special case where $D_1 = D_2 = D_3$ for $\lambda/D = 0.81$ and $\theta_{i2} \approx \theta_{d2} \approx \theta_{i3} \approx \theta_{d3} \approx 23.89°$, equation (74) becomes

$$\cos \theta_{d1} = \frac{\cos \theta_{i2}}{2} \tag{75}$$

and $\theta_{d1} = 62.8°$, corresponding to $\theta_{i1} = -4.55°$.

An equation for the scan-beam tracking error due to wavelength shift for each hologon rotation angle is achieved by differentiating equation (72) with respect to $d\lambda$ and setting $d\theta_{i3} = -d\theta_{d2}$, where equation (54) is used for the $d\theta_{d2}$ value.

By setting $d\theta_{d3} = 0$ in this equation, the condition for total correction of wavelength shift as a function of hologon rotation angle is achieved:

$$D_1 \cos \theta_{d1} = \frac{D_2 D_3 \cos \theta_{i2} \cos \theta_{i3}(K_2 \cos \theta_R - \sin \theta_{i2})}{D_3 \cos \theta_{i3}(K_2 - \sin \theta_{i2} \cos \theta_R) - D_2 \sin \theta_{d2} \cos \theta_{d2}} \tag{76}$$

Equation (76) can be solved so that total correction for wavelength shift occurs at the 0.71 scan-field position and, thereby, minimizes the scan tracking error associated with wavelength change for the total scan field.

5.6.4 Summary of System Performance for Improved PDG Hologon Disk Deflectors

Theoretical and experimental data have been presented that show how the bow in a scan-line generated by a PDG hologon disk deflector system can be minimized or eliminated by incorporating either scan-lens tilt or an auxiliary prism or grating element into the deflector geometry. These scan bow compensation techniques enable optimized hologon disk deflector system solutions to be constructed for K values less than 1.4142. When compared with the basic PDG disk deflector having $K = 1.4142$, the lower K value hologon systems have improved deflector performance with regard to scan-beam tracking error, system scan duty cycle, and radiometric scan uniformity and throughput efficiency for both S- and P-polarization states.

The improvements in performance for these lower-K-value PDG hologon disk deflectors enables them to better meet the requirements of graphic arts electronic imaging systems used for high-resolution reproduction of line art and halftone images. Typical uncorrected scan-beam tracking error performance between all facet scans for the deflector unit in Figure 5.33 is <2 arc seconds when $K = 1.0$, $\lambda = 0.6328$ μm, $\theta_i = 30.8°$, and $\theta_s = \pm 18°$. This scan tracking error performance is achieved for the hologon deflector mounted on a ball-bearing motor and used with an optical imaging system comprising only spherical lens elements. Radiometric throughput efficiency for the hologon element in this system is typically >85% for S- and P-polarized light when photoresist is used as the holographic recording medium. Variations in scan-to-scan and intrascan radiometric uniformity for this photoresist hologon deflector element are typically both $<\pm 3\%$.

These improved hologon disk deflector geometries still typically achieve only a maximum scan-beam tracking error insensitivity with regard to deflector wobble of 100 to 1, still have some measurable scan-line bow, and are restricted to usable scan-beam angles of $<\pm 20°$ for most graphic arts imaging applications. Major factors contributing to the restriction on scan angle are scan-line bow, scan-beam positional error sensitivity with regard to deflector wobble, and scan-beam image spot ellipticity. These factors increase as a function of scan-beam angle for disk hologon deflector geometries.

The major performance problem for these improved hologon disk deflector systems is still wavelength shift due to mode hopping in diode lasers. Diode lasers cannot be utilized with these deflector systems for graphic arts imaging applications unless the diode laser is stabilized with regard to operating wavelength or is modified so that the wavelength shifts gradually with time in comparison to the system scan rate.

The scan bow compensation techniques described in this section can be used to correct for the scan-line bow in galvanometer and rotating polygonal mirror systems [43]. It is sometimes advantageous in these mirror systems to bring in the laser beam at an angle to the mirror rotation plane. The reflected scan beam under these conditions will produce a bowed scan-line.

5.7 NONDISK PLANE GRATING HOLOGON DEFLECTORS

It is apparent from the preceding discussions of PDG disk hologon deflectors that the disk configuration suffers from being limited to scan angles of $< \pm 20°$ for most graphic arts imaging applications. This scan-angle limitation makes it impractical to utilize a hologon disk deflector for imaging applications requiring a scan addressability greater than approximately 22,000 dots per scan, based on $\lambda = 0.6328$ μm and a 50% intensity spot overlap. Other benefits derived from having a large scan-beam angle include improved scan system duty cycle, reduction in f-θ scan-lens focal length for a given format scan length, reduction in deflector size, reduction in scanning system package size, and reduction in the image plane magnitude of scan beam displacement errors associated with deflector scan-beam tracking errors, air turbulence, and system vibrations.

5.7.1 Nondisk Monograting Deflector for Internal Drum Imaging System

Very large scan-beam angles are only practical in deflector systems that possess total rotational symmetry with regard to imaging properties. Total image rotation symmetry is achieved with the internal drum laser beam deflector unit that is illustrated in Figure 5.40 [6, 7, 45]. This monofacet deflector utilizes a single transmission, plane grating facet orientated at approximately 45° to the deflector rotation axis. Following the PDG facet is a single element scan lens that rotates with the deflector unit and, therefore, is always used on its optical axis. The incident beam propagating parallel to the rotation axis is redirected after diffraction

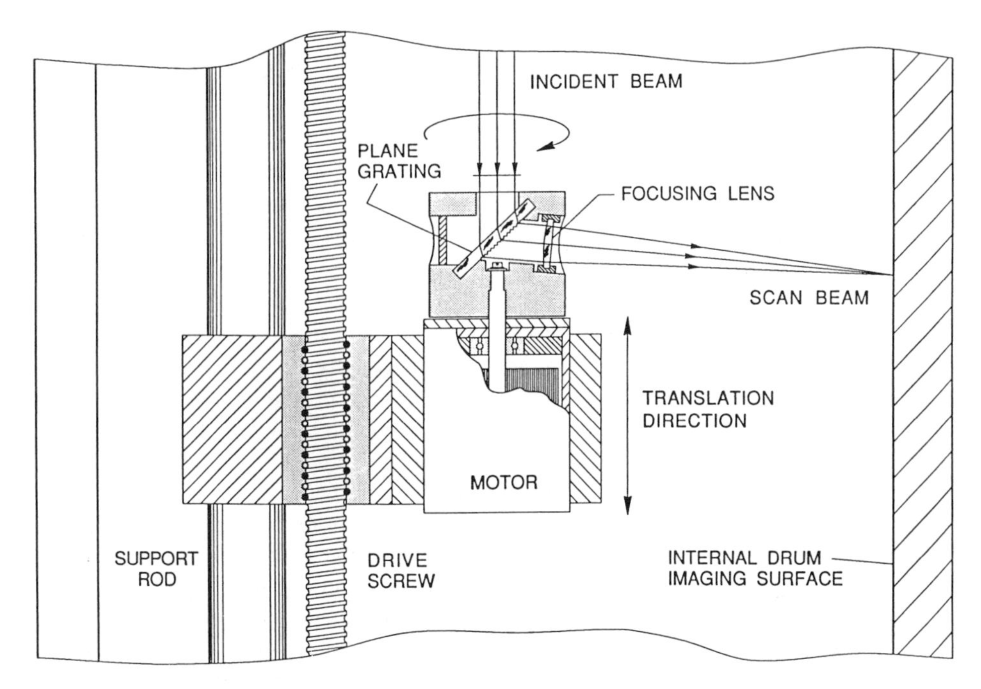

Figure 5.40 Sectional side view of a monograting deflector unit being used in an internal drum imaging system. (Courtesy of Holotek Ltd., Rochester, New York.)

by the grating facet so that it exits the deflector unit approximately perpendicular to the rotation axis, that is, $\theta_i = \theta_d \simeq 45°$. The deflected beam in Figure 5.40 is slightly offset from being perpendicular to the rotation axis so that the retro-reflected specular light from the internal drum image surface does not propagate back along the incident beam and cause ghost beams and laser intensity instability. For this nondisk PDG deflector system geometry, the scanning image spot size has both constant size and constant scan velocity along the entire image surface because the image surface is a cylinder having its axis collinear with the deflector rotation axis.

The PDG facet and focusing lens in the monograting deflector unit in Figure 5.40 could be replaced by a single, flat holographically formed facet that incorporates optical imaging power. For this case, the facet would both diffract and focus the input collimated beam to a point at the internal drum image surface. This approach is not as commercially desirable as the PDG facet case for the following reasons: it is presently easier to manufacture the PDG facet plus focusing lens than the single holographic facet lens element; putting optical power in the facet makes it difficult to make at one wavelength and use at a different wavelength; and the axial position of the present lens can be adjusted to compensate for small differences in either lens focal length or internal drum diameter.

Configuration of Internal Drum Imaging System

As indicated in Figure 5.40, the deflector unit is translated along the axis of the internal drum imaging surface while the recording film is held stationary to the internal drum surface by means of either gravity or vacuum assist. The laser light source, acoustooptic modulator, and beam expansion/collimating optics are typically mounted outside the drum with a fold mirror directing the collimated incident beam along the drum axis, as illustrated in Figure 5.41. Factors that limit the recording scan angle for this recording configuration are the deflector support and drive mechanisms, and the need to provide an opening along one axial section of the drum wall for film loading and unloading. Recording scan angle for this type of imaging system is typically between 180° and 270°.

The internal drum imaging approach trades the problems and cost associated with minimizing the field curvature of a wide angle, flat-field, f-θ scan lens for the problems and cost associated with fabricating a drum imaging surface that must be concentric with the deflector translation axis to within the depth of focus of the scanning image spot. The problems and cost associated with the mechanisms for loading, holding, and unloading the recording material from the internal drum surface have also to be compared with those of accurately translating film past a flat-field scanning image spot. Experience indicates that the internal drum scanning approach becomes more cost effective than flat-field scanning systems when greater than approximately 28,000 addressable dots per scan are required, based on $\lambda = 0.6328\ \mu m$ and a 50% intensity spot overlap.

Internal Drum Deflector System Imaging Properties

Rotational symmetry with respect to the deflector element is achieved in both the Cindrich-type hologon disk deflector in Figure 5.5 and in the nondisk monograting deflector in Figure 5.40 by having the collimated incident beam be parallel to the deflector rotation axis. Under these conditions, the incident beam maintains both a constant angle of incidence with regard to the facet surface normal and with the coordinate system associated with the rotating facet grating lines. Therefore, the diffraction angle of the scan beam remains constant with regard to the facet surface normal, $\theta_d = \theta_{do}$, and the scan-beam angle is equal to the deflector rotation angle, $\theta_s = \theta_R$.

The PDG facets for the nondisk monograting deflector and its multifacet version are individually fabricated and then placed in the cylindrically shaped deflector housing at approximately 45° to the deflector rotation axis. These grating facets operate at the minimum deviation condition for every deflector rotation angle, and, therefore, scan-beam tracking error is virtually totally insensitive with regard

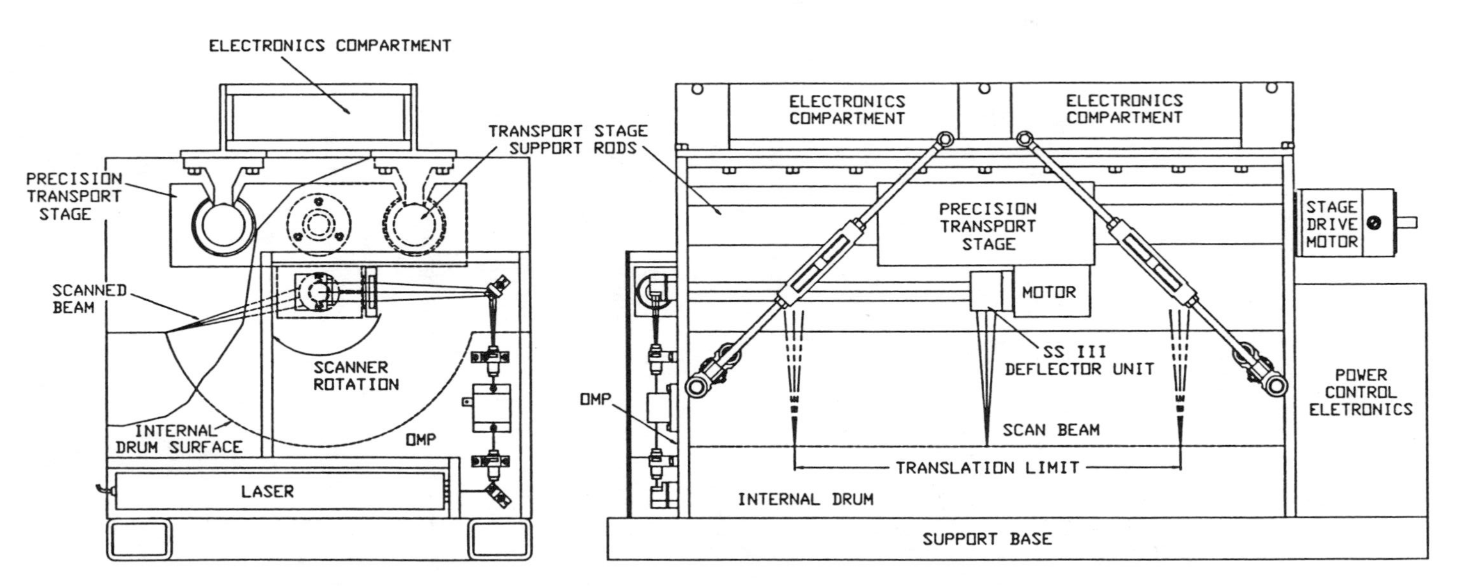

Figure 5.41 Schematic illustration of an internal drum imaging system that utilizes the monograting deflector unit in Figure 5.40 to scan and image the incident modulated laser beam along the internal drum surface.

to deflector wobble. These nondisk PDG deflectors do not have a problem with regard to the image spot becoming elliptically shaped as a function of scan-beam angle when underfilled facet operation is used. This is because the incident beam profile on the facet surface tracks with scan beam angle. A circular profile image spot is achieved for every scan angle when the incident beam to the deflector has a circular profile.

An angle of 42.5° is used for θ_i and θ_d in the monograting deflector in Figure 5.40 and, therefore, the principal scan ray exits the deflector unit at an angle of 85° with regard to the deflector rotation axis. For this internal drum imaging configuration, scan-line straightness and image spot velocity uniformity are independent of the angle that the scan beam makes with respect to the deflector rotation axis. These imaging parameters depend on the accuracy of the concentricity between the deflector rotation axis and the internal drum recording surface. A 5° deviation angle between the scan beam and image surface normal was selected because this ensured that retroreflected specular light from the image surface does not reenter the focusing lens, even for the largest designed scan-beam ray cone angle. Image resolution is degraded by utilizing too large of a deviation angle between the image surface normal and the scan-beam principal ray. This degradation occurs because the image spot becomes elliptically shaped in the cross scan direction and because of an increase in image flair associated with multiple reflection of the skewed incident scan beam within the recording medium.

Ray tracing calculations and system measurements have shown that when the single element focusing lens in the monograting deflector in Figure 5.40 has the optimum shape factor to minimize spherical aberration, the deflector will produce a diffraction-limited, $1/e^2$ intensity diameter spot size of 10 μm for $\lambda = 0.6328$ μm. This F/10 system has a maximum ray cone angle of approximately 4.6° when used with an untruncated incident Gaussian beam. Over 5000 dpi scan resolution can be achieved with this deflector system for an image format scan length that is limited only by the diameter of the internal drum. A drum diameter of 260 mm provides a 610 mm image format scan length when a recording scan angle of 270° is used, corresponding to a scan duty cycle of 75%. A scan resolution of 5000 dpi can be achieved for $\lambda = 0.6328$ μm and this 260 mm drum diameter by utilizing a monograting deflector having a clear aperture of only 16 mm.

Equation (5) states that when the incident Gaussian beam to the monograting deflector is truncated at the $1/e^2$ intensity points, the theoretical depth of focus for a 10% growth in image spot size is only about 166 μm for the 10 μm spot-size case. This depth of focus value indicates the required cumulative accuracy for the concentricity between the drum recording surface curvature and the deflector rotation axis, and for the tolerance on the focal length of the single-element focusing lens. The lens focal length tolerance is relaxed by the axial position adjustment provided by screwing the lens element into the deflector head. The cumulative accuracy for these system parameters is relaxed by a factor of 5 when the image spot size is increased to 22.5 μm for the $1/e^2$ intensity points. This scanning spot size is adequate for most high-resolution graphic arts imaging applications.

Scan-Beam Tracking Error Performance of Internal Drum Deflector

When the nondisk monograting deflector is hard mounted to the motor shaft, as depicted in Figure 5.40, it has a fixed wobble tilt angle with respect to the motor shaft of about 1 to 10 min of arc. This fixed deflector wobble angle does not cause a scan-beam tracking error problem for the monofacet deflector system since any resulting variation in scan-beam tracking angle is the same for each scan line. For the same reason, substrate wedge angle and grating periodicity error also do not cause a scan tracking error problem for the monograting deflector system. Relative changes in the fixed deflector wobble angle are what cause change in the scan-to-scan beam tracking error performance for a monofacet deflector system. Both random and periodic changes in the fixed deflector wobble angle can occur due to inaccuracies and/or vibrations within the rotor/bearing assembly of the deflector motor. These changes in the fixed deflector wobble angle have negligible influence on both halftone and line art reproduction image quality when the resulting changes in the scan-beam tracking error performance of the deflector system are minimized.

Neither random nor periodic changes in the fixed deflector wobble angle are expected to cause a problem with respect to the halftone and line art reproduction image quality performance of the monograting deflector system in Figure 5.40. This is because the scan-beam tracking error performance of this deflector unit is insensitive with regard to change in the cross scan angle of the unit. The grating facet in this deflector unit operates at the minimum deviation condition for every deflector rotation angle, and therefore, its scan-beam tracking error performance is essentially totally insensitive with regard to change in its cross scan angle. For example, if it is assumed that deflector motor deficiencies cause a maximum change in the deflector cross scan tilt angle of about 5 arc seconds, equation (42) predicts that the resulting change in cross-scan beam angle is 2.22×10^{-4} arc second for a grating facet having $\theta_i = \theta_d = 42.5°$. This change in cross-scan beam angle is not detectable and can be assumed to be zero. For this grating facet geometry, 10 arc minutes of fixed deflector wobble angle cause only 1.92 arc seconds of fixed scan-beam tilt angle.

Deflector wobble can cause changes in both the angular orientation and cross scan position of the focusing lens rotating with the monograting deflector unit in Figure 5.40. The angular direction of the rays leaving this focusing lens are essentially insensitive to changes in lens angular orientation. This is because the principal planes of a single-element lens are located near the center of the lens. If the deflector wobbles about a point displaced from the focusing lens center, the lens center and the image spot produced by the lens are both equally displaced.

For example, consider the case when deflector motor deficiencies produce a change in the fixed deflector cross scan tilt angle of 3 arc seconds about a point located on the deflector rotation axis. If the focusing lens element is located 10 mm from the deflector rotation axis, the lens center and corresponding scan image spot are both displaced in the cross scan direction by 0.24 μm. The magnitude of this beam displacement is small, particularly when compared with the associated change in deflector tilt angle. It is expected that deflector motor deficiencies will cause a maximum change in the fixed deflector wobble angle of about 2 to 3 arc seconds. For this range of angular change, the resulting image spot displacement due to change in position of the focusing lens rotating with the monofacet deflector unit in Figure 5.40 has negligible effect on the line art and halftone reproduction quality of the imaging system.

The preceding discussion on the correlation of focusing lens displacement with image spot displacement indicates that dynamic change in the axial runout of the deflector motor shaft also contributes to the scan-beam tracking error problem for internal drum imaging systems. This is true whether the focusing lens follows the deflector element, as in Figure 5.40, or precedes it, as is the case for most mirror-based internal drum imaging systems. Dynamic change in the axial runout of the deflector motor shaft does not contribute directly to the scan-beam tracking error in PDG hologon systems having stationary focusing lenses. This is because nonangular positional changes in a collimated incident beam to a stationary focusing lens do not alter the location of the imaging spot produced by that lens.

Dynamic change in axial runout in ball-bearing motors can be minimized by increasing the axial preload on the motor bearings. Unfortunately, increasing bearing preload reduces the life for the motor bearings and usually reduces the maximum operating speed of the motor. A major consideration in the design of an internal drum imaging system is the dynamic axial specification for the deflector motor bearing assembly.

As described earlier the major disadvantage of the grating deflector technology is that is must be utilized with a very monochromatic laser source. Wavelength shifts due to mode hopping in a diode laser cause abrupt, unacceptable scan beam tracking errors in grating deflector systems used for graphic arts imaging applications. Therefore, diode laser light sources cannot be used with grating deflectors for these imaging applications unless the diode laser is stabilized with regard to operating wavelength or is modified so that the wavelength shifts gradually with time in comparison to the system scan rate.

It is interesting to note that for a hologon deflector possessing rotational symmetry, such as the nondisk PDG deflector geometry of Figure 5.40, wavelength shift does not produce any in-scan positional beam error, only scan-beam tracking error. Unfortunately, when the compensation grating method described earlier is used to reduce the scan-beam tracking error for a rotational symmetric hologon deflector, it introduces in-scan positional beam error with wavelength shift. This problem does not occur if either a cylinder lens or a toroidal lens is used to correct for the scan tracking error caused by wavelength shift in a rotational symmetric hologon deflector. These anamorphic optical elements introduce their own problems.

Scan-Beam Jitter Performance of Internal Drum Deflector

As was the case for the scan-beam tracking error performance, fixed deflector wobble angle, substrate wedge angle, and grating periodicity error do not contribute to the scan jitter error for the nondisk monograting deflector system since any resulting variation in scan beam angle is the same for each scan line. A fixed in-scan deflector wobble angle causes a corresponding offset in the scan-beam angle for the nondisk PDG deflector geometry. This offset angle remains constant with scan position and, therefore, is readily compensated for by using a start-of-scan detector signal to synchronize the start of scan-line pixel data.

It follows from the preceding statement that for the nondisk PDG deflector geometry, an in-scan change in the fixed deflector wobble angle causes an equal corresponding angular change in both the facet element rotation angle and scan-beam angle. Changes in the scan jitter error associated with changes in the fixed deflector wobble angle are of greater concern in nondisk PDG deflector systems than in disk PDG deflector systems. This is because in a nondisk PDG deflector system there is a one-to-one relationship between these changes, whereas, the maximum change in scan-beam angle for a disk PDG deflector system is typically 20 times smaller than the in-scan change in deflector wobble angle, as calculated by equation (43). A one-to-one relationship between change in deflector wobble angle and change in scan-beam angle also applies to a monofacet deflector that uses a pyramidal mirror facet or a pentaprism facet [46, 47].

While changes in the fixed deflector wobble angle have only negligible influence on the scan-beam tracking error performance of a nondisk deflector system, these changes can contribute significantly to the scan jitter performance of the system. Therefore, care has to be taken in nondisk PDG deflector systems to minimize change in the deflector wobble angle. This is accomplished by reducing the vibrations in the deflector motor assembly that cause the change in the fixed deflector wobble angle. This vibration minimization is accomplished by precisely balancing the deflector motor assembly and specifying tight tolerances on the rotor/bearing assembly used in the motor. Reducing the mass of the deflector unit also contributes significantly to minimizing dynamic changes in deflector wobble angle.

If it is assumed that residual deflector motor deficiencies cause a maximum change in the in-scan deflector wobble angle of about 3 arc seconds, the corresponding in-scan positional beam error for a monofacet deflector would be 1.5

μm when the deflector is used in an internal drum imaging system having a radius of 100 mm. When this monofacet deflector is used in a flat-field imaging system, as described in Section 5.7.2 a 3-arc-second in-scan change in deflector wobble angle would produce a 7.5-μm in-scan positional beam error when the system incorporates an *f*-θ scan lens having a 500-mm focal length. While the magnitude of deflector wobble induced in-scan errors can be fairly large, they usually have only a small influence on recorded image quality due to their random nature and because they only change a limited number of pixel positions within a scan line.

Scan-beam jitter error associated with deflector motor hunting is normally more of a problem with monofacet deflectors than with their multifacet equivalents. This is because the monofacet usually has less inertia of rotation as a result of having a significantly smaller deflector diameter. Motor hunting causes a slow variation in the scan jitter that can be compensated for by using a variable pixel clock rate, as described in Section 5.2.3. Many monofacet deflector units incorporate a shaft encoder to measure the variation in deflector motor speed associated with motor hunting. Shaft encoder data can be used in conjunction with a start-of-scan detector signal to calculate the change in the pixel clock rate required to compensate for the motor hunting component of scan jitter.

Radiometric Performance of Internal Drum Deflector System

A linearly polarized incident laser beam should not be utilized with the monograting deflector in the internal drum imaging system in Figure 5.40. It is evident from equation (58) and the diffraction efficiency data in Figure 5.23 that if a linearly polarized beam were used the radiometric throughput efficiency of the deflector unit would vary between approximately 90% and 5% for a 90° deflector rotation angle. Uniform intrascan radiometric efficiency is achieved for this deflector geometry by using either a randomly or circularly polarized incident laser beam. Approximately 50% deflector radiometric throughput efficiency is achieved for these polarization states.

For halftone imaging applications, it is preferable to circularly polarize the output of a linearly polarized laser than to use a randomly polarized laser. This preference occurs because randomly polarized lasers can exhibit fairly large noise intensity fluctuations when used with linear polarizing elements, such as a grating element having $K = 1.4142$. Either a quarter-wave retardation plate or a Fresnel rhomb can be used to convert a linearly polarized beam to circular polarization.

Measured Performance for Internal Drum Deflector System

Performance data for two different sizes of nondisk monograting deflector units that are designed for internal drum imaging applications are reviewed. These deflector units have optical clear apertures of 12 mm and 24 mm and outer diameters of 46 mm and 55 mm, respectively. Each deflector unit is normally supplied hard mounted to a ball-bearing, hysteresis synchronous motor. For any angular orientation, the larger deflector unit can be operated at rotation rates between 2000 and 15,000 rpm, while the smaller deflector unit can be operated at rotation rates of 2000 to 20,000 rpm. Excessive deflector vibration occurs at some rotation rates between 15,000 and 20,000 rpm for the larger deflector unit, thereby causing degraded scan-beam tracking and scan jitter performance. Analysis indicates that these excessive vibrations can be minimized to acceptable levels by reducing the mass of the deflector unit and/or increasing the diameter of the rotor used in the deflector motor, or by only operating the deflector unit with its rotation axis parallel to the direction of gravity.

Scan imaging performance was measured for a 12-mm clear aperture monograting deflector unit incorporating a single-element focusing lens having a focal length of 95 mm. These deflector measurements were performed using an untruncated, collimated incident beam having a $1/e^2$ intensity diameter of 5.6 mm and a wavelength of 0.6328 μm. A measured image spot size of 14 μm for the $1/e^2$ intensity points was achieved for the deflector unit. This measured image spot size agrees closely with the theoretical value calculated using equation (2).

The average measured scan-beam tracking error performance for this 12-mm clear aperture deflector unit was less than 0.1 arc second, which was the limit of the measurement equipment [8]. Random peak-to-peak changes in this measured tracking error performance were typically less than 0.3 arc second. These random changes in scan tracking error performance were primarily due to vibrations in the measurement setup and air currents.

Measured radiometric intrascan uniformity of $\pm 0.5\%$ was achieved with this monograting deflector unit when a circularly polarized incident beam was used and the recording scan angle was 180°.

5.7.2 Nondisk Monograting Deflector for Flat-Field Imaging System

A flat-field imaging version of the nondisk monograting deflector unit is illustrated in Figure 5.42. This monofacet deflector unit possesses the same dynamic scanning beam properties as the monofacet deflector unit in Figure 5.40. These dynamic scan-beam performance properties are analyzed in Section 5.7.1.

A comparison of the internal drum deflector system in Figure 5.40 with the flat-field deflector system in Figure 5.42 reveals that scan-beam tracking error performance of the flat-field monograting deflector unit is far less sensitive to changes in both the deflector wobble angle and in the deflector axial position. This is because the rotating focusing lens in the internal drum deflector unit in Figure 5.40 is replaced in the flat-field system by a parallel plate window and a stationary scan lens. The rotational symmetry of the imaging properties of the Figure 5.42 deflector system are limited by the angular performance of the flat-field f-θ scan lens. A flat-field scan lens which covers a large angular field requires

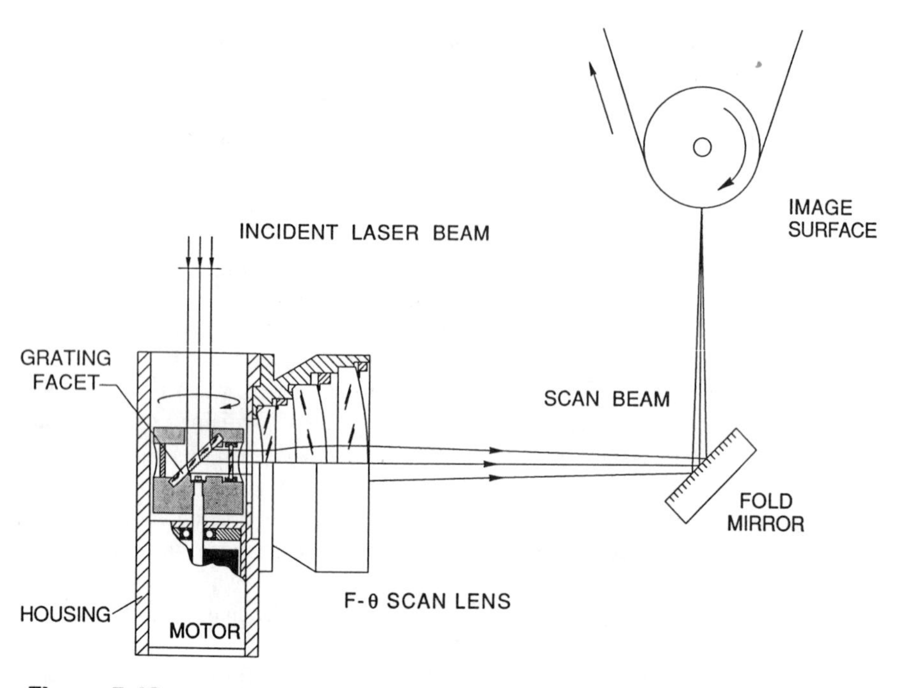

Figure 5.42 Sectional side view of a monograting deflector unit being used in a flat-field imaging system. (Courtesy of Holotek Ltd., Rochester, New York.)

more elements and is more difficult to manufacture and, hence, must be analyzed in terms of the cost trade-offs of the total deflector system. A flat-field f-θ scan lens having a half-scan-angle value in the range of about 24° to 28° usually provides a good compromise between system performance and system cost goals. A system scan duty cycle of 15% is achieved for the Figure 5.42 monograting deflector unit when the scan angle is $\pm 27°$.

If system scan duty cycle and scan speed are not of prime importance, the monograting deflector approach of Figure 5.42 provides the most cost-effective scanning solution for high-resolution, flat-field imaging. Deflector size, construction complexity, and cost are minimized for this monofacet deflector geometry because both the facet and incident beam are centered on the deflector rotation axis and, therefore, the facet need only be as large as the incident beam.

For example, if the incident Gaussian beam is truncated at the $1/e^2$ intensity points, then the incident beam diameter and facet clear aperture need only be 22.4 mm to achieve a $1/e^2$ intensity image spot size of 34 μm (1500 dpi) for a wavelength of 632.8 nm and a f-θ scan lens focal length of 647 mm. An image format scan-line length of 610 mm is achieved for this scan lens focal length when the scan angle is $\pm 27°$. Total deflector diameter for this case is about 50 mm. This deflector size is easily rotated to 15,000 rpm, thereby achieving a scan rate of 250 scans/s. An image plane translation rate of 10 in/min is required for this scan rate when a scan addressability of 1500 dpi is used. These scan imaging parameters require the raster image processor (RIP) to generate an average data rate of 9 Mpixels/s. A burst data rate of 60 Mpixels/s is required because the system scan duty cycle is only 15%.

Scan-Line Straightness for Flat-Field Monograting Deflector System

Scan-line straightness for the monograting deflector unit in Figure 5.42 is dependent on how perpendicular the scan beam is with respect to the deflector rotation axis. If the scan beam makes an angle θ_z with respect to the plane that is perpendicular to the deflector rotation axis, the scan line departs from a straight line by the value ΔZ, which is given by

$$\Delta Z = (\sqrt{x^2 + Y^2} - Y) \tan \theta_z \tag{77}$$

where X is the displacement of the scan beam from the center of the image plane and Y is the focal length of the focusing lens for the preobjective deflector case, or is the distance from the scanning facet to the center of the image plane when no lens follows the deflector. For the case where $Y = 647$ mm and $X = 305$ mm, corresponding to $\theta_s = \pm 27°$, θ_z must be $<0.2°$ in order to achieve the maximum allowed scan-line bow band value of <0.04% of scan-line length, as presented in Table 5.1 for noncomposite images.

If the deflector system is to be used to generate composite images, such as randomly ordered color separation films or circuit board artwork, it is desirable to reduce the scan-line bow band to 0.004% of the scan-line length, as presented in Table 5.1. To achieve this bow band value for the case where $\theta_s = \pm 27°$, θ_z must be $<0.02°$. This stringent angular alignment requirement is achieved by having the grating facet in Figure 5.42 be oriented at 45° to the deflector rotation axis, having $K = 1.4142$ for the facet, and having the f-θ scan lens mounting surface be designed to allow adjustment of the lens tilt angle in the cross scan direction so that the lens optical axis can be aligned to the central scan ray. The incident beam can be aligned to the deflector rotation axis by using the retrodiffraction beam technique described in Section 5.8.

Measured Performance for Flat-Field Monograting Deflector System

Performance data of three different sizes of nondisk monograting deflector units that are designed for flat-field imaging applications are reviewed. These deflector units have optical clear apertures of 12 mm, 24 mm, and 30 mm with corresponding outer diameters of 46 mm, 50 mm, and 58 mm. Each deflector unit is normally supplied hard mounted to a ball-bearing, hysteresis synchronous motor. For any angular orientation, the larger two deflector units can be operated at rotation rates between 2000 and 15,000 rpm, while the smaller deflector unit can be operated at rotation rates of 2000 to 20,000 rpm. Excessive deflector vibration occurs at some rotation rates between 15,000 and 20,000 rpm for the two larger deflector units, thereby causing degraded scan jitter performance. Analysis indicates that these excessive vibrations can be minimized to acceptable levels by reducing the mass of the deflector units and/or increasing the diameter of the rotor used in the deflector motors, or by only operating these deflector units with their rotation axis parallel to the direction of gravity.

Scan imaging performance was measured for a 30-mm clear aperture monograting deflector unit used with a flat-field f-θ scan lens having a designed focal length of 323 mm, a scan angle of $\pm 27°$, and an operating wavelength of 0.6328 μm. The recording scan-line length for this system was 305 mm. Measured image spot sizes at the $1/e^2$ intensity diameter were between 20 and 23 μm for the entire 305-mm scan length. These measured image spot sizes agreed closely with the theoretical value calculated using equation (2). Both theoretical and measured results were based on the collimated incident beam to the deflector unit having a truncation diameter of 20 mm and a $1/e^2$ intensity diameter of approximately 16.5 mm.

The average measured scan-beam tracking error performance for this 30-mm clear aperture deflector unit was less than 0.1 arc second, which was the limit of the measurement equipment [8]. Random peak-to-peak changes in this measured tracking error performance were typically less than 0.3 arc second. These random changes in scan tracking error performance were primarily due to vibrations in the measurement setup and air currents.

Measured radiometric scan uniformity data for this flat-field monograting deflector system agreed closely with theoretical calculated values. These calculations were performed using equation (58) and the data values presented in Figure 5.23 for the S- and P-polarized diffraction efficiencies of a grating having $K = 1.4142$. A measured radiometric throughput system efficiency of approximately 75% was achieved for the deflector unit at the center-of-scan position when the incident beam to the deflector was S-polarized. As predicted, the radiometric throughput efficiency decreased from the center scan peak value at a rate essentially equal to $\cos^2 \theta_s$. A radiometric throughput system efficiency of approximately 60% was measured for the end-of-scan position, which correlates with the $\pm 27°$ scan angle. These measured radiometric throughput system efficiency values include the combined loss for deflector head windows, grating facet, and f-θ scan lens.

Radiometric intrascan uniformity of $\pm 1\%$ was achieved with this flat-field monograting deflector system by synchronizing the intensity modulation level of the incident beam with deflector rotation angle. Under these conditions, measured radiometric throughput system efficiency was limited to approximately 60% for the deflector unit operating with an S-polarized incident beam. Radiometric intrascan uniformity of $\pm 0.5\%$ was achieved with the flat-field monograting deflector system by using a circularly polarized incident beam. Measured radiometric throughput system efficiency for this polarization state was approximately 40%. It is apparent from this data that a 50% increase in system radiometric throughput efficiency is achieved for this deflector system when S-polarized light and a complex beam intensity modulation technique is used to achieve scan radiometric uniformity. The variations in intrascan uniformity for S-polarized light need be measured only once and the corresponding digital correction values stored in a ROM chip.

While performing these radiometric scan-beam measurements, it was observed that the radiometric intrascan uniformity of the monograting deflector unit was strongly influenced by small angular misalignments between the incident beam and the deflector rotation axis. It was determined that the intrascan uniformity improved significantly when angular misalignment between the incident beam and deflector rotation axis was less than 2 arc minutes for the cross-scan direction and less than 1 arc minute for the in-scan direction. These radiometric related alignment tolerances are essentially equal to the alignment tolerances that have to be used to achieve the required scan-line straightness for composite image applications, such as circuit board artwork. The incident beam can be aligned to the deflector rotation axis by using the retrodiffraction beam technique described in Section 5.8 or by observing the radiometric intrascan uniformity.

5.7.3 Nondisk Dipole Monograting Deflector System

Scan duty cycle and scan rate can be doubled for the flat-field nondisk monograting deflector system by utilizing the nondisk dipole monograting deflector configuration in Figure 5.43. The through-borehole construction of the deflector unit in this figure enables two oppositely directed, collinear laser beams to be incident on the monograting facet. These two incident beams generate two collinear, oppositely propagating, diffracted beams that are sequentially scanned through the scan lens, and thereby, generate two scan lines for every complete rotation of the deflector head. The deflector head in Figure 5.43 is depicted as being supported on its outer surface by two bearing assemblies. Rotational drive to the deflector head can be by drive belt or by fabricating the deflector head as part of the rotor assembly for either a DC or AC motor, as illustrated in Figure 5.26 for the centered facet hologon disk deflector unit.

The two oppositely directed incident beams in Figure 5.43 are generated from a single incident laser beam by means of a beam splitter. Beam choppers are included in the two separated beam paths in Figure 5.43. The beam choppers' rotation rates are synchronized to that of the deflector head, so that each incident beam is only allowed to strike the grating facet during the deflector rotation cycle corresponding to it generating the scan line. In this operating mode, the beam

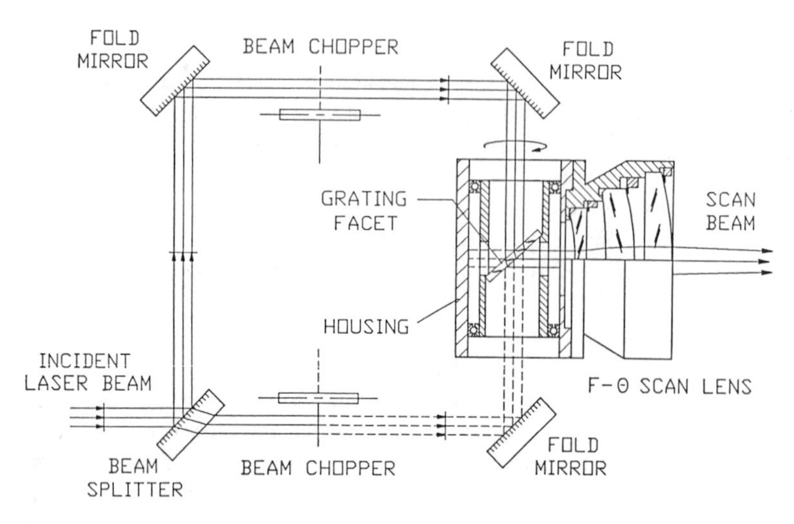

Figure 5.43 Schematic illustration of a nondisk dipole PDG deflector flat-field imaging system.

choppers block the undiffracted light from one incident beam from propagating back along the other incident beam path, thereby causing ghost beams and laser intensity instability. The mechanical beam choppers in Figure 5.43 could be replaced by electrooptic beam blockers or by magnetooptic beam isolators based on the Faraday effect. If the two oppositely directed incident beams are generated from two different laser sources, the beam choppers (blockers) can probably be eliminated.

Scan-Beam Tracking Error Performance for Nondisk Dipole Deflector

Periodic differences in scan-to-scan beam tracking error for the dipole monograting deflector in Figure 5.43 are caused by angular misalignment between the two oppositely directed incident beams, angular misalignment of the grating facet, error in grating facet K value, and substrate wedge angle in the grating facet and in windows used in the deflector head. These fixed potential error sources do not present a problem with regard to scan jitter error for this deflector configuration since their associated in-scan positional beam error can be easily corrected for by using a start-of-scan detector system. As previously noted, scan beam tracking error performance is virtually unaffected by dynamic change in deflector wobble angle for this monograting facet geometry.

A corresponding one-for-one change in scan-beam tracking angle for the nondisk dipole deflector is caused by both substrate wedge angle and misalignment error between the angle of the two incident beams. Error in the dipole grating facet K value causes an error in scan-beam tracking angle which is double the equation (46) calculated angular error in K, based on $\theta_i = \theta_d = 45°$. Scan beam tracking error caused by facet angular misalignment error in the dipole facet is reduced relative to the misalignment error by a factor calculated using equation (42).

Angular misalignment between the two oppositely directed incident beams is expected to be the major cause of scan-beam tracking error in the nondisk dipole deflector system because it is difficult to align and maintain the alignment of two beams to within 1 arc second. Better than 0.5 arc second, long-term alignment has been achieved between two beams by means of a feedback tracking system [48]. This system used the amplified output signal from a beam-position-sensing photodetector to control a piezoelectric-based minideflector mirror. This mini-deflector controlled the angular orientation of one of the beams relative to the other. The other fixed sources of scan-to-scan beam tracking error for the nondisk dipole deflector can be compensated for by adjusting the tilt angle of the grating facet. A technique for adjusting facet tilt is described in Section 5.7.5. for the pyramidal grating deflector unit.

Comparison of Nondisk Dipole PDG Deflector with Disk Dipole PDG Deflector

A comparison of the dipole disk hologon deflector system in Figure 5.26 with the dipole nondisk hologon deflector system in Figure 5.43 reveals that the Figure 5.26 system has an inherent advantage. This inherent advantage occurs because the centered disk facet in Figure 5.26 performs the dual function of deflector element and beam splitter element. In essence, the centered disk facet system in Figure 5.26 does not require the beam splitter, beam choppers and fold mirrors incorporated in the Figure 5.43 system. Also, the inherent dual function property of the facet in Figure 5.26 eliminates the requirement to maintain essentially perfect alignment between two oppositely directed incident beams. Expenses associated with the alignment and isolation requirements of the Figure 5.43 system can only be justified in system applications that require greater than $\pm 20°$ scan angle and/or a bowfree scan line.

5.7.4 Simultaneous Dual-Beam Scanning in a Hologon Deflector System

As previously noted, the nondisk monograting deflector approach provides the most cost-effective deflector solution for high-resolution imaging applications when scan duty cycle and scan rate are not of prime importance. While the dipole deflector configuration in Figure 5.43 can be used to double both the scan rate and scan duty of a monograting deflector system, it requires beam isolation and active feedback beam alignment components, as well as a through bore hole construction for the deflector unit. The major advantage of the sequential dipole scanning approach of Figure 5.43 is that it can be utilized for input (reading) scanning applications. Simultaneous multibeam scanning techniques [48] are usually not applicable for input scanning applications. This is because it is usually not feasible with multibeam scanning techniques to distinguish the individual reflected beam signal corresponding to an individual scan beam.

The simultaneous dual-beam scanning technique used with disk hologon deflector systems [48] to double the system scan rate for a given deflector rotation rate, can also be applied to nondisk hologon systems used for output recording applications. Reference 48 describes how orthogonal laser polarization states are used to achieve simultaneous dual beam scanning from a common hologon disk deflector facet. If lack of laser power is not a problem, then only a single laser source need be used instead of the dual laser source arrangement that is schematically illustrated in Ref. 48.

When a single laser source is used to generate dual simultaneous scanning beams, the laser is linearly polarized with its plane of polarization angularly orientated at 45° to the plane containing the predeflector optical components, as illustrated in Figure 5.44. For this angular orientation, the laser output power can be equally divided into S- and P-polarized beams by means of a polarization sensitive, beam splitter element. The separated S- and P-polarized beams are directed through separate acoustooptic modulator cells. These cells independently modulate the intensity of each beam. The independently modulated beams are recombined into a single beam by means of a polarization-sensitive, beam combiner element. Separating and combining the beams by means of orthogonal polarization states ensures high radiometric efficiency for the splitting and combining steps. This orthogonal polarization approach also ensures that the combined independently modulated beams do not coherently interfere with each other, thereby producing undesirable image plane intensity fluctuations.

The angular orientation between the combined orthogonally polarized (COP) beams can be accurately controlled by directing a small portion of the COP beams to a beam-position-sensing photodetector, such as a quadrature photodetector array element. Amplified output signals from this beam-position-sensing photodetector can be used to control a piezoelectric based minideflector mirror. This minideflector is used to control the angular orientation of one of the COP beams relative to the other.

Due to the polarization diffraction efficiency properties of the nondisk PDG hologon deflector, the COP beams would be passed through a quarter-wave retardation plate, thereby circularly polarizing the beams. Because the combined beams have orthogonal polarizations, their circular polarizations will be of opposite states (in opposite senses). After passing through the quarter-wave plate, the COP beams are expanded, collimated, and directed to the deflector unit.

If the simultaneous dual-beam scanning technique is used to double the scan

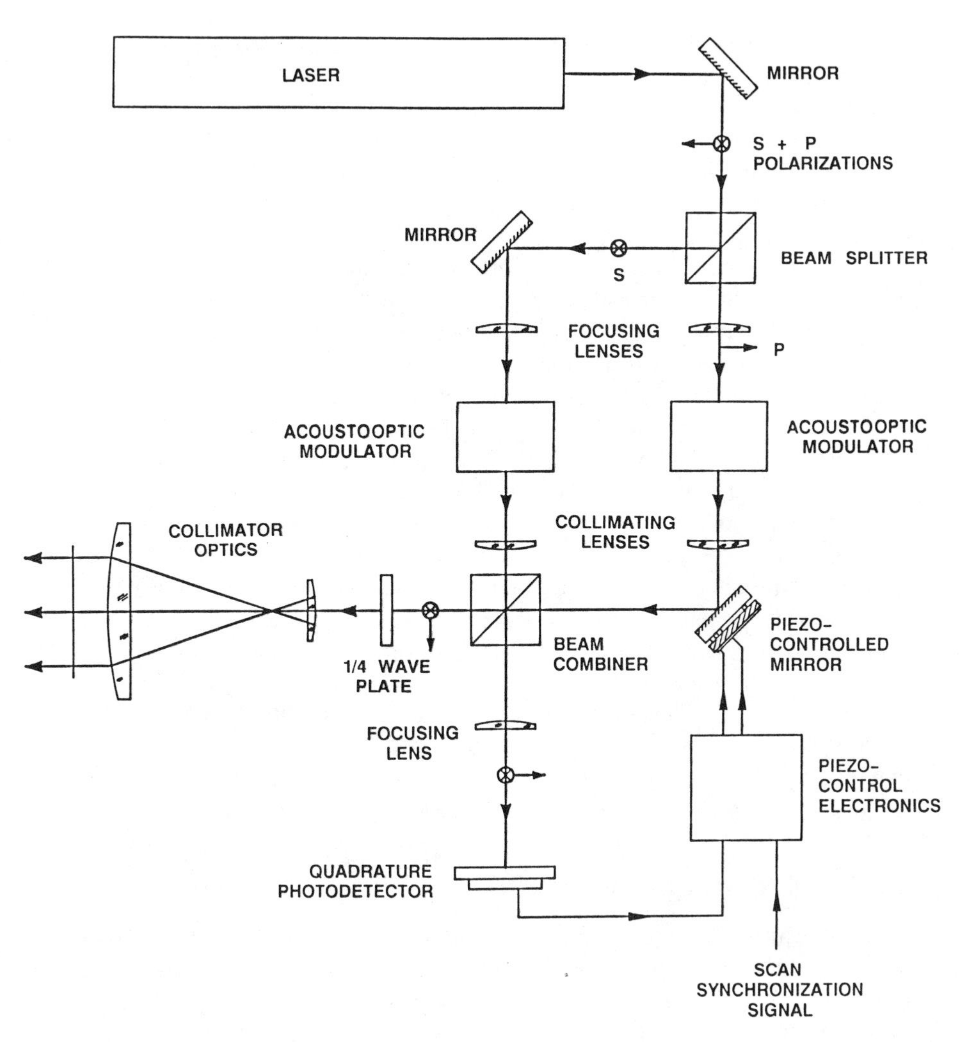

Figure 5.44 Optical system used to generate two orthogonal polarized incident laser beams to the hologon deflector unit.

rate of the flat-field monograting deflector system in Figure 5.42, the COP beams are angularly orientated so that the beams are parallel to each other in the plane that is perpendicular to the scan-line direction. In the cross scan direction, the COP beams are angularly separated by the angular spacing between adjacent scan lines. Under these conditions, two individually modulated, overlapping adjacent scan lines are generated for each rotation of the monograting deflector unit. Between 500 to 600 scan lines per second can be achieved with a nondisk monograting deflector if simultaneous dual beam scanning is used. Differential scan-line bow between scans generated with the dual-beam scanning technique is negligible when the total scan angle is less than about $\pm 27°$. Residual differential scan-line bow can be compensated for by using the minideflector control system to alter the cross scan angular separation between COP beams as a function of scan-field position.

Dual simultaneous beam scanning can be accomplished in an internal drum imaging system by dynamically changing the angular relationship between COP beams. This is achieved by utilizing a minideflector control system that allows the angular orientation between COP beams to be altered in both the cross scan and in-scan directions. Under these conditions, the minideflector element is driven so that one of the COP beams rotates about the other stationary COP beam. Feedback signals from both the beam-position-sensing photodetector array and from the deflector unit can be used so that the rotation rate of the one COP beam about the other COP beam is properly synchronized with the deflector angular position. The feedback signals from the deflector unit can be derived from either a start-of-scan detector or a deflector motor shaft encoder. This arrangement should enable two individually modulated, overlapping adjacent scan lines to be simultaneously generated on the internal drum surface at a rate of between 500 and 600 scan lines per second.

Dual simultaneous beam scanning doubles the scan rate of the deflector system, but it does increase the basic scan duty cycle of the system. While scan system duty cycle is not expected to be a problem for the monograting internal drum imaging system, it may be a concern in flat-field monograting scanning systems.

5.7.5 Multifacet Pyramidal Grating Deflector System

Schematically illustrated in Figures 5.45 and 5.46 is a multifacet pyramidal grating deflector system that retains essentially all of the scan-beam imaging advantages of the monograting deflector system in Figure 5.42, while providing significant improvement in scan system duty cycle for flat-field scanning applications. The multifacet pyramidal deflector unit possesses the same dynamic scanning beam properties as the monograting deflector units in Figures 5.40 and 42. These dynamic scan performance properties are analyzed in Sections 5.7.1 and 5.7.2.

The multifacet pyramidal deflector utilizes individual transmission PDG facets that are placed in metal frames. These frames are positioned in the rotating grating carrier at an angle of 45° to the rotation axis. The deflected beam exits the deflector unit perpendicular to the rotation axis, thereby generating a bow-free scan line. The tilt angle of each grating facet can be individually adjusted, thus allowing compensation for residual scan-beam tracking error problems associated with fixed deflector wobble, facet and window substrate wedge angle, and variation

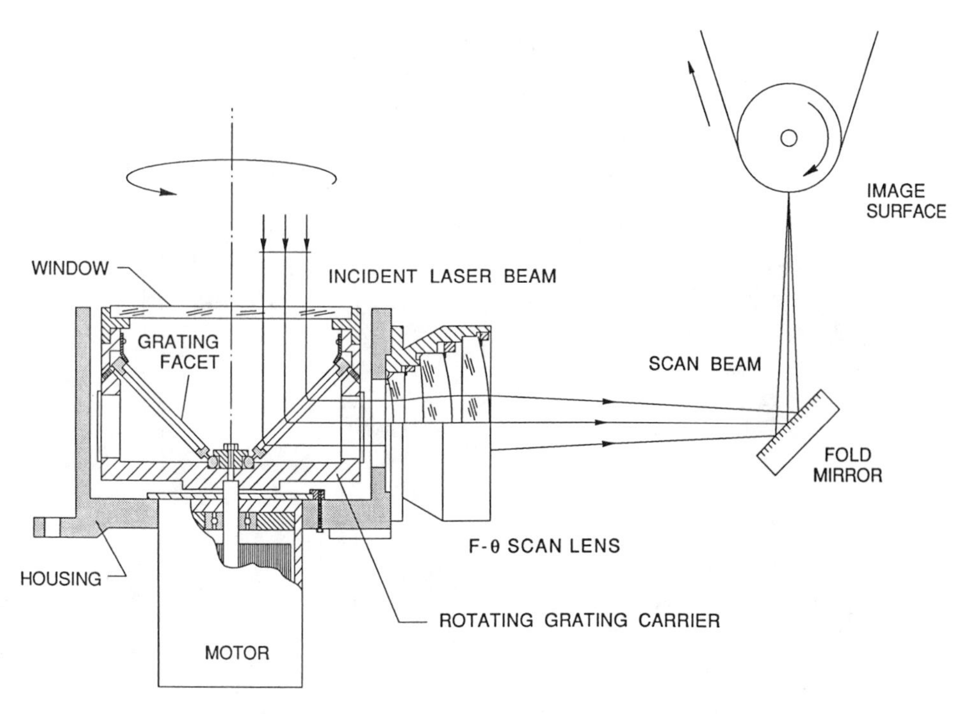

Figure 5.45 Sectional side view of a pyramidal grating laser beam deflector system. (Courtesy of Holotek Ltd., Rochester, New York.)

in facet-to-facet grating periodicity. This facet tilt angle adjustment does not affect the scan jitter performance of the deflector system since scan jitter associated with fixed deflector deficiencies can be easily corrected for by using a start-of-scan detector system.

Facet tilt angle adjustment is performed by a fine pitch setscrew located in the rotating grating carrier side wall. The outer part of the facet metal frame is held against the setscrew by a spring metal member when the deflector is stationary and by the spring metal member and centrifugal force during deflector rotation. The inner part of the facet metal frame is the rotating part of a ball joint. This mechanical arrangement ensures that tilt angle adjustments made when the deflector is stationary are maintained during deflector rotation. Periodic scan-beam tracking error between all facets of the pyramidal grating deflector is less than 1 arc second when the facet tilt angles are properly adjusted.

Similar facet tilt angle adjustment techniques have proven less successful when applied to polygonal mirror deflectors. It requires either a differential screw mechanism or ultra-fine adjustment of a very fine pitch screw to align mirror

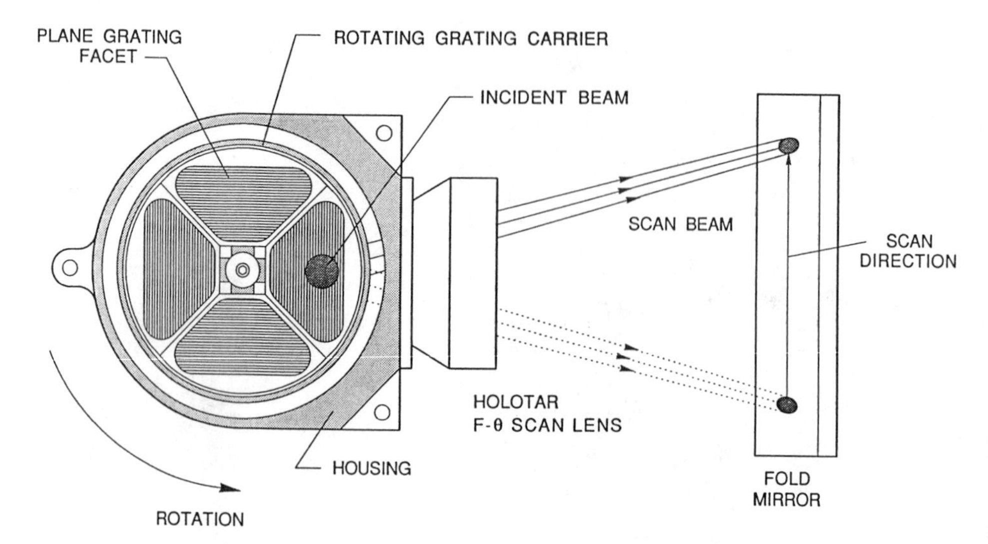

Figure 5.46 Top view of a pyramidal grating laser beam deflector unit with four facets. (Courtesy of Holotek Ltd., Rochester, New York.)

facets to the 0.5-arc-second tolerance required to achieve 1 arc second of scan-beam tracking error between all facets of a polygonal mirror deflector. This 1-arc-second level of scan-beam tracking error performance is usually maintained for only a short time period for the polygonal mirror deflector. This is because misalignment between facets occurs as a result of a change in rotor wobble tilt angle of the deflector motor and/or differential thermal expansion between deflector assembly components.

The problems of achieving initial facet angular alignment and maintaining that alignment are reduced in the multifacet pyramidal grating deflector by about 2000 times when compared with the polygonal mirror deflector case. This is because the pyramidal grating facets operate at the minimum deviation condition, and therefore, the change in cross scan beam angle is typically 1000 times smaller than the corresponding change in the facet cross scan tilt angle for this operating condition.

One major difference between the nondisk monograting deflector and its multifacet version is that the incident beam in Figure 5.45 is not collinear with the deflector rotation axis, but is parallel to it. This difference means that the facets translate through the incident beam. Therefore, the facet size in the scan direction must be at least twice as large as the incident beam if underfilled facet operation is used, as depicted in Figure 5.46. This increase in facet size requires a corresponding increase in deflector size and cost.

The following example illustrates the relative facet size required in a multifacet pyramidal deflector used in the underfilled facet mode. This example is based on the requirement that the multifacet deflector be able to reproduce the same imaging parameters that were examined for the flat-field monograting deflector system. These imaging parameters being a collimated Gaussian incident beam truncated at the $1/e^2$ intensity level and having a 22.4-mm diameter, $\lambda = 0.6328$ μm, a f-θ scan lens having a 647 mm focal length, and a $\pm 27°$ scan angle. The recording scan-line length corresponding to these imaging parameters is 610 mm. Theoretical calculated image spot for these imaging parameters is 34 μm for the $1/e^2$ intensity points, corresponding to a scan resolution of 1500 dpi. If the deflector has four facets, as illustrated in Figure 5.46, the angular duty cycle for the deflector is 60%. If the deflector mechanical duty cycle is set equal to its angular duty cycle, a mean facet width of approximately 56 mm is required in the direction of scan. The mean facet radius for this four-faceted deflector would be about 30 mm, corresponding to a total deflector diameter of approximately 120 mm.

A deflector of this size is not only expensive to manufacture but can present performance problems when operated at rotation rates greater than about 10,000 rpm. This size deflector can only be operated with its axis of rotation parallel to the direction of gravity for a cantilevered bearing assembly, as illustrated in Figure 5.45. This restriction on operating orientation also applies to large polygon mirror deflector units having the same bearing configuration.

5.7.6 Summary of System Performance for Nondisk PDG Deflectors

It has been demonstrated that a straight scan line and virtually no scan-beam tracking error are achieved for a nondisk PDG deflector system when the facets are oriented at 45° to the deflector rotation axis and the incident and diffracted beams are parallel and perpendicular to the deflector rotation axis, respectively. For this nondisk PDG deflector geometry: scan beam image spot ellipticity is independent of scan angle and can be virtually zero; scan angle is equal to deflector rotation angle; and scan-beam tracking error insensitivity with regard to deflector wobble is independent of deflector rotation angle. Essentially all of the scan imaging properties of this nondisk PDG deflector are independent of scan angle, thereby enabling the deflector system to utilize large scan angles for both non-composite and composite imaging applications. Being able to use large scan angles allows these deflector systems to achieve very high scan resolutions for large scan formats. Nondisk PDG deflectors can meet the performance requirements of graphic arts electronic imaging systems used for high-resolution reproduction of line art and halftone images when utilized with a ball-bearing motor and an optical imaging system comprising only spherical lens elements.

Deflector size and cost can be minimized by using the nondisk monograting deflector geometry if system scan duty cycle and scan rate are not of prime importance. Simultaneous dual beam scanning techniques can be used with either internal drum or flat-field monograting deflectors to double the system scan rate for a given deflector rotation rate. Scan duty cycle for the monograting deflector is significantly improved by incorporating the deflector into an internal drum imaging system. The internal drum imaging approach trades the problems and cost associated with minimizing the aberrations and the field curvature of a wide-angle flat-field f-θ scan lens for the problems and cost associated with fabricating a drum imaging surface that must be concentric with the deflector translation axis to within the depth of focus of the scanning image spot. Experience indicates that the internal drum scanning approach becomes more cost effective than flat-field scanning systems when greater than approximately 28,000 addressable dots per scan are required, based on $\lambda = 0.6328$ μm and a 50% intensity spot overlap.

The only major disadvantage of the nondisk PDG deflector technology is that it must be utilized with a very monochromatic laser source. Wavelength shifts due to mode hopping in a diode laser cause abrupt, unacceptable scan-beam tracking errors in systems used for graphic arts imaging applications. Therefore, diode laser light sources cannot be used with grating deflectors for these imaging applications unless the diode laser is stabilized with regard to operating wavelength or is modified so that the wavelength shifts gradually with time in comparison to the system scan rate.

Comparison of Nondisk PDG Deflectors to Previous Nondisk Hologon Deflectors

Neither of the nondisk hologon configurations in Figure 5.6, or other previous nondisk hologon geometries [13], provided sufficient advantages to offset drawbacks associated with not being a disk geometry. Therefore, these previous nondisk hologon systems have not been incorporated into commercial products. The nondisk PDG deflectors described in Section 5.7 are commercial products because they differ in a number of important ways from the previous nondisk hologon deflector configurations.

Fabrication of grating facets on the previous nondisk hologon deflectors was difficult because the facet surfaces had either spherical, cylindrical or conical curvature. Facets formed on a curved substrate must incorporate optical power which makes them difficult to make at one wavelength and use at a different wavelength. Since the facets in those previous devices were constructed on a continuous surface they could not be individually adjusted after fabrication to compensate for facet-to-facet scan differences caused by such factors as fixed deflector wobble angle, facet substrate wedge angle, differences in facet radiometric efficiency, and differences in facet grating periodicity. A major performance advantage is provided by the ability to compensate after fabrication for these types of facet-to-facet scan differences. This ability is incorporated into the multifacet pyramidal grating deflector that is described in Section 5.7.5. Also, none of the previous nondisk hologon deflectors were operated at the minimum deviation condition, and, therefore, their scan-beam tracking error was sensitive to deflector wobble.

Comparison of Nondisk PDG Deflectors to Disk PDG Deflectors

A comparison of PDG nondisk deflector flat-field systems with PDG disk deflector flat-field systems reveals that for the nondisk system, scan-line bow can be 10 times smaller for a $\pm 20°$ scan angle. In-scan and cross scan positional beam error sensitivity with regard to deflector wobble, as well as scan image spot shape are functions of the scan beam angle for a disk deflector system. These scan-beam properties are independent of scan angle for the nondisk deflector unit which means that this deflector geometry is not limited, like the disk deflector geometry, to scan angles of $< \pm 20°$ for most graphic arts imaging applications. Image performance consideration associated with the flat-field *f*-θ scan lens limit the scan angle of PDG nondisk deflector flat-field systems to approximately $\pm 30°$. Recording scan angle for the nondisk monograting deflector internal drum system is typically between 180° and 270°.

The $\pm 20°$ scan angle limitation makes it impractical to utilize a PDG disk deflector for imaging applications requiring a scan addressability greater than approximately 22,000 dots per scan, based on $\lambda = 0.6328$ μm and a 50% intensity spot overlap. For the same imaging criteria, scan addressabilities of 30,000 dots per scan and greater than 100,000 dots per scan can be achieved for the nondisk flat-field and internal drum PDG deflector systems, respectively. Other benefits derived from having a large scan-beam angle include improved scan system duty cycle, reduction in *f*-θ scan lens focal length for a given format scan length, reduction in deflector size, reduction in scanning system package size, and reduction in the image plane magnitude of scan-beam displacement errors associated with deflector scan-beam tracking errors, air turbulence, and system vibrations.

Improved scan-beam tracking error performance is achieved for both the monofacet and multifacet nondisk PDG deflectors when compared to the disk PDG version of these deflectors. This is because the nondisk deflectors operate at the minimum deviation condition for every deflector rotation angle, and, therefore, their scan-beam tracking error is typically 10 times less sensitive with regard to deflector wobble angle than that achieved for the disk deflector geometry. Individual facet tilt angle adjustment in the multifacet nondisk deflector allows for compensation of scan-beam tracking error problems associated with fixed deflector wobble, facet and window substrate wedge angle, and variation in facet-to-facet grating periodicity. Periodic scan-beam tracking error between all facet scans of a multifacet disk PDG deflector is typically 2 to 3 arc seconds, compared with less than 1 arc second for the multifacet PDG nondisk deflector unit.

The PDG disk and nondisk deflector systems have essentially equal scan jitter performance. Scan jitter associated with fixed deflector wobble angle is easily corrected for in the nondisk deflector system by means of a start-of-scan detector arrangement. A more complex pixel grating clock arrangement is required in the disk deflector system to totally correct for the scan jitter associated with fixed deflector wobble angle. Scan jitter associated with dynamic change in the fixed deflector wobble angle is approximately 20 times larger for the nondisk deflector system than for the disk deflector system. This dynamic component of scan jitter error can only be totally corrected for by using the pixel grating clock method.

When only a start-of-scan detector arrangement is used, fixed deflector wobble angle is expected to generate an uncorrected periodic scan jitter error of about 2 arc seconds for the PDG disk deflector system, while dynamic change in the fixed deflector wobble angle is expected to generate an uncorrected scan jitter error of about 2 to 3 arc seconds for the PDG nondisk deflector system. The dynamic scan jitter associated with the nondisk deflector system is expected, due to its random nature, to cause a smaller influence on recorded image quality than the periodic scan jitter associated with the disk deflector system.

Mechanical simplicity of the deflector structure is the major advantage that the disk deflector geometry has when compared with the nondisk deflector system. This mechanical simplicity enables both the centered monofacet and multifacet disk deflectors to operate at considerably higher rotation rates than the corresponding nondisk PDG versions of these deflectors. As a result of smoother surfaces, the disk hologon deflector produces less air turbulence than the nondisk deflectors. This reduction in turbulence contributes to a reduction in random scan-beam positional error associated with air currents, as well as a reduction in deflector windage noise and in the required deflector motor drive power.

Comparison of Nondisk PDG Deflectors with Mirror Deflectors

When both a PDG nondisk monofacet deflector element and a single mirror deflector element are mounted on similar ball-bearing motors, the grating deflector unit provides a significant advantage in scan-beam tracking error performance when compared with the mirror deflector. Scan-beam tracking error performance for the monomirror deflector suffers from changes in mirror wobble caused by motor bearing inaccuracies and/or vibrations. If it is assumed that these deflector motor deficiencies cause a maximum change in the cross scan deflector wobble angle of about 3 arc seconds, then the corresponding scan-beam tracking error would be 6 arc seconds for the monomirror deflector and approximately 0.003 arc second for the monograting deflector. The 6 arc seconds of scan beam tracking error is too large for most halftone image reproduction applications.

Scan-beam tracking error can be eliminated in a monofacet mirror system by the use of a pentaprism configuration [46, 47]. The pentaprism approach achieves insensitivity with regard to deflector wobble by using two tightly coupled reflecting surfaces to redirect an incident beam through a 90° angle. Drawbacks of the pentaprism approach include an asymmetrical optical and mass geometry that presents problems in a rotating environment, two reflecting surfaces which can degrade wavefront fidelity, and a fairly large deflector mass that can contribute significantly to inducing dynamic changes in deflector wobble angle. While the scan beam tracking error for both the pentaprism deflector and nondisk PDG monofacet deflector are essentially insensitive with regard to deflector wobble, the in-scan jitter error for these two deflectors is directly proportional to the in-scan component of change in deflector wobble angle. The relative lower mass of the nondisk monograting deflector contributes to lower in-scan jitter error for this deflector when compared to similar pentaprism deflector motor combinations.

Centrifugal-force-induced deflector element distortion is a problem in mirror deflectors used at high rotation rates. This is particularly true for the pentaprism deflector element due to the presence of the two reflecting surfaces and the asymmetric arrangement of those reflecting surfaces with respect to the deflector rotation axis. Since the PDG deflectors function in transmission, deflector facet distortion associated with rotation rates as high as 20,000 rpm have negligible effect on their scan-beam wavefront fidelity. The optical and mass symmetry of the nondisk monograting deflector unit enables it to be easily driven to very high rotation rates.

Similar scan-beam tracking error performance advantages are achieved with the multifacet pyramidal grating deflector when compared with polygonal mirror deflectors that are mounted on ball-bearing motors. The multifacet pyramidal PDG deflectors achieve scan-beam tracking error performance comparable to the highest-quality air-bearing polygonal mirror deflector systems, but at a fraction of the cost of air-bearing systems.

One advantage that mirror deflector systems have is that they can be used with a broadband light source. This advantage enables mirror deflectors to use diode lasers for many graphic arts imaging applications.

5.8 HOLOGON DEFLECTOR SYSTEM ALIGNMENT PROCEDURES

This section briefly describes procedures and equipment used to align a PDG hologon deflector unit to the other optical components in the scan imaging system. A number of these procedures are also useful for aligning mirror deflector scanning systems.

In addition to the normal care that is required when aligning any high-resolution imaging system, there are a few unique properties of a hologon deflector that have to be considered when incorporating it into a scan-beam imaging system. It is very important that the incident laser beam to the deflector unit be both highly monochromatic and collimated since the PDG facets have both large chromatic and anamorphic angular dispersive power. Care must be taken when choosing a red HeNe laser having an output power of 4 mW or larger since some manufacturers' lasers routinely output in addition to the desired 632.8 nm laser line, a low-level laser line at 640.1 nm. While this extra line usually causes no problem in a mirror deflector system, it generates a displaced shadow scan beam in the hologon deflector. Unwanted extra laser lines can also be a problem when using an argon laser. Scan lines associated with unwanted laser lines, as well as a large component of unwanted scattered system light, can be blocked from reaching the image plane by passing the scan beam through a slit aperture field stop located near the image plane.

Often the divergence angle of the input beam to a mirror deflector is altered in order to compensate for the difference in scan lens focal length associated with manufacturing tolerances. Unfortunately, this technique cannot be used with the PDG hologon deflector since divergence in the incident beam to the hologon produces astigmatism in the wavefront of the diffracted scan beam. Collimation of the incident beam to the deflector unit can be accomplished by the use of a high-quality mirror that is placed in the beam after the collimating lens. Collimation is achieved by adjusting the axial position of the collimating lens until the size of the retroreflected beam at a location prior to the collimator optics matches the size of the unexpanded input laser beam to those optics (autocollimation).

Beam collimation based on the parallel alignment of interference fringes to the roofline of a shear plate has proven to be faster and more accurate than the autocollimation method. The shear plate is placed in the incident beam at an orientation angle of approximately 45° to the beam propagating direction. Reflected laser light from the uncoated surfaces of the shear plate are intercepted by a white card used to observe their straight-line interference pattern. The angular orientation of the straight-line interference pattern changes as the axial position of the collimating lens is altered. Beam collimation is achieved when the straight-line fringes are parallel to the shear plate roofline which is usually indicated by an opaque bar that bisects the shear plate aperture. Collimating shear plate devices are sold by Diamond Electro-Optic, Inc.

When alignment of the deflector optical system is completed, very fine adjustment of the incident beam collimation state can be used to optimize the scan image spot performance of a flat-field hologon deflector system. This spot size optimization is performed by minimizing the measured system image spot size in the cross scan direction for the ± 0.7 scan-field positions. This minimization is determined by making very fine step and repeat adjustments to the axial position of the collimating lens and observing what the minimum image spot size is as the spot size measuring device is translated through focus. Adjustments in the axial position of the collimating lens as small as 0.1% of its focal length can be observed as changes in both the image spot size and in the location of minimum spot size. A similar collimation adjustment technique can be used to minimize image spot astigmatism between the cross scan and in-scan directions for the monograting internal drum deflector system. The beam scan device sold by Photon, Inc. is useful for making these image spot size measurements.

After collimation, the next task is adjusting the angular orientation and spatial position of the incident beam with regard to the hologon deflector element. It is important in any deflector system, but particularly in a multifacet deflector unit, to have the incident beam be centered on the facet in the in-scan direction for the center-of-scan field position. This beam spatial alignment condition ensures that truncation occurring in the scan beam by the f-θ scan lens barrel and truncation of the incident beam by the facet edges of a multifacet deflector element are both symmetric about the center-of-scan field position. An easy and accurate method of performing this beam positioning task is to place an alignment target on the entrance aperture of the deflector unit.

When initial adjustment of both the angular orientation and spatial position of the incident beam to the deflector unit has been accomplished, the following technique can be used to check and/or align the spatial position of the incident beam to the deflector facet. Evaluation and/or adjustment of the relative in-scan position of the incident beam with respect to the deflector element can be performed by observing the truncation in the scan beam at either the scan lens or image plane location as the deflector element is manually rotated. The cross scan position of the incident beam is easily and accurately adjusted by observing the relative change in truncation of the deflected beam when the deflector element is statically oriented for the center-of-scan field position.

It is important that the incident beam be perpendicular to the hologon deflector element in the in-scan direction and have the incident angle in the cross scan direction that minimizes both the scan-beam tracking error sensitivity with respect to deflector wobble and the scan-line bow band. When the incident beam is not perpendicular to the deflector element in the in-scan direction, the scan-line is shifted to one side and is angularly rotated. Initial adjustment of the in-scan incident beam angle consisted of making this angle perpendicular to the deflector unit alignment reference surface that is perpendicular to the deflector rotation axis. Alignment of the deflector unit reference surface to the incident beam is performed by placing a parallel plate mirror on the reference surface so that the mirror reflecting surface is parallel to this surface. The in-scan incident beam angle is adjusted so that the retroreflected incident beam from this mirror is parallel in the in-scan direction to the incident beam.

For the disk hologon deflector unit in Figure 5.33, the alignment reference surface is the baseplate to which the deflector motor is attached and through which the incident beam passes. For the nondisk flat-field deflector units in Figures 5.42 and 5.45, the alignment reference surface is the end housing surface that contains the opening through which the incident beam passes. For the StraightScan-3 monograting internal drum deflector unit in Figure 5.40, the alignment reference surface is either the top surface of the deflector head or a surface of the deflector unit support bracket that is perpendicular to the deflector rotation axis and through which the deflector motor passes.

A unique property of a diffraction grating is exploited when aligning the cross scan incident beam angle θ_i to the grating facet. Figure 5.47 schematically illustrates the angular relationship between the incident laser beam, the diffracted laser beam and the retrodiffracted laser beam for a surface-relief diffraction grating. Due to Fresnel reflection at the air-grating interface, a small fraction (~0.5%) of the incident beam power is retrodiffracted back in the direction of the incident beam. From the grating equation, equation (20), the diffracted and retrodiffracted beams must have the same angle with respect to the grating normal. If $\theta_i = 30.6°$ and $K = 1.0$, then both diffracted beams must be at 29.4° with respect to the deflector normal, and, therefore, the incident beam and retrodiffracted beam will be separated by an angular difference of 1.2°.

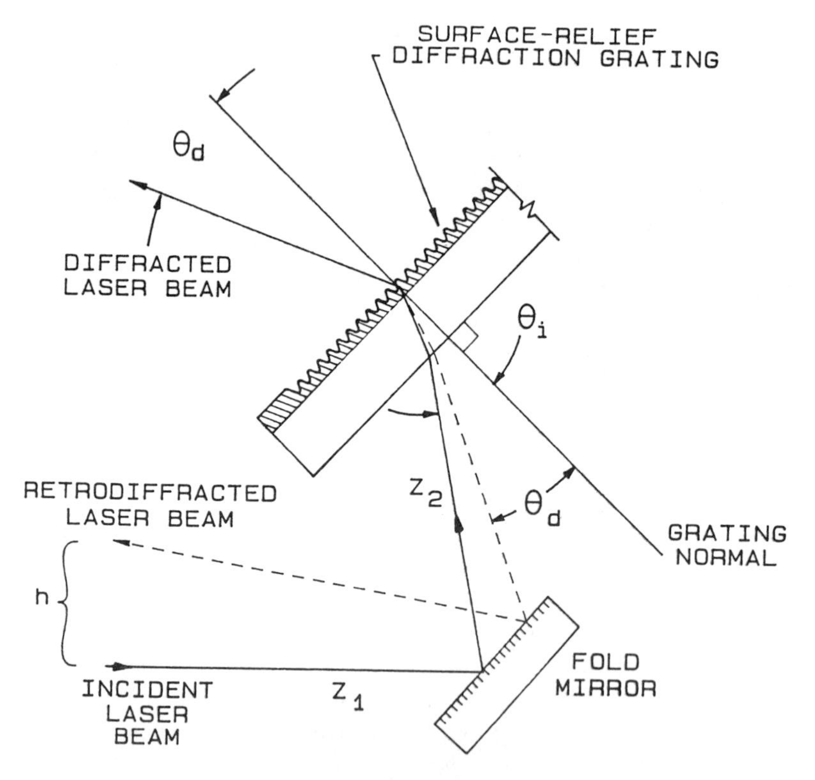

Figure 5.47 Diagram showing the angular relation between incident, diffracted, and retrodiffracted beams with respect to the grating normal.

Setting θ_i for a given angular value is accomplished by adjusting the cross scan incident beam angle to the grating facet so that the retrodiffracted beam has the correct angular separation with regard to the incident beam. Determination of the angular separation between these beams is accomplished by measuring the spatial separation between the beams at a known distance from the grating facet. This measurement is performed by manually rotating the hologon so that the retrodiffracted beam is parallel in the in-scan direction to the incident beam. Due to the low intensity of the retrodiffracted beam, it helps to view the beam in a darkened room using a light colored viewing screen. It also helps to aperture down the incident beam so only the central Gaussian portion of the incident beam is passed through a small opening in the viewing screen. By knowing the distance Z of the viewing screen from the grating facet and the spatial separation h between the incident and retrodiffracted beams on the viewing screen enables the angular separation between these beams to be easily calculated:

$$h = Z(\theta_i - \theta_d) \tag{78}$$

The retrodiffracted beam can be used to accurately align the incident beam parallel to the deflector rotation axis for the nondisk PDG deflector systems. If the nondisk PDG deflector element had no fixed misalignment tilt angle with respect to the deflector rotation axis, the incident beam angle in both the in-scan and cross scan directions would be adjusted so that the retrodiffracted beam propagates back along the incident beam. For this collinear beam condition, the incident beam is parallel to the deflector rotation axis when the grating facet is at 45° to the rotation axis and $K = 1.4142$ or when the grating facet is at 42.5° to the rotation axis and $K = 1.3512$. When the nondisk PDG deflector facet has a slight fixed angular misalignment with respect to the deflector rotation axis, the retrodiffracted beam rotates around the incident beam at the deflector rotation rate. Under these conditions, the incident beam is parallel to the deflector rotation axis when the circular pattern of the rotating retrodiffracted beam is centered on the incident beam.

After initial alignment of the disk deflector system is accomplished, the retrodiffracted beam technique can be used to adjust the perpendicularity of the in-scan incident beam angle with regard to the hologon disk surface. The center scan position for the hologon disk is found by manually rotating the hologon so that the in-scan retrodiffracted beam angle is parallel to the incident beam. The hologon disk center-of-scan position will coincide with the scan beam being located at the central scan-line position for the deflector system when the in-scan incident beam angle is perpendicular to the hologon disk. If these two center-of-scan positions do not coincide then both the in-scan incident beam angle and the hologon rotation angle are adjusted until these scan centers coincide. A further check of the perpendicularity of the in-scan incident beam angle with respect to the hologon disk is achieved by comparing the angular orientation of the scan-line relative to a plane that is perpendicular to the hologon rotation axis.

The described alignment techniques have not dealt with the alignment of deflector system optical components that precede the collimator optics. This is because there is nothing unique with regard to a hologon deflector system that would effect the alignment of these precollimator optical components.

5.9 DESIGN EXAMPLES FOR HOLOGON DEFLECTOR SYSTEMS

This section examines some of the trade-offs that have to be considered when designing hologon-based laser scanning systems for graphic arts imaging applications. These trade-offs include laser wavelength, scan spot size versus scan addressability, duty cycle, scan angle, deflector size, pulse imaging, facet tracking, etc. These system trade-offs will be analyzed with regard to the imaging requirements for a hologon disk deflector system used to record computer data on a microfilm format. Also briefly examined in this section is the integration of a hologon deflector with other optical and mechanical components to construct a laser imaging system that can be used to generate high-resolution images.

5.9.1 Computer Output Microfilm System

This example illustrates the trade-offs associated with designing a disk hologon deflector for either a computer output microfilm/microfiche (COM) system or for optically recording/reading binary digital data on a tape format. Desired system specifications for a COM system are presented in Table 5.2.

Trade-offs Associated with Laser Wavelength Selection

Optical wavelength is an important parameter in any recording application. Wavelength selection has greater importance in COM recording applications due to the relatively large numerical aperture required by these systems. Presented in equations (2) through (5) is the relation between laser wavelength and system truncation ratio, F-number, and usable depth of focus. The relationship between these imaging parameters becomes more evident by performing a series of system calculations. For these calculations it will be assumed that the incident beam is untruncated ($T = 0.5$), the image spot size S is 7 μm, and the scan-lens focal length, f, is 27 mm. Results for these calculations are presented in Table 5.3 for a number of important laser wavelengths, the $1/e^2$ intensity diameter required for an incident beam of that wavelength to achieve the image spot size, and the corresponding depth of focus for a 10% spot growth.

The linear relationship between the listed parameters in Table 5.3 indicates that deflector size is reduced while system alignment tolerances are loosened by

Table 5.2 Desired Deflector System Imaging Specifications

Operating wavelength	0.5435 μm
Scan-line length	7.41 mm
Image spot at $1/e^2$ points	5–10 μm
Residual scan-line bow	<2 μm
Scan-beam tracking error	<±0.5 μm
Scan system duty cycle	>66%
Scan rate	600 scan/s
Scan jitter	<±0.5 μm
System transmission efficiency	>50%

XY optical beam deflection is desirable for some COM applications so that film can be held stationary during the recording of a page format.

utilizing shorter wavelengths. Blue lasers are currently (1990) expensive relative to HeNe and diode lasers, and therefore, the 0.5435 *um* HeNe laser line provides a good trade-off between these system parameters and system costs.

Scan System Duty Cycle

Image spot size has a major effect on hologon deflector size and/or mechanical duty cycle of the deflector. Presented in Equations (7) through (9) is the relation between scan system duty cycle, deflector size and deflector system imaging parameters. Equations (11) and (15) show that the system burst bandwidth requirement and system radiometric efficiency are both improved by increasing scan system duty cycle. The number of facets for a hologon disk with tangential grating line orientation is given by equation (36) for the case where system scan duty cycle is optimized by having $U_m = U_a$. This equation can be rewritten as

$$N = \frac{2\pi MrSf}{WrS + 1.56\, M\lambda f^2} \tag{79}$$

Table 5.3 Imaging Properties versus Wavelength

Laser type	Wavelength (μm)	Input beam (mm)	Depth of focus (μm)
YAG	1.0640	5.33	32
HeNe	0.6328	3.17	54
HeNe	0.5435	2.72	63
YAG[a]	0.5320	2.67	64
Ar^+	0.4880	2.44	70
HeCd	0.4416	2.21	78

[a] Second harmonic

where k_c was reset to a value of 1.2 and θ_s set equal to W/f, where W is the f-θ recording scan-line length. Table 5.4 presents the relationship between scan system duty cycle, λ, and system optical magnification factor M, for a system having an optimum number of facets for the following parameters: $f = 27$ mm, $W = 7.41$ mm, $S = 7$ μm, and a fixed mean facet radius of $r = 40$ mm. It is evident from the data presented in Table 5.4 that scan system duty cycle is strongly dependent on λ and M.

Again, the green HeNe laser line provides a good trade-off between system scan duty cycle and system cost. Reducing M from 1.4142 to 1.0 increases scan system duty cycle while reducing the number of facets on the deflector. Deflector manufacturing cost is significantly less for the $M = 1.0$ case as a result of the reduction in both N and in grating periodicity for disk hologons.

Scan-Lens Design Considerations

Scan system duty cycle can be improved by going to larger scan angles. Larger scan angles are achieved by using either a shorter focal length scan lens or by increasing the length of the scan line. Increasing the scan-line length from 7.41 to 52.5 mm is recommended if one were designing a microfiche recording system instead of a microfilm system. This scan length recommendation is based on the opinion that an optimum microfiche deflector system would scan half the 105-mm width of the fiche and, thereby, minimize film translation within a fiche and reduce the deflector rotation rate requirement in relation to the system data transfer rate.

Required clearance distance between the deflector element and the scan lens imposes a limit on how short a focal length the scan lens can have without using an intermediary relay lens approach. This problem is compounded for the case where an XY raster scan is to be performed by having two deflectors positioned before the scan lens. The hologon deflector system in Figure 5.33 incorporates

Table 5.4 Deflector System Parameters versus Scan Duty Cycle

Wavelength (μm)	M	N	Duty cycle (%)
1.0640	1.000	14	61.1
1.0640	1.414	17	52.5
0.6328	1.000	17	74.2
0.6328	1.414	21	64.8
0.5435	1.000	17	74.2
0.5435	1.414	22	67.9
0.4416	1.000	18	78.6
0.4416	1.414	24	74.1

a bow compensation prism that limits the minimum distance between hologon and scan lens to about 34 mm. Scan lenses used for COM applications are designed to be telecentric, and, therefore, it is difficult to make the back focal length (BFL) of the scan lens longer than its effective focal length (EFL). A telecentric scan lens having a EFL of 27 to 30 mm and a BFL of 34 mm can be designed, but this BFL to EFL ratio does increase the design complexity of the scan lens.

Deflector System Configuration Trade-offs

If one were to use the deflector unit in Figure 5.33 without incorporating a relay lens, there would be no room between the prism and scan lens for a second deflector element. For this case, the film must be moved to achieve the second raster motion. There are advantages to moving the film to generate the second raster motion since the only suitable deflector element for producing the second raster scan is a mirror. This mirror deflector would either be a galvanometer or a rotating single-mirror facet. Angular positional errors of a mirror deflector are doubled and these errors would be further increased by the EFL throw distance of the scan lens. Velocity errors in a film transport system are reproduced directly in the recorded image without any magnification factor. It is possible to build a film transport system having the velocity accuracy required to record within a microfilm page.

A relay lens is required in order to use the Figure 5.33 deflector unit as part of a preobjective two-deflector, *XY* raster scan system. This relay lens is positioned between the prism and the mirror deflector as illustrated in Figure 5.48. The relay lens images the entrance pupil at the hologon deflector onto the surface of the mirror deflector. The scan lens images the relayed pupil at the mirror deflector onto the image plane. If a beam expanding telescope is incorporated as part of the relay lens, the scan duty cycle of the deflector system can be improved because the hologon deflector can be scanned to larger angles using a smaller input beam. Associated with this beam expansion is a corresponding decrease in the optical angular magnification factor, M_L, of the lens system following the hologon deflector. This decrease in M_L enables a corresponding increase in the hologon scan angle. Also, the scan lens for the beam expansion case can have a longer EFL and, thereby, increase the working distance between it and the mirror deflector.

For the deflector system shown in Figure 5.48, the relay lens is depicted as providing a 1.5*X* beam expansion. For these imaging conditions, the hologon deflector can achieve a scan duty cycle of 82.5% for the following system parameters: $\theta_s = 10.56°$, $f = 30$ mm, $W = 7.41$ mm, $\lambda = 0.5435$ μm, $S = 7$ μm, $M_d = 1.00$, $r = 40$ mm, and $N = 14$. The scan lens for this case should be designed as a flat-field F-tan θ lens to avoid the bowing of off-axis scan lines that occurs with f-θ designs, as described in Section 5.6.1. A long EFL for the scan lens is desirable for this case because it reduces the residual image distortion caused by the deflector elements. For a scan-lens EFL of 30 mm, the scan angle

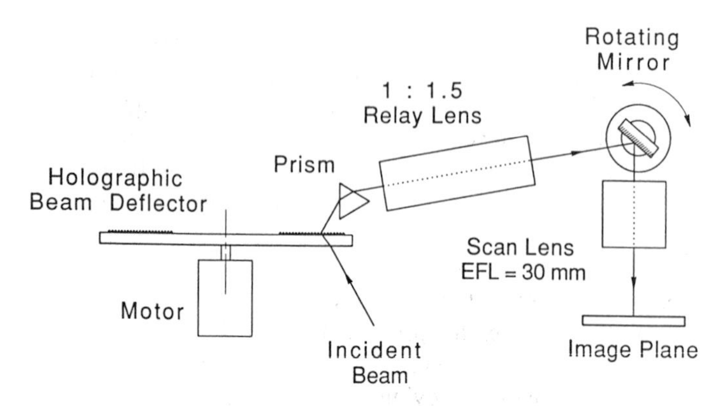

Figure 5.48 A hologon deflector system incorporating relay lens, second deflector element and F-tan θ scan lens. System designed for XY recording of COM data.

is ± 7.07 and the residual distortion at the edge of field is 0.5%. This level of distortion is not large enough to be perceived. If required, this distortion could be electronically corrected by altering the timing of the pixel clock as a function of θ_s [49].

A scan bow compensation prism element is not required for a hologon deflector having $K = 1.4142$ and $\theta_i \approx \theta_d \approx 45°$. Therefore, a deflector mirror can be positioned between the hologon and scan lens without the need for a relay lens. An XY raster deflector system based on this geometry is illustrated in Figure 5.49. This deflector system geometry is similar to what DatagraphiX Inc. utilizes in their Aries COM machine. For the case shown in Figure 5.49, the scan lens is depicted as having an EFL of 27 mm which corresponds to the calculations used to generate the data presented in Tables 5.3 and 5.4. Table 5.4 shows that for $\lambda = 0.5435$ um and $M = 1.4142$, a scan duty cycle of 67.9% is expected for the deflector system parameters assumed. While this system is optically simpler than the one illustrated in Figure 5.48, its scan duty cycle is 21% lower than that system.

Three alternative deflector system configurations have been described. All three could be used to achieve the deflector system imaging specifications listed in Table 5.2. These systems are summarized as follows.

Deflector System I. A hologon deflector system incorporating a bow compensation prism element and a f-θ scan lens, as illustrated in Figure 5.33. This deflector would provide 1 motion of raster scan, and the second raster scan would be performed by moving the film at a constant rate. This deflector system configuration may be utilized for optical digital data recording/reading applications and is useful for COM application.

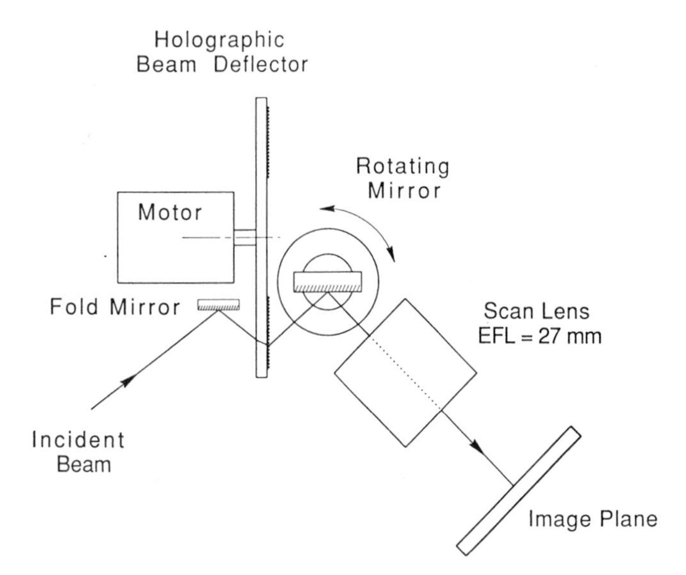

Figure 5.49 Basic PDG hologon deflector system incorporating a second deflector element and F-tan θ scan lens. System designed for XY recording of COM data.

Deflector System II. A hologon deflector system incorporating a bow compensation prism element, relay lens, second deflector element and F-tan θ scan lens, as illustrated in Figure 5.48. This deflector system configuration would provide an XY raster scan pattern and, therefore, is suitable for COM applications that desire the film to be held stationary during the recording of a page format.

Deflector System III. A hologon deflector having $K = 1.4142$ and $\theta_i \approx \theta_d \approx 45°$. This hologon deflector being part of a deflector system that includes a second deflector element and a F-tan θ scan lens, as illustrated in Figure 5.49. Like deflector system II, this deflector system would provide an XY raster scan pattern.

Table 5.5 compares the calculated performance expected by these three different deflector systems. When calculating the data presented in Table 5.5, it was assumed that $\lambda = 0.5435$ μm, $S = 7$ μm, $W = 7.41$ mm, and $r = 40$ mm. Transmission efficiency, as listed in this table, is the percentage of incident light transmitted by the deflector system including scan lens. For the XY deflector systems it includes losses associated with any relay lenses and reflectivity of second mirror deflector. All lens elements are assumed to be multilayer AR coated. Radiometric efficiency is the product of transmission efficiency and scan system duty cycle.

These three deflector systems have essentially equal performance properties except for scan system duty cycle. System II achieves the highest scan system

Table 5.5 Comparison of Expected System Performance

	System I	System II	System III
Scan-lens focal length	27 mm	30 mm	27 mm
Half-scan angle	7.86	7.07	7.86
Residual scan distortion	<0.2%	0.51%	0.63%
Number of hologon facets	14	12	17
Scan system duty cycle	74.2%	82.5%	67.9%
Residual scan-line bow	<1 μm	<1 μm	<1 μm
Scan tracking error	<0.5 μm	<0.5 μm	<0.5 μm
Scan jitter	<±0.5 μm	<±0.5 μm	<±0.5 μm
Transmission efficiency	>75%	>70%	>75%
Laser polarization state	*S* or *P*	*S* or *P*	*S*
Radiometric efficiency	>55%	>57%	>50%

duty cycle by incorporating a relay lens that adds complexity and cost to the system. The difference in scan system duty cycle between systems I and III is important but not critical in terms of system performance requirement. The major factor between these systems is the cost associated with fabricating deflector elements having K values of 1.0 versus 1.4142 for N values of 14 versus 17, respectively.

5.9.2 Hologon Deflector System for High-Resolution Imaging Applications

This subsection briefly examines the integration of a hologon deflector with other optical and mechanical components to construct a laser scanning system that can be used to generate high-resolution images on either photographic media or a xerographic photoreceptor. The optical component layout of the laser scan imaging system is schematically illustrated in Figure 5.50. The imaging medium in this figure is depicted as a rotating drum xerographic phtoreceptor but could also be film on a rotating drum capstan or film on a linear translating flat-bed image plane. The hologon deflector in Figure 5.50 is depicted as a PDG disk deflector but could also be one of the nondisk PDG deflectors described in Section 5.7.

The optical system in Figure 5.50 is based on a gas laser such as a helium-neon, helium-cadmium, or argon-ion. The output beam from the laser is focused by a single-element spherical lens into the acoustooptic (A-O) modulator cell used for turning the laser beam on and off. Commercial A-O modulators of this type can operate in either the digital or analog video mode at rates up to 100

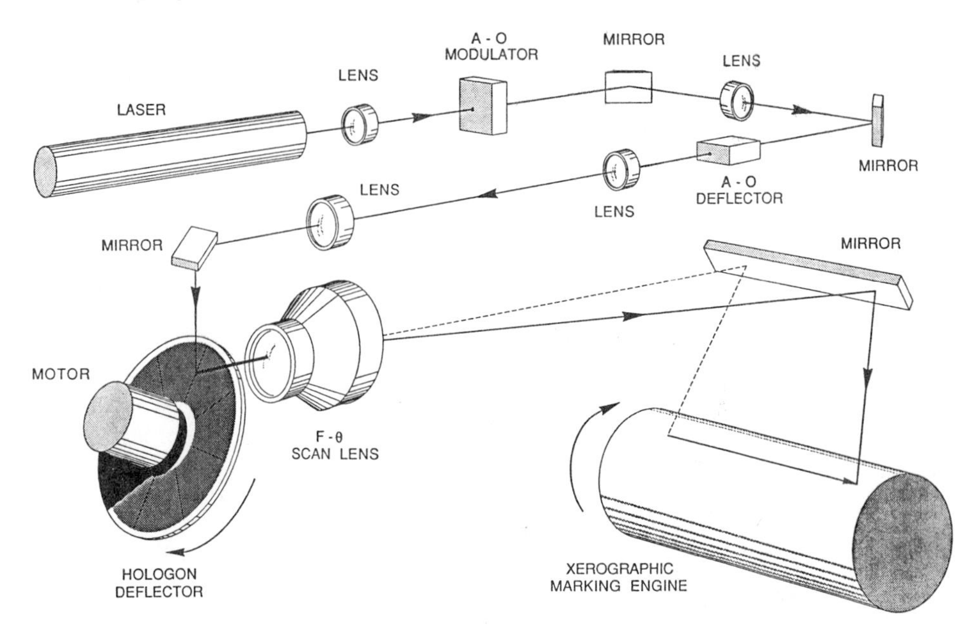

Figure 5.50 Isometric illustration of a laser scan imaging system that utilizes a hologon deflector element. (Courtesy of Holotek Ltd., Rochester, New York.)

MHz. A modulation rate of 50 MHz requires the focused laser beam in the modulator to have a spot size of about 50–60 μm for the $1/e^2$ intensity points. The first-order modulated laser beam from the A-O cell is recollimated by a single element spherical lens. The unmodulated zeroth order laser beam is blocked from entering this collimating lens by an aperture stop, not shown in Figure 5.50, that is located near the lens.

The collimated modulated beam is directed through an acoustooptic based minideflector. This minideflector is used to correct for periodic scan beam tracking errors in the deflector, as described in Section 5.5, "Active Technique for Reducing Scan-Beam Tracking Error," and for radiometric intensity variation associated with the deflector, as described in Section 5.6, "Technique for Improving Radiometric Scan Uniformity." This minideflector can also be used to compensate for minor positional errors in the image transport mechanism if used as part of a closed loop feedback system [50], as illustrated in Figure 5.51.

The xerographic marking engine in Figure 5.51 has a surface motion detector incorporated in it. This surface motion detector could consist of a shaft encoder on the xerographic drum rotation axis or an optical system that reads a structured pattern on the drum surface. Information about the drum velocity obtained with

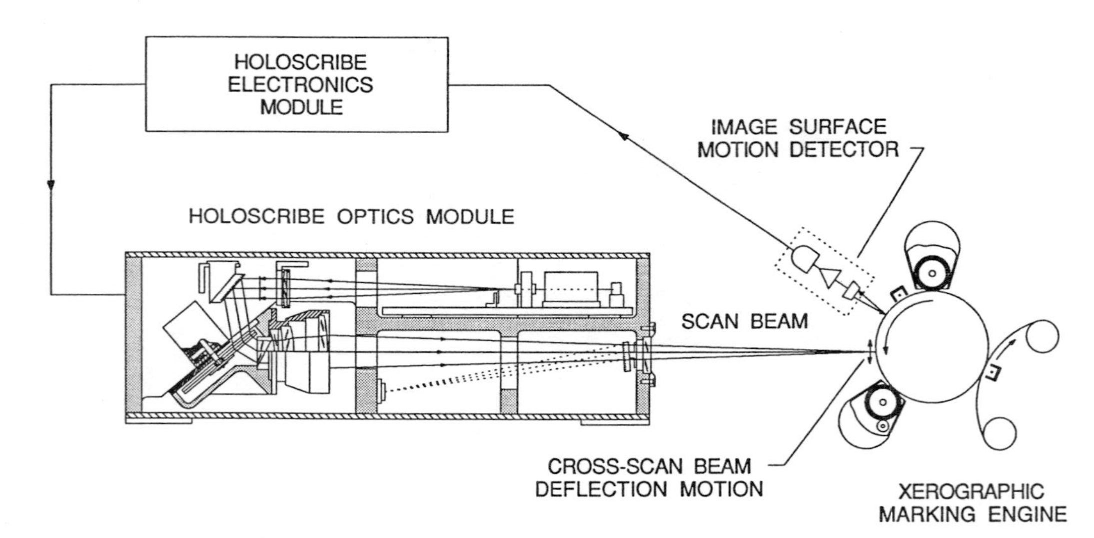

Figure 5.51 The active cross scan positional beam control mechanism that is incorporated in this imaging system can be used in conjunction with a surface motion detector to correct for errors in the speed or the recording imaging media. (Courtesy of Holotek Ltd., Rochester, New York.)

this motion detector is sent to the minideflector electronic control system. When the drum velocity varies, the minideflector is used to change the cross scan angle of the scan beam so that the beam tracks the velocity change in the drum surface. Since electrons, acoustic waves, and light waves are the components of this feedback tracking system, it can respond several thousand times faster than a mechanical-based corrective feedback system. Response time can be on the order of microseconds for this A-O minideflector system.

The change in the cross scan angle of the laser beam from the A-O minideflector is controlled by changing the frequency of the *RF* acoustic carrier which is altered by means of a voltage controlled oscillator. Change in the angle of the beam from the minideflector must be larger than that required at the image plane by the ratio of beam expansion produced in the beam expander/collimator optics following the minideflector. A collimated beam is used in the A-O minideflector to achieve the maximum change in beam angle for a given change in *RF* carrier frequency. Systematic radiometric variations are easily corrected for by varying the amplitude of the *RF* carrier of the A-O modulator and/or A-O minideflector elements.

Diffracted and undiffracted laser beams from the A-O minideflector are focused to aerial images by a single-element spherical lens. An aperture stop, not shown in Figure 5.50, that is located at the focal point of this lens and, thereby, functions as a spatial filter that stops all but the first diffracted-order beam from the A-O minideflector. This first order beam expands in diameter until striking the collimator lens. The collimated beam is directed by a fold mirror to the hologon disk deflector. The deflector beam from the hologon is imaged to a scanning spot at the image plane by the f-θ scan lens. The alignment procedures described in Section 5.8 can be used to adjust the collimator in this system and adjust the cross scan and in-scan beam angles to the hologon deflector unit.

The casting design of the optics module in Figure 5.51 effectively utilizes the transmission nature of the hologon deflector by angling the deflector unit so that the optical components before the deflector reside on one level of the casting while the scan-beam optics are on a second level. This bilevel, I-beam mechanical geometry provides a compact volume structure that is inherently rigid without the need for additional stiffening supports [51]. Also, the structure totally encloses the optical beam path up to the output window and thereby minimizes both beam positional errors caused by air turbulence and system image spot degradation caused by airborne contamination of optical system elements.

Just prior to the output window in Figure 5.51 are shown backward directed dashed lines. These dashed lines depict the start and the end of the scan beams which are intercepted by mirrors that direct these beams to start-of-scan and end-of-scan detectors. These detectors are used to measure the time of flight of the scan beam across the image plane. Time-of-flight information is used to correct in-scan positional errors associated with hologon deflector motor hunting. This correction is accomplished by means of a variable rate pixel data clock.

5.10 CONCLUSIONS

This chapter has described and analyzed three different categories of rotating, plane-grating-based laser beam deflection systems. These three categories comprise the following PDG hologon deflectors: the basic PDG hologon disk deflector of Section 5.5 which achieves a virtually straight scan-line by means of only the relationship between the incident beam angle and the grating facet periodicity value; the improved PDG hologon disk deflectors of Section 5.6 which incorporate additional optical elements after the hologon disk to compensate for the scan bow in the hologon deflected beam; and the nondisk PDG hologon deflectors of Section 5.7 which achieve a straight scan-line by having PDG facets orientated at 45° to the deflector rotation axis and having the incident and deflected beams be parallel and perpendicular to the deflector rotation axis, respectively. The advantages and problems associated with each of these categories of PDG hologon deflector systems are summarized at the end of the section that analyzes that deflector system. A comparison of the nondisk PDG hologon deflectors with other deflector geometries, including PDG disk hologons, is included as part of the summary of deflector system performance at the end of Section 5.7. Also described in this chapter are techniques for optically aligning the hologon deflectors to the other optical components in the scan imaging system and design examples for hologon deflector systems.

What all three categories of PDG hologon configurations have in common is that they function in transmission versus the reflection mode employed by rotating polygonal mirror and galvanometer deflectors. Surface errors in a reflective system, whether they are nonflatness or tilt in nature, produce an error in the optical wavefront exiting from the surface that is twice the magnitude of the corresponding surface error. Errors in transmissive elements, such as tilt and surface nonflatness, are generally reduced by a factor between 4 and 10,000, depending on the optical component and the nature of the error. Wobble in a rotating mirror deflector causes a scan-beam tracking error that is twice the corresponding deflector wobble angle. Wobble-induced tracking error in a transmissive hologon deflector may only be $^1/_{100}$ of the deflector wobble angle when the grating facets are used at the minimum deviation condition. All three categories of PDG hologon deflectors are operated at the minimum deviation condition, and therefore, achieve scan-beam tracking error performance comparable to high-quality air-bearing polygonal mirror deflector systems, but at a fraction of the cost of air-bearing systems. This performance is achieved for PDG deflector units incorporating ball-bearing motors and optical imaging systems comprised of only spherical lens elements. All of these PDG deflector systems achieve excellent scan-beam wavefront fidelity as a result of the flat transmission nature of the deflector element.

ACKNOWLEDGEMENTS

The author wishes to acknowledge the contributions made by Robert Hopkins to the optical design of the hologon deflector unit of Figure 5.33. The author also wishes to acknowledge the contributions and assistance of many of his colleagues at Holotek, in particular Mehdi Araghi and Joachim Ritter, who is now at Optical Gaging Products. The author wishes to acknowledge Henry Kelley of Agfa-Compugraphic Division for helpful discussions and constructive criticism during the early design stages of the hologon deflector unit of Figure 5.33.

REFERENCES

1. C. J. Kramer, Hologon laser-beam deflectors meet cost/performance criteria for graphic arts applications, *Laser Focus*, *24*:94 (1988).
2. F. Bestenreiner, V. Greis, J. Helmberger, and K. Stadler, Visibility and corrections of periodic interference structure in line-by-line recorded images, *J. Appl. Phot. Eng.*, 2:86 (1976).
3. U. C. Urbach, T. S. Fisli, and G. K. Starkweather, Laser scanning for electronic printing, *Proc. IEEE*, *70*:597 (1982).
4. A. Arimoto, S. Saitoh, T. Mochizuki, Y. Kikuchi, and K. Hatazawa, Dual beam laser diode scanning system for ultrahigh speed laser beam printers using a spot control method, *Appl. Opt.*, *26*:2554 (1987).
5. L. Beiser, Laser scanning systems. In *Laser Applications* (M. Ross, ed.), Academic Press, New York, 1974, Vol. 2, p. 55.
6. C. J. Kramer, Hologon deflectors for high-resolution internal drum and flat-field imaging, *SPIE Proc. Hard Copy Output*, *1079*:427 (1989).
7. C. J. Kramer, *Hologon Scanner System*, U.S. Patent 4,852,956, August 1989.
8. C. J. Kramer, Specification and acceptance test procedures for hologon laser scanner systems, *SPIE Proc. Metrology of Optroelectronic Systems*, *776*:81 (1987).
9. M. J. Buzawa, Lens systems for laser scanners, *Laser Focus*, *Vol. 16*, *82*: (1980).
10. *Laser Scan Lens Guide*, Melles Griot Optics, Rochester, N.Y., 1987.
11. L. C. DeBenedictis and R. V. Johnson, *Laser Scanning Utilizing Facet Tracking and Acoustic Pulse Imaging Techniques*, U.S. Patent 4,205,348, May 1980.
12. I. Cindrich, Image scanning by rotation of a hologram, *Appl. Opt.*, *6*:1531 (1967).
13. L. Beiser, *Holographic Scanning*, Wiley, New York, 1988.
14. D. H. McMahon, A. R. Franklin, and J. B. Thaxter, Light beam deflection using holographic scanning techniques, *Appl. Opt.*, *8*:399 (1969).
15. L. Beiser, Advances in holofacet laser scanning, *Proc. Electro-Optical Systems Design Conference* 1975, p. 333.
16. R. V. Pole and H. P. Wolenmann, Holographic laser beam deflector, *Appl. Opt.*, *14*:976 (1975).

17. C. S. Ih, Holographic laser beam scanners utilizing an auxiliary reflector, *Appl. Opt.*, *16*:2137 (1977).
18. C. Kramer, Holographic beam scanner with concentric reflector, *1979 Annual Meeting of the Optical Society of America*, Paper F13.
19. C. Kramer, *Holographic Scanner Insensitive to Mechanical Wobble*, U.S. Patent No. 4,243,293, Jan. 1981.
20. H. P. Herzig, *Holographic Optical Scanning Elements*, Doctoral Thesis at University of Neuchatel, Institute of Microtechnology, Switzerland, 1987.
21. H. P. Herzig and R. Dandiker, Holographic optical scanning elements with minimum aberrations, *App. Opt.*, *27*:4739 (1988).
22. C. J. Kramer, Holographic laser scanners for nonimpact printing, *Laser Focus*, *17*:70 (1981).
23. C. J. Kramer, *Optical Scanner Using Plane Linear Diffraction Gratings on a Rotating Spinner*, U.S. Patent 4,289,371, Sept. 1981.
24. M. V. Antipin and N. G. Kiselev, Laser beam deflector based on transmission holograms, *Tech. Kino Telev.*, *No. 6*:43 (1979) (in Russian).
25. C. J. Kramer, *Holographic Scanner for Reconstructing a Scanning Light Spot Insensitive to Mechanical Wobble*, U.S. Patent 4,239,326, Dec. 1980.
26. C. J. Kramer, *Multi-Wavelength Scanning System*, U.S. Patent 4,848,863, July 1989.
27. R. E. Brasier, *Linear Beam Scanning Apparatus Especially Suitable for Recording Data on Light Sensitive Film*, U.S. Patent 4,337,994, July 1982.
28. C. J. Kramer, Hologon deflectors incorporating dispersive optical elements for scan-line bow correction, *SPIE Proc. Holographic Optics: Design and Applications*, *883*:230 (1988).
29. C. J. Kramer, Hologon laser scanners for nonimpact printing, *SPIE Proc. High-Speed Read/Write Techniques for Advanced Printing and Data Handling*, *390*:165 (1983).
30. C. J. Kramer, *Holographic Scanning Spinner*, U.S. Patent 4,067,639, Jan. 1978.
31. C. J. Kramer, A. F. McCarroll, and T. S. Fisli, *Dynamic Mounting for Holographic Spinners*, U.S. Patent 4,353,615, October 1982.
32. C. J. Kramer, *Reflective Holographic Scanning System Insensitive to Spinner Wobble Effects*, U.S. Patent 4,304,459, Dec. 1981.
33. H. P. Brueggemann, *Scanning System with Two Reflections from Scanning Surface by Mirrors with Optical Power*, U.S. Patent 4,682,842, July 1987.
34. H. Ishikawa, *Light Beam Scanning Apparatus*, U.S. Patent 4,626,062, Dec. 1986.
35. D. B. Kay, *Optical Scanning System with Wavelength Shift Correction*, U.S. Patent 4,428,643, Jan. 1984.
36. H. Werlech, G. Sincerbox, and B. Yung, Fabrication of high efficiency surface relief holograms, *J. Imaging Tech.*, *10(3)*:105 (1984.
37. M. G. Maharam and T. K. Gaylord, Diffraction analysis of dielectric surface-relief gratings, *J. Opt. Soc. Am.*, *72*:1385 (1982).
38. E. G. Loewen, M. Neviere, and D. Maystre, Grating efficiency theory as it applies to blazed and holographic gratings, *Appl. Opt.*, *16*:2711 (1977).
39. H. H. Ryffel, ed., *Machinery's Handbook 22nd Edition*, Industrial Press Inc., New York, 1984, p. 233.

40. J. A. Ritter, M. N. Araghi, and C. J. Kramer, *Hologon Scanner System*, U.S. Patent Application filed March 1990.
41. C. J. Kramer, Preobjective Hologon Scanner System,'' U.S. Patent 4,583,816, April 1986.
42. Rudolf Kingslake, ed., *Applied Optics and Optical Engineering*. Vol. V, *Optical Instruments*. Part II, Academic Press, New York, 1969.
43. K. Matsumoto, N. Kawamura, and S. Minami, *Scanning Device*, U.S. Patent 4,176,907, Dec. 1979.
44. C. J. Kramer, Holographic grating for high resolution scanning, *1986 Lasers in Graphics Conference Proceedings*, vol. II, p. 106, Dec. 1986.
45. C. J. Kramer, A rotating pyramidal grating laser beam deflector for high resolution imaging, *Proc. SPSE 41st Annual Conf.*, May 1988, p. 339.
46. G. K. Starkweather, *Single Facet Wobble Free Scanner*, U.S. Patent 4,475,787, October 1984.
47. A. Stein and M. Nagler, *Internal Drum Plotter*, U.S. Patent 4,853,709, August 1989.
48. C. J. Kramer, *Hologon Scanner System*, U.S. Patent 4,786,126, Nov. 1988.
49. T. L. Whitman and M. N. Araghi, Electronic f-theta correction for hologon deflector systems, *SPIE Proc. Beam Deflection and Scanning Technologies*, 1454, (1991).
50. C. J. Kramer, *Hologon Scanner System*, U.S. Patent 4,826,268, May 1989.
51. J. A. Ritter and C. J. Kramer, *Integrated Laser Beam Scanning System*, U.S. Patent 4,779,944, October 1988.

6

Polygonal Scanners: Applications, Performance, and Design

Randy J. Sherman

Lincoln Laser Company
Phoenix, Arizona

6.1 INTRODUCTION

Polygonal mirrors are useful devices for scanning light beams within certain boundaries of performance. This section is intended to outline those boundaries as an aid to the system designer in selecting the most practical method of scanning in view of a particular need.

Table 6.1 lists a number of parameters to be considered when selecting a method of scanning. Opposite the parameters are the values associated with polygonal mirrors when used in conventional optical designs. A note of caution about the use of this table; it should not be assumed that arbitrary values of all the parameters shown (within the ranges given) can be achieved in the same device.

The primary advantages of polygonal mirrors are speed, the availability of wide scan angles, and velocity stability. They are usually rotated continuously in one direction at a fixed speed to provide repetitive unidirectional scans which are superimposed in the scan field or plane, as the case may be. A common method of producing rotation is to fasten the polygonal mirror directly to an electric motor shaft (see Figure 6.1).

The combined inertia of polygon and motor rotor contribute to rotational stability. In some cases an additional flywheel is added to further stabilize rotational rate. The relatively high inertia of polygons and drive motors, on the

Table 6.1 Practical Performance Boundaries for Polygon Mirrors

Light beam diameter	0–70 mm
Light beam wavelength	240–14,000 nm
Light beam power (continuous wave)	To several kilowatts
Scan angle	5–360°
Scan rate (unidirectional repetition rate)	30–20,000 Hz
Scan efficiency[a] $\left(\frac{\text{active scan time}}{\text{total time}}\right)$	To 100%
Light transmission efficiency	80–95%
Scan resolution (spots/line)	To 20,000+
Scan linearity[b]	0.02%

[a] 100% scan efficiency can be achieved by filling two polygon facets at a time with the input light beam.
[b] Scan linearity is defined as the percent deviation of angular scan velocity from the mean angular velocity of light reflected by polygon facets.

other hand, render them impractical for applications requiring rapid changes in scan velocity or start/stop formats.

Regular polygonal mirrors are available from the inventories of standard manufacture from several suppliers. Table 6.2 lists typical commercial-grade polygon dimensions and characteristics.

Rotational speeds up to 120,000 rpm are practical for alternating-current (AC) motors. For applications where rotational rates exceed the capability of electric motors, gas turbines provide an alternative. Rotational speeds in the range 90,000–1,000,000 rpm are achieved with such devices (see Figure 6.2). Turbine-driven rotating mirrors should be considered as an alternative only when short duty cycles (a few minutes) can be tolerated and relatively high costs can be afforded. Frictional heating of the rotor due to air drag at speeds approaching or exceeding Mach 1 at the periphery can determine the maximum practical run time.

6.2 TYPES OF POLYGONAL SCAN MIRRORS

Four common types of polygonal scan mirrors have been developed and put into use over the years:

1. Regular polygons
2. Irregular polygons
3. Inverted polygons
4. Pyramidal polygons

The purpose of this section is to define each of the types and give examples of their use to familiarize the reader with the various advantages and disadvantages they present to potential application.

Figure 6.1 Synchronous motor/polygon assembly. The polygon is fastened directly to the motor shaft, which is supported by ball bearings, one at each extreme end of the rotor. (Courtesy of Lincoln Laser Co.)

6.2.1 Regular Polygons

A regular polygonal scan mirror is defined as one having a number of plane mirror surfaces (facets) which are parallel to and facing away from a rotational axis, uniformly distant from the rotational axis, and located with equal polar

Table 6.2 Range of Dimensions and Characteristics Typical of Commercial-Grade Polygons

Feature	Range
Number of facets	1–60
Inscribed circle diameter	50.8–355.6 mm (2–14 in)
Pyramidal error	5 arcsec–2 arcmin
Facet flatness	$\lambda/2$–$\lambda/10$ at 550 nm
Facet-to-facet angular error	5–30 arcsec

coordinates (see Figure 6.3). The significant technical feature of a regular polygon is that it may be used to produce repetitively superimposed straight scans.

Regular polygons are the most common type of polygon found in scanning systems, since their geometry lends itself to mass production and rather low manufacturing cost. The major applications for regular polygons are

Nonimpact computer output printers
Laser inspection systems
Infrared viewing systems
Graphics facsimile systems

The symmetry of regular polygons enables the stacking of several polygons on an arbor for processing during manufacture (see Figure 6.4). Since the manufacturing cost of major processes involved in the fabrication of the stack is divided over several finished parts, substantial cost reduction is realized.

6.2.2 Irregular Polygons

An irregular polygon is defined as one having a number of plane mirror surfaces (facets) which are at a variety of angles with respect to, and face away from, the rotational axis. The polar location may also be irregular (a variety of angles) or regular (equiangular) (see Figure 6.5).

The significant technical feature of irregular polygons is that they produce nonsuperimposing repetitive scans (i.e., a raster).

Irregular polygons are commonly used in applications where a coarse (few lines) raster is of value, such as

Point-of-sale scanners for reading universal product code (UPC) symbols
Multiple-detector infrared scanners
Laser heat-treating systems
Intrusion-alarm scanning systems

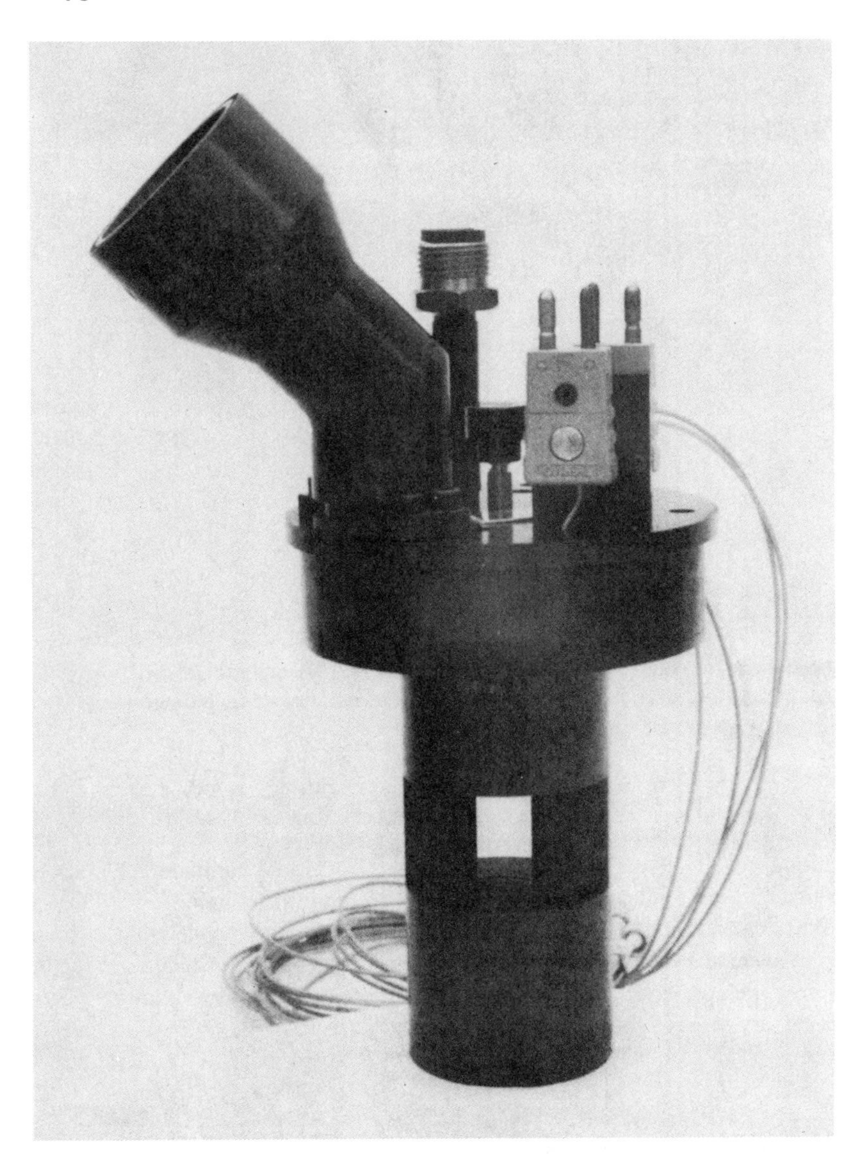

Figure 6.2 Compressed-air-driven rotating mirror assembly that operates at 498,000 rpm.

Figure 6.3 Regular 36-faceted polygon used in a production high-speed computer output printer. The facets on this polygon are nickel-plated and pitch-polished.

Because of their asymmetry, irregular polygons can seldom be stacked or grouped together during processing and therefore tend to have a higher manufacturing cost than that of regular polygons. The most common methods of manufacture are replication and single-point diamond machining.

When considering irregular polygons for use, it is recommended that a design employing diametrical symmetry be used, for the sake of dynamic balance. Figure 6.6 describes preferred and problem configurations. At high rotation speeds it is necessary to have the centers of gravity of diametrically opposed sections of the polygon in a plane at right angles to the rotational axis.

6.2.3 Inverted Polygons

An inverted polygon is defined as one having a number of plane mirror surfaces which are parallel to and facing toward a rotational axis, usually equidistant from the axis and located with equal polar coordinates (see Figure 6.7).

Irregular polygons are also inverted. The significant feature of inverted polygons is that they lend themselves to compact optical designs. Typical applications are scanning systems for missile guidance and reconnaissance.

Figure 6.4 Stack of polygons on a steel arbor. This group of eight individual polygons has undergone machining, lapping, plating, and polishing as a unit.

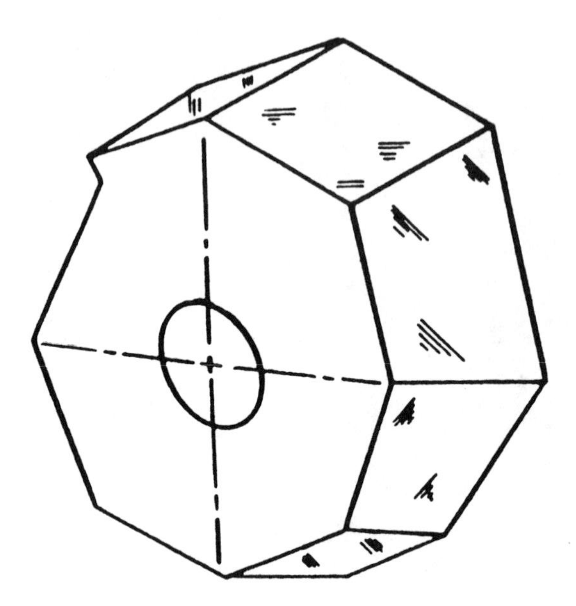

Figure 6.5 Irregular polygon.

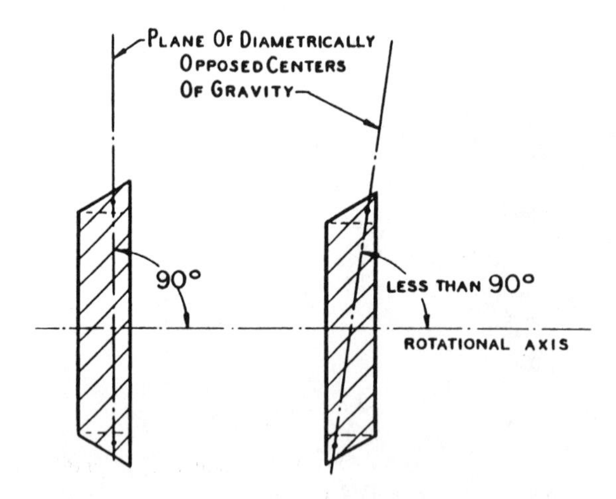

Figure 6.6 Dynamic balance considerations in irregular polygon construction.

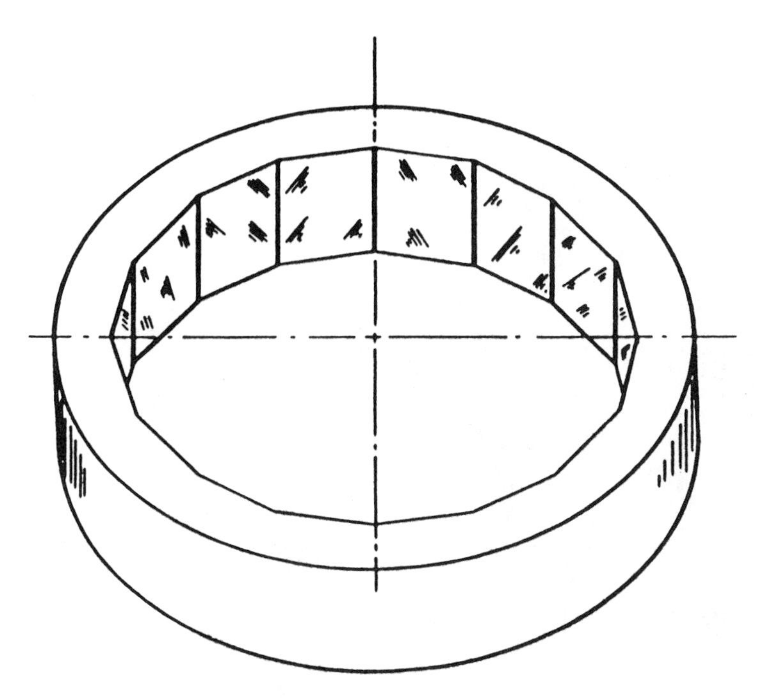

Figure 6.7 Inverted regular polygon.

Inverted polygons are limited to relatively low speed scanning because of their inherent lack of structural integrity and the necessity of using large bearings to support them. The most common fabrication techniques are replication and single-point diamond machining.

6.2.4 Pyramidal Polygons

Regular pyramidal polygons are defined as having a number of plane mirror facets at an angle (rather than parallel) to the rotational axis. In the most common variety the facet angle is 45° to the spin axis (see Figure 6.8).

Since the incident rays on a pyramidal mirror may be at other angles than normal (and still produce a straight scan), they are useful devices for producing smaller scan angles with fewer facets than are achievable with polygonal mirrors. For example, a four-sided, 45°, pyramidal scan mirror deflects incident rays (parallel to the rotational axis) through 90° for 90° of pyramid rotation. A four-sided regular polygon deflects incident rays (at right angles to the rotational axis) through 180° for 90° of polygon rotation.

Figure 6.8 Regular 45° pyramidal polygon.

6.3 APPLICATION TO SCANNING

6.3.1 Entrance Scanning (Preobjective)

Entrance scanning is a term used to describe the use of a polygon to deflect a ray bundle, which after deflection is finally imaged by a lens or curved mirror (see Figure 6.9). This method of scanning places the function of focal plane definition on the lens design rather than on the scanning facet. This is beneficial if the desired focal surface is to be flat.

Several desirable characteristics can be designed into the scan lens when employed in entrance scanning. An example is a lens design commonly referred to as an f-θ configuration. An f-θ lens provides the following features:

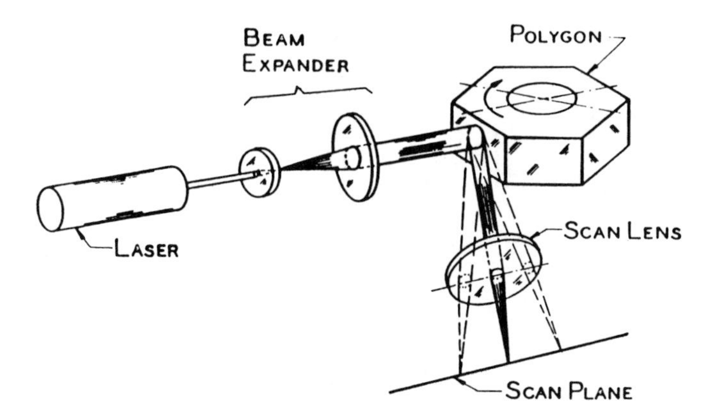

Figure 6.9 Schematic diagram of a typical entrance scanning application of a polygonal mirror.

A flat focal plane
Uniform spot diameter over the entire scan
Linear spot velocity at the focal plane (assuming constant angular velocity of the entrance beam)

Usually, it is desirable to have the scanning spot move with a highly accurate and constant velocity in the focal plane. Polygonal mirror deflectors provide angular velocity stability in the range 0.005 to 0.05%, depending on the speed and moment of inertia of the scanner. Without the aid of an f-θ lens, however, the spot velocity variation at a flat focal surface will be proportional to the tangent of the scan angle, which for systems involving several degrees of scan means several percent variation.

6.3.2 Exit Scanning (Postobjective)

Exit scanning is a term used to describe the use of a polygon to deflect a focusing ray bundle over a focal surface (see Figure 6.10). This method places the function of focal plane definition on the polygonal mirror, and the imaging (spot forming) lens is a relatively simple component.

The focal surface of an exit scanning system is curved. The radius of curvature is approximately the distance from the center of the scanning facet to the focal point of the lens. In scanning systems utilizing relatively small scan angles, $>20°$ and large f numbers, f/50 or greater, exit scanners may be used effectively, since the chord depth of the focal surface may be less than the depth of focus of the

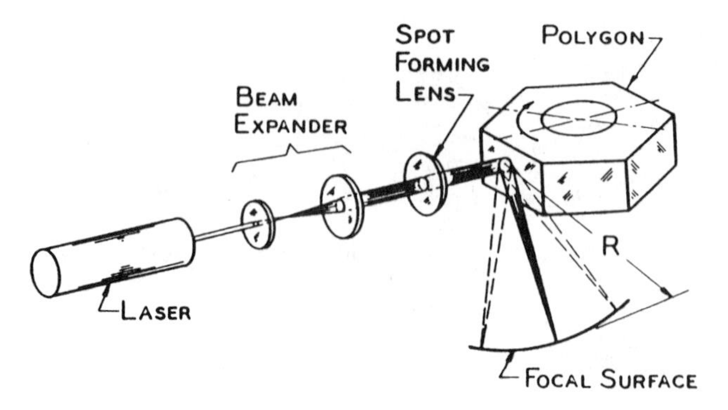

Figure 6.10 Schematic diagram of a typical exit scanning application of a polygonal mirror.

scanning beam, in which event spot size may be effectively maintained over a flat photoreceptor plane.

6.3.3 Scanning a Focused Spot along a Straight Line

Frequently a scanning system is required to scan a focused spot along a straight line on a planar photoreceptor surface. To some extent this is a special case, since there are a limited number of scan mirror configurations that produce straight-line scans in a plane.

If the line of focus of a scan is to be straight and in a plane, there are two scan mirror configurations that may be used. These are (1) a 45° pyramidal polygon and (2) a regular polygon. All other configurations produce conical scans from a single deflection by the rotating mirror surface.

If the photoreceptor surface can be curved, then a number of other configurations of scan mirror may be used, in principal matching the conical scan to an appropriate cylinder and thereby generating a scan line curved in one direction and in a plane in the other direction.

A rotating mirror has two effects on an incident ray. One effect is angular deflection. The degree of angular deflection is described by the first mirror law, which states that the angle of incidence is always equal to the angle of reflection. The angular path of a reflected ray in spherical space is described by the second mirror law, which states that the incident ray, reflected ray, and the mirror normal (a line at right angles to the mirror) always lie in a plane.

The second effect of a rotating mirror is translation, defined here as a parallel offset of the reflected ray. Only when a mirror surface lies on its axis of rotation does translation not occur.

The effects of angular deflection and of translation both are important to the

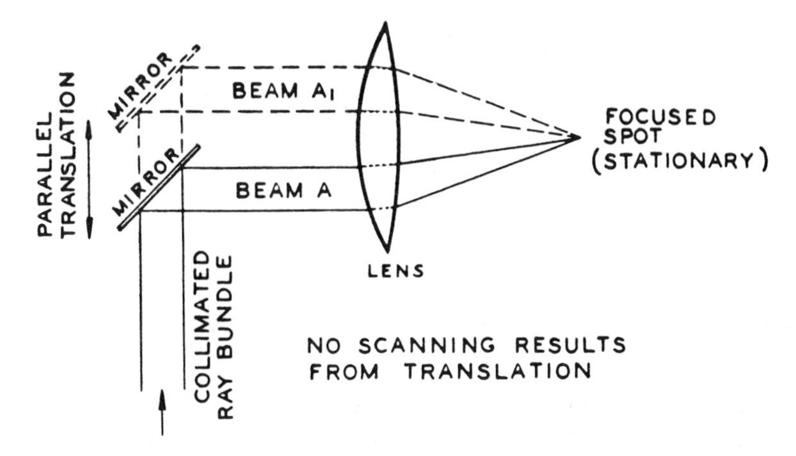

Figure 6.11 Beam translation effects on focal spot location.

behavior of mirrors and lenses used to scan. In general, beam translation caused by a mirror deflector in an entrance scanner configuration produces no motion of the scanning spot, provided the translation does not move the beam outside the usable lens aperture. Figure 6.11 describes this lens characteristic.

Figure 6.12 describes a rotating mirror and scan lens combination wherein

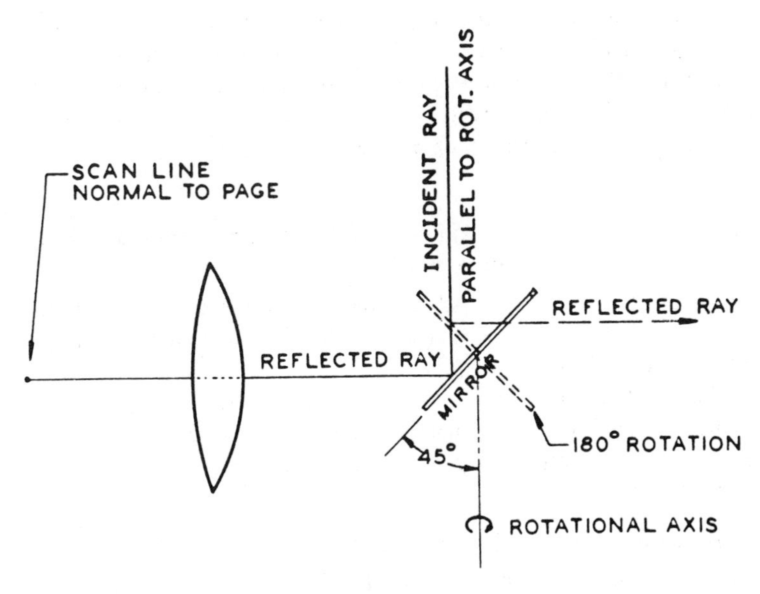

Figure 6.12 Simultaneous translation and deflection.

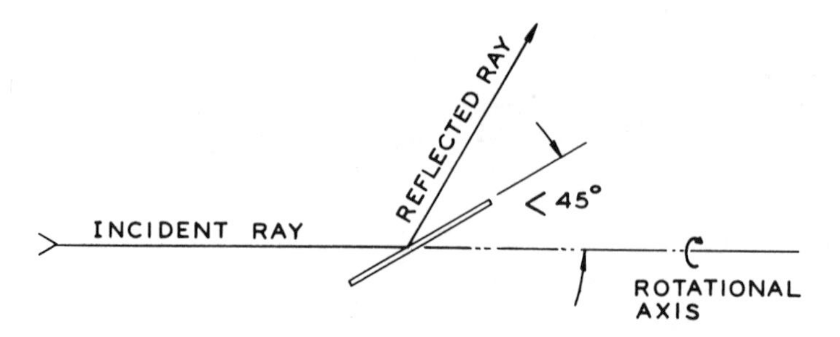

Figure 6.13 Conical scan from a mirror at less than 45° to rotational axis.

both translation and deflection occur. In this arrangement translation may be ignored, since over the limited useful scan angle the amount of translation is trivial and has little impact on the lens size required. Were this same arrangement to be employed in an exit scanner, a curved scan line would result in both principal planes, i.e., parallel to and at right angles to the rotational axis (see Figure 6.13).

6.3.4 Correcting Polygon-Induced Scanning Anomalies

Under circumstances where the combined errors of polygon and drive motor produce undesirable errors in the location of the resulting scans in the direction of scan (timing errors) or at right angles to the scan (tracking errors), some alternatives are available. These become particularly important when the requirements for scanning accuracy become difficult and/or impractical to achieve by placing tolerances of the order of a few arc seconds or less on the polygon/ motor assembly. The following correction schemes are those most commonly found in scanning systems requiring high-accuracy scans.

Sources of Tracking Errors

There are four major sources of tracking errors that may be contributed to a scanning system by a polygon/motor assembly:

1. Errors in the angular location of polygon facets with respect to the datum surface of the polygon. These errors are attributable to limitations in the manufacturing process of the polygon.
2. Errors in the fastening of the polygon to the motor shaft.
3. Errors in the rotation of the motor shaft attributable to bearings, lack of rotor stiffness, thermal gradients causing rotor warp, and dynamic unbalance.
4. Errors in facet flatness.

Active Methods of Correcting Tracking Errors

Methods for correcting tracking errors fall into two basic categories, active and passive. Active systems involve auxiliary small-angle deflectors located in the

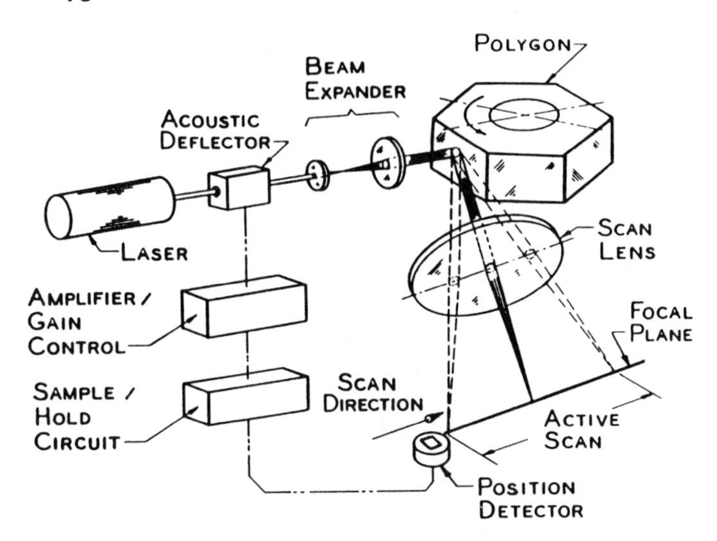

Figure 6.14 Schematic diagram of a small-angle correction scheme employing a scan position sensor and an acoustic deflector to compensate for polygon/motor-induced tracking errors.

scanner optical system to compensate for motor/polygon-induced errors in tracking. If all the errors to be corrected are repeatable, the auxiliary deflector may be programmed to make corrections. If nonrepeatable errors must be corrected, an error detector must be used to interrogate the scanning facet immediately prior to the active time of scan and produce an error signal on which the auxiliary deflector can act. However, to be effective, both of these active methods must rely on the integrity of a scanning facet during its active period of traverse. Errors resulting from items 3 and 4 may produce undesirable effects which are uncontrollable by programmed correctors or "feedback" correctors. Typically, corrections are made during the "dead time" between scans, rather than during the active period of a scan, hence the dependence on inherent stability of scans between correction periods.

Figure 6.14 depicts an active correction system for tracking errors that are polygon/motor-induced. In this case an acoustic beam deflector (with deflection capability of several arc minutes) is placed in the laser* beam path prior to the beam expander. The small aperture of the deflector dictates a location where the light beam is also small. In this location the acoustic deflector must be able to produce a correction angle which is twice the facet angle error multiplied by the

* Due to the fact that acoustooptic beam deflectors are diffractive devices, they are impractical for light sources other than laser.

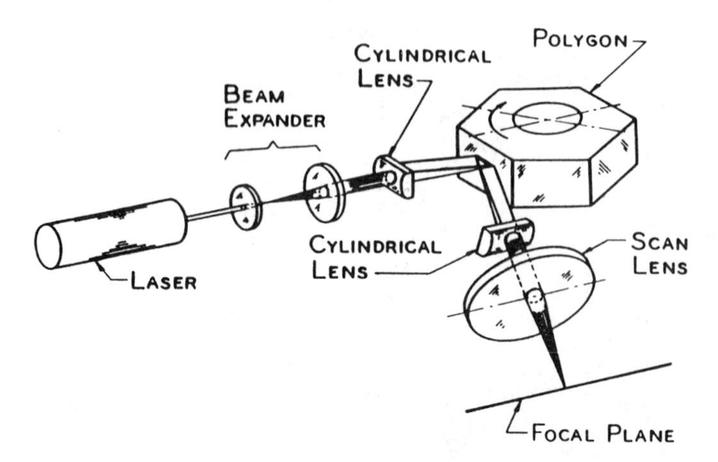

Figure 6.15 Schematic diagram of an optical system employing cylindrical lenses to correct for polygon/motor-induced tracking errors.

magnification of the beam-expanding telescope. The telescope will reduce the deflected beam angle in direct proportion to its power, and the scanned beam will have an angular error twice that of the facet angular error.

This scheme is used effectively to correct for facet errors in the range of an arc minute. Polygon/motor assemblies are readily available within this tolerance. A note of caution on this approach, however, is that acoustic deflection devices exhibit small changes in light energy transmitted as a function of scan angle. A difference of several percent (5–7%) can be expected from 0° to 10 arcmin of deflection. It is necessary to employ sample-and-hold electronic circuitry in controlling the deflector, since the correction made must be maintained over the period of the scan of the facet.

Passive Methods of Correcting Tracking Errors

Passive correction schemes usually utilize additional optical elements to reduce polygon-induced tracking errors. The most popular method of passive correction involves the use of cylindrical lenses in the scanning optical imaging system* (see Figure 6.15). These cylindrical lenses cause focusing in one direction of the parallel rays of light directed toward the polygonal scan mirror. This focusing in one direction creates a line image on the facet with the direction of the line running in the scan direction. The reflected rays from the polygon are recollimated by a second cylindrical lens. The result of this one-dimensional imaging is to

---

* John Fleischer, U.S. Patent #3,750,189.

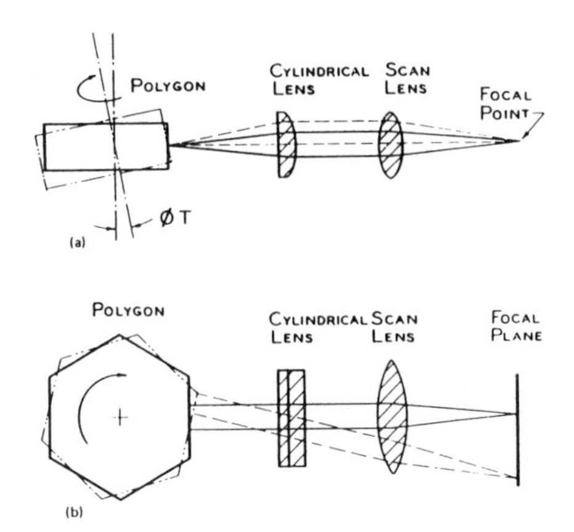

Figure 6.16 Mechanism of jitter correction in the system as shown in Figure 6.15. (a) A tilt of the polygon ϕT does not produce a displacement of the focused spot in the direction at right angles to the scan. (b) A rotation of the polygon around its axis produces a scan.

permit scanning in one direction, the direction of rotation of the polygon, but it prevents scanning in the direction at right angles to that. Figure 6.16 describes two views of the light energy emanating from a polygon facet, as shown in Figure 6.15, showing why a facet error or "wobble" of the rotational axis produces no undesired scanning.

Another novel passive approach involves the use of a retroreflecting prism which accepts the first scan beam from a polygon facet and returns those rays to the same facet. The second reflection will have an error identical to, but with the reverse sign of, the rays reflected in the first bounce. The result is a cancellation of the errors of the facet (see Figure 6.17).

The corrector prism method is simple and effective in correcting for facet dislocation errors. These errors result in a small lateral shift of the beam exiting the prism, but since these rays are collimated, this shift does not result in a motion of the spot in the focal plane.

In the direction of scan the corrector prism behaves as a plane mirror (except for the fixed offset between input and output), so it reflects light back in the direction of the facet. Of course, the input beam to the prism is coming from a moving facet, and so scanning occurs in the prism. The double bounce off the polygon facet causes velocity doubling of the final output beam. As a result, the angular velocity of the output is four times the polygon rotational velocity.

Since the beam returning to the facet is scanning, it migrates across the facet

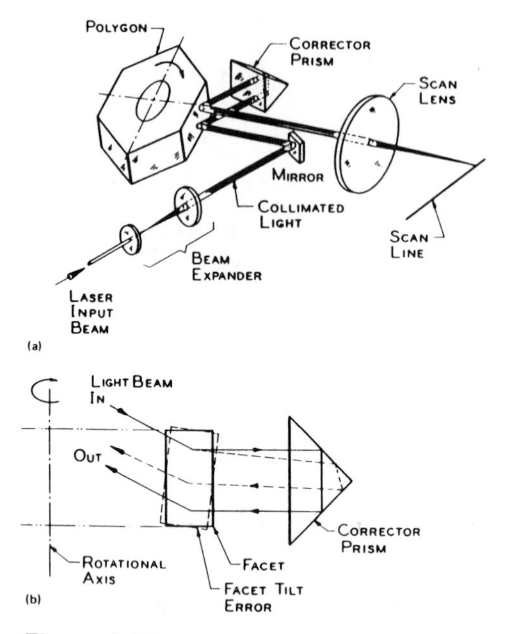

Figure 6.17 Schematic diagram of the use of a 90° corner prism (PORRO prism) to cancel polygon/motor-induced tracking errors. (a) General arrangement of components; (b) effect of corrector prism as facet tilt error is introduced.

(in the scan direction) (see Figure 6.18). The result of this migration in most configurations limits the scan efficiency (scan time/total time) to 50% or less; however, due to the velocity doubling, there is seldom a sacrifice in scan angle of the output.

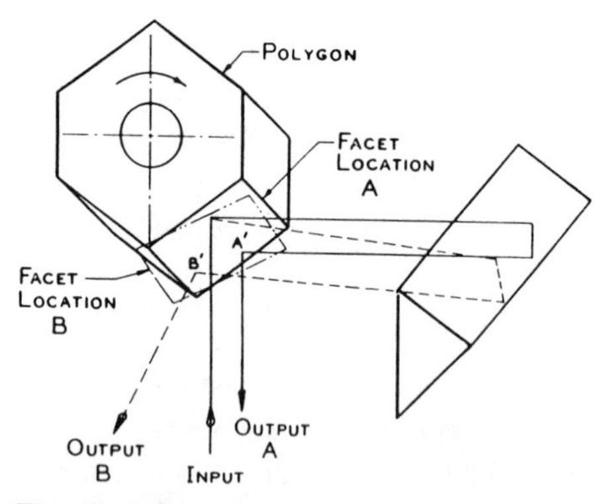

Figure 6.18 Schemtic diagram showing limited-use angle of the polygon when using a prism corrector due to beam migration along the facet. When the facet rotates from position A to B, the point of incidence of the output beam moves from A′ to B′.

6.3.5 Design Trade-Offs

Trading off design possibilities for a scanning system is to some extent an iterative process. A suggested guideline is to define the requirement at the scan focal plane (or focal surface) and work your way backward (up the photon stream) to the light source. Table 6.3 provides a rudimentary list of parameters that must be developed in order to determine a preferred approach to a scanning system employing a polygonal deflector. It also provides some comments on how the parameters of output relate directly to the polygon.

The key parameters to determining the basic polygon configuration are spot size, wavelength, and scan length. Having specified spot size (most commonly that diameter of spot at which intensity is down to 13% of the peak intensity, i.e., $1/e^2$) and wavelength, we may determine the length-to-diameter ratio of the cone of rays necessary to produce the spot. This ratio is commonly called $f\#$, and in the case where the input ray bundle is truncated at the $1/e^2$ diameter, is given by focused spot size/1.8λ (see Figure 6.19).

For an exit scanner configuration to be used with a planar photoreceptor, it is necessary to calculate the depth of focus of the focusing cone using the formula $D = (f\#)^2\lambda$ as a rule of thumb. This value becomes the permissible chord depth of the scan at the photoreceptor and thereby determines the scan angle and radius. Knowing the scan radius r, we can determine the width, the major dimension D', of the footprint of the ray bundle at the mirror facet by

Table 6.3 Design Parameters for a Polygonal Scanner System

Parameter	Relationship to polygon/motor performance
Spot size	Facet size and flatness
Scan line length/scan angle	Number of facets
Wavelength or wavelengths	Type of reflection enhancement coating
Line straightness	Average pyramidal error and flatness of facets
Scan rate	Motor speed
Scan cross-track accuracy	Polygon/motor tracking accuracy or corrector selection
Scan timing accuracy	Motor velocity stability and/or facet polar location accuracy
	Facet flatness
Scan efficiency	Facet length (scan direction)
Scatter losses permissible	Facet surface quality
Light transmission uniformity	Facet surface quality
	Coating uniformity
Mean time between failures	Facet durability (cleanability)
	Bearing type and quality
	Dynamic balance
Signal/noise ratio	Facet roll-off width

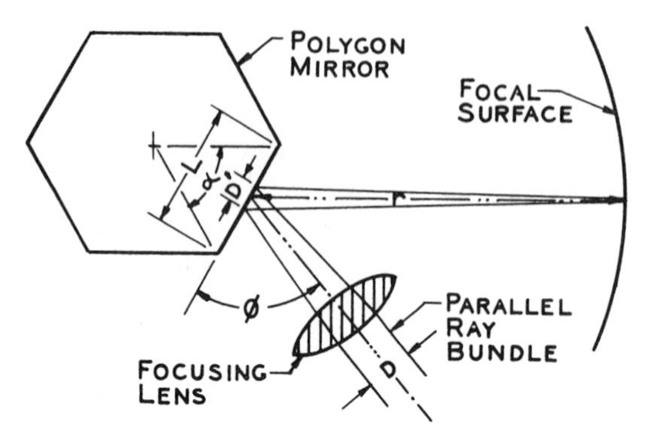

Figure 6.19 Schematic diagram of the important criteria for determining polygon facet size for an exit scanning system.

$$D' = \frac{r}{f\# \text{ Sin } \phi}$$

where ϕ is the tilt angle between the facet and the central ray of the focusing cone.

If the calculated facet length and/or the facet subtended angle is impractically long or small, respectively (numbers of facets exceeding 60 are impractical), it is an indication that an entrance-type scanning system is called for.

When determining the polygon characteristics for an entrance-type scanner, it is necessary to examine the required scan angle and beam diameter at the entrance pupil of the scan lens. Since the scan lens is designed to provide a flat field, the polygon dimensions need not be related to a scan radius sufficient to scan a chord depth within the beam depth of focus.

A particularly useful lens design for scanning is one referred to as an f-θ lens. The design provides a correction for the velocity change that would otherwise result from the tangent of the scan angle on a flat plane. The primary characteristics of an f-θ lens are

Flat focal plane
Uniformity of spot size over the scan length
Constant-scan-spot-velocity output for a constant angular scan input

Beyond sizing of the polygon there are more subtle concerns of tolerancing and practicality of manufacture. These are dealt with in Section 6.4.

6.4 CONSIDERATIONS OF QUALITY

6.4.1 Manufacturing Methods

There are numerous methods of fabricating multifaceted polygons. Five are discussed here:

1. Injection molding
2. Assembly construction
3. Single-point diamond machining
4. Replication
5. Conventional polishing

These various methods of manufacture have been refined over the years to rather sophisticated levels. Each manufacturer has proprietary tooling and/or procedures to apply to the task of making polygonal mirrors. These proprietary items, usually referred to as trade secrets, preclude any in-depth disclosure of the precise methods being employed. The intent here is to provide an overview of the most common techniques in use so as to put into perspective their merits in regard to the finished product.

Injection Molding

Injection molding has been successfully applied to the making of polygonal mirrors on an experimental basis, and there is some promise that this method will be refined over time to provide very low cost polygons in the medium to low end of the quality spectrum.

Special material formulations are necessary to mold an optical part.

Assembly Construction

This manufacturing method involves assembling a polygon with individual planar first surface mirrors fastened to the perimeter of a machined or cast support structure (see Figure 6.20). Assembly construction is particularly useful for fabricating irregular polygons.

The individual mirror segments may be made from glass or metal. Methods of fastening include bonding with epoxy adhesives and machine screws. Since the centrifugal stresses become great at high rotational speeds, careful attention must be paid to the potential for distortion of surface flatness and the possibility of fastener failure of assembly construction polygons that are to be used in relatively fast scanning applications.

Single-Point Diamond Machining

Single-point diamond machining is a process of material removal using a finely sharpened single-crystal diamond cutting tool. Diamond-machining machine tools are available in the forms of lathes and millers. The use of ultra-precise air-bearing spindles and table ways enable machining to optical-quality surface figure and finish. Figure 6.21 shows a diamond-machining system with a polygon in process.

Diamond machining is an efficient process for generating an optical surface, since it can be automated and the technique essentially creates the surface directly

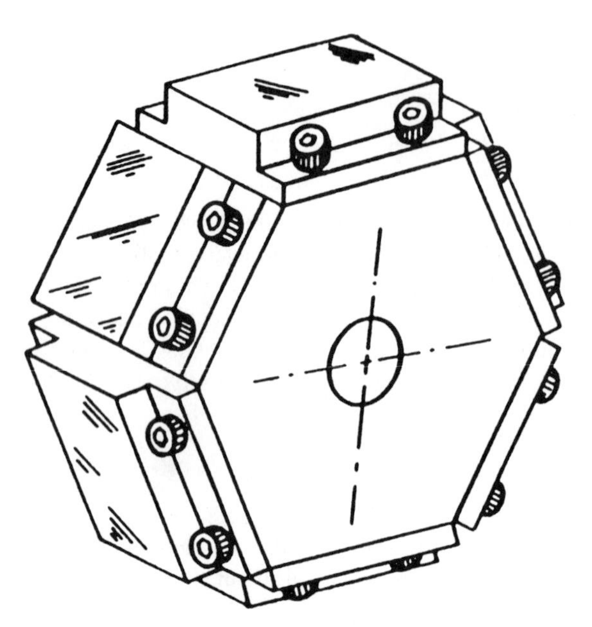

Figure 6.20 Polygon manufactured by assembly construction.

from the raw material. The most common substrate material used to make diamond-machined polygons is aluminum, although satisfactory results are achieved on brass, copper, and electroless-plated nickel. Bare aluminum mirror surfaces are fragile, and it is recommended that diamond-machined surfaces be overcoated with a harder dielectric layer (SiO) to enhance durability.

To the naked eye, diamond-machined surfaces appear to be perfect mirrors. With optical aids, however, it will be seen that the surfaces resemble a high-frequency grating. As a result, the specular reflectance of the surface degrades at short wavelengths (i.e., ultraviolet). Over the longer wavelengths of the visible spectrum and in the infrared, the grating effects are negligible to immeasurably small. Stacking of polygons can be done to increase yield and reduce cost in the diamond-machining process.

Replication

Replication is a process of impressing the pattern of a "master surface" onto the surface of a hardenable fluid. In recent times the fluid is epoxy resin. In the case of fabricating polygons by replication, the polygon substrate is machined to a dimension slightly smaller than the finished article and the space between the substrate and the master is filled up with resin. The master contains a microthin coating of release agent to prevent the resin from adhering to it (see Figure 6.22).

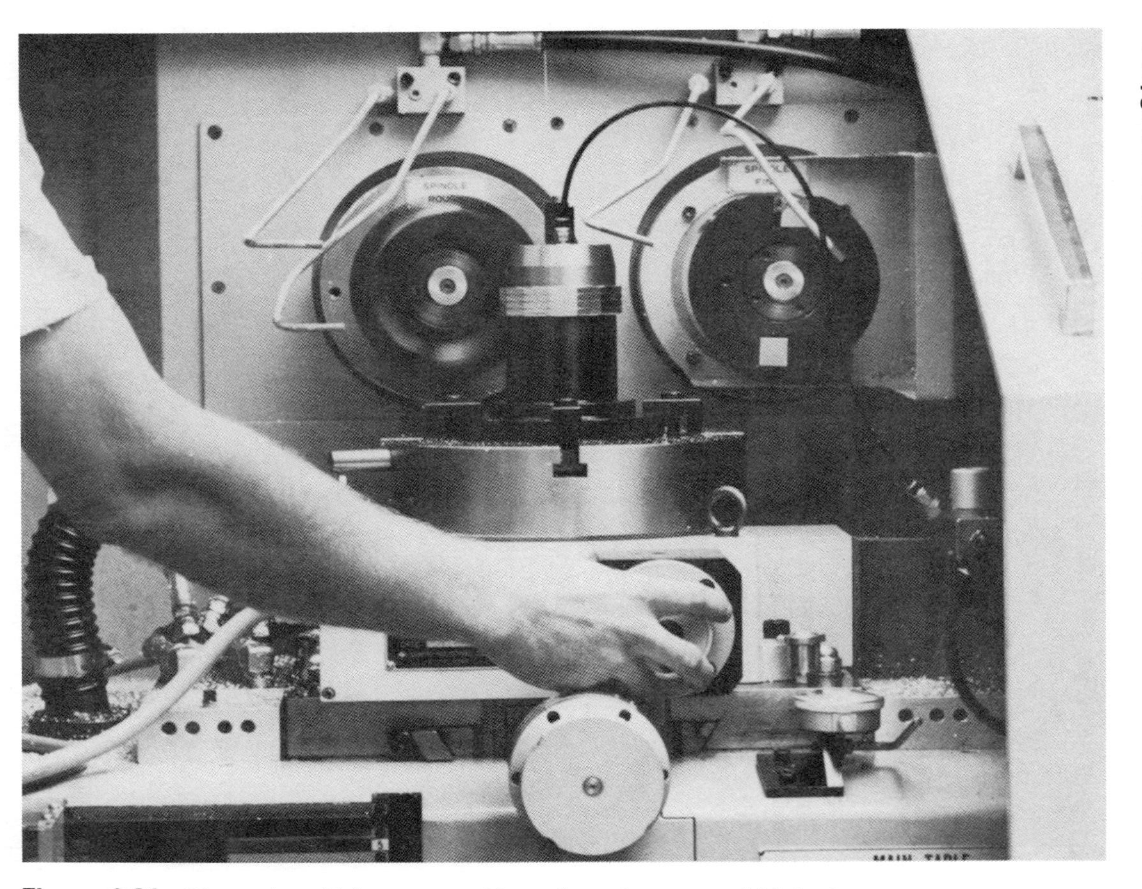

Figure 6.21 Diamond-machining system with a polygon in process of fabrication.

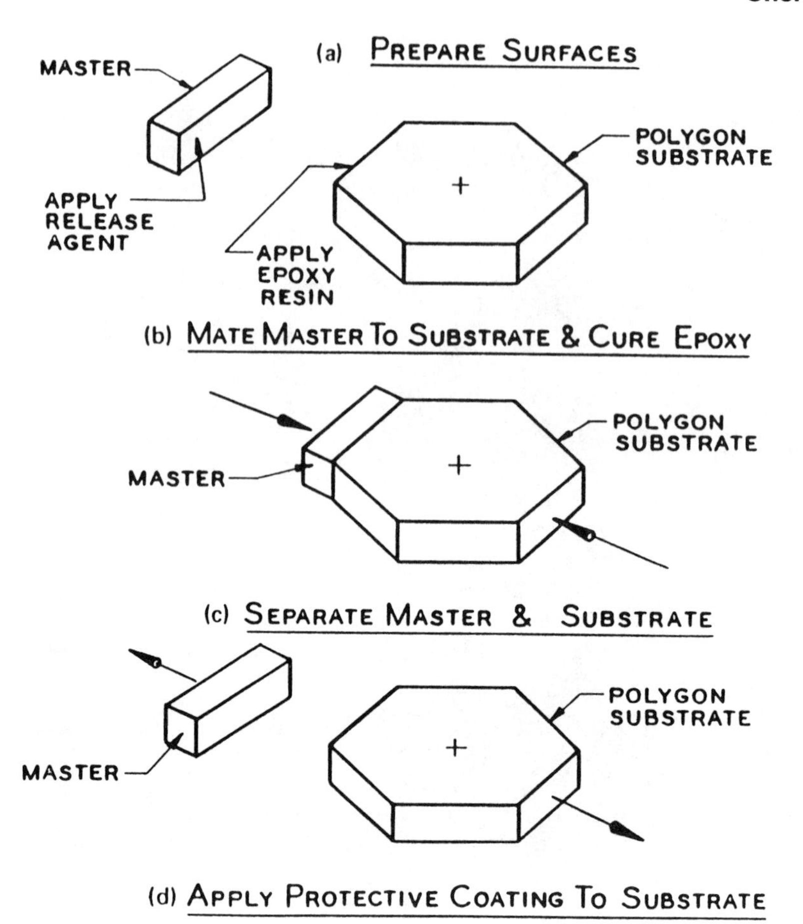

Figure 6.22 Schematic diagram of the fundamentals of replication.

The replication process is particularly useful for fabricating pyramidal polygons and types of irregular polygons, wherein the geometry permits the practical fabrication of masters. Due to the shrink characteristics of the best of resins, the replication is usually limited to producing surfaces of $\lambda/4$ or less flatness over relatively small apertures.

Conventional Polishing

Conventional polishing in this context is defined as pitch lapping, in much the same manner as glass lenses and prisms are polished. Pitch lapping can be used to produce high-quality surfaces on a number of polygon surfaces, including stainless steel, beryllium, chrome plating, electroless nickel plating, glass, and other glasslike materials (quartz, Pyrex, etc.).

Pitch-resin-faced polishing tools are stroked repetitively over the polygon facet in specially tooled polishing machines that utilize finely divided alumina and water as the polishing compound (see Figure 6.23). A large surround plate or "blocking tool" is placed around the facets being polished. This surround is held precisely in the plane of the facets to support the polishing tool and prevent rounding of facet edges, known as "roll-off."

Electroless nickel-plated aluminum is the most common polygon configuration. This combination provides the low cost and ease of machining of aluminum as the structural material with the superior polishing properties and durability of nickel.

6.4.2 Materials

Material selection for polygonal mirrors involves considerations of performance and cost. A rule of thumb for polygonal or pyramidal mirrors with solid cross sections (except for a central mounting hole) is that if the maximum operating peripheral velocity will not exceed 122 m/sec (400 ft/sec), the polygon facets can be expected to stay flat to $\lambda/10$ per inch. If the requirement exceeds that peripheral velocity, analysis will be necessary to determine whether a material with a superior stiffness/weight characteristic is necessary, such as beryllium. This rule of thumb has been derived as the result of a variety of tests conducted over a number of years on various polygon designs, as well as a few finite element analyses which correlated rather well with testing.

In situations where minor (or major) distortions in facet flatness can be tolerated and peripheral velocities exceeding 122 m/sec (400 ft/sec), are contemplated, safety must be considered. The speed at which the centrifugal stress will reach the yield strength (causing permanent distortion and dangerously close to the breaking speed) is found using the formula below. For an added margin of safety it is suggested that the circumscribed circle diameter of the polygon be used in the equation (tip diameter):

$$N = 57.3 \sqrt{\frac{32.16S}{(R^2 + Rr + r^2)W}}$$

where

N = limiting speed of the polygon (rpm)
S = yield strength of material (lb/in^2)
R = radius of polygon circumscribed circle
r = radius of the polygon bore
W = weight of the polygon material (lb/in^3)

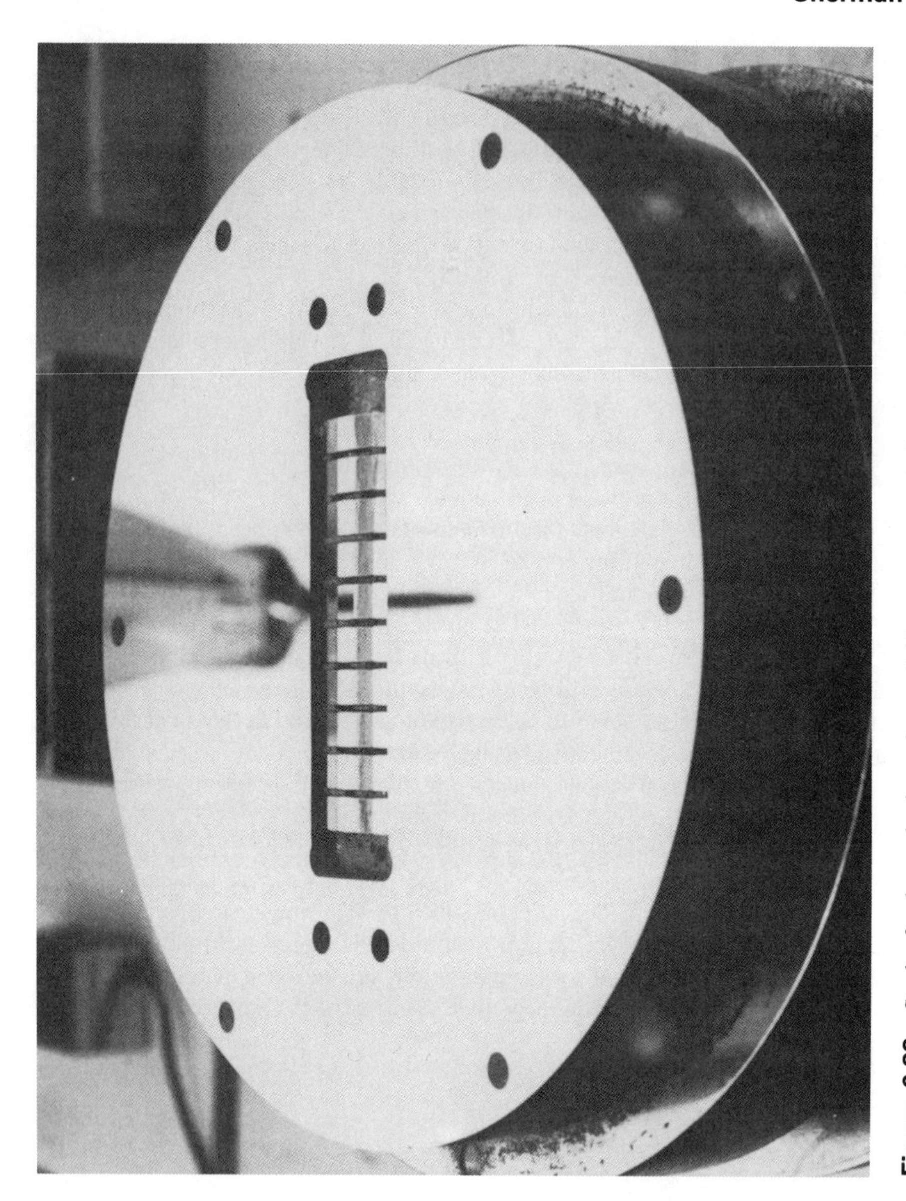

Figure 6.23 Stack of polygonal mirrors in a polishing support, ready for conventional pitch lap polishing.

6.4.3 Angular Tolerances and Measurement

The angular locations of facets on a polygonal mirror are usually specified in relation to two datum surfaces on the polygons. The most common datum surfaces are (1) one side of the polygon at right angles to its rotational axis, and (2) one or more facets (see Figure 6.24).

Facet-to-datum angles are measured using a measuring autocollimator and a flat surface plate. The polygon is placed datum surface down on the surface plate and rotated to align a facet in the rotational direction in the reticle of the autocollimator. The error is measured and the process repeated for the remaining facets. This technique is employed for angular tolerances of 10 arcsec or less. For tolerances greater than 10 arcsec, a ''spin test'' fixture is used, wherein the polygon is mounted on a precision spin platform and the facets are used to scan a focused laser beam by a horizontal knife edge with a photodetector behind it. The fixture is calibrated to read tracking error as a function of detector voltage variations produced by varying scan locations.

The polar location of facets is determined by fabrication accuracy and is usually measured one time after fabrication. Measurement of polar location is done using a precision rotary indexing head (angle reference) and a measuring autocollimator. The autocollimator is aligned to one facet to produce a ''zero'' reading. Then

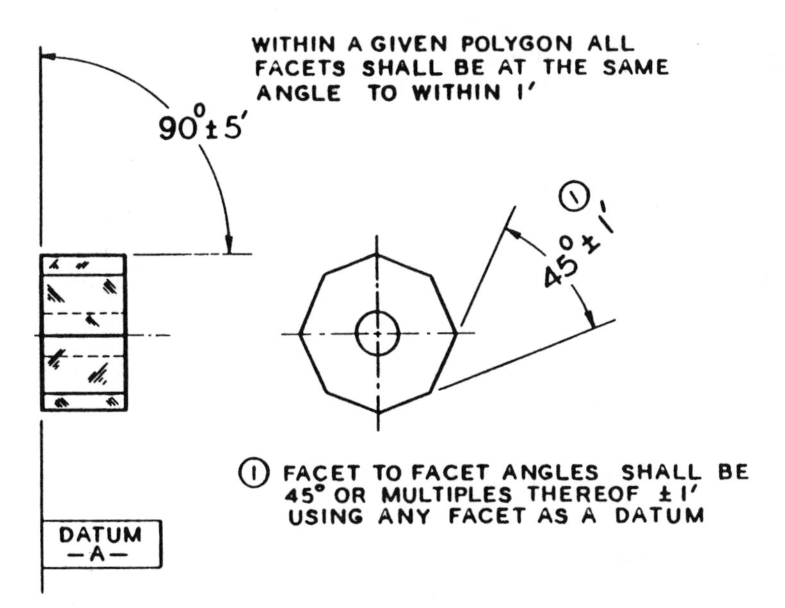

Figure 6.24 Typical method of specifying facet to datum angles and polar location angles on a polygonal mirror.

the indexing head is rotated to the desired location of the next facet, and the deviation of that facet is measured with the autocollimator (see Figure 6.25). This method provides true position information.

6.4.4 Facet Flatness: Static and Dynamic

Facet flatness is important to scanner system resolution. How flat a scan mirror facet must be to produce a given resolving power in a scanner system depends on wavelength, f number, and the range of incidence angles over which the facet will be used. Several factors influence the flatness of polygon facets:

Initial fabrication tolerances
Distortion due to mounting stresses
Distortion due to centrifugal stress when rotating at high speed
Distortions due to stress relief over long periods of time

Interferometry is used to measure static facet flatness. Flatness is specified in wavelengths (or fraction thereof) of light. A typical flatness specification will read: flatness–$\lambda/4$ at 632.8 nm. This specification requires that all points on a facet surface lie in a plane to within 632.8/4 nm.

Departure from flatness may have a variety of forms, depending on how the surface has been fabricated. For example, conventionally polished mirror surfaces tend to depart from flat in a regular spherical form, either concave or convex, but more usually convex. Diamond-machined surfaces usually depart from flat in a regular cylindrical form, either concave or convex. Most polygons used in reprographic applications are specified in the range $\lambda/4$–$\lambda/10$ in the wavelength of interest.

To preserve the facet flatness achieved during initial polygon fabrication, it is necessary to fasten the polygon to its drive motor with care, particularly if $\lambda/10$ flatness is required. The polygon/motor interface must be achieved without producing sufficient stresses to cause facet distortion.

A typical mounting scheme is shown in Figure 6.26. In this case the datum surface of the polygon and the locating annulus of the mounting hub are lapped to optical-quality surfaces so that when the two are firmly held together distortions do not occur. Equally important to accurate mating of datum and hub surfaces are cleanliness at assembly and appropriate torquing of the fastening screws. If the fastening screw location is close to the facet, it is desirable to counterbore the screw hole to achieve a minimum land thickness. This will reduce the amount of radiation of stress to the facet.

Polygons operated at particularly high rotational speeds may be subjected to sufficient centrifugal stress to degrade facet flatness. A polygon will stay flat while rotating at high speed if

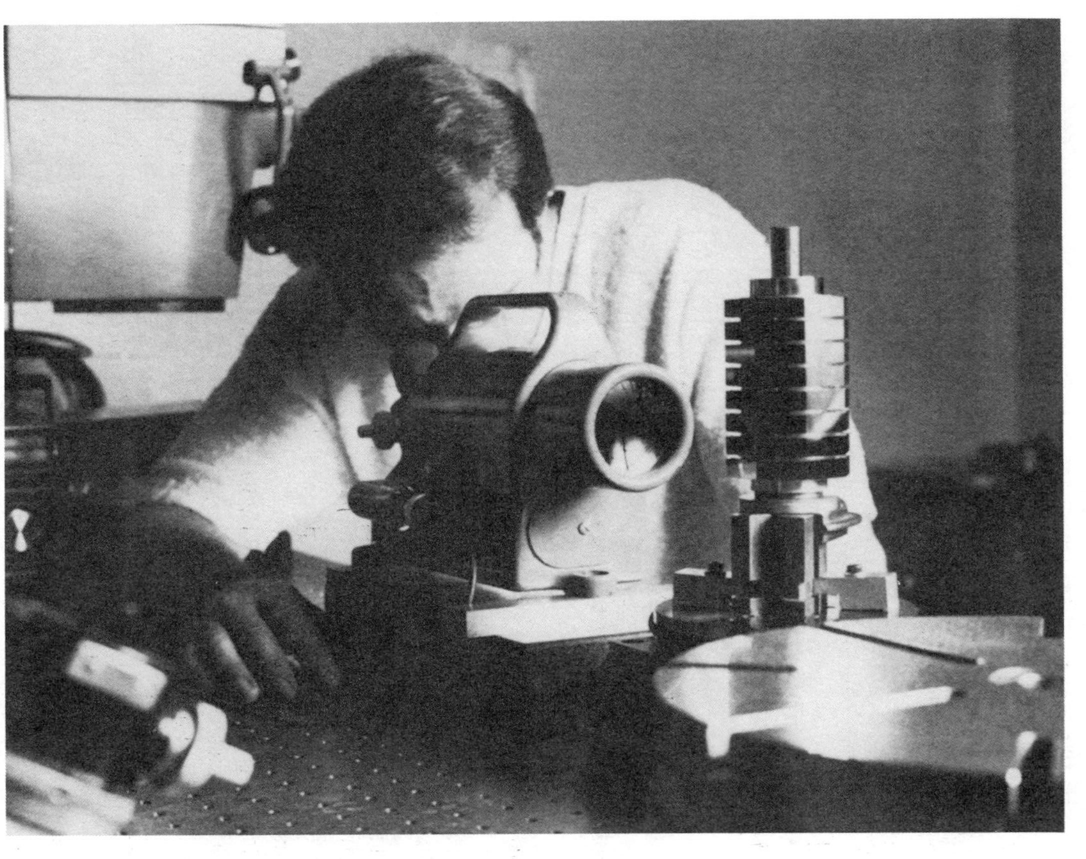

Figure 6.25 Measurement of polar location of polygon facets using reference indexing head and measuring autocollimator.

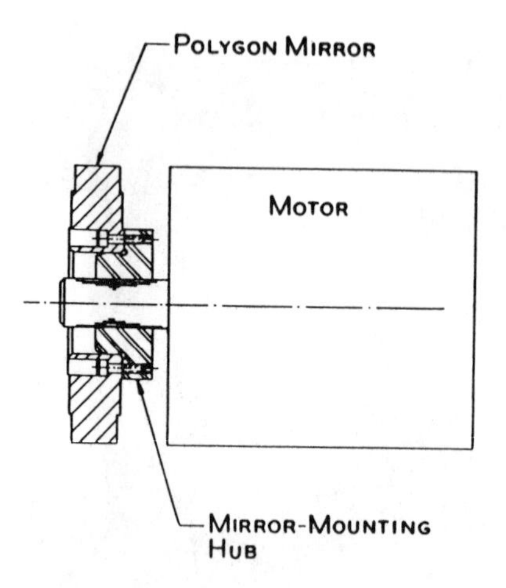

Figure 6.26 Typical mounting scheme for fastening a polygon to a motor shaft.

It has 12 or more facets.
It is made of aluminum or steel (tempered).
It has a solid radial cross section at least as thick as the facet is wide.
It has a central hole $\frac{3}{8}$ the diameter of the polygon-inscribed circle or smaller.

It may then safely operate up to a peripheral velocity (tip velocity) of 122 m/sec (400 f/sec) with facet distortions of less than $\lambda/10$ at 632 nm. This rule of thumb has iterated from a number of experimental results of testing polygons over the years.

Polygons that fall out of the boundaries of the rule of thumb should be dealt with by dynamic testing. Finite-element analysis may be used to determine close approximation of limiting speeds, but if the analysis shows a borderline case, a test may be in order.

There are a number of methods for measuring facet flatness of a high-speed rotating polygon, and they all involve rather sophisticated optical systems to attain accurate data. One of the simpler and easier-to-use methods is shown in Figure 6.27.

6.4.5 Surface Quality

In the context of first surface mirrors, a perfect surface will reflect all incident light rays at an angle precisely identical to the incidence angle. Any feature of the surface that causes other behavior, such as transmission through, absorption by, or scattering by the surface is considered a defect in surface quality. *Specular* is the term used to describe the perfect area of a mirror.

For some time in the past a system of quantifying mirror surface quality (and

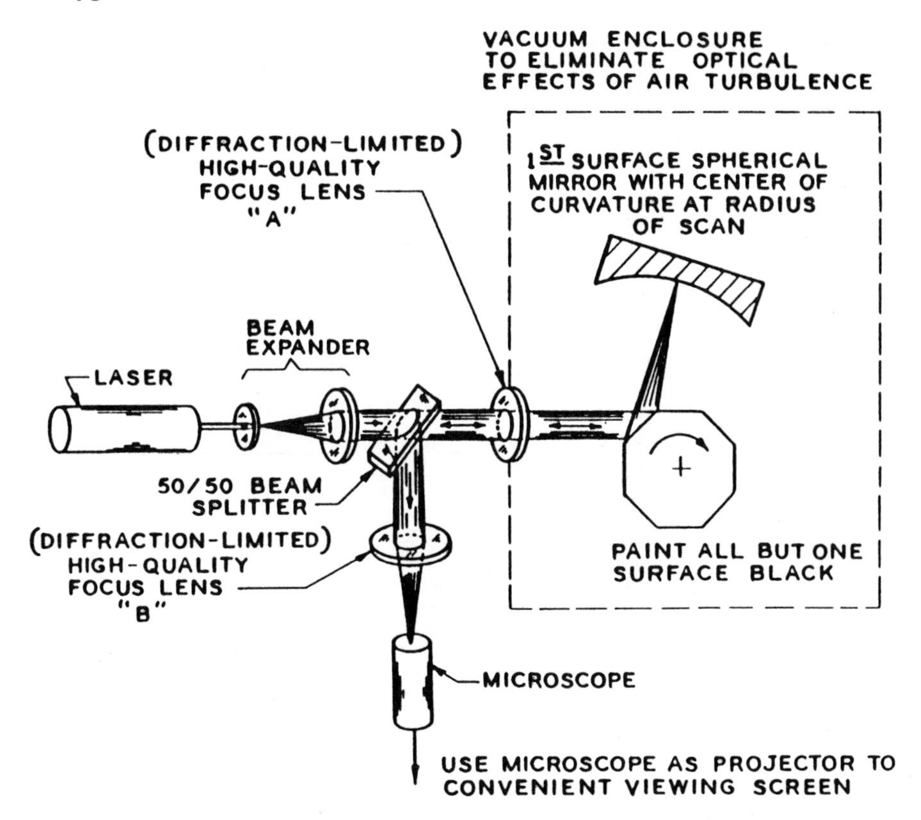

Figure 6.27 Dynamic facet flatness measuring optical setup. The scanning facet both scans and relays the scanned beam back through the scan lens. Wavefront distortion is measured as a growth or change of shape of the diffraction image formed by lens B. Static and dynamic cases are compared. Due to the high "strobe" frequency, the viewed image appears as steady state.

the quality of refractive elements such as lenses and prisms) known as a "scratch-and-dig" specification has sufficed. With the advent of machined optical surfaces, new methods have evolved. The U.S. military developed a scratch-and-dig specification which became documented as MIL 13508 (current revision letter is C) and which has become commonly used in the optics industry. This method of quality determination involves examination of a surface microscopically and identifying by a two-digit, hyphen, two-digit set of numbers the number of permissible scratches and digs (pits or holes) in a given unit of area. A typical high-quality conventionally polished polygonal mirror will rate 40–20 scratch and dig.

Table 6.4 Relationship Between RMS Roughness and Total Integrated Scatter of a Diamond-Machined Surface

Surface roughness (rms)		Wavelength	Total integrated scatter (%)	
microinches	angstroms	microns	$\theta = 0°$	$\theta = 45°$
1.0	250	0.6328	25.3	12.6
		10.6	9.0	4.5
0.8	200	0.6328	16.2	8.1
		10.6	5.8	2.9
0.6	150	0.6328	9.1	4.5
		10.6	3.2	1.6
0.4	100	0.6328	4.0	2.0
		10.6	1.4	0.72
0.2	50	0.6328	1.0	0.50
		10.6	0.36	0.18
0.1	25	0.6328	0.25	0.13
		10.6	0.09	0.04

TIS $\cong [4\pi\sigma(\cos\theta)/\lambda]^2$; σ = rms surface roughness; λ = wavelength of interest; θ = angle of incidence

Machined optical surfaces, on the other hand, are made up of a precise regular pattern of scratches (machine toolmarks) which are sufficiently high in frequency and low in peak-to-valley roughness as to behave aggregately in the manner of an acceptable "plane mirror" at certain wavelengths. The scratch-and-dig system is inappropriate here. A more useful method is that which specifies root-mean-square (rms) surface finish, since there is an approximate correlation between this value and the value of measured nonspecular reflectance from a machined surface (see Table 6.4).

6.4.6 Reflectance and Uniformity

The reflectance of a polygon facet and the uniformity of reflectance across that facet and other facets on the same wheel will be determined to the greatest extent by two factors. One is the specularity or quality of surface and the second is the enhancement coating used to improve the reflectance. Incident light on a mirror surface is either specularly reflected, scattered, absorbed, or transmitted by the surface. Surface reflectance is normally measured by monitoring the specularly reflected component. The scattered component is generally dealt with by specifying surface quality, and it is assumed that the remaining losses are by absorption. In the case where the polygon may be made out of glass, there may be a small component of transmitted light if the enhancement coating is not opaque.

The common measuring tools to determine reflectance are spectrophotometers and laser reflectometers. Most spectrophotometers with a reflectance measuring attachment are able to take only a given sample size, generally circular and approximately a 12–25 mm (½–1 in) diameter, to make a measurement. This fact usually precludes measuring the polygon itself but rather, measuring a witness sample which is supposed to be identical to the polygon facet itself. Using a spectrophotometer can be a reliable method of ascertaining the performance of a polygon facet if one has determined that the witness sample is an exact representation of the facet. If this exactness cannot be relied on, it is necessary to use another instrument for making the measurement. In a case where a laser wavelength is to be used in the scanning process with the polygon, the laser itself can be reflected from the surface and the reflected value compared to the incident value (see Figure 6.28).

Specific reflectance (i.e., the highest value attainable considering costs) is important from the standpoint of overall system efficiency. The polygon is one of several components in most systems that are optimized to make the most efficient use of the available light. This maximum reflectance value is normally determined by application of an enhancement coating on top of the micromachined or polished surface. Reflectance uniformity, on the other hand, can affect the uniformity of images read and/or recorded by the polygon, so it is necessary to achieve some value of constancy of reflection over the surface of any given facet and from one facet to any other facet on the polygon wheel. In the event that the tolerance on uniformity should be a fraction of 1%, it may be necessary to measure the uniformity of reflectance of a polygon before the enhancement coating is applied. In practice, the reflectance uniformity of rather large polygon facets (e.g., several mm or a few in square) is more difficult to achieve to high tolerances than if the facet is small (e.g., 12 mm or ½ in square). This phenomenon has to do with the length of time of processing of the facet and the degree of difficulty of achieving a number of characteristics simultaneously, such as reflectance uniformity, flatness, facet-to-datum angular tolerance, and facet polar location.

When measuring reflectance uniformity of polygon facets, it is good practice to use an optical aperture which reproduces that expected in the system. This will eliminate unwanted contributions by unused areas of the facet. In other words, instead of illuminating the entire facet with incident radiation, the incident radiation should be limited to the effective clear aperture of the facet. As a case in point, polygons used in correction optical systems that produce a line image on the facet have an effective clear aperture that is a very small fraction of the total facet area; this is really the area of interest. In this case a measurement of

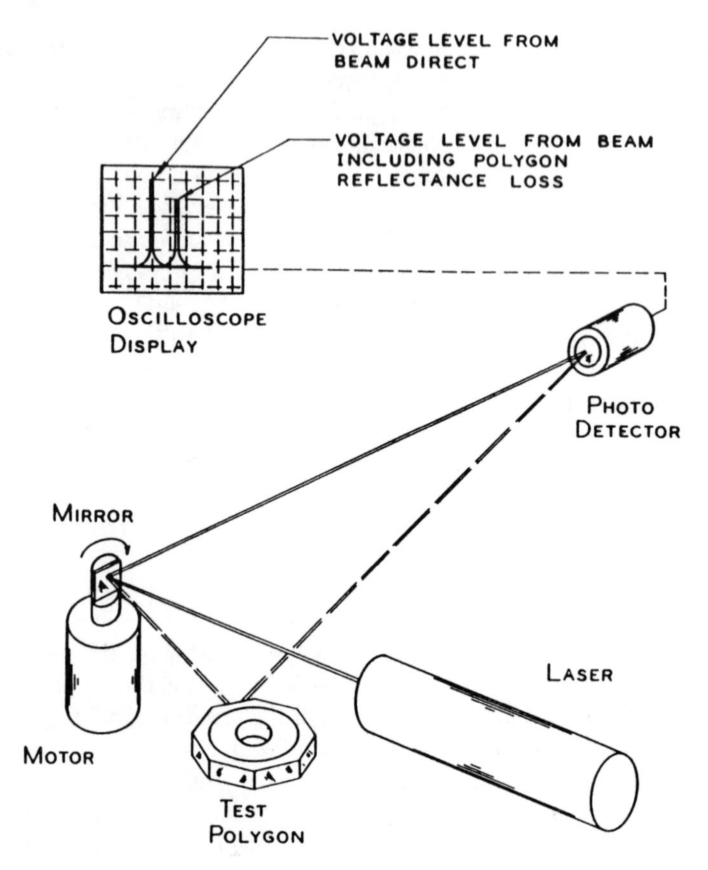

Figure 6.28 Reflectance test optical system.

the total facet may produce unsatisfactory results, whereas measurement of the useful clear aperture may produce satisfactory results.

The measurement of specific reflectance versus the measurement of comparative reflectance (i.e., uniformity) involves two different problems of measurement. In the case of measuring specific reflectance it is necessary to compare the measured reflectance against 100%. It has been the author's experience that sending the same sample to a number of laboratories for a measurement of the specific reflectance in a certain wavelength brings results that vary a few percent one from the other. If reflectance is a critically key element in the successful use of a polygon in a system, it is suggested that supplier and customer routinely and carefully calibrate their reflectance measuring equipment and frequently correlate supplier's and customer's measurements. In a production situation where a good deal of product is being manufactured, sold, and used, this is a good general rule to follow. Reflectance and uniformity are one matter of concern in the process of manufacture and acceptance. They are another matter over the long haul after the polygon has been installed in the system. In use the polygon may be subjected to contaminants from a variety of sources. Typical offenders are cigarette smoke, toner, bearing-oil vapors, paper dust, and fingerprints. Only careful handling and careful attention to the details of the environment in which the polygon operates will maintain high levels of reflectance and uniformity for reasonable periods of time.

It is important to mention that uniformity of reflectance of polygon facets and of mirrors in general is more difficult to achieve at shorter wavelengths than at long wavelengths. At wavelengths in the region 9–14 μm, for example, it is relatively easy to achieve extremely high levels of uniformity. This is due to the fact that the quality of surface necessary to appear specular at rather long wavelengths is easy to achieve. At wavelengths in the range 0.2–0.4 μm, it is a different matter altogether, and very small surface defects disturb the apparent specularity of the surface.

6.4.7 Enhancement Coatings

The matter of optical enhancement coatings is a subject unto its own, and we will deal with it here only with regard to the practical application of a few coatings to polygonal mirrors. The characteristics required of coatings to enhance the reflectance of polygonal mirrors are

High reflectance
Good uniformity
Good adhesion
Good life

There are two major functions of enhancement coatings on polygons: to improve the reflectance of the polished substrate and/or to improve its durability. In the case of diamond-machined polygons, the substrate is usually aluminum (in itself a good reflector over most of the visible and infrared spectrum). A $\lambda/2$-thick coating of silicon monoxide provides a protective "window" on the facet. In the case of replicated surface (epoxy), the enhancement coating adds both reflectance improvement and durability. Nickel-plated aluminum polygons are coated primarily to enhance reflectance, since nickel is a relatively durable surface in its own right.

In most cases the substrate material in a polygon is metal. In a number of cases the base material will be plated to accommodate polishing and to enhance durability. The ground rules on applying vacuum-deposited enhancement coatings to multifaceted metal and/or plated metal polygons are somewhat different from those that may be applied to coating glass or Pyrex or quartz optical elements.

Vacuum-deposited coatings for polygons are usually comprised of two or more layers of material, each a fraction of a wavelength of light thick. They are applied by evaporating or sublimating the materials in a vacuum chamber. The materials disperse in the chamber, adhering to everything that is clean and is in the line of sight of the source.

As a practical matter, coatings must be applied while the polygon rotates in the path of the vapor in order to expose all of it to the line of sight of the source and to ensure uniformity of application. Several polygons may be coated simultaneously on a single rotating shaft.

The first layer deposited is a metal, such as aluminum, silver, or gold. The remaining layer, or layers, is a dielectric material, such as silicon monoxide, magnesium fluoride, or thorium fluoride. The metallic layer is selected to provide the best reflectance at the wavelength of interest. The dielectric layers are added to protect (and in some cases, enhance) the underlying metal.

The index of refraction of some dielectric materials applied can be controlled usefully by controlling (elevating) the temperature of the polygon. A note of caution is in order that the polygon temperature not be elevated to an otherwise destructive temperature during coating. For example, nickel-plated aluminum polygons will distort (go unflat) and visibly "orange peel" if elevated beyond

Table 6.5 Contaminants to Be Alert for on Polygons to Be Coated, Together with Suggested Cleaning Method

Contamination source	Cleaning method
Water vapor on and in the surface (small water droplets resist evaporation for long periods of time at 1×10^{-6} torr vacuum)	Glow discharge vacuum chamber with polygon
Oil and debris from machining in tapped holes	Retap—ultrasonically clean in Freon TF solvent
Mineral residue from dried solvents or wafer	Clean with cotton[a] and distilled water or solution of 5 drops of liquid detergent in .0047 m^3 (1 pint) of distilled water; then wash with distilled water; rapidly remove residual water with alcohol on a cotton swab
Fingerprints (*note*: remove immediately, as etching may occur from some skin oils)	Cotton and acetone or denatured alcohol
Stains (*note*: examine microscopically to determine presence of etching, which cannot be removed except by resurfacing)	Cotton and acetone, kerosene, methyl ethyl ketone, or as above for mineral residues

[a] A good-quality, multiple-layer, soft toilet tissue is an acceptable alternative.

200°C after polishing. Some aluminum (unplated) alloys will detemper and distort at 310°C.

Silver exhibits very desirable reflectance characteristics over a broad spectrum and is frequently considered as a polygon coating. In practice, however, it is frequently a disappointing choice over the long haul. The slightest pinhole (or minute scratch from cleaning) will expose the silver to reactive contaminants from the atmosphere, which over time (several days or weeks) will diffuse into the silver, producing an ever-increasing size of blemish. Aluminum, which initially exhibits somewhat lower values of reflectance than silver, is far superior in terms of durability.

Crucial to all of the desired characteristics of the coating of a polygon is its cleanliness prior to and during coating. Irrespective of fabrication method, polygons will be handled prior to coating during inspection processes, transport, and installation into the coating chamber. Few coatings can be successfully removed if they are improperly applied or applied on a dirty surface.

The polygon must be cleaned thoroughly to remove foreign material that will degrade surface quality, prevent good coating adhesion, or outgas in the coating system. Typical offenders in achieving this level of cleanliness, and suggested methods of cleaning, are shown in Table 6.5.

6.4.8 Polygon Durability

In use, a polygon is a rotating part, usually running at several thousand rpm. As such it will have an exceptional opportunity to come into contact with the atmosphere surrounding it. This atmosphere generally contains varying degrees of such bothersome contaminants as cigarette smoke, paper toner, oil mist from the motor bearings, vapors, paper dust, ordinary dust, and others. Good design practice would dictate that the polygon and other delicate optical components in the scanning system be housed in an enclosure to minimize this exposure. Measurements of reflectance of polygon surfaces operated for periods as short as 50–60 h at several thousand rpm (unenclosed) in an ordinary office environment (including a modest number of cigarette smokers) show degradation as great as 20–30%.

With proper system design, polygons may be expected to provide 15,000–20,000 h of operating life over periods of five years or longer even at speeds of 25,000–30,000 rpm. Periodic maintenance of the system will include cleaning of the polygon facets. Maintenance and service personnel must be provided with training and procedures for cleaning the polygon without degrading its quality.

It is good practice to design the scanning system so that the polygon motor assembly may readily be removed (and replaced without necessity for optical alignment) for examination and cleaning of the facets. In systems that are less than desirably clean, the polygon will be the greatest contributor to light loss because of the high-velocity airstream at its perimeter.

The foregoing factors should be considered in selecting the type of polygon to use at the outset. There are trade-offs between cost and durability. In this case life-cycle cost should be traded off against initial cost.

6.5 METHODS OF ROTATING POLYGONS

6.5.1 Pneumatic Drives for High Speed

Much of today's scan mirror technology has evolved from the development of ultrahigh-speed polygon/turbine motors for the high-speed photography industry. Compressed air turbines continue to offer an attractive method of rotating a polygonal mirror at speeds beyond the capability of electric motors. The advantages of turbine drives are

Substantial horsepower can be delivered to the scan mirror to produce rapid acceleration and very high speed (up to 1,000,000 rpm).
They are compact in size and low in weight in proportion to delivered power.
Some versions are equipped with shaft seals so that the scan mirror can be used in a modest vacuum.

The disadvantages of turbine drives are

They require a compressed air source.
They are asynchronous devices.
They are costly.
They have a relatively short total running life.

Pneumatic drives are only recommended for short duty cycles and where ultrahigh speed is essential.

6.5.2 Motor Drives for Intermediate and Low Speeds

Four types of electric motors are commonly utilized for rotating polygonal mirrors:

1. Hysteresis synchronous
2. Induction synchronous
3. Induction slipped
4. Brushless DC

Most applications are in the fractional-horsepower category, so the motors fall into the class of "instrument" as opposed to "utility" motors. Hysteresis and induction synchronous motors synchronize with the AC-induced magnetic

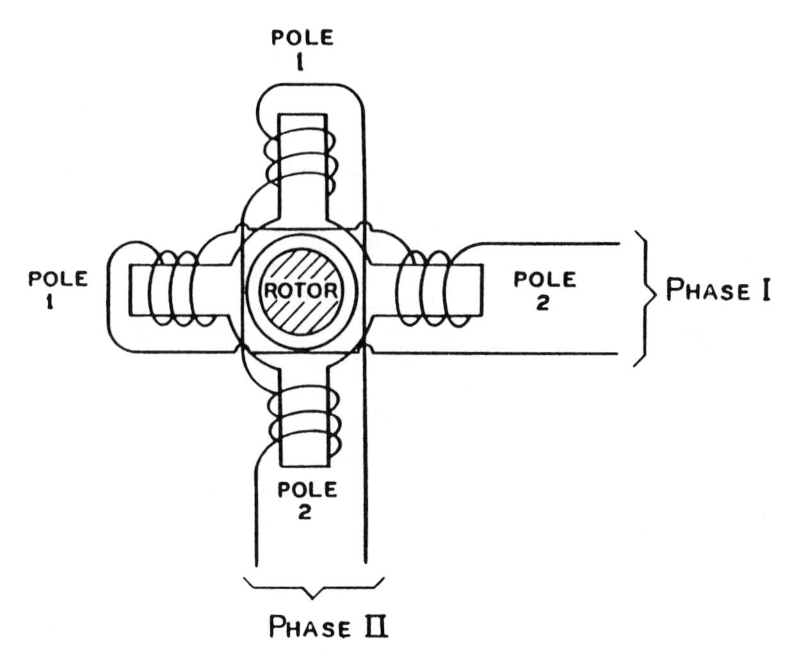

Figure 6.29 Schematic diagram of AC motor stator.

flux that drives them. Brushless direct-current (DC) motors run at that speed at which the drive torque (DC dependent) matches the resistance torque of the load they are driving. In the case of rotating polygons, the dominant resistance is air drag.

In all four types of motor the stator is essentially the same, consisting of a multipole laminated iron core wound with a number of coils of insulated copper wire (see Figure 6.29). The number of poles (terminated coils) ÷ 2 = the number of times the rotor will go around per cycle of current into the stator.* Therefore, four-pole motors (per phase) run at half the drive current frequency, six poles at one-third, and so on. Using more poles than eight poles is usually impractical. If you want to run the scan motor from the mains, you are limited to the frequency of the mains or fractions thereof of ½, ⅓, or ¼. If you wish to run at other speeds, an inverter must be used to create the desired frequency. An exception to this case is the brushless DC motor, wherein the excitor for commutation (current inversion) is on the rotor so that the stator frequency is always synchronous with (and leading in phase) the rotor frequency.

* This is also true of brushless DC motors when considering that an electronically commutated inverter is built into these motors to provide AC to the winding.

The rotors of the four motors are substantially different in form and function. The rotor of a hysteresis motor is usually fabricated from a single piece of hardened steel selected out of a group (predominantly alloyed with cobalt) that exhibit substantial hysteresis loss. This resistance to the movement of magnetic flux in the material imparts torque to a rotor out of sync with the drive current. This torque is responsible for the motor's ability to start rotation. When the rotor approaches the speed of the stator flux, it becomes permanently magnetized and ''locks in'' to synchronism with the drive. If the motor is turned off and restarted, the stator flux demagnetizes the rotor and hysteresis takes over again. The synchronous mode of function is more efficient than the hysteresis or start-up mode, and in many systems a sync detector is used to reduce drive current after the motor is locked in to save energy and minimize motor heating.

Induction slip motors contain conductor bars in the rotor. Magnetic flux from the stator induces current flow (and therefore magnetic flux) in the conductor bars. The stator flux attracts the rotor. The amount of torque produced is related to the amount of slip or velocity difference between stator field and rotor. Velocity servo systems are used to control the speed of these motors when high accuracy is required.

Induction synchronous motor rotors are more complex in their construction. They contain conductor loops or coils in which current flow is induced by the moving stator flux. This current flow, in turn, produces magnetic flux in the rotor. Torque is developed by magnetic attraction between stator and rotor. Since this torque diminishes as synchronism is approached (due to reduced flux motion in the rotor coil), the rotor must be equipped with permanent-magnet poles to achieve synchronism. Such a motor is a combination of induction slip and permanent-magnet sync.

The rotor of a brushless DC motor is constructed of a shaft, usually made of steel, with a permanent-magnet ring pressed on to it. Starting torque and running torque are imparted to the rotor by magnetic attraction.

As an aid to the system designer, the major significance of the design differences among these four motors is given in Table 6.6. References 1–3 deal with these design parameters in further detail.

Selection of a motor for a polygonal scanner application requires that a detailed performance requirement be generated and a trade-off study be done to weigh the advantages and disadvantages of available designs. The major issues of that trade-off will probably be speed stability and the mechanical rotational accuracy of the output shaft (hence the accuracy with which the polygon will be rotated relative to its datum surface). For systems employing correction optics to correct for polygon tracking errors (see Section 6.3.4), the trade-off may well change to one of motor/controller complexity and cost. For systems requiring accurate tracking (10 arcsec or better), the trade-off will be one of rotor integrity. Complex rotors, those made up of material of various thermal coefficients of expansion,

Table 6.6 Design Differences Between Scan Motor Types

Scan motor type	Prime power	Speed control[a]	Velocity stability	Rotor configuration[b]	Speed capability[c] (rpm)	MTBM[d]	Comparative cost
Hysteresis synchronous	AC: one-, two-, or three-phase from mains or separate inverter	Synchronous	Opened loop 0.5–0.01% with feedback externally supplied 0.01–0.002%	One-piece hardened steel	900–120,000	12,000 h at 24,000	1
Induction synchronous	AC: one-, two-, or three-phase from mains or separate inverter	Synchronous	Opened loop 0.5–0.02% with feedback externally supplied 0.01–0.005%	Complex: several materials	900–24,000	12,000 h at 16,000	1.3
Induction slip	AC: one-, two-, or three-phase from mains or inverter	External feedback loop	Opened loop: load dependent; closed loop: 0.5–0.005%	Laminated rotor with copper or aluminum bars	1000–36,000	Bearing-dependent	1.5
Brushless DC	DC	External feedback loop	Opened loop: load dependent; closed loop: 0.01–0.002%	Two-piece shaft/magnet	0–30,000	Various (dependent on electronic component reliability of internal inverter)	1.5

[a] See the text with regard to phase stability.
[b] Depends on tensile strength of rotor magnet.
[c] Independent of applied load.
[d] Mean time between maintenance; measured values.

are problematical in this accuracy range. For systems requiring a high degree of velocity stability of the polygon/motor assembly, it may be necessary to close a feedback loop around the motor/controller combination, utilizing a shaft encoder on the motor shaft.

Synchronous motors exhibit a characteristic called phase jitter (hunting). The rotors behave as though they were coupled to the drive waveform by a spring. Within synchronism the rotor springs forward and back in phase at a rate determined by the spring rate (flux density) and the torque/inertia ratio of the system. Typically, the frequency of this phase jitter is in the range 0.5 to 2 or 3 Hz, at an amplitude of a few degrees (1–6° peak to peak). Under perfect conditions this jitter damps to zero values of amplitude. However, perfection is seldom seen and continual recurrence of jitter may be expected, caused by electrical transients on the input, mechanical shock to the assembly, variable resistance torque of the motor bearings, and so on. For many systems the 0.5–0.01% velocity error contribution of phase jitter is acceptably small. For others it is not, and feedback should be employed to reach acceptable values. Figure 6.30 shows a typical feedback loop configuration for motor speed control.

As mentioned earlier, brushless DC motors, due to rotor-excited inversion of current to the stator (equivalent of brush commutation), have no synchronous relationship with the world outside their housing. Control of these motors is done by varying the input current (voltage). The speed, accuracy, and stability derived by simply applying a fixed DC input current are seldom satisfactory for scanning requirements. It is necessary to monitor the rotor speed and vary the input voltage to regulate speed. This feedback loop resembles the loop used to control AC motors in principle but is different in one major aspect. Input regulation is applied to DC rather than AC.

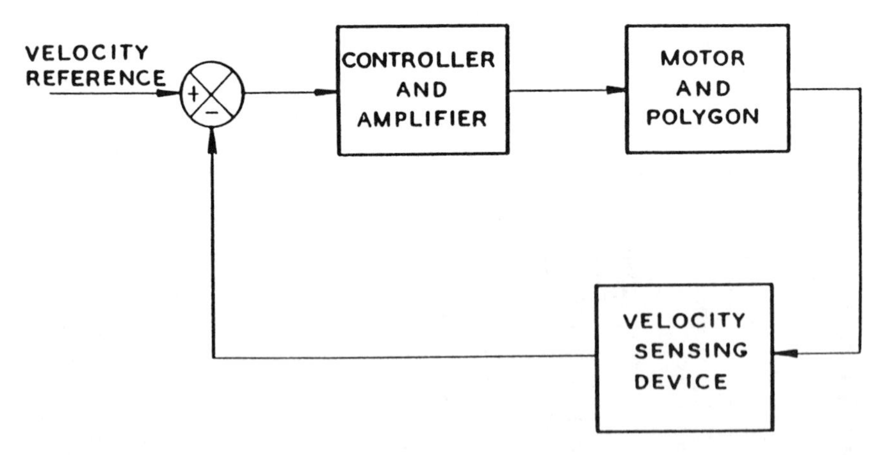

Figure 6.30 Block diagram of feedback loop for motor velocity control.

6.5.3 Bearing Systems for Accuracy and Longevity

The following is a brief discussion on scanner bearings based on the author's experience. For a more in-depth treatment on bearings, refer to Chapter 9.

Air Bearings

Many applications for rotating polygonal mirrors require a level of precision performance and lifetime that is difficult to achieve with ball bearing supported rotors. The difficulty arises primarily from two inherent features of ball bearings, which are

1. Ball bearings require lubricant, usually an oil or grease. This lubricant leaks and contaminates nearby optical elements at speeds beyond approximately 15,000 rpm. The evaporation rate of lubricant (particularly at elevated temperatures) directly affects bearing life. Additionally, lubricant affects the resistance torque of ball bearings, frequently and intermittently changing the torsional load on the rotor which perturbs rotor velocity stability.
2. Ball bearings are fragile components sensitive to shock and vibration and can be permanently damaged beyond repair by surprisingly low values of force.

As technology in printing and reading optical scanners has advanced to higher resolution and speed, requirements have developed for improved bearing support systems for the rotating scan element. One solution that has evolved is the hydrodynamic air bearing.

Gas lubricated bearings (air in this case) are not new technology. They have been around for many years and are commonly found in precision machine tool spindles, textile machinery, printed circuit drillers, etc. Most of these gas bearings are of the hydrostatic type, requiring well-filtered, uninterrupted, dry pressurized air from a compressor located nearby. Hydrostatic gas bearings are very useful for applications requiring the ultimate in stiffness and rotary precision. They are also expensive.

Hydrodynamic air bearings develop their own air pressure internally. The advantage is that no compressor is needed. The disadvantage is that the rotor and stator rub together during start-up and run-down. In the rotating mirror application where the load is light and constant, perfectly adequate stiffness and performance are achievable with hydrodynamic air bearings. Just as important, the air bearing is relatively immune to damage from shock and vibration since it exhibits large contact area at rest and large support area when operating.

How Air Bearings Work. When two surfaces are brought together with some force and then caused to slide against one another, wear occurs. Wear is the result of one surface removing material from the other. When wear continues long enough, something ''wears out''; in other words, enough material wears away to produce an undesirable effect.

Two things are done commonly to reduce wear when surfaces must bear load and motion. These are to use bearing materials which exhibit low friction and therefore lower rates of wearing, and also to use a lubricant to prevent actual contact of the bearing surfaces, which under perfect circumstances prevents wear altogether.

In its simplest form an air bearing is a plain journal bearing (a cylindrical shaft rotating in a cylindrical sleeve, or vice versa) made of materials that exhibit low friction and lubricated with air. In a regular journal, bearing oil or grease is used instead of air.

The primary difference between a journal bearing that is air-lubricated and one that is oil- or grease-lubricated is the size of the gap for lubricant. Oil and grease need more room than air does. The gap in an air bearing needs to be very small, about 2.5 Ωm to 3.5 Ωm (0.0001 to 0.00015 in). A little more than that is not good, and a little less than that is not good either.

Another difference between air and oil or grease as bearing lubricants is that the oil and grease ''wet'' the bearing surfaces and air does not. So in an air-lubricated bearing one needs a mechanism for distributing the air around to where it is needed. This mechanism can take the form of micro- or macrochannels in patterned or random arrangement to move air molecules to form an ''air wedge'' to keep rotor and stator material separated. This wedge or standing wave of air pressurized by the relative motion of bearing surfaces is what provides the lubrication or support for the load. In a well-designed air bearing this standing pressure wave develops at very low speed (a few hundred rpm or less) so that the amount of time that the bearing rubs is minimized, and, by the same token, the bearing surfaces are made of materials and finishes that are comfortable with this minimal rubbing at start-up or run-down.

Other key design parameters for a hydrodynamic air bearing for use in motor/polygon assemblies are

The control of thermal expansion coefficients of key components to provide satisfactory performance over a wide range of ambient temperatures

The use of materials and finishes that provide immunity to the effects of high-humidity environments

The control of critical tolerances and dimensions to ensure satisfactory performance over a range of ambient altitude and resultant air density variation

A design consistent with low manufacturing cost

Ball Bearings

Precision-instrument-class ball bearings (ABEC 5 and 7) are currently the most common bearings being utilized in production polygon/motor assemblies, where assembly tracking tolerances of 10 arcsec or greater are common and lifetimes of 10,000 h and more are specified. The successful use of ball bearings in a polygon/motor assembly (as in any precision instrument requiring optical-quality levels) demands meticulous attention to a number of details, including

Proper selection of bearing size, type, class, material, and supplier
Proper selection of lubricant and amount
Careful handling during and after installation
Proper installation and fit to shaft
Correct installation in housing
Correct selection and application of thrust preload
Correct attention to rotor dynamic balancing
Careful packaging and handling of completed assemblies

R3 and R4 ($\frac{3}{16}$ in ID × $\frac{1}{2}$ in OD and $\frac{1}{4}$ in ID × $\frac{5}{8}$ in OD, respectively) are the most commonly used sizes. This size range is an excellent trade-off of size versus ruggedness for high-quality bearings for the type of duty in a polygon/motor assembly. Full race bearings of 440C stainless steel in class ABEC 5 have a history of exceptional reliability.

It is noteworthy to mention that the only difference between ABEC 5 and ABEC 7 ball bearings is the amount of radial play between inner and outer races. In scan motor use the bearings are thrust-loaded to remove all radial play, and in practice ABEC 5 is preferred even though it has a greater radial play. The reason is its ability to "digest" larger particles of contamination without damage. In the occasional situation where preloading is impractical, ABEC 7 is suggested.

Two basic lubrication techniques are applicable to scan motor bearing use. One technique utilizes porous phenolic as the ball separator in the bearing. It is vacuum-impregnated with oil, becoming a reservoir of long-term supply. This technique is suggested for polygon/motor assemblies operating at 3600 rpm or less. At higher speeds the oil tends to leak out of the phenolic (due to centrifugal force) and then out of the bearing. The second lubrication technique is to use a grease such as Andock C.* This technique is much less susceptible to leakage out of the bearing and more satisfactory for extended high-speed use.

In very precise scanning applications the amount of radial play in the ball bearings (less thrust preload) may exceed the values desired and thrust load must be used. This is simply done by application of springs (wavy washers or Belleville springs) located axially between the bearing outer race and the assembly housing. This thrust load is parasitic and should be minimized. The mechanism that causes the rotating shaft to exercise radial play is dynamic imbalance. Therefore, the better balanced the rotor is, the less thrust preload must be applied to the bearings and the longer the bearing may be expected to last.

---

* Andock C is a trade name for a bearing grease manufactured by Humble Oil.

The polygon/motor application of bearings is a relatively gentle use. The operating environment is one of small load (primarily the gravity load of the rotor, i.e., a few ounces or pounds, or in high-speed cases the centrifugal ball load or imbalance load), usually clean, and hopefully low thermal condition. The major causes of failure, nevertheless, are (in order of importance)

1. Abuse (brinelling due to shock from mishandling)
2. Dirt (paper dust, toner, ordinary grit)
3. Heat

A polygon/motor assembly, like a lens, modulator, laser, or other precision optical/mechanical device, will seldom survive a drop to the floor without bearing (or other) damage. Shock isolation of the ball bearings using elastic O-rings between bearing and housing can minimize damage due to abuse, not eliminate it. Shock isolation in shipping containers will limit damage. A 1-in thickness of foam rubber surrounding a scanner assembly packed snugly with several others in a sturdy cardboard shipping box will prevent shock-induced damage in all but extreme cases of abuse.

Having survived shipment and installation into the scanning system, the polygon/motor assembly should provide several thousand hours of maintenance free service, except for an occasional cleaning of mirror facets. It is said that bearing life is lubricant-life-dependent, and that is largely the case at this point. For a bearing to last for 10,000 h or more of operation (plus many more hours at rest), the lubricant must stay in the bearing for that time. The two main reasons why lubricant disappears from the bearing are leakage and evaporation. In the two lubrication methods mentioned above, leakage normally occurs during the first few hours of operation, when excess lube is slung out. This will ordinarily occur during the test or "burn-in" period at the manufacturer's location. After that, evaporation is the enemy of life. Lubricant evaporation rate is a function of heat and ambient pressure. The vapor pressure of most bearing lubricants at 25°C is in the range 0.01–0.1 torr absolute. The vapor pressure rises exponentially with increase in temperature. As a result, it may be said that lubricant life (and therefore bearing life) is temperature-dependent.

A number of things can be done to minimize operating temperature in the polygon/motor assembly, among them:

Minimize the operating drive current to that necessary to achieve performance. All current beyond that will convert to heat in the assembly.

Provide good thermal coupling between the motor housing and an adequate heat sink (the aluminum base plate of the system, for example). Use thermal conductive grease if necessary.

In systems where the polygon/motor assembly must be operated in a vacuum to reduce windage and drive power, it is a good idea to configure the system to

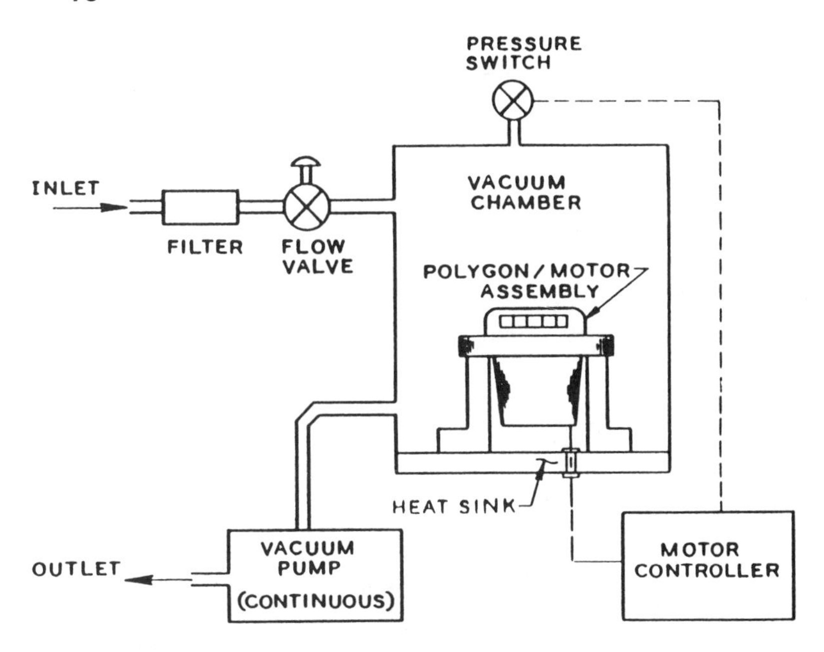

Figure 6.31 Schematic diagram showing method for providing cooling for vacuum operation of polygon/motor assembly.

provide cooling, as shown in Figure 6.31. A continuously operating vacuum pump maintains a reduced pressure in the vacuum housing. Ambient air is drawn into the chamber through a particle filter (1-μm sieve-type, replaceable cartridge) and then a flow valve. The flow valve is adjusted to throttle the inlet to control chamber pressure. Ambient air is cooled due to adiabatic expansion into the chamber, which in turn cools the scan motor. A pressure switch is suggested to disable the motor driver in the event that the pressure drops below a set point, normally caused by a clogged filter.

6.5.4 Polygon/Motor Interface

There are two basic mechanical design configurations of polygon/motor assembly. These are known as *cantilevered* and *captured* designs. These terms refer to the location of the polygon on the motor shaft relative to the bearings (see Figure 6.32).

The cantilevered approach is commonly used for relatively low speeds (10,000 rpm or less) and/or relaxed tracking tolerances (1 arc minute or greater). The primary advantage of this design is its adaptability to the use of standard com-

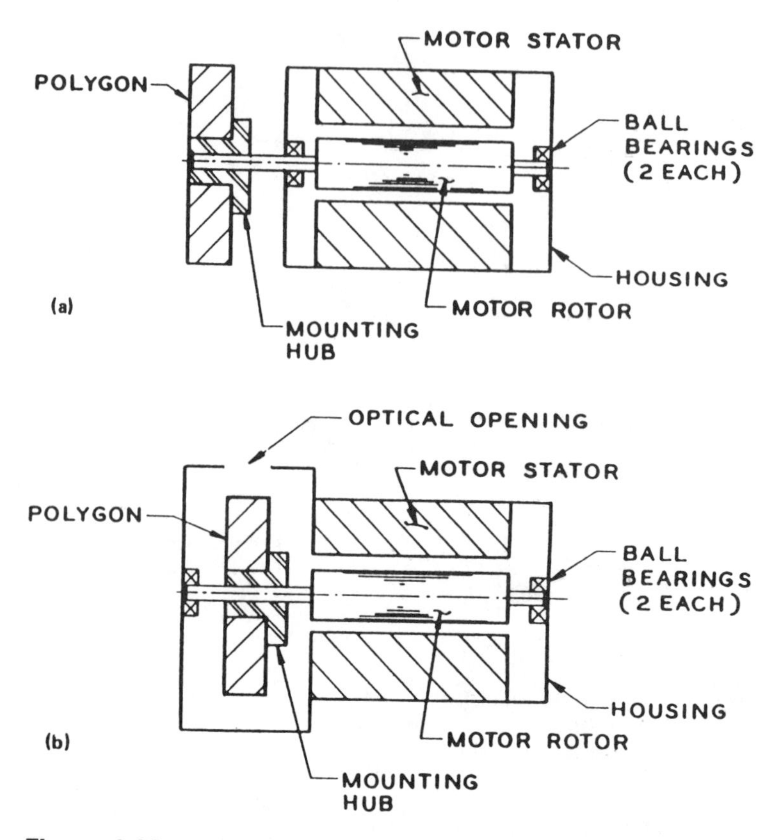

Figure 6.32 Cross-section showing cantilevered and captured polygon/motor construction: (a) cantilevered construction; (b) captured construction.

mercial motors. Additionally, it is more convenient to those who elect to purchase motors and polygon mirrors from different suppliers, since the motor may be adapted for use without disassembly. The main disadvantages of this design type are:

Limited stiffness. The output shaft must be small enough to go through the bearing.
Difficult bearing replacement. The polygon mounting hub is usually interference-fitted and must be removed to replace the bearing behind it.
Limited dynamic balancing capacity. The motor end of the rotor is inaccessible after the polygon is installed.

The captured approach is recommended for applications requiring high speed and/or high tracking accuracy. In this approach the center-to-center distance

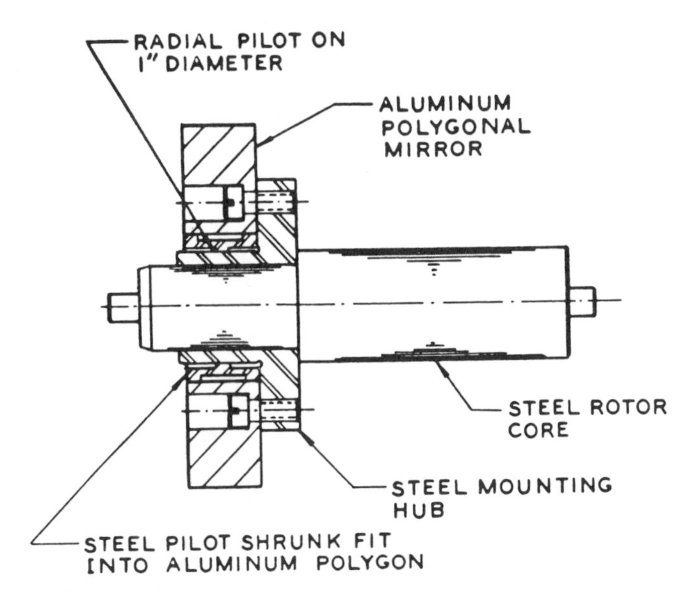

Figure 6.33 Thermally stable mounting scheme for polygonal mirrors.

between bearings is increased, thereby reducing the contribution of bearing runout to shaft angular runout. Also important is the fact that the motor rotor may have a larger diameter near the polygon location (bearing ID not being a limitation), providing greater stiffness and stability.

A major benefit of the captured design is that the entire rotor assembly and bearings may be assembled outside the housing and two-plane dynamically balanced. Additionally, bearings may be replaced without having to remove the polygon mounting hub. The major disadvantages of the captured design are

Lack of adaptability to standard commercial motors.
Rotor must be removed to install or remove polygon.

6.5.5 Dynamic Balance

Accurate dynamic balancing of polygon/motor assemblies is important for several reasons:

To enhance bearing life
To enable accurate polygon tracking
To prevent vibrations which may adversely affect other components of the optical system
To minimize acoustic noise

When free to do so, all rotating bodies rotate about their center of mass. If the center of mass does not coincide with the mechanically defined rotational axis, an unbalance force is created. This is a force of restraint coupled through the rotor bearings to the bearing holders, in this case the polygon/motor housing. It is known as dynamic imbalance and is expressed in terms of weight/radial distance (i.e., oz/in or g/cm). This force increases with speed. The relationship of the factors influencing imbalance is given by

$$\text{Force exerted against the shaft} = \frac{W\omega^2 r}{g}$$

where W = weight of the rotating body
ω = angular velocity, rad/s
r = distance from the rotational axis to the center of gravity
g = acceleration due to gravity = 980 cm/s^2 (32.16 ft/s^2)

Polygon/motor assemblies are dynamically balanced on a two-plane balancing machine similar to that shown in Figure 6.35. This machine supports the opposite ends of the rotor to be balanced on two separate motion transducers whose outputs are amplified and filtered and read out on meters on the machine control. The unit is precalibrated for rotor weight and speed.

Signal filtering is employed to eliminate frequencies of vibration occurring at other than rotor speed. The output of the motion transducer is also used to trigger a strobe light to provide visual cues to the operator as to the phase relationship of the measured unbalance load.

Having measured the amplitude and polar location of the unbalance load at each end of the rotor, weight is added or subtracted from indicated locations to alter the location of the mass axis to more nearly coincide with the rotational axis. Grinding or drilling away small amounts of material from the ''heavy side'' of the rotor, although a common technique employed in motor manufacture over the years, is not recommended. The opportunity for bearing contamination is too great with this method. A preferred approach is to provide a number of equally spaced, radially drilled and threaded holes at each rotor end. Small machine screws are inserted in appropriate locations and numbers to achieve balance.

The range of dynamic imbalance maximums most commonly specified for polygon/motor rotors is 3.5×10^{-6}–3.5×10^{-7} Nm (0.005–0.0005 oz/in) at n rotations per minute, where n is in the range 3600–24,000 rpm. Since the effect of imbalance increases as the square of the rotor speed, the importance of careful balancing increases by the same ratio. Attempts at perfection in balancing at high speed are frustrated by lack of rotor stiffness and thermal inhomogeneities in the rotor materials; further reason for design simplicity and attention to symmetry.

In the design of rotor assemblies, particularly for high-speed operation, it is important to provide high rotor stiffness. Good design practice dictates that the

Figure 6.34 Polygon/motor assembly being dynamically balanced on a two-plane balancing machine.

rotational frequency of a rotor should be significantly lower than the rotor's natural resonant frequency (0.8 or less). When unavoidable constraints dictate that the rotational speed exceed the rotor's natural resonant frequency, then some rules of thumb are useful in achieving an acceptable result. These are:

1. Design so that the operating speed (when higher than shaft resonance) be an odd integer of the resonant frequency.
2. Spend as little time at resonance as possible (accelerate through it rapidly) to minimize fatigue of parts.
3. Provide dampening where possible, such as O-ring mounting of bearings.

Designing for Dynamic Balance

In order for a rotating body to be in perfect dynamic balance, it is necessary that every plane through the body at right angles to and everywhere along the rotational axis have its center of mass precisely on the rotational axis. When such a condition exists, no unbalance coupling forces occur in the rotor during rotation. This is a perfect case and is seldom if ever achieved. Satisfactory results may be achieved nevertheless, even though it may not be possible to add or subtract mass in the exact plane where a mismatch occurs between the center of mass and the rotational axis. Typically, mass is added or subtracted in a position of convenience, dictated by design, and, in fact, imbalance coupling forces almost always exist in a restrained rotor. This condition is acceptable provided that the coupling forces are sufficiently small compared to rotor stiffness so that bending or distortion does not occur at speed.

Rotating mirror assemblies differ in an important way from most other high-speed rotating machinery, in that the polygonal mirror provides an excellent measuring tool to aid in characterizing rotor behavior at speed with a high degree of sensitivity. By examining the positions of scanning spots in the scanner focus line or plane at a variety of speeds, one may determine how the changes in imbalance coupling forces are effecting the integrity of the rotor assembly. ''Strobing'' or intensity-modulating the light source being scanned in synchronism with one or more facets removes the confusion over which scan is related to which rotor position.

Typically, a scanner rotor assembly is assembled with a number of component parts fabricated from a variety of materials, piloted one to the other with precision at room temperature and at thermal equilibrium. Then the rotor assembly is installed in its housing and balanced dynamically in a relatively short time at room temperature. Next the complete unit is installed in a system and operated for extensive periods, usually at elevated temperature. Unless care is taken, the performance measured during the manufacturing process may differ significantly from that measured in actual use.

Materials expanding or contracting due to a change in their temperature do so with a force (against restraint) equal to their compressive or tensile strength, respectively. This is a key point to remember when designing rotor assemblies that will operate at elevated temperature and high speed, since the various component parts may have differing expansion rates due to thermal change. An example of what may occur under such conditions is described as follows:

Assembly condition (aluminum polygon on steel shaft)

1. Radial pilot—line-to-line fit to 5 Ωm (0.0002 in) clearance
2. Temperature 24°C (75°F).

Operating condition

1. Radial pilot—5 to 10 Ωm (0.0002–0.0004 in) clearance
2. Temperature 60°C (140°F).

Due to the temperature difference between assembly and operating conditions, the pilot diameter of the polygonal mirror bore will increase by 0.0004 in more than the pilot diameter of the mounting hub. As luck would have it, the polygonal mirror will decenter on the hub by the amount of increase in radial clearance, in this case 0.0002 in. In a high-speed rotating assembly this will create a significant amount of unbalance.

The reason that decentering occurs in the example given is that one or more of the fastening screws will create more clamping force between the polygonal mirror and the mounting hub than the others. That superior clamping force location will become an anchor point about which the mirror will move relative to the hub. When the parts return to a lower temperature, the amount of decentering will be reduced, but it can only return to its original position if the pilot was, and still is, a line-to-line fit at the temperature returned to.

An alternative design solution for operation over a broad temperature range is suggested in Figure 6.33. The significant difference between this method and that previously described is the addition of a shrink-fit steel pilot ring to the polygonal mirror bore. The pilot ring thermal coefficient of expansion matches that of the hub and rotor core so that when the assembly temperature is elevated, no change in radial clearance occurs between these parts.

The shrink fit between the aluminum polygonal mirror bore and the steel pilot ring is toleranced so that compressive force exists between these parts over the anticipated temperature difference, thereby allowing no radial play between them. This approach is quite practical over temperature differences of up to 27°C (80°F) and pilot diameters up to 38.1 mm (1.5 in).

Occasionally, and for a variety of reasons, the mass distribution in a polygonal mirror structure may not be symmetrical in a series of planes about the center of its pilot diameter (the desired rotational axis). Some of the reasons for this condition may be the

1. Asymmetry of the machined geometry of the part. This can be measured and, if not correctable, at least identified.
2. Inhomogeneity of the material of the part, i.e., variations in material density from one place to another in the same part due to voids, impurities, alloy concentrations, plating thickness variations, and/or other anomalies more difficult to find by relatively simple measuring techniques.

These asymmetries can produce imbalance coupling forces to the rotor assembly which are difficult to compensate, since it may be impractical to add or remove mass in the plane in which the problem exists. Particularly bothersome in this regard are polygons which have diameters as great or greater than the separation distance between the bearings on their support shaft. When balancing such rotors, the shaft end opposite the polygon end may appear to be out of balance when in fact it is simply nutating in response to a coupling force introduced by the polygon. In this case it helps to remove the polygon from the assembly, balance the shaft on its own, replace the polygon and limit any further adjustments to the polygon only.

6.5.6 Electronic Controls for Speed

Figure 6.35 shows a schematic of a typical two-phase AC synchronous motor controller and a time diagram of its output voltage waveform. Scanner motors are typically fractional-horsepower, low-voltage devices, operating in the range 10–70 V peak to peak and drawing current in the range 0.5–5 A rms per phase. These operating ranges are well within the capability of semiconductor driver amplifiers operating in the square-wave mode.

The primary component of reactance in a synchronous motor is inductance, usually in the range 10–50 mH. Table 6.7 lists the reactive components of a typical hysteresis scan motor over a range of operating speeds. The significance of the inductive load characteristic is twofold. First, the drive current will lag the voltage delivered to the motor winding by the driver, resulting in a power factor of less than 1. As a result, more current must be delivered to the motor to produce torque than if unity power factor were maintained. For the case where the motor is to be driven at a fixed speed, capacitance may be put in parallel with the driver output to the motor to tune the circuit at operating frequency and improve power factor.

Second, the inductive motor load will distort (round off) the current waveform, particularly at higher frequencies (120 Hz and greater), reducing losses due to stator heating. Similar current and voltage waveform distortions occur in sine wave drivers in this use, thereby diminishing their desirability compared to rectangular wave systems.

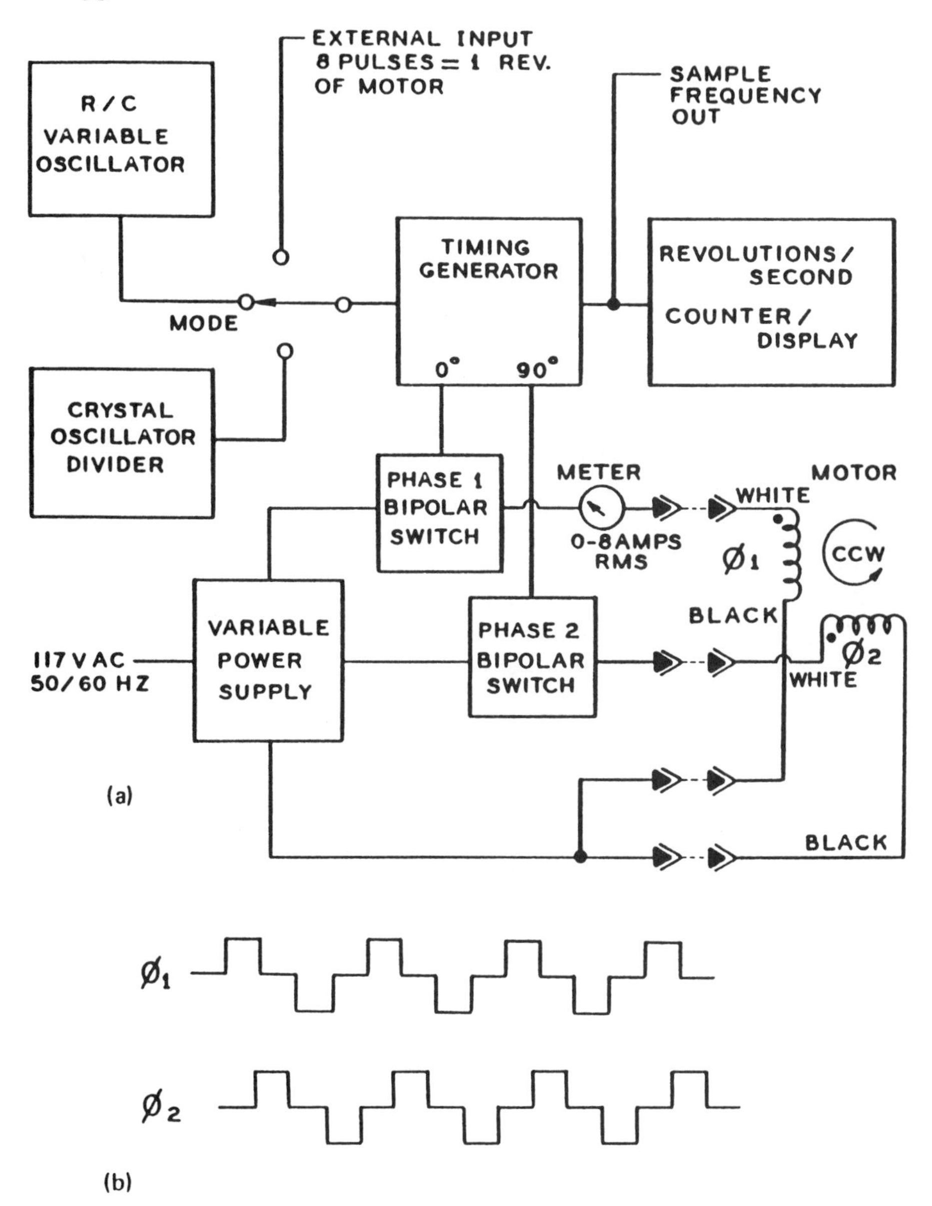

Figure 6.35 Schematic diagram of a two-phase synchronous motor controller circuit and output diagram: (a) circuit schematic; (b) output diagram—voltage waveform.

Table 6.7 Reactive Components of a Typical Synchronous Scanner Motor

Typical Design Characteristics—Hysteresis Synchronous Motor

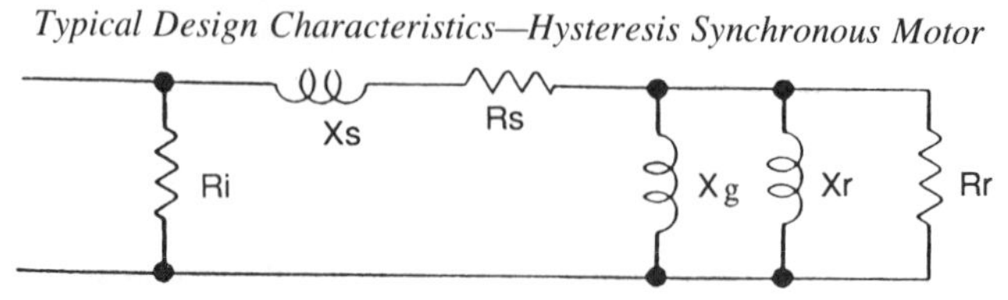

Equivalent Circuit of Hysteresis Synchronous Motor

R_r is the power component of current drawn by the ring. Figuratively this resistor has a shaft sticking out of it, and I^2R_r is horsepower, not heat.

X_r and X_g account for the current drawn to magnetize the ring and the gap, respectively. Values given are ohms reactance, not henries inductance, and therefore vary with frequency.

X_s is reactance from flux that links the stator windings but never gets to the rotor. Small but not negligible. Also ohms reactive, varying with frequency.

R_s is the resistance of the stator copper. It is the only element that can be measured directly with a meter, and the only one that does not vary with frequency.

R_i represents eddy-current and hysteresis losses in the stator iron. It is not negligible, 8.5 W at 25,000 rpm, compared with 1.9 W in the stator copper.

R_r varies with frequency, just as the reactances do. If you hold torque constant but vary speed, you must hold I constant and vary the value of I^2R_r. If torque also varies with speed, R_r will vary as (speed)^2.

R_i varies as (speed)$^{1.5}$ with current held constant.

All these can be combined into two parallel components, X_p and R_p, with values as tabulated below.

Rpm	25,000	15,882	10,002	6,300	3969.6	2,500
Frequency, Hz	416.7	264.7	166.7	105.00	66.16	41.67
Parallel resistance	12.57	7.450	4.307	2.454	1.369	0.750
Parallel reactance	5.847	3.734	2.374	1.520	0.985	0.648
Impedance	5.300	3.338	2.080	1.292	0.800	0.490
Apparent inductance (millihenry)	2.23	2.25	2.27	2.30	2.37	2.47

The above values are for just before lock-in, when motor is delivering rated torque at nearly rated speed. This is the only place they can be computed, as after lock-in, impedance is a function of load, past history of load, and past history of excitation.

6.6 TESTING POLYGONAL SCANNERS FOR PEFORMANCE

Component-level measurement and testing were discussed in Sections 6.4.3–6.4.6. Assembly-level testing may be approached in one of two ways:

Figure 6.36 Polygon/motor final test setup.

1. A series of individual and combined tests to confirm conformance to a detailed set of specifications, item by item
2. A single "in system" test to verify acceptability on the basis of use

Each method has merit in certain areas of application. In some instances, both types of testing are necessary to establish the acceptability of a polygon/motor assembly in order to find assemblies which, although correct in each item of component detail, may embody tolerance "stack-ups" as an assembly that render it "out of spec."

It is very beneficial for supplier and purchaser to agree on and implement identical procedures and test equipment at each other's facilities to verify compliance with performance specifications. This holds true for polygon/motor assemblies as well as any precision component or subassembly.

When a rigorous, effective, well-documented quality assurance system is employed to verify specification compliance at key locations and times of in-process manufacture of polygon and motor, final inspection should have only to confirm

1. Motor behavior
   a. Current draw for specified voltage
   b. Achievement of synchronism
   c. Velocity stability
   d. Rotation direction
2. Polygon behavior: tracking accuracy

Figure 6.36 shows a test setup for measuring the polygon/motor characteristics as given above.

An efficient means of qualifying polygon/motor assemblies frequently is simply to install them in the system and examine the total system output. For example, in a computer output printer, one may rather rapidly "inspect" polygon/motor assemblies by altering a single printing machine to permit the rapid exchange of scanners, so an actual test pattern may be printed with each one. This is a rapid "acid test" of the scanner. Often defects in the printout may serve as a diagnosis of a problem.

REFERENCES

1. F. G. Spreadbury, *Fractional HP Electric Motors.*
2. B. R. Teare, Jr., Theory of hysteresis motor torque, *AIEE Trans.*, *59*:907 (1940).
3. C. G. Veinott, *Fractional and Subfractional Horsepower Electric Motors.*

7

High-Performance Polygonal Scanners, Motors, and Control Systems

Gerald A. Rynkowski

Speedring Systems, Inc.
Rochester Hills, Michigan

7.1 INTRODUCTION

7.1.1 Applications

Polygonal scanners have been designed, developed, and manufactured in all shapes and configurations during the past 20 years. These devices have been employed in military reconnaissance and earth resources studies, high-density computer storage, TV displays, laser printers and video records, flight simulators, and supermarket UPC (universal product code) readers, to name a few of the well-known applications.

In general, three types of scanner mirror configurations are popularly utilized in collimated or convergent, passive or laser scanning optical systems. These rotating mirror spinners are at the center of the electrooptical system. They direct incoming optical signals as well as outgoing modulated laser beams.

Utilized at the center of the optical system, the spinner's key specifications interrelate with the system performance parameters, and as such must be optimized with regard to optical characteristics, accuracies, dynamic performance, bearing and material selection, and fabrication processes. Technical and cost-effective trade-offs are possible where communications between scanner manufacturer and system design engineers are maintained.

7.1.2 Scanner Configurations

The three most utilized scanner beam deflector configurations are the regular polygon, pyramidal, and the single-faceted cantilever design. The regular polygonal scanner is generally the most popular with system designers and can be utilized with either the collimated-beam or the convergent-beam scanning configurations.

Figures 7.1 and 7.2 illustrate the two configurations using two six-sided polygons having the spin axis projected into the page. In Figure 7.1 the facets are illuminated with a collimated beam and reflected to a concave mirror that focuses at a curved focal plane. Figure 7.2 shows a lens system that focuses the collimated beam prior to being reflected at the scanner facets, and then converging at a focus. Comparing the two configurations, it is obvious that both focal image surfaces are curved. This presents a problem to the system designer, but is usually corrected optically with a suitable field-flattening correction lens system, or perhaps the recording surface is curved to conform to the focal surface. Another difference with regard to the polygon is that the Figure 7.2 convergent-beam bundle uses less area of the facet, all else being equal, than the other configuration. However, maximum utilization of the facet area is desirable because the facet surface flatness irregularities tend to be averaged out, and, therefore, minimize modulation of the exiting-beam scanning angles. Most precision polygonal scanning systems use the collimating-beam scanning configuration which utilizes a larger proportion of the area of the facets.

Figure 7.3 illustrates a pyramidal mirror scanner commonly used with the rotational and optical axes parallel, but not coincident. Note that either the collimated or convergent configuration can be utilized with this design.

Figure 7.4 illustrates a regular polygonal scanner in which the optical axis is normal to the rotational axis, or where the angle is acute to normal. Note that

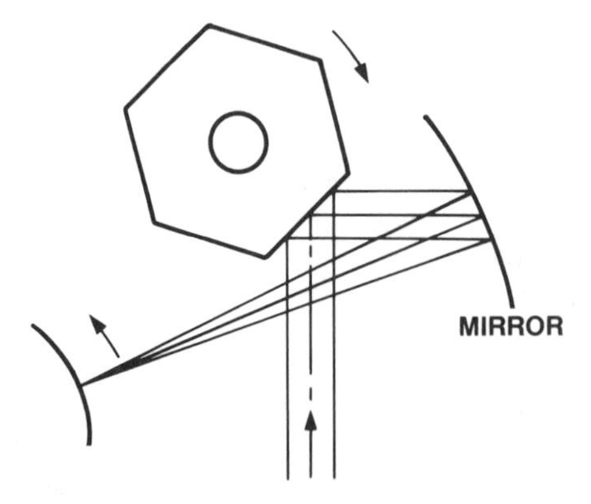

Figure 7.1 Collimated beam scanning [3].

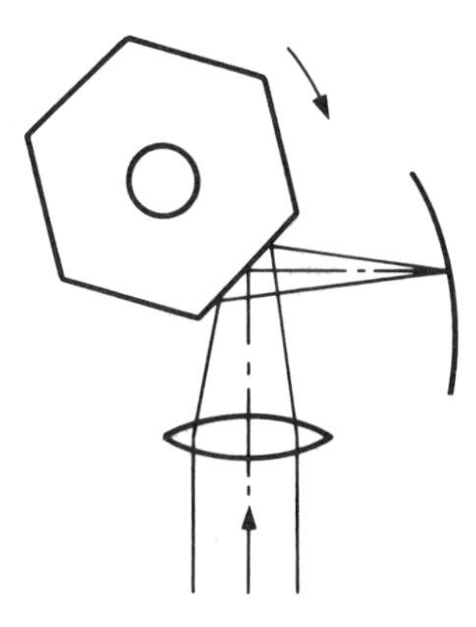

Figure 7.2 Convergent beam scanning.

when the two axes, optical and rotational, are normal, the beam can reflect back upon itself.

At this point, it is noteworthy to realize the relation between the facet angle and scan angle. The facet angle is defined as 360°/N (number of facets). The optical scan angle may be expressed as

$$\text{Scan angle (degrees)} = \frac{720}{N} \qquad \text{for } N \geqslant 2$$

Observe that for $N \geqslant 2$, the optical scan angle is two times the shaft angle. This angle-doubling effect must obviously be considered when relating the shaft and

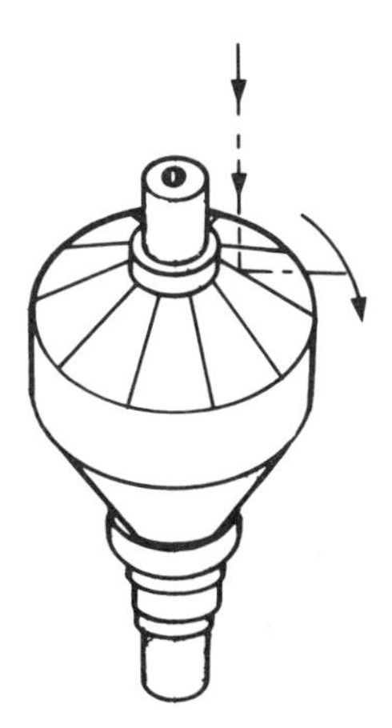

Figure 7.3 Parallel, but not coincident, optical and rotational axes [3].

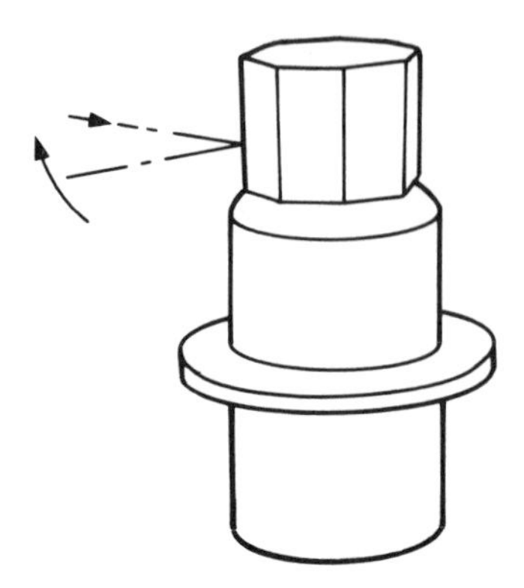

Figure 7.4 Optical and rotational axes normal.

facet parameters and their effects on the angular position of the focused spot at the focal plane.

The scanner rotational speed may be expressed as:

$$\text{rpm} = \frac{60\,W}{N}$$

where W = line scan/second

N = number of facets

An increase in the number of facets reduces the motor speed requirements as well as the maximum scan angle. However, the usable scan angle may in some cases also be limited by aperture size and the allowable vignette effect.

Shown in Figure 7.5 is a single-faceted cantilevered scanner with the beam and rotational axes coincident and reflecting from a 45° facet, and thereby generating a continuous 360° scan angle and a circular focused scan line. This configuration is used in passive infrared scanning systems having long focal length and requiring a large aperture. Nine-inch, clear-aperture scanners have been manufactured for high collection efficiency.

7.1.3 Reference System

A film recording scanning system has been selected as a reference subsystem for purposes of discussing the related spinner and optical performance parameters as well as the dynamic performance requirements.

Figure 7.6 depicts a laser recording system capable of recording high-resolution video or digital data on film. The spinner generates a line scan at the film plane using a focused intensity modulated laser beam. Line-to-line scan is accomplished

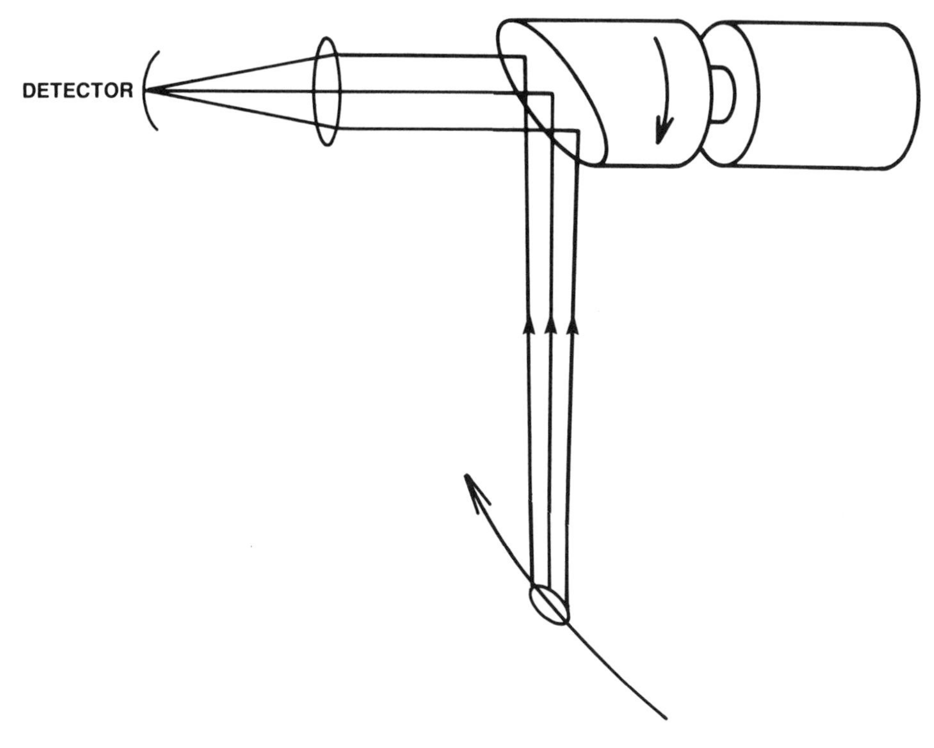

Figure 7.5 Parallel and coincident optical and rotational axis.

by moving the photographic film at constant velocity and recording a continuous corridor of data limited only by the length of film. The film controller provides the required frequency/phase lock control of film velocity. The expanded and collimated laser beam is intensity-modulated with video or digital data and then scanned by the facets of the spinner to the film plane. A field correction lens (F-θ) is used to focus the beam, to linearize the scan line with respect to the scan angle, and thereby provide a uniform spot size along the line at the film plane.

The scanner assembly contains a 12-faceted beam deflector as required to perform the optical scan function. The rotating mirror, its drive motor rotor, and a precision optical tachometer are supported radially and axially by externally pressurized gas bearings.

The electronic controller provides precise motor speed control and synchronization between the reference frequency sync generator and the high-density data track of the optical encoder. The encoder also supplies an index pulse used for facet identification and derivation of the synchronized field frequency that is

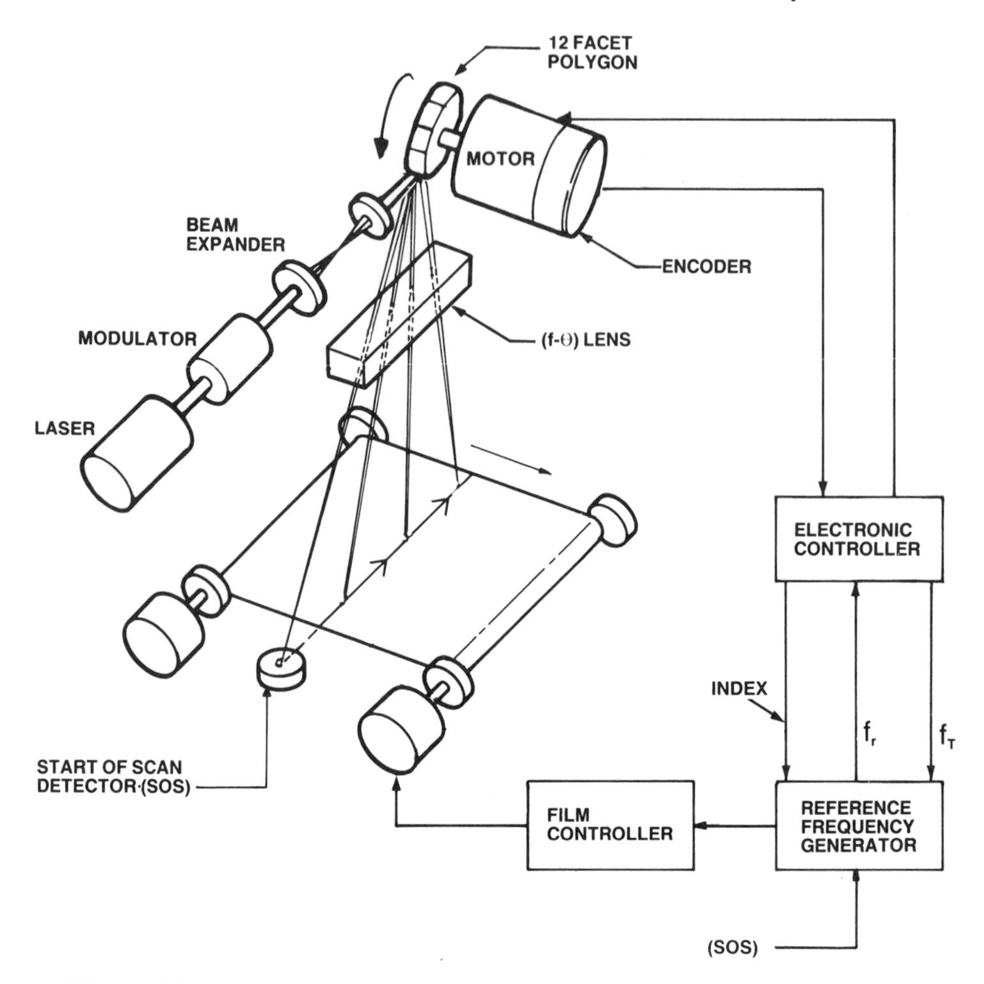

Figure 7.6 Film recording system.

required for some raster scanning systems, as well as pixel registration and control of the film drive motor.

7.2 BASIC CONSIDERATIONS

7.2.1 System Performance Requirements

The system performance parameters for our "strawman" digital film recorder are discussed in the following paragraphs. These parameters are summarized in

Table 7.1 Summary of System Performance Requirements

Line resolution (both directions)	10,000 pixels/line
Line scan length	5.5 in
Scan rate	1200 lines/sc
Film speed	1.668 cm/sec
Pixel frequency (clock)	12 MHz
Pixel diameter	10 μm at $1/e^2$
Pixel-to-pixel spacing	13.97 μm

Table 7.1 and are typical and representative of a recently manufactured film recording system.

The system resolution requirements are at the film plane. The application requires 10,000 pixels of digital data per line at the scan plane using a 10-μm spot diameter measured at $1/e^2$ irradiance level. One can calculate that for a 5.5-in line scan the pixel spacing, center to center, must be 13.97 μm as must the spacing between scan lines. The line spacing variance is selected to be less than ±5 μm (half the nominal spot size).

The application dictates a scan rate requirement of 1200 lines per second. From this information the calculated pixel frequency is 12 MHz, and the film speed needed is 1.668 cm/sec.

7.2.2 Spinner Parameters

The spinner, the beam deflector, specification is determined and driven by the form, fit, and functional performances set by the optical requirements of the system. At this point, the optical engineer must optimize the design of the optical elements, which includes specifying the polygon type, facet number, facet width and height, inscribed diameter of the polygon, facet flatness and reflectance; speed, quality, and tolerance of variations. These variations are both spatial and temporal and do interact with each other as a function of temperature, balance, bearing and material selection, cost, and manufacturing process. Due to the parametric specification interactions and the high-accuracy requirements, the spinner is addressed as a scanner subsystem to allow the scanner designer to make trade-offs within the agreed-upon limits imposed by the optical and system performance requirements.

The scanner subsystem consists of a one-piece beryllium scan mirror and shaft, suspended on hydrostatic gas bearings, which is driven by a servocontrolled motor. Depicted in Table 7.2 is a summary of the subsystem requirements.

The F-θ lens is designed to function with a 60° optical scan entrance angle which dictates a 30° facet angle specification. The number of facets is therefore

Table 7.2 Summary of Scanner Subsystem Requirements

Number of facets	12
Facet angle	30°
Inscribed diameter	4.0 in
Facet height	0.5 in
Facet reflectance	89–95%
Facet flatness	λ/20
Facet quality	Scratch and digs (MIL-F-48616)
Scan rate	1200 scans/s
Rotational speed	6000 rpm

calculated to be 12. The film sensitivity, optical system transmittance laser power, facet reflectance, and spot shape and size enable the facet area and inscribed diameter of the polygon to be specified.

A motor speed of 100 rev/s, or 6000 rpm, ensures 1200 scan lines per second.

7.2.3 Scanner Tolerance Requirements

Scanner tolerance is determined by the permissible static and dynamic pixel position errors acceptable at the film plane. These worst-case errors are referenced back through the optical system and scanner subsystem to be distributed and budgeted between the operational elements and reference datums. The tolerable variances are often specified as a percentage of pixel to pixel angle, pixel diameter, pixel-to-pixel spacing, or the motor speed regulation (stability) over one or more

Table 7.3 Subsystem Characteristics and Tolerances

Characteristic	Tolerance	Comments
Number of facets	N.A.	Determined by scan angle
Facet angle	±10 arc sec	One pixel-pixel angle
Diameter	N.A.	Controlled by facet width dimension
Facet width	1.035 in min.	0.020-in roll-off
Facet height	0.5 in min.	0.020-in roll-off
Flatness	λ/20 max.	Spot control
Reflectance	± 3%	Tolerance
Apex angle	1.00 arc sec	Total variation – 10% of line-line angle
Speed regulation:		
(1 revolution)	±10 ppm	±1.08 arc sec/line
(long term)	±50 ppm	±5.40 arc sec/line

Note: Scan error for any 12 scans ≤ ±12.96 arc sec

revolutions. The conversion of these variances to meaningful and quantifiable units is necessary for manufacturing, measurement, inspection, and testing.

7.2.4 Need for Speed Control

Speed control is required for accurate pixel positioning, repeatability, and linearity, as well as line-to-line pixel registration and synchronization.

In order to accurately position data pixels in a line at the film plane, the system must generate precision pulses spatially related to the facet angles. These pulses, occurring on a one per pixel basis, are used to gate in and turn off the intensity-modulated video being projected to the light sensitive film surface. Since there are 120,000 pixels (12 times 10,000) per revolution, one clock pulse would be required for every 10 arc sec of shaft rotation.

Optical encoders are well suited to the task of generating accurately timed and positioned pulses. However, incremental, high-density data track encoders are expensive, large in diameter, and difficult to mechanically interface with an integral spinner/motor/shaft assembly. To overcome this problem, a smaller and inexpensive, low-density optical encoder (6000 pulses/rev, ppr) was designed into the system. The required 120,000-ppr pixel clock pulses are obtained by electronically multiplying the encoder data track frequency (6000 ppr) by a factor of 20. Scanner speed control is accomplished by frequency/phase locking the encoder data track (600,000 Hz) to an accurate and stable crystal oscillator.

An index pulse is accurately positioned at the normal of a facet on a second encoder track, thereby providing start of scan (refer to Figure 7.6, SOS detector) synchronization and pixel registration.

This system of generating pixel clock pulses places a heavy burden on the performance accuracy of the rotating shaft assembly, the encoder design and

Table 7.4 High-Performance Polygonal Scanner Summary

Characteristics	Reference system	State-of-art system
Facet number	12	20
Facet tolerance	±10 arc sec	±1 arc sec
Apex angle error	±0.4 arc sec	±0.2 arc sec
Speed	6000 rpm	28,800 rpm
Scan rate	1200 scans/s	9600 scans/s
Regulation/rev	< 10 ppm	<1 ppm
Pixels/scan	10,000	50,000
Pixel/jitter/rev	< ±25 ns	< ±2 ns
Pixel clock	12 MHz	480 MHz

adjustment, and the stability of the crystal oscillator. Nevertheless, the net speed control and jitter performance has been optically measured to be less than ±10 parts per million for one revolution of the scanner.

7.2.5 High Performance, Defined

Table 7.4 depicts the summary performance of the film recorder reference system in comparison to a state-of-the-art recording system considered by many to have the highest resolution and fastest system manufactured to date.

7.3 MANUFACTURING AND ENGINEERING DESIGN CONSIDERATIONS

7.3.1 Facilities and Disciplines

The successful engineering and manufacture of a cost-effective scanner subsystem requires an organization staffed with a dedicated team of people having many disciplines and able to communicate with each other at all levels.

Figure 7.7 depicts the in-house disciplines and facilities needed to engineer and manufacture a typical scanner system. Shown are some popular applications that require a high-performance polygonal scanner subsystem. The basic components that comprise a typical scanner subsystem are shown as a subset consisting of the optics and spinner, bearings, motor, housing, encoder, and electronics drive and control. Overall engineering and manufacturing considerations are given to the system requirements with regard to the essential metrology capability, and precision assembly facilities (e.g., clean rooms).

7.3.2 Spinner/Optics Design

To obtain the highest performance and state-of-the-art accuracy, the shaft and polygon are manufactured as an integral assembly using selected beryllium as a substrate material.

Table 7.5 details typical characteristics of four applicable metals. It can be seen from this table that beryllium offers the highest strength-to-weight ratios and an order of magnitude difference in Poisson's ratio. Beryllium's low Poisson's ratio, combined with its very low density, yields mirror distortions much smaller than any other metal. Also, the distortion factor of beryllium is 67 times better than type 410 stainless steel.

Electroless nickel is generally deposited on the beryllium substrate to obtain

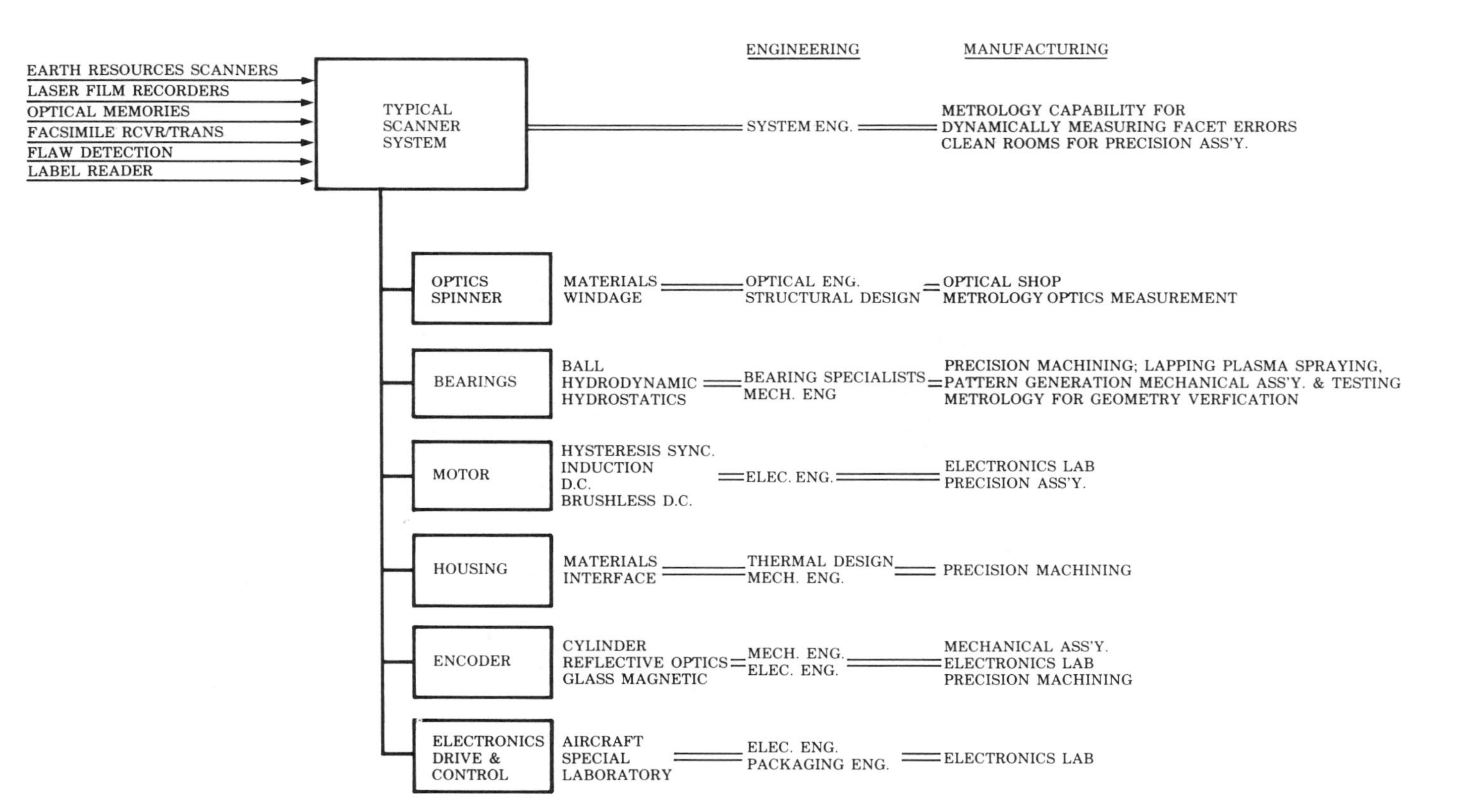

Figure 7.7 Disciplines (in-house) or special facilities [3].

Table 7.5 Characteristics of Typical Substrate Metals

	2024-T86 Aluminum	S-350 Beryllium	410 Stainless steel	ZK51A Magnesium
ρ Density (lb/in)	0.10	0.067	0.29	0.0656
α Coef. of linear thermal expansion (in/in/°F)	12.6×10^{-6}	6.5×10^{-6}	5.5×10^{-6}	14.5×10^{-6}
κ Thermal conductivity (Btu/hr/ft^2/ft/°F)	70.0	10.4	14.4	63
E Modulus of elasticity (lb/in^2)	10.6×10^{6}	44×10^{6}	29×10^{6}	6.5×10^{6}
E/ρ Modulus to density ratio (in)	106×10^{6}	657×10^{6}	100×10^{6}	99×10^{6}
γ Poisson's ratio	0.33	0.03	0.3	0.35
$\gamma\rho/E$ Distortion factor (1/in)	3.11×10^{-9}	0.045×10^{-9}	3.0×10^{-9}	3.54×10^{-9}

a finely polished surface. Although polished beryllium has a 52–55% maximum reflectivity at 550-nm wavelength, the surface is highly abrasive and not easily polished. Hence, the use of electroless nickel enables a high degree of control of both surface temper and polishing quality.

A variety of materials are selected for vacuum deposition on the scanner mirror surfaces to improve reflectivity. Aluminum is most popular, although silver and gold are also used. Protection is provided by overcoating with hard quartz, silicon monoxide, or magnesium fluoride. In some instances, in addition to aluminizing, further enhancement is possible for selected wavelength regions by multiple stacks of proprietary coatings to provide reflectivity approaching 95–96% over large incident angles. All coating techniques mentioned can meet or exceed the military aluminizing specifications of MIL-M-13508.

7.3.3 Bearing Selection

The basic bearing selection considerations for high-performance scanners are dictated by the precise and stable apex-angle requirements, trouble-free operation (high MTBF), ease of manufacture, and low cost.

Ball bearing scanners are generally limited to low-accuracy systems having apex-angle errors from 10 to 60 sec of arc. Operation at speeds greater than 24,000 rpm is obtainable only with a decrease of MTBF and life. The geometric errors due to manufacturing limitations, loss of lubrication by migration and outgassing, torque perturbations, and contamination are a few of the reasons for limited performance and life.

Self-acting or hydrodynamic gas bearings are serious candidates for high-performance scanners. These bearings meet the accuracy and stability requirements and have much greater life expectancy than ball bearings at high speeds. Operation at speeds up to 120,000 rpm have been achieved. High rotational accuracies are accomplished using an error-averaging principal, whereas the shaft is coaxially suspended on a pressurized gas flow inside a cylinder separated by a small clearance. Figure 7.8 illustrates the machining errors of the cylinder and shaft that are averaged out, as the shaft rotates, thereby reducing their influence on the radial positioning (wobble) of the shaft. Self-acting gas bearings are so called because they develop the load supporting gas pressure (internally) as the shaft rotates. Figure 7.9 shows some of the popular configurations used to support scanner mirrors. Note the grooved sections that are etched into the bearing surfaces and act as miniature pumps that generate the load-supporting pressure. These bearings have no load capacity unless they are rotating. Bearing surfaces remain in contact until sufficient pressure is developed for the rotating shaft to ''lift off'' and be gas-supported. Depending on the number of start and stops, the life of the bearing is limited by the surface wear characteristics and the resulting dimensional changes and contamination that can eventually terminate bearing life.

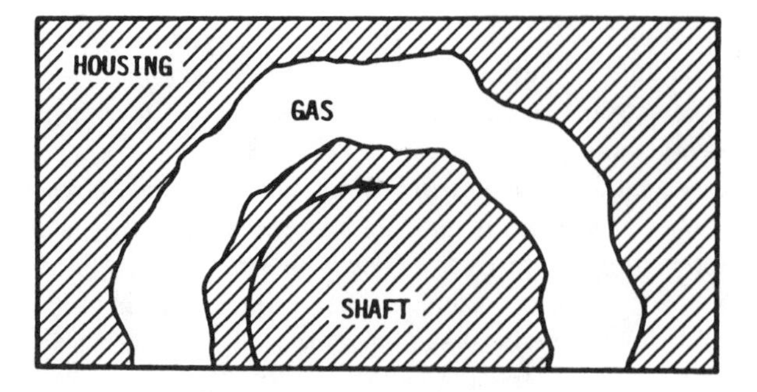

Figure 7.8 Simplified gas bearing illustration [3].

Another limitation of self-acting bearings for scanners is their load capacity. In order to develop and maintain sufficient gas pressure to lift heavy loads, large surface bearing areas and pumping groove densities are required. These design requirements reduce the MTBF (mean time between failure), impose limits on the radial stiffness of the bearing, increase the cost and manufacturing complexity, and place an impractical burden on the power requirements of the motor to provide "lift off" and maintain pumping power at speed. However, these limitations are

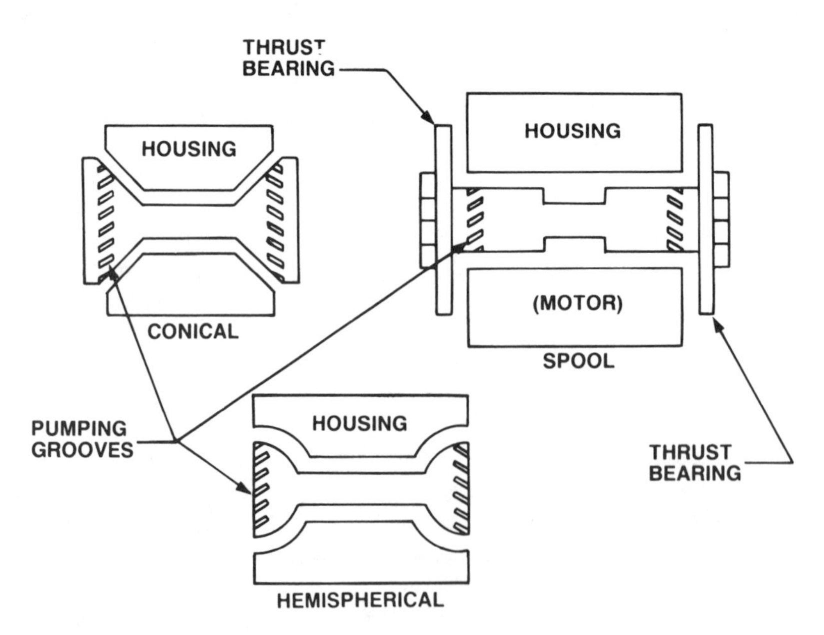

Figure 7.9 Self-acting gas bearing configurations [3].

overcome by utilizing the externally pressurized gas bearing, which is also known as a hydrostatic air bearing.

The externally pressurized bearing, shown in Figure 7.10, can also be made in the configurations shown in Figure 7.9. The grooved areas are omitted, and air inlet jets are suitably positioned in the stator. The design of these bearings is more straightforward than that of the self-acting types. Calculated load capacity is only slightly dependent on speed. Also, this type of bearing is much more resistant to instability and can be considered free from half-speed whirl (an instability, generally associated with hyrodynamic bearings) for most purposes. In designing rotor bearing systems with these bearings, classical critical speeds become important, just as with ball bearings; the main difference is that the air bearing has little inherent damping, and so care has to be taken if the bearing has to run up through a critical speed. These bearings have been successfully used at speeds of 250,000 rpm. As far as materials are concerned, there is only need for particular care if an overload condition is foreseen. For a self-acting spool bearing which is suitable for an optical scanner, the bearing dimensions and tolerances are given in Table 7.6. Also, for comparative purposes, data are shown for an externally pressured bearing in Table 7.7.

Tables 7.6 and 7.7 show that the self-acting bearing has much smaller clearances than the externally pressured bearing, resulting in tighter tolerances. The externally pressurized bearing has more parts, and the manufacture of smaller jets can be a problem. Table 7.8 shows a comparison of the features of the three types of bearings considered.

7.3.4 Motor Requirements

The precision of the integral polygon/shaft and gas bearing assembly, as well as the sophisticated material used, places a heavy burden on motor rotor selection.

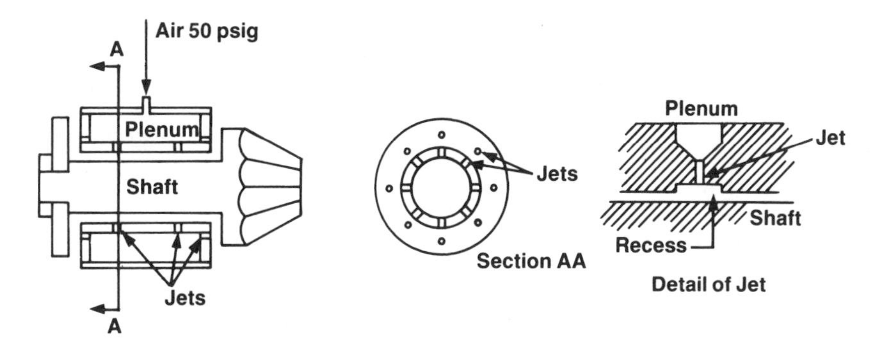

Figure 7.10 Hydrostatic air bearing design.

Table 7.6 Typical Self-Acting Bearing Dimensions [3]

Feature	Dimensions	Tolerance
Journal diameter (in)	0.5	0.005
clearance (μin)	50	5
groove depth (μin)	70	15
Thrust outer diameter (in)	0.8	0.005
clearance (μin)	50	5
groove depth (μin)	125	25
Squareness of shaft (arc sec)	1.0	max. error
Surface finish (μin)	3.0	max. error

Any rotor attached to the shaft must have very stable and predictable characteristics with regard to strength and temperature; and if possible, the rotor material should be homogeneous. If the rotor is of a complex mechanical configuration and consists of laminations and windings, the assembly may not maintain a precision balance (less than 20 μin oz) when operating at high speeds. Additionally, thermal expansion and high centrifugal forces may shift and reposition the rotor and perhaps cause a catastrophic seizure of the air bearing. Two motor designs are considered for use with high- and low-speed air bearing scanners; respectively, they are the hysteresis synchronous and the DC brushless.

Hysteresis Synchronous Motor

The difficult characteristic rotor requirements as previously stated are easily obtained with the use of a hysteresis synchronous motor (see Figure 7.11). The hysteresis rotor is uncommonly simple in design and consists of a cylinder of hardened cobalt steel that is heat-shrunk onto the shaft. Careful calculations are

Table 7.7 Typical Externally Pressurized Bearing Dimensions [3]

Features	Dimension	Tolerance
Journal diameter (in)	0.5	0.005
clearance (μin)	400	40
jet diameter (mils)	2.5	0.5
Thrust diameter (in)	0.75	0.005
clearance (μin)	400	40
jet diameter (mils)	20	2
Squareness of shaft (arc sec)	10	max. error
Surface finish (μin)	16	max. error

Table 7.8 Summary of Bearing Features [3]

Feature	Ball bearing	Gas bearing: Self-acting	Gas bearing: Externally pressurized
Rotational accuracy	10 μin	Better than 1 μin	Better than 1 μin
Lifetime	Limited: 250 to 1000 hours for very high speeds	Depends on materials and start-stop frequency	Unlimited by bearing: lasts as long as air supply
Load capacity	Higher than gas bearing, depends on size, number of balls	Depends on speed, gas, ambient pressure, and size	Can be high: depends on supply, pressure, size
Power demand equivalent bearings	Lower than gas bearing	Higher, goes as square of speed	Lower than self-acting bearing
Starting torque requirement	Medium	High	Low
Torque noise or ''hash''	Pronounced	None	None
Thermal conductivity	Poor	Medium	Excellent

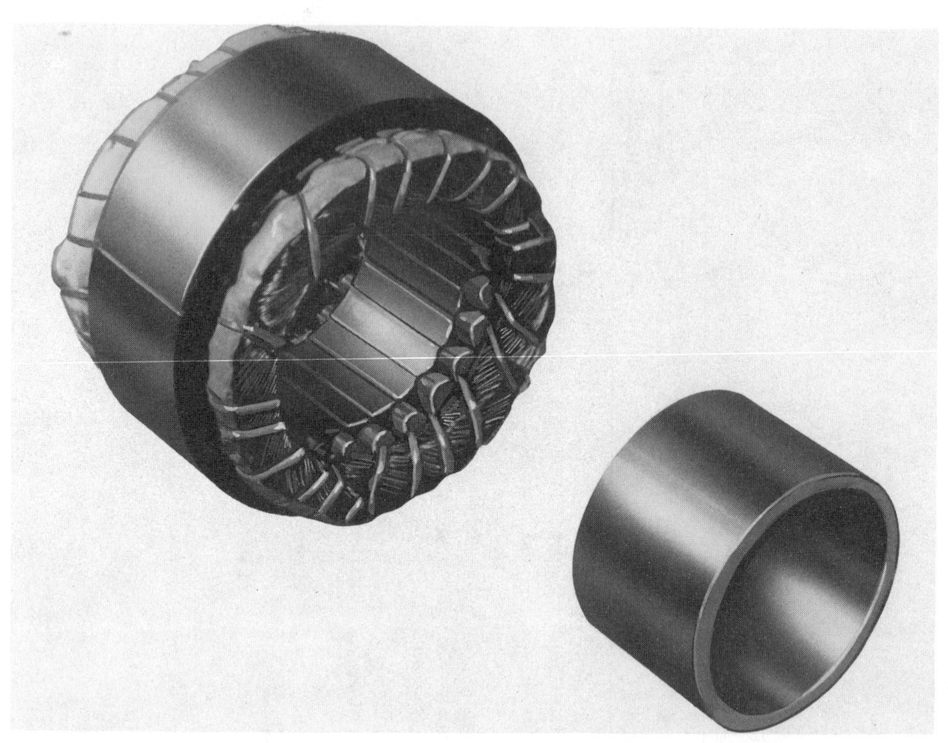

Figure 7.11 Hysteresis synchronous motor, stator, and rotor [4].

made with regard to the centrifugal forces and thermal expansion influences on the precision-fit requirements for safe and reliable operation.

The motor is well suited for operation at speeds ranging from 1000 to 120,000 rpm. Motors as large as 3 hp have been successfully used on large-aperture IR scanning systems operating at 6000 rpm.

The principle operation of a hysteresis synchronous motor relies on the magnetic hysteresis characteristics of the rotor material. As the magnetizing force from a suitably powered stator (not unlike that used with reluctance type motors) is applied to a cobalt steel rotor ring or cylinder, the induced rotor magnetic flux density will follow the stator coil current, as illustrated in Figure 7.12. The sinusoidal current is shown to increase from zero along the initial magnetization curve to point (a), thereby magnetizing the material to a corresponding flux level at the peak of the sinusoid. As the current decreases to zero, the rotor remains magnetized at point (b). If the current at this point in time were to remain at zero, the rotor would be permanently magnetized at the point (b) flux level. However, as the current reverses direction, the flux reduces to zero at some

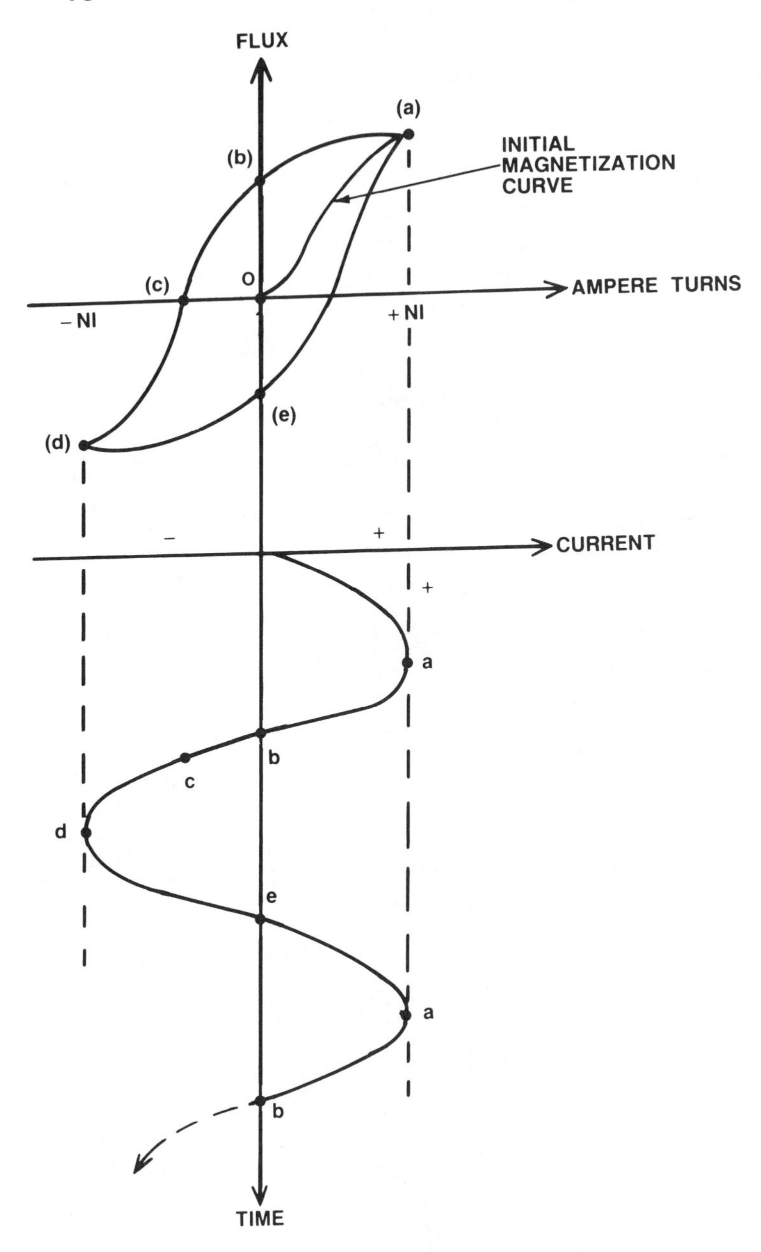

Figure 7.12 Magnetic hysteresis curve [2].

negative value of current as shown at point (c). Further decreases in current (negative direction) reverses the direction of flux as shown at point (d), corresponding with the negative peak of the current. The process continues to point (e) and back to point (a), completing the loop for one cycle of current. The figure generated is called a magnetic hysteresis loop. In physics, hysteresis is defined as a lag in the magnetization behind a varying magnetizing force. By analogy, as the axis of the magnetizing force rotates, the axis of the lagging force of the rotor will accelerate the rotor in the same direction as the rotating field. As the rotor accelerates, its speed will increase until it reaches the synchronous rotating frequency of the field. At this point, the rotor becomes permanently magnetized and follows the rotating field in synchronism. The synchronous speed of the rotor can be calculated with the following expression:

$$\text{rpm} = \frac{120f}{N}$$

where f = line frequency (Hz)
N = number of poles

Figure 7.13 is a typical speed versus torque characteristic curve for a hysteresis synchronous motor. As the fixed line voltage and frequency are applied to the

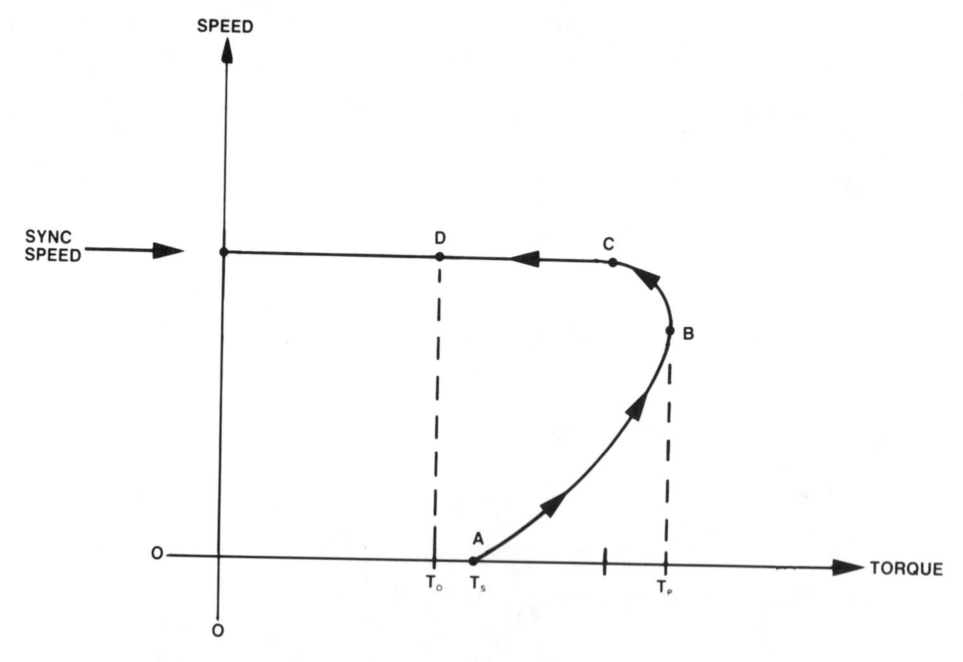

Figure 7.13 Speed/torque performance curve [2].

stator winding of the motor, an accelerating torque is developed equal to the starting torque T_s shown at point A. As the speed of the rotor increases, the operating point on the curve moves through the maximum torque developed at point B and through point C, at which time synchronous speed is reached. The final operating point D is determined by the operating load torque presented to the shaft at torque level T_0. Note that if the operating load torque is greater than the in-sync torque, T_{sync}, synchronous speed will not be reached.

Figure 7.14 is a vectorial representation of the rotating magnetizing field and the magnetized rotor field while in synchronism. Note that the rotor field vector lags the magnetizing field by an angle α. The operating torque (as developed by the motor in synchronism) is in proportion to the sine of the angle α in electrical degrees. If the load torque and stator frequency are absolutely constant, their frequencies will be precisely equal. However, should the torque angle be modulated sinusoidally, as indicated in Figure 7.13 (with a torque variance of $\pm\beta$), the rotor vector will advance and retard as indicated about the average angle α. The long-term average speed will be as constant as the applied stator source frequency, and the instantaneous speed will follow the derivative of the sine wave on a one-to-one basis. Torque perturbations, that are not attenuated by system inertia, will also modulate the shaft speed accordingly.

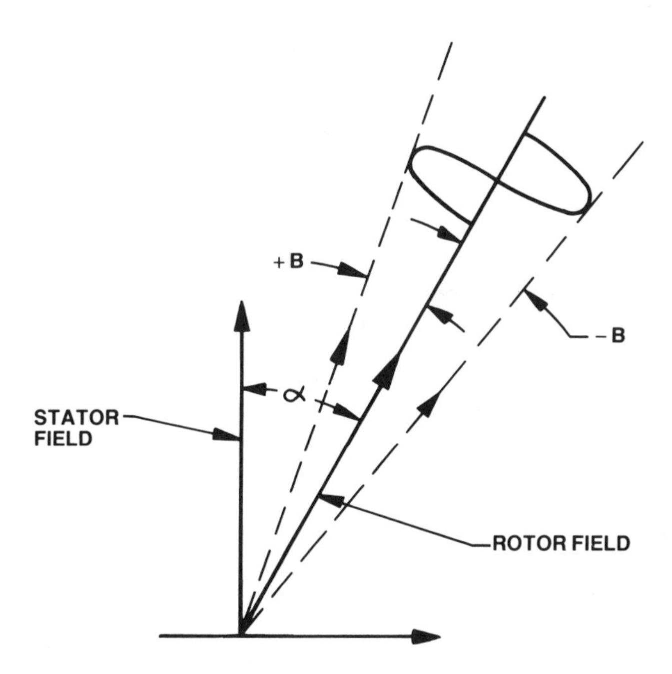

Figure 7.14 Stator/rotor field vectors [2].

A characteristic of hysteresis synchronous motors (and other second-order devices and systems) called "hunting" may be observed when operating the motor in a system having low losses and damping factor. The motor rotor will oscillate in a sine wave fashion, not unlike that depicted in Figure 7.14, if perturbed by applied forces internal or external to the system. If the perturbations are sustained, the oscillations will also be sustained; however, if the perturbations are not sustained, the oscillations will diminish in amplitude to essentially zero. The internal damping factor of the motor can be influenced by the rotor resistivity, rotor-to-stator coupling coefficient, and driver source and stator impedances. In a typical open-loop operation (no external velocity or position feedback) the oscillations may not be predictable and, therefore, may suddenly appear due to an unknown source or sources of pertubing forces. The amplitude of the oscillations in shaft degrees can typically range from 1 to 10°, and, as a practical matter, is very difficult to calculate; however, the oscillating frequency can be calculated as follows:

$$W_n = \sqrt{\frac{K}{I}}$$

where W_n = natural resonant frequency (rad/s)
K = motor stiffness (oz-in/rad)
I = shaft moment of inertia (oz-in s^2)

The maximum instantaneous speed is determined by setting the derivative of the sine function to zero, and then calculating the maximum positional rate of change in radians per second:

$$\text{Change in speed (rad/s)} = \pm A_p W_n$$

where A_p = $\pm\beta$ (Figure 7.14, rad)
W_n = natural resonant frequency (rad/s)

The maximum change in speed is often expressed as a percentage change relative to the nominal operating speed of the motor, as follows:

$$\text{\% Velocity regulation} = \left(\frac{A_p W_n}{W_s}\right) 100$$

where W_s = nominal operating speed (rad/s)

Figure 7.15 shows two curves of percent velocity regulation versus speed in rpm, for a typical open-loop scanning system having peak angular displacements of 1 and 5°. A four-pole motor with a peak torque of 10-oz in and having a total inertia of 0.076 oz-in s^2 was used for the calculations in this figure. Note that for peak angular displacements of 1°, 0.05% velocity regulation could be claimed for all speeds greater than 3000 rpm; however, should the peak angular dis-

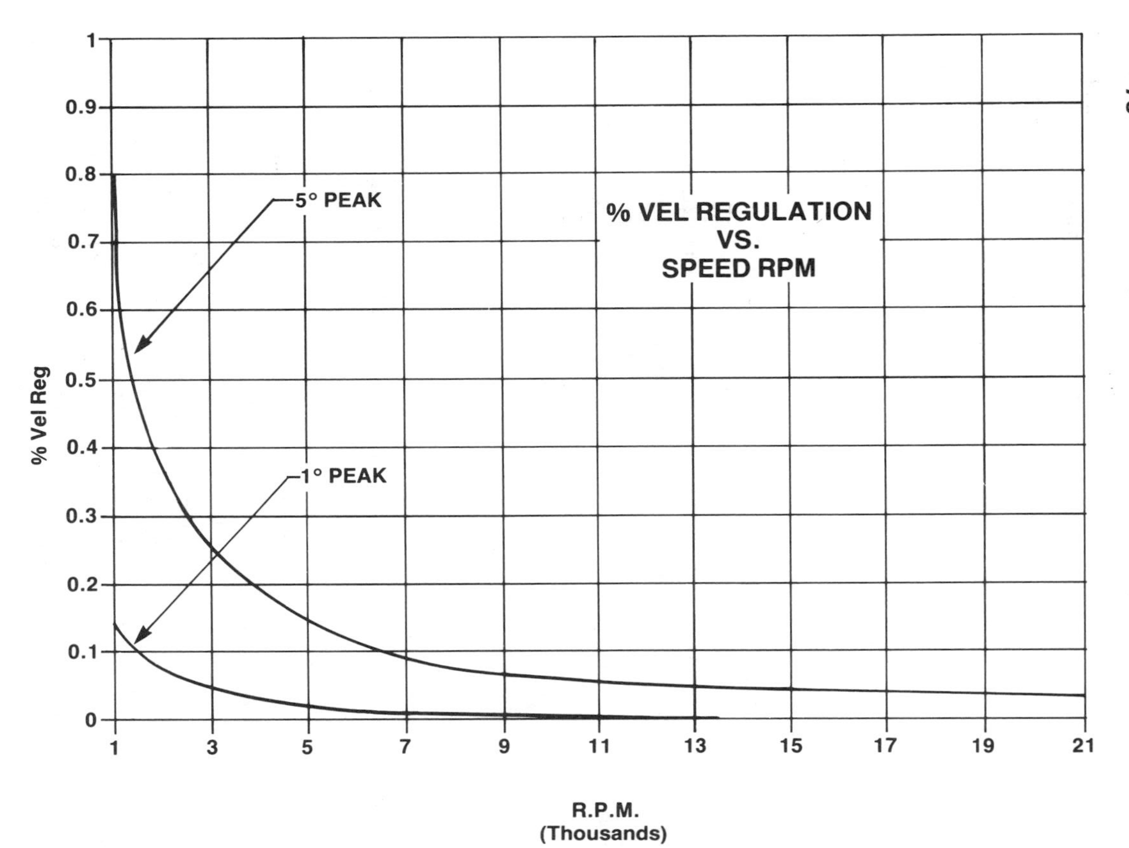

Figure 7.15 Percent velocity regulation versus speed rpm.

placements increase to 5°, then 0.05% velocity regulation is only obtainable at speeds greater than 14,000 rpm. Figure 7.15 illustrates that, for a typical open-loop scanning system operating at speeds above 3000 rpm, the system designer can expect variance in velocity ranging from less than 0.05% to as high as 0.25%. In conclusion, if system speed regulation requirements must be guaranteed to be less than 0.05% (500 parts per million), the closed-loop configuration using position feedback must be incorporated.

Another important reason for using hysteresis synchronous motors in precision air bearing scanners is the near absence of rotor eddy currents when the rotor is synchronized with the rotating stator field. These currents can produce increased I^2R rotor losses that can cause rotor/shaft distortions due to generated temperature gradients. The primary source of rotor eddy current losses is from the spurious flux changes that occur as the rotor passes the stator slots. These parasitic losses are often referred to as *slot effect losses* and can be very significant at high speeds, rendering the device very inefficient as is often noted in archaic conventional designs. Careful stator design can minimize these losses to the extent that the primary source of rotor/shaft heating is through the air bearing gap from the stator. The stator losses are further diverted away from the bearing/rotor system by water or air cooling the stator housing to minimize the rotor/shaft temperature rise.

Brushless DC Motor

The brushless DC motor is well suited for speeds ranging from near 0 rpm to as high as 30,000 rpm. These motors exhibit the same characteristic as brush commutating types and can therefore be used in the same applications. They are also suited for velocity and position servo applications since they have a near ideal linear transfer characteristic. The elimination of brushes and commutating bars provides reduced electromagnetic interference, higher operating speeds and reliability, and no brush material debris from brush wear. The commutating switching function is accomplished by using magnetic or optical rotor position switches that control the electronic commutating logic switching sequence. In actual operation, a DC current is applied to the stator windings which generates a magnetic field that attracts the permanent magnets of the rotor, causing rotation. As the rotor magnetic field aligns with the stator field, the field currents are switched, thereby rotating the stator field and the rotor magnets follow accordingly. The rotor will continue to accelerate until the back electromotive force (BEMF) generates a voltage equal to the stator supply voltage minus the DC winding resistance voltage drop. At this point, the rotor speed reaches an equilibrium level as determined by the BEMF motor constant. The open-loop speed stability and regulation under controlled supply and temperature conditions is usually 1 to 5%, so the device is typically used with closed-loop feedback control. In the closed-loop mode of operation, short-term speed stability of 1 ppm is obtainable. However, on a long-term basis, the stability and accuracy is as good as the reference crystal, which is readily specified at 50 ppm or less.

The brushless DC rotors most often used for air bearing scanners consist of rare-earth/cobalt permanent magnets that are contained with a rigid ring or cup for the outer rotor configurations, but are usually glued (epoxied) in place with the inner rotor configurations. The inner rotor configuration is generally used at the higher speeds because of the lower centrifugal forces resulting from a reduced rotor diameter. However, due to the elastic characteristics of epoxy and other adhesives, and the need for stable and reliable precision balancing, the operating speed of these precision spinner rotors is conservatively restricted to less than 15,000 rpm for precision polygonal scanners.

7.3.5 Dynamic Control Requirements

The basic control requirements for a precision polygonal spinner are to provide synchronization and velocity control for precise pixel registration. To this end, the principles of feedback control are utilized for synchronization/velocity and position (shaft angle) control.

Velocity control of the hysteresis synchronous motor is intrinsic to its design; that is, the long-term speed is as accurate as the applied frequency. With reference to Figure 7.14, the stator and rotor fields rotate together (at an integer submultiple of the applied frequency) with the rotor lagging by the torque angle α and with possible modulations of $\pm\beta$, as was previously discussed. The control systems' task is to fix the rotor vector position, and therefore eliminate hunting and other speed variances. To implement position control, the shaft position is measured with a shaft-mounted incremental encoder. The encoder pulses are frequency- and phase-compared with the reference frequency using a frequency/phase comparator, as shown in Figure 7.16. Since the motor is at synchronous speed, the tachometer frequency will equal the reference frequency, and the frequency/phase comparator will operate in the phase comparator mode. In this mode of operation, the output of the phase comparator is an analog voltage proportional to the phase difference between f_T and f_R. At zero frequency and phase differences, the two signals are edge locked, and the phase comparator output voltage is zero. Should the shaft advance or retard for any reason, the phase comparator error voltage will be a proportional measure of the phase difference within ± 360 electrical degrees maximum of the reference frequency.

With reference to Figure 7.16, the phase comparator error is processed through a proportional-integral-derivative (PID) controller compensation scheme and onto the control input of the phase modulator. The phase-error-corrected phase modulator output frequency, f_M, is applied to the motor, thereby completing the position control loop back through the encoder to the frequency/phase comparator.

The open-loop DC gain of the system is primarily determined by the product

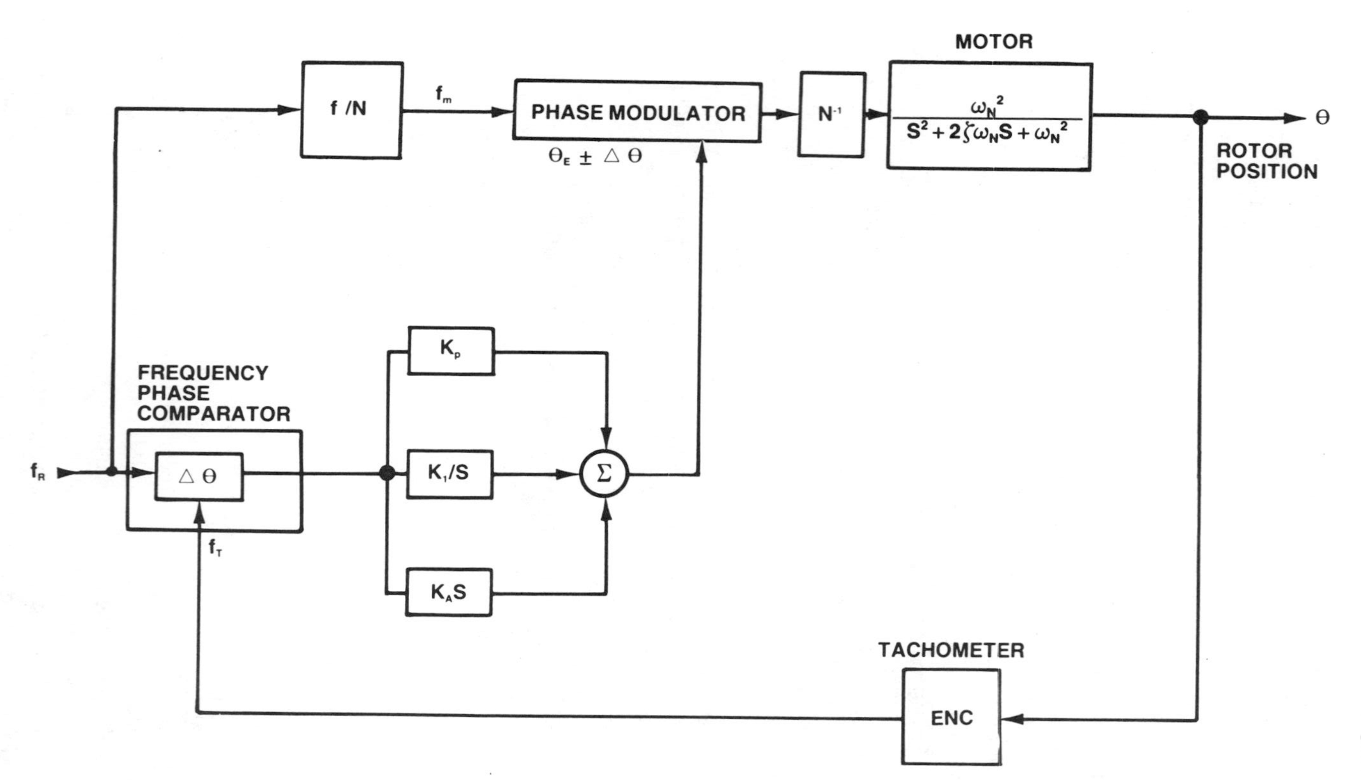

Figure 7.16 Position control block diagram (hysteresis synchronous motor).

of the encoder, frequency/phase comparator, integrator, and phase modulator gains. The high DC gain of the integrator (100 dB) reduces the phase error between f_R and f_T to zero, resulting in near perfect synchronization. The differentiation gain constant provides sufficient damping to eliminate ''hunting'' and improve the overall dynamic performance and speed regulation to less than 1 ppm.

The velocity control of a brushless DC motor differs from that of hysteresis synchronous in that the brushless motor speed is a function of applied motor voltage, as opposed to the frequency/phase as the driving function of the latter. The same principles of feedback velocity/position control are utilized in a similar fashion and are depicted in Figure 7.17. The elements within the closed-loop block diagram are essentially the same, with the exception of the motor transfer function and the addition of a DC power amplifier. For simplification purposes, the commutating and pulsewidth modulation circuits have been omitted, but will be covered in a later discussion. As before, to implement position control, the shaft frequency and position are measured with a shaft-mounted incremental encoder. The encoder pulses are frequency/phase compared with the frequency phase comparator, which generates an error signal that is processed through a PID controller compensator, similar to that used in Figure 7.16. As the system is turned on, the error signal is power-amplified, causing the motor to accelerate to a speed at which f_T exceeds (overshoots) f_R. At this point, the error signal reverses polarity at the motor (through the PID differentiator) reducing the speed until f_T equals f_R. Ultimately, a point of equilibrium is reached at which time the frequency/phase comparator error voltage is zero, and the integrator output voltage regulates the speed of the motor. Furthermore, the high DC gain of the integrator maintains a zero phase difference between f_T and f_R, resulting in edge lock synchronization.

7.4 SCANNER SUBSYSTEMS DESIGN

7.4.1 Host/Subsystem Interface

For long-term operation, the basic scanner subsystem requires the installation of two reliable support subsystems consisting of the air bearing gas supply and a continuous-duty vacuum pump. Other interface requirements include an electric power source and a means for communicating the synchronization and control signals to and from the controller. The host systems' optical elements and the polygon must also be mechanically and optically interfaced.

Figure 7.18 shows the typical host system/subsystem interface requirements for an air bearing scanning system. Note that a 135 psig pressure source is applied through a check valve, reserve accumulator tank, and on through a 0.5-micron

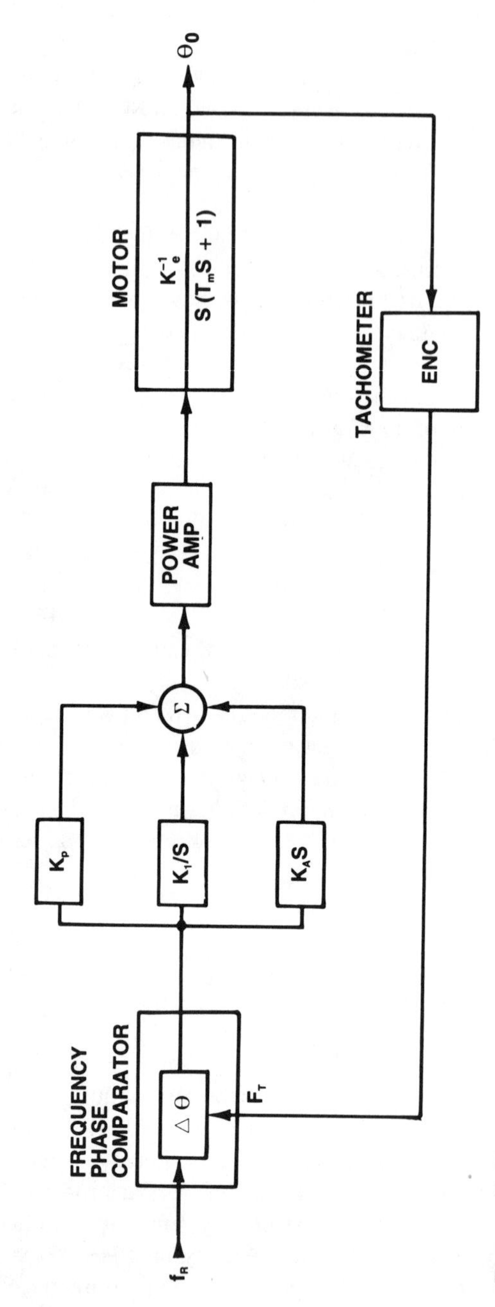

Figure 7.17 Position control block diagram (brushless DC motor).

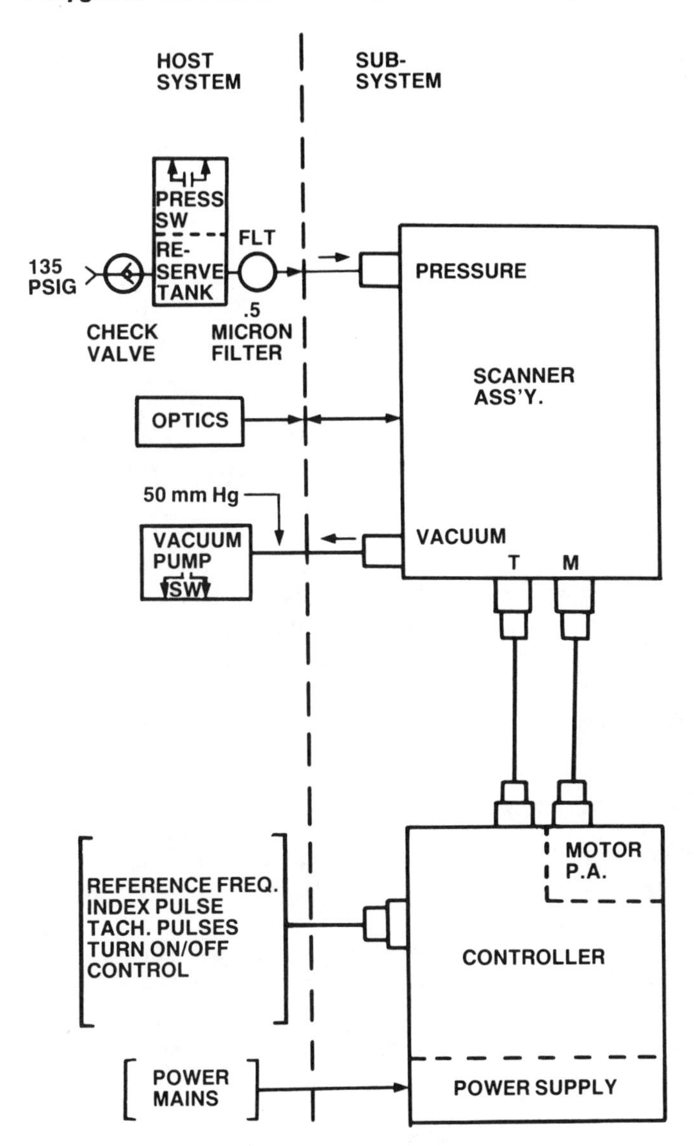

Figure 7.18 Typical host/subsystem interface.

filter to the pressure input port of the scanner assembly. The check valve is necessary to prevent the reserve tank from dumping air back into the supply in the event of a supply pressure drop. A large 10-gal reserve accumulator tank provides sufficient pressure and airflow to the scanner assembly in the event of a pump failure or pressure reduction below 130 psig. The supply pump is equipped with a switch (not shown) that disables the scanner assembly, enabling the spinner to come to rest before the reserve tank is discharged. The accumulator is also equipped with a similar pressure switch for the same reason. The 0.5-micron filter (internal in some scanner assemblies) prevents particle contamination of the air bearing, necessary for obvious reasons. A vacuum pump that maintains less than 50 mmHg pressure is shown connected to the vacuum port of the scanner. The vacuum pump is used to evacuate the air surrounding the polygon, thereby reducing the windage and motor power requirements. Additionally, the vacuum port provides a return path (not shown) for the bearing airflow which is vented at the polygon. A great deal of emphasis is placed on the quality, reliability, and cleanliness of these pneumatic support systems. Special emphasis is placed on the quality of material and installation of the tubing and fittings that connect the reserve tank to the high-pressure port. A failure of this line would result in an instant loss of pressure and possible catastrophic failure of the air bearing.

The speed control, regulation, and synchronization of the scanner assembly are the primary functions of the controller. The host system must provide the primary power, reference frequency, and control signals as well as receive the scanner positioning pulses for further processing of the pixel clock and frame synchronization frequencies. The power for the subsystem is provided through a separate connector at the controller. All other electrical interfacing functions are connected to the controller interface connector as shown in Figure 7.18.

7.4.2 Pneumatic and Vacuum Requirements

The following subsystem pneumatic and vacuum requirements are taken from the operation and maintenance manual for a typical scanner subsystem, similar to the film recorder previously described.

During in-house testing over several years, it has been concluded that standard industrial-grade equipment for these systems is quite satisfactory if care is given to selecting minimum maintenance versions of filter and rotary components. The bearing air supply train should optimally consist of the following series-connected components:

1. A continuous-receiver-rated, two-stage compressor and 20-gal receiver capable of delivering 2.0 scfm at 145 psig. Pressure switch range approximately 150 to 160 psig.
2. An automatic drain oil/water trap.
3. A self-renewing type air dryer, 0°C dew point.
4. An oil-vapor-removing filter with automatic drain.
5. An oil-odor-removing filter with automatic drain (0.5 micron).
6. A 250-psig pressure regulator set to 135 psig.

7. A high-quality, in-line, low-differential-pressure check valve.
8. A diaphragm or bellows-type pressure-sensing switch with low current contacts. Pressure switch setting is 130 psig. Tee switch into line.
9. A reserve air receiver, approximately 10-gal capacity.
10. A duplicate pressure switch to the above, series connected. Tee switch into line. This switch should be provided with a separate set of contacts to pilot-start the vacuum supply system.
11. A ¼-in minimum-inner-diameter, armored, high-pressure air line with swaged fittings leading to the scanner assembly.

The vacuum supply must be timed to start operation after the gas bearing pressure is applied. It should also be the first of the two support subsystems to be shut off when scanner operation is discontinued. The supply can consist of a single-stage, belt or direct-drive laboratory-type vacuum pump of 5 scfm minimum capacity, connected to the scanner vacuum port by any length of vacuum tubing, so long as the scanner vacuum port pressure does not exceed 50 mmHg. The line should be equipped with a vacuum-sensing interlock switch to verify correct pump operation, and must also be fitted with a low-differential, pop-off check valve to allow release of bearing exhaust air if the vacuum supply is not operating.

It is imperative that the gas supply be fed to the scanner assembly before the motor power is applied. Application of motor power to the scanner without prior gas flow into the scanner assembly can cause extensive damage to the gas bearing surfaces. A gas supply "interlock" controls the motor power application to prevent operation without gas flow.

7.4.3 Controller Functions

With reference to Figure 7.19, the scanner assembly is shown mechanically coupled to the mirror and three encoder tracks designated T_1, T_2, and T_3. These encoder tracks generate the motor commutating signals, the tachometer data pulses, and the one pulse/revolution index, respectively.

The brushless DC motor used in this system is driven by a three-phase electronic inverter which derives its pulse timing from the commutating logic, line receivers, encoder circuits, and a three-phase, 2-pulse/rev optical position encoder, T_1. A 20-kHz fixed-frequency carrier signal is utilized to pulsewidth-modulate the three-phase commutating encoder signals in linear proportion to the DC input of the pulsewidth modulator (PWM). The rotational direction of the motor field is controlled by the polarity of the PWM control input. At speeds from motor start-up to 400 rpm, the rotational direction is dictated by the direction relay. For speeds greater than 400 rpm, the servo loop is automatically closed, and the input reference frequency controls the speed of the scanner rotor. This method of motor control simulates a near-ideal DC motor transfer characteristic. The

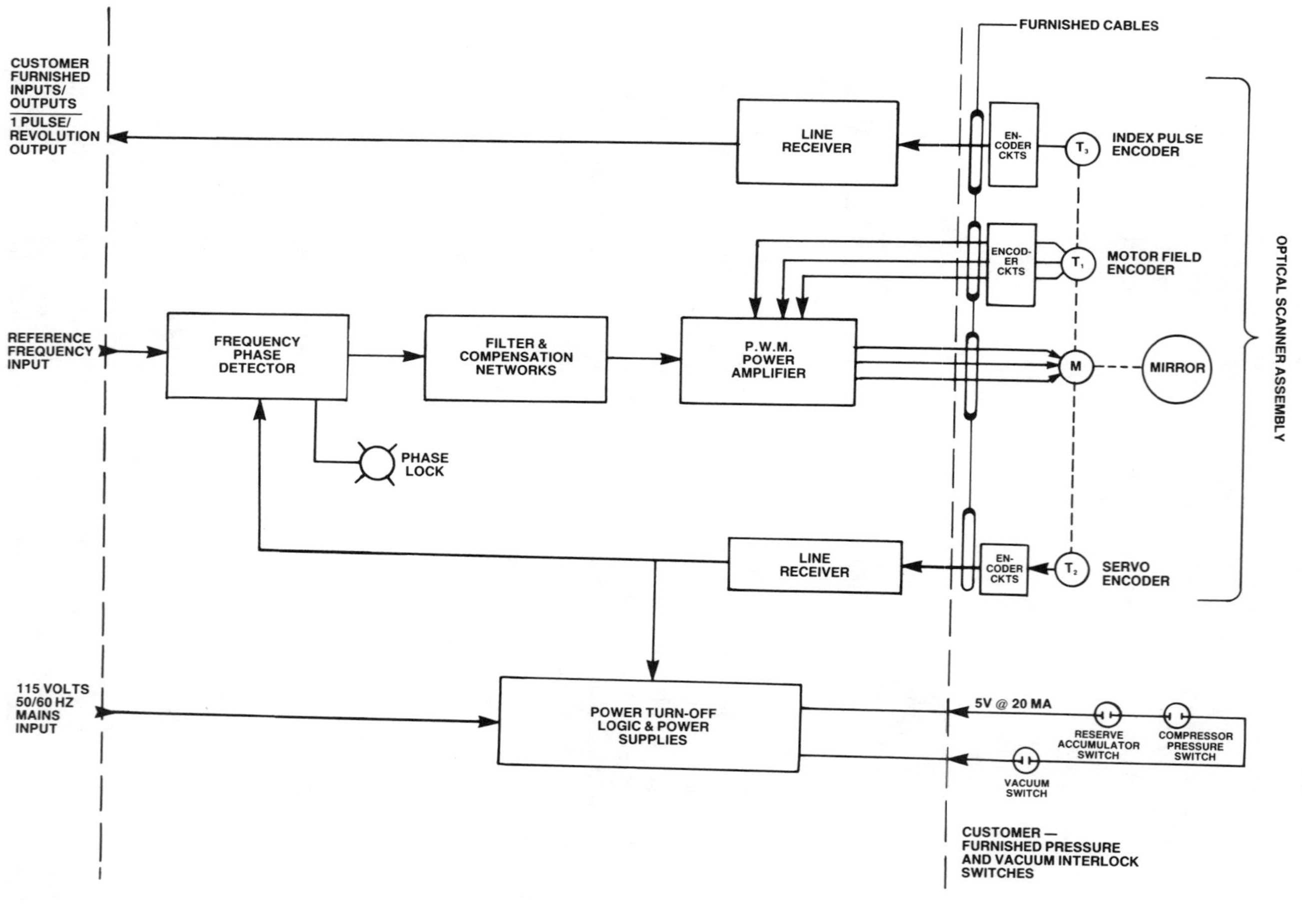

Figure 7.19 Controller functional block diagram.

6000 pulses/rev incremental encoder signal (T_2) is processed through the encoder circuits, line receiver, and the phase locked loop (PLL) frequency-phase detector. Similarly, the reference frequency signal from the host system is applied to the PLL frequency-phase detector for frequency-phase comparison with the signal from T_2. The PLL frequency-phase detector derives a shaft position signal by comparing the leading edges of the reference and encoder signals. This analog signal is processed through a low-pass carrier filter, lag-lead stabilization and damping network, and then to the input of the PWM. The PWM will advance or retard the motor torque angle to maintain a phase lock condition between the encoder signal T_2 and the reference frequency signal. An analog signal proportional to the reference frequency is derived using a D/A converter. This signal is used as a DC offset to automatically bias the motor operating torque throughout the operating speed range, and is also used to provide a loss-of-reference system turn-off signal.

7.4.4 Controller Ancillary Functions

The ancillary controller functions include overspeed and underspeed detection, motor power turn-on and turn-off, dynamic braking, monitor outputs, and system status indicators. The overspeed/underspeed detection circuits monitor the commutating encoder frequency and develop a latched inhibit signal, turning off the motor DC power supply if the scanner speed equals or is greater than 7000 rpm. The motor field DC power supply will remain off until the motor/reset front panel switch is reset. These detector circuits also energize the direction input command relay if the motor speed is equal to or less than 400 rpm. This is necessary to establish the desired direction of rotation when the system is first turned on.

The motor power supply is a current-limited, voltage-regulated DC supply. This voltage is processed through the shaft brake relay contacts to power the motor via the three-phase power inverter. The shaft brake relay is deenergized with a loss of reference frequency, loss of compressor or reserve tank via pressure, loss of vacuum, scanner overspeed, or if the front panel motor power/reset switch is turned off. The deenergized shaft brake relay contacts provide a return path to ground for the motor-BEMF-generated current, braking the motor to a complete stop at any one of its four encoder positions.

7.4.5 Operation

Remote speed selection is accomplished by providing a reference frequency to the TTL differential interface input connector, using the following relation to determine the frequency:

f_r (Hz) = 100 × RPM

The turn-on sequence begins with the motor on/off reset switch in the off position (refer to Figure 7.20). Power is applied by operating the system power mains switch. The system status lamps will indicate as follows:

Air interlock	ON, indicating the accumulator switch is closed at 130 psig
Motor overspeed	ON, indicating the scanner motor DC supply is latched off
Input reference	ON, indicating the presence of the input reference frequency
Servo sync	OFF, indicating an out-of-sync condition

Turning the scanner motor on by operating the motor on switch will turn off the overspeed indicator lamp, and turn on the motor on indicator lamp. The scanner will accelerate to operating speed in approximately 20 s. The servo sync lamp will then indicate when the shaft encoder frequency is properly synchronized with the input reference frequency.

The system performance can be observed by connecting an oscilloscope to the monitor output signal connectors located on the front panel.

7.5 TYPICAL DESIGNS

7.5.1 Film Recording System

The polygonal scanner shown in Figure 7.6 is an oversimplification of the manufactured scanner subassembly used in the film recording system. Figure 7.21

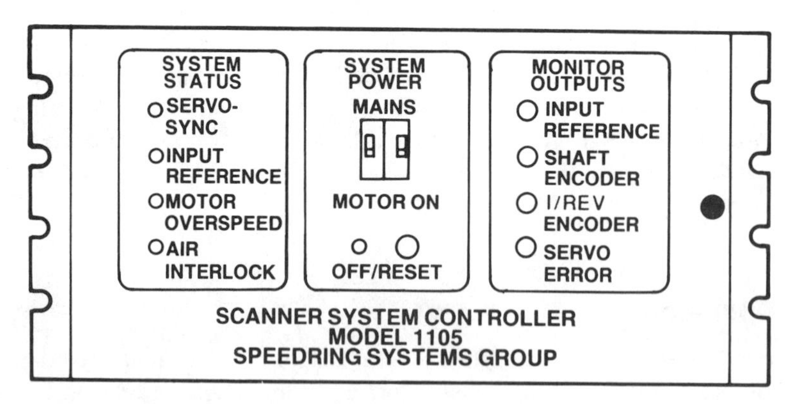

Figure 7.20 Controller front panel.

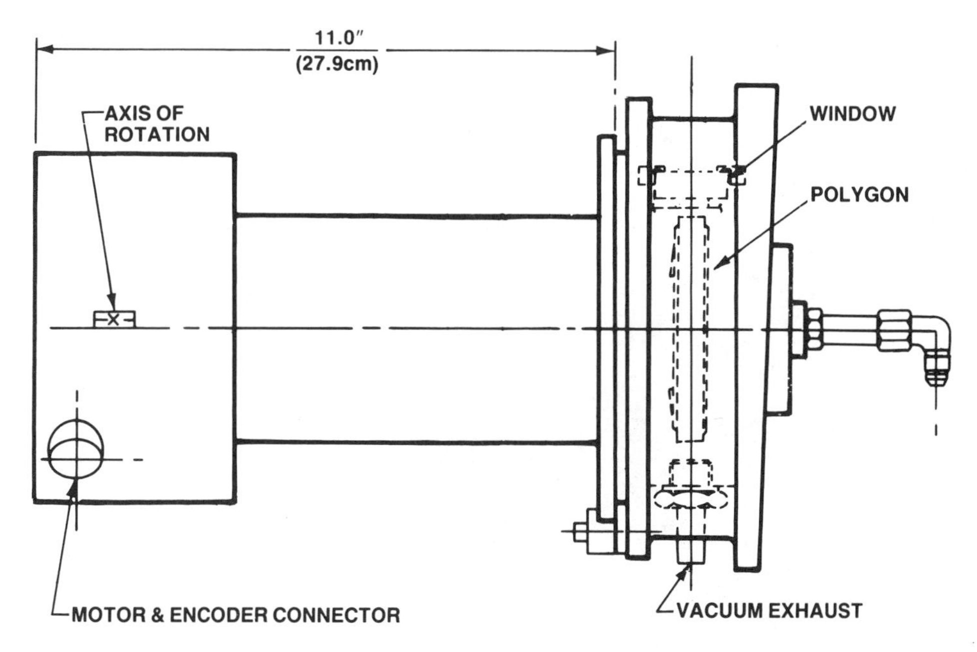

Figure 7.21 Film recording scanner.

depicts the actual scanner assembly used for this system and is representative of a typical design for a high-performance, low-speed scanner subsystem.

The scanner assembly contains a 12-faceted truncated pyramidal scan mirror that is driven with a brushless DC motor (not shown) as required to perform the optical beam scanning function. The rotating mirror and its drive motor rotor and a precision optical tachometer (enclosed at the opposite end of the scan mirror) are supported radially and axially by externally pressurized gas bearings. Inlet pressure and vacuum fittings are provided to interface with the pneumatic and vacuum support subsystems. An electrical connector is also provided at the encoder end of the assembly to interface the motor and encoder electronics with the electronic controller assembly, which is usually remotely located. Note that the scanner facets and the vacuum enclosure/mount are tilted to conform with the optical (f-θ lens and laser) interface requirements, as specified by the system designer.

Due to the high operating efficiencies of the low-speed brushless DC motor, no special cooling jackets or cooling fins are required with this design, and low-temperature-rise operation is ensured by thermal conduction of the stator power losses through the mounting flange.

Table 7.9 is a summary of the film recorder subsystem performance characteristics.

Table 7.9 Film Recorder Subsystem Characteristics

Characteristics	Value or type
Power input requirements	105–132 VAC, 55–65 Hz, 15-A, 2-cycle inrush current, 0.4-A operating current
Gas bearing supply	Customer-furnished pump and filters: 135 ± 5 psig dry atmospheric air, 0°C dew point, filtered to 0.5 micron, <2 scfm
Vacuum environment	Customer-furnished pump: 50 mmHg; maximum pressure, 5 actual cfm min
Controller input speed command and sync requirements	600 kHz, TTL twisted pair, differential line receiver, type DS8820
Interface input format	TTL twisted pair into differential line receiver, type, DS8820
Interface output	1 pulse/rev, 6.5 ± 2μs, TTL transmitter, type DS8830
Controller front panel	10 TTL loads: input reference frequency, shaft encoder, 1/rev encoder, and servo error (analog)
Controller panel indicators	Servo sync, input reference, motor overspeed, and air interlock
System performance (maximum specified values)	Time to speed, 0 to 6000 rpm: <20 s Run-down time, 6000 to 150 rpm: <30 s Speed control (long term) <50 ppm Scanner speed jitter: <10 ppm (1 rev) <0.05% (>2 rev)
Scan mirror (maximum specified values)	Number of facets: 12 Tip-to-tip facet diameter 4.045 in Height: 0.500 in Flatness: λ/20 at 488 nm Surface quality: 60–40 (MIL-O-13830) Reflectance: >89% at 514.5 nm; all facets within 2% at 5° Facet-to-facet angle: 30° ± 10 arc sec Absolute pyramid angle: 5 ± 0.006° Pyramid angle matching: ± 0.40 arc s
Operating ambient	Temperature: 85 ± 5°F Relative humidity: 30–60%

7.5.2 Reticle Writer Scanner

A high-speed, laser beam ''reticle writer'' scanner assembly, used by the semiconductor industry to design and manufacture integrated circuit chips, is shown in Figure 7.22. The system enables reticles to be engraved to accuracies of less than 150 nm.

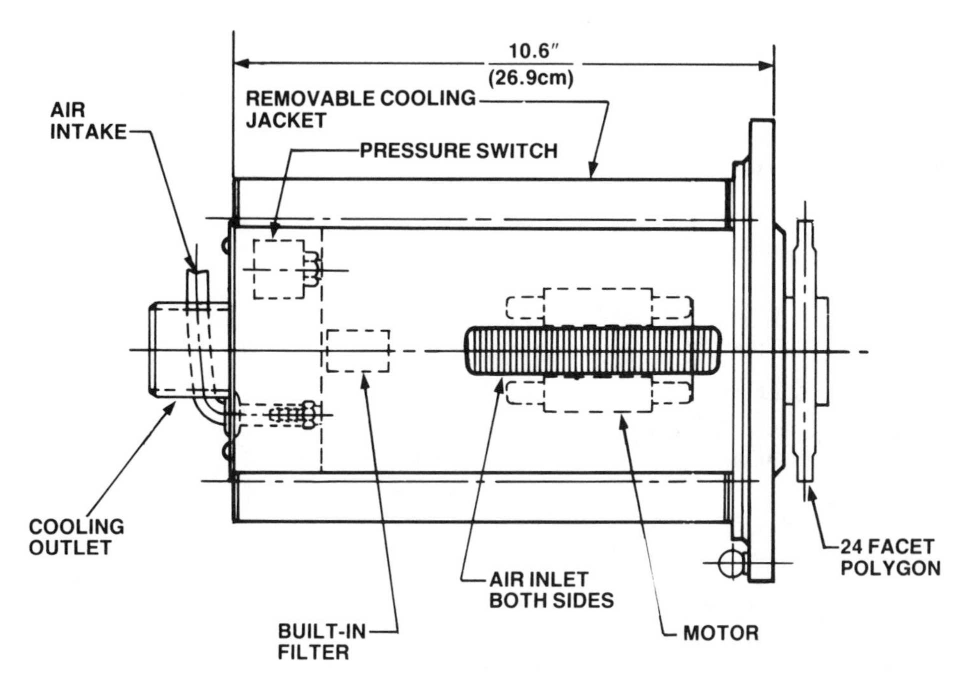

Figure 7.22 Reticle writer scanner.

The scanner assembly contains a 24-faceted cantilevered normal polygon scan mirror which is driven by a hysteresis synchronous motor. The polygon/motor rotor/shaft assembly is radially and axially supported by externally pressurized gas bearings (not shown), and included in this assembly is a built-in air filter and pressure switch. The assembly operates at a user-selectable speed between 16,000 to 20,000 rpm; however, due to the stator power losses at higher operating speeds, a forced-air cooling jacket is provided that surrounds the housing to maintain a low-temperature rise. A remotely located blower, coupled to the air outlet fitting, removes the heated air from the jacket that was drawn into the system at the air inlets.

Table 7.10 is a summary of the reticle writer subsystem performance characteristics.

7.5.3 Flight Simulator Scanner

Flight simulators provide an enhanced visual simulation of aircraft flight for purposes of training commercial airline and military pilots. A high-speed scanner is used to laser-illuminate a terrain model through an optical probe. The probe

Table 7.10 Reticle Writer Subsystem Characteristics

Characteristics	Value or type
Power input requirements	220 VAC ± 5%, 50–60 Hz
Gas bearing supply	Customer-furnished pump and filters: 90 ± 5 psig dry atmospheric air, 0°C dew point, filtered to 0.5 micron, <1 scfm
Controller input speed command and sync requirements	8 kHz/20 krpm, ECL twisted pair, differential receiver
Interface input format	ECL twisted pair into differential line receiver; ECL input feedback facet sync pulses
Controller front panel	10 TTL loads: input reference frequency
Controller panel indicators	Servo sync, input reference, and air interlock
System performance	Time to speed, 0 to 20,000 rpm <60 s
	Run-down time, 20,000 to 0 rpm: <60 s
	Speed control (long term) <50 ppm
	Scanner speed jitter: <10 ppm (1 rev) <0.05% (>2 rev)
Scan mirror (maximum specified values)	Number of facets: 24
	Radius to facet flats, 61.2 + 0.5, −0.0 mm
	Height: 7.62 ± 0.5 mm
Scan mirror (continued)	Flatness: λ/20 at 663 nm
	Surface: 60–40 (MIL-O-13830)
	Reflectance: >84% at 364 nm; all facets within 2%
	Facet-to-facet angle: 15°, 10 arc sec
	Absolute pyramid angle: <1 arc sec
	Pyramid angle matching: ±0.50 arc sec
Operating ambient	Temperature: case temperature rise <5°C
	Relative humidity: 30–60%

is servo-position- and velocity-controlled by the pilot to simulate the flight of the aircraft. The probe also collects the model's terrain image which is then reproduced and viewed by the pilot on a high-resolution display in real time.

A high-speed flight simulator line scanner is shown in Figures 7.23 and 7.24. Note that the polygon is integral to the shaft assembly, which is supported radially and axially on hydrostatic air bearings. A membrane filter is internally provided as well as a high-speed magnetic encoder, which is mounted on the opposite end of the shaft. Two removable cooling jackets are provided for water cooling the assembly to miminize the effects of temperature rise on the shaft/polygon geometry. The scanner is powered by a hysteresis synchronous motor, and is operated at 76,725 RPM and provides a 30,690 lines per second scan rate.

The scanner/controller subsystem is characterized by Table 7.11.

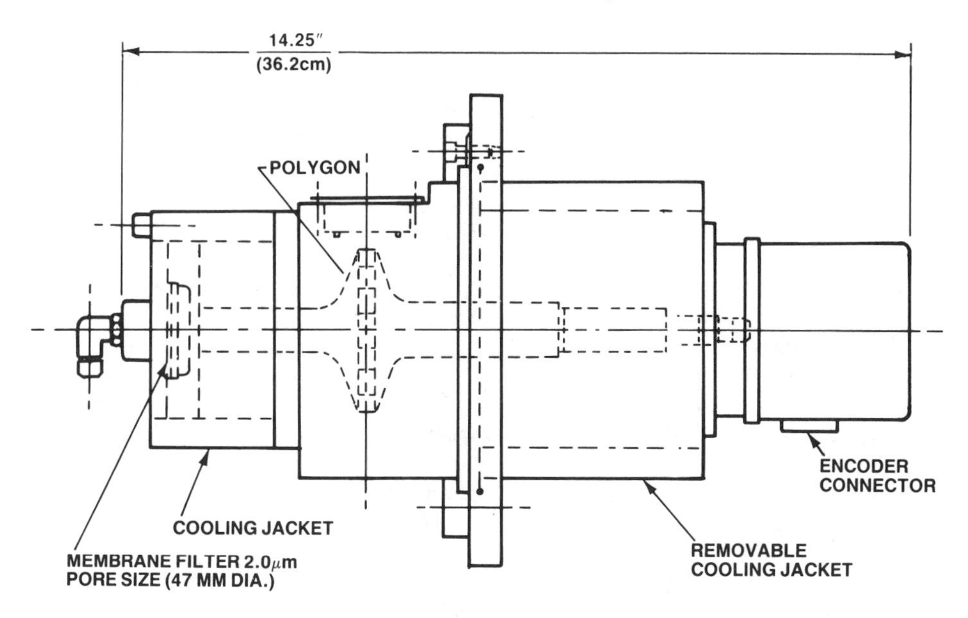

Figure 7.23 Flight simulator scanner.

Figure 7.24 Flight simulator scanner.

Table 7.11 Flight Simulator Subsystem Characteristics

Characteristics	Value or type
Power input requirements	120 VAC, 60 Hz, 1 phase
Gas bearing supply	Customer-furnished pump and filters: 160 ± 5 psig dry atmospheric air, 15.5–37.8°C, approx. 0.944 L/s (120 scfh), minimum dew point 0°C through a 500-nm in-line filter
Vacuum environment	50 mmHg maximum pressure, 2.36 L/s (5 scfm) minimum pumping capacity
Controller input speed command and sync requirements	491 kHz, ECL twisted pair, differential receiver
System performance	Time to speed, 0 to 76,725 rpm <30 s Run-down time, 76,725 to 0 rpm: <180 s Scanner speed: 76,725 rpm servo accuracy: ±2 ns (facet output) at speed
Scan mirror (maximum specified values)	Number of facets: 24 Radius-to-facet flats, 61.2 + 0.5, −0.0 mm Height: 7.62 ± 0.5 mm Flatness: $\lambda/20$ at 663 nm Surface quality: 60–40 (MIL-O-13830) Reflectance: >84% at 364 nm; all facets within 1% Facet-to-facet angle: 15°, <1 arc sec Absolute pyramid angle: <5 arc sec
Magnetic shaft encoder scanner output	1 pulse/rev ECL logic, differential line driver 384 pulses/rev (ECL)
Water cooling input	22.8 ± 2.8°C 0.044 to 0.063 L/s (0.7 to 1.0 gpm)

7.5.4 Hemispherical Gas Bearing Scanner

Lithographic scanning applications such as digitizing and printing newspapers and photographs, as well as engraving copper plates and printed circuit boards, utilize the cost-effective hemispherical gas bearing design shown in Figure 7.25. A similar scanner, currently in production, is shown in Figure 7.26.

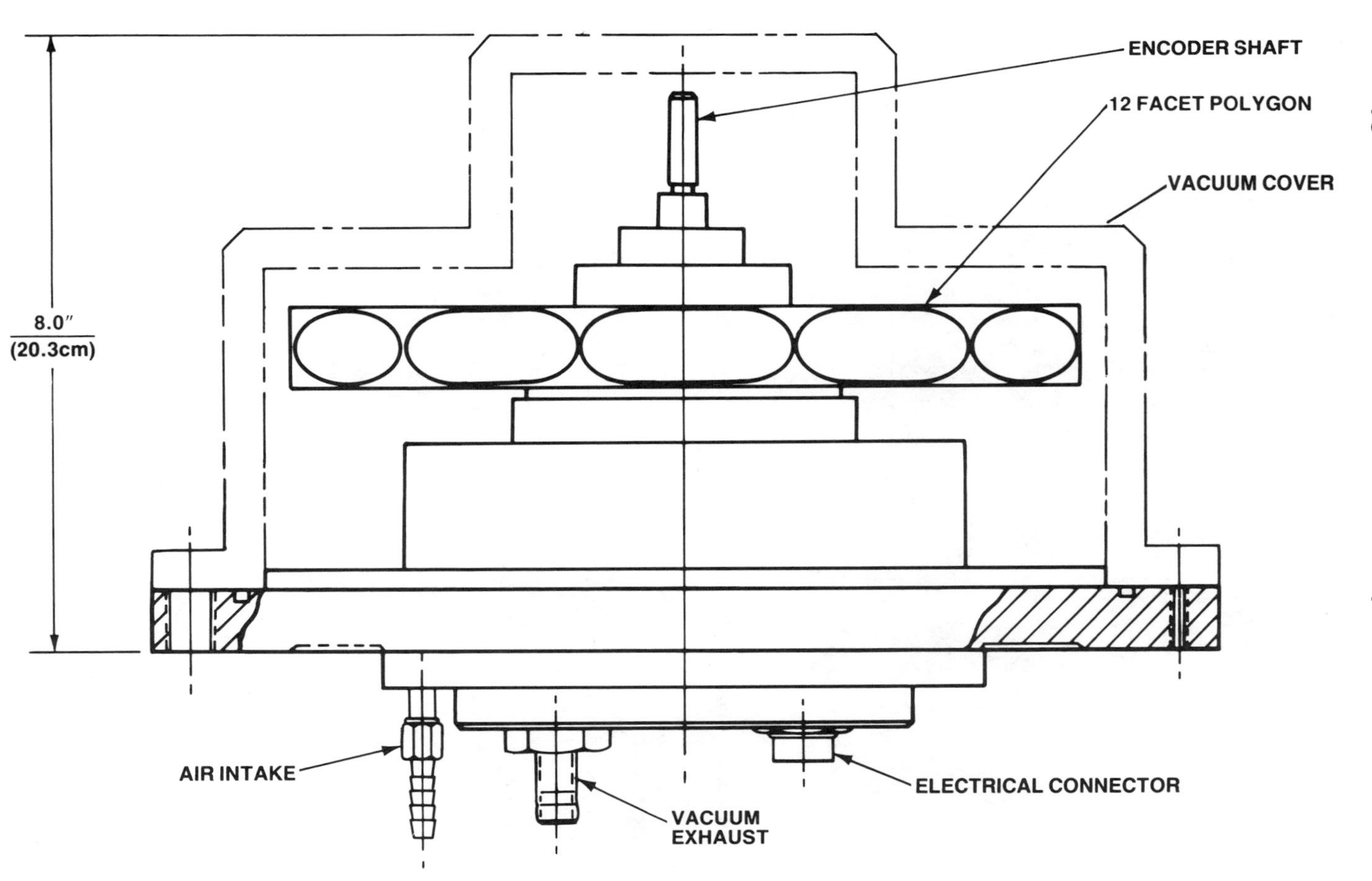

Figure 7.25 Hemispherical gas bearing scanner.

Figure 7.26 Lithographic scanner assembly.

REFERENCES

1. R. W. Landee, D. C. Davis, and A. P. Albrecht, *Electronic Designer Handbook*, 1st ed., McGraw-Hill, New York, 1957.
2. T. C. Lloyd, *Electric Motors and Their Applications*, Wiley, New York, 1969.
3. Speedring Systems Group, Rochester Hills, MI, Technical Bulletins, *Ultra Precise Bearing for High Speed Use* 102-1, *Gas Bearing Design Considerations* 102:2, *Gas Bearings Manufacturing Considerations 102-3, and Rotating Mirror Scanners* 101-1, 101-2, and 101-3.
4. H. C. Rotors, *The Hysteresis Motor-Advances Which Permit Economical Fractional Horsepower Ratings*, AIEE Technical Paper 47-218, 1947.

8

Windage of Rotating Polygons

Joseph Shepherd*

Ferranti International, Silverknowes, Edinburgh, Scotland

8.1 INTRODUCTION

Rotary mirror scanners are often purchased as "standard" motorized spindles to whose shafts a range of mirror-faceted polygons can be fixed. This approach usually offers low unit costs and has proved satisfactory for many optical scanning systems, but for more exacting requirements these units can be wasteful of energy and space and can have inadequate performance.

Many advanced systems require features such as more compact designs, higher scan speeds, more complex rotors, smaller power consumption, and smaller clearances from other optics. To incorporate such features successfully, the design of the rotating scanner must be integrated with that of the complete optical scanning system. This can be achieved only if the performance of any shape of rotor, rotating at any likely speed in any surroundings, can be predicted. Such information will allow an optical system designer to consider more fully the effects of possible optical scanning variables and to optimize the design for a particular requirement, reducing the design risk and increasing market potential.

It is the purpose of this chapter to look at the windage of rotating polygons, an aspect of performance normally difficult to estimate, and to describe a proven method of predicting the windage for any rotating scanner.

* *Current affiliation:* Consultant, Edinburgh, Scotland

8.2 WINDAGE

8.2.1 What Is Windage?

Windage (refer to Figure 8.1) is the rotational equivalent of linear wind resistance or drag. It is the torque (T_w) required to overcome the counterrotational forces induced on any rotor spinning in a fluid medium. It is often useful to consider the windage power (P_w), which is the product of the windage torque and the rotational speed (ω).

The rotational power losses in a rotary mirror scanner are those associated with the drag of the spin axis bearings and the windage of the rotating polygon. In addition, there will be further losses in the drive motor. Of these losses, windage is not only often the largest component but the most difficult to predict.

There are many advantages to be gained from minimizing windage losses, such as reduced total power input, smaller motor and drive supply requirements, and lower running temperatures, which in turn have their own clearly beneficial consequences. It is also important to be able to predict the magnitude of the windage power loss, especially when considering the motor specification for a proposed scanner design.

Very little information relating directly to the analysis of windage for rotating polygons was found in a literature search, with the exception of a paper by Hayosh [1]. The majority of work published which is in any way related deals either with rotating disks [2] or rotating cylinders [3]. It has therefore been necessary to base this chapter mainly on unpublished work carried out at Ferranti. This has been concerned with polygonal rotors having facets parallel to the spin axis. However, the principal and method described to evaluate windage can also be

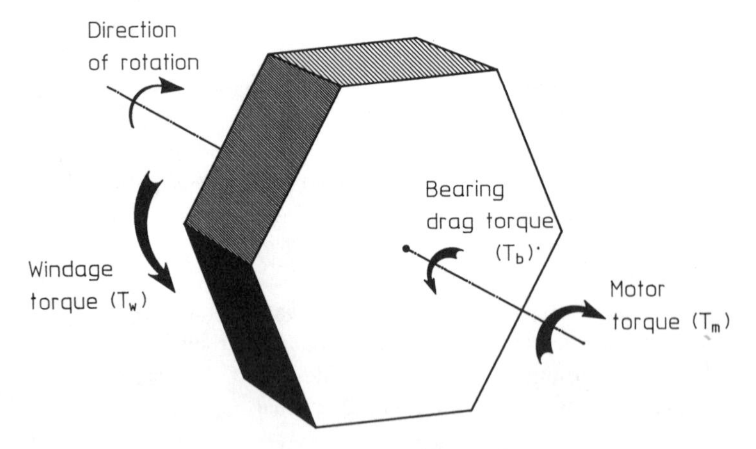

Figure 8.1 Torques acting on a polygon spinner in steady rotation.

applied to pyramidal and irregular polygons or to angled single mirror scanners. Constant rotational speed has been assumed when analysing windage. If transient information is required it can be derived from constant speed results.

8.2.2 Sources of Windage

For any object moving through a fluid there are two possible sources of drag: skin friction and form drag. Skin friction is the result of shear forces induced by the velocity gradient in the fluid boundary layer surrounding the rotor. The fluid in contact with the rotor has the same velocity as the rotor's periphery and the fluid in contact with the housing, or at a reasonable distance from the rotor, is stationary. Factors affecting the magnitude of the skin friction drag torque acting on a given rotor include fluid medium viscosity, velocity gradient, and flow regime (i.e., laminar or turbulent).

Form drag torques are induced as a result of boundary layer separation and wake generation in the fluid surrounding a noncylindrical rotor. As a result of energy dissipated by the highly turbulent motion in the wake, the pressure there is reduced and the form drag (or pressure drag) on the rotor is thus increased. The magnitude of this type of drag torque depends very much on the size of the wakes, which in turn depends upon the position of boundary layer separation. For polygonal rotors the magnitude of this type of drag torque is affected by factors such as number of facets, facet size, and fluid density.

The overall windage torque acting on a given polygonal rotor normally contains elements of both skin friction and form drag, although often one or the other will predominate. This is one of the reasons for the windage characteristics being so difficult to predict. The influence of both sources of drag is illustrated in Figure 8.2, which shows schematically the relationship between windage torque, rotational speed, and number of facets. At high speeds the flow separates at the leading edge of the facets for all practical numbers of sides, and so form drag will predominate over most of the range. The magnitude of the drag decreases as the number of facets increases since the total volume of wake decreases. There is a slight increase as the polygon tends to a cylinder since the flow will no longer separate and skin friction starts to take effect. At lower speeds skin friction starts to dominate much sooner (at around eight-sided polygons) since the regions of flow separation begin to decrease, reducing the volume of wake but increasing the surface area which generates skin friction drag.

8.3 DIRECT MEASUREMENT OF WINDAGE

There are two basic methods for making direct measurements of scanner windage torque: reaction torque testing and acceleration torque testing. Some of the relative merits of each are discussed below.

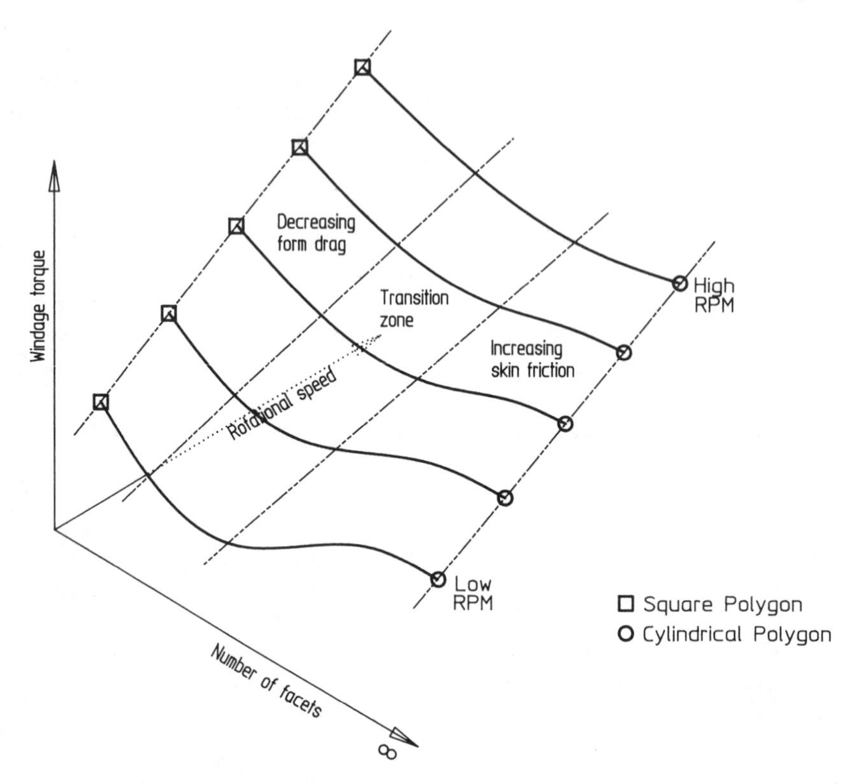

Figure 8.2 Typical relationship between torque, rotational speed, and number of facets.

8.3.1 Reaction Torque Testing

The reaction torque from a scanner is the torque produced on the stationary components (stator) as a reaction to the torques acting on the rotating components (rotor). By measuring the value of this torque, it is possible to deduce a value for the windage drag torque acting on the rotor.

The reaction torque may be measured by isolating the scanner using some type of low friction bearing (e.g., a rotary air bearing) and then measuring the torque as the scanner is run up to the speed of interest then down again. The rotor speed can either be measured independently or calculated by a process of integration on the relationship

$$T = I\alpha \qquad (1)$$

where T = reaction torque
I = polar moment of inertia of all rotating parts
α = angular acceleration of the rotor

Hence, the torque/speed relationship may be determined.

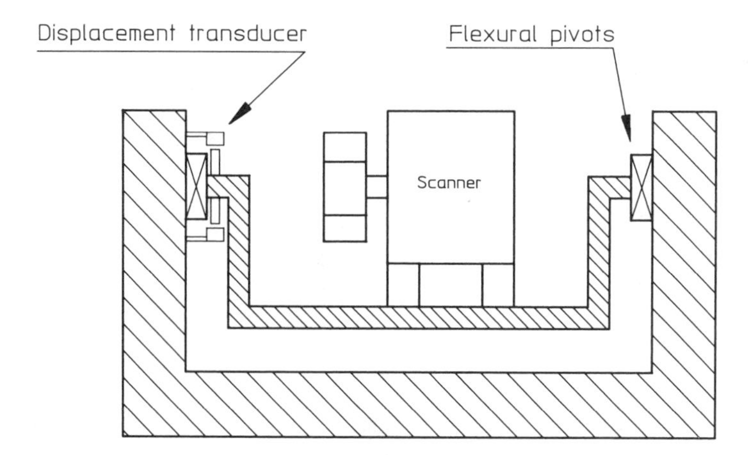

Figure 8.3 Schematic representation of a torque tester.

For the direct measurements of scanner windage torque reported in this chapter a reaction torque test method was used. A schematic diagram of the apparatus is shown in Figure 8.3. This shows a scanner held in a cradle supported between two cross-flexural pivots, which perform as low-friction rotational bearings. Any reaction torque from the scanner deflects the pivots, and the resulting movement (proportional to the magnitude of the torque) is measured using a high-sensitivity angular displacement transducer, which has been calibrated using known values of applied torque.

For this type of testing there are two distinct configurations possible for the scanner. The rotor can either spin in the open, as in Figure 8.4a, or in a housing which is part of the stator fixture (Figure 8.4b). The form of the results from the two situations have important differences, which may be explained by considering the separate torque components acting on different parts of the scanner.

For both cases, when the motor is running at synchronous speed, the motor torque (T_m) must be equal to the torque on the rotor (see Figure 8.1):

$$T_m = T_w + T_b \tag{2}$$

From Figure 8.4a it may be seen that the torque on the stator (T_s) is given by

$$T_s = T_m - T_b \tag{3}$$

as the reaction to the windage torque does not act on the stator.

Substituting (2) in (3) gives

$$T_s = T_w \tag{4}$$

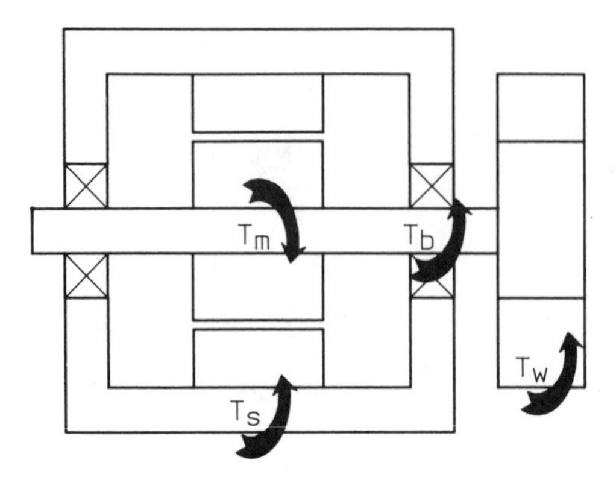

Figure 8.4a Torques acting on a scanner with the rotor running in the open.

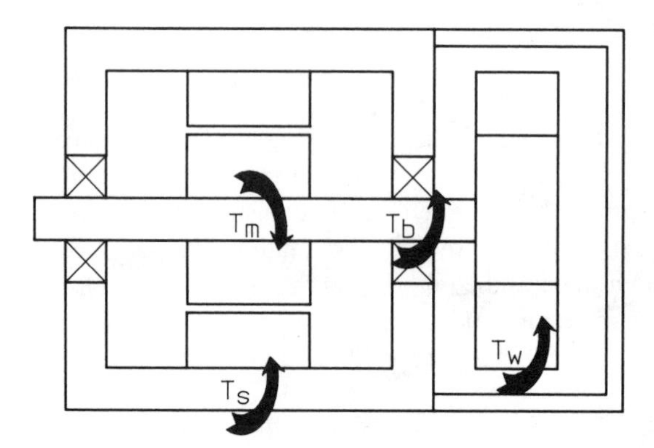

Figure 8.4b Torques acting on a scanner with the rotor running in an enclosed housing.

Therefore, for the case of the rotor running at constant speed in the open (or in any enclosure isolated from the stator), the torque on the stator, and hence the measured reaction torque, is equal to the windage torque at that speed. If the motor is then switched off (i.e., $T_m = 0$ in equation (3)) a negative reaction torque is produced due solely to the bearing drag. A typical example of the resulting torque/speed relationship for this situation is shown in Figure 8.5a. This approach was used in the direct measurement of windage to verify the results obtained from the model-testing experiments discussed later in this chapter.

If the scanner design has the rotor enclosed within a housing which is an integral part of the stator, as shown in Figure 8.4b, then the reaction torque acting on the stator is modified. The windage torque is transferred to the stator via the housing, giving an expression for the torque on the stator as

$$T_s = T_m - T_b - T_w \tag{5}$$

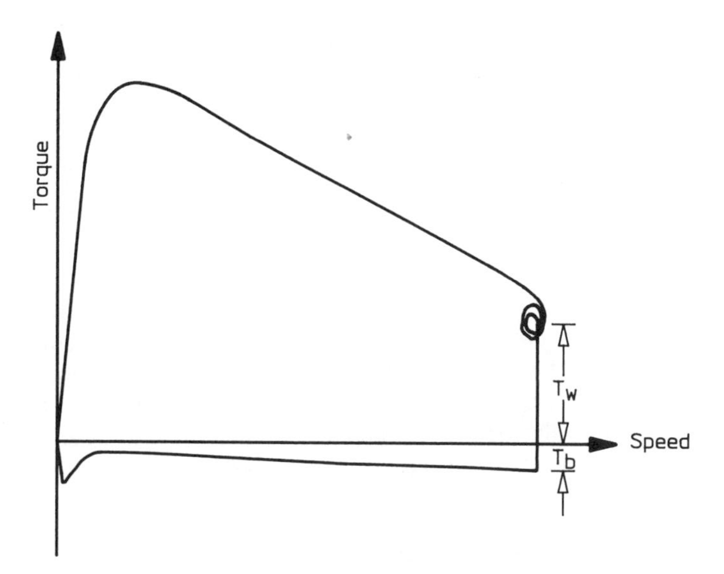

Figure 8.5a Torque-versus-speed curve for scanner with the rotor running in the open.

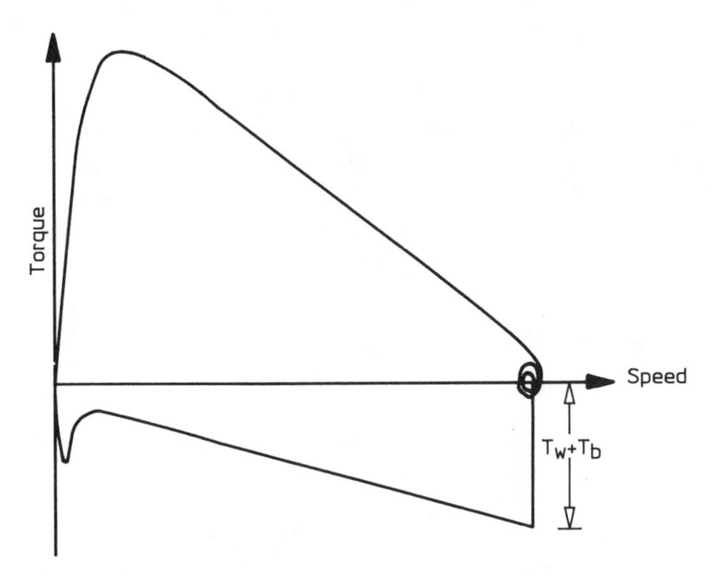

Figure 8.5b Torque-versus-speed curve for scanner with the rotor running in an enclosed housing.

Equation (2) still holds and substituting this in equation (5) gives

$$T_s = 0 \qquad (6)$$

Therefore, for the scanner running at constant speed the reaction torque is zero. When the motor is then switched off, a negative reaction torque is produced equal to the sum of the windage and bearing drag (i.e., substituting $T_m = 0$ into equation (5)). The type of torque speed relationship produced by this method is shown in Figure 8.5b, and it is obviously not possible to distinguish between bearing and windage drag contributions. Therefore, an estimate of the bearing drag is necessary in order to determine the magnitude of the windage component.

Reaction torque testing is a very useful technique, but can have limitations when applied to large bulky scanners due mainly to problems involved with providing a suitable rotationally frictionless support. An alternative approach, which is not subject to such problems, is acceleration torque testing.

8.3.2 Acceleration Torque Testing

This method of windage torque measurement may be applied to any rotor system where an accurate measurement of speed can be made (normally using some type of noncontact method, e.g., capacitive or inductive probe). In the case of laser beam scanners the obvious method of speed measurement is an optical one, as there are accurate reflective surfaces available on the rotor. A high-speed counter measures the interval between successive laser scans, allowing the rotor speed and acceleration to be calculated.

If the rotor polar moment of inertia is known, either from calculation or direct measurement, the torque may be calculated using the relationship in equation (1). The torque/speed relationship obtained using this method is of the same form as that shown in Figure 8.5b, so it only contains information on the sum of windage and bearing drag torques. This is not normally a serious problem since bearing drag characteristics tend to be well documented.

8.4 METHODS OF PREDICTING WINDAGE

8.4.1 Predicting from Previous Results

Basing predictions for windage on results for previous scanners is most useful if the speed, size, and configuration of the proposed design are reasonably close to the previous designs measured. Often the use of a "windage equation," derived from dimensional analysis or some other source, may be helpful. For example, such an equation would be useful when considering a new design which is of a similar configuration but a different scale from an existing one.

This method is obviously limited by the range of previous scanner designs for which data are available. Even where a new design appears to be very similar to an existing one, it is known that seemingly unimportant differences can lead to significant changes in the windage characteristics. It is preferable to have results for a series of scanners of similar design to the proposed one so that the likely effect of differences in configuration may be assessed. It would not normally be practical to build full-speed prototype scanners in order to establish the effect of the various parameters in windage prediction. This would involve time-consuming and expensive experimentation with the variables involved.

8.4.2 Derivation of Windage Equations

Many attempts have been made to develop an equation-based solution for windage calculation, both analytically and empirically. Empirical solutions tend to apply to a limited range of conditions and require a great deal of experimentation in their construction. More general equations have been derived analytically, notably that presented by Hayosh [1]. The derivation of such an equation is based upon certain assumptions about the nature of the mechanics of windage drag, and the following example shows how an equation of the type proposed by Hayosh may be arrived at.

This procedure is based upon the definition for the drag force on a body placed in a fluid flow of linear velocity U and density ρ with a frontal area A_f. The drag force F_d acting on the body in the direction of the fluid flow is given by the expression

$$F_d = Cd\left(\frac{\rho U^2}{2}\right)A_f \tag{7}$$

where

$\rho U^2/2$ = fluid dynamic pressure [4]
Cd = drag coefficient (a function of Reynolds number Re and boundary conditions)

Now, referring to Figure 8.6, this expression is related to the facet-projected area for an N-sided polygon. Taking the polygon average radius R_a (i.e., $R_t + R_f)/2$) as the position at which the force on each facet acts, then the torque acting on a polygon of facet projected area

$$A_f = L(R_t - R_f)$$

and rotating at speed ω will be

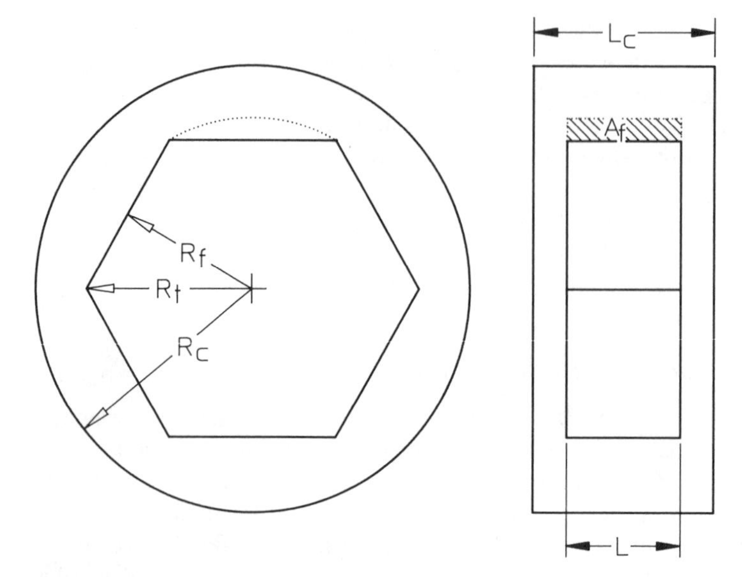

Figure 8.6 Dimensions of a typical polygon.

$$Tw = Cd\left(\frac{\rho N\omega^2}{2}\right)R_a^{\,3}A_f \tag{8}$$

Here Cd is the drag coefficient for the complete rotor; Hayosh suggests that a value of 0.5 should give satisfactory results for most cases and that Cd is dependent upon Reynolds number and boundary conditions, but that it is independent of features affecting the scanner design (e.g., N, L, and R_t).

However, experimental results indicate that the relationship between Cd and Re does in fact vary with changes in the rotor design parameters as well as the housing design parameters. This limits the equation's usefulness for windage prediction unless the dependence of Cd on Re, for the particular situation of interest, has been established either from existing results or using the methods outlined in this chapter.

It is clear that the assumptions used for this derivation have been inadequate to describe the flow regime at the polygon periphery. For example, equating the facet movement through the fluid environment to the movement of a body in a uniform fluid flow is unjustified, since each facet is moving into the wake left by the preceding one, which will normally be a highly disturbed flow.

Equations of this type and those derived empirically can prove useful in certain situations, but the complicated nature of the windage mechanisms for polygonal scanners makes a comprehensive solution extremely difficult to achieve.

8.4.3 Windage Modeling

Model testing is an alternative to these two approaches, which is strongly recommended as a relatively simple and inexpensive method where it is desired to

1. Build up a databank of information on a wide range of polygons and surroundings.
2. Minimize the windage of a particular design or to take account of complex configurations.

Model testing also allows additional rotor features, such as bearing parts or balancing rings, to be taken into account, which in certain cases can be significant.

8.5 MODEL TESTING

8.5.1 Dimensional Analysis

Model testing, which is one of the most useful techniques of fluid mechanics, is based on the principles of dimensional analysis as discussed by Taylor [5]. The basis for this approach is the assumption that if a list can be made of all the variables that describe some fluid mechanical situation, be it a ship moving through water or a steel ball falling through glycerine, the complexity of their interrelationships can be reduced by gathering the variables together into "dimensionless groups." These are composite variables organized in such a way that the dimensions (in this case mass, length, and time) of their constituent parts all cancel out.

Considering the case of a polygon rotating in some sort of housing immersed in a fluid, and making the fundamental assumption that the fluid is incompressible the windage will depend on

1. Rotor size
2. Rotor shape
3. Speed of rotation
4. Number of facets
5. Housing shape and size
6. Viscosity of the fluid in which it rotates
7. Density of the fluid in which it rotates, which can be characterized by a relationship of the form

$$T_w = f(\rho, \omega, r, \mu) \tag{9}$$

where T_w = windage torque
ρ = fluid density
r = a representative dimension (e.g., the polygon tip radius)

μ = dynamic viscosity of the fluid
ω = angular frequency of rotation of the rotor

Dimensional analysis allows this to be rearranged into the form

$$\frac{T_w}{\rho\omega^2 r^5/2} = f\left(\frac{\rho\omega r^2}{\mu}\right) \tag{10}$$

where a torque coefficient (Cm) analogous to the drag coefficient Cd is defined to be

$$Cm = \frac{T_w}{\rho\omega^2 r^5/2}$$

and where the Reynolds number is

$$Re = \frac{\rho\omega r^2}{\mu}$$

Therefore,

$$T_w = Cm\left(\frac{\rho\omega^2 r^5}{2}\right) \tag{11}$$

and

$$Cm = f(Re)$$

which is a relationship between two dimensionless quantities that can be plotted on one graph. This is a powerful simplification, which from dimensional analyses, is valid for all geometrically similar models. Note that for each configuration, (see Section 8.9.1) the length variable r in effect stands for all the lengths in the system, both rotor and housing.

What has been established here is that if a model of a proposed spinner is made, and tested in such a way that the Reynolds number is the same as that for the "real" rotor, the calculated value of torque coefficient will be the same as that in the "real" case. The assumption of an incompressible fluid is valid for windage, as the gas pressures generated are low and compressibility effects can be neglected.

8.5.2 Model Testing of Spinners

The main benefit of model testing in liquid, as applied to high-speed spinners, is that for the same Reynolds number a model can be run at much lower speeds (e.g., a spinner operating at 9000 rpm in 1 atm of air at 25°C can be modeled

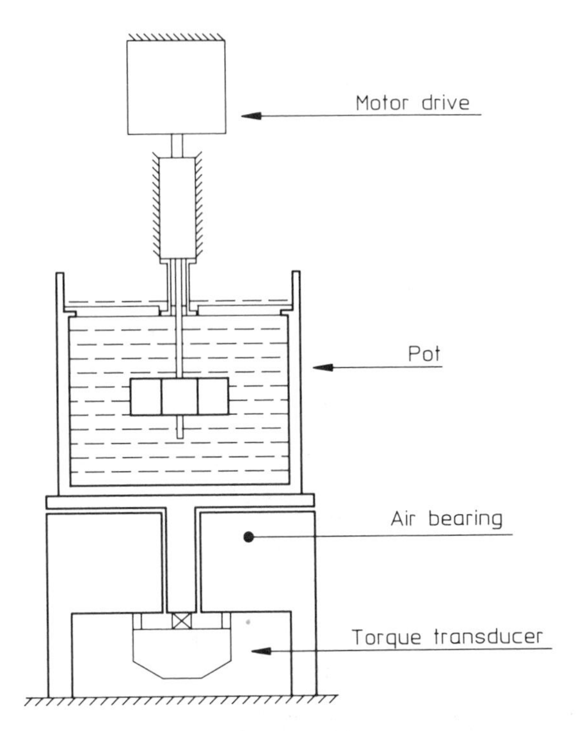

Figure 8.7 Schematic diagram of windage-modeling apparatus.

at 550 rpm in water at 23°C). The resulting advantages simplify and speed up the gathering of information, as follows:

1. The reduced dynamic unbalance forces allow larger tolerances on parts and the use of longer reach spindles to hold the rotor, simplifying the test rig design.
2. The range of speeds required can be obtained without using special high-speed motors and drives.
3. The difficulties of instrumenting rotors running at high speeds in gas environments are overcome. Also the problem of differentiating between bearing and windage drag is avoided.
4. Rotor and housing configuration changes may be made with ease.

A model testing rig which has been used successfully is shown in Figure 8.7. This is fairly self-explanatory, but note that the windage reacts back on the housing (labeled Pot) and thence is applied directly to the torque transducer. The rig used has been linked to a microcomputer to automate the control of the motor speed and the collection and processing of data (i.e., measuring drag torque and

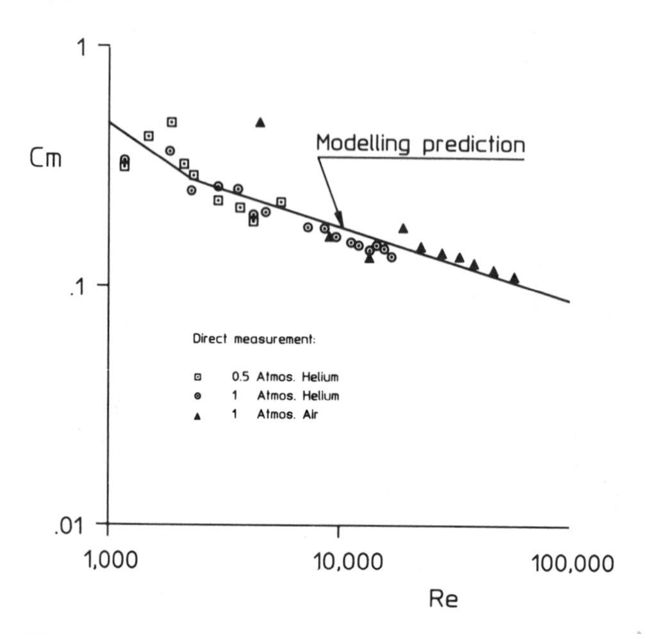

Figure 8.8 Comparison of model test with direct measurement on a type 203-001 scanner in a cylindrical housing with $R_c - R_t = 0.3R_t$ and $L_c = 3.2R_t$.

fluid temperature giving values for viscosity and density and then calculating and displaying graphically Reynolds numbers and torque coefficients), although this may also be achieved manually.

In order to ensure the validity of the model-testing approach, a number of direct measurements were made, at Ferranti, on "real" spinners using the method outlined in Section 8.3. These measurements were compared with predictions from model tests. Agreement was obtained to an accuracy of $\pm$ 12% in the majority of cases. Figure 8.8 shows an example of this using the rotor shown in Figure 8.9.

Figure 8.9 Type 203-001 scanner.

The use of model testing to predict whether the windage of any proposed rotary scanner will fall within acceptable limits can be done in two ways. The first to to model the rotor and housing as exactly as possible, and the second is to do a number of tests modeling different rotor and housing features, to establish a database from which to infer the likely windage of a new spinner and housing design.

The first approach is valid for one particular configuration and will give an accurate prediction of windage. However, a scanner design is rarely fixed at an early stage, and where there is likely to be an optimization process the second approach is probably the more useful, in that it predicts trends as well. The latter approach has been the basis of the information presented in this chapter. There is, unfortunately, not enough space to give more than a flavor of the subject and to point out a few trends that may be of interest to designers wishing to keep rotor windage to a minimum.

8.6 RESULTS OF MODEL TESTS PERFORMED ON ROTATING POLYGONS

These tests involved only regular polygonal rotors with between 4 and 12 facets, parallel to the spin axis, but since only the broad trends of the results can be given here, the information given may be applicable to other rotor types as well.

8.6.1 Shape of Curves of *Cm* against *Re*

Curves of *Cm* against *Re* were typically (although there were some exceptions) found to have the form

$$\ln(Cm) = \ln(A) - B\ln(Re) \tag{12}$$

where ln indicates the natural logarithm. The constant *B* was found to have values between 0 and 0.83, the lower values of *B* indicating a relatively more disturbed

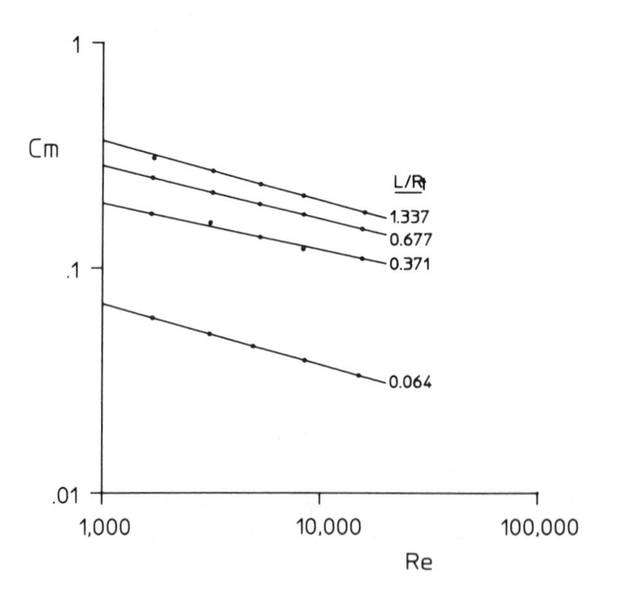

Figure 8.10 Typical model test results for a four-sided polygon in a cylindrical cavity with $R_c - R_t = 0.066R_t$ and $L_c = 2.7R_t$.

flow (i.e., windage increasing more rapidly with speed). This applies over the range of Reynolds numbers used in the tests (2000–20,000), and in those cases where high values of Re (up to 100,000) were tested the equation remains valid to $\pm 10\%$ approximately. Figure 8.10 shows an example, a test on a four-sided polygon rotating in a plain cylindrical cavity.

Equation (12) can be rearranged to bring it into a more useful form:

$$Cm = A\,Re^{-B} \tag{13}$$

Substituting for the dimensionless groups

$$\frac{T_w}{\rho\omega^2 r^5/2} = A\left(\frac{\rho\omega r^2}{\mu}\right)^{-B}$$

and rearranging, we obtain the windage torque

$$T_w = \frac{1}{2}A[\rho^{(1-B)}][\omega^{(2-B)}][r^{(5-2B)}]\mu^B \tag{14}$$

or the windage power

$$\begin{aligned} P_w &= \omega T_w \\ &= \frac{1}{2}A[\rho^{(1-B)}][\omega^{(3-B)}][r^{(5-2B)}]\mu^B \end{aligned} \tag{15}$$

It was found that A and B vary in a complicated manner with the dimensional parameters of the system. Table 8.1 gives examples of the values of A and B

Table 8.1 Values of the Parameters A and B for a Number of Different Polygons Rotating in Plain Cylindrical Housings

Number of facets	Length-to-radius ratio, L/R_t	Radial clearance, $R_c - R_t$	Curve parameter A	Curve parameter B
4	0.094	$0.066R_t$	0.46	0.272
4	0.371	$0.066R_t$	0.90	0.220
4	0.677	$0.066R_t$	1.62	0.255
4	1.337	$0.066R_t$	2.68	0.287
5	0.367	$0.062R_t$	0.83	0.228
6	0.374	$0.066R_t$	0.74	0.226
7	0.373	$0.065R_t$	0.75	0.251
8	0.425	$0.109R_t$	1.03	0.288
12	0.366	$0.066R_t$	3.3	0.454
∞ (Cylinder)	1.337	$0.026R_t$	252	0.829

for various rotors in cylindrical housings, but it should be noted that the values given will be completely upset by features causing gross disturbances to the fluid flow, such as window holes, obstructions, and baffles. Indeed, in one series of tests, on a five-sided polygon at $Re = 20{,}000$, a variation in windage of over 500% from the best to the worst case was obtained merely by changing the housing shape and by using side cheeks (disks attached to the end faces of the rotor) to smooth the flow. The results of these tests are summarized in Table 8.2. This shows that not only do the actual values of windage vary very widely for the same rotor at the same speed, but the slope of the graph Cm against Re is highly variable. This means that the higher the value of Reynolds number, the more the windage is affected by obstructions to the smooth flow of fluid around the rotor.

8.7 DISCUSSION

As a result of the research and experimental work that has been carried out on the subject of windage of rotating polygon scanners, an insight has been gained into the often complex nature of the windage characteristics. The following discussion is not intended to define any hard and fast rules for those concerned with this problem, but rather to indicate general trends and suggest some measures to take in order to keep windage to a minimum. The discussion is followed by an example of a typical scanner in order to demonstrate the potential effects of bad or good design features, from the point of view of windage, with a view to design optimization.

8.7.1 Important Parameters Affecting Windage

Size

Given a fixed scanning rate in facets per second, efforts should be made to reduce the facet size as much as possible. This is because, for any one configuration, windage is approximately proportional to size raised to the power 4.5 (see equation (14) and Table 8.1); that is, if a spinner is scaled down by 10%, windage is reduced by nearly 40%.

Housing Design

Housing design is very important, since seemingly small changes in the housing configuration often have a most significant effect on the rotor windage characteristics. Experimental results indicate that the optimum housing to minimize windage in any situation is a plain cylinder with an inside radius (R_c) of approximately 1.15 times the polygon tip radius (R_t), so an open housing with large clearance around the rotor does not necessarily give the minimum windage. A

Table 8.2 Change in Windage with the Changes to Rotor and Housing Illustrated in Figure 8.12 and Figure 8.13

Number of facets	Length-to-radius ratio, L/R_t	Radial clearance, $R_c - R_t$	Comments	Curve parameter		Windage at $Re = 20{,}000$
				A	B	
5	0.68	$0.023R_t$	Cylindrical housing	1.90	0.298	100 (nominal)
5	0.68	$0.023R_t$	Side cheeks	8.3	0.494	59
5	0.68	$0.038R_t$	Hole in side of housing	0.88	0.156	168
5	0.68	$0.369R_t$	Rectangular obstruction	0.66	0.110	204
5	0.68	$0.2.34R_t$	Radial baffle	0.70	0.070	318

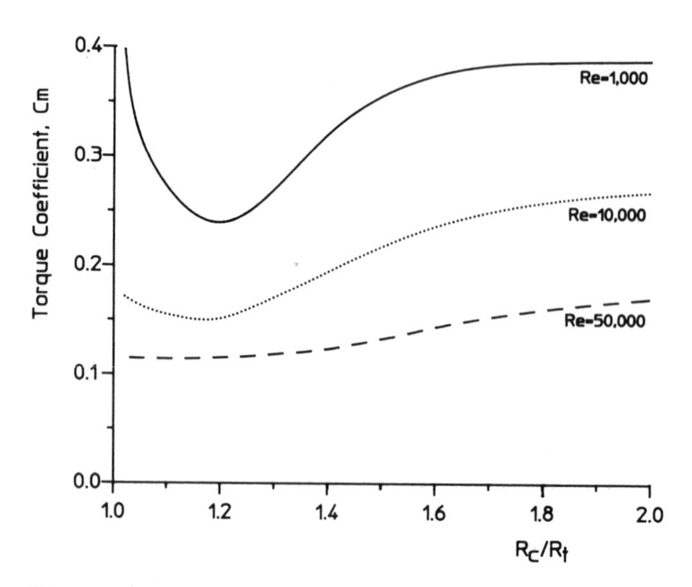

Figure 8.11 Typical relationship between torque coefficient and radial clearance for a six-sided polygon in a cylindrical housing with $L_c = 2.7R_t$ and $L = 1.3R_t$.

typical relationship between windage and radial clearance is shown in Figure 8.11. Real housings will normally depart from this ideal, however, and any additions to the housing in the form of window cutouts, subsidiary mirrors, or other obstructions (see Fig. 8.12) may cause large increases in windage. The potential severity of these effects may be alleviated by careful design: for example, keeping obstructions and baffles as far away from the rotor as possible, smoothing window cutouts, and keeping the cutouts as small as possible.

Atmosphere

The gas environment in which the rotor spins has an important effect on windage. The choice of gas may be largely a consequence of the structure of the scanner; consider three likely classes of seal between the polygon rotor and the outside world:

1. Dust seal only; ambient air (1 atm) must be used.
2. Polygon rotor running without a special sealed housing inside the machine, but with good sealing between the scanner and the outside world; this will allow the use of an inert-gas atmosphere, e.g., helium at 1 atm pressure.
3. Polygon rotor in separate sealed housing; a reduced atmosphere, e.g., ½ atm helium, can be used.

Class 1 is very severe from the windage point of view. In this case the designer has to be careful to take every step to reduce windage by smoothing the airflow and by making the polygon facets as small as possible. In classes 2 and 3, the

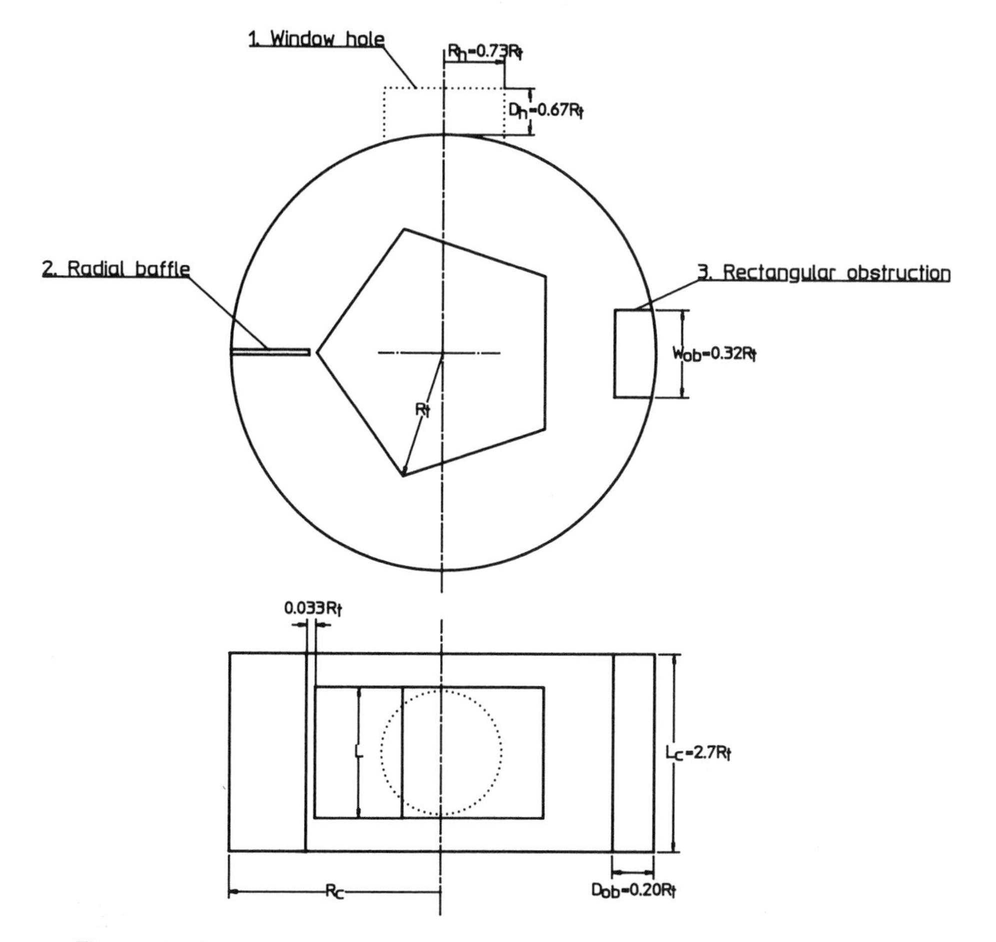

Figure 8.12 Important dimensions of polygon with a housing showing the features whose effects are analyzed in Table 8.2; i.e., a window hole, a radial baffle, and a rectangular obstruction.

use of helium, with its lower density, typically decreases the windage power by a factor of 4.5 for 1 atm pressure or 7.3 for ½ atm pressure.

Polygon Modifications

In certain cases significant reductions in windage may be achieved by modifying the polygonal rotor either by the use of side cheeks or by rounding off the corners (see Figures 8.13 and 8.14). Such measures tend to be most effective for polygons with large facet size, running at high speed and especially when obstructions to the flow are present in the housing. In such cases modifying the polygon can lead to reductions in windage power of over 50%. These modifications can prove to be optically restrictive, however, limiting their suitability in certain cases.

8.7.2 Examples of Design Optimization

In order to demonstrate clearly the possible improvements in design which may be achieved, consider the following case: a five-faceted polygon of diameter 34 mm and length 11.5 mm rotating at 50,000 rpm in 1 atm air at 25°C. Assume that the housing has a bad design feature in the form of an optical mirror positioned in such a way that it acts as a radial baffle (see Figure 8.12), giving a windage power loss for the scanner of 40 W. The use of side cheeks on the rotor, as illustrated in Figure 8.13, would achieve an immediate reduction in this power loss of over 40% to 23 W. If the housing was redesigned to remove the mirror and give a housing diameter of 19.5 mm, then, without any modifications to the polygon, the windage power would be 8.5 W. Using side cheeks would reduce this even further to only 3.6 W, less than one-tenth of the worst case. Any further significant reductions could only be achieved by a change in the operating en-

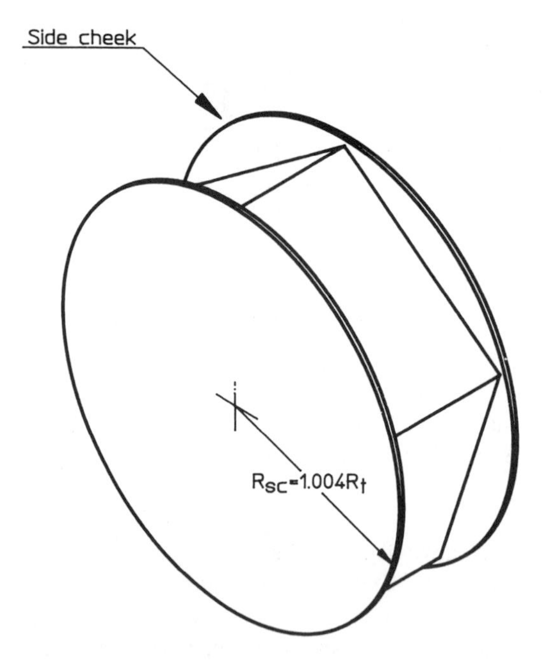

Figure 8.13 A polygon rotor with side cheeks.

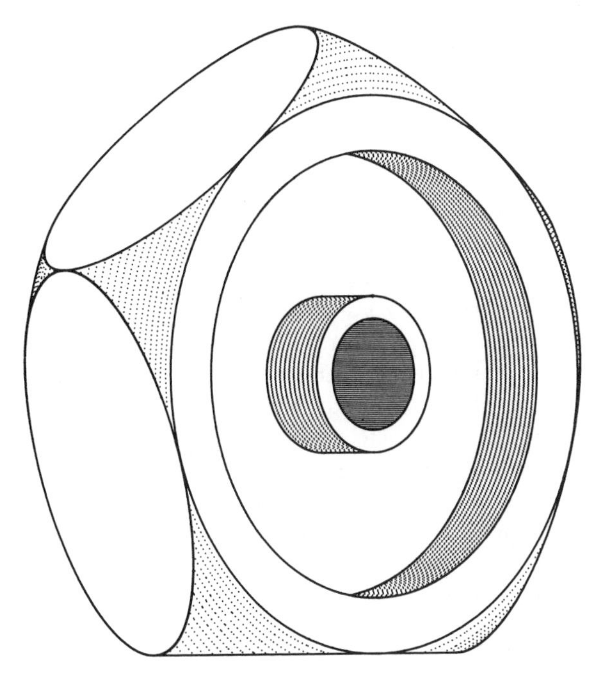

Figure 8.14 Schematic diagram of a polygon with rounded facets.

vironment, which may be limited by housing design constraints, as discussed in Section 8.7.1. If it were possible, and 1 atm He used, the windage power would be as little as 1.75 W without any polygon modifications. Using a $\frac{1}{2}$-atm He environment and side cheeks would reduce the power loss to under 1 W, less than one-fortieth of the worst case!

A less dramatic example, but one which highlights an important point, is to observe the effect of altering the dimensions of just one housing feature. Consider a six-faceted polygon of tip diameter 40 mm, length 13.5 mm, and rotating at 40,000 rpm in 1 atm air at 25°C. A window is required in the housing wall, and a circular hole (see Figure 8.12) of diameter 40 mm has been chosen which gives a windage loss of over 20 W. If the hole diameter could be reduced to 19 mm, then there would be a saving in power of over 50%.

It is clear that the influence of windage is a very important consideration in scanner design optimization. Any decrease in power loss reduces the risks from thermal effects as well as the required supply power. In many cases the potential reductions will be so great that the overall scanner design can be improved. For example, a large decrease in the power requirement could allow the use of a smaller motor with decreased mass. This in turn should reduce the bearing size and lead to an overall reduction in the scanner size and mass with obvious improvements in its market potential.

8.8 CONCLUSION

The windage of rotary scanners is an important consideration during system design from the point of view of motor specification, power requirements, and thermal effects. The windage characteristics of polygon rotors follow complex relationships, and estimation of windage values at the early stages of design can be difficult. No suitably accurate and comprehensive equation-based solution is available for the wide range of scanner configurations potentially of interest. Building and testing full-speed prototypes can produce useful results, but would normally be too expensive and time-consuming to contemplate. Model testing of prototypes in liquid at low speeds is a proven means of obtaining results which is relatively quick and cheap and sufficiently accurate for most applications. Modeling can be used to (1) determine the windage characteristics of a particular design or (2) indicate the relationship between windage and changes in the scanner configuration, information which may then be used in an optimization process to minimize windage.

The quantity of information which would be required to cover the majority of likely scanner configurations is far too great to be presented in a short chapter. The test results included in this chapter will only be of use in making first approximations of windage. However, most engineering institutions should be able to generate the necessary detailed information for their own designs using the modeling techniques outlined. A proper analysis of windage during the scanner design stages can lead to substantial savings in power consumption and heat dissipation and to improved performance, such as higher speeds or better speed stability. These factors serve to decrease the risks with prototype scanners and greatly enhance the market potential of scanner systems.

8.9 APPENDIX

8.9.1 Definitions

Configuration: A scanner design in which the overall size of the machine may be scaled up or down without the configuration being changed. The restriction is that all the dimensions of the system must remain in a fixed relation to one another. This scaling property is fundamental to the validity of model testing.

Environment: The type of gas, its temperature, and pressure that the rotor or scanner operates in.

Rotor: In this chapter, either identical to "spinner" or a rotating polygon that models a spinner for the purpose of model testing.

Scanner: The mechanical part of an optical scanning system, including the spinner and its housing.

Spinner: The rotating part of an optical scanner.

8.9.2 Nomenclature

Symbol	Meaning	Dimensions	SI units
A_f	Facet projected area	L^2	m^2
A	Constant	—	—
B	Constant	—	—
Cd	Drag coefficient	—	—
Cm	Torque coefficient $= T_w/(\rho\omega^2 r^5/2)$	—	—
I	Polar moment of inertia	ML^2	kg m^2
L	Length of polygon rotor	L	m
L_c	Length of housing cavity	L	m
N	Number of facets	—	—
P_w	Windage power	ML^2/T^3	W
r	Representative dimension	L	m
R_a	Polygon average radius	L	m
R_f	Polygon inscribed radius	L	m
R_t	Tip radius of polygon	L	m
R_c	Housing internal radius	L	m
Re	Reynolds number $(= \rho\omega r^2/\mu)$	—	—
R_f	Radius-to-facet face	L	m
T_b	Bearing drag torque	ML^2/T^2	$N\ m$
T_m	Motor torque	ML^2/T^2	$N\ m$
T_s	Stator reaction torque	ML^2/T^2	$N\ m$
T_w	Windage torque	ML^2/T^2	$N\ m$
U	Linear velocity	L/T	m/s
α	Angular acceleration	$1/T^2$	rad/s^2
μ	Dynamic viscosity	M/LT	$N\text{-}s/m^2$
ω	Rotational speed	$1/T$	rad/s
ρ	Fluid density	M/L^3	kg/m^3

REFERENCES

1. T. D. Hayosh, *Motors and Control Systems for Rotating Mirror Deflectors*, Laser Scanning Components and Techniques, Proc. SPIE 84, pp. 97–108, 1976.
2. J. W. Daily and R. C. Nece, Chamber dimension effects on induced flow and frictional resistance of enclosed rotating disks, *J. Basic Eng., Trans. ASME*, *82*:218 (1960).

3. L. A. Dorfman, *Hydrodynamic Resistance and the Heat Loss of Rotating Solids*, 1st English ed., Oliver & Boyd, London, 1963.
4. B. S. Massey, *Mechanics of Fluids*, 5th ed., Van Nostrand Reinhold, New York, 1984.
5. E. S. Taylor, *Dimensional Analysis for Engineers*, Oxford University Press, London, 1974.

9

Bearings for Rotary Scanners

Joseph Shepherd*

Ferranti International, Silverknowes, Edinburgh, Scotland

9.1 INTRODUCTION

A rotary scanner consists of a mirror-faceted rotor caused to spin about an axis defined by bearings. The bearings chosen have a dominant influence on the accuracy of this spin axis and therefore on the optical scan accuracy achievable. They must be capable of repeating this accuracy within acceptably low limits of deviation, throughout many start-ups, for the required life, and while sustaining specified static and dynamic loads. Instrument life is normally that of the bearings and especially for high scan-accuracy or high speed, they have a major influence on cost.

Rotary scanning applications range from low accuracy, with scanners costing a few hundred dollars, to extremely high accuracy and speed, where costs can be tens of thousands of dollars each. Throughout this range, bearings should be considered as part of the overall design optimization if cost and performance targets are to be met.

A rotary scanner bearing specification will depend on the scanning system requirements. For example, speeds up to 500,000 rpm, rotational center accuracy to millionths of an inch, a wide temperature range, low cost, silence, and cleanliness may be required. Fortunately, the extremes of these are not often all required at once, and different types of bearings give special advantages.

It is a tribute to the work of many engineers and scientists that high-accuracy

* *Current affiliation:* Consultant, Edinburgh, Scotland

mechanical scanning is at all possible. Their work over the last two decades has improved the reliability and accuracy of both gas and ball bearings. Many of the advances in bearing technology have been achieved through experience gained in the design of gyroscopes for inertial navigation systems; in such systems the need for the ultimate in spin axis accuracy and reliability, freedom from contamination, and long life has much in common with that of high-performance rotary scanners.

In most types of bearings the principal aim is to produce true fluid film separation of the bearing surfaces when rotating, thereby achieving contactless running which will ensure long life. It is hoped that the designer of systems incorporating rotary scanners, who has little previous knowledge of bearings, will gain from this chapter an appreciation of the advantages and limitations of the various types of spin axis bearings available. While not intended to be a guide to bearing design, a designer should be able to approach the manufacturers of bearings and rotary scanners with a working knowledge of their products.

The bearing types most likely to be considered for rotary scanners are ball, pressurized gas, and self-acting gas. These have all been covered as fully as possible in this chapter. Further information on ball bearings is well documented and readily available, whereas information on gas bearings, especially self-acting types, is much more restricted. To help to redress this, more illustrations and design references have therefore been included for gas bearings than for ball bearings.

9.1.1 Historical Background

A bearing carries or supports the load of an object while allowing relative movement, preferably with low friction. There are two general classes, linear bearings and rotating bearings.

Many forms of bearings occur in nature: for example, animal joints, which are beautiful examples of sealed low-friction rotating bearings, or snails slithering on their mucus, and glaciers sliding on ice water melted by pressure, both of which are examples of linear bearings. Early man, having slid on scree or slipped on fat or perhaps even a banana skin, soon realized that by placing a "bearing" under a heavy object it became easier to move. Early examples from history are the use of tallow or oil to lubricate the runners of sledges used to convey the massive stones used in the pyramids (fluid film bearings) or a number of round tree trunks placed under heavy weights to transport them (rolling element bearings). The wheel is often regarded as one of the most important advances in history, but a wheel needs a bearing and, preferably, lubrication to reduce friction.

Plain journal bearings such as those used on early wooden wheels worked much better when lubricated; however, this was usually boundary lubrication without true separation of the bearing surfaces. Such bearings existed, with little change, until advances in manufacturing during the nineteenth century led to more accurate bearing surfaces, which greatly reduced friction and wear of plain cylindrical (journal) bearings. The reduction in friction, with better surface accuracy, was not, however, seen in flat thrust bearings. This was a puzzle, especially to naval designers working on the development of ships' propellers, as this was being held back because of the high losses on the shaft thrust bearings.

The first understanding of a possible reason for this was by Tower [1], who in 1883 published data showing that the pressure distribution in a plain journal bearing was not uniform around the side of the bearing opposing a load, but increased considerably at points near the convergence in the bearing gap. The importance of this work was recognized by Reynolds [2], who in 1886 was able to explain Tower's results when he published his now famous theory for hydrodynamic fluid film lubrication.

Air is a fluid, although with some significant differences from liquids, as described later, and journal bearings with air as the lubricant were demonstrated within about 10 years of Reynolds's work.

As mentioned previously, rolling element bearings utilizing round logs were used by early man, as were smooth spherical stones, which were the forerunners of ball bearings. There is evidence of a type of ball bearing used by the Romans in the form of a thrust bearing incorporating spherical wheels. However, it was many centuries later, during the Industrial Revolution, that ball bearings began to be produced more regularly as machining standards improved. The first true instrument-size ball bearings available commercially were probably produced by the Swiss firm RMB in 1936, as a diversification from precision watchmaking.

9.2 GAS BEARINGS

In many forms of bearing, surface separation results from the pressure generated, by viscous effects, when a fluid is induced to flow in a narrow gap. When the fluid is a gas, the effect is much less because of its relatively low viscosity—which has some disadvantages, but also some unique advantages, as described later. Because of those advantages, gas bearings are now used in a wide range of fields from ultra-precision self-acting bearings for inertial navigation gyroscopes to pressurized bearings used to support elements of machine tools weighing several tons.

Gas bearings are particularly suited to low-noise, high-speed, high-accuracy bearing applications, which makes them attractive for many rotary mirror scanners. In many instances it is the availability of gas bearings, providing a high-accuracy, high-speed spin axis, which has broadened the scope for mirror scanning applications.

They can generally be divided into two groups, self-acting bearings, which can generate their own internal pressure by virtue of the relative motion of converging surfaces, and externally pressurized bearings, in which the pressure is generated by an external source and the bearing needs no relative motion for lift. There is an intermediate group, hybrid bearings, which are a cross between the other two and are basically externally pressurized bearings which have been designed to take account of their self-acting capability when running at speed, thus producing improved load capacity or stability. All types can be designed to take axial or radial loads.

9.2.1 Background of Gas Bearings

Gas bearings were a fairly natural derivative of the early work on hydrodynamic fluid film bearings, when it was realized that air could carry loads like any other fluid. The first known experimental self-acting air bearing was produced by Kingsbury [3] in 1897. Other early workers in this field were de Ferranti [4], who in 1909 patented an air bearing spindle for textile machinery, and Harrison [5], who in 1913 put forward an approximate theory for induced fluid film lubrication which took into account the compressibility of gases.

Those early designs were not very successful because of their low load capacity, and it was only in the early 1950s that concentrated effort was applied to improving the design of gas bearings, when it was realized that bearing problems in several advanced fields could be best solved by utilizing the unique advantages of gas bearings. Some of the requirements were for bearings to run at greater extremes of temperature, higher speeds, and better accuracies than was possible with the bearings then available. Other needs were for bearings that did not cause lubricant contamination of the local environment, could withstand nuclear irradiation, and had minimum drag. Much of this work was necessary for the gas-cooled nuclear reactor program in the United Kingdom and for high-speed power plant and inertial gyroscope development in the United States. The importance of these programs stimulated a period of intense theoretical and experimental activity on gas bearings, which has been summarized in several textbooks published in the early 1960s. Those by Gross [6] and Grassam and Powell [7] cover the period well and cite comprehensive references. Work on refining the theory and improving the technology of gas bearings has continued, until the present, when they can be designed and used with confidence; useful references are the Mechanical Technology, Inc., manuals, *Design of Gas Bearings*, edited by Wilcock [8].

A range of self-acting gas bearing rotary scanners for thermal imaging and laser scanning is shown in Figure 9.1. These have been run at speeds up to 90,000 rpm and with facet-to-spin axis accuracy of less than 10 arc seconds.

Figure 9.1 Gas bearing rotary scanners.

9.2.2 Unique Attributes of Gas as a Lubricant

The drag power of a plain journal bearing is approximately given by

$$P_D = K_1 \frac{\mu\omega^2 R^3 l}{c} \tag{1}$$

and the load capacity by

$$L = K_2 \frac{\mu\omega R l}{c^2} \tag{2}$$

where μ = dynamic viscosity
ω = speed
R = outer radius
l = length
c = bearing clearance (gap)
K_1, K_2 = constants of proportionality

Gas bearings can be run at much higher speeds than liquid bearings because of the very low viscosity of gas compared to that of liquids. The dynamic viscosity of a typical instrument bearing lubricant oil is several thousand times that of air at 20°C. From equation (2), for a similar load capacity, a gas bearing would therefore have to be much larger than a liquid bearing, the gap would have to be smaller, or it would have to be run at higher speed. A larger bearing is not necessarily a disadvantage, however, as it allows larger-diameter stiffer shafts to be used, which can be beneficial in increasing the shaft critical speed, often a limiting factor in high-speed operation.

Dividing equation (2) by equation (1) shows the load capacity per unit of drag power as some measure of bearing efficiency as

$$\eta \propto \frac{1}{\omega c}$$

which indicates that at first sight, this factor is independent of fluid viscosity, improves as gap decreases, and worsens as speed increases. For a gas bearing, there is a practical lower limit on the gap size, as this in turn is limited by the accuracy attainable on the bearing components, but this is steadily improving.

Another important and exclusive advantage of gases as lubricants is their ability to accommodate wide variations or extremes of temperature. Under normal conditions, gases, unlike liquids, do not change state, or degrade with temperature and their viscosity varies only slightly over the normal range of ambient conditions experienced. This is illustrated in Figure 9.2.

Finally, and importantly, a gas bearing is surrounded by its lubricant, the atmosphere. The gas is drawn or pumped from this readily available source into

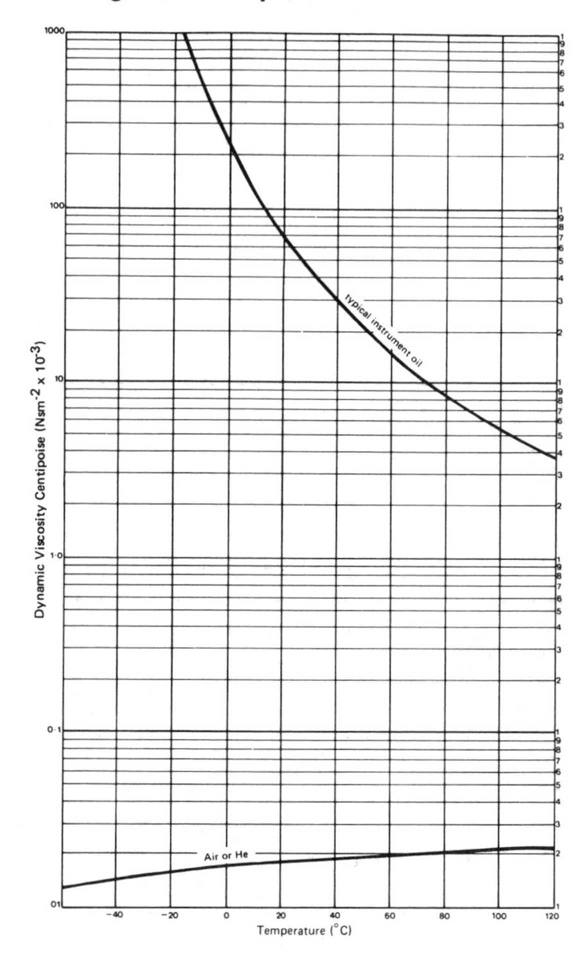

Figure 9.2 Comparison of gas and oil viscosity change with temperature.

the bearing and returned, normally without contamination of the surroundings. If the atmosphere is clean, there will also be no contamination of the bearing. There is therefore no need to seal the bearing, which may be necessary in liquid or ball bearings to avoid contamination of adjacent optical surfaces by traces of lubricant.

9.2.3 Self-Acting Bearings

Self-acting bearings are also known as self-pressurized, aerodynamic, or hydrodynamic bearings. Their load capacity comes from pressure generated by the

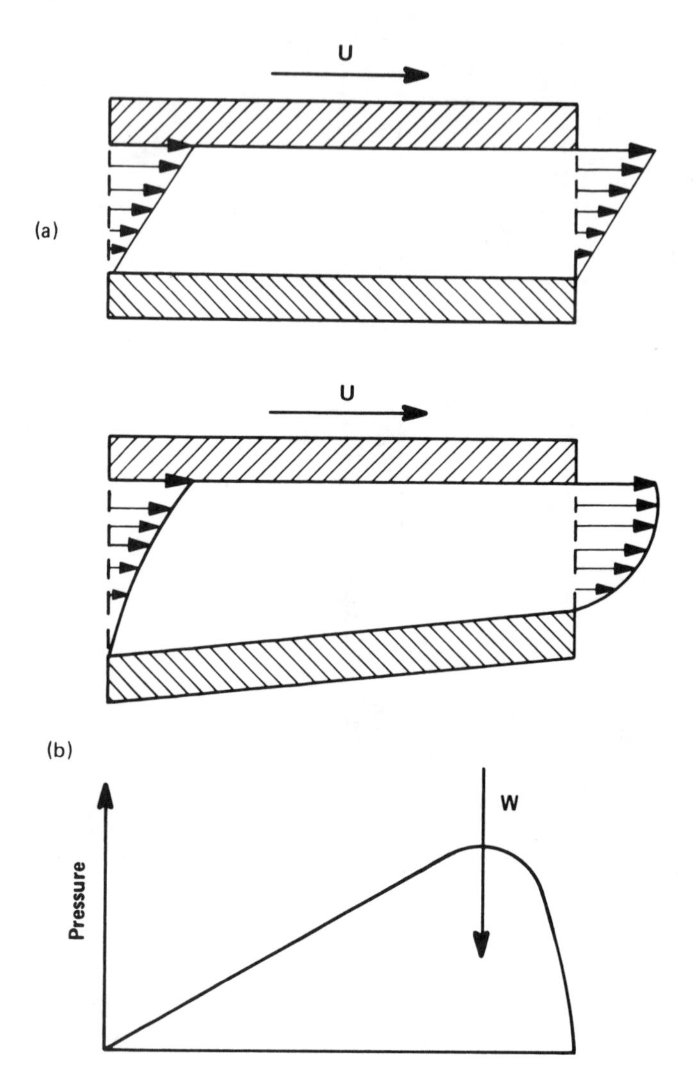

Figure 9.3 Induced flow between flat surfaces.

''wedge effect'' produced when gas is induced to flow through a convergence in the bearing gap, by the motion of the bearing surfaces.

An understanding of this effect can be obtained by considering two parallel plates in a fluid. If one of the plates is moving with velocity U, flow will be induced between the plates by viscous shear, the velocity varying linearly from U to zero (Figure 9.3a). When one of the plates is tilted, continuity of mass flow through the bearing dictates that the velocity must increase as the gap narrows,

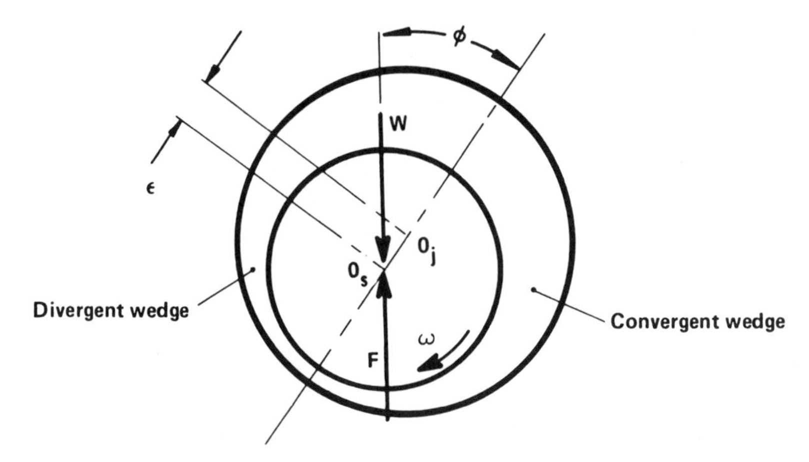

Figure 9.4 Journal bearing: wedge effect with rotation.

so that the area of the velocity profile is constant (Figure 9.3b). This increased velocity must be accompanied by an increase in pressure above ambient, which produces load capacity in a pressure profile as shown.

If the fluid is a gas, the pressure generated will also compress the gas, thereby increasing its density. Consequently, the velocity increase in the constriction will not be as great, resulting in some loss in load capacity. This effect will be slight at low speed but will become more significant as speed is increased.

A plain journal bearing (Figure 9.4) develops a wedge effect only when the shaft is deflected from its central position by an applied load. It will deflect until the pressure rise balances out the load. This will be accompanied by a sideways movement as a result of pressure rise above ambient in the convergent wedge and a pressure drop below ambient in the divergent wedge. The shaft thus takes up a position at an angle ϕ to the load direction, known as the attitude angle, with an eccentricity ϵ. When running centrally, there is no wedge effect in the bearing, and since no restraining forces are generated it is basically unstable.

A plain thrust bearing with its surfaces normal to the axis of rotation has no wedge effect, even when deflected. This was the problem seen in early ships' thrust bearings described previously.

To overcome these faults, the surfaces must deviate from true surfaces of revolution, in some way, to achieve a built-in wedge effect. There are two methods of achieving this:

1. By allowing one of the surfaces to move or deflect. (Types include pivoted pad bearings, resilient conforming bearings, and foil bearings.) These bearings have an advantage in their ability to remain stable over a wide speed range because they adjust their shape as speed and load changes. But this

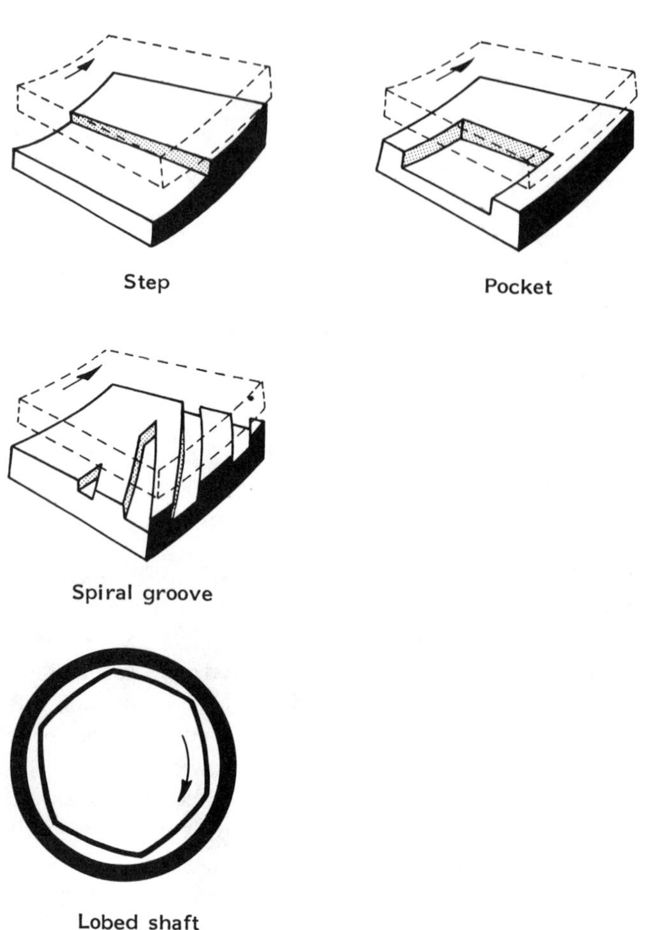

Figure 9.5 Built-in wedge effect features.

adjustment can also be a disadvantage in rotary scanners, where as little movement as possible of the optical axis is required. These types are also generally less efficient as load carriers than fixed geometry bearings.

2. By fixed geometry bearings where one of the mating surfaces is permanently altered from a surface of revolution to produce a built-in wedge compression effect. This effect can be achieved by adding surface steps, pockets, or grooves, which can be applied to both thrust and journal bearings, or by adding lobing to journal bearings (Figure 9.5). Of these, spiral groove bearings have several advantages in performance, which make them the most widely used self-acting bearings for precision spin axis applications.

Spiral Groove Bearings

In spiral groove bearings one of the surfaces is grooved such that when the surfaces move relative to each other, gas is drawn into the grooves by viscous shear. When the flow thus induced is throttled at the end of the grooves, a pressure rise is produced to give load-carrying capacity.

Much of the original work on the bearings was carried out at the U.K. Atomic Energy Authority, and Whipple [9] published the first theoretical analysis of

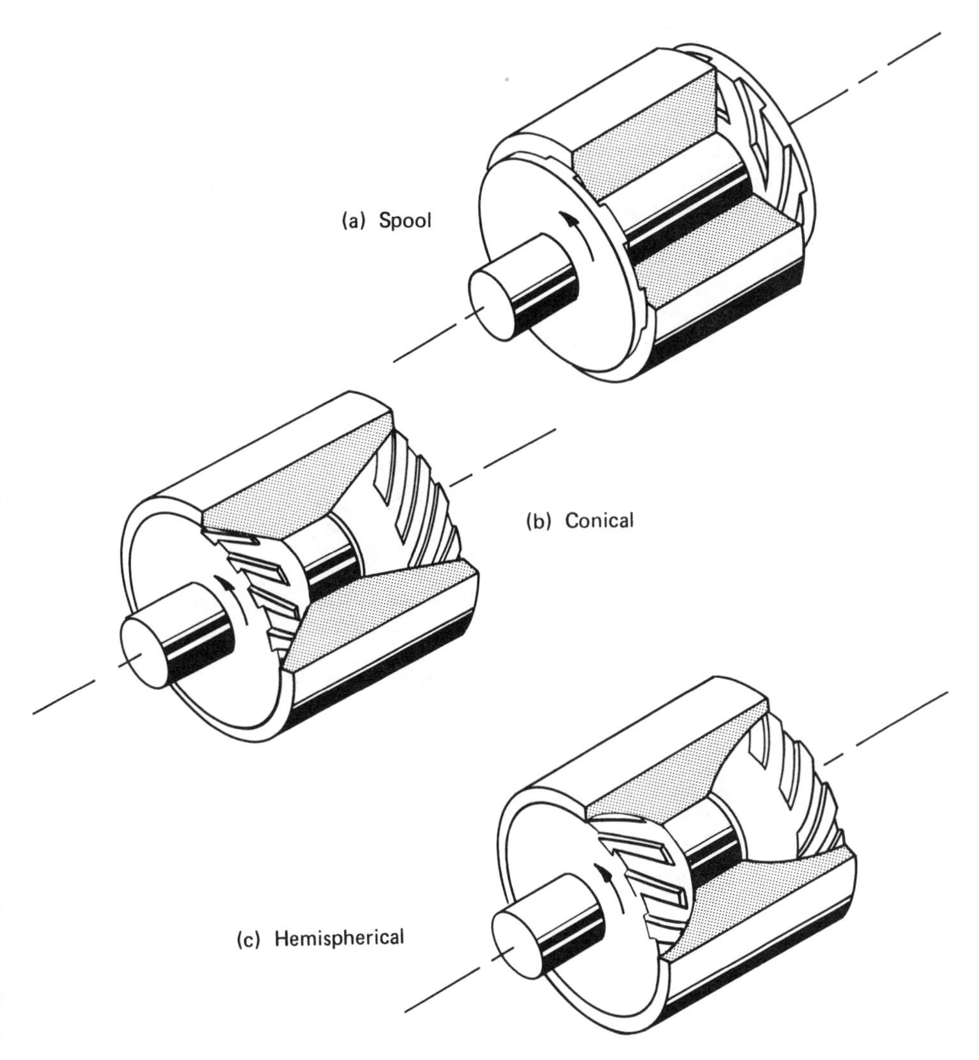

Figure 9.6 Spiral groove bearing configurations.

them in 1949; this and other reports from the same source were not published until much later because of government restrictions. They were later adopted as the ideal spin axis bearing for precision gyroscopes, and a general analysis on the application of spiral groove bearings to gyroscopes was carried out by Vohr and Pan [10], followed by Faddy [11], who in 1969 produced a design analysis for a conical configuration.

Configurations of Spiral Groove Bearings

The spool or H-form bearing (Figure 9.6a) was the first type to be produced commercially because the techniques of manufacture (finishing cylinders and flats) were already well established in other fields. The thrust plates have spiral grooves, and the journal bearing can be plain, lobed, or grooved.

Two other types available are the opposed conical (Figure 9.6b) and the opposed hemispherical (Figure 9.6c) bearing. These have more demanding manufacturing problems, but their advantages have encouraged a small number of manufacturers to persevere, and scanners based on these are now also available commercially. Although both types need more specialized techniques to produce accurate geometry, they have the advantage of being able to take radial and thrust loads on

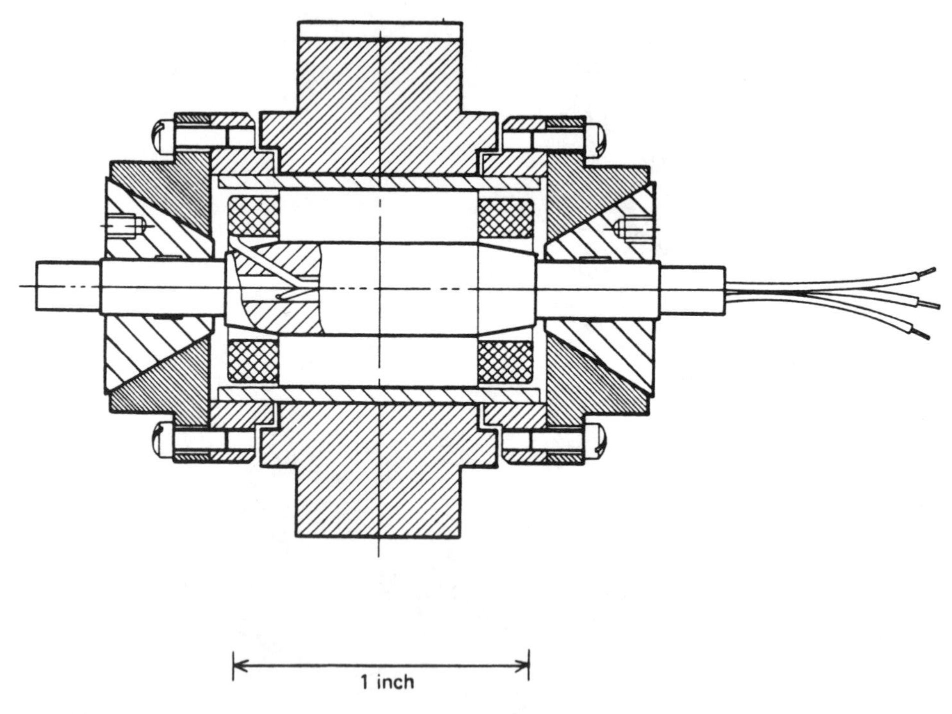

Figure 9.7 Scanner with conical gas bearings.

one surface. It is thus possible to produce compact standard designs (Figure 9.7) which are extremely adaptable for rotary scanner applications. Hemispherical bearings also have the advantage of being inherently self-aligning, which can greatly simplify the required accuracy of assembly.

If the projected apexes of opposed conical bearings coincide (Figure 9.8), materials of differing coefficient of expansion can be used for the rotor and the stator without any change in the bearing gap with temperature. This can be a major advantage, as lightweight rotor materials are often used to reduce mass or inertia effects together with heavier and stiffer stator materials. The gap of conical bearings can also be easily adjusted by moving one of the cones axially; thus the same standard bearings can often be used for a range of loads and speeds. For the same maximum diameter, minimum diameter, and length, the theoretical bearing performance characteristics of both conical and spherical types have been shown to be similar by Vohr and Pan in 1968 [10].

Other Characteristics of Self-Acting Gas Bearings

A self-acting gas bearing requires relative movement of the surfaces to generate pressure, and rubbing of the surfaces takes place at start-up. To minimize wear, bearing surfaces are normally manufactured from, or coated with, hard materials such as ceramics or cermets; these materials are also more easily finished to the extremely high accuracies required for self-acting gas bearings. Boundary lubrication with fatty acids has also been employed to reduce starting friction and wear. The use of these techniques has been highly successful, so that there should be no real problem in achieving many thousands of starts and stops, without damage.

In a plain journal gas bearing, load capacity initially increases linearly with bearing speed (see equation (2)); however, at high speeds the linear increase no longer applies and there is some loss of load capacity, as described previously (see Section 9.2.3). It has been shown by Faddy [11] that this effect is not present in spiral groove bearings (Figure 9.9) within the range of compressibility numbers

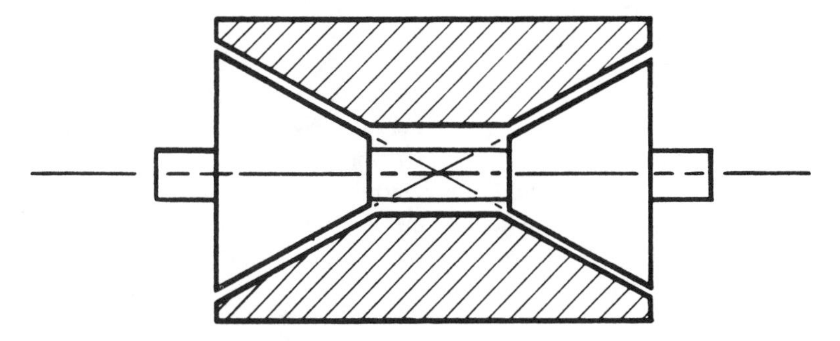

Figure 9.8 Coapex conical bearing.

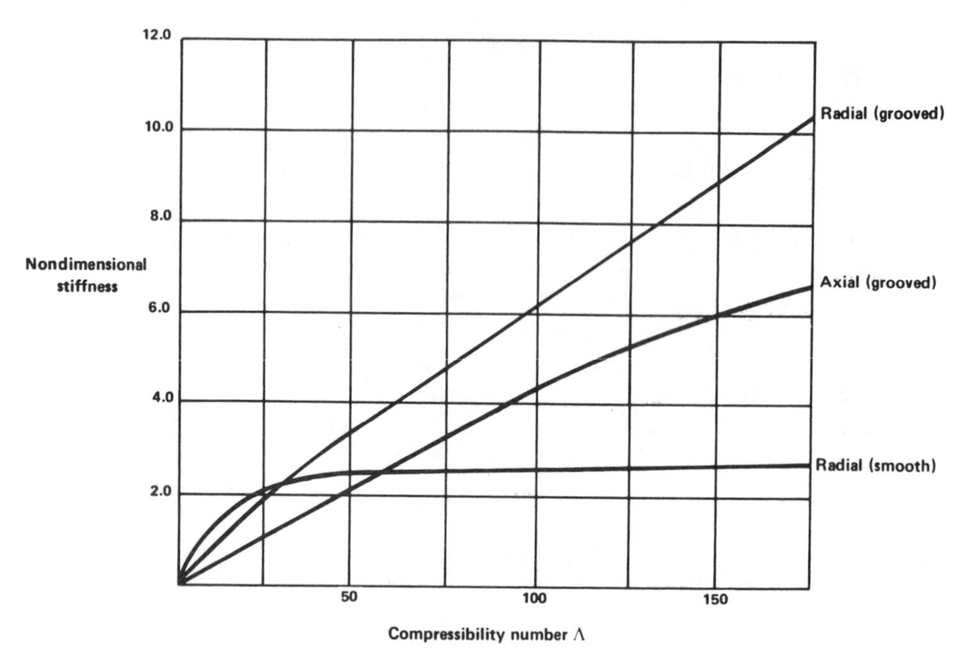

Figure 9.9 Comparison of grooved and smooth gas bearing stiffness.

likely for rotary scanners. The compressibility number Λ is the dimensionless group

$$\frac{6\mu\omega R^2}{P_a c^2}$$

where μ = dynamic viscosity
ω = speed
R = radius
P_a = ambient pressure
c = bearing clearance

This number is a measure of the pressure generated in the bearing relative to ambient. A dimensionless group is a gathering of interdependent variables into groups whose dimensions (length, mass, and time) cancel out. In a complex situation, such groups greatly simplify the interrelationship of the variables.

Two further factors that can influence the performance of self-acting gas bearings, and whose effect should be minimized at the design state are

1. Centrifugal effects, which can cause distortion of the bearing surfaces when running at high speed, can become a significant proportion of the small gaps used in self-acting bearings.

2. A loss of load capacity, which was observed by Burgdorfer [12] in 1959, occurs when the film thickness of a self-acting gas bearing becomes less than about 100 times the magnitude of the mean free path of the ambient gas. This is especially relevant when running at reduced pressure or when using low-density gases to reduce windage on the polygon.

Instability of Self-Acting Gas Bearings

Instability in self-acting gas bearings can be a complex phenomenon. The following is a simplified description of two of the main factors.

1. Mass/stiffness resonance. This is the well-known type of resonance which occurs in every spring-mass system. However, in self-acting gas bearings, stiffness will change with speed and will not necessarily be linear with deflection. Bearings should therefore be designed such that this resonance lies outside the frequency range of any likely external vibration and the speed range is below this resonance. If this is not possible, sufficient damping must be built in to limit the vibration amplitude.
2. Half-speed whirl. The average induced flow in a plain journal bearing running with no eccentricity is in the same direction as the rotation but at half the speed. If the shaft is deflected from its central position, it will experience a centering force from the wedge effect and a backward torque from shear forces. These forces will cause the shaft to whirl in a direction opposite to the direction of rotation. When this whirl speed reaches the speed of flow around the bearing (i.e., half speed), there will be no relative flow through the gap and therefore lubrication forces will collapse. Bearings with a built-in wedge effect, such as spiral grooving, are not usually troubled by half-speed whirl, as the effect is much reduced: however, it will still be there and should be considered in any design.

   Obviously, if a bearing is run at twice the speed of its resonant frequency, it will tend to whirl at its resonant frequency, which is almost certain to cause bearing failure if it is held at this speed; but if damped, it is often possible to run successfully through this resonance to higher speeds.

For a pair of bearings, instabilities can occur in cylindrical and conical modes. External vibration at specific frequencies, or poor balance, can stimulate these instabilities. The rotor should therefore be balanced as well as possible and external vibrations taken into account.

9.2.4 Externally Pressurized Gas Bearings

Other names for externally pressurized gas bearings are hydrostatic or aerostatic bearings. In these, the pressure to produce load support is provided from an external source, avoiding any need for relative motion of the bearing surfaces. They can therefore be used to define an accurate axis from zero up to very high speeds. At zero and low speed they have virtually zero friction torque, and many instrument designs make use of this. They have the advantages of the use of gas as a lubricant, described earlier (see Section 9.2.2).

The bearing gap is generally about an order of magnitude greater than that for self-acting bearings, so manufacturing and assembly tolerances can be more relaxed. These tolerances can be met by most precision machine shops, and this has led to fairly wide use and availability of pressurized bearings.

Pressurized bearings should have long life in spin axis applications, because no wear should occur since the bearings are pressurized to separate the surfaces prior to running. However, it is still desirable to have wear resistance surfaces, to withstand accidental turning when unpressurized, or overload touchdown.

The need for an external pressure source is often a major disadvantage; if this is not already available, a local compressor will be necessary, which may be bulkier and noisier than permissible. There is also a need to exhaust the gas supplied, which precludes their use in sealed systems. If these conditions are acceptable, the pressurized system has an advantage, since if the supply is filtered and dried, continuous flow through the bearings can keep them cool and clean even when working in relatively dirty atmospheres. It is necessary to use dry air, as the pressure drop across restrictors (see the section "Types of Restrictor") can cause ice crystals to form in moist air.

Early pressurized bearings were made with relatively large clearances, which made them easy to manufacture, but also made them comparatively inefficient, with large gas consumption and low stiffness, and also more liable to pneumatic hammer instability (see the section "Instability of Pressurized Bearings"). As the demands on these bearings have increased and the design knowledge improved, the clearances have been reduced. This reduced clearance also means that when the bearings are used for high-speed applications, there is an appreciable self-acting effect, which takes them into the hybrid bearing class; both effects must be allowed for in any design.

Journal Bearings

The principle of operation of pressurized journal bearings can be explained with reference to the bearing as shown (Figure 9.10). Supply pressure P_s feeds a chamber around the bearing. From this, gas passes onto the shaft through restrictors, which limit the gas flow, causing a pressure drop across the restrictor, to pressure P_g in the gap. There will be a further pressure drop to ambient as gas flows to the ends of the journal. With the shaft central and no load applied, the pressure P_g in the gap will be equal around the shaft and will vary along the shaft as illustrated (Figure 9.10a). If a load is now applied to the shaft, it will be displaced in the direction of the load. Because the gap has increased on the

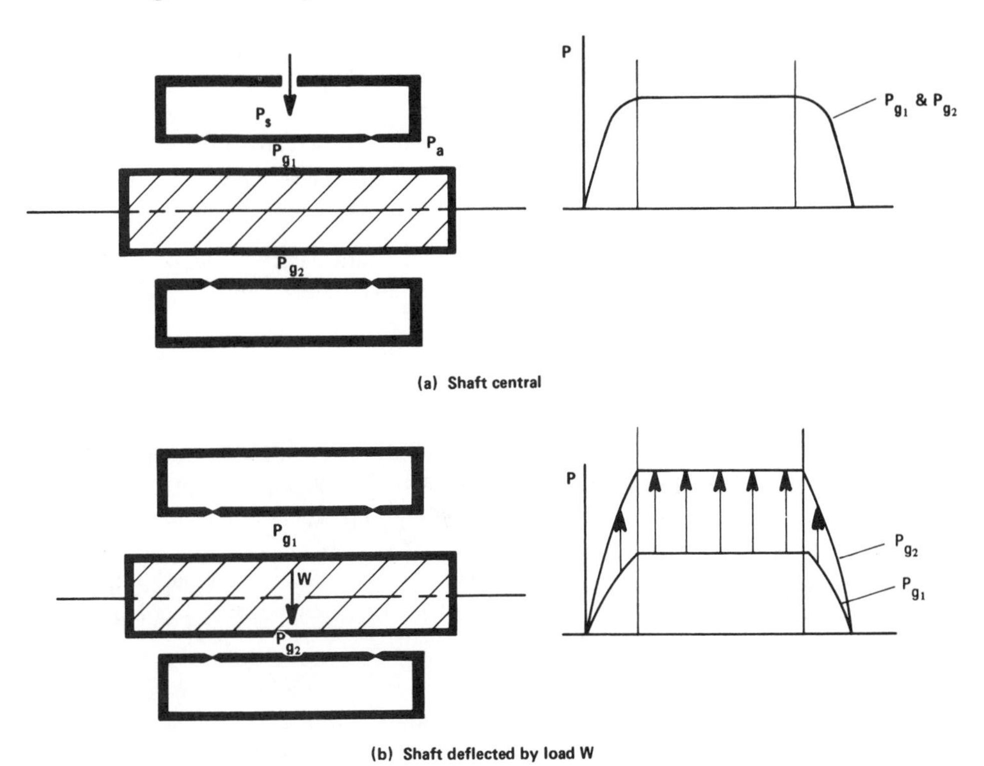

Figure 9.10 Externally pressurized journal bearing.

load side, flow will increase through the restrictors and the pressure drop will increase, lowering the pressure. Similarly, the decreased gap on the opposite side will produce an increase in pressure, and the resultant pressure difference will balance the load.

The lifting efficiency of a pressurized bearing is normally measured by its load coefficient

$$C_L = \frac{W_{\max}}{P_s A}$$

where P_s = supply pressure above ambient
$W_{\max}$ = maximum load
A = bearing projected area

The load coefficient varies with bearing deflection. At a deflection of half the bearing clearance the load coefficient is normally about $\frac{1}{3}$; however, this can be increased, up to a maximum of about ⅔, for high-efficiency bearings, but usually only with additional manufacturing complication.

As described earlier, the lifting capacity of a journal bearing is determined by the pressure difference created between the two sides of the shaft when it is deflected under load. Any factor that reduces this difference will also reduce the lifting capacity. Typical factors are

1. The flow of gas circumferentially around the bearing from the high-pressure side to the low-pressure side, limiting the area of the high-pressure zone (Figure 9.11a).
2. The effect of dispersion of the gas as it flows from discrete feedpoints (Figure 9.11b), resulting in local reductions in pressure; this can be reduced by increasing the number of outlet points.
3. The length of the reduced pressure zone between the feedpoints and the ends of the bearing; this length can be decreased by moving the outlet points nearer to the ends of the bearing, but only at the expense of increased gas consumption.
4. The restrictor characteristics, which control the maximum pressure differential generated across the shaft when deflected. These characteristics depend on the restrictor type and dimensions.

There are several basic types of restrictors, with many variations on these. New designs are regularly being introduced with claimed advantages in some aspects of performance. Some of the basic types are outlined below.

Types of Restrictor

1. Jet or orifice. There are two main types, the plain jet and the pocketed jet, illustrated in Figure 9.12. In the plain jet, flow is regulated by the area of the annulus surrounding the jet πdh (if $\pi dh < \pi d^2/4$); this is often described as annular orifice regulation. For the pocketed jet, flow is regulated directly by the area of the orifice $\pi d^2/4$, known as simple orifice regulation. The regulation of flow is important in limiting gas consumption and providing control over the bearing response to load fluctuations. Generally, pocketed jets have higher stiffness than plain jets, because of reduced dispersion losses, while plain jets have better stability characteristics.

2. Porous pad. In these, the jet is replaced by a porous pad or, more commonly, a complete porous journal is fitted (Figure 9.13). With a complete journal, an almost infinite number of point source outlets will be created, so that dispersion effects will be decreased and load capacity increased. Porous bearings are also more tolerant of contamination in the supply, which can have a disastrous effect on a jet-fed bearing if one or more jets become plugged.

The main disadvantages are that special microporous materials are needed to obtain the correct flow characteristics; these materials must machine cleanly to

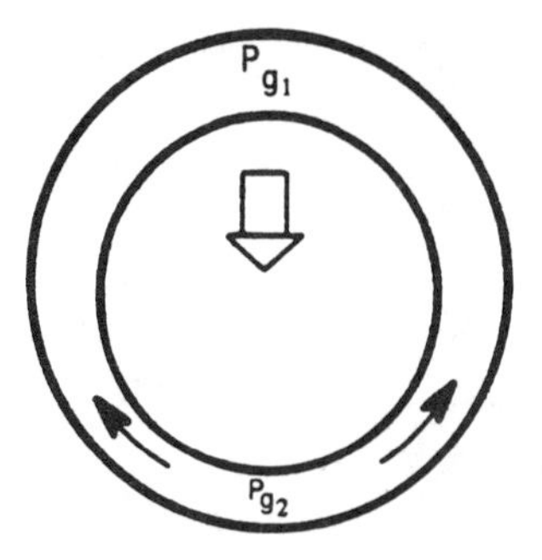

(a) Circumferential flow

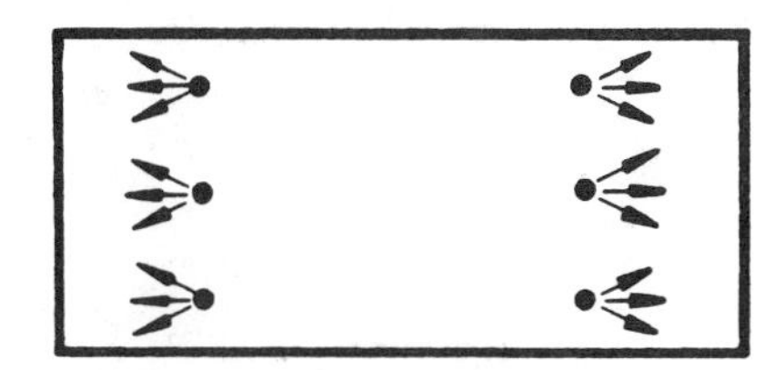

(b) Dispersion effect at jets

Figure 9.11 Factors reducing load capacity in externally pressurized bearings. a) Circumferential flow. b) Dispersion effect at jets.

tight tolerances and should have excellent rubbing qualities, without smearing, to withstand accidental damage before lift-off or during overload touchdown.

3. Slot entry. These normally have a circumferential slot, often around the entire circumference of the bearing (Figure 9.14). While these have a lower

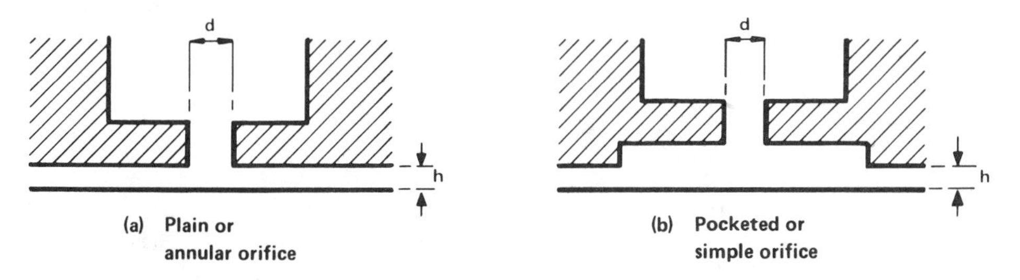

Figure 9.12 Types of jet restrictors. a) Plain or annular orifice. b) Pocketed or simple orifice.

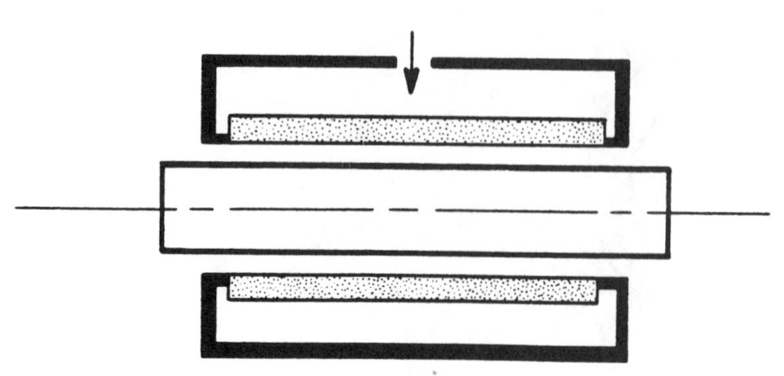

Figure 9.13 Porous journal bearing.

theoretical stiffness and load capacity than similar jet-fed bearings, the line source of supply eliminates dispersion effects which reduce actual performance in jet-fed bearings. Slot entry bearings therefore have an advantage for short, low length-to-diameter bearings, where flow dispersion has the greatest effect on jet-fed types. In addition, the optimum design for slot entry bearings is independent of the fluid characteristics and supply pressure. This feature makes them extremely versatile as "standard" manufactured bearings, which will give good performance at any available supply pressure, and which can even be used with liquids.

Other Design Features

For a mirror scanner of given mass and inertia, the bearing load capacity limits the maximum input rates that can be applied before touchdown of the bearing surfaces. To leave a margin of safety it is desirable to design for sufficient load capacity when the bearing gap is closed by only about half. If this is not possible, the surfaces should be hard coated as a precaution against high-speed touchdown.

Spin axis positional displacement under load will be determined by the bearing stiffness, which can be increased by reducing the mean radial clearance; this also greatly reduces gas consumption. Angular displacement will also be controlled by bearing stiffness and the distance between bearing support points.

A journal bearing of cylindrical form will support only radial loads. If axial load capacity is also required, it is usual to provide this on a flat circular or annular thrust bearing surface, normal to the spin axis, which is provided with restrictor outlets. If these axial loads are likely to be small, there are advantages, in simplicity, in using types of combined journal and thrust bearing where no outlets are present on the thrust face. Some axial load capacity is produced by allowing the pressurized exhaust from the journal to pass through the thrust faces to ambient pressure.

Conical or hemispherical pressurized bearings are also sometimes used in

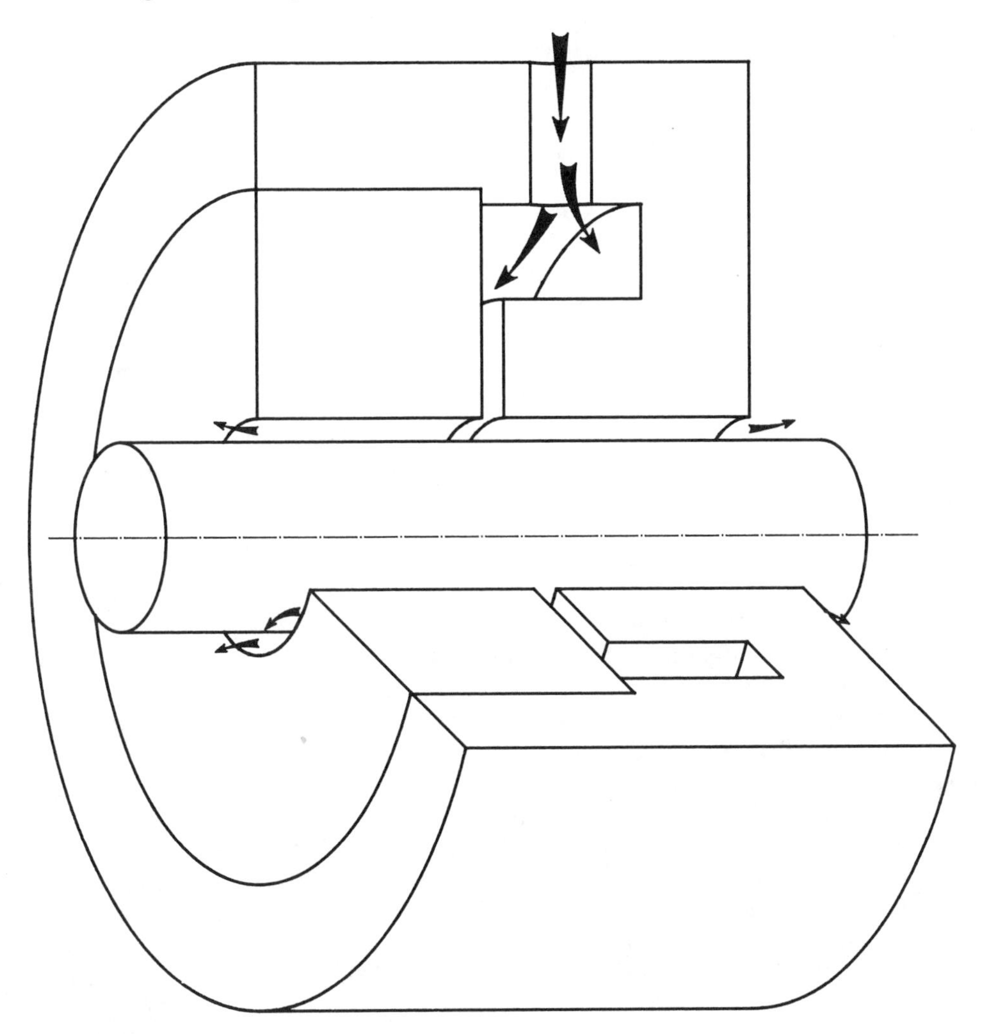

Figure 9.14 Slot entry bearing.

special designs, where the advantages of these configurations are considered worth the probable extra cost of manufacture. Their advantages are the ability to take axial and radial loads on one surface and the other attributes described previously for these configurations (see Section 9.2.3).

Instability of Pressurized Bearings

The instabilities associated with externally pressurized bearings are caused primarily by the resonance and half-speed whirl effects already outlined for self-acting bearings. Where instabilities do occur within the speed range required, a well-established method of overcoming the problem is to mount the stationary member on soft rubber O-rings, which contribute sufficient damping to limit the amplitude. This may be a disadvantage in some accurate spin axis applications, as soft mounting will contribute to the total axis deflection under load. If this is important, other methods of damping out the oscillations will have to be built in, such as the use of stabilizing cavities, as described by Sixsmith in 1959 [13].

The other instability associated with externally pressurized bearings is that known as pneumatic or air hammer, where the rotor vibrates in a self-sustained oscillation similar in principle to that of a pneumatic hammer. This condition occurs independently of any rotation of the bearing and can often be heard as a low-frequency moan at standstill. It is more likely to occur in thrust bearings than in journal bearings. An analysis of the problem of self-excited vibrations and suggestions for overcoming it were presented by Licht and Elrod in 1960 [14].

9.3 BALL BEARINGS

Small instrument ball bearings were first produced commercially about 50 years ago. They were immediately successful because of their low friction and were widely used on military and commercial equipment. With their increased usage more specialized machinery was introduced to improve output and standards. Even with these improved standards, many bearings suffered from unexplained failures and limited life. The problem continued until the late 1950s, when a theoretical understanding of the true nature of the lubrication process was sought and research and development on the influence of ball bearing components was carried out. Much of this work was initiated at the Charles Stark Draper Laboratory (then the Instrument Laboratory of MIT), stimulated by the demand for better performance from spin axis bearings for inertial navigation gyroscopes.

The theoretical work on elastohydrodynamic (EHD) lubrication by Archard and Cowking [15], published in 1965, showed that the lubricant film thickness in a typical instrument bearing was less than 10 microinches (250 nm). This knowledge highlighted the need to control all factors that could affect this extremely thin film; these factors include complete raceway geometry and finish, ball standards, cage material and type, cage geometry, lubricant type and quantity, steel quality, preloading, environment, and freedom from contamination. Control over all these factors is neither easy, nor inexpensive, but if attained it will ensure extremely good consistency and accuracy, together with long life.

Ball bearing manufacturers and others have steadily improved the standard of bearing components, so that today geometry, finish and materials are all significantly better than a decade ago; even for the same nominal grade of bearing.

In addition several recent technology advances show great promise of further performance improvements (see Section 9.3.8).

Standard instrument bearings up to ABEC5 quality standard (see Section 9.3.2) are available at competitive prices since these bearings are widely used in industry. As accuracy increases above this, choice becomes progressively less and price can increase considerably.

A few manufacturers concentrate on high accuracy and "special" bearings—i.e., bearings with some feature designed and manufactured specifically to customer requirements, this can range from simple modifications to a completely unique design. These are likely to be used in applications where the bearings are integrated into the overall design and where this gives life and performance advantages to justify this extra cost.

One important form of special bearing is where the bearing raceways are integral with either the inner shaft or the outer component of an assembly. This ensures better control over raceway alignment resulting in greater spin axis accuracy and reduced bearing noise. Some manufacturers will produce complete spindle assemblies with integral bearings. As these are assembled and lubricated in controlled conditions, good accuracy and life can be expected.

9.3.1 Description and Terminology

Instrument ball bearings are usually either radial deep groove bearings or angular contact bearings. For accurate spin axis applications, such rotary as scanners, free movement (play) in the bearings should be eliminated by using them in pairs loaded against each other (preloaded pairs). The load displaces the balls to the side of the race into an angular position (contact angle), which determines the axial and radial stiffnesses and therefore controls the dynamic response of the rotor. Preloading is also beneficial in reducing noise, and, in general, more predictable bearing performance is obtained. Radial bearings can be used in this manner, but there are some advantages in using the more specialized angular contact bearings described later.

A ball bearing consists of an inner race and an outer race, separated by a set of balls held in position by a cage (retainer, separator) (Figure 9.15). The outer and inner races have grooves (raceways) machined in them in which the balls run. Generally, these raceways are finished to very high geometric accuracies relative to their locating faces by grinding and honing, often finished by lapping and polishing to improve the surface finish further.

The type of bearing shown is known as a radial deep groove bearing. The balls have a slight clearance in the raceways, which allows them to deflect to the sides of the groove to sustain axial loads. This bearing clearance is usually measured and quoted as radial play; it is directly related to axial play by the bearing geometry. The greater the radial play quoted, the greater the contact angle and the thrust load when deflected. Radial play is *not* related to quality or the ABEC tolerance class. Radial loads push the balls to the center of the groove, the load being taken by an arc of balls opposing the load. Radial bearings can be run with these clearances, but for high-accuracy spin axis applications the normal practice is to take up the clearance with an axial preload between a pair of bearings. They are assembled by displacing the inner race to one side within the outer and filling this gap with as many balls as can be accommodated. The balls are then spaced even round the race, and a snap-on or two-piece riveted cage is placed over them. This procedure has certain disadvantages in that the raceways can be only about half-filled with balls, limiting the load capacity and the cage must be one-sided or two-piece, limiting high-speed performance.

These disadvantages can be overcome by using angular contact bearings, which are commonly used for accurate, high-performance spin axis applications. Angular contact bearings (Figure 9.16) have one side of either the outer or inner race machined away, allowing more balls and more rigid cages to be fitted, which

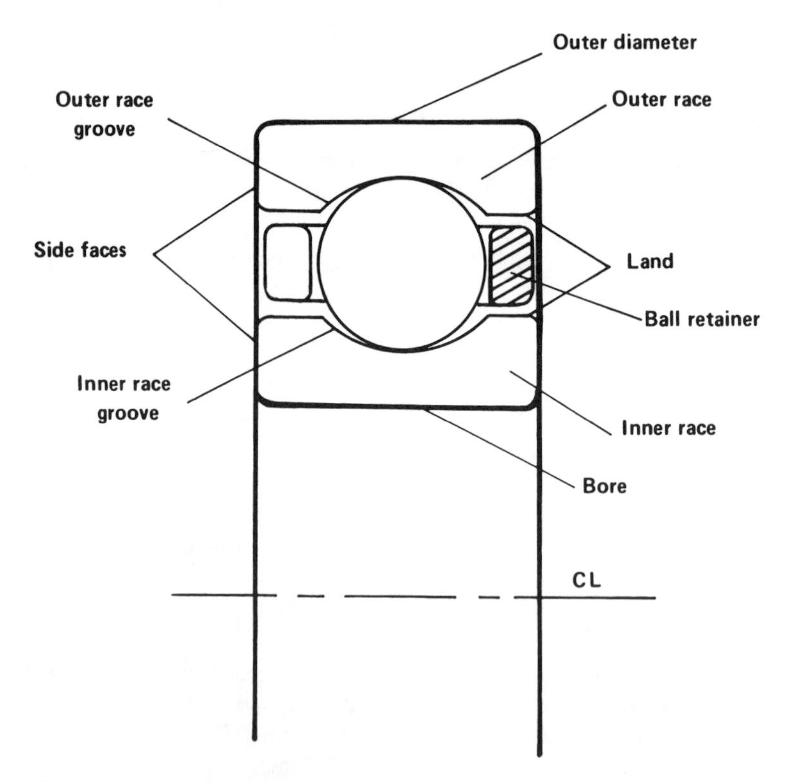

Figure 9.15 Ball bearing nomenclature.

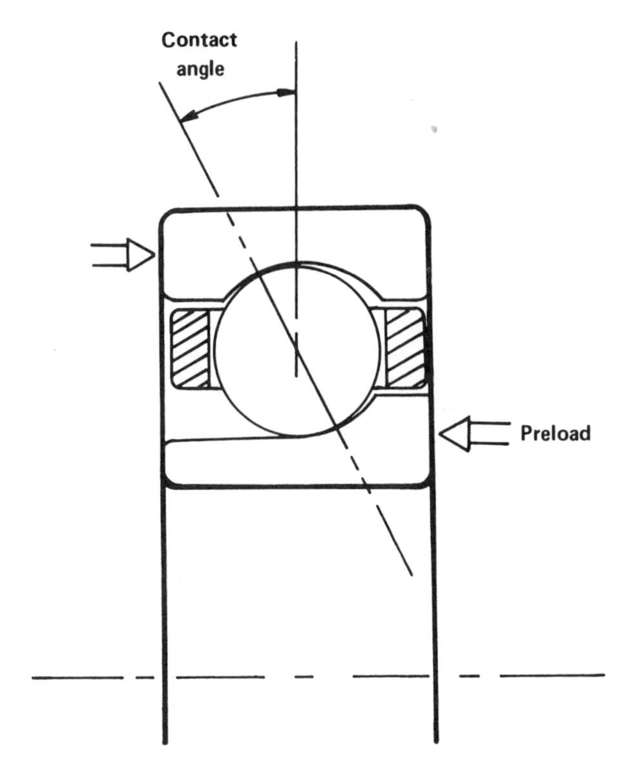

Figure 9.16 Separable angular contact bearing.

improves their performance. If only enough of the shoulder is removed to allow the balls to clear when the outer race is heated and the inner race is cooled, a nonseparable angular contact bearing is produced, or if one side is removed completely this gives a separable angular contact bearing.

In these bearings, the contact angle is the deviation of the contact line from the radial position. An increase in contact angle gives greater end load or thrust capacity. At large contact angles the life of the bearing can be affected, because of the greater deviation from true rolling contact to that with a component of spinning of the balls. The maximum satisfactory contact angle for high speed will be just above 30° when axial and radial stiffness are about equal.

For elimination of shaft play and to produce an accurate spin axis, bearings must be used as preloaded pairs. The preload (Figure 9.16) can be introduced by an axial spring or by solid preloading using differing length spacers between the two outer and the two inner races, to produce fixed compression. For critical applications the preload should be measured and adjusted to give optimum performance and life.

9.3.2 Ball Bearing Standards

Ball bearings bought as separate items from a manufacturer will usually be supplied to ABEC (Annular Bearing Engineers Committee) or ISO (International Standards Organization) standards. These give limits for the errors measured on the locating faces of either race, when it is rotated, while the other race locating faces are held in a fixed position. Typical figures are shown in Table 9.1.

Bearing tolerance standards are intended to produce uniformity and constancy in bearing quality which is independent of the manufacturer. They have had some success in this aim and can be taken as benchmarks for bearing selection. These standards should not however be taken as an absolute measure of bearing accuracy or quality, as they omit many important factors which can vary with different manufacturers and sometimes even with separate batches from the same manufacturers. All high-quality instrument bearing manufacturers have their own wider-ranging internal standards and quality checks to control geometry and other important components such as lubricants, cages, and steels, which are not covered in ABEC standards. These controls are often indicated in the manufacturers catalogues, and consultation of them is always beneficial, rather than simply purchasing on cost alone to ABEC standards.

Cost competition restricts the standards and controls on off-the-shelf bearings, so that where the highest standards of accuracy, reliability, performance, and life are necessary, these can only be ensured by (1) buying "specials" (i.e., bearings custom made against a design specification), or (2) quality checking and, if necessary, finishing all bearings immediately prior to instrument assembly by methods similar to those described by Holmes and Shepherd in 1973 [16].

The geometrical errors that must be controlled include

1. Balls: diameter, sphericity, surface finish, and diameter matching of the ball set.

Table 9.1 Tolerance Errors for Ball Bearings

Precision ball bearing tolerance[a] (in $\times\ 10^{-4}$) [2.54 μm]

	Bearing class	Radial runout outer	Radial runout inner	Bore out of round	Outside diameter out of round	Size bore	Outside diameter
Normal	ABEC 5P	2.0	1.5	1.0	1.0	2.0	2.0
Precision Bearings	ABEC 7P	1.5	1.0	1.0	1.0	2.0	2.0
Special	ABEC 9P	0.5	0.5	0.5	0.5	1.0	1.0

[a] For shielded bearings, tolerances are worse.

2. Outer and inner raceways: raceway pitch circle diameter, roundness, cross-track profile (groove radius accuracy), concentricity and squareness of raceway to mounting diameter, squareness of the end face to the mounting diameter, and surface finish.

For the highest accuracy, all these parameters should be controlled on every bearing, and this has become general practice on ultra-precision spin axis instruments such as gyroscopes for inertial navigation. This adds cost but the benefits in accuracy, reliability, and life can be considerable. With ball bearings assemblies controlled in this manner, rotary scanners with total scan errors of only a few arc seconds can be consistently produced.

9.3.3 Cages

Cages are also referred to as retainers or separators. The purpose of the cage is to keep the balls in their correct angular position and to provide lubricant as required. They may be located in three ways: (1) on the balls (ball riding), (2) their outer diameter may ride on the outer land (outer riding), or (3) the inner diameter may locate on the inner land (inner riding).

For radial bearings, simple pressed steel cages locating on the balls are used for slower speeds, usually in conjunction with grease lubrication. One-piece plastic cages which snap onto the balls from one side are normal for higher speeds, again using a grease lubricant.

One-piece symmetrical cages of phenolic-cotton or phenolic-paper material are usual for angular contact bearings. These can be vacuum impregnated to give a slight increase in oil retention. Although often referred to as ''porous,'' these materials retain oil only at their surface where machining has opened the paper or cotton laminating material to local wicking. Porous nylon or porous polyimide cages which have a completely microporous structure are available from some manufacturers as nonstandard items. These can be partly filled with lubricant which circulates within the cage and bearing when running, providing a continuous lubricant supply and acting as a filter to trap degradation products in the pores. They are more expensive, but as they increase bearing life they can usually be justified, especially at higher speed where greater lubrication degradation occurs.

While raceways and balls are finished to microinch accuracy, the machining accuracy of plastic cages is necessarily several orders lower. Machining parameters are important however, since geometric or surface finish variations inherent in plastic cages are often the source of inconsistent performance in bearings.

To avoid migration of the lubricant to adjacent optical surfaces, the cages should be vacuum impregnated and then centrifuged to an acceleration level higher than they are likely to sustain while running. This step is also employed to reduce the risk of uneven distribution of the lubricant, which would cause running problems.

9.3.4 Lubricants

The purpose of a lubricant in a ball bearing is to provide a fluid film between the bearing elements. If this film thickness is greater than the sum of the peak surface roughnesses, then no asperity contact will take place, eliminating wear and minimizing lubricant breakdown. Film thickness can be calculated from classic elastohydrodynamic theory and, for a given geometry, increases with speed and the oil viscosity in the ball contact zone.

A bearing with perfect geometry, running at constant speed, and with a lubricant of sufficient viscosity should have almost infinite life, and this has been demonstrated with life tests of over 10 years running without failure. But most standard instrument bearings are not perfect, may have been subject to some damage or contamination, have insufficient oil to form a proper film, and must change speed at least to stop. For this reason almost all instrument oils have additives to enhance their ability to withstand oil film breakdown and prevent lubricant degradation. Additive packages are usually proprietary to either oil companies, specialist oil stockists, or sometimes ball bearing manufacturers, and often include oxidation inhibitors, viscosity index improvers, extreme pressure boundary additives, and boundary lubricants.

No lubricant is universally good in all conditions, and this, together with the many formulations and types available, makes selection extremely difficult. Bearing manufacturers are a sound source of advice and will supply tables of commonly used lubricants and their characteristics. Selection on this basis will limit the choice of contenders, and these must then be instrument tested for the particular application until a suitable one is found. Only then can life and performance be guaranteed.

Some of the properties which need to be considered in selecting a lubricant are listed here.

1. Volatility (or vapor pressure at a specific temperature) will affect the condensation of vapor on adjacent optical surfaces. Low-volatility lubricants are available, but, even with these, vapor can still be released if lubricant degradation occurs.
2. Viscosity and its associated viscosity indexes (temperature and pressure) determine the film thickness and therefore influence the running life. The change in viscosity with temperature is important if wide variations in temperature are expected; the pressure viscosity index also affects the film thickness because of the extremely high pressures at the ball contact zone.
3. Oxidation resistance is directly related to the effectiveness of lubricants in that failures of lubricants in bearings are commonly caused by oxidation of the lubricant. Oxidation produces resins, sludges, and acids, which cause an increase in temperature when running and therefore more oxidation. Partial control is possible by using oxidation inhibitors or by using an inert at-

mosphere (N_2, He), or by changing to an oil with greater oxidation resistance (see also Section 9.3.7).

4. Lubricity is important, but, like a good whiskey, it is difficult to analyze what makes a good lubricant. While running with a true fluid film separation of the surfaces, only the viscosity affects performance, but under adverse conditions, such as touchdown or shock loads when boundary lubrication takes places, good lubricants protect bearing surfaces and themselves usually by beneficial chemical or physical reaction with the surfaces, forming protective layers. This action is usually too complex for full analysis, and the best way to prove a lubricant is to test run it in instrument bearings. Additives to improve the performance are also often tried by empirical testing.
5. Hydrolysis is the reaction of a lubricant with water. Moisture is present in air and can react with lubricant molecules, causing decomposition products (sludges, acids). Hydrolysis increases with temperature, as can the possible moisture content of air.
6. Surface tension controls the spread of a lubricant over surfaces. The optimum surface tension is a compromise between the need to ''wet'' the bearing surface, to ensure a continuous oil film, and the need to keep the lubricant in the bearing. Surface tension depends on both the lubricant and the surface condition (cleanliness). Barrier films have been used to control the spreading of lubricants to nonbearing surfaces; these should be used with great care to avoid contamination of the bearings surfaces, making them nonwettable.
7. Pour point is the lowest temperature at which flow will occur, and this is generally taken as the lower operational limit. Bearings can be run from slightly below this temperature if sufficient torque is available to overcome the initial drag as the lubricant will soon heat up. Storage below the pour point is acceptable.

Types of Lubricant

The choice of lubricant for precision instrument bearings is probably limited to mineral (petroleum), synthetic diester, or synthetic hydrocarbon (SHC). The poor lubricating properties of silicone oils and greases probably rules them out except where their ability to withstand extremes of temperature or their extremely low volatility are necessary, and limited life is acceptable.

Mineral oils have been used with great success in instrument bearings and have excellent lubricating properties. Their limitations are their low viscosity indexes and their high pour-point temperatures. This restricts their operating temperature range. Synthetic oils have lower volatility, less change in viscosity with temperature, inherent oxidation resistance, and can be used over a wider temperature range. These improved properties could make them attractive for rotary scanners.

Each class of synthetic comprises a wide range of lubricants with differing properties. In general, SHCs have better lubrication properties than diesters and are much more compatible with rubbers and elastomers.

Grease consists of an oil with a thickener; thus it can be packed into a bearing to make more oil available. Common lubricating oils are also often available in grease form. They have some advantages when running in dirtier atmospheres, as the grease also acts as a seal, but they are less suitable at higher speeds because of higher friction losses.

9.3.5 Raceway Materials

The standard steel for instrument bearings running in air is AISI 440C stainless steel. As supplied, it has high hardness, good stability and excellent corrosion resistance in normal atmosphere. It is less reactive than AISI 52100, which helps to prevent lubrication degradation during the wear process. To further enhance its corrosion resistance, 440C is often supplied ''passivated'' after finishing.

AISI 52100 (often referred to as chrome steel) is a high-hardness carbon chromium steel with extremely uniform microstructure, which gives greater wear resistance and the potential for improved surface finish compared to 440C. Its corrosion resistance is relatively poor, and because of this it is mainly selected for use where an inert atmosphere can be provided.

The microcleanliness of all bearing steels is of great importance as inclusions at the bearing surface can cause local collapse and breakdown of the lubricant film. For applications where performance and life is critical, most bearing manufacturers offer superclean steels on request at some extra cost.

9.3.6 Seals and Shields

Seals and shields are often collectively referred to as closures and have three principal purposes: (1) to prevent contamination entering the bearing during service, (2) to give protection when the bearing is handled during assembly, and (3) to minimize leakage of lubricant from the bearing. Closures are normally attached to both sides of the outer bearing ring, and if they contact the inner ring they are referred to as seals; otherwise they are known as shields.

As seals rub on the inner ring, they increase the bearing drag torque, although this is usually minimized by the use of low-friction materials such as Teflon. They give maximum protection against contamination. Shields are normally stamped stainless steel, and because they clear the inner ring they have little effect on drag torque. However, this clearance reduces the protection they afford against contamination and leakage.

Radial bearings are available fitted with closures, usually shields, as standard catalogue items. Their use protects the bearings during handling and assembly in lower-quality facilities and allows bearings to be run in environments with some contamination. Where bearings are built and used with care in clean conditions, and where the lubricant is properly selected and controlled, closures can be omitted.

An alternative to mechanical seals is the use of ferrofluid seals. In these a magnetic field holds a ferromagnetic fluid in a small gap between the rotating and stationary parts to form a seal. The advantage is complete sealing with relatively low drag, and they have been widely used to isolate Winchester disks from possible contamination from their ball bearing spindles. There are some limitations due to the large changes in their drag with temperature and speed and because of the low pressure difference they can sustain across a single fluid seal.

9.3.7 Life

The life of a ball bearing depends on the establishment and continued existence of the thin EHD fluid film to separate and protect the bearing surfaces. A great many factors can affect this, such as preload, cage type, temperature, atmosphere, lubricant type and quantity, contamination, loads applied (including vibration and shock), raceway geometrical accuracy, raceway surface finish, and speed. It is only by carefully selecting and controlling all such factors and building up application experience that life can be confidently guaranteed.

Fatigue life is usually quoted in bearing manufacturer's literature, but this is seldom a factor in precision instrument bearings because of the relatively low Hertzian stresses (i.e., the stresses beneath the contact ellipse). In the author's experience of many thousands of precision gyroscope bearings, fatigue has never been the cause of failure, even after life tests exceeding 100,000 hrs running.

Bearing failure is often seen as lubricant degradation (or polymerization); this is an increasing thickening and darkening of the lubricant. The breakdown is often progressive, as the thickening causes increased drag, higher temperatures, and so further degradation, often ending in complete failure of the bearing to run. Failure can usually be traced to

1. Inadequate specification or control of the bearings or lubricant, resulting in poor geometry or surface finish, wrong bearing type, incorrect contact angle, unsuitable cage, unsuitable lubricant, and so on.
2. Mishandling or poor assembly methods, resulting in contamination of lubricant or bearing surfaces, excessive preloading, brinelling, surface damage, and so on.
3. External influences, such as extremes of temperature, shock, slew, vibration, and so on.
4. The cage, which needs special mention, as it is the most difficult component to control adequately (see Section 9.3.4): lubricant failure is often initiated at the cage rubbing surfaces.

Despite this catalog of potential troubles, where ball bearings are properly selected, handled, and installed, there is no reason why true fluid film separation of surfaces should not be achieved, ensuring long life and consistent performance. There will be a few where statistical chance works against this, and it is advisable to test run all bearing assemblies for a time to weed these out. Tests could vary from simply measuring the constancy of the time to run down from a fixed speed, to more complex tests for the incremental stability of torque, power, or speed over a period. Speed stability is especially relevant in most scanner applications.

Bearing life decreases as speed increases above a certain level, where dynamic forces on the cage, ball set, and lubricant cause such increased frictional drag that overheating and lubricant breakdown occur.

9.3.8 Recent Advances in Ball Bearing Technology

The surge in activity in the development and the understanding of instrument ball bearings which occurred in the decade to the early 1970s was brought about largely by the critical requirements of spin axis bearings for gyroscopes. This was followed by a period of steady progress with gradual improvements in the quality and accuracy of components taking place rather than step changes in performance. Possibly the only exception to this was the development of synthetic hydrocarbon oils.

During the last few years, this pattern has changed, and several exciting new technological developments have been reported as producing significant improvements in ball bearing performance, although generally with some increase in unit costs. Some of these advances, which were presented at the 1987 Instrument Ball Bearing Symposium, are highlighted, together with other known developments which could be of significance to instrument bearings.

Titanium Carbide (TiC) Coated Balls

If the ball and raceway surface finishes are not quite good enough to allow full EHD fluid film separation of the surfaces or if starved lubrication occurs, then surface asperity contact will occur causing microwelding and fracturing of asperity peaks. The instantaneous high temperature generated in these contacts will cause lubricant breakdown, wear particles will be generated and the freshly exposed metal can act as a catalyst to further breakdown.

By replacing conventional balls with balls coated with titanium carbide (TiC), this breakdown process has been shown to be much reduced or eliminated entirely. The factors producing this change are the smoother surface finish possible with this hard coating compared to bearing steel and the greatly reduced prospect of microwelding between TiC and steel. Boving et al. [17] has suggested that, as an additional factor, the raceway steel asperities are plastically deformed when they come into contact with the hard TiC surface, thus further reducing the possibility of microwelds.

Field experience which demonstrates reduced lubricant breakdown and improved general performance with the use of TiC-coated balls in otherwise normal ball bearings is now fairly extensive. The improvements demonstrated with TiC coated balls in high-performance spin axis bearings should encourage their wider use and, as costs come down, their more general availability in "standard" bearings.

Ion Implantation

In ion implantation, high-energy ions are directed at a material's surface. Some of these ions enter into the surface, which can change the chemical composition and physical properties of the surface layer. Most elements can be ionized and implanted to form alloys with those already in the bulk material or with another implanted element. This gives wide possibilities for the alloys and, therefore, the enhanced surface properties which can be produced, even on low-cost bulk materials.

For ball bearing steels, ion implantation of titanium followed by carbon forms titanium carbide precipitates at the surface, greatly increasing the wear resistance. Alternatively, implanting tantalum has been claimed by Sioshansi [18] to give improvements to both wear and corrosion resistance.

The surface layer is typically less than a micron thick, the implanted ion species filling spaces in the near-surface crystal lattice network, so that no changes in dimensions or surface finish occur. Finished bearings can thus be treated without affecting their geometry. Some heating occurs in the process, but by mounting the components on water-cooled fixtures their temperature can be kept below 150°C, which prevents changes in the bearing steel properties. Bearings finished by relatively cheap conventional means can thus be treated and used without the need for further finishing, but, conversely, no work is possible to improve the surface finish to higher standards as has been reported for TiC balls.

At present the application of ion implantation to instrument-size ball bearings is limited. Trials with balls, implanted with carbon followed by titanium, running in conventional steel raceways, were reported by Hanson [19] to reduce wear and lubricant degradation.

Silicon Nitride (Si_3N_4)

Silicon nitride has been widely and successfully tested in large rolling element bearings such as jet engine main shaft bearings. Ceramic bearings were investigated for such applications because of their excellent high-temperature properties and because their low densities reduced the ball centrifugal forces at the high engine speeds. Other advantages of ceramics are high hardness, inertness, low friction coefficient, and excellent wear resistance.

The principal disadvantage is the need for expensive manufacturing techniques, normally using diamond abrasives for finishing. For this reason it is unlikely that complete ceramic bearings will be used in instrument bearings at present. Limited trials, showing similar advantages to TiC balls in preventing degradation, were reported by Hanson, using Si_3N_4 balls with steel raceways in a ''hybrid'' bearing, which limits the cost premium.

This course has some other disadvantages however; Si_3N_4 has a much lower coefficient of expansion and a higher stiffness than raceway steels, which will affect the performance characteristics. The higher stiffness is especially important, as this will result in smaller deflections, a correspondingly smaller ball/race contact ellipse, and therefore higher stresses in this area. This can be compensated for by making the ball radius closer in size to the raceway cross-track radius (tighter conformity), but this change will itself change bearing characteristics.

These disadvantages are likely to limit the use of solid ceramics in instrument bearings at present, especially as most of the advantages are also available with the better-established and less radical change of using TiC-coated balls.

Dry (Solid) Lubricants

The tendency for liquid lubricants in ball bearings to migrate due to gravity or centrifugal forces, or for some of their constituents to vaporize during service, can produce lubrication problems and especially for scanner applications can cause contamination of adjacent optical surfaces. For this reason dry lubricants have been suggested.

Dry lubricants divide into two main classes: laminar soft materials whose low shear allows surfaces to slip over each other and hard materials which run with minimal wear. Much of the development work on dry lubrication has been carried out for space applications where bearings must run in vacuum and temperature extremes. In these difficult circumstances ball bearings with dry lubricants have been successfully employed for applications with restricted angular rotation, but for spin axis bearings performance is more limited and less predictable than that of bearings where fluid film lubricants can be employed.

Where ball bearings must be run in extreme conditions then ball bearings with solid lubricants could be considered, although there are usually alternative approaches where bearings with proven performance can be used. Gas bearings can be utilized for temperature extremes, and by running in a reduced atmosphere of helium (or hydrogen) the polygon windage can usually be kept sufficiently low (see Chapter 8) to obviate any advantage of running in vacuum.

9.4 OTHER BEARING TYPES

9.4.1 Liquid Bearings

Plain oil bearings were available long before either ball or gas bearings, but it is only since the emergence of the spiral groove bearing in the early 1960s that self-acting bearings using oil have become contenders for precision spin axis applications. Spiral groove oil bearings are similar to those using gas, described in Section 9.2.3, but because of the much higher viscosity of oil, they will be much smaller for the same load capacity. This makes them especially suitable for small-diameter, high-load-capacity instruments.

Unlike gas bearings, spiral groove oil bearings do not normally have the advantage of being surrounded by their lubricant supply, and the oil must therefore be retained within the bearing. When rotating, the pumping action of the spiral grooves does this, but when stationary the oil would normally leak out; this has been a major problem in the development of these bearings. Another problem is that voids can occur in the fluid because of cavitation or air entrapment at the inlet.

Methods of preventing leakage when stationary include (1) the use of barrier film coatings; (2) using ferrofluids, which are held in the bearing gap by magnetic force; and (3) using special greases. Good results on the use of special greases to overcome both leakage and voids have been reported by Muijderman et al. in 1980 [20] on the application of spiral groove journal bearings to video equipment spindles. Another method to overcome the problems of these bearings described by Molyneaux in 1982 [21] is to completely surround the bearing with a chamber, attached to the rotor, filled with a grease developed to stay in position but to have low shear viscosity.

All these methods, while successful, have some limitations (e.g., in size, speed, or design adaptability). Further development is probably required before spiral groove oil (or grease) bearings become generally available for use in rotary scanners.

Spiral groove oil bearings have only two bearing surfaces which, when rotating, will be separated by a relatively thick oil film. This oil film will average out the roundness errors of the bearing surfaces, to produce a spin axis accuracy, similar to gas bearings. Their shock load capacity should be better than for ball bearings, as the load will be distributed over a larger area. The use of oil will limit the temperature and speed ranges and will give potential problems with oil outgassing and migration.

Within their range of application, they should cost less to manufacture than spiral groove gas bearings, as the larger bearing gap will ease tolerances. They are also likely to be inherently stable, because of the grooves and the damping of the oil (see Section 9.2.3).

While most work with self-acting liquid bearings has been with the spiral groove design, other types such as tilting pad and stepped pocket have also been used for instrument bearings. These will have similar problems in retaining the liquid within the bearing. One interesting development uses a ferrofluid liquid with a stepped pocket design, the fluid being contained by magnetic seals.

9.4.2 Magnetic Bearings

With magnetic bearings the rotor is suspended entirely by magnetic forces, producing very low drag and allowing operation from stationary up to very high speed. An active electromagnetic control system is normally used to improve stiffness and stability which adds complexity and cost. Active control is especially useful for high-speed operation, as resonances can be changed and controlled electronically.

It is unlikely that magnetic bearings will be used for laser scanning applications at present because of cost and complexity, except possibly where high-speed rotation in vacuum is essential. The recent advances in high-temperature superconductivity may eventually change this situation, and simple passive superconducting bearings could become a practical possibility within the next decade.

9.5 PROCUREMENT OF ROTARY SCANNERS/BEARINGS

There are many possible approaches to the provision of scanners and their bearings. The choice is necessarily wide because of the extensive range of applications, each with its own individual requirements in cost, accuracy, life, speed, size, etc. Other considerations such as the motor type, whether the optical rotor should be overhung or symmetrical between the bearings, the option of running in an enclosed housing with a low-density atmosphere and environmental conditions can all affect the choice of bearings and design. Even for a particular bearing type there can be a wide variation in standards. This is especially true of ball bearings where the quality and performance range is extremely wide. The following are some of the more general approaches to obtaining scanners.

1. Select a commercially available ''standard'' motor and fit an optical rotor to it. For ball bearing motors, this is a common approach for low-cost and low-accuracy scanners and has the advantage of being easy and cheap to try with a wide choice of motors available. Disadvantages are that life and performance will probably be unknown or variable. The design will also not be optimized so that power, weight, and size may be greater than necessary. For large quantities a special motor, designed to a requirement specification, could be obtained which should ensure improved performance.

Externally pressurized bearing motors (motorized air bearing spindles) which are available commercially have also been used. These should have high accuracy and stiffness and have predictable performance, but could be larger and possibly more expensive than required.

2. Purchase a complete spindle from a specialist manufacturer and fit a motor and optical rotor to it. High-accuracy ball bearing spindles can be obtained, often with integral bearings (see Section 9.3), and if this has been properly specified and assembled in controlled conditions then performance should be predictable.

The need for an external motor could produce a larger overall size than necessary.

Externally pressurized spindles are available with similar characteristics to the motorized types outlined in the first approach.

3. Build the scanner assembly in-house. The facilities needed for this would depend upon the performance required. It is likely that this course can only be justified where facilities already exist for other comparable quality instruments, some specialist knowledge or experience is available, or where large throughput is contemplated. Some large companies produce scanners exclusively for use in their own optical systems.

4. Purchase a complete rotary scanner from a specialist manufacturer. Manufacturers can vary greatly in the optical accuracy, range of performance, and design options they can provide.

Designs based on standard ball bearings are readily available, often off the shelf. At the lowest level these may be put together as described in approach 1, or they may be special proprietary designs utilizing standard bearings, usually ABEC5 or 7. Such assemblies will normally run in a mixed lubrication regime, i.e., without full separation of the bearing surfaces by an oil film producing intermittent asperity contact. Lubrication degradation will gradually occur limiting life and performance. This limitation could be improved by the use of TiC balls or other measures as described in Section 9.3.8.

The highest-quality ball bearing assemblies will use "special" bearings or in-house finishing techniques (see Section 9.3.2) to improve the bearing geometry and so ensure the true fluid film separation of the surfaces associated with long life and reliability. Because of the higher accuracy of the components of these bearings, random axis "wobble" and noise will also be reduced.

Commercial scanners based on externally pressurized bearings are now fairly widely available. Many of these designs derive from those produced for the machine tool industry and are often bulkier than other types. If properly designed, they provide an extremely accurate rotational axis with high load capacity, and are therefore especially suited to large accurate polygon assemblies, generally for slow to medium speeds. As speed increases, availability becomes less and price will rise. This is because of the special design techniques and better manufacturing and assembly standards required to overcome high-speed dynamic effects. Extremely high speeds are possible.

More compact designs of externally pressurized gas bearings are becoming available. These would probably utilize higher accuracy components to improve touchdown load capacity and to reduce air consumption.

Self-acting gas bearing scanners are only available from a few suppliers, as highly specialized techniques of manufacture and assembly are necessary to achieve the extremely high bearing geometric accuracy required. This accuracy is largely offset by the basic simplicity of the bearing components, so that highly competitive designs, for many applications, are possible. They are particularly suited to high-performance units where the bearings are incorporated into the complete scanner assembly as part of an optimized design approach, to ensure that the benefits of the bearings are maximized.

Compared to the long historical developments of precision ball bearing technology, self-acting gas bearings are in their relative infancy and have therefore much greater potential for further development than the other types outlined here. Advanced materials, designs, and manufacturing methods are being developed which could give increased output at lower cost while maintaining high geometric accuracy. This will widen the range of application and availability of these bearings.

9.6 BEARING-RELATED OPTOMECHANICAL ERRORS

Geometrical errors of the bearings (Section 9.6.1) and those of the polygon (Section 9.6.2) combine to produce errors in the parallelism of the facets to the spin axis, thus causing scan repeatability errors (weave). This applies equally to regular polygons and those with deliberately irregular facets where a repeatable scan pattern is still required. Monogons are a special case as some of the errors will be seen merely as a small fixed change of facet angle which may not be critical.

9.6.1 Bearing Errors

Errors can be divided into (1) those regular once per revolution, (2) irregular, i.e., not once per revolution, and (3) coning. The first two are caused by geometrical errors inherent in the bearings or distortion during mounting of the bearings. Such errors will be greatest for ball bearings, much less for liquid bearings, and least for gas bearings where the accuracy of components necessary for performance plus the averaging effect of a compressible gas will ensure that the bearing contribution to spin axis weave is very small, typically less than 1 arc sec.

Bearing runout errors in standard ABEC ball bearings were given in Table 9.1. In the case of ABEC 5P and 7P the inner runout, as would be seen in a shaft rotating assembly, is slightly less than the outer runout, as with the polygon mounted external to the bearings. These radial runouts will cause a spin axis wobble which will have a regular (once per revolution) component due to the track roundness error of the rotating bearing and an irregular component caused by the errors in other bearing components coupled by the ratio of the speed of the ball set to the spin speed. This ratio, known as the epicyclic ratio, is almost invariably an irregular fraction.

If standard bearings were used without any adjustment to reduce these bearing errors (see Section 9.6.3), the resulting spin axis wobble, for a 2-in (5.08 cm) distance between the bearings, could be 30 arc seconds, 20 arc seconds, and 10 arc seconds for ABEC 5P, 7P and 9P, respectively.

Coning is caused by eccentricity errors between an axis referenced by the polygon mounting surface and the centers of the bearings. This can probably best be visualized in the case of a perfect cylindrical shaft mounted eccentrically on perfect bearings. The shaft center will develop a conical motion on rotation resulting in a sinusoidal scan pattern for a perfect polygon mounted on it. Of course, both ends can be eccentric and in different planes, producing two superimposed sinusoidal scanning errors.

Eccentricity errors are always present because of (1) tolerance runouts on the bearing mounting surfaces, (2) any clearance fits used, and (3) inaccuracy between the bearing mounting surfaces and the polygon mounting surface.

9.6.2 Polygon Errors

There are two main sources of error: (1) the variation of the polygon from perfect shape (pyramidal and random errors) and (2) polygon mounting surface errors (axes tilt).

Pyramidal

Pyramidal error is the part of the angle between the facet and the polygon axis which is constant for all facets. As it is constant, it can usually be allowed for.

Random

Random errors are likely to be caused by errors in finishing the facets, either due to faults in the master indexer rotational axis or in the method of finishing. These errors are likely to be small.

Tilt Errors (Sinusoidal)

Tilt errors are caused by misalignment of the polygon axis to the spin axis. This can occur due to errors in (1) finishing the polygon relative to its mounting surface or (2) mounting the polygon relative to the spin axis. Because of the number of manufacturing steps involved, these errors are likely to be the largest source of polygon weave errors. To hold a weave of, say, 10 arc seconds on a 1-in (2.54 cm)-diameter mounting surface would mean that the *total* buildup of mounting errors during manufacture and assembly must be less than 50 μin (1.27 μm).

9.6.3 Some Special Techniques to Reduce Errors

1. Weave resulting from bearing errors is inversely proportional to the distance between the bearings. Bearings should therefore be as far apart as possible within the limits of sensible design. Designs with the polygon mounted between the bearings have some advantage.

2. If the regular bearing runout errors are measured and the high spots noted, they can be matched at both ends to reduce weave from this source.
3. The surface of the rotor assembly, where the polygon is to be mounted, can be finish machined while it is rotated on its own bearings. This can be highly effective, as it can remove all coning errors, which are often the largest component of weave. Disadvantages are possible bearing damage or contamination, as machining must be done on a completed assembly, and extra costs and design constraints.
4. By making the polygon mounting position adjustable, sinusoidal errors from both the bearing assembly and the polygon can be reduced or eliminated entirely.

   The simplest method is to rotate the polygon on its mounting to back off the errors of one against the other; this will be most effective if the errors of both are the same; however mounting errors are usually greater because of bearing assembly tolerance buildups.

   Another method is to introduce some flexibility into the polygon mounting surface so that it can be tilted with the polygon in position to eliminate sinusoidal weave. This is effective for low-speed applications, but could give problems at higher speeds because of shifts caused by the higher forces.

An alternative for outer rotating scanner assemblies where the outer bearing is flange-mounted is to allow for radial adjustment of the bearing after checking the polygon weave. This radial adjustment tilts the spin axis to bring it in line with the optical axis (see Figure 9.17). It will also slightly tilt the outer bearing to the spin axis, but as the correction necessary is likely to be less than 20 arc seconds no detrimental effects will be seen especially on ball bearings. This method has been shown to be highly effective, and ball bearing scanners of this type have been built with total scan weaves of less than 4 arc seconds.

All the preceding methods can be used to reduce or eliminate polygon tilt or sinusoidal weave errors. Random or irregular bearing errors will still be present but have been shown to be almost negligible for gas bearings and very small for liquid bearings. For ball bearings, random errors depend upon the roundness accuracy of individual components, and where these are completely checked and selected random errors of a few arc seconds are possible.

9.7 PERFORMANCE COMPARISON

Any comparison of the types of bearing covered (gas, ball, liquid) must necessarily be generalized because of the large number of variables outlined in previous sections. There are general characteristics, however, which make one type more suitable for certain applications. Table 9.2 summarizes some of the most likely requirements, together with the suitability of the types of bearing considered to meet these requirements.

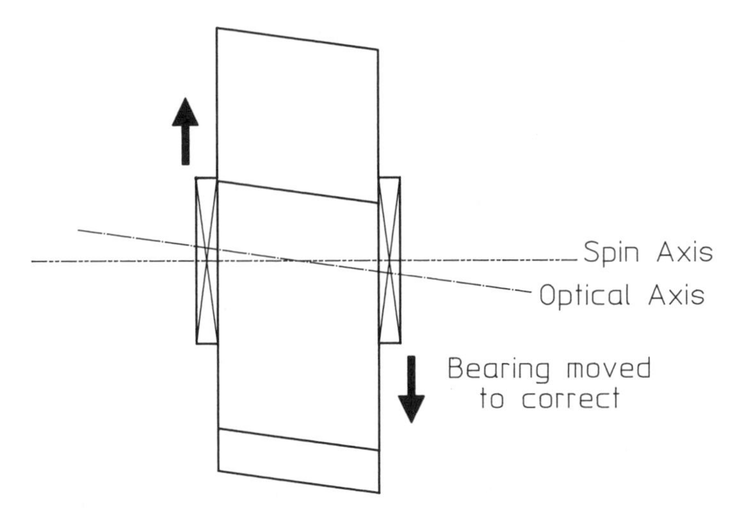

Figure 9.17 Axis alignment by moving bearings.

It is only where more than one type would be suitable for a particular set of requirements that a choice is necessary. To aid this choice, the advantages and disadvantages of all four types are listed in more detail in Section 9.7.1.

Spiral groove liquid bearings are not generally available, and only limited information has been published on their performance. They are most likely to compete with ball or gas bearings in the lower-speed, higher-accuracy range.

Externally pressurized bearings are most likely to be used in less compact applications, where their excellent performance over a wide speed range, or their ability to work in dirtier atmospheres, is needed and where an external pressure supply can be provided.

The most likely area of contention is the choice between angular contact ball bearings and spiral groove gas bearings, for compact, medium-speed (up to 25,000 rpm) applications. A more detailed discussion on the merits of these two types is therefore given in Section 9.7.2.

This comparison cannot be comprehensive over such a wide possible field; magnetic, squeeze film, and ferrofluid bearings could also have been considered, as could hybrids between these and the other types. Magnetic and squeeze film bearings are more specialized and complicated than any of the others, and would probably be considered only if none of these were suitable. Ferrofluid bearings, as such, have low load capacities and are most likely to be used, as seals, to extend the performance of the other types, for example, to prevent oil migration from ball bearings, or to allow self-acting gas bearings to be used in dirty atmospheres, or to hold lubricant in place in self-acting oil bearings. Finally, it

Table 9.2 Summary of the Performance of Likely Bearing Types[a,b]

Feature	Bearing type			
	Ball	Pressure gas	Self-acting gas	Self-acting liquid
Speed (rpm)				
0–1000	L	G	P	L
1000–20,000	G	G	G	G
20,000–30,000	L	G	G	L
above 30,000	P	G	G	P
Spin axis error (arc sec)				
$<\pm 10$	L	G	G	G
$>\pm 10$	G	G	G	G
Life (h)				
0–1000	G	G	G	G
1000–20,000	L	G	G	G
above 20,000	L	G	G	L
Low noise	L	L	G	G
Noncontaminating	L	L	G	L
Use in dirty atmosphere	L	G	P	L

[a] This is a general comparative guide only; special designs could improve on these ratings.
[b] G, good performance; L, limited performance; P, poor or no performance.

should be emphasized that this comparison is for guidance only, as all the bearing types considered are capable of improvement, by special design or manufacturing, to take them outside the limits of performance suggested.

9.7.1 General Comparison

Ball Bearings

Advantages

- Can be run in either direction
- Will usually continue to run after shock loads (but performance and life may be affected)
- Can be run in relatively dirty atmosphere
- Inexpensive for low-performance scanners
- Readily available
- More compact than gas bearings
- Lower starting torque than self-acting gas bearings

Disadvantages

Limit on spin axis accuracy
Possibility of oil migration onto optics
Limit on speed
Limit on life (depends upon quality and speed)
Susceptible to damage by severe shock loads especially when stationary

Self-Acting Liquid Bearings (Spiral Groove)

Advantages

High-accuracy spin axis possible
Very compact
Should have long life (limited information available)
Low noise
Fairly easy manufacturing tolerances

Disadvantages

Possibility of oil migration onto optics
Very limited availability
Limit on speed
Small size may limit shaft stiffness

Self-Acting Gas Bearings (Spiral Groove)

Advantages

Extremely high spin axis accuracy possible
High speed
Almost unlimited life in clean atmospheres
No contamination of optics
Compact
Wide temperature range
Very low noise

Disadvantages

Run in only one direction
Need reasonably clean atmosphere
Less widely available than ball bearings
Higher starting torque than ball bearings
Tight machining tolerances

Externally Pressurized Gas Bearings

Advantages

Extremely high spin axis accuracy possible
Good performance from zero to high speed
Very high stiffness possible
Long life with clean air supply
Wide temperature range
Fairly easy manufacturing tolerances

Disadvantages

Need separate air supply
Fairly bulky
Can be liable to instabilities
Sealed system impossible because of exhaust gas

9.7.2 Choosing between Ball Bearings and Self-Acting Gas Bearings

Generally, ball bearings have been used in lower-speed, lower-accuracy systems, while for higher-speed, higher-accuracy systems it is necessary to use gas bearings. It is in the medium-accuracy, medium-speed (up to 25,000 rpm) range that the choice is not so clear cut and would depend on the particular specification. Some of the factors likely to be considered are reviewed here.

Bearing Power

Figure 9.18 shows a comparison between measured running power for spiral groove conical bearings and bearing manufacturers' data for ball bearing dynamic lubricant drag. The ball bearings shown are R2 [3/8 in (9.525 mm) nominal OD], R3 [1/2 in (12.7 mm) nominal OD], and R4 [5/8 in (15.875 mm) nominal OD], which cover the likely range of sizes for rotary scanners. The gas bearings considered are of roughly similar load capacity.

The effect of lubricant viscosity dominates ball bearing drag torque, as illustrated in Table 9.3, where a 40°C fall in lubricant temperature produces a fourfold increase in drag. In practice, a viscosity should be chosen to have sufficient elastohydrodynamic film thickness [say, 6–12 μin (150–300 nm)] to ensure a safety margin over raceway surface roughness at maximum operating temperature. Gas bearing running power is proportional to gas viscosity, which changes little with temperature.

Spin Axis Errors

Gas bearings are inherently more accurate because they have only two surfaces separated by a gas film, whose compressibility averages out roundness errors of these surfaces. Errors achievable should therefore be about an order of magnitude lower than for ball bearings. This is usually only important for extremely high accuracy, as polygon and mounting errors are usually more significant.

Life

The life of an instrument ball bearing is usually terminated by lubricant degradation (Section 9.3.4), which depends on factors such as speed, load, atmosphere, bearing quality, and lubricant. The life of a gas bearing, if it is run in a clean atmosphere, is almost certainly superior to that of a ball bearing.

Shelf Life

Ball and gas bearings should both have adequate shelf life under suitable storage conditions.

Vibration and Shock

1. Bearings stationary. This is the worst condition for a preloaded ball bearing because of the high local stresses (hertzian stresses) already present in the ball contact zone and the absence of an oil film between the surfaces. On the other hand, a gas bearing will take higher loads because of the larger areas and close conformity of bearing surfaces.

2. Bearings rotating. In this case the elastohydrodynamic film and the distribution of the load stresses lead to a large increase in the ability of ball bearings to withstand vibration. When running, gas bearings have been shown to withstand similar levels of vibration to ball bearings.

Contamination from Bearings

With ball bearings, the lubricant may be gradually centrifuged from the bearing, especially at higher speeds, or may migrate as a vapor as temperature increases. Suitable shields or seals may be required to protect adjacent optical surfaces. Gas bearings produce no significant contamination.

Number of Starts

There is no normal limit to the number of start-ups with a ball bearing. For a gas bearing, where surface rubbing occurs during start-up, some wear occurs, and special hard (antiwear) surfaces are used. With suitable surfaces, gas bearings have been shown to be capable of tens of thousands of starts/stops without degradation in bearing performance.

Speed

It is probable that the top speed of ball bearing scanners should be between 25,000 and 30,000 rpm. Higher speeds can be achieved but only at the expense of life, even with the highest-quality bearings. In this case the cost would be similar to gas bearings, but their performance would be worse.

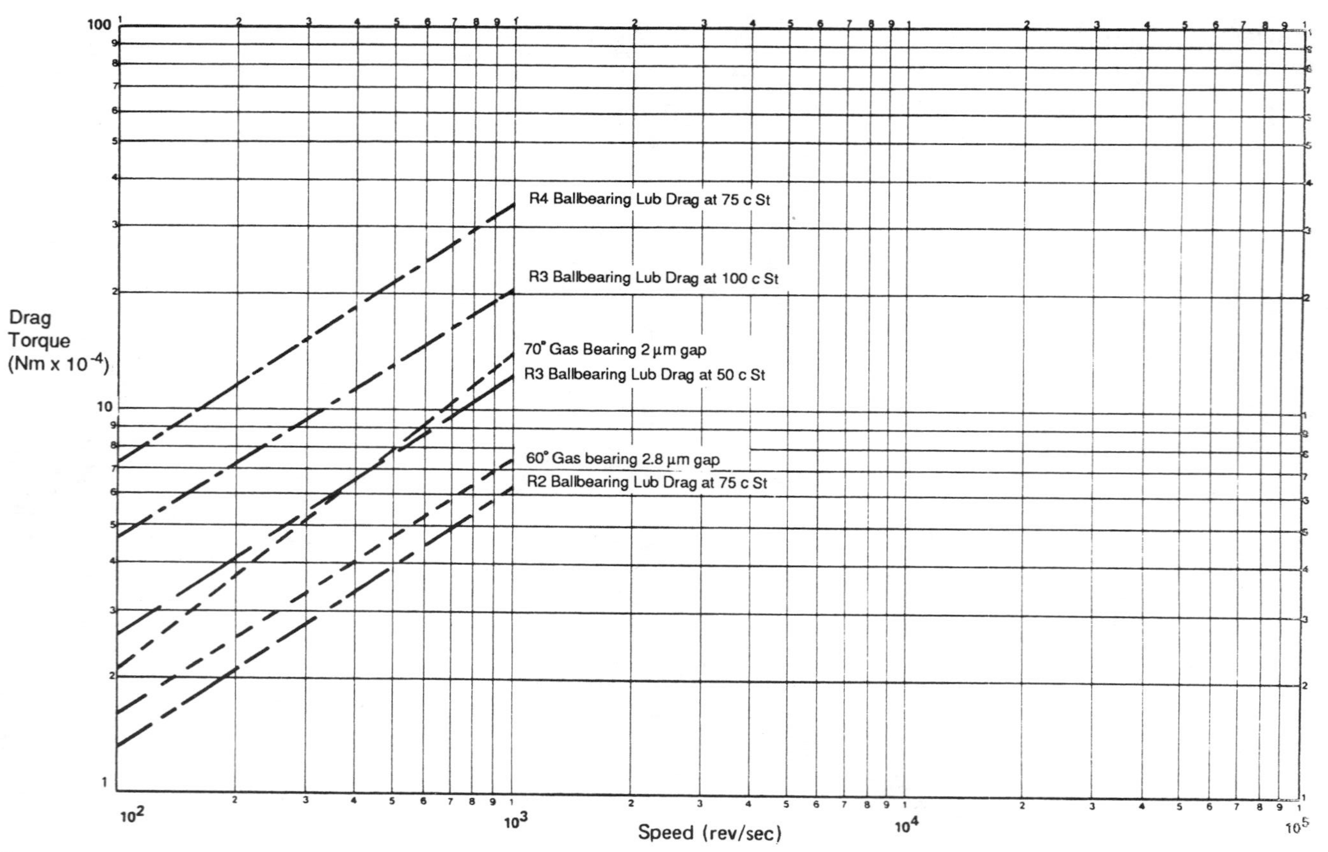

Figure 9.18 Comparison of ball bearings and self-acting gas bearing: Drag torque versus speed.

Table 9.3 Effect of Lubricant Viscosity on Ball Bearing Drag and EHL Film Thickness[a]

Lubricant		Minimum EHL film[b] thickness (m × 10^{-6})		Drag torque (N-m × 10^{-4})	
Temp. (°C)	Viscosity (cst)	100 rev/s	250 rev/s	100 rev/s	250 rev/s
63	50	0.1	0.2	3	5
47	100	0.18	0.33	4	8
21	430	0.48	1.0	11	20

[a] Typical results for an R3 bearing.
[b] Minimum elastohydrodynamic film thickness between balls and inner race.

Self-acting gas bearing designs have been run successfully at speeds from about 1000 to 500,000 rpm. In practice, the upper speed limit will normally depend on overall design considerations, such as the mass and inertia of the rotor, dynamic forces generated, and external forces applied.

Cost

Experience has shown that for high-accuracy systems the cost of ownership of gas bearing rotary scanners can be similar to or less than that for equivalent ball bearing versions. For low-accuracy, low-speed systems ball bearing scanners should cost less.

Atmosphere

Ball bearings can be run in very low pressure atmospheres with special lubricants. They will have longer life in a clean inert-gas atmosphere. Gas bearings require a clean dry gas atmosphere for maximum life. Some designs have been run in an office and laboratory atmosphere with good results.

REFERENCES

1. Tower, B. (1883). First report on friction experiments. *Proc. Inst. Mech. Eng.*, (a) 1883, pp. 632–659; (b) 1884, pp. 29–35. Also, Second report on friction experiments. *Proc. Inst. Mech. Eng.*, 1885, pp. 58–70.
2. Reynolds, O. (1886). On the theory of lubrication and its application to Mr. Beauchamp Tower's experiments including an experimental determination of the viscosity of olive oil. *Philos. Trans. R. Soc. London, 177*, 157–234.
3. Kingsbury, A. (1897). Experiments with an air lubricated journal. *J. Am. Soc. Nav. Eng.*, *9*, 267–292.
4. de Ferranti, S. Z. (1909). *Air Bearing for High Speeds*. U.S. Patent 930,851.

5. Harrison, W. J. (1913). The hydrodynamical theory of lubrication with special reference to air as a lubricant. *Trans. Camb. Philos. Soc.*, *22*, 39–59.
6. Gross, W. A. (1962). *Gas Film Lubrication*, Wiley, New York.
7. Grassam, N. S., and Powell, J. W. (1964). *Gas Lubricated Bearings*, Butterworth, London.
8. Wilcock, D. F., ed. (1972). *Design of Gas Bearings*, Vols. 1 and 2, Mechanical Technology, Inc., New York.
9. Whipple, R. T. P. (1949). Herringbone pattern thrust bearing. *U.K. Atomic Energy Authority Report AERE-T/M-29.*
10. Vohr, J. H., and Pan, C. T. H. (1968). Design data of gaslubricated spin axis bearings for gyroscopes. *Mechanical Technology, Inc. (New York), Technical Report M.T.I.-68 TR 29.*
11. Faddy, D. (1969). The theoretical performance of statically loaded pairs of conical gas bearings. *U.K. Royal Aircraft Establishment Report 69174.*
12. Burghdorfer, A. (1959). The influence of the molecular mean free path on the performance of hydrodynamic gas lubricated bearings. *J. Basic Eng., Trans. ASME*, *81D*, 94–99.
13. Sixsmith, H. (1959). The theory and design of a gas lubricated bearing of high stability. *First Int. Symp. Gas-Lubricated Bearings*, Washington, D.C., pp. 418–434.
14. Licht, L., and Elrod, H. (1960). A study of the stability of externally pressurized gas bearings. *J. Appl. Mech., Trans. ASME*, *82*, 250–258.
15. Archard, J. F. and Cowking E. W. (1965) Elastohydrodynamic lubrication at point contacts. *Proc. Inst. Mech. Eng.*, *180 Pt. 3B*, 47–56.
16. Holmes, J., and Shepherd, J. (1973). Control of track quality. *International Ball Bearing Symp.*, Charles Stark Draper Lab., Inc., Cambridge, Mass.
17. Boving, H. J., Hänni, W., Hintermann, H. E., Flühmann, F. and Waelti, M. (1987). High precision coated steel balls. *Bearing Conference Proceedings*, Orlando, Florida.
18. Sioshansi, P. (1987). Ion implantation for wear and corrosion resistant bearings. *Bearing Conference Proceedings*, Orlando, Florida.
19. Hanson, R. A. (1987). The effect of advanced materials on wear in instrument bearings. *Bearing Conference Proceedings*, Orlando, Florida.
20. Muijderman, E. A., Remmers, G., and Tielmans, L. P. M. (1980). Grease lubricated spiral groove bearings. *Philips Tech. Rev.*, *39*(6–7), 184–198.
21. Molyneaux, A. K. (1982). Theoretical and empirical evaluation of spiral grooved grease bearings. *Proc. U.S./U.K. Fluid-Film Bearing Meet.* Southampton University, U.K., pp. 62–76.

10

Galvanometric and Resonant Low-Inertia Scanners

Jean I. Montagu

General Scanning, Inc.
Watertown, Massachusetts

10.1 INTRODUCTION

After defining the two types of low-inertia scanner (LIS), this chapter describes typical applications and then presents the critical issues underlying LIS system design. The concluding sections review the factors affecting selection and design of each scanner element, and describe performances of specific commercial LIS products.

Low-inertia scanners have long been used by science and industry in optical scanning. These scanners were used in the early electrocardiograph recorder and in the first oscilloscope, as well as in the making of film sound track. Advances in computer and laser technology have stimulated their evolution, and they are now found in complex precision instruments.

The term *low-inertia scanner* designates any scanner that deflects light with the aid of a moving mirror and which can achieve full scanning speed in less than about 1 s. There are two fundamentally different types of low-inertia scanners: galvanometric and resonant.

Galvanometric scanners can produce a steady-state deflection and follow with considerable fidelity sinusoidal as well as nonsinusoidal wave forms such as sawtooth, triangular, and random access. Resonant scanners can operate at only one frequency or its immediate vicinity, and they can oscillate only in harmonic motion: they are of interest because much higher scanning angles can be achieved at high frequencies by taking advantage of resonant amplitude magnification.

10.2 APPLICATIONS OF LOW-INERTIA SCANNERS

Like most industrial products, the LIS has evolved to meet the needs of its users. Literally hundreds of different applications have appeared in industry, medicine, military hardware, and communications. Some of these are artificial horizon displays for aircraft cockpits, scanning laser acoustic microscopes, silicon wafer annealers, bar code readers, birthmark erasers, lead bonders, microfilm equipment for computer output, color separators for printing, component insertion designators, 3-D lithography, laser communication systems, image digitizers, dye laser tuners, facsimile transmitters and receivers, film thickness monitors, surface flaw detectors, wire-harness wire-path designators, interferometers, infrared scanners, laser trimmers, laser cutting machines, laser markers, laser printers, liquid crystal large-scale displays, light and sound shows, 35mm movie converters for television, night vision systems, laser ophthalmoscopes, printed circuit board exposure and inspection equipment, phototypesetters, prepress proofers, Q-switches, ring lasers, range finders, scanning infrared microscopes, optical disk readers, robotic vision systems, war game simulators, and bacteria colony counters.

Although the market for each application is different, applications can be grouped in categories according to common technological disciplines.

10.2.1 Visual Communication

Visual communication is one of the most demanding applications of the LIS. The human eye has not only an astounding resolution, but also very fast and accurate autocorrelation capabilities which permit the detection of minute image imperfections. Physiologists have shown that the human visual system readily detects pixel position errors as small as 1/1000 in and that it is even 10 times more sensitive to the "periodic scanning artifacts," which consist of periodic shading variations, generally called a venetian blind defect [1,2].

Raster scanning is the preferred imaging technique for visual communication. Most applications in this group need gray scale presentation capability and the most demanding also require color. The following classifications are representative of these applications:

1. Print plate exposure
2. Laser printing for electronic publishing
3. Photocomposition

4. Phototransmission
5. Photoplotters
6. Computer output microfilm
7. Large-scale display

Print Plate Exposure

The use of the hot-lead linotype process begin to diminish shortly after World War II when the first phototypesetters were introduced. At present, newspaper composition is largely automated. Text, headlines, line art, photographs, and advertisements are paginated and edited into final form via computerized systems.

As computer input devices for image data, or digitizers, scanners convert text and graphics into picture data elements which can be manipulated digitally. Capabilities range from processing letter-sized documents to full newspaper pages with monochromatic or full-color content. As output devices, or plotters, scanners are capable of scanning a full printing plate (typically 20 in × 24 in) as well as a single photograph or poster for enlargement or reduction. The scanners must be fast, reliable, versatile, and economical, with resolution compatible with the industry. Originally drum scanners were used. Flatbed scanners are now accepted because they are simpler, faster to operate, and can handle nonflexible paste-ups.

Flatbed scanners used as input devices or output devices share a common layout. The work to be digitized or the photographic film to be sensitized is translated at constant speed under a beam scanned along the other axis. Optical elements are usually located near the work, ensuring field flatness and normal incidence at all points. Pixel position is usually determined with the help of a grating. In order to achieve small spot size, the scanning mirror is greater than 1 in in diameter.

A flatbed scanner built as a digitizer normally has a scanned focused laser beam which is absorbed or reflected by the work, and the reflected light is collected by a wedge or bundle of fiber optics located in close proximity to the trace and then guided to a photomultiplier for further processing. As a plotter, a modulator interrupts the beam to create the desired image. In order to correct for line straightness, a small high-speed linear scanner corrects the laser beam prior to its entering the beam expander. Figure 10.1 is a schematic presentation of such a unit built around the SX resonant scanner of General Scanning Inc. (GSI).

Table 10.1 shows the range of the performance needed for system continuity in the printing industry.

Printing quality varies as widely as applications do. The requirements shown in Table 10.1 are typical of those for scanners and plotters used in the printing industry. The reader unfamiliar with graphic art reproduction may find these unnecessarily demanding; however, the production of sharp, clean pictures can

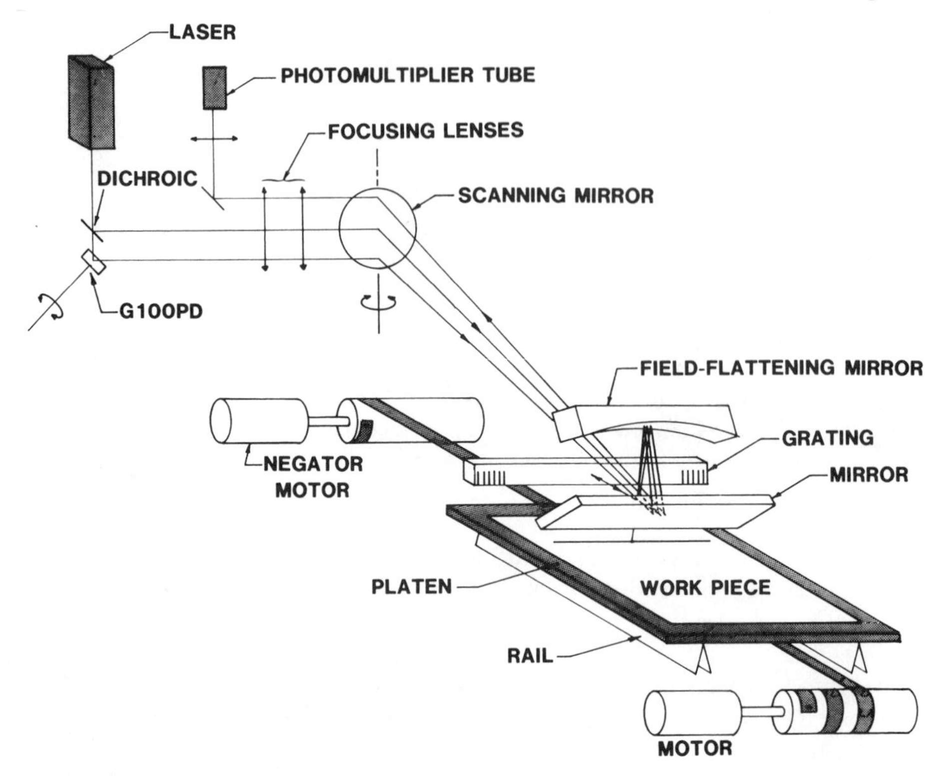

Figure 10.1 Flatbed scanner system.

only be obtained with resolution accuracies of the magnitude cited in Table 10.1. These dictate in turn the use of large mirrors and high-speed scanners with extremely low wobble as indicated in Table 10.2.

Laser Printers

The current generation of computer printers combines electrophotographic and laser printing technologies. These are used by distribution centers and are relatively high in cost. They include such printers as the IBM 3800, Model 3 Siemens ND-3, and Xerox 9790. Their output format, letter size 8.5 in × 11 in, replaces the perforated computer printout at speeds greater than 200 copies per minute.

The comparatively low resolution printers of this type use multifaceted polygon rotating mirrors and offer resolutions of 240 and 300 dots per inch. Higher-resolution printers with fewer copies per minute are also available from 400 to 1200 dots per inch using both rotating polygon and LIS technologies. Their print quality is comparable to that of daisy wheel machines, and they can raster create

Table 10.1 Flatbed Scanner System Performances

Half-tone images	80–200 dots/in
Scan lines per inch	800–2000/in
Line spacing position tolerance	1–2 μm
Resolution	12,000–25,000 pixels/scan
Maximum plate size	20 in × 24 in
System speed	1–2 min/plate

Table 10.2 Typical Resonant Scanner Specifications for Flatbed Scanners

Beam rotation	32°
Resonant frequency	200 Hz
Mirror dimension	35mm diameter
Mirror inertia	39 g-cm^2
Repeatability, scan/scan	1 μrad typical
Wobble, scan left to right/ scan right to left	5 μrad
Life	10^{11} cycles minimum

line art, graphic, and halftone pictures with as many as 16 shades of gray. Electronic digital control systems provide great flexibility in choice of fonts and formatting.

These printers are plain-paper drum machines or coated-paper continuous feed machines, depending on the particular photosensitive or ion deposition technology used.

Laser printers are produced today in large volume at low cost. However, the office environment in which these lightweight machines must operate demands that the scanners be capable of operation under moderate vibration. Table 10.3 shows expected system performance, and Table 10.4 shows typical scanner performance requirements for such an application.

Table 10.3 Laser Printer Performances

Resolution	300–1200 lines/in
Half-tone images	240–600 dots/in
Spot overlap	50%
Spot size	3.5–8 × 10^{-3} in
Speed	3–10 s/page

Table 10.4 LIS Performances

Mechanical	
oscillating frequency	600–1200 Hz
mechanical deflection	20–30°
mirror size	10–20 mm
wobble	5 μrad
Environmental	
vibration	1 g, DC-100 Hz
temperature	20–100°C

Photocompositor

The printing preparation technology is at this date in a state of flux. Software and computer work stations now available make complex pagination and composition increasingly easy and inexpensive, putting both proofing and production capability directly into the hands of the writer. Such systems are known as desktop publishing systems and many are available in the current marketplace.

The new generation of photocompositors—able to compose graphic images as well as text—have performances approaching those of plate making equipment described in the section on print plate exposure. They print on film sensitive to laser diode or helium-neon wavelength. Figure 10.2 is a schematic presentation of a modern "pell box" imagesetter from ECRM using a low-inertia scanner. Table 10.5 lists the scanner performances called for by this application.

Phototransmission

In electronic printing or facsimile transmission, images are digitized at one location and transmitted via telephone lines or satellite to another location, where

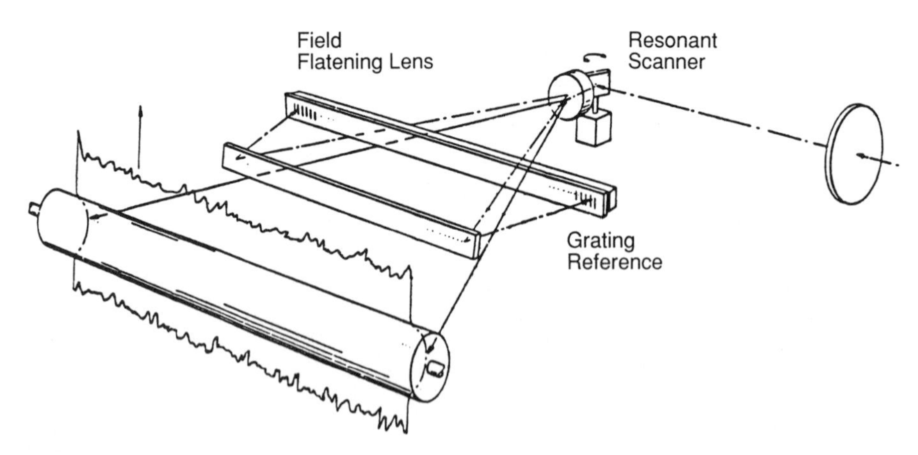

Figure 10.2 Pell box optical schematic.

Table 10.5 Photocompositor Performances

Line length	310 mm
Line resolution	
low	7,250 pixels
high	31,000 pixels
Line spacing	
low	40 μm
high	10 μm
Speed (time to set an A4 page)	
low	20 s
high	150 s
Linearity error, max.	0.3 mm
Accuracy	
Scan repeatability (scanner jitter and drift)	
adjacent line, max.	8 μm
within 100 lines, max.	16 μm
within 1000 lines, max.	32 μm
over 1000 lines, max.	40 μm
Scan spacing (scanner wobble and transport errors)	
adjacent lines, max.	5 μm
within 10 lines, max.	10 μm
within 1000 lines, max.	32 μm

they are reconstructed. Phototransmission handles continuous tone information as well as text and line art. Color images are sent as a set of color separations. Unattended operation at the receiving end is possible. The data content per image is considerable because of the gray tone information attached to each pixel.

Laser scanning systems printing on dry silver, heat-developed photographic paper, or film are the only systems now used for newspaper phototransmission. A standard 8 in × 10 in picture with a resolution of five lines per millimeter requires 4 or 5 min for image transmission. Reproduction speed is low because it is not commercially justifiable to use transmission lines with a greater capacity than the 4800-Hz bandwidth of the telephone lines. Police or military systems which operate without communication bandwidth limitation transmit at a rate of 60–200 scan lines/s for both transmission and reception.

Since such pictures are intended for viewing rather than scaling, avoiding image discontinuities is important, because the human eye readily detects a pixel 0.1 mm out of position. Critical scanner features are repeatability, low wobble, and low jitter. The transmission of color images demands a high degree of address repeatability (rather than accuracy) to ensure proper color registration.

Linearity and temperature stability are sometimes required by special applications and environmental conditions.

Photoplotter

Photoplotters are conceptually similar to phototransmission receivers, but the required output quality as well as other factors lead to a totally different imaging technology and different output media.

Photoplotters create from digitally acquired modalities film transparencies with gray scale and resolution equivalent to those of standard chest x-ray negatives. They are used to supply permanent records for medical computerized tomography (CT) and magnetic resonance (MR) imaging, as well as prints or transparencies of computer manipulated graphic imagery.

The required image specifications are presented in Table 10.6. It should be noted that though the image quality required is extremely high, as exemplified by the pixel placement variation tolerance of 8 microns (one part in 54,000 of frame scan length) and a capability of 1024 shades of gray, the true dimensional accuracy is not very demanding at ½% of full scale.

The scanning system is a postobjective scanner (scanning mirror after imaging objective lens) diagrammatically shown in Figures 10.3 and 10.4. It is exceptional with respect to the scanner and lens translator selection. The gray scale resolution and the small spot size lead to a large optical pupil, while the fast scan rate of 200 Hz and the large image size require a fast lens translator with a long displacement.

The photoplotter represented in Figures 10.3 and 10.4 has a resonant line scanner as well as a resonant lens translator, both operating at 200 Hz. A judicious phase and amplitude definition and control of the two resonant units yield a flat-field focused spot. The frame scanner is a high-accuracy galvanometric scanner with performances similar to those described in Table 10.13a.

Both resonant scanners are frequency tunable to accommodate frequency drift due to temperature, construction or other perturbations. They also can be synchronized to the rate of the data source. Their specifications are described in Tables 10.7 and 10.8.

Computer Output on Microfilm

The output of computer data directly on microfilm makes it possible to produce microfiches with user-readable information. Two machines now commercially available are capable of printing over 200 pages a minute, with 64 lines of 132 characters, as well as producing graphics.

The "Komstar," shown schematically in Figure 10.5, uses a line scanner to position characters on a line, and a page scanner to position lines on a page. Each line scanner moves an array of nine beams. The characters are composed by turning each beam on or off at any of the 1500 possible positions on a line. The nine-beam array is created from a single laser passing through an acoustooptic modulator operating at nine different frequencies between 40 and 50 MHz. Each frequency is controlled in amplitude to ensure individual beam output independent

Table 10.6 Photoplotter Performances

Film, format, resolution	
Film sizes	14″ × 17″, 11″ × 14″, 8″ × 10″
Film type	red sensitive
Number of addressable pixels per line	4446
pixel width (spacing)	80 μm
pixels per horiz. dimension	12.5/mm (318/in)
Number of lines per frame	5398
pixel height	80 μm
pixels per vert. dim. (line spacing)	12.5/mm (318/in)
Total pixels	23,999,508
Spot size	Beam overlap at half power points
Spacing between horizontal lines	
Line-to-line spacing variations	±1% of nominal (max.)
True position line placement variation	±0.5% of full scale (0.5% × 17″ = 0.085″)
Spacing between vertical lines	
Nonrepeatability (jitter)	Less than 8 μm for 14″ × 17″ image Less than 6.5 μm for 11″ × 14″ image Less than 4.6 μm for 8″ × 10″ image
True position pixel variation	±0.5% full scale (0.5% × 14″ = 0.070″)
Image density	
Density uniformity (solid gray, exposed at half digital full scale intensity, ptp variation anywhere on frame will not exceed 0.2 OD)	Less than or = 0.2 OD ptp variation
Gray scale range	1024 (10 bits) levels, 2.8 OD max.
"Gamma" settable for	16 film types
Throughput	
X scanning frequency	200 Hz
Frame time	27 s to expose a full image
Laser type	HeNe at 633 nm
Data transfer rate	1.8 Mbytes/s avg.
Physical	
Size	Within 22″ H × 22″ W × 32″ D
Weight (with electronics)	150 lb

Note: values shown refer to the 14″ × 17″ image size, unless otherwise indicated.

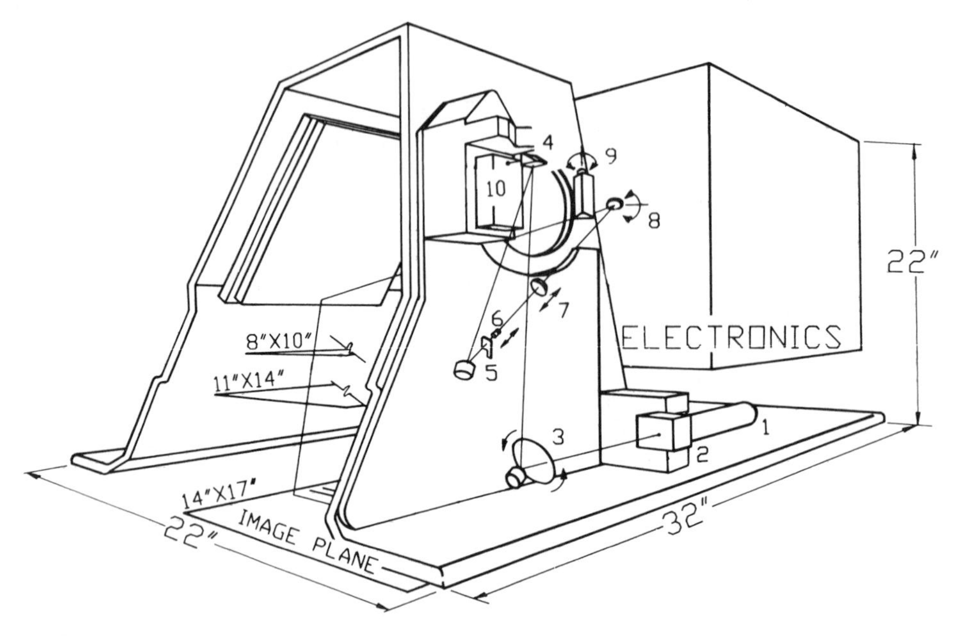

Figure 10.3 Photoplotter scanning system (physical layout).

of the number of beams sharing the same source at any instant. Of the 4.8-ms cycle of the line scanner, 1 ms is reserved for retrace and settling. The scanner operates with optical position feedback and locates characters with an accuracy of 1 part in 10,000 of line length. The cross-axis wobble of the two scanners and the interaction between them must be below 5 arc sec to produce consistent accuracy.

The ERIS II, manufactured by Datagraphix, Inc., has similar performance characteristics but uses a holographic rotating scanner to position a four-beam cluster to generate symbols. A number of line scans is required to complete a line of characters. Page scan is achieved with an LIS. The scan duration is 150 msec, with an accuracy of 30 arc sec and less than 5-arc sec wobble.

Large-Scale Image Projection

Images with large and changing information content are often displayed by large-scale projection. These types of displays are used in applications such as public address panels; traffic control displays for air, city, and harbor activity; interactive computer-aided design/computer-aided manufacturing (CAD/CAM) output for the design of complex products, such as very large scale integrated (VLSI) chips or hydraulic systems for aircrafts or power plants; CAD/CAM microfiche or microfilm generation; and military information distribution centers.

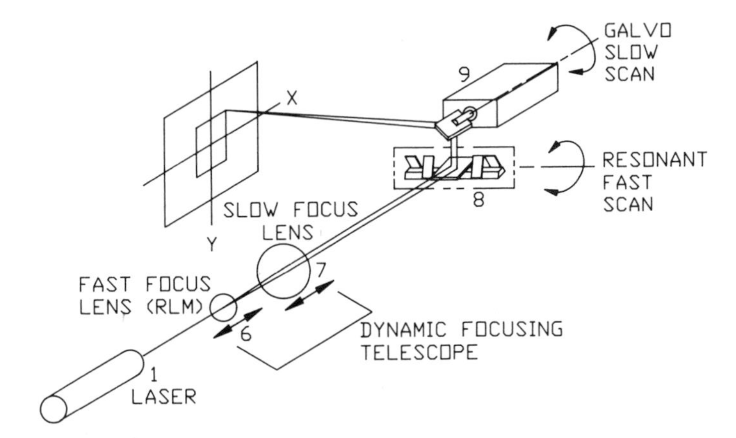

Figure 10.4(a) Photoplotter scanning system (optical path).

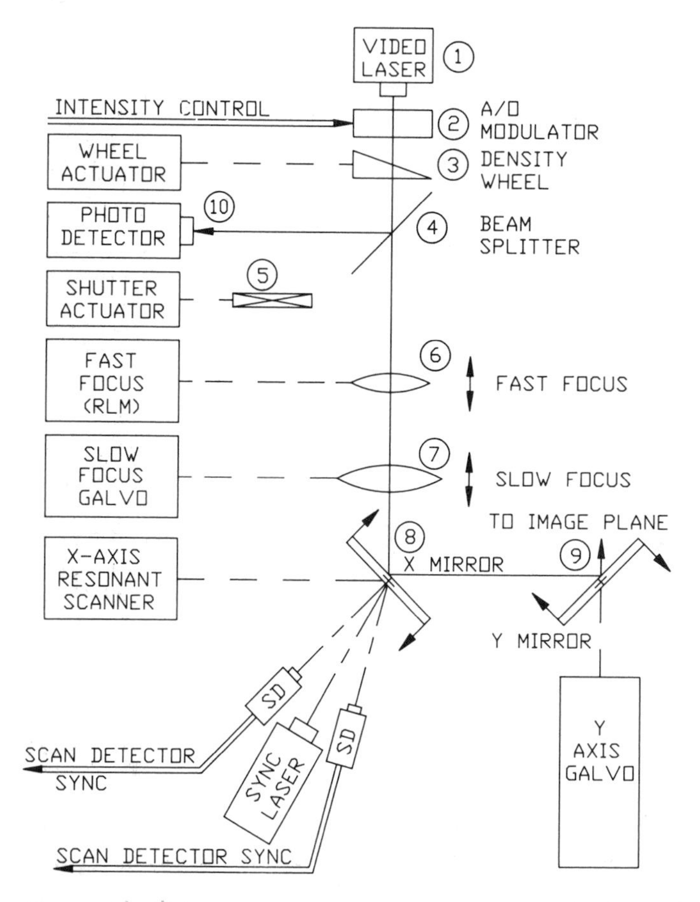

Figure 10.4(b) Photoplotter scanning system (symbolic description).

Table 10.7 Parameters of Photoplotter Tunable Resonant Scanner

Parameters	Units	
Natural frequency	Hz	200
Beam rotation	radians	0.84
Wobble, max.	μrad	3
Spring rate	g-cm/rad	5300
Mirror diameter	cm	1.7
Mirror flatness	HeNe wave	¼
Tuning range	Hz	1.6
Tuning bandwidth	Hz	20
Tuning spring rate	g-cm/rad/amp	100
Tuning power, max.	watts	4.5

Table 10.8 Parameters of Photoplotter Tunable Resonant Lens Translator

Parameters	Units	
Natural frequency	Hz	200
Excursion	cm	0.3
Wobble, max.	μrad	2
Clear aperture	cm	0.4
Tuning range	Hz	20
Tuning bandwidth	Hz	0.2
Tuning power, max	watts	1.0

A high-contrast image is created on a 2-cm to 10-cm square intermediate substrate with a modulated laser scanned in vector, raster, or spiral mode. It is then projected enlarged 10–50 times onto a viewing screen by conventional methods. The intermediate substrate, called a light valve, is either a photochromic film or smectic liquid crystal cell.

The resolution of these displays exceeds that of the best cathode ray tube screens, with densities of 2000–64,000 points per axis. Refresh rates often exceed 30 s; although slow, the refresh speed is consistent with other limitations such as data handling and audience information assimilation. Figures 10.6 and 10.7 are schematic presentations of two such displays.

Scanner resolution, wobble, and jitter are critical parameters for these applications. Most systems use LIS for both axes; in those where vector generation is not necessary, rotating polygons may be used for one axis.

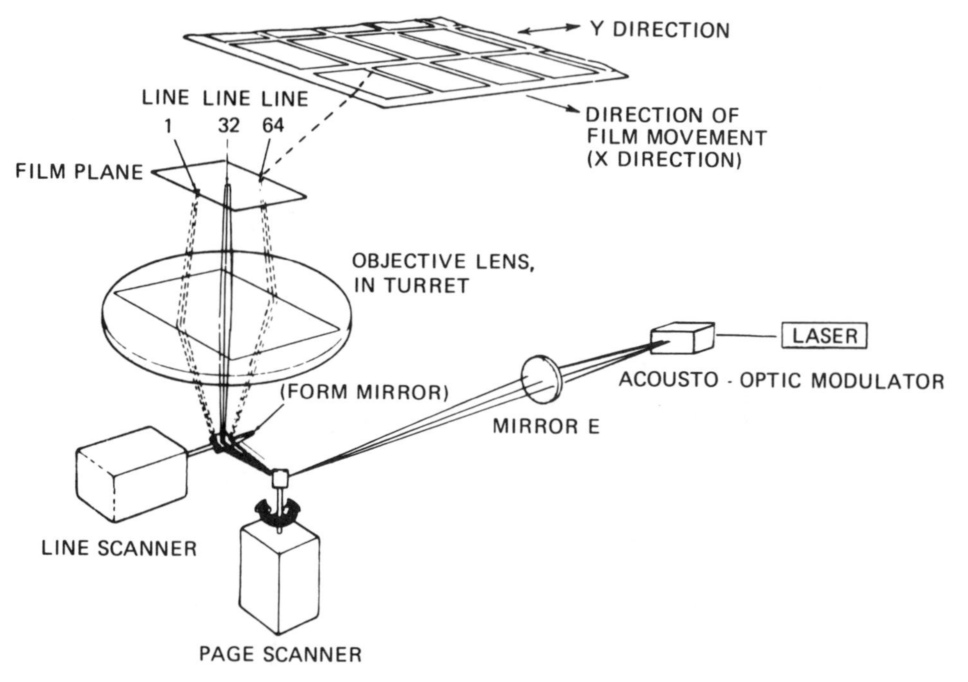

Figure 10.5 "Komstar."

10.2.2 Artificial Extensions of Vision

Scanning systems which extend vision serve three major purposes:

1. To achieve greater accuracy, flexibility, or resolution than conventional techniques. Examples are scanning spectrometers, laser gages, and laser readers for 3D metrology.
2. To extend our vision beyond the reach of our senses—for example, to the infrared spectrum and ultrasound or microscopic field. Usually the information is displayed on a TV monitor; it is also available for numerical analysis.
3. To extend our reach. Examples are satellite communication and satellite docking; there are also a number of military applications, such as heads-up displays for aircraft.

Scanning Grating Spectrometer

A conventional spectrometer design can be dramatically improved with a high-precision LIS to support and scan the grating. As each scan is calibrated with a known mercury or other source, long-term drifts can be compensated for. Com-

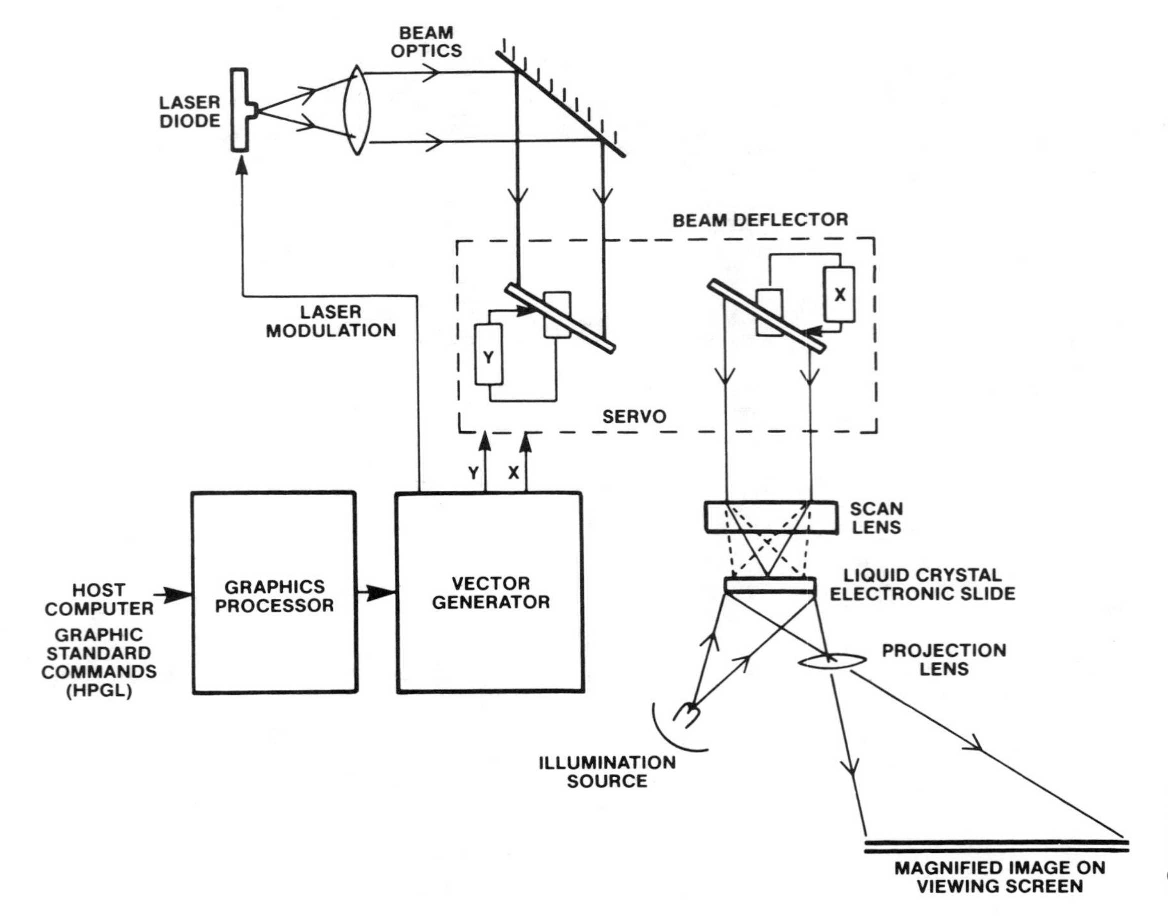

Figure 10.6(a) "SOFTPLOT(tm)" image projection system.

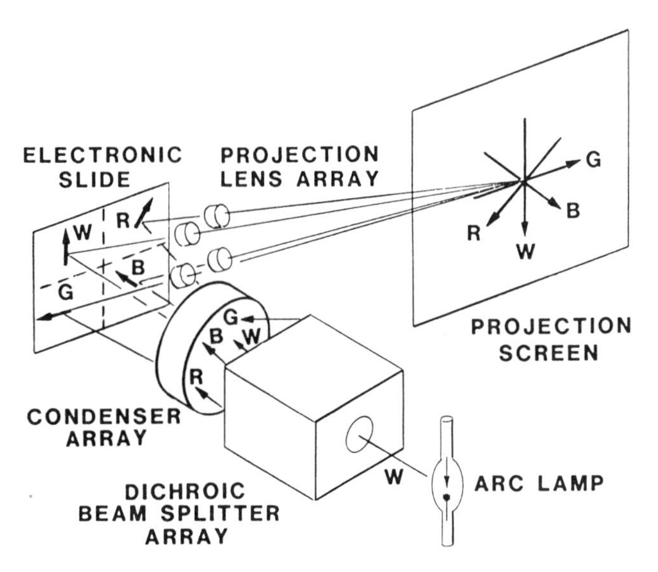

Figure 10.6(b) "SOFTPLOT(tm)" color picture generation.

monly achieved LIS angular resolutions of 1 part in 262,000 correspond to a spectral resolution smaller than $\frac{1}{40}$ Å in the visible range.

Forward-Looking Infrared Sensors

Forward-looking infrared sensors (FLIRS) and night vision systems are made for both civilian and military applications. The civilian sector includes applications such as heat loss surveys for homes and pole-mounted transformers, forest fire control, and security. In the military FLIRS are used to safeguard foot soldiers, as well as tanks, aircraft, and ships.

All objects emit electromagnetic radiation (light). By measuring the intensity and color of this radiation, a thermal image of a field of view can be created. A fairly good representation of the scene is possible even through smoke and vapor, however, the quality of the image is strongly dependent upon the thermal resolution of the system.

A modern FLIRS uses a single or double HgCdTe detector, cooled at liquid nitrogen temperatures to receive a narrow light bundle of limited spectral range. A scanner, acting as a television camera in a raster mode, aims the detector to a sequence of points in the object field. Since commercial TV raster scan rates are used, this equipment interfaces with conventional TV equipment.

Figure 10.8 is a schematic view of a FLIRS using two LIS. The horizontal trace is obtained by a resonant scanner generating 7866 scan lines per second with a viewing angle of 28°. A dual-element detector obtains conventional TV

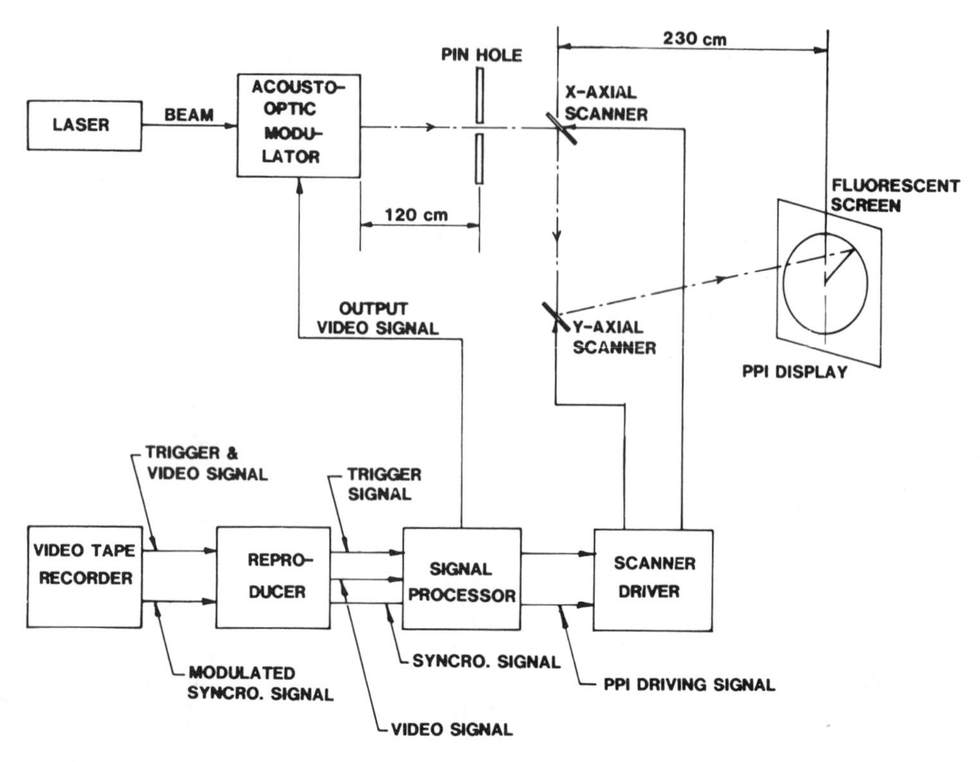

Figure 10.7 Radar imaging system.

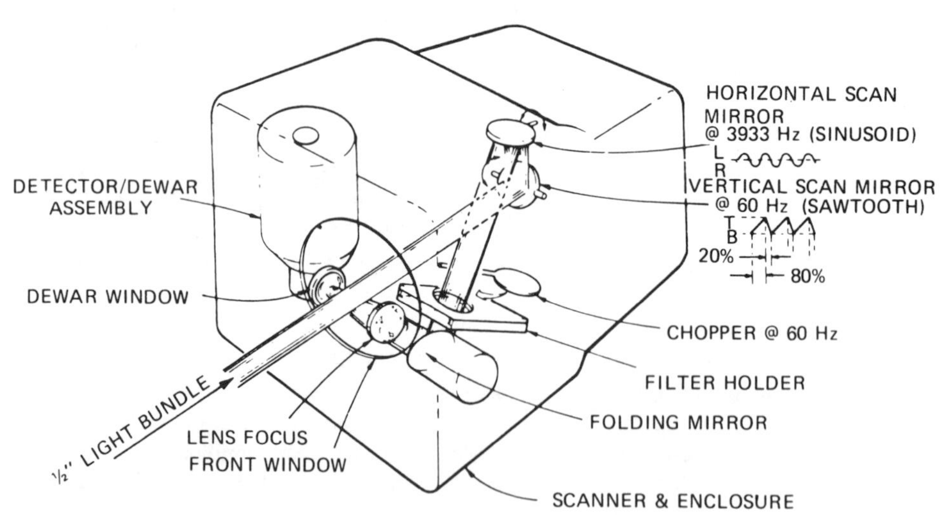

Figure 10.8 Forward-looking infrared sensors.

raster density; a position sensor, closed-loop galvanometric LIS generates the frame scan. The thermal resolution of the system is 0.1°C, and image resolution is comparable to that of commercial TV. Temperature is color coded for ease of interpretation.

Tunable resonant scanners are being explored as a replacement for fixed-frequency units. A cascade of two resonant scanners, the first unit operating at the fundamental frequency at full scan amplitude, and the second unit operating at the third harmonic at 10% amplitude, is the mechanical equivalent of the two first terms of the Taylor expansion of a triangular signal. This structure simulates a linear scan with better than 99% fidelity and 95% time efficiency.

This technique is only practical with tunable resonant scanners, as the two units must be synchronized in phase to each other and preferably also to the monitor display clock.

Such triangular wave scanning permits a uniform energy collection and the use of multidetector signal integration techniques or SPRITE detectors.

Because of their long procurement cycles, FLIRS presently used by the military incorporate traditional rotating polygon scanners. The advantages of resonant scanners, that is, greater reliability, compactness, less power consumption, and freedom from gyroscopic effect, are now being recognized, particularly by military procurers outside the United States.

Satellite Communication

Laser beams are used to ensure privacy of communication between satellites. The divergence of laser can be held down to 10 μrad, making snooping virtually impossible (and latching and communication extremely difficult).

Figure 10.9 presents the relationship of two geosynchronous satellites with the earth. The travel time of a light signal originating from satellite 1 to reach satellite 2 and return, assuming no communication delay, is 0.56 s. During this time satellite 1 has traveled a total of 1.67 km and 1.43 km in the direction normal to the original line of sight, or 17 μad.

This gives credibility to the requirements of millisecond response and 5 μrad beam pointing accuracy in two dimensions under all environmental conditions of outer space.

Laser Scanning Microscope

Conventional optical microscopes are limited in resolution by the wavelength of light, and their depth of focus at high magnification is inconveniently small. Laser scanning microscopes have similar limitations, but, because images are created by accumulating discrete picture elements, they can be assembled from different focus planes by a tracking dynamic focusing scheme. Even in applications where an image is not required, significant advantage is obtained by employing laser scanning techniques. For example, it is possible to detect light

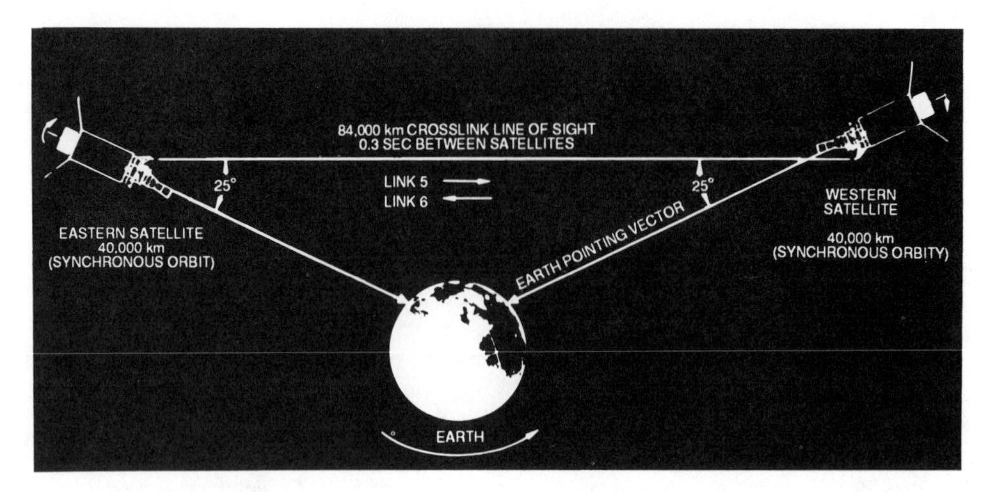

Figure 10.9 DSP cross-link satellite configuration.

diffusely scattered from surface features much smaller than the focused beam spot. Dust or defect detection scanners for silicon wafers or masks use this property.

An important application in the semiconductor industry is the infrared microimaging capability of this technology. Figure 10.10 is a schematic description of such a unit with a conventional microscope used for alignment.

Scanning Laser Imaging System

Scanning imaging systems are applicable to rapid quantitative image capture and digitization where there are special requirements for field size, speed of acquisition, spatial resolution, dynamic range, and low light sensitivity. Conventional optical imaging instruments or the type of scanning microscope described in the previous subsection meet many other imaging system requirements. The same unit can give densitometric imaging information and/or forward laser scattered and fluorescence emission quantitative imaging measurements.

A scanning laser imaging system is a laser flying spot scanner equipped with a translatable focusing objective lens as described in the section entitled "Focus Errors in Postobjective Scanning, page 563." The large field address places strict demands on the scanner's accuracy and resolution as well as the design of the light gathering process.

The scanning laser imaging system of American Matrix Inc. and Ohio State University, represented symbolically in Figure 10.11, uses an optical-fiber-based detector assembly having a numerical aperture value between 0.58 and 0.95 which, combined with proper filters, can capture from 14% to 32% of fluorescence

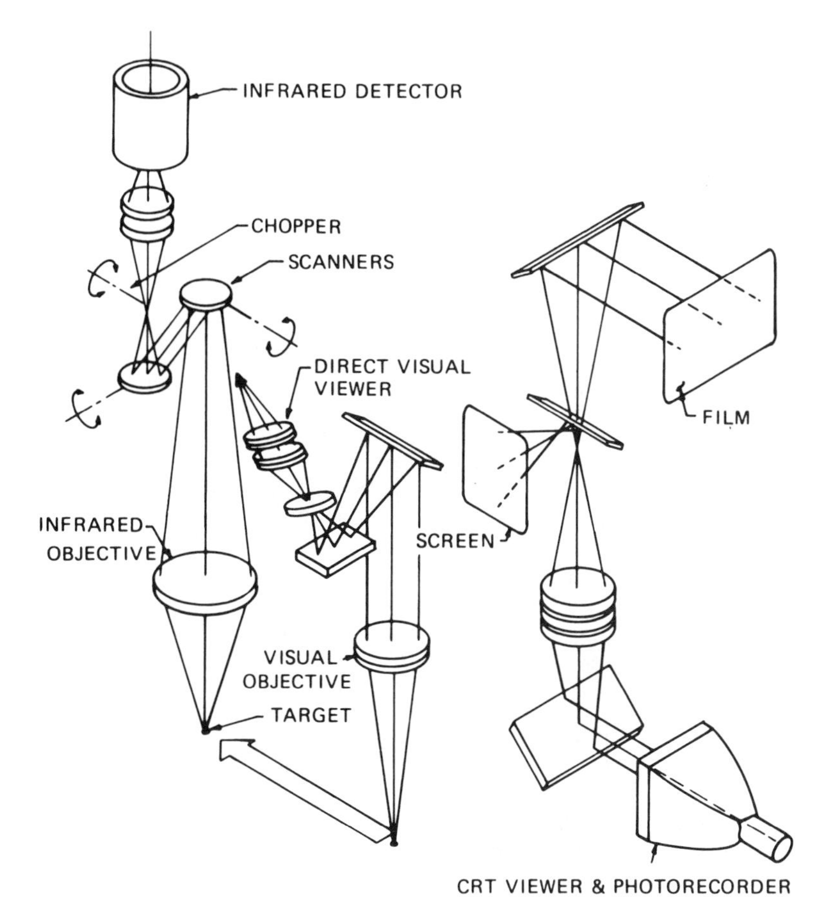

Figure 10.10 Infrared microscope.

emission. Light collection through the optical path rarely can offer similar performances.

This instrument, built with a helium-cadmium laser, is used primarily as a cell or bacteria colony counter. It can address microscope slide preparations as well as agar cultures with substantial depth. It is capable of image capture with a 10-μm spot at 100 scans/s, and yields 4096 × 4096 × 16 bit images.

Scanning Laser Acoustic Microscope

A scanning laser acoustic microscope is a specialized application of the scanning microscope. The work is subjected to standing acoustic waves which create surface deformation conditions representative of subsurface density variations. The scan-

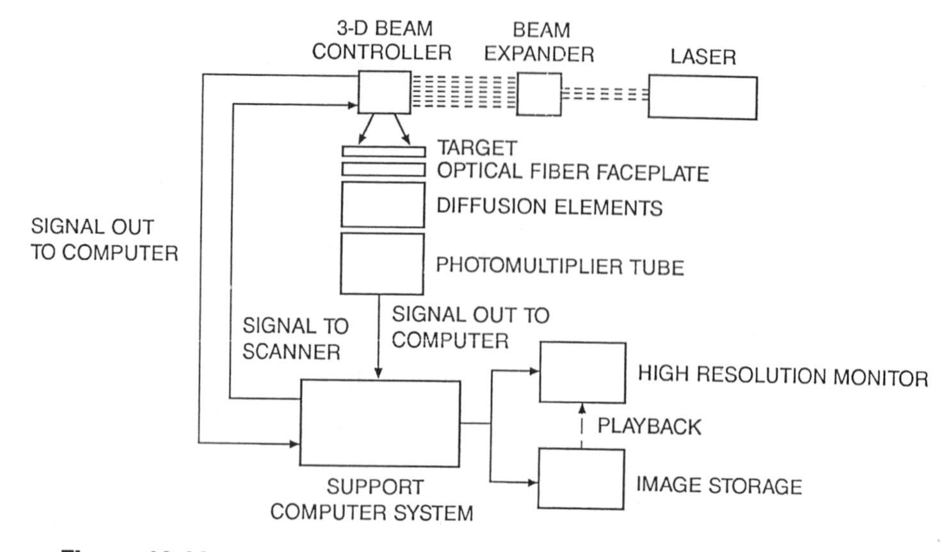

Figure 10.11 Scanning laser imaging system.

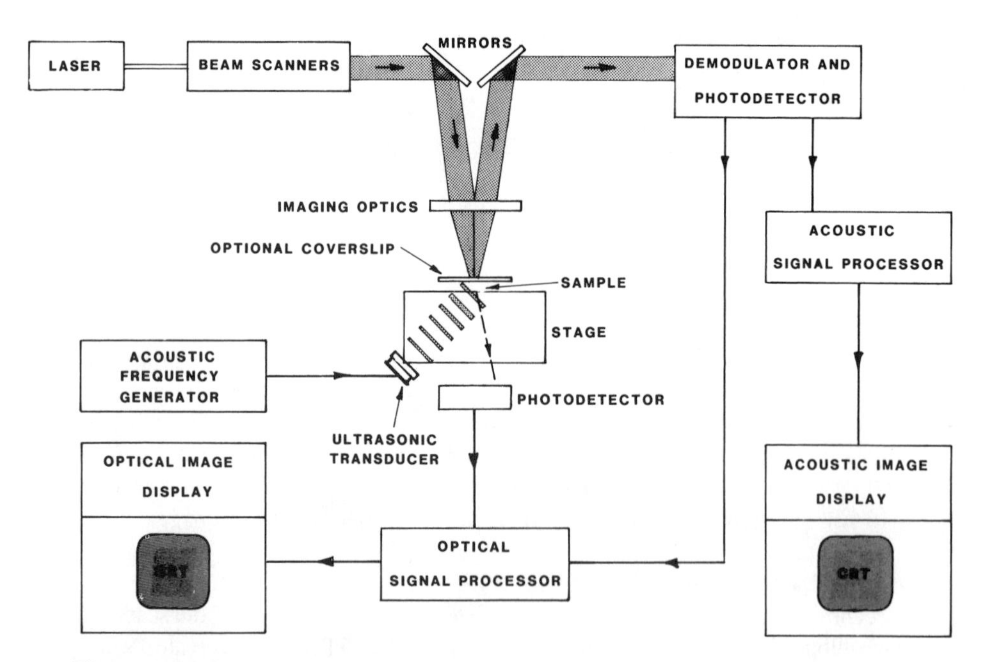

Figure 10.12 Scanning laser acoustic microscope.

ning microscope digitizes this surface information so that it can be interpreted and displayed on a monitor. Figure 10.12 shows a system where the line scan is obtained from an acoustooptic scanner and the frame rate with an LIS. Acoustooptic scanners will not be dicussed here.

10.2.3 Medical Applications

Though numerous medical applications of scanning techniques have been described or reported, it is unclear how prevalent some of these are or may ever be. Port wine, birthmark, and tattoo removal are the most frequently mentioned applications, but numerous others have been considered.

LIS have been incorporated into systems designed to search out the surface of organs for stained cancerous cells and destroy them with a burst of higher-power light. LIS three-axis scanning systems have been considered for sculpting the cornea in order to correct vision defects; a focused UV laser could be used to ablate the cornea with total dimensional control at the micron level. Dentists have also described the use of a scanning system to distribute focused laser power over a larger area than could be done practically by hand.

Most commercial medical applications of LIS are found in ophthalmology.

Computerized Electrooculographic System Diagnostic Eye-Tracking Instruments

Computerized electrooculographic system techniques are used to gain information about patients with disturbed sensations of balance. Such information is acquired by observing the patient's eye movement during certain kinds of visual and vestibular stimuli. A two-dimensional LIS scanning system is used to present such stimuli in the patient's field of vision.

Similar instruments are used to diagnose eye tracking dysfunction, as in screening for reading disabilities. Miniaturized versions are used for pilot training.

Scanning Laser Ophthalmoscope

The scanning laser ophthalmoscope is used to view the fundus oculi, the retina, and its supporting structures, including blood vessels, nerve bundles, and underlying layers. The instrument, built by R. H. Webb and G. W. Hughes at the Retina Foundation in Boston, is a TV ophthalmoscope. It uses flying spot scanner to achieve high-light-collection efficiency and reduce the light needed for viewing and recording. A resonant scanner provides the horizontal scan at the commercial TV rate and a galvanometric scanner provides the 60-Hz frame rate. The viewing field is 55° peak to peak.

Figure 10.13 is a schematic description of the scanning laser ophthalmoscope's optical path. The instrument uses 2% of the pupil opening to create the raster scan on the retina, and 98% for collection of reflected light. It resolves 10 μm

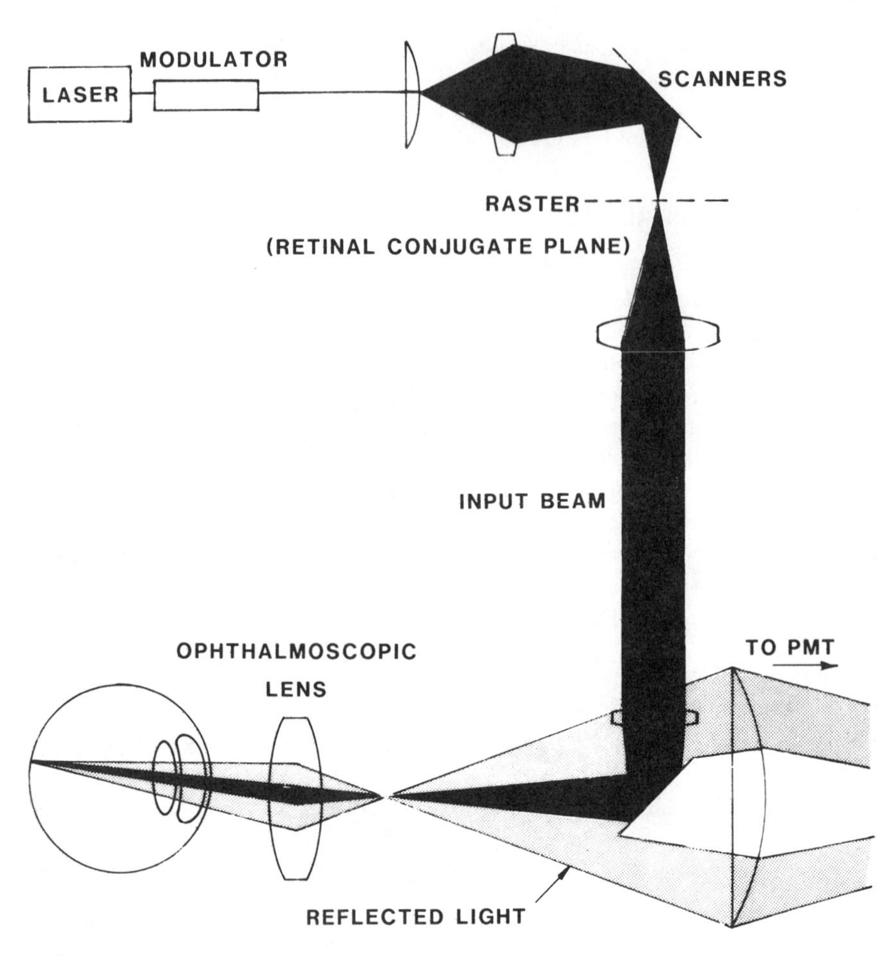

Figure 10.13 Scanning laser opthalmoscope.

on the retina with 5 mW of laser power. A HeNe and an ArKr laser are used to cover the visual spectrum. The power density on the retina does not exceed 50 $\mu W/cm^2$. Continuing developments will lead to greater resolution, and eventually performing laser surgery while viewing the retina may be possible.

10.2.4 Material Processing

The last decade has witnessed a surprising acceptance of the laser as a working tool with both low-power and high-power applications. The beam steering capabilities of galvanometric LIS offer new applications for those lasers. They interface very well with programmable and computerized manufacturing and vision systems. They also offer random addressing capabilities, speed, precision, and the ability to handle large amounts of power. Many industrial applications exist, mostly in the clean and technology driven industries. Among the most common are welding, cutting, stitching, annealing, drilling, marking, metrology, robotics, and laser-induced chemical reactions.

As the reliability of power lasers has improved, and their operating cost decreased, they have penetrated new markets and the demand on scanners has been for higher speed and greater accuracy.

Laser Trimming

Laser trimmers were first designed to adjust by ablation the value of resistors formed by thin-film or thick-film technology. With the advance of the semiconductor industry, requirements for system speed, accuracy, and reliability have grown more demanding, and the LIS has become the standard solution to beam positioning. The main optical element of these systems is a very precise thermally controlled laser, flying spot scanner, as shown in Figure 10.14. In order to satisfy demand for greater excursion, accuracy and speed, scanning heads are built to ensure fine scanner alignment. Such a unit is shown in Figure 10.15.

Figure 10.14 Laser trim system.

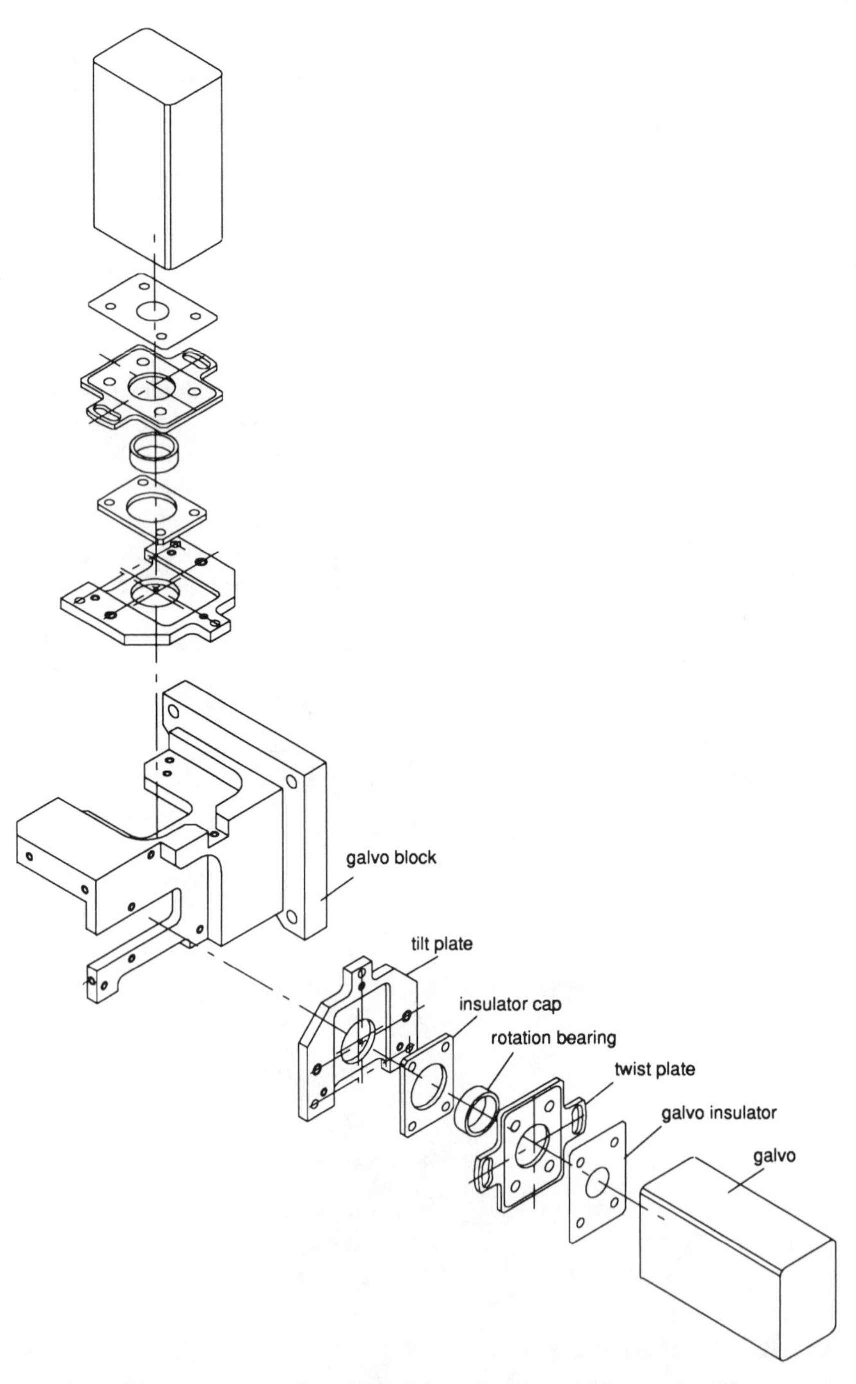

Figure 10.15 Precision two-axis scanning head for laser trimmer.

A YAG laser is used for cutting, and a colinear HeNe laser, with a TV camera for observation, is used for system programming. A high quality multielement, flat-field focusing lens ensures uniform spot size over the entire work area. System calibration, including lens nonlinearity correction, is kept in memory so that the major requirement is for high repeatability. Performances of such a beam positioning head are shown in Table 10.9.

The success of these trimmers has led to the recent development of more precise machines capable of operating directly on integrated circuit chips in order to correct faulty elements or to customize devices. Redundant memory designs for high-density memory devices add extra memory locations to replace defective locations. The redundant memory repair is expected to fix 95% of the defective devices up to and including 16 Meg. memory chips. Table 10.9 also shows the performances of these scanners.

Laser Marking

Marking products with identifying codes or serial numbers can be a difficult manufacturing task, and laser scanning systems have proven to have important advantages over mechanical and inking processes, especially when the marking must be indelible. The first laser markers were introduced more than a decade ago as a natural extension of laser resistor trimming. Now thousands of systems have been installed. Scanning laser markers are best suited for engraving metal parts, for situations requiring rapid change of marking, as in serialization, and for the marking of large symbols on nonflat surfaces such as the ends of compressed-air cylinders.

Marking applications cover a wide variety of parts, from automobile engine blocks to hypodermic needles. One of the larger-volume applications is wafer, chip, and package identification or serialization in the semiconductor industry. Figure 10.16 shows typical YAG laser marking patterns.

Table 10.9 Beam Positioning Head Performances

Application	Thick film/Thin film	VLSI Chips
Work area	150 × 100 mm	30 × 30 mm
Repeatability	20 microns	0.3 micron
Resolution	5 microns	0.1 micron
Peak acceleration	800 G's	800 G's
Settling time, full excursion	5 ms	10 ms
Spot size	25 microns	3–5 microns
Depth of focus	0.55 mm	5 microns
Temperature tolerance	N.A.	± 1°C

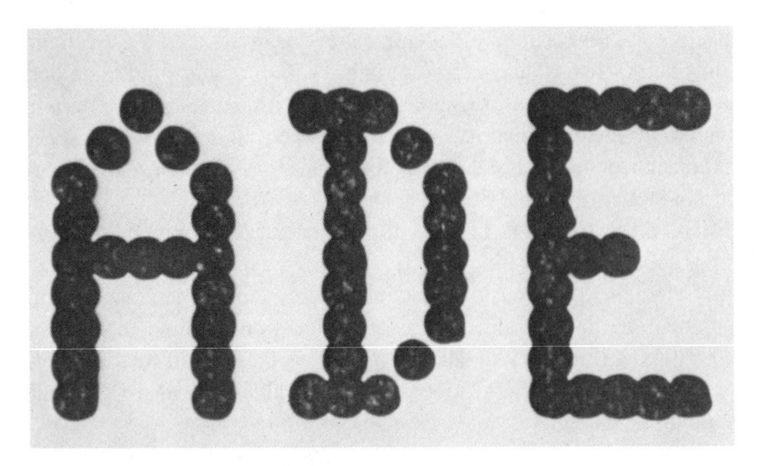

Figure 10.16 Laser marking patterns.

The optical structure of a typical laser marker is that of a preobjective scanner, as shown in Figure 10.25. The marking system of Figure 10.17 addresses a flat work surface of 24 × 24 inches. It is used mostly to generate serialized precision bar-coded tags and short-run instrument panels on a specially treated metal base.

As marking is expected to be machine-readable, location and type must be quite accurate. The optical system has a low F number to reduce the required power and to minimize ablation debris; therefore, field-flattening lenses or focusing systems are desirable to provide a constant spot size.

Laser marking is increasingly used in the production process to incorporate features which either simplify or permit flexible computer-aided manufacturing. For example, a keyboard can be assembled with blanks, and the keys marked in place without fear of assembly errors. Promotional objects as well as signs are also customized in this fashion.

These applications demand faster markers with higher precision than code marking. Table 10.10 lists the performance parameters of laser markers in both identification and manufacturing applications.

Laser-Induced Chemical Reaction (3D Lithography)

This process is used to fabricate quickly out of plastic an ''actual'' full scale model of a 3D CAD ''virtual'' model, as either a solid or a shell. Applications exist mostly in the manufacture of prototypes of objects whose final forms are not limited to plastic.

The material used is a liquid plastic which polymerizes when irradiated with violet or UV light. Numerous photocurable polymers exist, one of the most common being polymethylmetacrylate (PMMA). Presumably this technique will be extended to low-cost high-power CO_2 lasers and heat curable plastics.

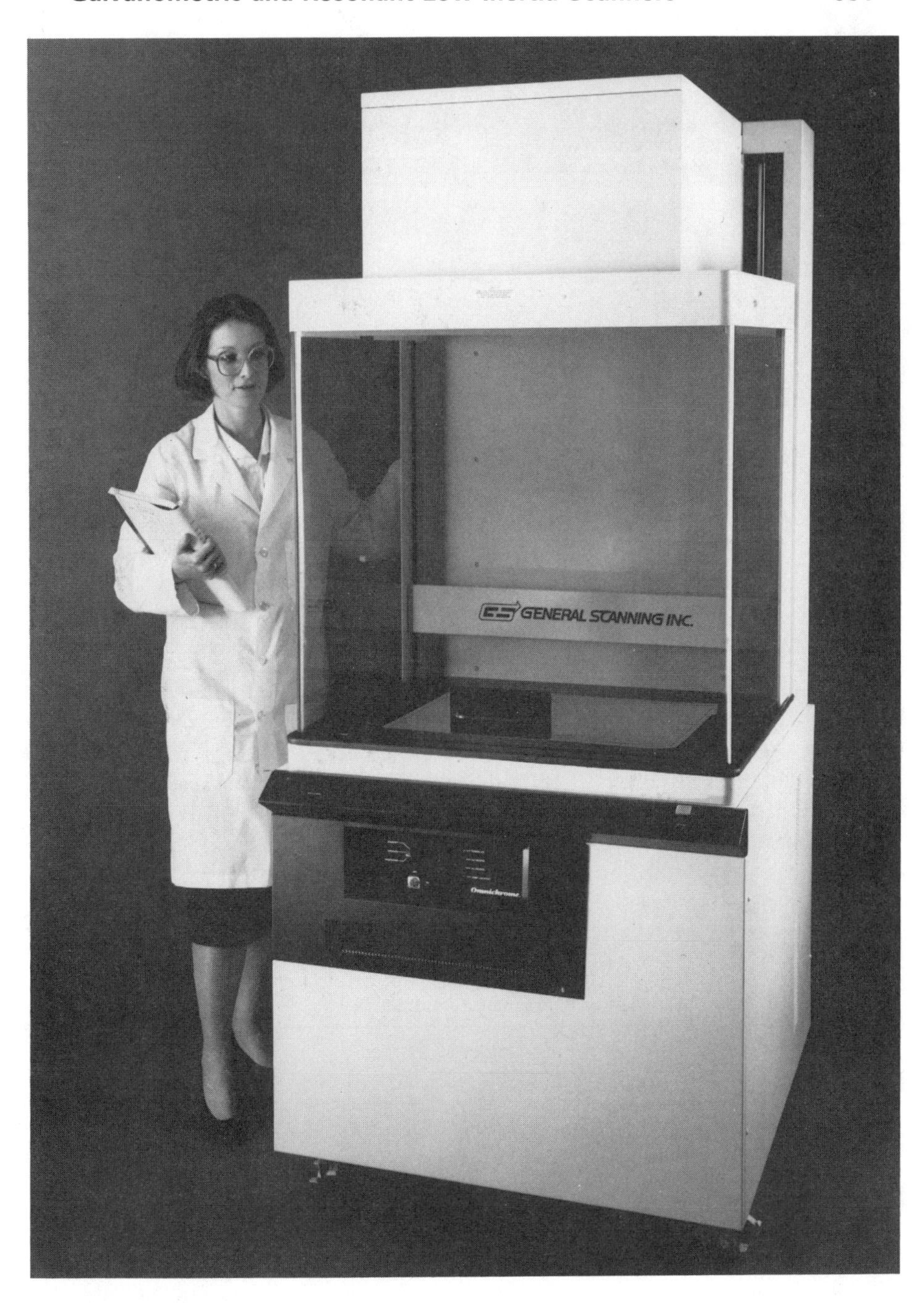

Figure 10.17(a) Scanning laser marker.

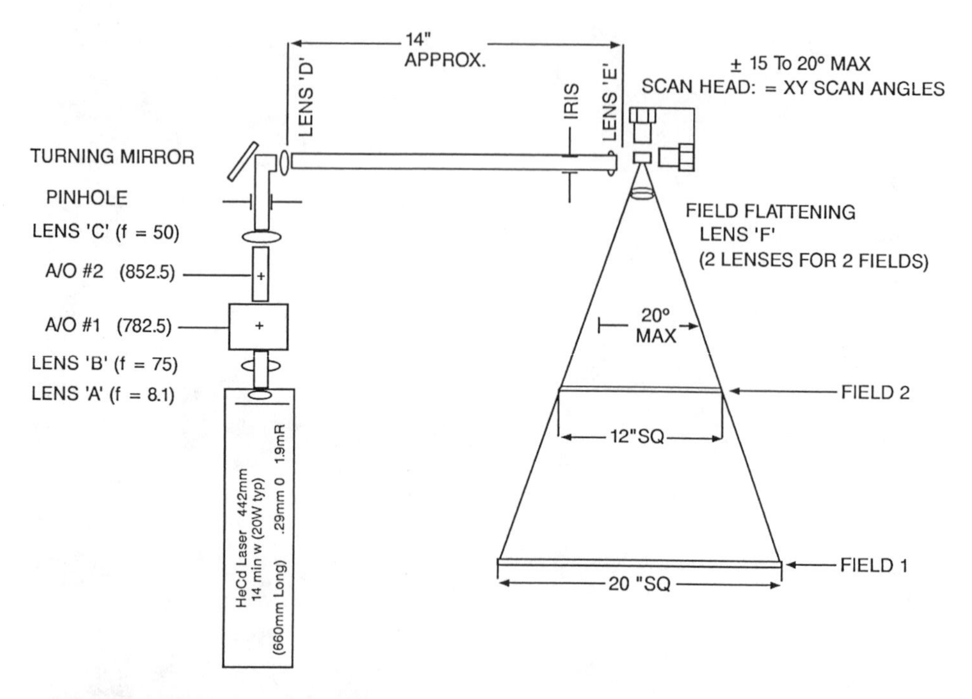

Figure 10.17(b) Scanning laser marker—schematic.

As described in Figure 10.18, a movable platform is located just below the surface of a pool of plastic. The laser draws on the surface of the liquid the bottommost cross section of the prototype. As the platform is lowered, the model is progressively formed by successive layers of polymerized plastic.

A solid or surface model in a CAD computer contains the information from

Table 10.10 Laser Marker Performance Parameters

Application	Identification	Bar Code
Optical beam size	20 mm diameter	20 mm diameter
Scanner response time	5 ms	2.5 ms
Beam excursion	0.55 rad	0.7 rad
Marking field	12 mm × 45 mm	610 × 610 mm
Marking rate	12 char/s	24 char/s
Accuracy		
Intracharacters	±0.005 mm	±0.02 mm
Intercharacters	±0.012 mm	±0.2 mm

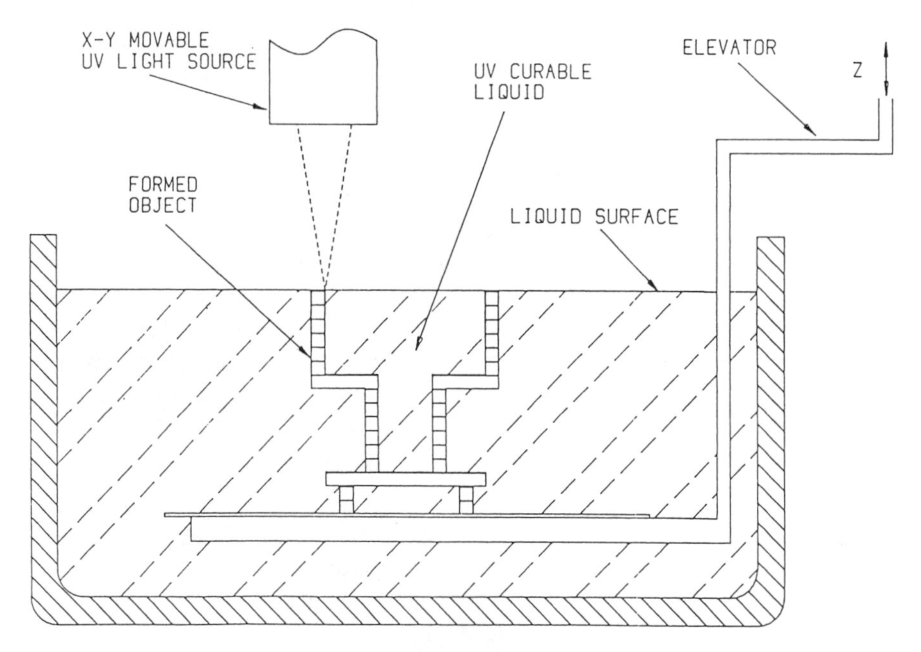

Figure 10.18 Stereolithography apparatus.

which the set of layered cross sections is extracted. The laser must always be kept focused on the surface of the liquid plastic.

The dimensional fidelity of the model is dependent on the accuracy of the scanner. In order to meet industrial expectations, the scanner must be able to locate the laser focal point within 0.001 or 0.002 of an inch on a 10-in scan anywhere on the liquid surface plane.

10.2.5 Entertainment

Scanners can be found in entertainment applications in projection television and "laser light shows."

Laser TV projectors were pursued very early by one of the laser inventors, Ted Maiman at Hughes. Commercial and military use of the idea was delayed more than a decade, and only a few installations exist. Though the image quality of present projectors is exceptionally good, cost and complexity are deterrents.

The use of lasers for entertainment began in the late 1960s, and scanners were incorporated soon after. The first scanners were loudspeakers with mirrors glued to the cone. Later on, diffraction devices, mirrored spheres, and moving coil galvanometers were used. In the early 1970s laser displays, in which modulated

Lissajous figures and mandalas typified the esthetics, used all the capabilities of the open-loop scanners then available. In the fall of 1975 lasers with scanners made graphic images. The November cover of *Industrial Research and Development* was a photograph of the magazine's initials laser scanned on the clouds over Boston, Massachusetts. Computer-generated animation, which used the advances in hardware, software, and laser technologies, soon followed.

Scanned laser graphic projections are best known for the planetrium shows and animated "son et lumière." They also are effective in enhancing meetings and promotional presentations, as well as advertisements and art.

The image is produced by aiming a laser beam at a screen and repetitively deflecting the beam to retrace a controlled path. When the time period is shorter than that of vision decay, the viewer perceives the path as a persistent line. A flicker-free image requires a minimum retrace rate of 15 to 25 Hz, depending on the color, light intensity, and ambient illumination.

Studio Projector

The first entertainment shows were designed as visual representations of music. Suitable signals were filtered out of the sound track and used as source data to drive scanners directly or to simulate auxiliary drive patterns; they were analog systems. Modern laser entertainment studios create specialized graphic generation systems adapted to the amplitude–bandwidth constraints of low-inertia scanners.

Scanners and amplifier dynamic responses determine the complexity and often the message content of the imagery. All other parameters have wide tolerances. The small-angle (2°) response of 100 μs is commonly demonstrated with the G120D scanner of Table 10.13a and 10.13b. This is possible because the components, the system, or the dynamic distortions are compensated by the graphic artist as he or she views the imagery being generated. Images with the equivalent complexity of 20 detached cursive letters can be presented in vector mode and flicker with only one pair of scanners.

The studio has a computer-based system with many editing features which are used to develop animation through scan-to-scan dynamic image deformation. Figure 10.19 presents the elements of a studio with full graphic capability. A typical program is composed of elements held on magnetic tape, disks, or programmable read-only memories (PROMs) and then stored on sets of PROMs for later use. The playback unit or projector also has manually controlled functions such as interpolation, pan, zoom, rotation, as well as color hue and intensity. The operator can, if needed, follow actors or speakers on stage.

Programming

The process of programming is best described by one of its practitioners, Jennifer Morris of the Image Engineering Corporation [3]:

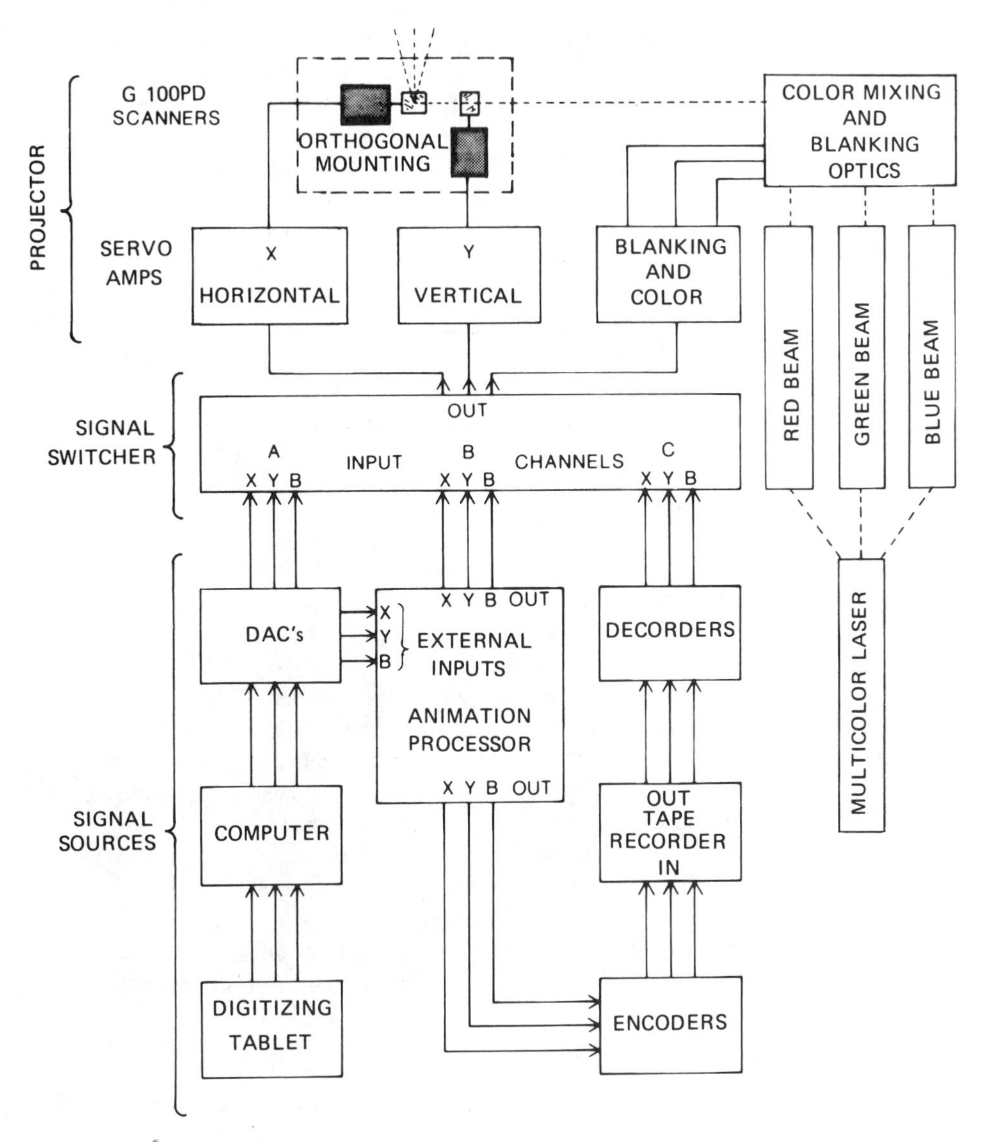

Figure 10.19 Studio projector functions.

The most familiar laser graphics effect, widely used in light shows and discotheques, is the Lissajous figure. More complex abstract designs, with circular motif, can be achieved by an analogue rotator.

Representational imagery, such as cartoons, messages, logos and other illustrations, are generated by means of an electronic data tablet interfaced to a microcomputer. An artist may then trace a line drawing and have it translated into the X–Y coordinate points stored in the computer. As the images are played back, they may be animated by using special circuits. Animated sequences can also be stored on magnetic tape with FM encoding.

A third signal is required to generate blanking of the laser beam. This makes it possible to display images with discontinuous lines. Blanking may also be used with Lissajous pattern generators to create more intricate dynamic designs.

In a full-spectrum color projector, three blanking signals are required, one each for red, blue, and green. These signals also contain intensity level information. With this system, it is possible to create images with precise, discrete colors, from soft pastels to vibrant primaries, at specific points within the image.

10.3 CRITICAL DESIGN ISSUES

The purpose of this section is to bring to the attention of the optical designer the particular requirements and features of LIS in order to assist him or her to best take advantage of them. The designer would be well advised to pursue the study of applicable items in a more extensive text.

Any one specific defect of a scanning system may originate from a number of sources. My first intention was to create a troubleshooting tree, but it became apparent that, for most defects, all the elements of the system could be construed as responsible. This led to a compilation of subjects which need to be analyzed in order to design a scanning system with low inertia scanners.

This section is not intended as an exhaustive study of scanning system design issues. Only those subjects which relate specifically to galvanometric and resonant scanners are addressed. All the subjects reviewed are not pertinent to all designs.

In this section it is assumed that the designer is building an advanced scanning system and needs to obtain the optimum performances out of every element.

10.3.1 Optical Constraints

Some reasonable optical choices can have disastrous dynamic consequences. The optical engineer must become familiar with the specific application.

On the other hand, it is undisputed that precision high-speed scanning is always facilitated when the mirror's inertia and mass are minimized. This section addresses possible compromises and design alternatives.

Gaussian Beams

The desire to minimize the size of a scanner's mirror has to be mitigated by the image quality or spot size consequences.

In the first chapter of this book Gerald Marshall gives a description of gaussian beams. O'Shea [4] treats gaussian beams in their more general applications. It is preferable to assume all beams to be gaussian because these are the only ones for which a manageable theoretical base exists. Also, lasers naturally produce gaussian beams or beams which can readily be made gaussian.

The minimum spot size d obtainable by focusing a monochromatic cylindrical parallel beam of wavelength λ and uniform energy density and diameter D with a perfect lens of focal distance L is given by the Rayleigh criterion:

$$d = \frac{2.44\lambda L}{D} \tag{1}$$

This is an incomplete definition, because not all of the energy is confined to the focus spot; the spot is surrounded by a number of concentric rings, called Airy rings, which contain 16% of the total energy. This is represented in Figure 10.20.

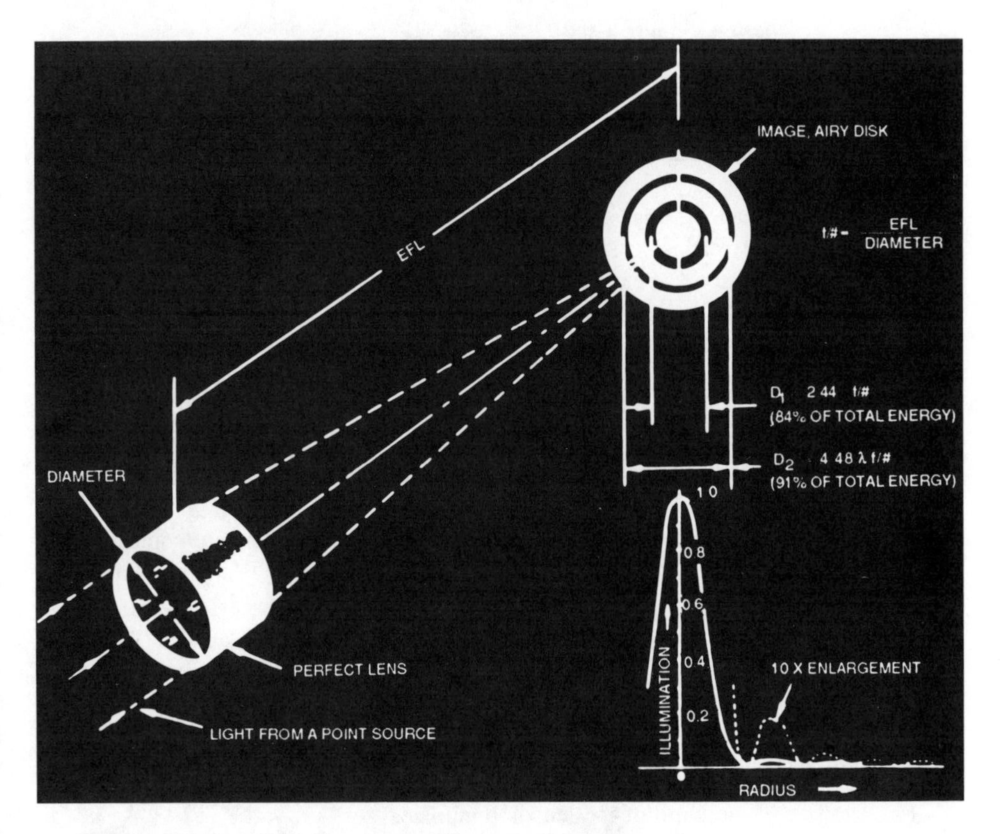

Figure 10.20 The distribution of illumination in the airy disk.

The Rayleigh criterion is commonly used because a truncated gaussian beam exhibits behavior similar to that of a beam of uniform energy density. In practice, the pupil as well as the scanning mirror can be defined only with knowledge of the threshold sensitivity of the target and the desired effect. It is evident that the important parameters of the mirror of a laser welder are quite different from those of a photocompositor.

The information in Chapter 1 should be followed to define the mirror size. This is frequently such a critical value that it is strongly recommended to verify experimentally that the smallest possible mirror has been determined.

It is not uncommon in graphic art scanners that the pupil diameter must be two or three times larger than indicated by the Rayleigh criterion.

Field Flattening and F-θ Lenses

F-θ lenses were conceived and developed for motor-driven polygons which scan at constant angular velocity. They have two functions:

1. To provide a flat field at the image plane of the scan
2. To force the image velocity to be proportional to the angular velocity of the polygons.

Low-inertia scanners, on the other hand, only need field-flattening optics; the control of the image velocity is best achieved electronically. When using a galvanometric scanner, this is done by suitable shaping of the drive signal. When using a resonant scanner, the image velocity is derived from a look-up table which can be created as needed.

Ordinary field-flattening lenses may be simpler or more economical to build than field-flattening F-θ lenses. Also, all other parameters being equal, the span of an imaging system using a simple field-flattening lens is always greater than that of one built with an F-θ field-flattening lens. This can be readily derived, since the tangent of an angle always has a greater value than the angle (expressed in radians). Conversely, in a given application, the use of a simple field-flattening lens will necessitate a lesser scanner mirror rotation.

The reach of a simple field-flattening lens and of an F-θ lens are compared in Figure 10.21 for a scanner rotation of $\pm 15°$. The difference is approximately 10%.

Angle of Incidence

It is advantageous to bring the lens as close as possible to the scanning mirror. This is best achieved when the incident beam enters the mirror as close to the normal rest position as possible. Two benefits are derived:

1. The size of the pupil of the lens is minimized.
2. The size and inertia of the scan mirror are minimized.

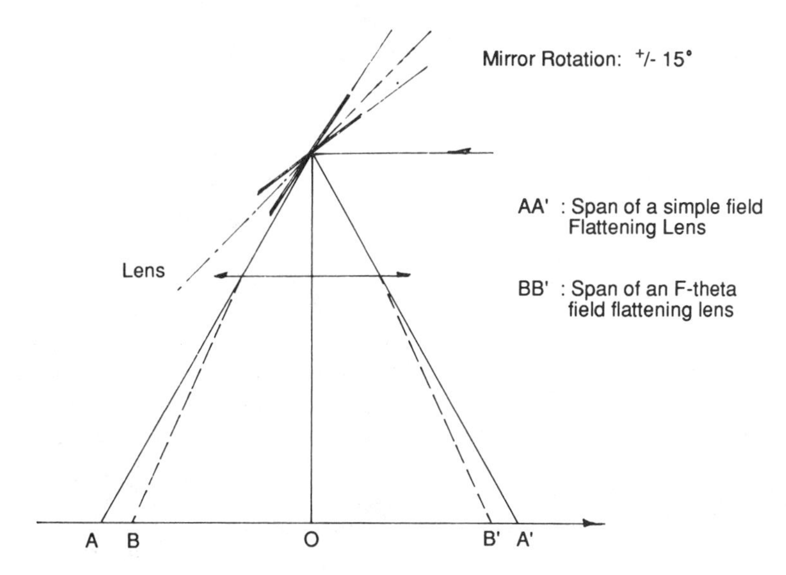

Figure 10.21 Comparison of beam path of F-θ and field-flattening lenses.

The optimum geometry is when the incident beam enters through the field-flattening lens. Table 10.11 compares three designs for a 10-mm incident beam with an active beam deflection of ±21°.

10.3.2 Image Distortions

This section reviews the most common scanner-related image distortions. While it is important to identify the source of any error, only the most common imperfections deriving from scanner sources will be discussed here.

Index of Refraction of Air

The index of refraction n for air is strongly dependent on the local pressure p, density d, and absolute temperature T. The pertinent relations are

Table 10.11 Effect of Angle of Incidence on Mirror Size and Inertia

Incident angle (deg)	Max. mirror dim. (mm)	Relative inertia (%)
0	11.7	100
35	17.2	320
45	21	580

$$\frac{p}{d} = RT \tag{2}$$

$$\frac{n - 1}{d} = K \tag{3}$$

where R is the gas constant and K is an empirical value known as the Gladstone-Dale constant.

One should keep in mind that the local air density is effectively proportional to its local velocity. In practice this forces the designer to enclose the optical bench and prevent air motion.

Excessive wobble of single axis or bending of two axis scanning systems operating at slow speed (a few scans per second) is frequently due to air motion and can be eliminated by proper baffling and occasional vigorous air stirring, such as with a fan.

Microphonic perturbations are also a consequence of the variability of the index of refraction of air.

Air Dynamics

As we have seen, air is not the ideal medium for light to travel through. Air also adds to scanning difficulties due to the damping effect of its viscosity and the buffeting perturbations caused by its turbulence. These disturbances are pertinent for systems of high speed and high precision. At present they are frequently encountered with high-performance resonant scanners which are selected for their high-frequency and large mirror capabilities. Many advanced systems being contemplated operate the scanners in partial vacuum or helium in order to minimize these disturbances.

There is no literature for LIS equivalent to that of Lawler and Shepherd [5] for polygons. Aerodynamic effects for LIS are complex due to the extremely low inertia and stored energy of the moving element compared to the effects of the aerodynamic forces generated. The Reynolds number has been used by Brosens for the purpose of evaluation [6].

The Reynolds number Re is a dimensionless quantity function of the fluid density d, viscosity v, velocity V, and the mirror radius r.

$$Re = \frac{dVr}{v} \tag{4}$$

A practical expression for air at standard atmospheric conditions in the MKS system is

$$Re = (Vr)\,6.7 \times 10^{-4} \tag{5}$$

At Reynolds numbers above 2000 the pressure forces proportional to mirror tip velocity add to the viscous losses and are the dominant cause of low Q for resonant scanners. This is the region where laminar flow changes to turbulent.

It is also these turbulences which induce jitter in resonant scanners which can exceed 5 μrad. This more than any other effect limits the dimensions and operating frequency of resonant scanners in air.

Mirror Surface Off Axis

For dynamic reasons, the reflecting surface of a scanner mirror is normally offset from the axis of rotation. This offset T causes an additional scan nonlinearity error. In order to minimize this effect the beam should be centered below the axis of rotation by an amount K as shown in Figure 10.22. This error E function is

$$E = \frac{T - K \sin \alpha}{\cos \alpha} \tag{6}$$

where α is the angle of the mirror to the normal of the work plane.

Figure 10.23 is a graph of typical single-axis, flat-field scan error for a mirror with a unity surface offset. The scan angles are beam rotations. The beam scan angle is the rotation added to the reference angles of 37° and 45°, respectively.

Beam Path Distortions

Beam path distortion (BPD) may come from the scan head or imaging system. It is caused by path length variations for different portions of the beam. Portions of the image may be blurred or may focus before or after the image plane. Alternately, portions of the image beam may be directed to an incorrect position in the image plane.

In focused laser systems, BPD frequently appears as elongated or distorted

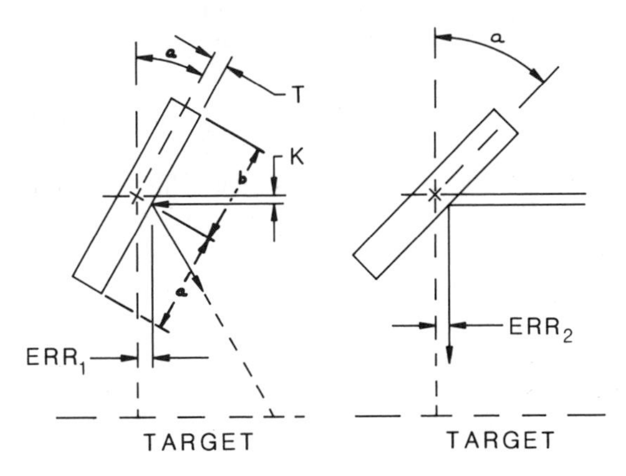

Figure 10.22 Compensation for mirror surface off-axis error.

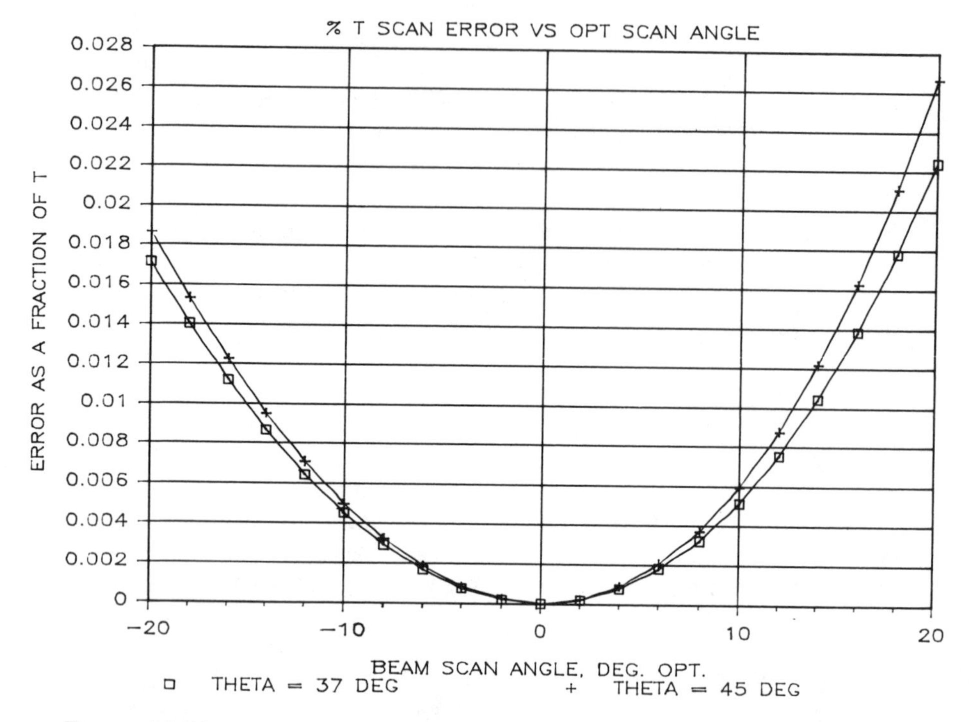

Figure 10.23 Scan error versus scan angle.

spots. These defects appear in different axes in the near and far focus about the optimum focus or as "lobes" of energy projecting from the focused spot.

In vision systems, the image is not always diffraction limited, particularly with the larger apertures. As much as $\frac{1}{2}$ or even 1 wave of distortion may be acceptable.

Mirror nonflatness in imaging systems is a common cause of BPD and is likely to impair the image astigmatically.

Vertical and horizontal axes do not focus at the same plane. This is seen in the two views of the image beam in Figure 10.24. The incorrect imaging position of certain bundles of light typically reduces contrast in the image.

Figure 10.24 shows the front and side views of an image beam reflected from a cylindrically deformed mirror. In the front view, because the beam's convergence is reduced, it focuses at a farther point than the undisturbed focus cone in the side view. The spot diagrams show that, in this case, the spot takes on an oval or enlarged form; it never attains the correct size and shape, shown on the right.

Lens system defects, such as decentering, stress, and other manufacturing-

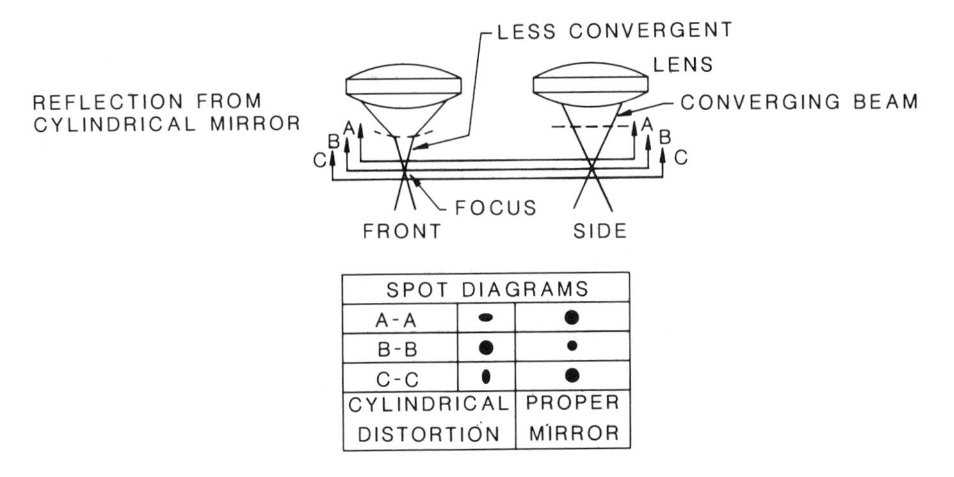

Figure 10.24 Astigmatism caused by cylindrical mirror surface.

related problems, can also cause image deformities. If image quality does not meet expectations, test the lens system without the scanner.

F-θ lenses commonly exhibit more than ½% nonlinearity. Field-flattening lenses also have performance tolerances which need to be specified.

Focus Errors in Postobjective Scanning

Postobjective scanning refers to systems in which the scanning elements act on the laser beam after, rather than before, the imaging optics. Diagrams of both preobjective and postobjective systems are shown in Figures 10.25 and 10.26.

Figure 10.27 shows how the focus error for two-mirror, two-axis flat-field scanning is derived. In this configuration, a is the center of the X mirror, b is the center of the Y mirror, and c is the point at coordinates $(0,Y_i)$; d is the length

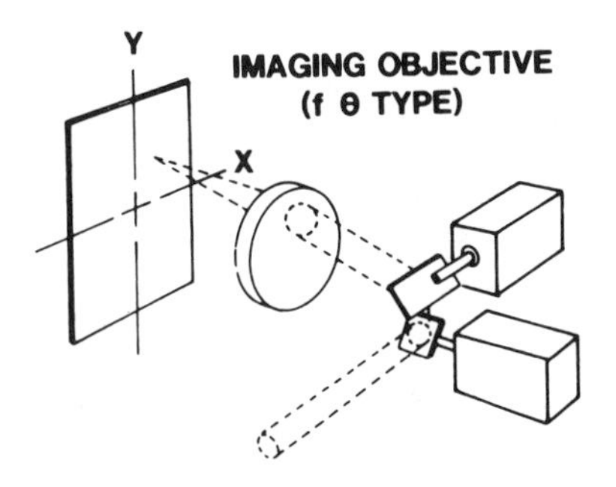

Figure 10.25 Preobjective scanning.

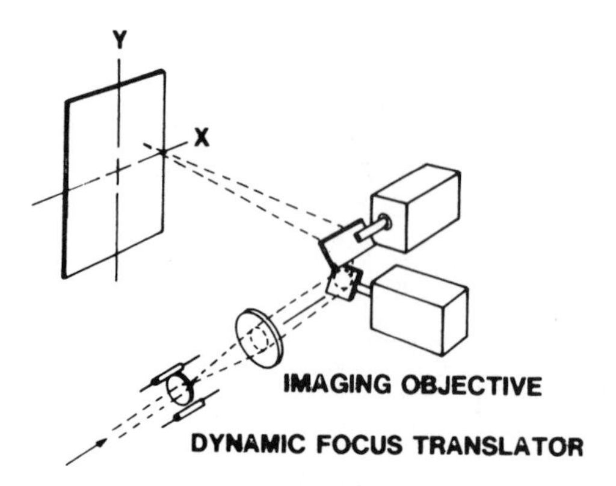

Figure 10.26 Postobjective scanning.

from b to (0,0) and e is the length from a to b. The optical scanner angles are θ_x and θ_y, and the coordinate (X_i, Y_i) is any point on the target fields. It can be seen that when $X_i = Y_i = 0$, then $\theta_x = \theta_y = 0°$. The equation that relates Y_i to θ_y is derived from the triangle of points (0,0), $(0,Y_i)$, and d. Solving for the length (0,0) to $(0,Y_i)$, which equals Y_i, we get

$$Y_i = d \tan \theta_y \tag{7}$$

The determination of the X equation is somewhat more complex and is best illustrated by projecting the target image onto the virtual image position of the

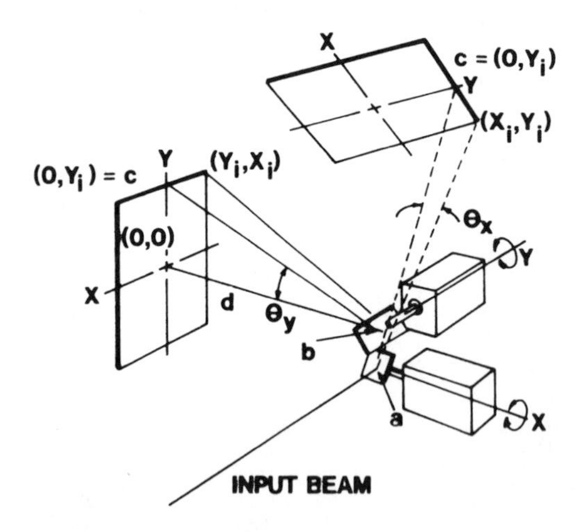

Figure 10.27 Two-mirror, two-axis flat-field scanning.

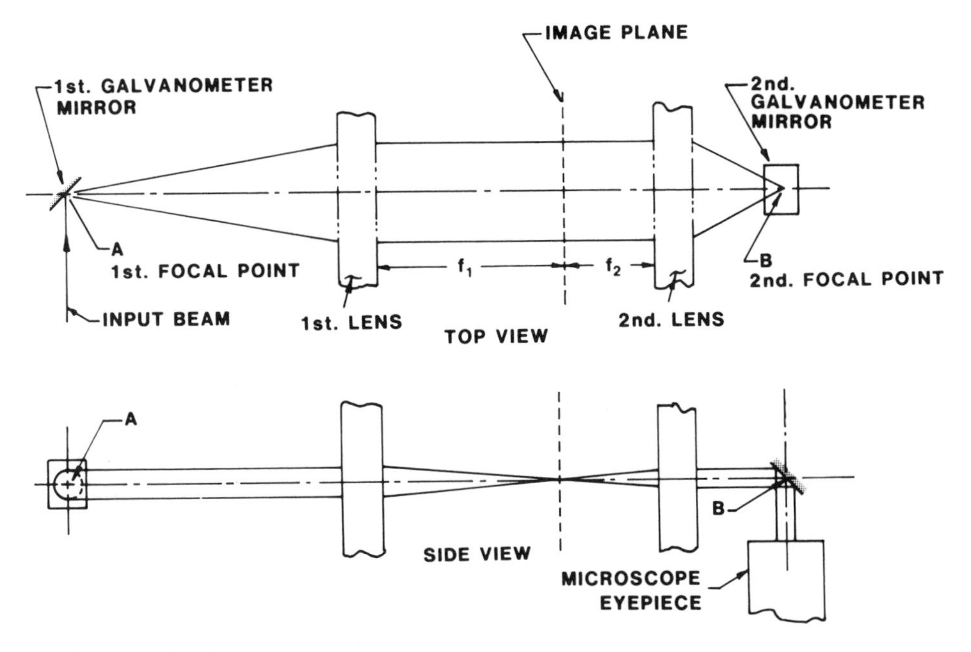

Figure 10.28 Relay lens configuration.

Y mirror, as shown by the phantom lines and phantom coordinates $(0,Y_i)$, (X_i,Y_i), and a in Figure 10.27. By solving the triangle of points a, $(0,Y_i)$, (X_i,Y_i) for the length $(0,Y_i)$ to (X_iY_i), which equals X_i, we have

$$X_i = ac \tan \theta_x \tag{8}$$

Since $ac = \sqrt{d^2 + Y_i^2} + e$, where $e = ab$, the solution is

$$X_i = (\sqrt{d^2 + Y_i^2} + e) \tan \theta_x \tag{9}$$

If we solve for the length from a to (X_i,Y_i), we will get the equation for the focus length:

$$f_i = \sqrt{(\sqrt{d^2 + Y_i^2} + e)^2 + X_i^2} \tag{10}$$

The resulting change in focus length for (X_i,Y_i) is

$$\Delta f = \sqrt{(\sqrt{d^2 + Y_i^2} + e)^2 + X_i^2} - (d + e) \tag{11}$$

Pincushion Errors in Two-Dimensional Scanning

Two-dimensional scanning can be accomplished with simple two-mirror systems as illustrated in Figure 10.27, with the use of relay lenses as illustrated in Figure 10.28, or with a paddle two-mirror system as illustrated in Figure 10.31.

Pincushion Errors in Relay Lens Scanning. Relay lens scanning makes it possible for the intersection of the scanning beam with the optical axis, or scan pivot point, to be a fixed point in space for all scan angles. This configuration is shown in Figure 10.28. The scan axis of the first scanner intersects the optical axis at point A, the focal point of the first relay lens with focal length f_1. A second relay lens, of focal length f_2, is located at a distance of $f_1 + f_2$ from the first. The scan axis of the second scanner is oriented orthogonally to the axis of the first scanner, and it intersects the scan axis at point B. For the case where a fixed beam enters at A and exits at B, the input beam and the scanner are oriented so that the first scanner reflects the input beam into the first lens at all scan angles. The second scanner is oriented so that its mirror scans the desired image area. If a collimated input beam is used and the distance between the relay lenses is maintained at $f_1 + f_2$, the exit beam is also collimated.

Figure 10.29 illustrates the scanning errors introduced by this type of system. Here θ_1 is the angle of incidence at mirror B and is produced by the deflection of mirror A. As mirror B rotates, the reflected beam describes a conical surface in space, with a half angle at the apex of $90° - \theta_1$. However, for any fixed position of mirror B, the rotation of mirror A causes the incident and reflected beams at B to sweep within a plane. As a result, if the beam reflected from B is projected onto a flat plane perpendicular to the beam position when both mirrors

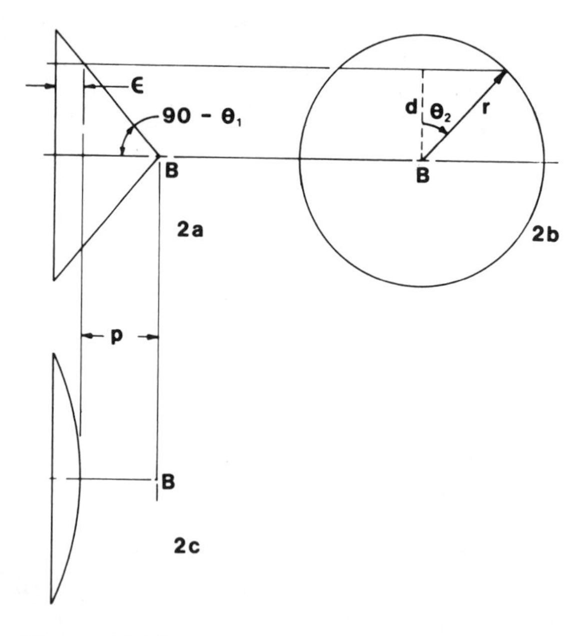

Figure 10.29 Relay lens scanner pincushion error.

are in their neutral position, the scan traces produced by rotation of B while holding A fixed are hyperbolic lines. On the other hand, the scan traces produced by A while holding B fixed are straight parallel lines. These errors are referred to as one-dimensional pincushion errors.

The magnitude of the scan pincushion error is calculated using the geometric views presented in Figure 10.29. The error is defined as the ratio of the maximum deviation from straightness e to the peak-to-peak amplitude $2p$. From Figure 10.29 it is evident that

$$\frac{e + p}{p} = \frac{r}{d} \tag{12}$$

If θ_2 designates the beam deflection produced by mirror B rotation, then

$$\frac{r}{d} = \frac{1}{\cos \theta_2} \tag{13}$$

from which it follows that the pincushion distortion is given by

$$\frac{e}{2p} = \frac{1 - \cos \theta_2}{2 \cos \theta_2} \tag{14}$$

Pincushion Errors in Simple Two-Mirror Systems. For the simpler system illustrated in Figure 10.30, the derivation of pincushion errors is more complicated. Using the terminology introduced with Figure 10.27, the equation derived earlier,

$$X_i = [(d^2 + Y_i^2)^{1/2} + e] \tan \theta_x \tag{15}$$

can be combined with $Y_i = d \tan \theta_y$ to yield

$$X_i = (d/\cos \theta_y + e)\tan \theta_x \tag{16}$$

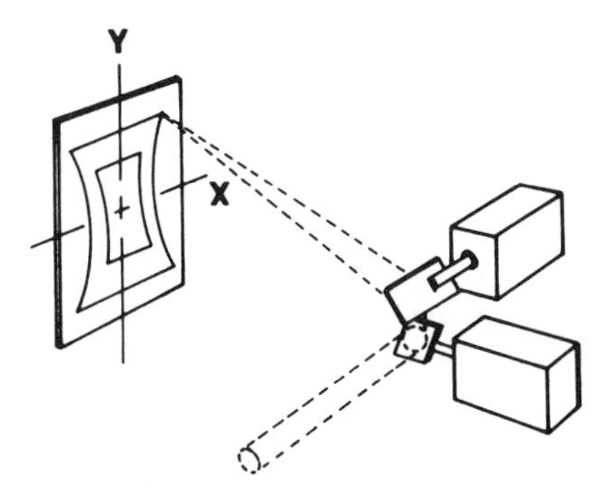

Figure 10.30 Simple two-mirror system pincushion error.

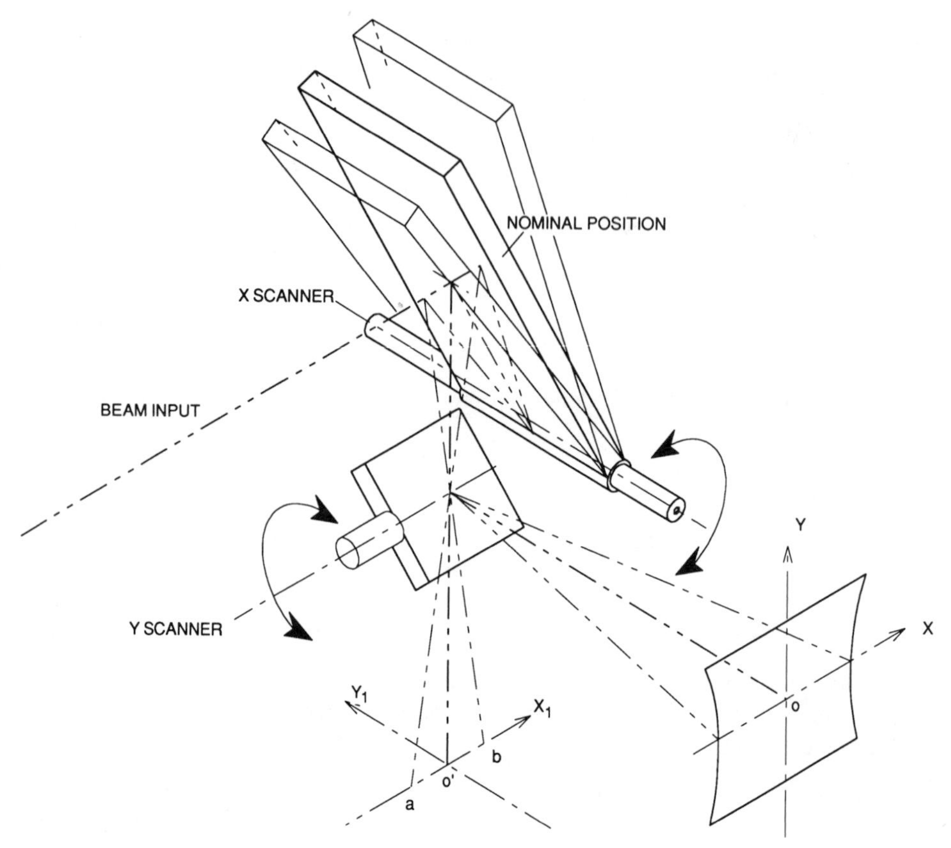

Figure 10.31 Paddle two-mirror scanning system.

The pincushion error ϵ is the ratio of the change in the value of X_i as θ_y changes from zero to a specified value, to the peak-to-peak amplitude $2X_i$ at $\theta_y = 0$:

$$\epsilon = \frac{X_{i\theta_y} - X_{i0}}{2X_{i0}} = \frac{1 - \cos\theta_y}{2(1 + e/d)\cos\theta_y} \tag{17}$$

Pincushion Error in Paddle Two-Mirror System. The paddle two-mirror scanning configuration is an alternative to the relay lens method when the limitations of the lens system are objectionable. This configuration is shown in Figure 10.31. The two scan axes are orthogonal and coplanar, forming plane x-y. The first mirror, x, is rotated from its extreme edge and, when centered, the plane of the mirror intersects the x-y plane at a 45° angle. With the incoming beam parallel to the y axis as a first-order approximation for small scan angles, the reflection off the x mirror intercepts the y mirror at the same location on mirror x. The plane of the y mirror intercepts the x-y plane at a 45° angle, and the y mirror rotates on its bisector. The output beam is orthogonal to the input beam and, when scanned, appears to pivot on the y mirror in two dimensions, as in a relay lens scanner. To a first approximation the same errors are found to be present as in a simple two-mirror system but rotated 90°.

Other Image Distortions

Distortions that vary as the cosine or cosine squared are common to numerous optical systems and have been analyzed by Warren J. Smith [7]. Their importance for a flying spot scanner is usually minimal, because scan angles are usually small, and can readily be calculated and compensated. A scanner used as a designator encounters a different set of requirements when projecting an off-axis image on a screen. These difficulties relate to the image size and the depth of focus of the system.

Intensity modulation is often traceable to improper aperture definition or alignment such that the scanner mirror or mirrors vignette the beam. This is quite evident on a single-axis scanner but more difficult to trace on a two-axis system.

The reflectance from mirrors can vary a few percent with angle of incidence. It is greater when the incidence is high and when light is polarized normal to the incident plane.

When a single-axis scan does not produce the expected straight line but instead a "smile" or a "frown," this is due to improper alignment: the incident beam is not normal to the axis of rotation and/or the mirror is not parallel to it. As shown by Schreiber [8], it is possible to obtain a straight line with a judicious optical path design even when incident and reflected beams do not define a plane normal to the axis of rotation of the mirror.

10.3.3 System Dynamics

The conventional measure of the speed of response of a galvanometric scanner system is its torque-to-inertia ratio. This is a necessary but not sufficient condition. Similarly resonant scanners upon careful scrutiny have multiple resonant frequencies, all with a high Q.

The position, amplitude, phase, timing with frequency controls, and mirror design are integral elements in the design of a low-inertia scanner system. Environmental conditions as well as mounting structures must be specified. The mechanical structure of a scanner must be designed with full consideration of all system elements.

Dynamic Considerations

The dynamic behavior of a resonant LIS is predictable [9], and manufacturers' data should be used for design purposes. Special consideration needs to be given to the ability of a scanning element to exchange energy with the frame.

The dynamic performance of a nonresonant LIS can be described by the bandwidth and step response of a second-order system. If it is assumed that drive power and control are of secondary importance, Newton's law applies and the limiting condition is as follows:

$$T = I\frac{d^2\theta}{dt^2} \tag{18}$$

where T is the torque required to impart the angular acceleration $d^2\theta/dt^2$ and I is the moment of inertia. To optimize a reciprocating motion system, equal energy and time must be allowed for both acceleration and deceleration of the mirror and armature. The maximum potential mechanical energy W which can be given to a rotating system is expressed as

$$W = \frac{1}{2}T\beta \tag{19}$$

where β is the total angular displacement (peak to peak). This energy can be expressed as equivalent dynamic energy by the kinetic-energy equation as follows:

$$W = \frac{1}{2}I\left(\frac{d\theta}{dt}\right)^2 \tag{20}$$

Solving the above three equations for time yields the expression for the minimum stepping time or step response:

$$t = 2\sqrt{\frac{I\beta}{T}} \tag{21}$$

Dynamic Imbalance

A rotating body can be balanced, but never perfectly. It is necessary to qualify and quantify resulting imbalance forces in order to assess their consequences and judge if they are acceptable for the application.

Bearings are built with some degree of radial play. When they are subjected to periodic eccentric forces which exceed the constraints of their preload, radial or axial, damage follows which can result in catastrophic failure.

Figure 10.32 is a schematic representation of a ball bearing mounted rotor from a galvanometer with an unbalanced load m. The total system inertia is J and a drive torque T imparts an acceleration $d^2\theta/dt^2$. The rotor is also subjected

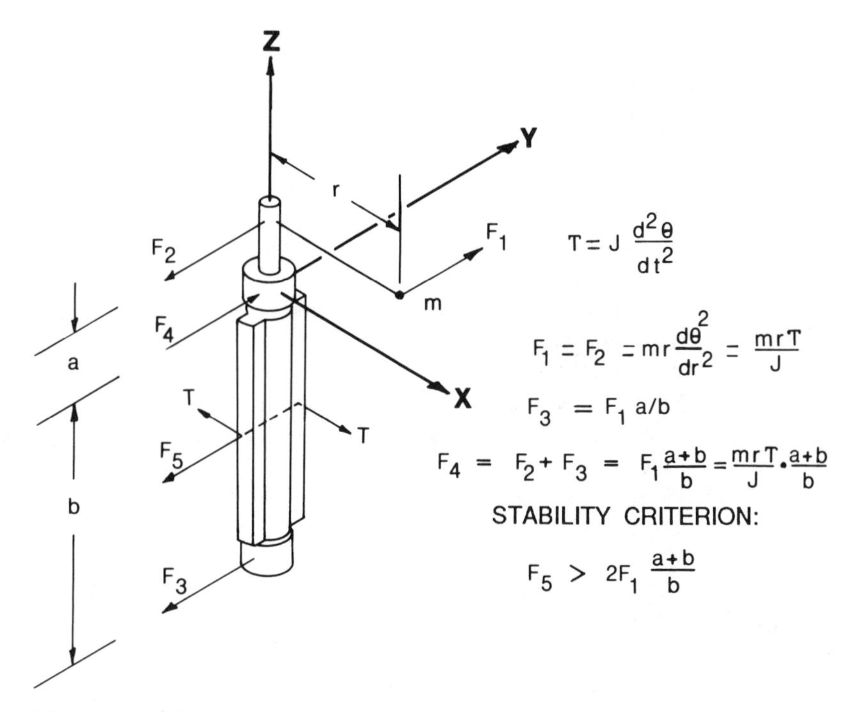

Figure 10.32 Dynamic forces on rotor.

in the middle to a radial force F_5, which is the bearing preload. The derivation uses the symbols of the figure.

All the torques and forces on the rotor must balance; consequently, the following conditions must be satisfied.

$$T = mr^2 \frac{d^2\theta}{dt^2} \tag{22}$$

$$F_1 = F_2 = mr \frac{d^2\theta}{dt^2} \tag{23}$$

$$F_3 = \frac{F_1 a}{b} \tag{24}$$

$$F_4 = F_1 + F_3 = \frac{F_1(a + b)}{b} \tag{25}$$

The stability criterion is for the preload to be greater than the period eccentric forces. If it is assumed, according to the figure, that the eccentric mass represents the effect of the mirror, the front bearing is the most vulnerable. The following relationship must be satisfied:

$$F_4 = \frac{(a + b)}{b} mr \frac{d^2\theta}{dt^2} < \frac{F_5}{2} \tag{26}$$

The most common mirror mounting technique is a mass balanced assembly with the reflecting surface forward of the axis of rotation. Lateral mass balance is equally imperative.

The advantages of a magnetic radial preload, as opposed to axial preload, are described in the section on ball bearings. Typical preload values on the front bearing range from 50 to 500 g, depending on the scanner size and design.

Armature imbalance of resonant scanners causes wobble and excites the instrument's chassis. Unacceptable audio coupling to the chassis can be minimized with massive construction or soft mounting of the scanner. Both are costly and undesirable as compared to a properly balanced armature.

Mechanical Resonances

A perfectly balanced armature-mirror assembly can still produce unacceptable oscillations. These are caused by the excitation of any possible natural frequency of one or more of the elements of the armature or occasionally the stator. These structures commonly have no damping and can be excited by a magnetic imbalance or external shocks and vibrations. Such oscillations are usually sensed by the control circuitry and amplified to cause system instabilities. It is necessary to design the armature so that its first resonance in any mode is substantially beyond the cutoff frequency of the amplifier or is properly damped.

This is the most common limiting factor to the speed of response of a small mirror galvanometric scanner. Two familiar modes of oscillation are reviewed here.

1. Mirror on a limb: The mirror is overhung at the end of the shaft and behaves like a freely supported beam. With reference to Figure 10.33, it has a deflection angle θ expressed by

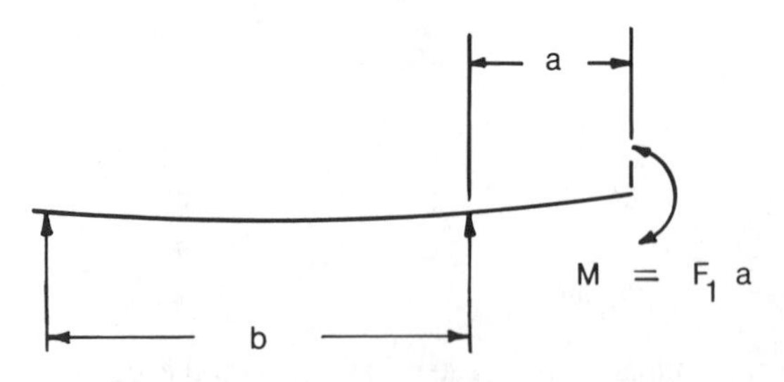

Figure 10.33 Bending forces on rotor.

$$\theta = \frac{Mb}{3EI} \tag{27}$$

where E is the Young modulus of the shaft's material and I its moment of inertia. For a heavy mirror of mass m the first cross-axis resonant frequency ω is expressed by

$$\omega = \left(\frac{3EI}{ma^3}\right)^{1/2} \tag{28}$$

For a small mirror, the first cross-axis resonant frequency is the rotor resonance. This has been analyzed by Den Hartog [10]. If the rotor can be represented by an iron cylinder, its resonance can be expressed by

$$\omega = 500{,}000\,\frac{d}{b^2} \tag{29}$$

where ω is in radians per second, d is an approximate value of the diameter of the rotor, and b is its length between bearings, as shown in Figure 10.33. (d and b are both in inches here.) Exciting this resonance will cause the mirror to wobble and/or render the servo unstable.

2. Torsional resonances: The rotor and mirror are two freely supported inertias connected by a shaft. The resonant frequency of such a system, with reference to Figure 10.34, is

$$\omega = \sqrt{\frac{K(J_1 + J_2)}{J_1 J_2}} \tag{30}$$

This will cause perturbations in the scan axis. Periodic compression and decompression of data will occur. If such a scanner is used to generate gray tone images, they will show lighter and darker waves normal to the raster lines.

10.3.4 Servo Control of Galvanometric Scanners

The servo controller of a scanner, also called the amplifier or simply "the electronics," provides operational control. In its simplest form it is a negative position

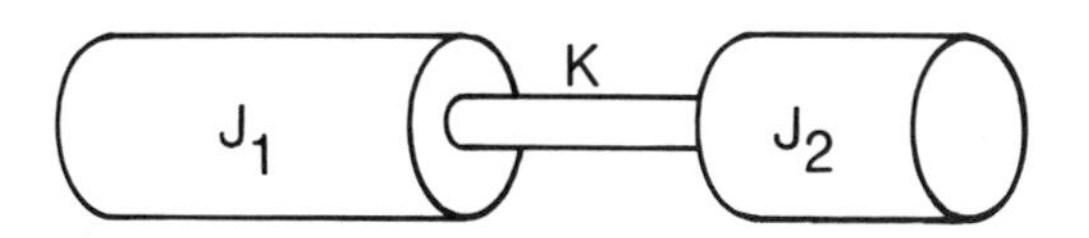

Figure 10.34 Angular rotor-mirror presentation.

feedback servo amplifier with limited bandwidth and accuracy. At its most complex it is a high-speed, 32-bit computer controllable digital filter/amplifier with variable parameters. Every intermediate possibility also exists. The choice is dictated by the application, available electronic and logic devices, and cost.

Galvanometric applications fall into two major application categories: (1) vector scanning and (2) tracking and pointing. Each group can be subdivided according to performance parameters. These are

1. Speed
2. Precision
   a. Absolute position
   b. Jitter
   c. Wobble
3. Stability
   a. Dynamic
   b. Gain drift
   c. Null drift
4. Repeatability
5. Life, definition of failure
6. Multiple scanner synchronized performance
7. Environmental immunity

When high-performance standards are applied, these conventional requirements can lead to complex design problems.

Performances on most of these parameters can be improved by slowing down the system. When this is not acceptable, the next best thing is to minimize accelerations and prevent any electronic saturation. Shaping the drive signal prior to its entering the amplifier is the recommended procedure.

The servo designer has to confront mechanical, electrical, and servo constraints whose elements and consequences are tightly interwoven. Compensation is required for the imperfections of the other components in the system—mirror, magnetic driver, and position transducer.

The servo is the interactive link for all system elements with each other. The performance demands placed upon them are justified.

Servo Control Constraints

A galvanometer can be simply modeled by a second-order differential equation such as

$$KI = K_s\theta + D\frac{\partial\theta}{\partial t} + J\frac{\partial^2\theta}{\partial t^2} \tag{31}$$

where I is the drive current, K_s is the spring constant, D is the damping constant, J is the rotor and load inertia, θ is the shaft rotation, and K is the torque constant.

When driven by an amplifier with position error and velocity feedback, the transfer function is

$$\frac{\theta}{e_{in}} = \frac{KA}{JS^2 + (D + KK_D K_\theta)S + (K_s + KK_\theta A)} \tag{32}$$

where K_D is the scanner damping constant and K_θ is the transducer position constant. There is a finite error in the response to any nonzero excitation. Magnetic hysteresis and frictional losses will also result in position errors. The effect of the closed-loop amplifier is to extend the bandwidth of the galvanometer with the addition of servo stiffness and to improve its static linearity to that of the position detector. The dynamic performance of the system is second order only as long as there are no additional perturbations.

For those applications requiring the utmost in position accuracy, the simple servo can be modified to include error integration. Then, given enough time, all errors associated with the internal spring, the hysteresis losses, the magnetic torque variations, and the frictional losses are removed, and the nonlinearity of the position sensor is the sole remaining contributor to position error. Unfortunately, dynamic behavior is slowed by the addition of error integration, and no improvement in dynamic errors results, as the achieved servo stiffness is no greater than without error integration. One benefit associated with the introduction of error integration is tracking a ramp input with no error after decay of the initial velocity transient.

Mechanical Constraints

The dynamic performance requirements of a second order servo assume that all the components are linear within the bandwidth of interest. Since the secondary resonances or the nonlinearities of the components will define the system's speed and accuracy, it is necessary to incorporate only components whose actual performances match the requirements of the application.

In practice, each component should preferably have a first resonance one order of magnitude beyond the system's bandwidth. If necessary, it is possible, at great cost, to attain a system bandwidth as high as 50% of the lowest first resonance.

Real mechanical components have multiple resonances, frequently with no damping. These resonances exist normally along degrees of freedom other than the axis of rotation, as in bending, compression, and twist. They all cross-couple within each component and also through the feedback position sensor, which is commonly slightly sensitive to radial and tilt motions. All the components of the scanning system should be carefully analyzed: the mirror, the driver's armature, the driver's stator, the position transducer, the electronic components, and the chassis. Every interface should be considered.

Real components also have nonlinearities. The transfer function equation in the previous section shows that the coefficient of every component contributes to the damping bandwidth criterion. If these coefficients have variable values, resulting performances are lower than those possible with a constant value, even the lowest. The linear relation of torque, current, and rotation must be calculated along with the transfer function of the sensor.

For high-speed and/or high-accuracy servo systems, requirements of the dynamic system override those of the torque to inertia ratio analyzed in the dynamic considerations section. This justifies the emphasis placed on the design of the scanner elements.

Other mechanical factors also deserving consideration include the damping properties of the material used, the bearing lubricant, the thermal regulation, and the inertial matching for two-axis systems.

Electrical Constraints

The general considerations for the design of closed-loop driver amplifiers to drive scanning galvanometers are similar to those of other precision servo designs. Engineering considerations start with defining the necessary power supply stability to minimize induced drifts. The use of low-noise, wide-bandwidth, low-drift operational amplifiers throughout the amplifier is needed when the application is demanding in the area of speed or stability. Careful attention must be given to isolation of the position signal from the output driver to eliminate spurious oscillations due to coupling modes. Unwanted ground paths must be studiously avoided, particularly in the presence of case-grounded position feedback sensors. ''Zeroing'' potentiometers should be avoided owing to their excessive drift characteristics. The use of temperature-stable components throughout the design, including resistors and capacitors, is highly advisable.

As all moving iron galvanometers present a significant inductive load to the driver amplifier, particular attention needs to be given to the output stage. Typically the configuration of the output amplifier looks like a transconductance amplifier, so as to remove the unwanted pole caused by the load inductance and give a second-order response. Since the maximum rate of rise of current in the load is limited by the ratio of drive voltage to inductance, significant increases in servo response can only be achieved by raising the driver voltage substantially. Under these conditions it is necessary to check that safe operating area restrictions on the bipolar output transistors are met at all times.

Provisions to limit the rotational excursion should be incorporated in the design if damage can occur to either the mirror or the galvanometer when overdriven. Overdriving is most likely during the turn-on/turn-off transient when power supplies are not properly sequenced. As current rise times are improved and servo bandwidths extended, care must be taken to guarantee that I^2R losses in the galvanometer do not exceed the steady-state limit. It may be desirable to provide thermal shutdown protection of the load by means of a thermistor or thermal switch mounted in the galvanometer.

10.3.5 Control of Resonant Scanners

Resonant scanners require low power and are possibly the simplest to drive. They are not, however, the simplest to control, when placed in high-precision systems. Resonant scanners require the same tolerance standards as galvanometric scanners, listed in Section 10.3.4. The elements of a resonant system are strongly mechanically coupled. Their performance is dictated by scanner design and construction as well as by the system design.

Resonant scanners can be viewed as galvanometric scanners with extremely low damping, or as high-Q filters. As such, they reproduce the drive signal with all its low- and high-frequency "noise" filtered out but not eliminated. A high resolution systems requires a "clean" drive signal for satisfactory performance.

System Considerations

The mirror of a resonant scanner follows a sinusoidal motion; unfortunately most applications require a linear motion of a spot on a flat plane. This is achieved by synchronizing the scanner to external drive electronics while a pixel clock strobes data either in or out of the user hardware. Low-cost microprocessors and high-speed complex logic elements make this easy and affordable. Numerous logic conversion systems have been published (e.g., Refs. 11–13). They all have to satisfy certain operational requirements.

A scanner operating at mechanical resonance executes a motion described by the following equation, where θ is the scanner shaft angle, θ_s is the maximum angle of excursion, t is time, and T_s is the period of the scanner:

$$\theta = \theta_s \sin\left(\frac{2\pi t}{T_s}\right) \tag{33}$$

It therefore follows that

$$t = \frac{T_s}{2\pi} \sin^{-1}\left(\frac{\theta}{\theta_s}\right) \tag{34}$$

and that

$$\frac{d\theta}{dt} = \frac{2\pi}{T_s}\theta_s \sqrt{1 - \left(\frac{\theta}{\theta_s}\right)^2} \tag{35}$$

Normalizing the velocity of the scan to the maximum at $\theta_s = 0$ yields

$$\frac{d\theta/dt}{2\pi\theta_s/T_s} = \sqrt{1 - \left(\frac{\theta}{\theta_s}\right)^2} \tag{36}$$

It can be seen that this function has a maximum value of unity at $\theta = 0$ and a minimum of zero at $\theta = \theta_s$. Taking a desired scan angle efficiency of 95% of the rated excursion as an arbitrary example yields a velocity ratio of 0.312. Thus the maximum velocity is 3.2 times the minimum. The clocks need to have periods in this ratio, and the light source or automatic gain control systems need to compensate their amplitudes by the same amount. Clearly, the velocity ratio decreases dramatically with reduced usable scanner angle and makes the clocking requirements easier to meet.

When mapping to or from a flat image plane, an additional nonlinearity is encountered due to the tangent error. The equations below describe the motion in the flat image plane; x is the displacement in the image plane and S is the distance from the scanner mirror to the image plane:

$$x = S\tan(2\theta) \tag{37}$$

$$x = \frac{\tan\theta}{1 - \tan^2\theta}2S \tag{38}$$

$$\frac{dx}{dt} = 2S(1 + \tan^2 2\theta)\frac{d\theta}{dt} \tag{39}$$

$$\frac{dS}{dt} \equiv 0 \tag{40}$$

It is thus possible to modify the scanner $d\theta$ in order to yield a constant displacement in the image plane.

Mechanical Constraints

High-performance resonant scanners are expected to display parameter repeatability and predictability, but a few potential problems still persist.

The effects of air turbulence described in the section on air dynamics change the effective air path and can be noticeable at low-frequency scanning. A change in scan amplitude, also, can change the tip speed of the mirror from a laminar flow region to a turbulent one and introduce jitter as well as a change in damping.

An unbalanced armature induces bending forces of the type analyzed in the section ''Dynamic Imbalance.'' These will be transferred to the chassis at the mounting, location and excite any element within the instrument with the same natural frequency.

Scanners commonly exhibit a small change in resonance with a change in temperature and amplitude. Use of a tunable resonant scanner can solve this problem.

Finally, to avoid scanner perturbances it is necessary to verify that no cross-resonance exists in the vicinity of the torsional resonance.

10.4 SCANNER ELEMENTS, SELECTION AND DESIGN

Scanning systems are similar to all other complex systems in that control of all the interrelated elements determines the maximum possible performances. The incorporation of elements with higher-performance capabilities does not correct for deficiencies caused by weaker elements. One is led to examine all the elements of a scanner with respect to the expected system performances.

In the previous section it was explained how the optical/geometric choices, the mechanical construction of a scanner, and its control electronics interact to affect the quality of the scan. In this next section the design choices for each element are reviewed. These are mirror, suspension, driver, transducer, and electronics.

10.4.1 Mirrors

Many scanning problems can be attributed to mirror problems. If mirrors could be made infinitely rigid, flat, and reflective and have negligible inertia, the design and operation of scanners would be extremely simple and there would be less demand for this text. No material able to fulfill these ideal requirements comes to mind; however, available materials do offer creative design solutions.

The facets of a polygon scanner are often viewed as the last link in the chain of components of a scanning system and burdened with all the system's faults. The mirror of a galvanometric or resonant scanner, however, is rarely perceived in the same fashion and receives comparatively little attention. Actually, they both must meet the same requirements for reflectivity, balance, thermal and dynamic deformation, mounting, etc.

The condition of the LIS mirror is of concern for that portion of its movement in which optical data is transmitted. Control of dynamic deformations is commonly associated with resonant scanners (and raster scanning). Fabrication and installation requirements are commonly associated with galvanometric scanners (and vector scanning). Thermal conductivity becomes an issue with use of high-power lasers (and galvanometric scanners), which tend to induce temperature gradients.

In LIS systems the design and installation of the mirror must, under all operating and storage conditions, be able to preserve the essential system features. Low wobble, low jitter, precise pointing or scanning accuracy, and long life are typical requirements that have to be addressed.

Mirror Construction and Mounting

Several guidelines for mirror design and mounting associated with scanning performances have been mentioned previously. These are

The mirror mass must be a minimum.
The mirror inertia must be a minimum.

The mirror must be mounted as close as possible to the front bearing of the scanner.

All moments of inertia with respect to the axis of rotation must be balanced in order to minimize wobble induced by angular acceleration and by environmental perturbations.

Balancing is most imperative for resonant scanners with torsion bar suspension.

Three other performance issues associated with mirror design and installation—alignment, mirror bonding, and mirror clamping/mounting—must be addressed.

Alignment. One consequence of mirror misalignment is beam positioning error. This can be easily compensated for a line scan, but a precise area scanner, or designator, may require precise mirror angular positioning and compensating gimbal scanner mounts. Such mounts may necessitate each scanner to be electrically as well as thermally isolated.

Alignment or balancing must be verified for each mirror along both axes to prevent imbalance and wobble described earlier.

An effective way to minimize problems with both alignment and bonding is to precision-machine both the mirror's reflective surface and its shaft mounting hole from a single piece of metal to make an integral mirror. Mirror design criteria for high-resonance frequency and mounting stress isolation must be considered.

Mirror Bonding. It is extremely difficult to bond a mirror to a mount and have it aligned with an accuracy of 1 mrad. For mirrors smaller than 1 cm the alignment tolerance can be as high as 5 mrad unless optical autocollimation methods and great care in bonding are used.

An improper bond process or mount design can cause mirror deformation when the adhesive cures or temperature changes during shipping or use with higher-power lasers. These thermal stresses may also cause the mirror to break or become unbonded.

The elasticity of the bond can cause dynamic pointing errors as well as undesirable resonances which could reduce the system's bandwidth.

The rigidity of the bond can cause mirror stresses and deformation when mounting to the shaft of the scanner.

Mirror Clamping and Mounting. Reiss [14] gives a brief overview of recommended mounting procedures.

> Clamping mirror mounting allows for repositioning and possible removal for replacement. Both integral and bonding mirror mounting are somewhat permanent conditions.
>
> The most successful removable clamping incorporates some form of collet clamp which provides isolation of the mirror substrate from clamping stress. A disadvantage of the collet approach is potential loosening of the collet clamp forces, with a resulting drift of mirror position.

Collet clamps, if not over-tightened, induce only compressive forces that produce no bending movement and thus no distortions in the mirror facesheet. One collet clamp technique is to mechanically isolate the shaft by relieved regions so that the distortions imposed by the clamping screw are not transmitted.

Fastening a mirror mount onto a shaft with set screws is not recommended, as they can deform the shaft. When set screws are properly fitted, removing them is nearly impossible. When not fitted properly, they act as a hinge, allowing wobble excitations. Set screws can fatigue and loosen.

Dynamic Deformations

The accelerating torques imparted to a scanning mirror can produce significant mirror surface distortions. This is particularly true in scanners driven with a sawtooth wave form and in high-frequency resonant scanners. Brosens' [15] analysis of the deformations induced by accelerating torques yields the approximate formula

$$f = 0.065\left(\frac{s^2T}{Eh^3l}\right) \tag{41}$$

where f is the maximum deflection from the original mirror shape, s is the width measured across the axis of rotation, E is the Young's modulus of the mirror material, h is its thickness, l is its length in the axial direction, and T is the applied torque, which is related to the angular acceleration a by

$$T = \frac{hs^3lda}{12} \tag{42}$$

with d the density of the mirror material. Combining the two equations, we get

$$f = 0.0055a\left(\frac{ds^5}{Eh^2}\right) \tag{43}$$

This expression points to the desirability of keeping mirrors as narrow as possible. For a glass mirror 1 cm in diameter and 1 mm thick, the resulting deflection at an acceleration of 10^6 rad/s^2 is about $\frac{1}{25}$ of the sodium D-line wavelength. Since the inertia of such a mirror is 0.011 g-cm^2, the above acceleration corresponds to a torque of only 11,000 dyne-cm. The actual mirror deformation may be smaller when a substantial portion of the width of the mirror is cemented to a mount.

Thermal Deformations

When a scanning mirror is exposed to radiation, a substantial part of the radiation that the reflecting surface absorbs is transferred in the form of heat to the rear surface. This heat is discharged by conduction to the mount and by radiation and convection to the surrounding atmosphere. The conduction of heat to the rear surface causes differential expansion and, if the incident radiation is particularly intense, significant distortions can occur. Such distortions can be estimated by assuming one-dimensional heat transfer to the rear surface.

The radius of curvature caused by a uniform temperature gradient is

$$R = \frac{1}{\sqrt{a(dt/dx)}} \tag{44}$$

where a is the coefficient of linear thermal expansion, and dt/dx is the temperature gradient in the material. From Fourier's law of conduction,

$$\frac{dt}{dx} = \frac{q}{kA} \tag{45}$$

where q is the heat transfer rate, k is the thermal conductivity, and A is the cross-sectional area. The camber assumed by a plate of width s, when it curves with a radius of curvature R, is given to a first-order approximation by

$$e = \frac{s^2}{2R} \tag{46}$$

Combining this equation with the previous expressions, we obtain

$$e = \frac{aqs^2}{2kA} \tag{47}$$

For a glass mirror of width 1 cm conducting 0.1 W/cm^2 to the back surface, the resulting camber is 0.5 μm, or about 1 wavelength.

Erosion

When a scanner mirror is moved through air at high speed, the collision of dust particles with its surface can cause gradual erosion of its reflective coating. Experience shows that for any coating the process of erosion does not occur below a critical impact velocity. It is believed that surface erosion occurs when the stress developed at the impact interface between the coating and the dust particle exceeds a value that is characteristic of the coating.

The stress developed by the impact of a rigid body against an elastic mass was analyzed by Timoshenko and Goodier [16]. The stress wave generated by impact is given by the formula

$$S = E\left(\frac{V}{c}\right) \tag{48}$$

where E is the Young's modulus of the substrate, V is the relative speed at impact, and c is the velocity of wave propagation (sound velocity) in the substrate.

Experimental evidence shows that AlSiO coatings on fused silica degrade through erosion at all points where the speed of motion exceeds 3 m/s at any instant during the scan cycle.

Users of high-speed scanners should take precautions to minimize the presence of suspended dust particles near scanning mirrors. Where such protection cannot be provided, hard coatings should be used in combination with substrates of low Young's modulus.

Material Selection

All scanner applications do not have the same performance requirements, so there is no optimum mirror material. The selection of a substrate is application dependent and has to satisfy some or all the performance requirements reviewed earlier.

Table 10.12 lists the properties of materials suitable for mirror substrate as well as mounts. The figure of merit for resonant scanners E/d^3 has been derived by Brosens and Vudler [17]. This is to be used as a comparative guide for a given geometry. One should keep in mind that design, construction, heat treatment, coating, and installation can each have a dominant influence on the performance of a mirror. The figure of merit for galvanometric scanner mirror design is E/d. This is based on fabrication requirements only. (For discussion, see Yoder [18].)

Cost, ease of fabrication, stability with time and environmental conditions (such as temperature and cyclic stresses), bonding capability, and mirror surface finishing are extremely important to the selection of a substrate material. Fatigue and yield strength are normally irrelevant.

Mirror Surface Finish

Definitions and specifications of available surface finishes and coatings for glass mirrors can be obtained from numerous sources, including government specifications, and will not be reviewed here.

For metal mirrors, difficulties begin with the form of the metal stock and the machining process. Each case is different and presents its own problems. After a blank has been machined to finished dimensions, it has to be stress relieved and stabilized. Dynamic stressing can also be used. Thermal stabilization can be achieved with three or four cycles of processing from liquid nitrogen to boiling water.

The following processes are used for surface finishing when polishing the substrate is not acceptable.

Plating and Polishing. All the surfaces of the mirror can be plated with equal thickness of hard nickel (typically 0.002–0.005 in) to avoid thermal deformation. Any material removal during polishing may need to be allowed for when plating,

Table 10.12. Mechanical and Thermal Properties of Substrates

Material	Density g/cc	Coef. T-exp. 10^{-6}/°C	Therm cond. W/cm–°C	E, Young mod kg/cm^2 × 10^5	Fig/merit E/d^3 × 10^5	Fig/merit E/d × 10^5
BK7	2.53	8.9	0.010	8.22	0.50	3.2
Fused silica	2.20	0.51	0.014	7.10	0.66	3.2
Fused quartz	2.20	0.51	0.014	7.10	0.66	3.2
Pyrex	2.23	3.3	0.011	6.67	0.54	3.0
Silicon	2.32	3.0	0.835	11.2	0.89	5.0
Aluminum	2.7	25	2.37	7.03	0.35	2.6
Iron alloys	7.86	0–20	0.1–0.8	13–21	0.03–0.04	<2.5
Al oxide	3.88	7.0	0.08	36.0	0.61	9.3
Titanium	4.3	8.5	0.20	11.2	0.14	2.6
Beryllium	1.8	12.0	2.10	30.8	5.2	17.0
Magnesium	1.7	26.0	1.59	4.2	0.80	2.5
Diamond	3.5	0.7	10–25	120.0	2.6	34.0
Silicon carbide	2.92	2.6	1.56	31.5	1.4	11.1
SXA	2.96	10.8	1.2	14.5	0.56	4.9
Tungsten carbide	15.3	5.94	0.5	68.5	0.02	4.5
Miralloy	2.10	6.3	1.1	20	2.1	9.5

and additional thermal stabilization may be required. Nickel can be ground and polished to a high-quality surface and then finished by conventional means.

Replication. This is a process in which a reflective surface and one or more coatings are formed by successive evaporation in reverse order onto a master, and then transferred together onto the substrate and bonded. The bonding agent is typically an epoxy layer with a viscosity of 100 centipoise and a thickness of a few micrometers. If thickness approaches 25 μm, the process introduces alignment errors of 0.1 to 1 mrad or greater. There is limited usable temperature range due to "bimetallic deformation." Weissman [19] shows that a 1 fringe curl per 25°C is to be expected for a disk with a 25/1 aspect ratio.

Additional limitations may be introduced by the power to be reflected from the mirror. Brosens and Vudler [17] calculate that the heat transfer coefficient of epoxy could limit replicated optics to energy pulsed under 0.0017 J/cm^2 or four orders of magnitude lower than polished copper optics at YAG and CO_2 wavelengths.

Diamond Machining. This technology has proven to be exceptionally successful in the manufacture of high-volume low-cost aluminum polygon mirrors with sagittal tolerance <50 μrad. Diamond machining is less attractive for beryllium and coated steel substrates, or when reflectivity requirements necessitate additional processing.

Glass Laminate. A very thin glass layer (typically 0.1 to 0.3 mm thick) can be bonded to an aluminum substrate with a suitable thickness of adhesive such that the composite has the same coefficient of thermal expansion as aluminum.

Strains occur in the epoxy and are due to the thermal expansion of the material, reduced by the stresses induced by the glass. The converse applies to the glass. All effects are temperature-dependent, and a judicious selection of the Young modulus for materials, the thermal coefficient of expansion, and the thickness of both the glass and the adhesive can yield a laminate stable over a wide temperature range.

This author [20] describes mirrors of this construction which remained flat within ¼ wave through a temperature range of 5 to 55°C.

This construction allows surface finishing typical for conventional glass mirrors.

10.4.2 Armature Suspension

An entire chapter of this book (Chapter 9) is devoted to bearing suspension designs for rotary scanners. Suspension is very critical to their performance. The armature suspension of a galvanometric or a resonant scanner is likewise critical.

The two technologies (rotary and LIS) share the same manufacturing standards and system performance requirements, but these are frequently achieved by different means. Ball bearings are common to both technologies, but fluid bearings have only been successfully implemented in rotary designs. Cross-flexures and torsion bars, on the other hand, are only applicable to periodic movement designs.

The suspension of a resonant scanner serves additional functions to that of a rotary scanner. It is an integral part of the dynamic system and must store and exchange large amounts of energy with substantially no losses. It must be compliant in one axis and rigid in all others. The compliance must be stable with time and environmental conditions and, for tunable scanners, variable and controllable.

The bearings of a galvanometric scanner must establish an extremely accurate axis of rotation under variable static and dynamic loads. Repeated smooth start-ups must be possible without stickiness, and an extremely long life must be obtained under conditions which may include fretting corrosion. Such requirements far exceed those for bearings of polygon spinners.

LIS are frequently cost effective where wobble and jitter requirements are extremely demanding and should be considered when system tolerances are under 10 μrad. The small mass and inertia of the armature and low-driving-power requirements for many applications provide significant LIS performance advantages for many system designs.

Ball Bearings

Bearing Wobble. A typical ball-bearing-supported spindle, as for a polygon, exhibits 50 to 100 μrad of wobble normal to scan. This represents spindle-only errors and does not include any sagittal errors of the polygon reflective surface. A typical galvanometric scanner using the same ball bearings on the same spacing will be an order of magnitude better, or 5 to 10 μrad. Selected units from production for both types of scanner can have wobble values lower by a factor of 5.

The components of a bearing, the inner and outer races as well as the balls, have specified tolerances. The same goes for the bearing seats and the shaft. If error in excess of 1 μm is associated with each interface, the cumulation of all imperfections is the source of spindle wobble. Add to this any polygon mirror errors as well.

The armature construction of a well-designed galvanometric scanner has the same tolerances and imperfections as any spindle movement. The mirror surface, however, is forced to keep a constant periodic relationship to all the bearing and other components, so the mirror repeats its sagittal behavior scan after scan and virtually no wobble is present.

The design of galvanometric scanners and mirror systems takes this periodicity as critical to their implementation. The sections ''Dynamic Imbalance'' and ''Mechanical Resonances'' provide information for analysis of dynamic radial imbalances and confirm the importance of armature-mirror balancing.

Bearing Preload. The conventional method of achieving radial rigidity of a spindle is to have axial preload for the bearings as described in Chapter 9, Section 3.

High-performance scanners preferably use radial preload only and not axial preload. Techniques to implement this have been described by Montagu [21–23] and Brosens [24]. Radial preload offers the advantage of minimum possible friction and the added benefit that friction is constant at all temperatures. The differential expansion of rotor and stator has no effect on the bearing preloading condition.

Galvonometric scanners designed for unbalanced load, heavy load, or operation in a vibrating environment (such as a helicopter) are built with axial preload for the best average performance.

The selection of the proper scanner bearings must address the following considerations:

Radial play and ABEC tolerance number
Lubricant
Installation and preload
Acceleration (acceleration greater than 500g results in sliding ball condition and early failure.)
Thermal expansion, axial and radial
Bearing rigidity

Cross-Flexure Bearings

Figure 10.35 is a section of a cross-flexure bearing unit built inside a cylindrical housing. These commercially available units come close to the size of a ball bearing and are used in some galvanometric scanners.

Figure 10.36 shows a resonant scanner with an armature supported on cross-flexure bearings. The unit is designed to capitalize on a feature of cross-axis flexure pivots, a suspension totally free of jitter. Flexures are for the LIS technology what air bearings are for polygons but at a cost lower by at least one order of magnitude. Both flexures and air bearings have a common weakness, however: low radial stiffness.

The advantages of flexure bearings are

Freedom from wobble
Nearly unlimited life in noncorrosive atmospheres
Operation in a vacuum (non-out-gassing)
No contamination of optics (versus lubricated bearings)
Compact
Wide temperature range
Very low noise
Very low damping losses

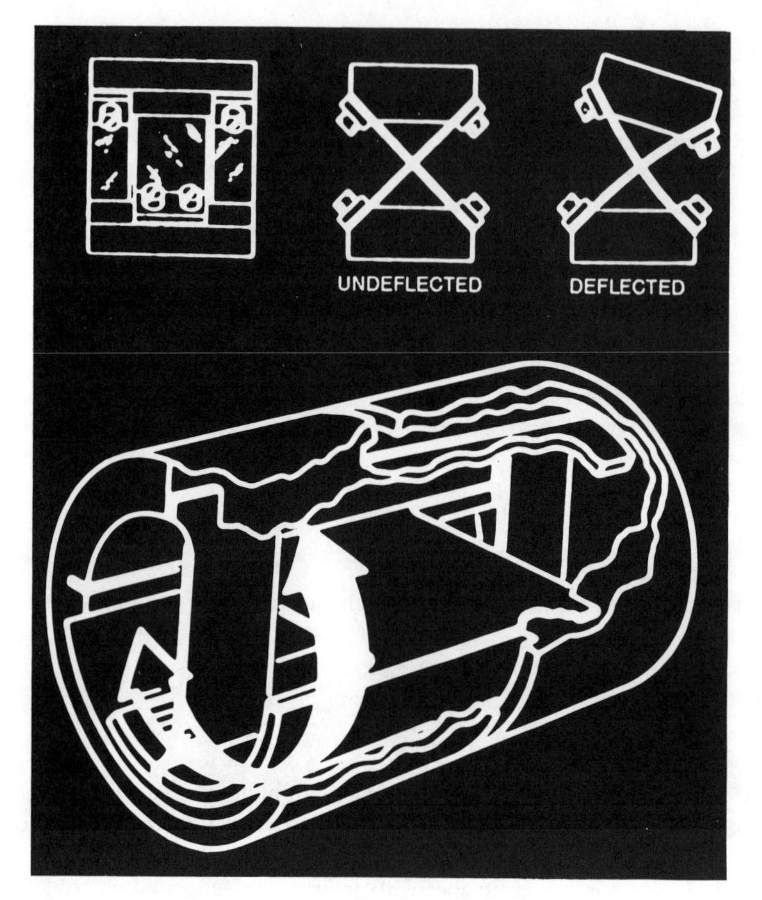

Figure 10.35 Free-flex flexural pivot.

The disadvantages are

Low radial stiffness
Shift of the axis of rotation with angle
Coupling of torsional and radial stiffness
Multiple elastic modes with low damping
Difficult installation
Intolerance of axial loading
Amplitude-dependent rigidity
Temperature-dependent rigidity

Fortunately, most of the shortcomings of flexure pivots can be circumvented by trade-offs in scanner design and installation procedures.

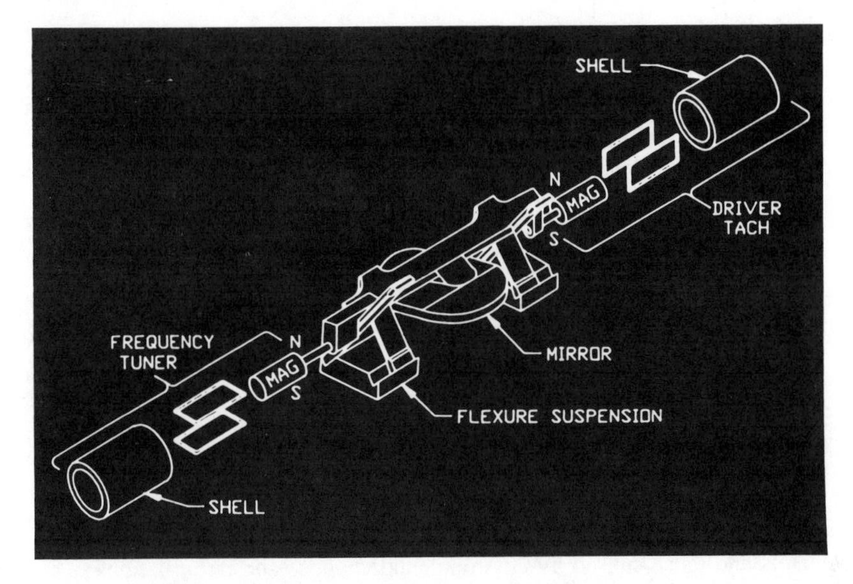

Figure 10.36 Tunable resonant scanner with magnetic spring.

Galvanometric scanners with flexure bearings have shorter bandwidth and less torque than their ball bearing equivalents. This is due to radial resonances and radial compliance constraints. Torque as well as sensor linearity is frequently affected by the radial shift of the axis of rotation.

Resonant scanners, on the other hand, are readily built to circumvent the flexure pivot's deficiencies and actually can benefit from some of them. High-performance resonant scanners require a high torsional rigidity and therefore achieve a desirable high radial rigidity. The high mechanical Q of this suspension is one of the major reasons for its selection. Rigidity variations due to amplitude and temperature changes can be compensated for in tunable scanners of sufficient bandwidth.

Torsion Bar Suspensions

Torsion bars have been used to support scanner armatures since the first moving coil optical galvanometers around the turn of the century. They are still used, but their extremely low radial rigidity restricts their application to resonant scanners. Figure 10.40 shows a unit of this type.

The single torsion bar construction with a mirror at the end is a very convenient geometry and is successfully incorporated into high-frequency resonant scanners, over 2.5 kHz, where the cross-resonance can be isolated from environmental perturbations. Numerous imaginative designs for low-frequency, vibration insensitive, single-torsion-bar scanners have been offered by Dostal [25], Montagu [26], and Burke [27]. They are not currently available commercially.

Taut double- and triple-torsion-bar (or band) constructions are used at lower frequencies, from 2500 down to 20 Hz. Figure 10.40 shows an exploded view of an induction-driven resonant scanner of this construction where the leaf spring keep the armature and torsion bar assembly in tension.

The cross-axis wobble of torsional elements due to environmental excitations can be minimized or neutralized with a thoroughly symmetrical and balanced construction. Armature translations cannot be eliminated, but introduce only negligible effects.

10.4.3 Electromagnetic Drivers

The mechanical transducer of a scanner should be selected for its ability to integrate with the other elements of the scanner, the mirror, the position sensor, and the electronic driver/controller. It must also support the dynamic performance requirements and those caused by environmental changes and perturbations.

The mechanical transducers or drivers analyzed in the following sections are all permanent magnet devices. Other technologies can also be used, such as electrostatics (electret, bimorph, piezo, etc.), pneumatics, hydraulics, and mechanics.

The list of features of an ideal magnetic driver is long and frequently a compromise is reached where some necessary system properties are obtained through other means. Environmental control and electronic compensation schemes have become standard features of high-performance scanners.

The ideal galvanometric or resonant scanner driver would have the following properties:

High torque-to-inertia ratio
Linear relationship between torque, current, and angular position
Freedom from cross-axis forces or excitations
No hysteresis or discontinuities
No elastic restraint
Some mechanical damping, constant and uniform
Very high rigidity in torsion and bending
A balanced armature
Low power consumption
Immunity to thermal expansion constraints
Good heat dissipation
Demagnetization protection
Simplicity of installation and use
Freedom from RF and other environmental perturbations

Immunity to external perturbations
Freedom from self-induced perturbations
Infinite life with stable parameters

Additionally, for a resonant scanner, damping properties should be minimum along the axis of rotation and high for all other degrees of freedom.

Moving Coil Driver

The most common magnetic driver is the moving coil or d'Arsonval galvanometer. The basic design has not changed since its invention a century ago, but availability of high-energy rare-earth magnets and high-saturation magnetic material has made practical some special-purpose designs.

Designs have been developed to satisfy most of the requirements listed in Section 10.4.3 except for thermal effects. Consequently, moving coil scanners should be considered for systems with a comparatively low duty factor. One should keep in mind that it is imperative to prevent coil deformation or fatigue due to periodic thermal expansion and contraction. Heat removal with magnetic fluids has not proven to be a long-term solution as it causes a temperature-dependent damping and is nearly impossible to contain totally when subjected to step motions.

Position servo scanners built with a moving coil driver compensate for their low structural rigidity by locating the sensor at the output shaft. It must then be elongated to accept it. The diameter of the shaft must also increase to yield proper rigidity. This unfortunately exacerbates heat conduction to the position sensor, which is the main sink for the heat generated in the coil.

Figure 10.37 is an example of a moving coil designed with a highly rigid and balanced ''shell'' armature incorporating a tachometer winding.

Though instruments of this type are available from half a dozen sources in the United States alone (strip chart recorder manufacturers among them), their limited appeal is due to their high cost, large bulk, and delicate construction.

Moving Iron Driver

The moving iron magnetic circuit of Figure 10.38 can be designed to satisfy all items listed in Section 10.4.3. The torque calculations are described in the first edition of this text [30], with the linearization technique.

The magnetic circuit is analogous to an electric bridge circuit. A polarizing flux is created at all four air gaps by the two permanent magnets. The remaining parts of the magnetic circuit are made from soft magnetic iron. The control flux is created in the same air gaps by the two control coils. The torque nonlinearities prevalent at high current are compensated by the introduction of alternate flux paths as described by Montagu [28]. The resulting torque expression within the

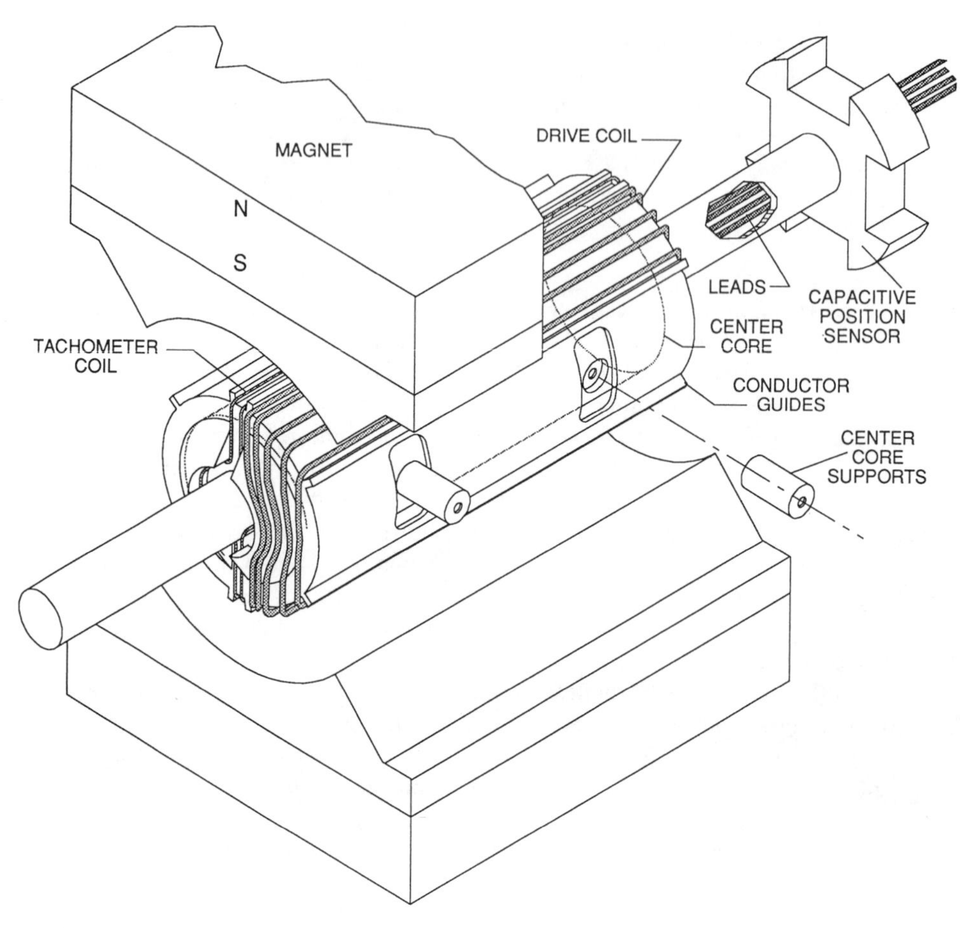

Figure 10.37 Shell armature—moving coil scanner.

rotation and current levels of interest is linear and that of a d'Arsonval movement. It is

$$T = BLNID \tag{49}$$

where T is the torque $(N - m)$, B is the air gap flux density (T), L is the rotor length (m), D is the rotor diameter (m), N is the number of coil turns, and I is the coil current (A).

Figure 10.39 shows the angular torque of four galvanometers with a ½-in bore, illustrating the evolution of the linearization techniques.

The torque is generated on a laminated moving iron vane structured to have

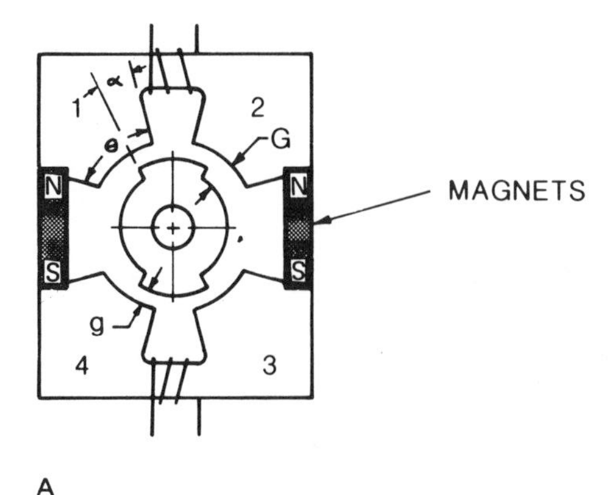

Figure 10.38 Moving iron galvanometer.

both high rigidity, to prevent bending, and low Q. The coils are enclosed in the stator, which permits good heat dissipation and eliminates the need for electrical connection to the rotor. All these support a wide servo bandwidth.

Induced Moving Coil

Two of the deficiencies of moving coil devices, lack of rigidity and moving electrical connections, can be bypassed by having a single turn coil energized by induction [29,39]. This technology is applicable to both galvanometric and resonant scanners. A resonant scanner implementation is illustrated in Figure 10.40. One of the most novel resonant LIS to have been introduced is the balanced torsion bar design with an induction torque driver. The mirror is suspended by two torsion bars in a fully symmetrical arrangement designed to be mass balanced, so that accelerations along the three principal axes of translation will not cause torsional excitation of the mirror.

The drive coil and the armature are magnetically linked in a transformer-like fashion by a soft iron core. The armature is a single-turn, rectangular drive loop with an edge colinear with the torsion current in the drive loop. This current interacts with the return path of the flux of a permanent magnet to create the drive torque:

$$T = \frac{\mu ANIBlr}{LR} \tag{50}$$

where μ is the permeability of the iron core, A is the cross section of the iron core, N is the number of turns of the drive coils, I is the driver coil current amplitude, B is the field of the permanent magnet, l is the length of the drive loop in the magnetic field, r is the drive loop acting radius, L is the length of

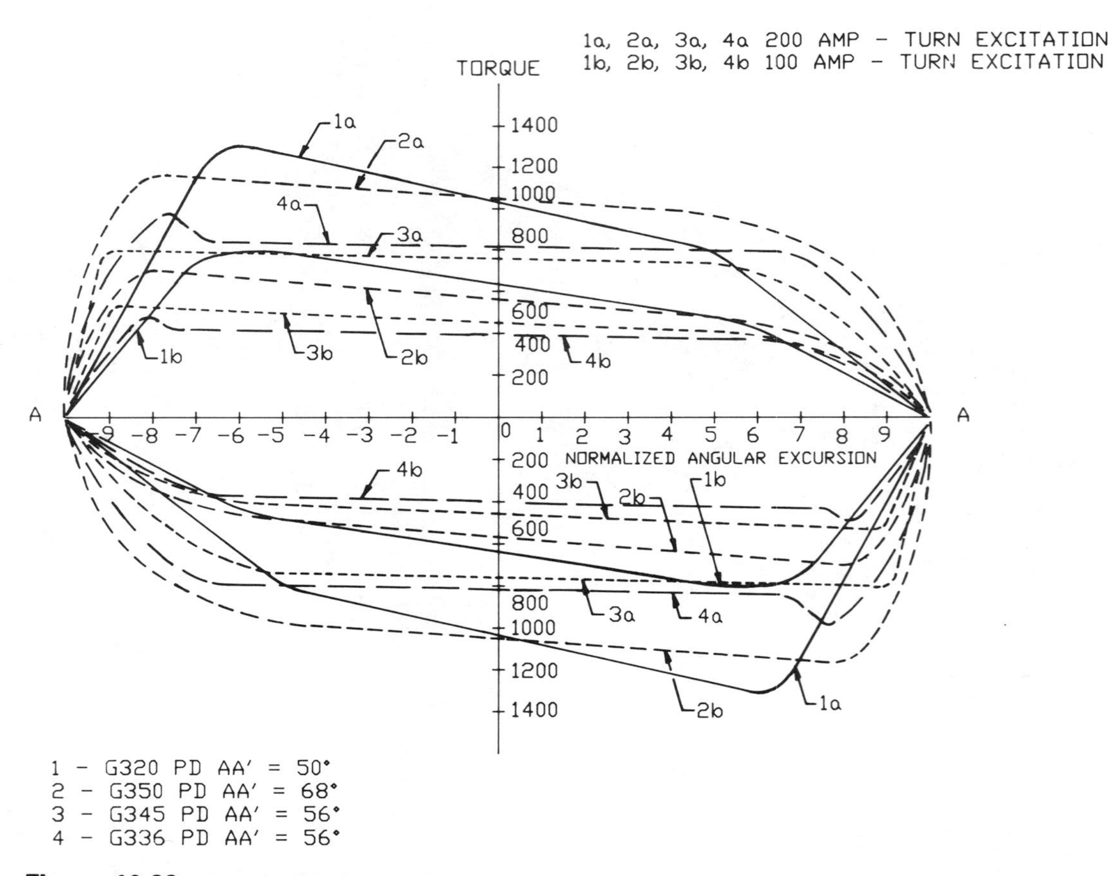

Figure 10.39 Comparative linearization curves.

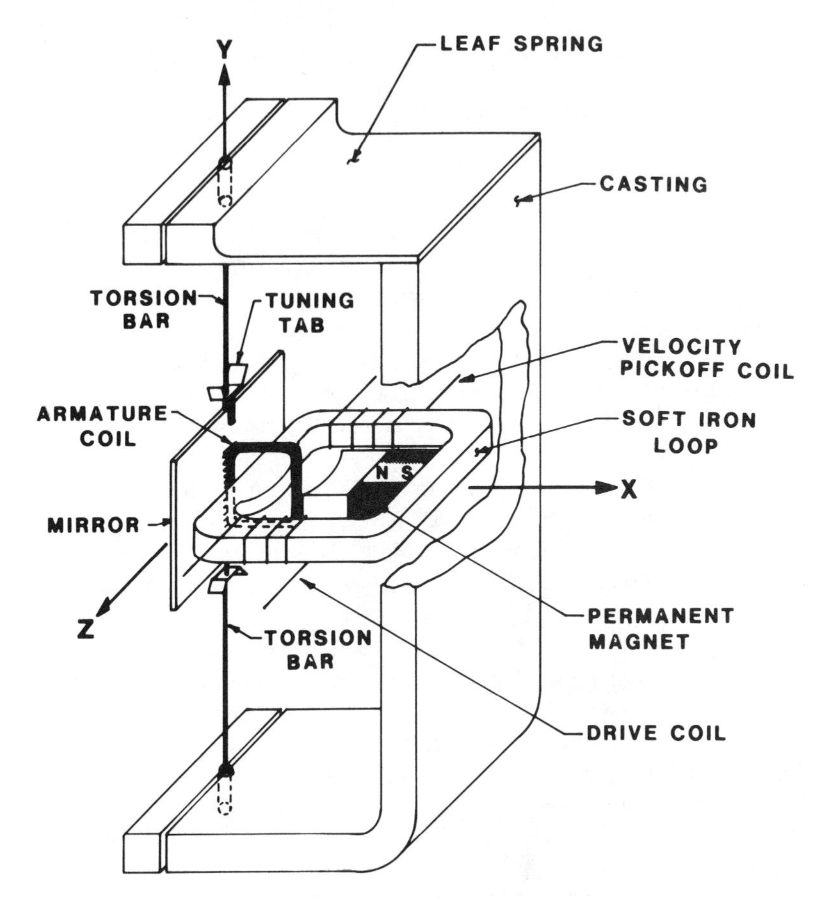

Figure 10.40 Resonant LIS.

the iron core path, and R is the resistance of the drive loop. In this manner a moving coil driver is obtained without having to provide leads or brush contacts to the moving armature. The motion of the armature induces a voltage and current which are sensed by a pick-off coil. The actual voltage measured is the sum of this induced voltage, which is proportional to velocity, and of the portion of the drive coil voltage induced by transformer couplings. It is expressed as

$$E = \mu \frac{AIN^2}{L} \omega \cos \omega t + NBLr \frac{\partial \theta}{\partial t} \tag{51}$$

where ω is the resonant frequency. The transformer coupling component is easily subtracted electronically. This velocity sensor permits extremely good, simple, and external amplitude control with drift below 100 ppm/°C of the peak-to-peak excursion. The resonant frequency is stable to -160 ppm/°C. A theoretical derivation of the pertinent parameter is presented in the appendix to Section 5 of the first edition of this text (Montagu [30]).

Moving Magnet Drivers

The high energy of rare-earth magnets makes them attractive driving elements of torque transducers. Moving magnet devices are practically free of radial forces. They can be incorporated into torque producing drivers which have most of the desired properties listed in Section 10.4.3. They are large-air-gap devices with comparatively low inductance, and simple to interface with electrically. Because they cannot achieve the torque-to-inertia ratio obtainable with moving iron torquers, they are found in slow systems and resonant scanners, as illustrated in Figure 10.36.

The driving stage is a conventional inside-out d'Arsonal movement, as shown in Figure 10.41. The torque can be calculated as the interaction of two fields or the effect of a field on a current. We shall follow the latter method.

For the purpose of determining the forces, let us assume that the magnet and the shell are stationary and calculate the torque T on the set of coils.

We know that to every torque there is an equal and opposite reaction. If the set of coils is used as a reference, the same torque acts on the magnet and shell combination.

As the shell is concentric to the magnet, unpolarized and anisotropic, all the torque appears at the magnet. The shell as well as the coils are stationary.

The torque T at a coil segment such as u, is

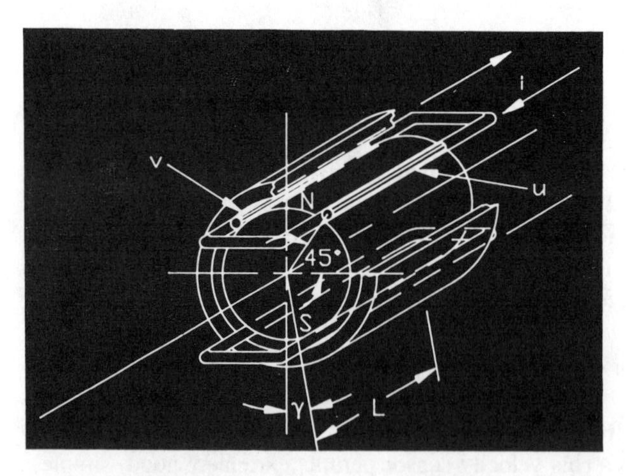

Figure 10.41 Inverted d'Arsonval movement.

$$T = \frac{BLNiD}{2} \tag{52}$$

where B is the magnetic field where the N conductors of the coil segment are located on a diameter D. The active length of the magnet is L.

For a highly isotropic material such as that of the rare-earth magnets, the magnetic field B between the magnet and the shell can be described as

$$B = KB_r \cos\theta \tag{53}$$

where θ is the angle between the axis of the magnet and the location of each segment of the coil. K is a constant defined by the magnet diameter and the air gap. B_r is the coercive force of the material.

If the coils are at 45°, a rotation γ at location u gives the field value

$$B_u = 0.707KB_r(\cos\gamma + \sin\gamma) \tag{54}$$

At location v the field is

$$B_v = 0.707\,KB_r(\cos\gamma - \sin\gamma) \tag{55}$$

The resultant torque T on the two coils is

$$T = 1.4\,KB_rLNiD\cos\gamma) \tag{56}$$

For values of γ under 10°, it approximates to better than 1%:

$$T = 1.4\,KB_rLNiD \tag{57}$$

10.4.4 Position Transducers

The position sensor is a crucial element in the performance of a position servo system. It must accurately convert the angular position of the armature into a convenient electric signal to be processed by the control electronics. Also, the derivative of the position signal must meet the damping requirements of analog servo controllers. The sensor may have an analog or a digital output signal, depending on the overall system design.

An ideal sensor would have the following characteristics:

Very high resolution, analog or digital
Stability equal to the resolution
A flat bandwidth extending to several orders of magnitude beyond the desired servo response
Freedom from hysteresis in the transfer function
A high signal-to-noise ratio
Insensitivity to wobble or radial motions
No phase delay

A linear transfer function
No coupling to the electromagnetic driver
Minimum inertia
Mechanical balance and freedom from resonances
Insensitivity to environmental/temperature factors
Easy electrical and mechanical interface
No variation with time, and infinite life
Low cost

All position transducers fall short of the ideal one. In the practical world of design engineering, the final selection is a compromise between the system requirements and the available techniques. Some of the design approaches used will now be described.

Inductive Encoders

There are three basic types of magnetic inductive encoders. First are the rotary variable differential transformers or RVDTs [31]; second are the Microsyns [32]; third are the Metresites [33]. They all incorporate some magnetic material within their structure to define the active region. This is one of the major limiting factors to their applications, since ferromagnetic properties terminate at between 20 and 100 kHz. Though ferrites are usable into the megahertz region, they are very temperature sensitive and difficult to machine.

RVDTs consist of a number of large air core windings. A primary winding is flanked with two identical secondary windings on either side. An excitation is applied to the primary and the motion of a magnetic core varies the mutual inductance of each secondary to the primary.

The Microsyn is a magnetic equivalent of the Wheatstone bridge. The motion of a magnetic vane changes the coupling of four legs of a magnetic stator.

The Metrisite is an iron core differential transformer (illustrated in Figure 10.42) where the imbalance is controlled by the position of a shorted turn.

Eddy-current displacement transducers are inductive devices with megahertz carrier frequency. The motion of a conductive core alters the mutual inductance of two coils from which the desired information is derived.

Inductive encoders have not found wide acceptance in high-performance scanning applications. They are characterized by the following:

Advantages

High signal-to-noise ratio
High resolution
Good linearity
Low inertia and freedom of self-resonances
Moderate sensor bandwidth
High reliability

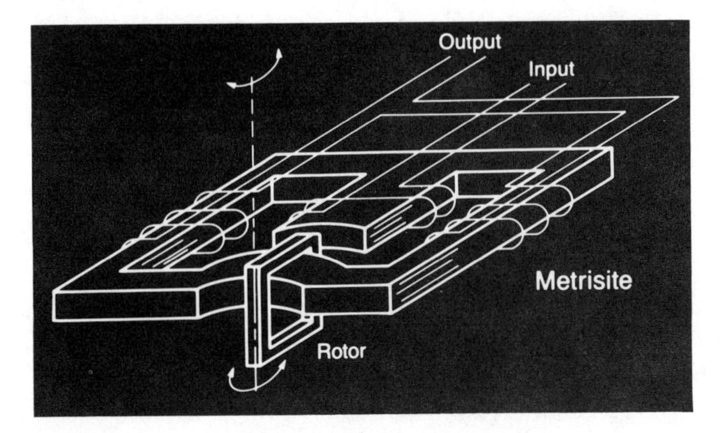

Figure 10.42 Inductive position sensors.

Disadvantages

Limited thermal stability
Coupling to magnetic driver
Limited bandwidth or RF noise generators.

Capacitive Position Sensor

Variable capacitance sensing represents a good compromise technique for rotation position sensing in high-speed scanning applications. Figure 10.43 presents an exploded view of such a compact device with protection guard and dimensionally/thermally stable moving and stationary elements.

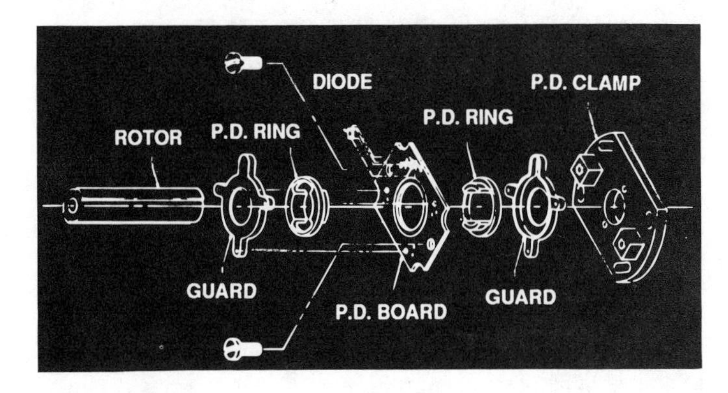

Figure 10.43 Capacitive position sensor—exploded view.

The sensing of capacitance can be done by any of a variety of methods including direct impedance measurement, the length of time to integrate to a known level, or the frequency shift of an oscillator. The need for high speed, resolution, and stability complicates the electrical interface, but it is well within the state of the art.

Advantages

High resolution
High repeatability
Wide bandwidth, in excess of 50 KHz
High signal-to-noise ratio
Very low hysteresis, under 1 μrad
Good linearity
Minimal coupling to the driver
Simple mechanical integration
Infinite life in clean environment
Moderate cost

Disadvantages

Gain drift with temperature
Null drift with temperature
Limited to analog metrology capability
RF noise generation

The temperature sensitivity of these sensors (as low as 10 ppm/°C of gain drift and 2 μ/°C null drift) can be virtually eliminated by proper system calibration techniques. A calibration protocol searches periodically for two reference points and sets new gain and null values. These points can either be located in the field of vision of an optical path or take the form of mechanical stops.

Optical Position Sensing

Optical sensing covers a wide range of techniques and finds applications at both ends of the precision spectrum. The common attractions are the noncontact feature and long life.

A number of simple medium-performance techniques have been implemented. They are usually variations of energy-collection schemes where the light from an LED is modulated by a rotating mirror or a shadowing or polarizing vane. They are easy to interface with electrically and comparatively low in cost. Unfortunately, they suffer from thermal stability problems associated with both the sensor and the light source. The best schemes have a 1% long-term accuracy.

Grating encoders rely for their accuracy on the precision and stability of all optical elements as well as the sensing devices. An encoder disk mounted directly on the scanner shaft offers an attractive package but penalizes dynamic performances. Canon [34] describes a small encoder of this type with an inertia of 5 g-cm^2 and a 77-μrad resolution without electronic interpolation.

When cost has not been a significant constraint the location sensing of a dedicated beam via the use of an optical grating in or near the work plane has found wide acceptance. Figure 10.1 illustrates this design. For line scan applications this scheme offers, inherently, perhaps the best of all performance with respect to system stability, as it can directly measure the location of the working light beam after most of the path-altering elements have been passed. A 10-μrad beam steering accuracy can be readily achieved. For surface (*XY*) scanning it is, however, expensive and cumbersome, as the rear of the mirrors must be used for metrology.

Interferometry offers the most accurate position sensing technology, down to submicroradians (with a 0.35-rad range), and with data transfer at 1 or 2 MHz. Interferometers have a digital output signal and easily interface with computer logic. In fact, the resolution and stability obtainable are under 10 parts per million, and substantially beyond what can be achieved with analog circuitry. Digital amplifiers or controllers are necessary to justify the use of this technology.

Interferometric measurements have intrinsic uncertainties (sources of possible error) due to the fact that the index of refraction of air (and consequently the wavelength of light) changes with the variations of temperature, pressure and humidity of the environment as described in the section titled "Index of Refraction of Air."

Temperature: For an interferometer with a HeNe laser, the error due to air temperature variation is approximately 1 μm/m of air path for a 1°C temperature variation.

Pressure: For an interferometer with a HeNe laser, the error due to air pressure variation is approximately 1 μm/m of air path for a pressure change of 2.8 mm of mercury.

Humidity: Similarly, 1 μm of error is derived from a relative change of 18 mm of Hg of water pressure. As humidity can be well controlled, this variable can be neglected in this application.

These sources of error associated with interferometric measurements are well understood and compensation techniques exist. The most common are:

1. Control of the environmental parameters, temperature, pressure, and humidity.
2. Measurement of the environmental parameters and appropriate correction of the measurement data.

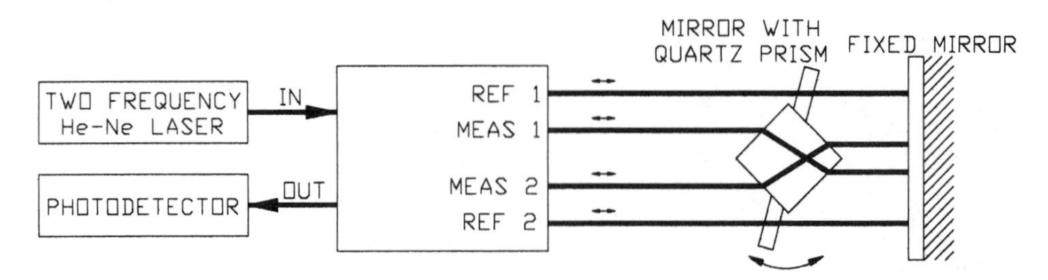

Figure 10.44 Zygo's large-angle measuring interferometer (LAMI).

3. Design of an alternate calibration path which shares the same environment.
4. Minimization of the air paths.

It is also necessary to keep in mind that interferometers measure the wavelength of light, which is a linear quantity. Angles, which are of interest here, are the ratio of two lengths, and this introduces a number of additional sources of error which must be controlled.

Two types of angular interferometers are used:

Heterodyne Interferometers. Heterodyne interferometers have two laser beams (usually coaxial) whose wavelengths are separated by a few megahertz. They are made to interfere and create a time-varying fringe pattern with a period defined by the period separation. Figure 10.44 is a symbolic diagram of a commercial angular measuring interferometer of this type. A solid prism mounted on the

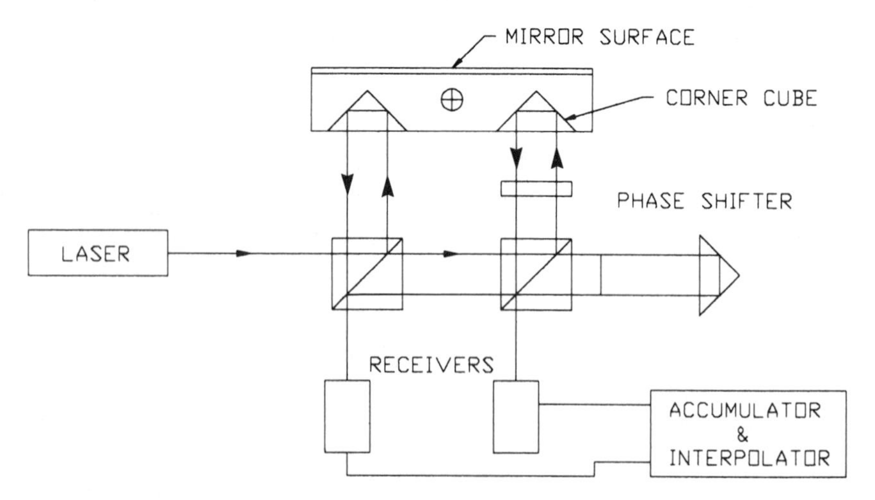

Figure 10.45 Corner cube Michelson interferometer.

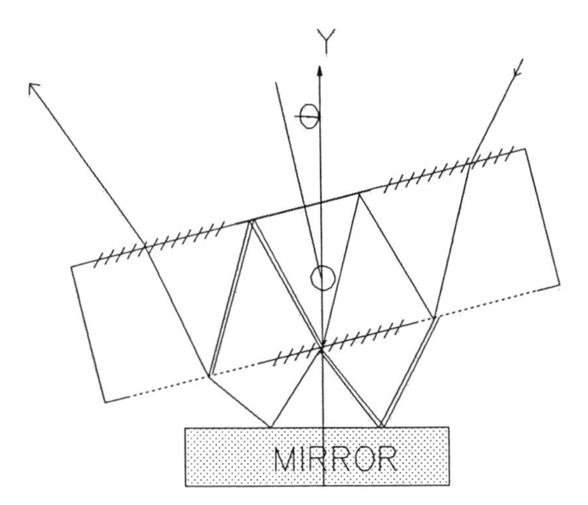

Figure 10.46 Young rotational interferometer.

scanner shaft next to the mirror alters the optical path lengths of the two measuring beams as it rotates. There are also two reference beams. Each beam operates as an independent heterodyne interferometer. Incident and reflected beams are separated with polarizing and rotating optical elements (shared among the four beams). The reference beams are used to calibrate the air path to null out the perturbations due to temperature, pressure, and humidity variations as well as to define phase and frequency references. Rotation detection is converted to frequency discrimination and the angular position is obtained by phase comparison. The signal to noise ratio ultimately limits the resolution. Cost is high.

A number of other rotary heterodyne interferometers have also been implemented [35].

Dual-Path Interferometers. Dual-path equal length interferometers have a much older ancestry and have been recently much revisited by Ledger [36], Young [37], and others. Two interferometers are required, 90° out of phase, in order to eliminate ambiguities of direction of motion. Figure 10.45 shows a Michelson design where the translation mirrors have been replaced by prism reflectors. Figure 10.46 shows the Young rotational interferometer, which can be traced to the Jamin refractometer—the second phase-shifted path is omitted. These designs have the advantage of being simple, and when compactly built the air paths can be so short that their susceptibility to atmospheric conditions can be neglected.

10.4.5 Electronic Drivers and Signal Conditioners

Electronic drivers for galvanometric and resonant scanners are diagrammed in Figures 10.47 and 10.48, respectively. The galvanometric scanner driver is a

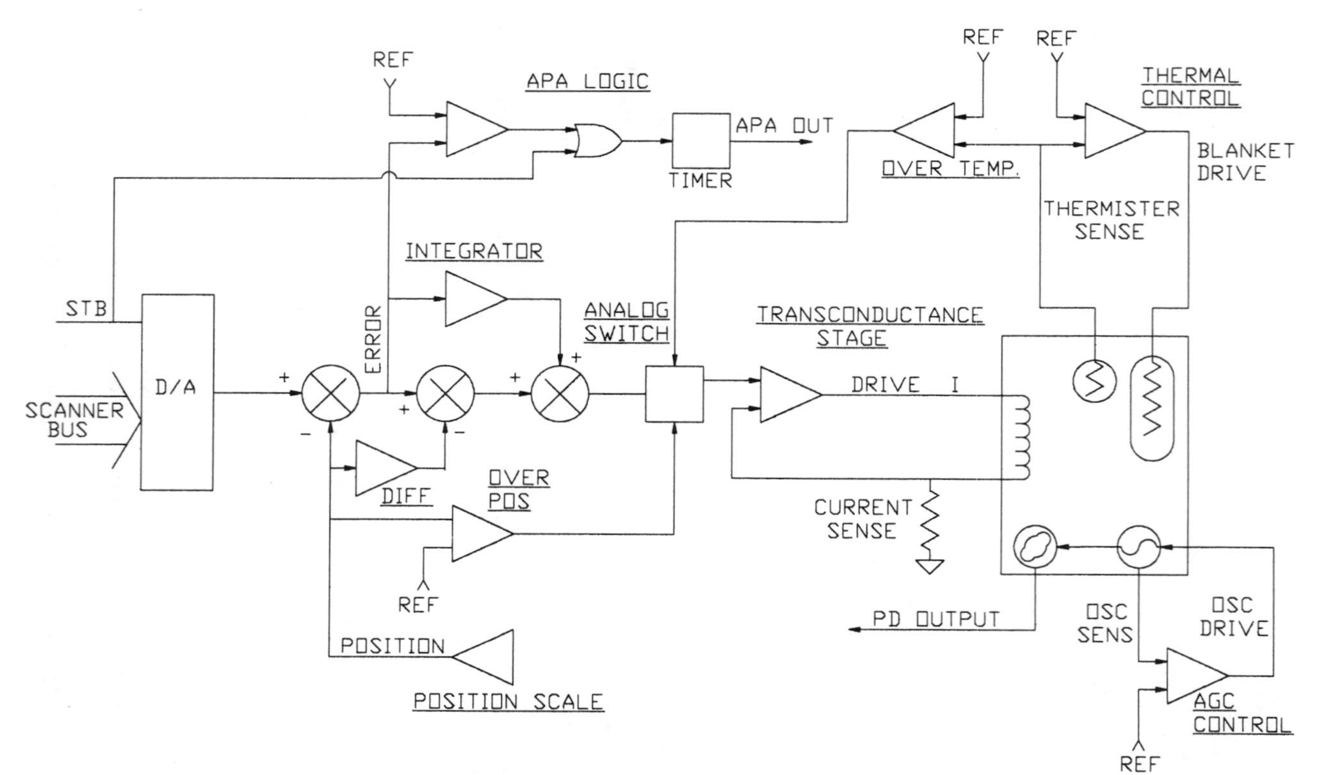

Figure 10.47 DX series servo amp—galvanometric scanners.

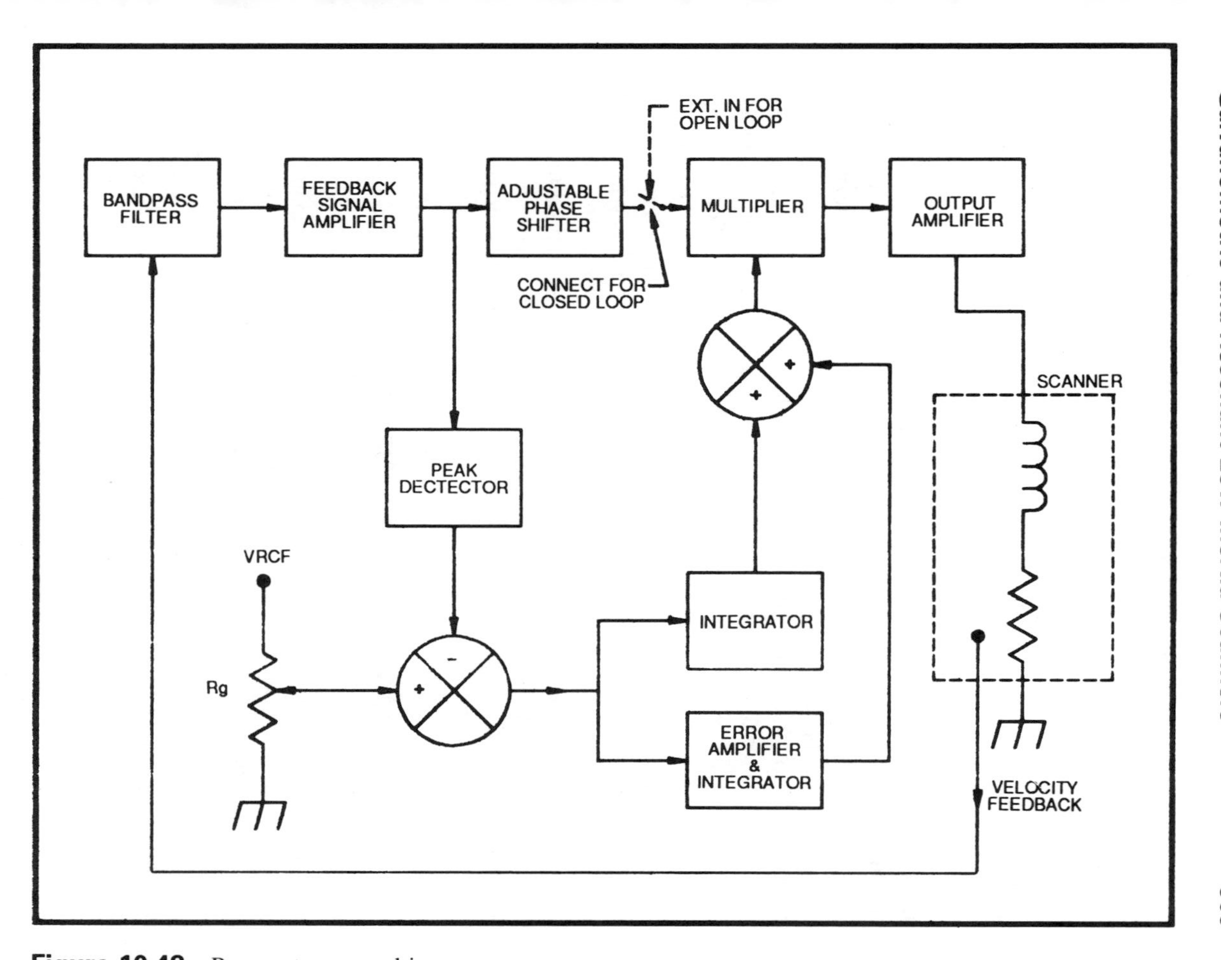

Figure 10.48 Resonant scanner drive.

conventional design which becomes complex in order to obtain higher speed and precision. It has provision for thermal control and thermal protection, and is intended to interface with a host computer.

It is extremely important to minimize the maximum acceleration the scanner will experience. Imbalances are always present and may cause radial and wobble motions or excite cross-resonances (see sections on dynamic imbalance and mechanical resonance).

In the resonant scanner driver block diagram (Figure 10.48), a transconductance amplifier with an electrically variable gain drives the scanner. A sinusoidal voltage of sufficient amplitude and of the correct frequency applied to the external input causes the scanner to resonate. The velocity transducer accomplishes automatic gain control by providing a signal whose amplitude is proportional to oscillation amplitude. This signal is filtered, amplified, and applied to a peak detector to form a DC voltage proportional to oscillation amplitude. This DC voltage is compared simultaneously to an adjustable reference voltage, and the difference is applied to an amplifier and an integrator to provide control.

The outputs are then summed and applied to the variable-gain element. This element adjusts the gain of the power amplifier so that any tendency to change the amplitude of oscillation is opposed. In the self-resonant mode, the external frequency source is removed, and the filtered gain- and phase-compensated feedback signal (velocity or position) is fed into the input of the power amplifier. Automatic gain control operation is the same. A TTL-compatible sync signal on the front panel is derived from the velocity/position signal.

10.5 PERFORMANCES OF LOW-INERTIA SCANNERS

It is not practical to give thorough general-purpose specifications of low-inertia scanners as electromagnetic components for use in high-performance applications. The list of parameters would be too long. Describing entire systems in detail is equally impractical because applications are so varied. The data listed in Tables 10.13 to 10.16 present performances of a few standard commercial products driven with general-purpose amplifiers. These data are intended as a guide to possible performance. The designer should obtain the physical parameters of the desired unit and do a thorough system analysis before performance can be predicted.

10.5.1 Performance of Galvanometric Low-Inertia Scanners

A concise presentation of the characteristics relating to the accuracy of servo controlled moving iron scanners has been offered by Brosens [38]; Tables 10.13 and 10.14 give the same information but includes more recent models.

Table 10.13a Step Response of Moving Iron Galvanometric Scanners with Closed-Loop Integrating Amplifier—Shaft or Mirror Movements

Scanner type	GSI G120D		GSI G400			GSI M3B			GSI G3B			GSI G325	
Mirror													
Face (mm)	7 × 7		15 × 15			24 × 38			24 × 38			26 × 26	
Thickness (mm)	1		1.5			3.2			3.2			3.2	
Inertia, with mount (g-cm^2)	0.016		0.40			6.5			6.5			3.3	
Rotor inertia (g-cm^2)	0.023		0.20			5.0			6.0			3.7	
First resonance (kHz)	4.0		7.0			5.0			3.0			1.0	
Stator													
Height (mm)	33		75			98			87			79	
Cross section (mm)	23 × 33		30 dia.			51 dia.			54 dia.			32 × 45	
Weight (g)	107		240			300			670			435	
Life (cycles)	10^{10}		10^{10}			10^{10}			10^{10}			10^{10}	
Bearing noise (arc sec)	10		<2			<2			<2			5	
Wobble/jitter (μrad)	25/50		8/15			4/12			4/12			12/25	
Step response (ms) for angular step	a	b	a	b	c	a	b	c	a	b	c	a	b
1°	0.50	0.5	1.5	—	2.5	1.8	—	2.0	2.5	3.8	3.6	3.4	2.8
5°	0.55	0.57	1.6	1.7	—	1.9	2.2	3.2	3.0	—	4.7	3.5	3.4
10°	0.63	0.75	1.6	1.9	3.3	2.5	3.6	5.6	3.1	5.0	6.5	3.6	3.7
15°	0.70	0.95	1.8	2.0	—	2.8	—	—	—	—	—	4.1	4.4
20°	0.80	1.1	2.0	2.5	4.3	3.6	4.3	7.0	5.4	7.0	9.2	4.6	4.8

Conditions: Closed-loop integration amplifier

Power supply: +/− 15 volts, 3 amps

a. Settled accuracy or error: 1% of step angle
b. Settled accuracy or error: 0.1% of step angle
c. Settled accuracy or error: 35 μrad

Table 10.13b Comparative Performance of Scanning Galvanometers

Model	G120PD	G400PD	6300	G3B	M3B	6650
Armature (moving)	Iron	Iron	Coil	Iron	Magnet	Coil
Resistance (ohm)	2.0	2.5	4.0	2.0	4.5	4.5
Torque constant (gc-cm/A)	60	100	115	500	500	570
Inertia (g-cm^2)	0.03	0.20	0.13	6.0	5.0	8.8
Max. torque (g-cm)	240	400	128	2000	1600	1800
Max. power (watts)	30	40	5[1]	32	45	45[1]
Acceleration (rad/s^2 10^6 ×)	8.0	2.0	1.0	0.33	0.32	0.20
Weight (gm)	107	240	540	670	300	750

[1] Burst mode only—all others specified at continuous power dissipation by manufacturer.
Source: G120PD, G400PD G3B, and M3B from General Scanning data sheets; 6300 and 6650 from Cambridge Technology data sheets and measurements.

This data represents only basic information. Additional parameters come into play to determine performance of a system. Listed below are some of the parameters to be considered.

Wobble
Jitter
Null drift
Gain drift
Repeatability
Settling time
On-axis first resonance
Cross-axis first resonance
Sensitivity to external vibration
Suspension rigidity and imbalance
Linearity of sensor
RF radiation
Mirror dynamic rigidity
Mirror temperature deformation

Mirror mounting rigidity and alignment (in two axis)
Air turbulence effect on the mirror
Accuracy of the mounting surface
Coupling of signal and power ground
Sensitivity to electrical noise
Sensitivity to magnetic noise
Heat generation
Heat dissipation
Effect of temperature gradients
Long-term mechanical stability
Outgasing properties
Sensitivity to dust and vibration

Table 10.14 Drift of Capacitive Scanner Transducers

Scanner type	GSI G120DC	GSI G3B	GSI M3B	GSI G400	State of the art
Test range (°C)	26–46	10–50	10–50	20–60	10–50
Gain drift (%/°C)	0.07	0.005	0.005	0.004	0.001
Zero drift (μrad/°C)	150	10	10	6	2

10.5.2 Performances of Resonant Low-Inertia Scanners

Resonant scanners have had limited use in laboratories as a simple way to obtain large scan angles with large beam diameters. When practical solutions were found to the problems caused by vibrational and temperature instability and when high repeatability and long life were achieved, resonant scanners appeared commercially.

High-strength steel and aluminum, CNC machining centers, and low-cost CAD equipment have helped solve the mechanical problems. High-speed logic has made it possible to implement practical solutions for the timing and interface requirements.

The data presented in Table 10.15 should be used as a general guide only. The ultimate performance of a system incorporating a resonant scanner will be strongly influenced by the mounting techniques as well as the sensing and driving electronics. The list of parameters to specify the scanner as a component is the same as that given in Section 10.5.1 for galvanometric units.

10.5.3 Tunable Resonant Scanners

Tuning of resonant scanners is effected by altering the spring rate [23]. Two types of tunable scanners exist.

Table 10.15 Typical Performance Characteristics of Resonant Scanners

Scanner suspension	Taut band	S-Flexure	Blade	Torsion bar	IDS	Double node	Torsion bar	X-Flexure	S-Flexure
Performance range: resonant frequency/ mirror diameter/ angle (Hz/mm/deg)	10/26/30 to 800/5/2	10/25/15 to 350/10/10	20/25/15 to 250/20/10	750/10/7 to 16,000/2/1.5	600/10/30 to 4000/7/10	4000/15/15 to 18,000/3/4	500/25/30 700/10/20	170/37/24 to 200/25/28	75/25/25 to 250/25/10
Model & number	L50	URS 100	1B20–100	L45	IDS 103010	IFS	CRS	IPX 4260	FLS 30
Typical performance									
Resonant frequency (Hz)	100	100	100	1000	1000	4000	650	200	200
Temperature stability (ppm/°C)	1000	200	300	500	200	200	−170	170	200
Mechanical scan angle (deg)	15	15	15	6	30	15	21	28	15
Temperature stability (ppm/°C)	NS	400	NS	NS	100	NS	NA	10	1000
Mirror dimension (mm)	15 × 16	20 × 20	20 × 20	10 × 10	10, diameter	15, diameter	13, diameter	25 dia.	25 dia.
flatness (λ/cm)	½	½	½	NS	½	½	¼	¼	¼
Wobble (μrad)	NS	10	10	NS	35	NS	<2	NS	NS
Repeatability (μrad)	NS	10	10	NS	5	5	10	<1	<5
Power (mW)	40	15	18	300	100	100	2000	1000	1000
Operating temperature (°C)	−20–40	0–60	0–60	−40–40	0–77	NS	10–32	15–60	0–60
Weight (g)	21	20	18	56	100	50	260	320	150
Vibration sensitivity	Poor	Good	Good	Good	Excellent	Very good	Excellent	Excellent	Good

[a] Not specified.

Source: L50 and L45 data from FCD data sheets; IB, IDS, IFS, CRS and IPX data from General Scanning, Inc.; data sheets. URS and FLS data from Laser Scanning product data sheets.

Table 10.16 Tuning Performances of Mini-ISX Electromagnetically Tunable Resonant Scanner

Natural frequency	200 Hz
Tuning range, total	1.6 Hz
Bandwidth, min.	20 Hz
Spring rate, mechanical	5.3 × 10 g-cm-rad
Spring rate, magnetic	100 g-cm/rad/amp
Tuning coil, resistance	26.6 ohms
Tuning coil, inductance	1.2 mH
Tuning torque time rise	0.1 ms
Rotor inertia	3.31 g-cm^2

Table 10.16 shows the performances of the unit in Figure 10.49, which is an ISX model from General Scanning, Inc. with a magnetic tuner built on axis. (A schematic description is given in Figure 10.36.) The frequency tuning section consists of a permanent magnet similar to that of the driving section. Its effect can be derived by noting that the coils are rotated 90° from those of the driver. The resulting torque calculates therefore as the difference of equations (56) and (57). Using the same symbols, it is:

Figure 10.49 Mini-ISX tunable resonant scanner.

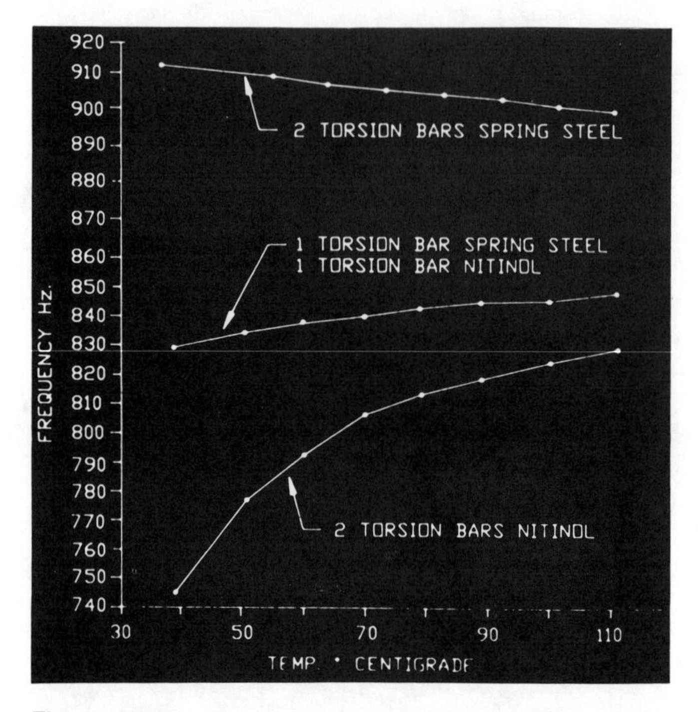

Figure 10.50 Tuning performances of IDS temperature-tunable resonant scanner.

$$T = 1.4KB_rLNiDy \tag{58}$$

which is in accord with the tabulated data.

The other method of controlling the spring rate of a scanner torsion bar is to control the environmental temperature of the unit. Tunability is much enhanced when the torsion bars are made of Nitinol, whose Young's modulus can be changed by as much as 10% without affecting the performance life of the scanner. Figure 10.50 compares the performances of an IDS scanner fitted with a choice of torsion bars.

REFERENCES

1. P. Schubert, *Appl. Opt.*, *25*:101–104 (1986).
2. F. Bestenreiner, *J. Appl. Photog. Eng.*, *2*:86–92 (1976).
3. J. Morris, personal communication.
4. D. C. O'Shea, W. T. Callen, and W. J. Rhodes, *An Introduction to Lasers and Their Applications*, Addison-Wesley, Reading, MA (1977).
5. A. Lawler and J. Shepherd, *Laser Beam Scanning* (G. F. Marshall, ed.), Marcel Dekker, pp. 125–147, New York (1985).

6. P. J. Brosens, personal communication.
7. W. J. Smith, *Modern Optical Engineering*, pp. 132–133, McGraw-Hill, New York (1966).
8. W. Schreiber, *Ann. New York Acad. Sci.*, *267*:469–476 (1976).
9. J. D. Zook, *Appl. Opt.*, *13*:875 (1974).
10. J. P. Den Hartog, *Mechanical Vibrations*, p. 396, McGraw-Hill, New York (1956).
11. D. Tweed, *Laser & Appl.*, *Aug.*, *1985*:65–69 (1985).
12. G. J. Burrer, U.S. Patent #3,978,281 (1976).
13. L. Burburry, U.S. Patent #4,482,902 (1974).
14. R. S. Reiss, *OE Reports*, *May*, *1989*:8. Quoted with permission of the author, 1989.
15. P. J. Brosens, *Appl. Opt.*, *2*:2988–2989 (1972).
16. S. Timoshenko and J. N. Goodier, *Theory of Elasticity*, McGraw-Hill, New York (1951).
17. P. J. Brosens and V. Vudler, P. *Opt. Eng.*, *28*:61–65 (1989).
18. P. Yoder, *Opto-Mechanical System Design*, pp. 71–77, Marcel Dekker, New York (1986).
19. H. Weissman, *Opt. Eng.*, *15*:435–441 (1976).
20. J. I. Montagu, *SPIE Proceedings*, *896*:40–44 (1988).
21. J. I. Montagu, U.S. Patent #3,624,574 (1971).
22. J. I. Montagu, U.S. Patent #4,694,212 (1987a).
23. J. I. Montagu, *SPIE Proceedings*, *817*:2–7 (1987b).
24. P. J. Brosens, U.S. Patent #4,135,119 (1977).
25. F. Dostal, U.S. Patent #3,609,485 (1971).
26. J. I. Montagu, U.S. Patent #3,959,673 (1976).
27. E. Burke, U.S. Patent #4,364,000 (1982).
28. J. I. Montagu, U.S. Patent #4,528,533 (1985b).
29. J. I. Montagu, *Electro Optics*, *May*, *1983*:51–56 (1983).
30. J. I. Montagu, *Laser Beam Scanning* (G. F. Marshall, ed.), pp. 283–286, Marcel Dekker, New York (1985a).
31. E. Herceg, *Handbook of Measurement and Control*, pp. 9-12–9-15, Shaevitz Engineering, Pennsauken, NJ (1976).
32. F. Mueller, U.S. Patent #2,488,734 (1949).
33. G. Smith, U.S. Patent #3,034,036 (1962).
34. Canon, Inc. *Motion Magazine*, *Sept.–Oct. 1986*:12–18 (1986).
35. S. Holly, *Opt. Eng.*, *15*:146 (1976).
36. A. Ledger, *Electro-Optical Systems Design*, *7*:32–36 (1975).
37. N. Young, U.S. Patent #3,348,446 (1967).
38. P. J. Brosens, *Opt. Eng.*, *15*:95–98 (1976).
39. J. I. Montagu, U.S. Patent #4,502,752 (1985c).

11

Acoustooptic Scanners and Modulators

Milton Gottlieb

Science and Technology Center, Westinghouse Electric Corporation, Pittsburgh, Pennsylvania

11.1 INTRODUCTION

It will be apparent to the reader of this book that there is a great variety of applications of lasers for which scanning devices are required, and that these applications include a wide range of performance requirements on the scanner. The basic specifications include speed, resolution, and random access time, and the choice of a scanner will be determined by these parameters. Acoustooptic scanners are best suited to those systems which are of moderate cost, since the cost of AO Bragg cells and the associated drive electronics are by no means trivial, and for which the resolution requirement is about 1000 spots. In addition, AO technology is most appropriate where random access times on the order of 10 μs are needed, or where it may be desired to perform intensity modulation on the laser beam, as in image recording. There are currently many systems employing AO scanners, perhaps the most familiar being laser printers, in which the scanner capability is an excellent match to the system requirements. Large-area television display was one of the first applications considered for AO scanners, and it performs this function very well, although such display systems are relatively uncommon. These as well as other applications of acoustooptic scanners, will be described in detail in a later section.

The interaction of light waves with sound waves has in recent years been the basis of a large number of devices related to various laser systems for display, information handling, optical signal processing, and numerous other applications requiring the spatial or temporal modulation of coherent light. The phenomena underlying these interactions were largely understood as long ago as the mid-1930s, but remained as scientific curiosities, having no practical significance, until the 1960s. During this period several technologies were developing rapidly, at the same time that many applications of the laser were being suggested which require high-speed, high-resolution scanning methods. These new technologies gave rise to high-efficiency, wideband acoustic transducers capable of operation to several gigahertz, high-power wideband solid-state amplifiers to drive such transducers, and the development of a number of new, synthetic acoustooptic crystals with very large figure of merit (low-drive-power requirements) and low acoustic losses at high frequencies. This combination of properties makes acoustooptics the method of choice for many systems, and is very often the only approach to satisfy demanding requirements. In this chapter, the underlying principles of acoustooptic interactions will be reviewed, and this will be followed by a description of the materials considerations and the relevant acoustic technology. Acoustooptic scanning devices will be described in some detail, including the important features of optical design for various types of systems.

11.2 ACOUSTOOPTIC INTERACTIONS

11.2.1 The Photoelastic effect

The underlying mechanism of all acoustooptic interactions is very simply the change induced in the refractive index of an optical medium due to the presence of an acoustic wave. An acoustic wave is a traveling pressure disturbance which produces regions of compression and rarefaction in the material, and the refractive index is related to the density, for the case of an ideal gas, by the Lorentz-Lorenz relation

$$\frac{n^2 - 1}{n^2 + 2} = \text{constant} \times \rho \tag{1}$$

where n is the refractive index and ρ is the density. In fact, this relation is adhered to remarkably well for most simple solid materials as well. The elastooptic coefficient is obtained directly by differentiation of (1):

$$\rho \frac{\partial n}{\partial \rho} = \frac{(n^2 - 1)(n^2 + 2)}{6n} \tag{2}$$

where it is understood that the derivative is taken under isentropic conditions. This is generally the case for ultrasonic waves, in which the flow of energy by thermal conduction is slow compared with the rate at which density changes within a volume smaller than an acoustic wavelength. The fundamental quantity given by equation (2), also known as the photoelastic constant p, can be easily related to the pressure applied, with the result

$$p = \frac{l}{\beta} \frac{\partial n}{\partial P} \tag{3}$$

where P is the applied pressure and β is the compressibility of the material. The photoelastic constant of an ideal material with refractive index of 1.5 is 0.59. It will be seen later that the photoelastic constants of a wide variety of materials lie in the range from about 0.1 to 0.6, so that this simple theory gives a reasonably good approximation to measured values.

The relation in equation (3) follow from the usual definition of the photoelastic constant:

$$\Delta\left(\frac{1}{\epsilon}\right) = \Delta\left(\frac{1}{n^2}\right) = pe \tag{4}$$

where ϵ is the dielectric constant ($\epsilon = n^2$) and e is the strain amplitude produced by the acoustic wave. From equation (4) it is easily seen that the change in refractive index, Δn, produced by the strain is

$$\Delta n = -\frac{1}{2}n^3 pe \tag{5}$$

where e is of the form, $e_0 \exp(i\Omega t)$ for an acoustic wave of frequency Ω. The magnitude of the changes in refractive index that are typical for acoustooptic devices are not large. Strain amplitudes lie in the range 10^{-8}–10^{-5}, so that using the above expressions for Δn and p gives for Δn about 10^{-8} to 10^{-5} (for n = 1.5). It may be somewhat surprising, then, that devices based upon such a small change in refractive index are capable of generating large effects, but it will be seen that this comes about because these devices are configured in a way that can produce large phase changes at optical wavelengths.

The relation defining the photoelastic interaction has been written in equation (5) as a scalar relation, in which the photoelastic constant is independent of the directional properties of the material. In fact, even for an isotropic material such as glass, longitudinal acoustic waves and transverse (shear) acoustic waves cause the photoelastic interaction to assume different parameters. A complete description of the interaction, particularly for anisotropic materials, requires a tensor relation between the dielectric properties, the elastic strain, and the photoelastic coefficient. This may be represented by the tensor equation

$$\Delta\left(\frac{1}{n^2}\right)_{ij} = \sum_{kl} p_{ijkl} e_{kl} \tag{6}$$

where $(1/n^2)_{ij}$ is a component of the optical index ellipsoid, e_{kl} are the cartesian strain components, and p_{ijkl} are the components of the photoelastic tensor. The crystal symmetry of any particular material determines which of the components of the photoelastic tensor may be nonzero, and also which components are related to others. This may be useful in determining whether some crystal, based only upon its symmetry, may even be considered for certain applications.

11.2.2 Diffraction by Acoustic Waves

The most useful photoelastic effect is the ability of acoustic waves to diffract a light beam. There are several ways to understand how diffraction comes about; the acoustic wave may be thought of as a diffraction grating, made up of periodic changes in optical phase, rather than transparency, and moving at sonic velocity rather than being stationary. Thus, it is possible to analyze the diffraction as resulting from a moving phase grating. Alternatively, the light and sound may be thought of as particles, photons and phonons, undergoing collisions in which energy and momentum are conserved. Either of these descriptions may be used to obtain all the important diffraction effects, but some are more easily understood on the basis of one or the other. It will be useful, then, to outline both of these approaches.

To examine the simplest case of plane acoustic waves interacting with plane light waves, consider Figure 11.1. Suppose the light wave, of frequency ω and wavelength λ, is incident from the left into a delay line with an acoustic wave of frequency Ω and wavelength Λ. If the refractive index of the delay medium

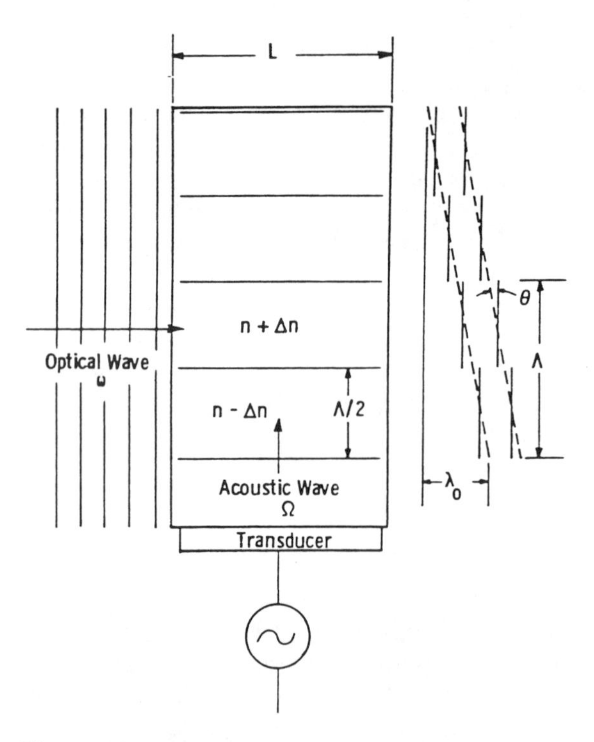

Figure 11.1. Tilting of optical wavefronts caused by upward-traveling acoustic wave.

is $n + \Delta n$ in the presence of the acoustic wave, the phase of the optical wave will be changed by an amount

$$\Delta\phi = 2\pi \frac{L}{\lambda} \Delta n \tag{7}$$

if the width of the delay line is L. Some typical values of $\Delta\phi$ can be obtained by assuming $L = 2.5$ cm (1 in) and $\lambda = 0.5\ \mu m$, with Δn reaching a peak value of 10^{-5}. This yields a phase change of π rad, which is, of course, quite large. It is large because L/λ, the number of optical wavelengths, is 50,000, so that a very small Δn can still produce a sizable $\Delta\phi$. If the electric field incident on the delay line is represented by

$$E = E_0 e^{i\omega t} \tag{8}$$

then the field of the phase-modulated emerging light will be

$$E = E_0 e^{i(\omega t + \Delta\phi)} = e^{i\omega t} e^{i2\pi(L/\lambda)(a_0 sin\Omega t)} \tag{9}$$

We shall not give a detailed derivation of the resulting temporal and spatial distribution of the light field, but we can use intuition and analogy with radio-wave modulation to arrive at the resultant fields. It is well known from RF engineering that the spectrum of a phase-modulated carrier of frequency ω consists of components separated by multiples of the modulation frequency Ω, as shown in Figure 11.2. There is a multiplicity of sidebands about the carrier frequency, such that the frequency of the nth sideband is $\omega + n\Omega$, where n is both positive and negative. The amplitudes of each of the sidebands are proportional to the Bessel function of order equal to the sideband number, and whose argument is

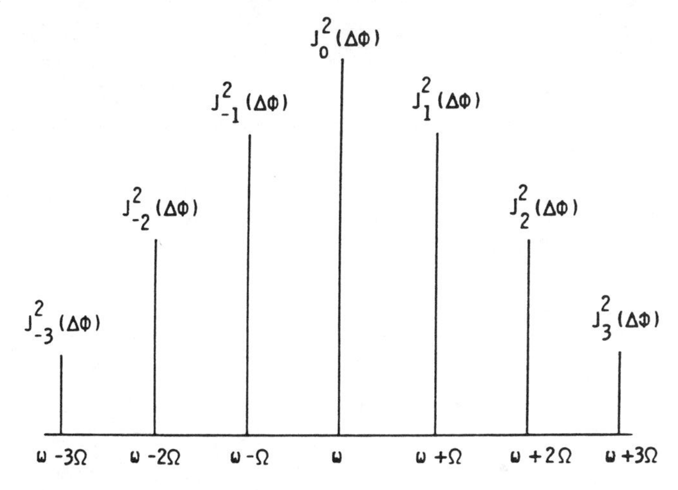

Figure 11.2. Intensity of diffracted orders due to Raman-Nath interaction.

the modulation index $\Delta\phi$. Although not shown by Figure 11.2, note that the odd-numbered negative orders are 180° out of phase with the others. The light emerging from the delay line is composed of a number of light waves whose frequencies have been shifted by $n\Omega$ from the frequency ω of the incident light. The relative amplitudes will be determined by the peak change in the refractive index.

In order to understand the diffraction of the light by the acoustic wave, consider the optical wavefronts in Figure 11.1. Since the velocity of light is about five orders of magnitude greater than the velocity of sound, it is a good approximation to assume that the acoustic wave is stationary in the time that it takes the optical wave to traverse the delay line. Suppose that during this instant the half-wavelength region labeled $n + \Delta n$ is under compression and $n - \Delta n$ is under rarefaction. Then the part of the optical plane wave passing through the compression will be slowed (relative to the undisturbed material of index n) while the part passing through the rarefaction will be speeded up. In this rough picture, the emerging wavefront will be "corrugated," so that if the corrugations are joined by a continuous plane its direction is tilted relative to that of the incident light wavefronts. Since the optical phase changes by 2π for each acoustic wavelength Λ along the acoustic beam direction, the tilt angle will be given by $\theta \cong \lambda/\Lambda$. The direction normal to the tilted plane is the direction of optical power flow and represents the diffracted light beam. Note that the corrugated wavefront could just as well have been connected by a tilted wavefront at an angle given by $\theta \cong -\lambda/\Lambda$. This corresponds to the first negative order, the other to the first positive order. At this point we will note that an important consideration in the operation of AO Bragg cells, or for that matter of most ultrasonic devices, is the ratio of the acoustic wavelength Λ to the transducer length L. The assumption that the acoustic energy propagates as a plane wave is valid when this ratio is very small or when there is little diffraction of the wave. However, when this ratio is not large, the acoustic propagation is more properly described in terms of the sum of plane waves, the angular spectrum of such plane waves increasing as the ratio increases. If we consider that partial wave which is propagating at an angle λ/Λ to the forward direction, then we see that the light which has been diffracted into the first order may be diffracted a second time by this partial wave into an angle $2\theta = 2\lambda/\Lambda$, and that the frequency of this light will once again be upshifted, for a total frequency shift of 2ω. If the angular spectrum of acoustic waves contains sufficient power of still higher orders, then this process can be repeated again, so that light will be multiply diffracted into higher-order angles, $n\theta = n\lambda/\Lambda$ each with a frequency shift $n\omega$. A similar argument holds for the negative orders, so that a complete set of diffracted light beams will appear as shown in Figure 11.3, where the angular deflection corresponding to the nth order is given by $\theta_n \cong \pm n\,\lambda/\Lambda$ and the frequency of the light deflected into the nth order is $\omega \pm n\Omega$. The intensity of the carrier wave, or zeroth order, will be

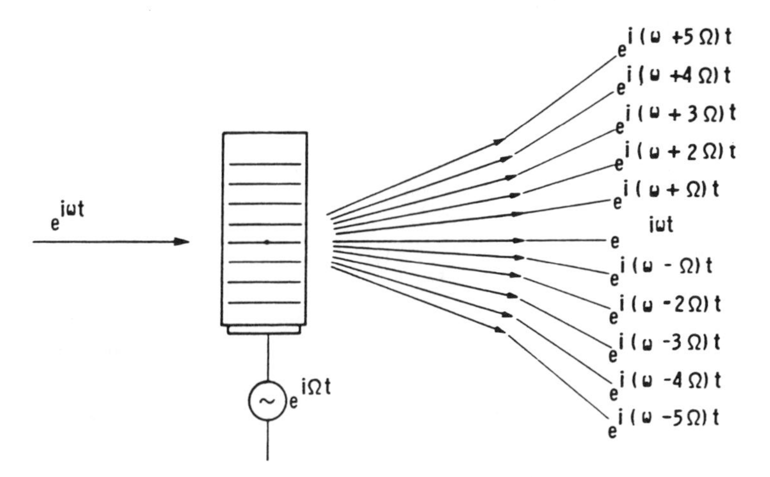

Figure 11.3. Raman-Nath diffraction of light into multiple orders.

zero when the modulation index $\Delta\phi$ is equal to 2.4. The generally important first order will be a maximum for $\Delta\phi = 1.8$, decreasing for higher modulation. These phenomena were described by Debye and Sears [1] and are often referred to as Debye-Sears diffraction. Similar observations were published almost simultaneously by Lucas and Biquard [2]. An extensive theoretical analysis of the effect was given by Raman and Nath [3], and so it is alternatively referred to as Raman-Nath diffraction. A distinctive feature of this type of diffraction is that it is limited to low acoustic frequencies (or relatively long wavelengths). The origin of this limitation lies in the diffraction spreading of the light beam as it traverses apertures formed by the columns of compression and rarefaction in the acoustic beam. If the length of the acoustic beam along the light propagation direction is large enough, the diffraction spread of the light between adjacent compression and rarefaction regions will overlap, so that there is some maximum length of interaction region beyond which the Debye-Sears effect smears out. To estimate this maximum length, suppose the compression and rarefaction apertures are one-half an acoustic wavelength, $\Lambda/2$, so that the angular diffraction spread of the light is $\delta\phi \approx 2\lambda/\Lambda$. Then l_{max} can be defined as that interaction length for which the aperture diffraction spreads the light by one-half an acoustic wavelength,

$$l_{max}\,\delta\phi = \frac{\Lambda}{2} \tag{10}$$

or

$$l_{max} = \frac{\Lambda^2}{4\lambda} \tag{11}$$

This can be expressed as

$$Q \equiv \frac{4l\lambda}{\Lambda^2} < 1$$

where the quantity Q, known as the Raman-Nath parameter,* relates l, λ, and Λ for which the "thin grating" approximation is valid. For typical values of $l = 1$ cm and $\lambda = 6.33 \times 10^{-5}$ cm, $Q = 1$ for $\Lambda = 0.0159$ cm (0.006 in), which corresponds to a frequency of 31.4 MHz for a material whose acoustic velocity is 5×10^5 cm/s (2×10^5 in/s).

For values of the interaction length $l > l_{max}$, for which the thin grating approximation no longer holds, a different regime of operation takes effect. If the incident light beam is normal to the sound beam propagation direction, the higher diffraction orders interfere destructively beyond l_{max}, eventually completely wiping out the diffraction pattern. In order for constructive interference to take place, the angle of incidence must be tilted with respect to the acoustic beam direction. To better understand what conditions must be satisfied for this, it is easier to think of the light and sound waves as colliding particles, photons and phonons. In this description, the light and sound take on the attributes of particles, and the dynamics of their collisions are governed by the laws of conservation of energy and conservation of momentum. The magnitudes of the momenta of the light and sound waves are given by the well-known expressions

$$|\boldsymbol{k}| = \frac{\omega n}{c} = \frac{2\pi n}{\lambda_0} \tag{12}$$

and

$$|\boldsymbol{K}| = \frac{\Omega}{v} = \frac{2\pi}{\Lambda} \tag{13}$$

respectively. In the latter equation, v is the velocity of sound in the delay medium, $v = 2\pi\Omega\Lambda$. Conservation of momentum is expressed by the vector relation

$$\boldsymbol{k}_i + \boldsymbol{K} = \boldsymbol{k}_d \tag{14}$$

the diagram for which is shown in Figure 11.4a, where $\boldsymbol{k}_i$ and $\boldsymbol{k}_d$ represent the momentum of the incident photon and the diffracted photon, respectively. The process may be thought of as one in which the acoustic phonon is absorbed by the incident photon to form the diffracted photon. Thus, conservation of energy requires that

$$h\omega_0 = h\omega_1 + h\Omega \tag{15}$$

or

---

* Often referred to as the Klein-Cook parameter.

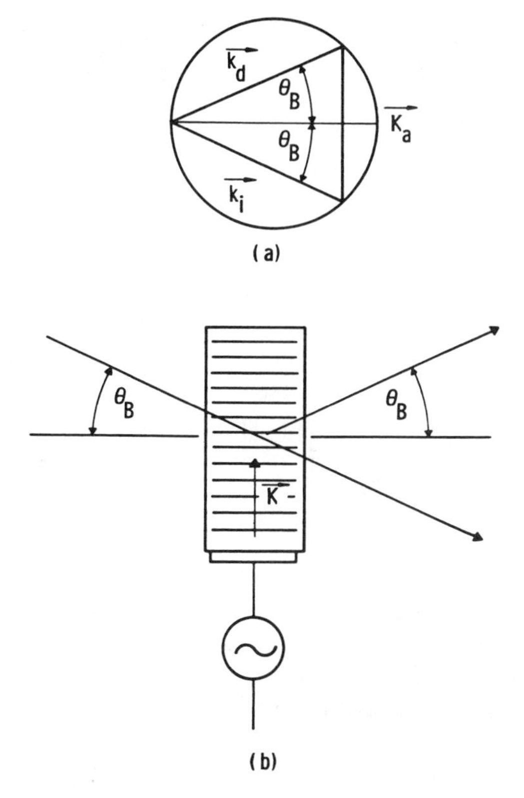

Figure 11.4. Bragg diffraction in isotropic medium.

$$\omega_d = \omega_i + \Omega$$

in which h is Planck's constant.

Since ω_i lies in the optical frequency range, and Ω will typically lie in the RF or microwave range, $\Omega << \omega_i$ so that $\omega_d \approx \omega_i$. This results in the magnitudes of $\boldsymbol{k}_i$ and $\boldsymbol{k}_d$ being almost equal, so that the momentum triangle of Figure 11.4a is isosceles, and the angle of incidence (with respect to the normal to K) is equal to the angle of diffraction. This angle is easily obtained from Figure 11.4a as

$$\sin \theta_B = \frac{1}{2}\frac{K}{k} = \frac{1}{2}\frac{\lambda}{\Lambda} \tag{16}$$

It is called the Bragg angle because of its similarity to the angle of diffraction of x-rays from the regularly spaced planes of atoms in crystals. The configuration of these vectors in relation to the delay line is shown in Figure 11.4b. In order for diffraction to take place, the light must be incident at the angle θ_B, and the diffracted beam will appear only at this same angle. In contrast to the Debye-

Sears regime, there are no higher-order diffracted beams. In the full mathematical treatment of the Bragg limit ($Q = 4l\lambda/\Lambda^2 >> 1$), light energy may appear at the higher orders, but the probability of its doing so is extremely small so that the intensity at higher orders is essentially zero. The diagrams in Figure 11.4 show the interaction in which the diffracted photon is higher in energy than that of the incident photon, but the reverse can also take place. If the sense of the vector $\boldsymbol{K}$ is reversed with respect to $\boldsymbol{k}_i$, then $\omega_d = \omega_i - \Omega$, and the diffracted negative first order results.

It is important to understand that the Debye-Sears effect and Bragg diffraction are not different phenomena, but are the limits of the same mechanism. The Raman-Nath parameter Q determines which is the appropriate limit for a given set of values λ, Λ, and l. Quite commonly in practice, these values will be chosen such that neither limit applies, and $Q \approx 1$. In this case, the mathematical treatment is quite complex, and experimentally it is found that one of the two first-order diffracted beams may be favored, but that higher orders will be present.

11.2.3 Diffraction Efficiency

Having obtained the angular behavior of light diffracted by acoustic waves, the next most important characteristic is the intensity of the diffracted beam. Again, the full mathematical treatment is beyond the scope of this book, but a very good intuitive calculation leads to results that are useful. Referring to the spectrum of a phase-modulated wave shown in Figure 11.2, we can see that the ratio of the intensity in the first order to that in the zero order is

$$\frac{I_1}{I_0} = \left[\frac{J_1(\Delta\phi)}{J_0(\Delta\phi)}\right]^2 \tag{17}$$

We shall now show in detail how this result comes about for acoustooptically diffracted light. The acoustic power flow is given by

$$P = \frac{1}{2}cve^2 \tag{18}$$

where c is the elastic stiffness constant. The elastic stiffness constant is related to the bulk modulus β and the density and acoustic velocity through the well-known expression

$$c = \frac{1}{\beta} = \rho v^2 \tag{19}$$

Thus, the acoustic power density is

$$P_A = \frac{1}{2}\rho v^3 e^2 \tag{20}$$

We can express the phase modulation depth in terms of the acoustic power density, using equation (5) for Δn and (7) for $\Delta\phi$, with the result

$$\Delta\phi = 2\pi \frac{L}{\lambda} \Delta n = -\pi \frac{L}{\lambda} n^3 p \left(\frac{2P_A}{\rho v^3}\right)^{1/2} \tag{21}$$

For small modulation index, the zero-order and first-order Bessel functions can be approximated by

$$J_0(\Delta\phi) \approx \cos(\Delta\phi) \approx 1 - \Delta\phi$$

and (22)

$$J_1(\Delta\phi) \approx \sin(\Delta\phi) \approx \Delta\phi$$

so that the small-signal approximation to the diffracted light is, from equation (17),

$$\frac{I_1}{I_0} \approx (\Delta\phi)^2 = \frac{\pi^2}{2}\left(\frac{L}{\lambda}\right)^2 \left(\frac{n^6 p^2}{\rho v^3}\right) P_A \tag{23}$$

This efficiency may be expressed in terms of the total acoustic power P,

$$P = P_A(L\,H) \tag{24}$$

where H is the height of the transducer, and

$$\frac{I_1}{I_0} = \frac{\pi^2}{2}\frac{L}{H}\left(\frac{n^6 p^2}{\rho v^3}\right)\frac{P}{\lambda^2} \tag{25}$$

The quantity in parentheses depends only upon the intrinsic properties of the acoustooptic material, while the other parameters depend upon external factors. It is therefore defined as the figure of merit of the material,

$$M_2 = \left(\frac{n^6 p^2}{\rho v^3}\right) \tag{26}$$

from which it can be seen that, in general, the most important factors leading to high acoustooptic efficiency will be a high refractive index and a low acoustic velocity. This does not guarantee a large figure of merit, since the photoelastic constant may be very small, or even zero.

The other factors in equation (25) have the following effect on the diffraction efficiency. The efficiency decreases quadratically with increasing wavelength, so that the power requirements for operation in the IR may be hundreds of times that required for the visible. For high efficiency, it will be desirable to have a large aspect ratio, L/H, leading to a configuration as shown in Figure 11.5. It is difficult to make conventional bulk devices with H much less than 1 mm, so

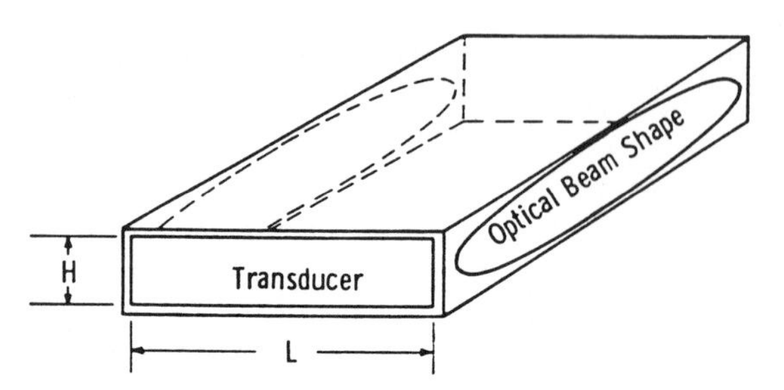

Figure 11.5. Transducer and optical beam shapes for optimization of acoustooptic diffraction.

that aspect ratios up to about 50 can be achieved. Much higher aspect ratios can be reached in guided optical wave devices. A more exact calculation of the diffraction efficiency in the Bragg regime [4] yields the result

$$\frac{I_1}{I_0} = \sin^2 \left[\frac{\pi^2}{2} \frac{L}{H} M_2 \frac{P}{\lambda^2} \right]^{1/2} \tag{27}$$

For low signal levels equation (27) reduces to the same expression as in (25). To obtain an order of magnitude for the power requirements of an acoustooptic deflector, let us assume a material with $n = 1.5$, $\rho = 3$, $v = 5 \times 10^5$ cm/s, and a photoelastic constant calculated from the Lorentz-Lorenz expression, $p \cong 0.6$, so that $M \simeq 1.1. \times 10^{-17}$ s^3/g. If the remaining parameters are $L = 1$ cm and $\lambda = 0.6$ μm, then by assuming a maximum acoustic power density for CW operation of 1 W/cm^2 (10^7 ergs/cm^2-s), the maximum obtainable efficiency is 15%. We shall see later, however, that materials and designs are available which are capable of realizing higher efficiencies with lower power levels.

11.2.4 Anisotropic Diffraction

Optical materials such as glass, or crystals with cubic structure, are isotropic with respect to their optical properties; i.e., they do not vary with direction. Many crystals, on the other hand, are of such structure, or symmetry, that their optical properties depend on the direction of polarization of the light in relation to the crystal axes. They are birefringent; i.e., the refractive index is different for different direction of light polarization.

The theory of diffraction of light thus far presented has assumed that the optical medium is isotropic, or at least that it is not birefringent. A number of important acoustooptic devices make use of the properties of birefringent materials, so a brief description of the important characteristics of anisotropic diffraction will be given here. The essential difference from diffraction in isotropic media is that the momentum of the light,

$$k = \frac{2\pi}{\lambda} = \frac{2\pi n}{\lambda_0} \tag{28}$$

will in general, be different for different light polarization directions. Thus, the vector diagram representing conservation of momentum will no longer be the simple isosceles triangle of Figure 11.4a. The momentum vectors for light that is ordinary polarized will terminate on a circle, as shown in Figure 11.6, while those for light that is extraordinary polarized will terminate on the ellipse of Figure 11.6.

To understand the effect of anisotropy on diffraction, it is necessary to mention another phenomenon which occurs when light interacts with shear acoustic waves, i.e., waves in which the displacement of matter is perpendicular to the direction of propagation of the acoustic wave. A shear acoustic wave may cause the direction of polarization of the diffracted light to be rotated by 90°. The underlying reason for this is that the shear disturbance induces a birefringence which acts upon the incident light as does a birefringent plate; i.e., it causes the plane of polarization

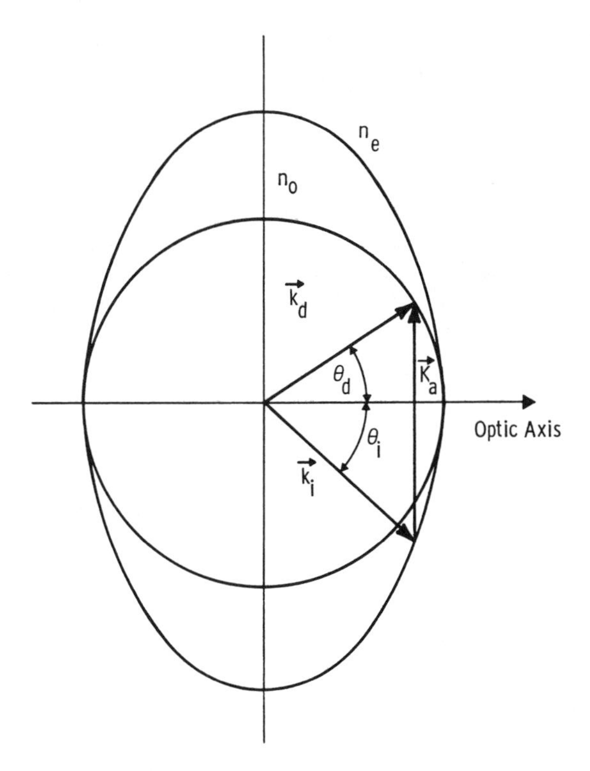

Figure 11.6. Vector diagram for diffraction in birefringent medium.

to be rotated. This phenomenon occurs in isotropic materials as well as in anisotropic materials; however, in isotropic materials the momentum vector, $k = 2\pi n/\lambda_0$, will be the same for both polarizations, so there is no effect on the diffraction relations. Suppose, instead, that the interaction occurs in a birefringent crystal in a plane containing the optic axis. Let us choose the example as shown in the index surfaces in Figure 11.6, in which the incident light is an extraordinary ray and the diffracted light is an ordinary ray. For this example,

$$k_i = \frac{2\pi n_e}{\lambda_0} \quad \text{and} \quad k_d = \frac{2\pi n_0}{\lambda_0} \tag{29}$$

and the angles of incidence, θ_i, and diffraction, θ_d, are in general not equal. The theory of anisotropic diffraction was developed by Dixon [5], in whose work the expressions for the anisotropic Bragg angles were derived as

$$\sin \theta_i = \frac{1}{2n_i} \frac{\lambda_0 f}{v} \left[1 + \left(\frac{v}{\lambda_0 f} \right)^2 (n_i^2 - n_d^2) \right] \tag{30}$$

$$\sin \theta_d = \frac{1}{2n_d} \frac{\lambda_0 f}{v} \left[1 - \left(\frac{v}{\lambda_0 f} \right)^2 (n_i^2 - n_d^2) \right] \tag{31}$$

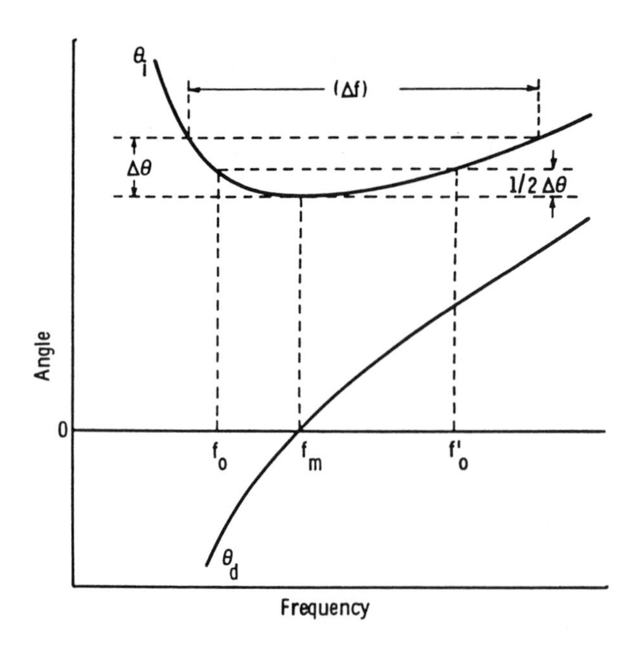

Figure 11.7. Angles of incidence and diffraction for anisotropic birefringent diffraction.

where n_i and n_d are the refractive indices corresponding to the incident and the diffracted light polarizations, and f is the acoustic frequency,

$$f = \frac{v}{\Lambda} \tag{32}$$

These angles are plotted in Figure 11.7 about that frequency f_m, for which there is a minimum in the angle of incidence. These curves, the general shapes of which are similar for all birefringent crystals, have a number of interesting characteristics which are useful for several types of acoustooptic devices. The minimum frequency for which an interaction may take place corresponds to $\theta_i = 90°$ and $\theta_d = -90°$, for which all three vectors will be collinear, as shown in Figure 11.8. It is easily shown that for this case, since the vector equation for conservation of momentum can be written as a scalar equation,

$$|\boldsymbol{k}_i| + |K| = |\boldsymbol{k}_d| \tag{33}$$

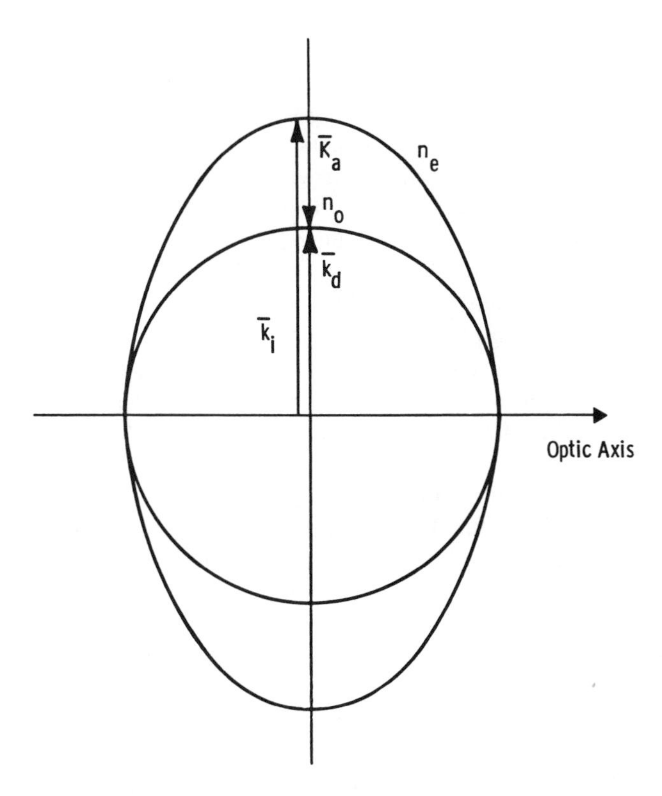

Figure 11.8. Vector diagram for collinear diffraction in birefringent medium.

the frequency for which collinear diffraction takes place is

$$f = \frac{v(n_i - n_d)}{\lambda_0} \tag{34}$$

Such collinear phase matching has been used as the basis of an important device, the electronically tunable acoustooptic filter [6]. Note that if the incident light had been chosen as ordinary rather than extraordinary polarized, the sense of the acoustic vector K would be reversed. In fact, the roles of the two curves in Figure 11.7 would be reversed by interchanging n_i and n_d.

Another interesting region of anisotropic diffraction occurs at the minimum value in the curve representing θ_i, at which frequency $\theta_d = 0$. This frequency, f_m, is obtained by setting the quantity in brackets in equation (31) equal to zero:

$$f = \frac{v}{\lambda_0}\sqrt{n_i^2 - n_d^2} \tag{35}$$

The significance of this point is that the angle of incidence of a scanned beam is relatively insensitive to change over a very broad range of frequencies. This frequency has important implications for the design of scanners because the bandwidth can be much greater than for a comparable isotropic scanner. Since the incident beam angle reaches a minimum value while the diffracted beam angle passes through zero at this point, and increases approximately linearly with frequency, the Bragg angle matching can be maintained over a large range, as will be described later. It will be seen that it is very difficult to achieve an interaction bandwidth this large by any other method.

The description of the interaction of light with sound we have given above is perhaps the simplest in terms of giving an intuitive understanding of the phenomena. Other descriptions, with totally different mathematical formalisms, have been carried out, and these lead to many details and subtleties in the behavior of acoustooptic systems that are beyond the scope of this book. Exact calculations have been carried out to extend the range of validity [7] from the limits allowed by the Raman-Nath theory [8], and this has been experimentally investigated [9]. Other studies have also been carried out to give accurate numerical results for the intensity distribution of light in the various diffraction orders [10]. The diffraction process has been reviewed and analyzed by Klein and Cook [11], using a coupled mode formulation, and there is continuing recent interest in refining the plane-wave scattering theory to give explicit results for intermediate cases [12, 13]. Finally, the acoustooptic interaction can be viewed as a parametric process in which the incident optic wave mixes with the acoustic wave to generate polarization waves at sum and difference frequencies, leading to new optical frequencies; this approach has been reviewed by Chang [14].

11.2.5 Resolution and Bandwidth Considerations

Resolution, bandwidth, and speed are the important characteristics of acoustooptic scanners, shared by all types of scanning devices. Depending upon the application, only one, or all of these characteristics may have to be optimized; in this section, we will examine which acoustooptic design parameters are involved in the determination of resolution, bandwidth, and speed. Consider an acoustooptic scanner with a collimated incident beam of width D, diffracted to an angle θ_0 at the center of its bandwidth Δf. If the diffracted beam is focused onto a plane by a lens, or lens combination, at the scanner, the diffraction spread of the optical beam will be

$$\delta x = F\delta\phi \cong F - \frac{\lambda}{D} \tag{36}$$

where F is the focal length of the lens. The light intensity will be distributed in the focal plane as illustrated in Figure 11.9. As an example for diffraction-limited optics, the spot size for a 25-mm (1-in)-wide light beam of wavelength 6.33 μm at a distance of 30 cm (12 in) from the delay line is 7.6 μm (3×10^{-4} in). There are, however, aberrations which prevent this from being fully realized, as will be discussed later.

The number of resolvable spots will be the angular scan range divided by the angular diffraction spread,

$$N = \frac{\Delta\theta}{\delta\phi} \tag{37}$$

where $\Delta\theta$ is the range of the angular scan. Differentiating the Bragg angle formula yields

$$\Delta\theta = \frac{\lambda}{v \cos\theta_0}\Delta f \tag{38}$$

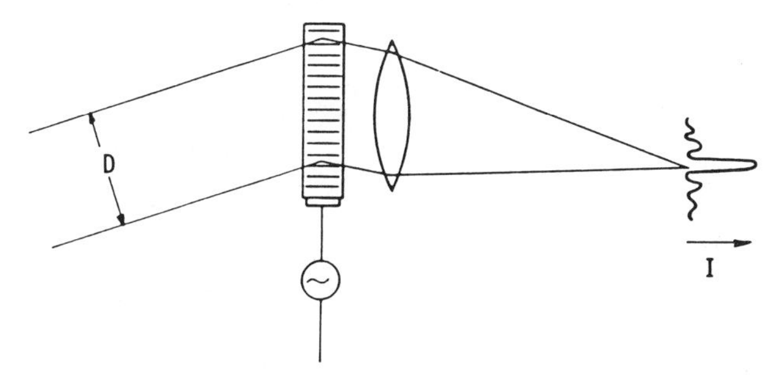

Figure 11.9. Distribution of light intensity due to diffraction by acoustic field.

and

$$N = (\Delta f)\left(\frac{D}{v \cos \theta_0}\right) = \Delta f \tau \tag{39}$$

where τ is the time that it takes the acoustic wave to cross the optical aperture. The resulting expression is the time-bandwidth product of the acoustooptic scanner, a concept applied to a variety of electronic devices as a measure of information handling capacity. The time-bandwidth product of an acoustooptic Bragg cell is equivalent to the number of bits of information which may be instantaneously processed by the system.

In order to maximize the number of resolution elements, it is desirable to have as large a bandwidth as possible (i.e., large frequency range) and also as large an aperture delay time as possible. There are two factors limiting the bandwidth of an acoustooptic device; the bandwidth of the transducer structure (discussed later), and acoustic absorption in the delay medium. The acoustic absorption increases with increasing frequency; for high-purity single crystals the increase generally goes with the square of the frequency. For glassy materials, on the other hand, the attenuation will increase more slowly with frequency, often approaching a linear function. The maximum frequency is generally taken as that for which the attenuation of the acoustic wave across the optical aperture is equal to 3 dB. A reasonable approximation of the maximum attainable bandwidth is $\Delta f = 0.7 f_{max}$, so that we may derive some relationships for the maximum number of resolution elements.

For a material with a quadratic dependence of attenuation on frequency,

$$\alpha(f) = \Gamma f^2 \tag{40}$$

and the maximum aperture for 3-dB loss is

$$D = \frac{3}{\Gamma f^2} \tag{41}$$

Using these results, the maximum number of resolution elements is

$$N_{max} \simeq \sqrt{\frac{1.5D}{v^2 \Gamma}} \tag{42}$$

from which it can be seen that, in principle, it is always advantageous to make the delay line as long as possible. In practice, the aperture will be limited by the largest crystals that can be prepared, or ultimately by the size of the optical system. For a glassy material for which the attenuation increases linearly with frequency,

$$\alpha(f) = \Gamma' f \tag{43}$$

and the maximum number of resolvable spots will be

$$N_{\text{max}} \simeq \frac{2}{\Gamma' v} \tag{44}$$

which is independent of the size of the aperture, being determined only by the material attenuation constant and the acoustic velocity.

In the next section we will review material considerations in some detail, and see what the performance limits are of currently available acoustooptic materials. As a numerical example, however, the highest-quality fused quartz has an attenuation of about 3 dB/cm at 500 MHz and an acoustic velocity of 5.96×10^5 cm/s (2.35×10^5 in/s) (for longitudinal waves), leading to $N_{\text{max}} = 560$.

11.2.6 Interaction Bandwidth

The number of resolution elements will be determined by the frequency bandwidth of the transducer and delay line, but a number of other bandwidth considerations are also of importance for the operation of a scanning system. While a large value of τ leads to a large value of N, the speed of the device is just equal to $1/\tau$. That is, the position of a spot cannot be changed randomly in a time less than τ. If the acoustic cell is being used to temporally modulate the light as well as to scan, then obviously the modulation bandwidth will similarly be limited by the travel time of the acoustic wave across the optical aperture. In order to increase the modulation bandwidth, the light beam must be focused to a small width, w, in the acoustic field. The 3-dB modulation bandwidth is approximately

$$\Delta f = \frac{0.75}{\tau} = \frac{0.75v}{w} \tag{45}$$

and the diffraction limited beam waist (the $1/e^2$ power points) of a gaussian beam is

$$w_0 = \frac{2\lambda_0 F}{\pi D} \tag{46}$$

where D is the incident beam diameter and F is the focal length of the lens. With this value of beam waist, the maximum modulation bandwidth is

$$\Delta f = 0.36\pi \frac{vD}{\lambda_0 F} \tag{47}$$

It can be seen from Equation (47) that the modulation bandwidth for a diffraction-limited focused gaussian beam can be very high; for example, for a material of acoustic velocity 5×10^5 cm/s(2×10^5 in/s) the bandwidth of a 0.633-μm light beam focused with an *f*/10 lens is about 1 GHz. Such a system, however, is practically useless, because the diffraction efficiency would be extremely small.

In order for the Bragg interaction bandwidth to be large, there must be a large spread of either the acoustic or the optical beam directions, $\delta\theta_a$ and $\delta\theta_0$ respectively, or both. This spread may occur either by focusing, which in the case of the acoustic beam is achieved by curving the plane of the transducer, or it may be due simply to the aperture diffraction for both beams. It follows from fairly simple arguments that the optimum configuration for the most efficient utilization of optical and acoustic energy corresponds to approximately equal angular spreading, $\delta\theta_0 \cong \delta\theta_a$, as illustrated in Figure 11.10.

In order to maximize the time-bandwidth product, the angular spread of the acoustic beam should be made large enough to match Bragg diffraction over the frequency range of the transducer-driving circuit bandwidth. As mentioned previously, this will result in some reduction in efficiency. To examine the relationship between bandwidth and the efficiency, we must first state another well-known result of acoustically diffracted light. This is, as shown by Cohen and Gordon [15], that the angular distribution of the diffracted light will represent the Fourier transform of the spatial distribution of the acoustic beam. This Fourier transform pair is illustrated in Figure 11.11 for the usual case of the rectangular acoustic beam profile. It seems intuitively obvious for this simple case, in which the diffraction spread of the incident optical beam is ignored, that there will be components in the diffracted light corresponding to the acoustic field sidelobes. It is shown in Ref. 15 that the Fourier transform relationship holds for an arbitrary acoustic beam profile.

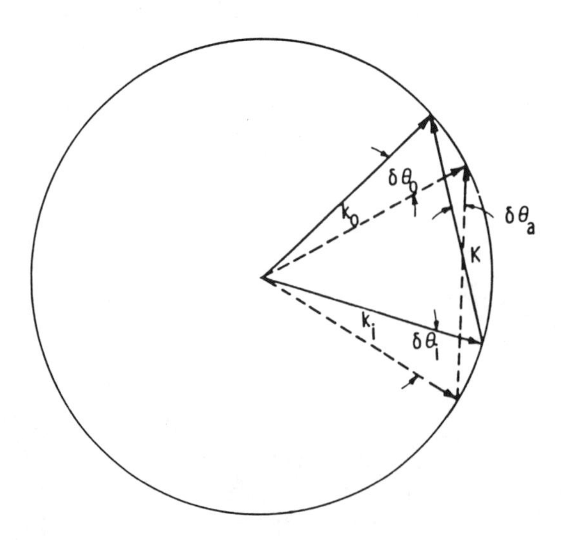

Figure 11.10. Vector diagram for Bragg diffraction in isotropic medium with angular spread of acoustic beam direction.

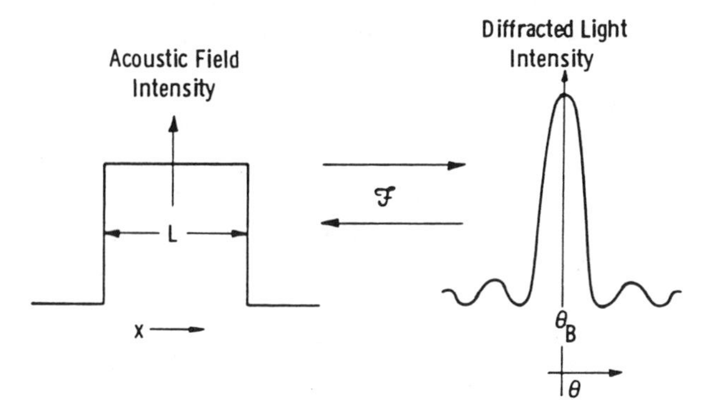

Figure 11.11. Fourier transform relationship between acoustic field intensity and diffracted light intensity.

For the rectangular profile, the angular dependence of the diffracted light, illustrated in Figure 11.11, is

$$\frac{I(\theta)}{I_0} \alpha \left[\frac{\sin \frac{1}{2} KL(\theta - \theta_B)}{\frac{1}{2} KL(\theta - \theta_B)} \right]^2 \tag{48}$$

for which the -3-dB points occur at

$$\frac{1}{2} KL(\Delta\theta)_{1/2} \simeq \pm 0.45\pi \tag{49}$$

where $(\Delta\theta)_{1/2}$ is the value of $\theta - \theta_B$ at the half-power points. This yields a value for the angular width of the optical beam, just equal to the diffraction spread of the acoustic beam, namely

$$2(\Delta\theta)_{1/2} \simeq \frac{1.8\pi}{KL} \tag{50}$$

The frequency bandwidth is obtained by equating this result to the differential of the Bragg condition:

$$\delta\theta = \frac{\lambda_0 \Delta f}{nv \cos \theta_B} \tag{51}$$

This result is

$$\Delta f = \frac{1.8nv^2 \cos \theta_B}{Lf_0\lambda_0} \tag{52}$$

For acoustooptic scanning devices in which the bandwidth as well as the diffraction efficiency is of importance, a more relevant figure of merit may be the product of the bandwidth with the efficiency. By combining equations (52) and (25), this product is

$$2f_0\,\Delta f \cdot \frac{I_1}{I_0} = \frac{1.8\pi^2}{\lambda_0^{\,3} H \cos\theta_B}\left[\frac{n^7 p^2}{\rho v}\right] P \tag{53}$$

The quantity in parentheses can be regarded as the figure of merit of the material when the efficiency-bandwidth product is the important criterion, and is designated as

$$M_1 = \frac{n^7 p^2}{\rho v} \tag{54}$$

Other methods of achieving a large interaction bandwidth include transducer designs that steer the acoustic beam in direction in order to track the Bragg angle as it changes with frequency. A description of beam steering will be included in the section on transducers.

Still another figure of merit was introduced by Dixon [16] in connection with wideband acoustooptic devices. Since the power requirements decrease as the transducer height H decreases, it is advantageous to make H as small as possible. If there are no limitations on the minimum size of H, it can be as small as the optical beam waist in the region of the interaction, h_{min}. The modulation bandwidth is determined by the travel time of the acoustic wave across this beam waist,

$$\tau \approx \frac{1}{\Delta f} = \frac{h_{min}}{v} \tag{55}$$

so that

$$h_{min} = \frac{v}{\Delta f} \tag{56}$$

Substitution of this value for H in equation (51) results in the relation

$$2f_0 \frac{I_1}{I_0} = \frac{1.8\pi^2}{\lambda_0^{\,3}\cos\theta_B}\left[\frac{n^7 p^2}{\rho v^2}\right] \tag{57}$$

and the appropriate figure of merit for this situation is the quantity in parentheses:

$$M_3 = \left[\frac{n^7 p^2}{\rho v^2}\right] \tag{58}$$

Note that the optical wavelength appears as $\lambda_0^{\,3}$ in both equations (53) and (57), so that operation at long wavelengths is relatively more difficult in terms of power requirements for configurations optimizing bandwidth as well as efficiency.

11.2.7 Bragg Cell Design Procedure

The useful optical aperture of a Bragg cell is usually considered to be that length across which the difference in acoustic attenuation between the highest and the lowest frequencies within the operating bandwidth of the cell is 3 dB. A particular application may dictate either a bandwidth or a resolution, i.e., a time-bandwidth product. In general, an optimized Bragg cell design will maximize the number of resolvable spots, as well as other transducer structure parameters.

The number of resolvable spots, or the time-bandwidth product, will be determined by three key factors [17]: the acoustic attenuation Γ, the optical aperture of the acoustooptic crystal D, the angular beam spreading of the acoustic wave, which is determined by the transducer length L, and the acoustic wavelength. The constraints placed upon the number of resolvable spots N by these three factors is given by the relations [17]

$$N \leqslant \frac{1.5\Lambda_c}{\Gamma\Lambda_1^{\,2}} \tag{59}$$

$$N \leqslant \frac{D}{2\Lambda_c} \tag{60}$$

$$N \leqslant \left(\frac{L}{2\Lambda_c}\right)^2 \tag{61}$$

Where Λ_c is the acoustic wavelength at the center frequency, Λ_1 is the acoustic wavelength at 1 GHz, and Γ is the acoustic attenuation in dB per unit length, normalized to 1 GHz (under the usual assumption that the attenuation increases quadratically with frequency). Note that equation (59) allows for a 3-dB attenuation. Once the center frequency and the bandwidth of the cell have been determined, the transducer structure must be designed. This will include the electrode length L and height H. The length must be chosen so that it is small enough to allow sufficient beam spread to satisfy the Bragg angle matching requirements over the desired bandwidth (for a fixed angle of incidence of the optical beam). At the same time, the diffraction efficiency will decrease as L decreases, so that we will want L to be as large as possible within the interaction bandwidth constraint.

11.3 MATERIALS FOR ACOUSTOOPTIC SCANNING

11.3.1 General Considerations

We have seen in the preceding section that two important criteria for choosing materials for acoustooptic scanning systems are the acoustooptic figure of merit and the high-frequency acoustic loss characteristics. Other properties that determine the usefulness of a material are its optical transmission range, optical quality, availability in suitable sizes, mechanical and handling characteristics as they may pertain to polishing and fabrication procedures, and chemical stability under normal conditions. As with most components, cost will be an important factor, even when all the other factors may be positive, if competing techniques are available.

One of the limitations on the use of acoustooptic scanners before the late 1960s was the availability of materials with reasonably high figure of merit. As we have seen, fused quartz, which is used as the standard for comparison, has a figure of merit so low that only a few percent diffraction efficiency can be obtained for scanners of typical dimensions, and with RF powers that can be applied without causing damage to transducer structures. Water is a fairly efficient material, with a figure of merit about 100 times larger than fused quartz, and has actually found use in some scanning systems. Like most liquids, it cannot be used at frequencies higher than about 50 MHz, so that large numbers of resolution elements cannot be achieved. Since the late 1960s, many new materials have been synthesized and existing ones were found to have excellent properties. Materials can now be found for most scanning applications from the UV through the intermediate IR where high bandwidth is required.

The selection of a material for any particular device will be dictated by the type of operation under consideration. In general, it is desirable to select a material with low-drive-power requirements, suggesting those with large refractive index and low density and acoustic velocity. If, however, high-speed modulation is of paramount importance, then a low acoustic velocity may lead to slower than required speeds. In the following section, we will review the factors and tradeoffs involved in the selection of materials for various acoustooptic applications. Whatever the particular material requirements may be, there are also a number of practical considerations which dictate several generally important material properties whatever the application: (1) the optical quality must be high so that not only absorption but scattering and large-scale inhomogeneities are small; (2) good chemical stability is required so that protective enclosures are not needed to maintain integrity; (3) good mechanical properties are required so that the device can be cut and polished without extraordinary procedures and can be adjusted and used with normal handling techniques; (4) the availability of crystal growth methods for obtaining suitably large, high-quality boules with reasonable cost is needed; and (5) a low-temperature coefficient of velocity is required to avoid drift of scan properties.

11.3.2 Theoretical Guidelines

There is no simple microscopic theory of the photoelastic effect in crystals. Therefore it is not possible to predict the magnitude of the photoelastic constants from first principles. However, Pinnow [18] has suggested the use of certain empirical relationships between the various physical properties in order to systematize and group acoustooptic materials. It is well known that such relations exist, for example, for the refractive index and the acoustic velocity for such groups as the alkali halides, the mineral oxides, and the III-IV compounds.

A large amount of data has been collected on the refractive indices of crystals, and generally good agreement is found with the GladstoneDale [19] equation

$$\frac{n-1}{\rho} = \sum_i q_i R_i \tag{62}$$

in which R_i is the specific refraction of the ith component and q_i is percentage by weight. Reliable values of R_i have been determined from mineralogical data over many years.

From the expression for the acoustooptic figure of merit, it is apparent that a high value of refractive index is desirable for achieving high diffraction efficiency. It is not, however, possible simply to select for consideration those materials with high refractive index, as even a casual survey shows that such materials tend to be opaque at shorter wavelengths. This trend was examined in great detail by Wemple and DiDomenico [20], who found that the refractive index is simply related to the energy band gap. The semiempirical relation for oxide materials is

$$n^2 = 1 + \frac{15}{E_g} \tag{63}$$

where E_g is the energy gap (expressed in electron volts). For other classes of materials the energy gap constant will be different, but the same form holds. It can be seen from equation (63) that the largest refractive index for an oxide material transparent over the entire visible (cutoff wavelength at 0.4 μm) is 2.44. Higher refractive indices can be chosen only by sacrificing transparency at short wavelengths.

Pinnow [18] has found that a good approximation to the acoustic velocity for a wide range of materials is obtained with the relation

$$\log\left(\frac{v}{\rho}\right) = -b\bar{M} + d \tag{64}$$

where $\bar{M}$ is the mean atomic weight, defined as the total molecular weight divided by the number of atoms per molecule, and b and d are constants. Large values of d are generally associated with harder materials, while b does not vary greatly for oxides. Thus, in general, low acoustic velocities tend to be found in materials of high density, as is intuitively expected. Another useful velocity relationship has been pointed out by Uchida and Niizeki [21]; this is the Lindemann formula relating the melting temperature T_m and the mean acoustic velocity v_m,

$$v_m^2 = \frac{cT_m}{\bar{M}} \tag{65}$$

in which c is a constant dependent upon the material class. This relation suggests that high-efficiency materials would likely be found among those with large mean atomic weight and low melting temperature, i.e., dense, soft materials.

In order for an acoustooptic material to be useful for wideband applications, the ultrasonic attenuation must be small at high frequencies. An attenuation that is often taken as an upper limit is 1 dB/μs (so that the useful aperture will depend upon the velocity). Many materials that might be highly efficient and otherwise suitable are excessively lossy at high frequency. A microscopic treatment of ultrasonic attenuation was done by Woodruff and Ehrenreich [22]. Their formula for the ultrasonic attenuation is

$$\alpha = \frac{\gamma^2\Omega^2\kappa T}{\rho v^5} \tag{66}$$

where Ω is the radian frequency, γ is the Grünneisen constant, κ is the thermal conductivity, and T is the absolute temperature. This formula would suggest that the requirement of low acoustic velocity and low attenuation conflict with each other, since $\alpha \sim v^{-5}$; it is quite unusual for materials with low acoustic velocity to not also have a high absorption, at least for the low-velocity modes.

The determination of the photoelastic constants of materials is essentially an empirical study, although a microscopic theory of Mueller [23], developed for cubic and amorphous structures, is still referenced. For both ionic and covalent bonded materials the photoelastic effect derives from two mechanisms: the change of refractive index with density, and the change in index with polarizability under the strain. Both of these effects may have the same or opposite sign under a given strain, and one or the other may be the larger. It is for this reason that the magnitude or even the sign of the photoelastic constant cannot be predicted, since the effects may completely cancel each other. It is possible, however, to estimate the maximum constants for groups of materials. This has been done for three important groups with the result

$$|P_{\max}| = \begin{cases} 0.21 & \text{(water insoluble oxides)} \\ 0.35 & \text{(water soluble oxides)} \\ 0.20 & \text{(alkali halides)} \end{cases}$$

In general, the photoelastic tensor components corresponding to shear strain will be less than those corresponding to compressional strain because there is no change, to first order, of density with shear; only the polarizability effect will

Table 11.1 Maximum Photoelastic Coefficients[a]

Material	$\lvert p_{max} \rvert_{measured}$
$LiNbO_3$	0.20
TiO_2	0.17
Al_2O_3	0.25
$PbMoO_4$	0.28
TeO_2	0.23
$Sr_{.5}Ba_{.5}Nb_2O_6$	0.23
SiO_2	0.27
YIG	0.07
$Ba(NO_3)_2$	0.35
α-HIO_3	0.50
$Pb(NO_3)_2$	0.60
ADP	0.30
CdS	0.14
GaAs	0.16
As_2S_3	0.30

[a] From Ref. 9.

be present. It is always possible that exceptionally large values of shear-related photoelastic coefficients may be found, but in no case could they be expected to be larger than the estimated value of $\lvert p_{max} \rvert$. The maximum values of photoelastic constant are shown in Table 11.1 for a number of important oxides and other materials.

11.3.3 Selected Materials for Acoustooptic Scanners

Among older materials, those that have been shown useful for acoustooptic applications are fused quartz, because of its excellent optical quality and low cost for large sizes, and sapphire and lithium niobate, because of their exceptionally low acoustic losses at microwave frequencies. For infrared applications germanium [24] has proven very useful, as has arsenic trisulfide glass, where bandwidth requirements are not high. Among the newer crystal materials, very good acoustooptic performance has been obtained in the visible with GaP [25] and $PbMoO_4$ [26,27]. One of the most interesting new materials to be developed within the past several years is TeO_2 [28], which along with $PbMoO_4$ has found wide use in commercially available acoustooptic scanners. More design details for devices employing this material will be given later. Among the new materials that have been developed for infrared applications, very high performance has been reached with several chalcogenide crystals [29]. Particularly important members of this group of materials include Tl_3AsS_4 [30] and Tl_3PSe_4 [31]. The compound Tl_3AsSe_3 [32] is particularly interesting beyond its possible use as an infrared acoustooptic modulator material. Since Tl_3AsSe_3 belongs to the crystal class 3m, its symmetry permits it to possess a nonzero p_{41} photoelastic coefficient, and it is suitable for use as a collinear tunable acoustooptic filter, a device first realized by Harris [33], using lithium niobate.

Tables 11.2–11.4 summarize the properties of some of the materials that have been studied for acoustooptic applications. The acoustic attenuation constant in these tables is defined as

$$\Gamma = \frac{\alpha}{f^2} \tag{67}$$

which supposes that the attenuation increases quadratically with frequency. This will be the case for good-quality single crystals, but not for polycrystalline, highly impure, or amorphous materials. For the latter, the constant given in the tables is a rough estimate, based on measurements at the higher frequencies. The light polarization direction is designated as parallel or perpendicular according as the light polarization is parallel or perpendicular to the acoustic beam direction. Table 11.2 lists some of the more important amorphous materials, which may be useful if large sizes are desired or very low cost is required, but none of which can be used at frequencies much above 30 MHz. Table 11.3 lists the most important class of materials, crystals that are transparent throughout the visible with very low acoustic losses. Table 11.4 lists high-efficiency crystal materials that are transparent in the infrared and have reasonably low acoustic losses.

An overall summary of a few outstanding (in one or another respect) selected acoustooptic materials presented in these tables is shown in Fig. 11.12. Using figure of merit and acoustic attenuation as criteria of quality, it is clear that a tradeoff between these two parameters exists, and that the selection of the optimum material will be determined by the system requirements.

11.3.4 Anisotropic Diffraction in Tellurium Dioxide

One of the most remarkable materials to have appeared recently for acoustooptic applications is paratellurite (TeO_2) [34]. It has a unique combination of properties which leads to an extraordinarily high figure of merit for a shear-wave interaction in a convenient RF range. It will be recalled that the anisotropic Bragg relations, equations(30) and (31), led to a particular frequency, given by equation (35), for which the angle of incidence is a minimum and therefore satisfies the Bragg condition over a wide frequency range. However, typical values of birefringence place this frequency around 1 GHz or higher. Of particular interest in TeO_2 is its optical activity for light propagating along the c-axis, or (001) direction; the

Table 11.2 Acoustooptic Properties of Amorphous Materials

Material	Trans range (μm)	Acoustic mode	v (cm/s × 10^5)	Γ (dB/cm-GHz^2)	Opt. Pol. Dir.	n (0.633 μm)	M_1 (cm^2-s/g × 10^{-7})	M_2 (s^3/g × 10^{-18})	M_3 (cm-s^2/g × 10^{-12})
Water	0.2–0.9	L	1.49	2400	$\parallel$ or $\perp$	1.33	37.2	126	25
Fused quartz	0.2–4.5	L	5.96	12	$\perp$	1.46	8.05	1.56	1.35
SF-4	0.38–1.8	L	3.63	220	$\perp$	1.62	1.83	4.51	3.97
SF-59	0.46–2.5	L	3.20	1200	$\parallel$ or $\perp$	1.95	39	19	12
SF-58		L	3.26	1200	$\parallel$ or $\perp$	1.91	18.2	9	5.6
SF-57		L	3.41	500	$\parallel$	1.84	19.3	9	5.65
SF-6		L	3.51	500	$\parallel$ or $\perp$	1.80	15.5	7	4.42
As_2S_3	0.6–11	L	2.6	170	$\parallel$	2.61	762	433	293
As_2S_5	0.5–10	L	2.22			2.2	278	256 (est.)	125

Table 11.3 Acoustooptic Properties of Crystals for the Visible

Material	Trans range (μm)	Acoustic mode & prop. dir.	v (cm/s × 10^5)	Γ (dB/cm-GHz^2)	Opt. pol. dir.	n (0.633 μm)	M_1 (cm^2-s/g × 10^{-7})	M_2 (s^3/g × 10^{-18})	M_3 (cm-s^2/g × 10^{-12})
$LiNbO_3$	0.04–4.5	L[100]	6.57	0.15		2.20	66.5	7.0	10.1
		S[001]	3.59	2.6	$\perp$	2.29	9.2	2.92	2.4
Al_2O_3	0.15–6.5	L[100]	11.0	0.2	$\parallel$	1.77	7.7	0.36	0.7
YAG	0.3–5.5	L[110]	8.60	0.25	$\perp$	1.83	0.98	0.073	0.114
		S[100]	5.03	1.1	$\parallel$ or $\perp$	1.83	1.1	0.25	0.23
TiO_2	0.45–6	L[001]	10.3	0.55	$\perp$	2.58	44	1.52	4
SiO_2	0.12–4.5	L[001]	6.32	2.1	$\perp$	1.54	9.11	1.48	1.44
		L[100]	5.72	3.0	[001]	1.55	12.1	2.38	2.11
α-HIO_3	0.3–1.8	L[001]	2.44	10	[100]	1.99	103	86	42
$PbMoO_4$		L[001]	3.63	15	$\parallel$	2.62	108	36.3	29.8
TeO_2	0.35–5	L[001]	4.20	15	$\perp$	2.26	138	34.5	32.8
		S[110]	0.616	90	Circ [001]	2.26	68.0	793	110
Pb_2MoO_5	0.4–5	L a-axis	2.96	25	b-axis	2.183	242	127	82

Table 11.4 Acoustooptic Properties of Infrared Crystals

Material	Transmission range (μm)	Acoustic mode & prop. div.	v (cm/s × 10^5)	Γ (dB/cm-GHz^2)	Opt. pol. dir.	λ (μm)	n	M_1 (cm^2-s/g × 10^{-7})	M_2 (s^3/g × 10^{-18})	M_3 (cm-s^2/g × 10^{-12})
Ge	2–20	L[111]	5.50	30	$\parallel$	10.6	4.00	10,200	840	1850
		S[100]	3.51	9	$\parallel$ or $\perp$	10.6	4.00	1,430	290	400
Tl_3AsS_4	0.6–12	L[001]	2.5	29	$\parallel$	1.15	2.63	620	510	290
GaAs	1–11	L[110]	5.15	30	$\parallel$	1.15	3.37	925	104	179
		S[100]	3.32		$\parallel$ or $\perp$	1.15	3.37	155	46	49
Ag_3AsS_3	0.6–13.5	L[001]	2.65	800	$\parallel$	.633	2.98	816	390	308
Tl_3AsSe_3	1.25–18	L[100]	2.15	314	$\perp$	3.39	3.15	654	445	303
Tl_3PSe_4	0.85–9	L[100]	2.0	150	$\parallel$	1.15	2.9	2,866	2069	1288
$TlGaSe_2$	0.6–20	L[001]	2.67	240	$\parallel$	.633	2.9	430	393	161
CdS	0.5–11	L[100]	4.17	90	$\parallel$	.633	2.44	52	12	12
ZnTe	0.55–20	L[110]	3.37	130	$\parallel$	1.15	2.77	75	18	19
GaP	0.6–10	L[110]	6.32	6.0	$\parallel$	.633	3.31	75	30	71
ZnS	0.4–12	L[001]	5.82	27	$\parallel$	.633	2.35	27	3.4	4.7
		S[001]	2.63	130		.633	2.35	14	8.4	5.2
Te	5–20	L[100]	2.2	60	$\parallel$	10.6	4.8	10,200	4400	4640

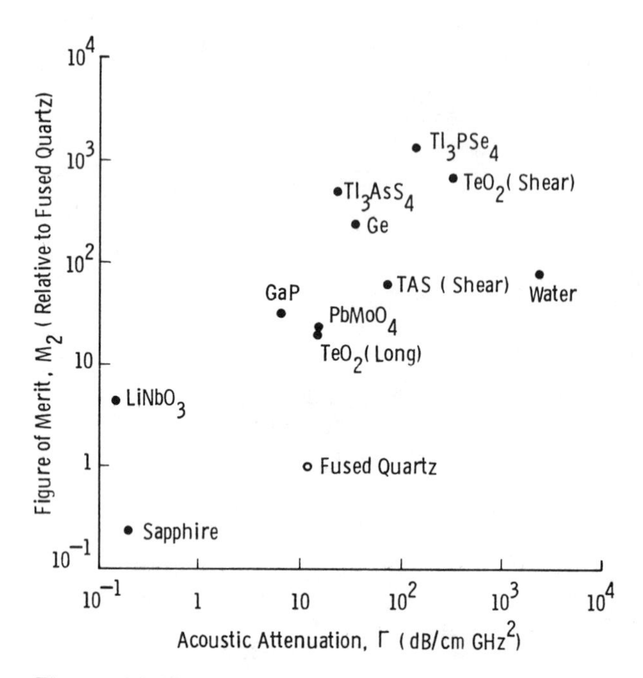

Figure 11.12. Figure of merit versus acoustic attenuation.

indices of refraction for left- and right-hand circularly polarized light are different, so that plane polarized light undergoes a rotation of its plane of polarization by an amount

$$R = \frac{2n_0}{\lambda}\delta \tag{68}$$

where δ is the index splitting between left- and right-hand polarized light,

$$\delta = \frac{n_\ell - n_r}{2n_0} \tag{69}$$

Just as acoustic shear waves can phase-match two linearly polarized light waves, they can also phase-match two oppositely circularly polarized light waves. Thus, shear waves propagating in the (110) direction, with shear polarization in the $(1\bar{1}0)$ direction, will diffract left- or right-hand polarized light propagating along the (001) direction, one into the other. The anisotropic Bragg relations apply to crystals with optical activity, where the birefringence is interpreted as

$$\Delta n = n_\ell - n_r = 2n_0\delta \tag{70}$$

and the value of δ obtained from specific rotation is wavelength-dependent. For the light and sound wave propagation directions described above, the acoustic velocity is 0.62×10^5 cm/s (0.24×10^5 in/s) and the figure of merit, M_2, is 515 relative to fused quartz. The frequency for which the Bragg angle of incidence is a minimum, as evaluated from equation (35) for $\lambda = 0.633$ μm, is $f = 42$ MHz, a very convenient frequency. For other important wavelengths, the minima occur at 36 MHz for $\lambda = 0.85$ μm and at 22 MHz for $\lambda = 1.15$ μm.

The application of the anisotropic Bragg equations to optically active crystals was discussed in detail by Warner et al.[35] They showed that near the optic axis the indices of refraction are approximated by the relations (for right-handed crystals, $n_r < n_\ell$)

$$\frac{n_r^2(\theta)\cos^2\theta}{n_0^2(1-\delta)^2} + \frac{n_r^2(\theta)\sin^2\theta}{n_e^2} = 1 \tag{71}$$

and

$$\frac{n_\ell^2(\theta)\cos^2\theta}{n_0^2(1+\delta)^2} + \frac{n_\ell^2(\theta)\sin^2\theta}{n_0^2} = 1 \tag{72}$$

For incident angles near zero with respect to the optic axis and for small values of δ,

$$n_r^2 = n_0^2\left(1 - 2\delta + \frac{n_e^2 - n_0^2}{n_e^2}\sin^2\theta\right) \tag{73}$$

and

$$n_\ell^2 = n_0^2\,(1 + 2\delta\cos^2\theta) \tag{74}$$

For light incident exactly along the optic axis the two refractive indices are simply

$$n_r = n_0(1-\delta) \tag{75}$$

and

$$n_\ell = n_0(1+\delta) \tag{76}$$

The anisotropic Bragg equations for optically active crystals are obtained by substitution of equations (73) and (74) into equations (30) and (31) for n_i and n_d. By ignoring the higher-order terms, this results in

$$\sin\theta_i \cong \frac{\lambda f}{2n_0 v}\left[1 + \frac{4n_0^2v^2}{\lambda^2 f^2}\delta + \frac{\sin^2\theta_r n_0^2}{\lambda^2 f^2}\left(\frac{n_e^2 - n_0^2}{n_e^2}\right)\right] \tag{77}$$

and

$$\sin\theta_d = \frac{\lambda f}{2n_0 v}\left[1 - \frac{4n_0^2v^2}{\lambda^2 f^2}\delta - \frac{\sin^2\theta_e n_0^2}{\lambda^2 f^2}\left(\frac{n_e^2 - n_0^2}{n_0^2}\right)\right] \tag{78}$$

The anisotropic Bragg angles (as measured external to the crystal) are shown in Figure 11.13 for TeO_2 at $\lambda = 0.6328\ \mu m$. It is obvious that for frequencies around the minimum, it will be possible to achieve a much larger bandwidth for a given interaction length than is possible with normal Bragg diffraction; a one-octave bandwidth corresponds to a variation in angle of incidence for perfect phase matching of only 0.16°. A useful advantage of such operation is that large bandwidths are compatible with large interaction lengths, which assumes the avoidance of higher diffraction orders from Raman-Nath effects. For normal Bragg diffraction, on the other hand, large bandwiths can only be reached with interaction lengths that are so small that significant higher-order diffraction occurs. This decreases the efficiency with which light can be directed to the desired first

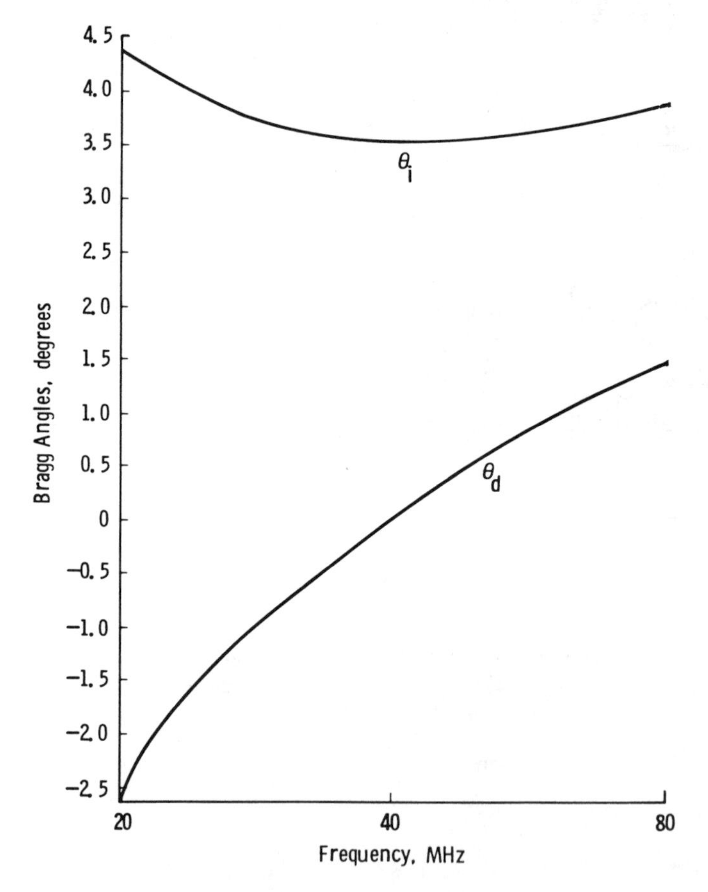

Figure 11.13. Bragg angles of incidence and diffraction (external) for anisotropic TeO_2 scanner, $\lambda = 0.6328\ \mu m$.

order, and also limits the bandwidth to less than one octave in order to avoid overlapping low-frequency second-order with higher-frequency first-order diffracted light. However, tellurium dioxide operating in the anisotropic mode can always be made to diffract in the Bragg mode with no bandwidth limitation on length, so that this feature combined with the extraordinary high figure of merit leads to deflector operation with very low drive powers.

An important degeneracy occurs for anisotropic Bragg diffraction which causes a pronounced dip in the diffracted light intensity at the midband frequency, where θ_i has its minimum. This degeneracy was explained by Warner et al. [35], and is easily understood by referring to the diagram in Figure 11.14. Two sets of curves are shown in the figure; the solid pair represent θ_i and θ_d when the incident light momentum vector has a positive component along the acoustic momentum vector, and the dotted pair represent these angles when the incident light momentum vector has a negative component along the acoustic vector. In the former case, the frequency of the diffracted light is upshifted, and in the latter it is downshifted. The vector diagram for this process is shown in Figure 11.15. Light is incident to the acoustic wave of frequency f_0 at an angle θ_0, and is diffracted as a frequency upshifted beam, $(\nu + f_0)$, normal to the acoustic wave. This light, in turn, may be rediffracted; referring to Figure 11.14, it can be seen that for a

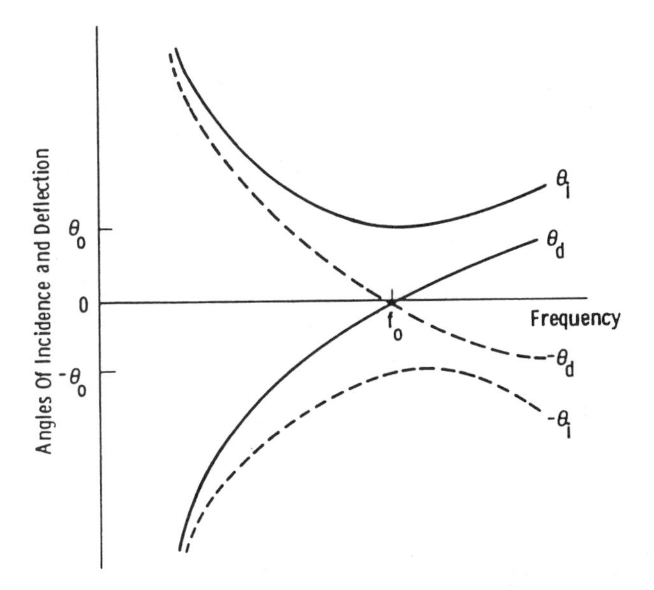

Figure 11.14. Angles of incidence and diffraction for anisotropic diffraction. Solid curves are for incident light having a component in same direction as acoustic wave, and dotted curves are for incident light having a component in opposite direction.

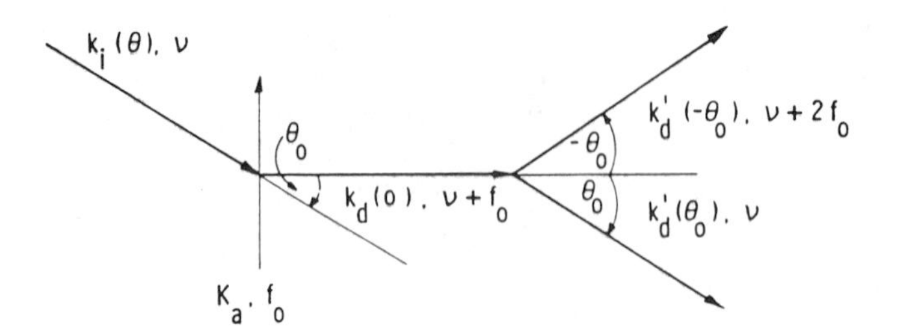

Figure 11.15. Vector diagram for midband degeneracy of Bragg diffraction in birefringent medium.

frequency f_0, light that is incident at $\theta = 0°$ can be rediffracted to either θ_0 or $-\theta_0$. In the former case, the light will be downshifted to the original incident light frequency ν, and in the latter case it is upshifted to $\nu + 2f_0$. Note that this degeneracy can only occur at the frequency f_0 where light incident normal to the acoustic wave is phase-matched for diffraction into both θ_0 and $-\theta_0$. How the light is distributed in intensity between the three modes will depend upon the interaction length and the acoustic power level. The exact solution to this is found by setting up the coupled mode propagation equations under phase-matched conditions. The result of this is that maximum efficiency for deflection into the desired mode at f_0 is 50%. At low acoustic power, the deflection of light into the undesired mode is negligible; at high powers the unwanted deflection increases so that, for example, if the efficiency is 50% for frequencies away from f_0, it will be 40% at f_0. The theoretical response of such a deflector is shown in Figure 11.16, and is in excellent agreement with experimental results.

11.3.5 Infrared Laser Scanning

Acoustooptic beam scanners for use with infrared lasers have been under consideration in recent years by the aerospace industry in connection with laser radar and optical communications systems. Where system requirements place excessive demands on mechanical scanning methods, various electronic approaches become attractive. In general, the carbon dioxide laser, with wavelengths from 9 to 11 μm, is the most common one for long-wavelength operation. There are a number of electronic approaches to infrared beam scanning besides acoustooptic, and all of them are quite difficult to implement, usually for reasons related to the long interaction length needed to achieve large optical phase excursion. For acoustooptic diffraction, we have seen that the RF power needed for a given diffraction efficiency increases quadratically with wavelength. At 10.6 μm therefore, 280 times the power for 0.633 μm is necessary. Clearly, there will be severe constraints on the available materials for such devices, and on their performance. Referring

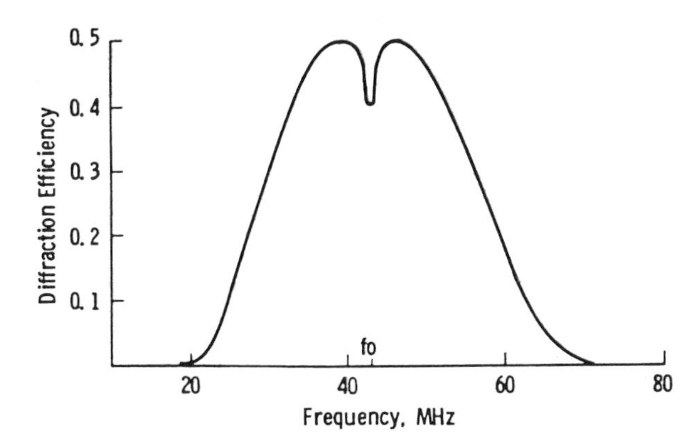

Figure 11.16. Effect of midband degeneracy on diffraction efficiency, for a maximum efficiency of 50%.

to Table 11.4, we can see that only a few materials that transmit to 11 μm and have large acoustooptic figure of merit have been identified. The most common of these is germanium, which can be purchased in very large single crystals of excellent optical quality; germanium acoustooptic scanners have been commercially available for a number of years. They operate best in the isotropic mode, typically near 100 MHz RF. Another favorable infrared material for use in the 9- to 11-μm range is thallium arsenic selenide, which has very recently become available on a commercial basis. This crystal is best used in an anisotropic mode, as described in a previous section. The acoustooptic figure of merit is very high due to the low value of shear acoustic wave velocity. The low velocity produces another result that may simplify the design of scanner optics. The scan angles at infrared wavelengths are quite large; the angular dispersion for 10.6μm carbon dioxide wavelength for this Bragg cell is shown in Figure 11.17. For an RF bandwidth of 30% around 110 MHz center frequency, a scan-angle range of 16° is reached. For many applications, no magnification of the scan angle will be needed, as may be the case for acoustooptic scanners in the visible.

One of the major problems associated with carbon dioxide laser beam scanners is heating, due both to absorption of optical energy and RF power heating. If very high laser beam powers are used, then even a small absorption coefficient in the scanner may cause unacceptable heating. For germanium and thallium arsenic selenide, the absorption coefficients at 10.6 μm are 0.032 and 0.015 cm^{-1}, respectively. The thermal conductivity of germanium is much higher than that of thallium arsenic selenide, but design considerations may favor one or the other, depending upon the detailed effects of thermal gradients. High RF power operation is limited by heating at the transducer, which will eventually damage

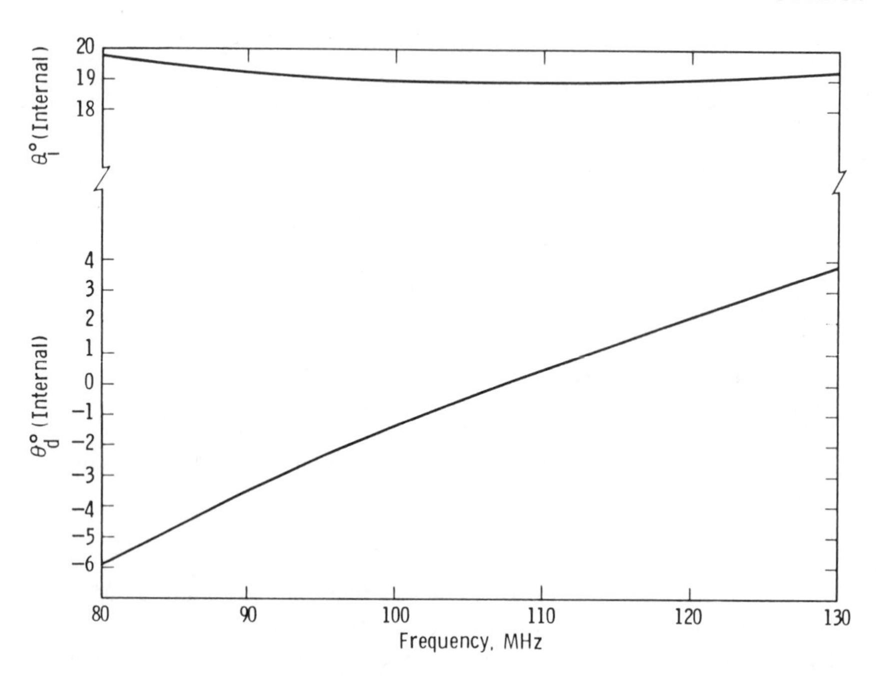

Figure 11.17. Anisotropic Bragg diffraction in TAS at $\lambda = 10.6\ \mu m$.

the transducer bond. Such thermal effects can be reduced by water- or air-cooling the RF mount and by heat-sinking the transducer; a photograph of a high-power transducer with a matching sapphire heat sink, which also serves to make electrical contact, is shown in Figure 11.18. The resolution of such infrared acoustooptic scanners will, in general, be limited by RF heating if high diffraction efficiency is to be obtained. This comes about because a large interaction bandwidth requires a small interaction length, so that high efficiency can only be achieved by high power. The relationship between resolution and acoustic power for the infrared scanner materials is shown in Figure 11.19. If CW operation is required, it can be seen that no more than a few hundred spots can be obtained for an aperture of 1 or 2 cm.

11.4 DESIGN AND FABRICATION OF TRANSDUCERS

11.4.1 Transducer Characteristics

The second key component of the acoustooptic scanner after the optical medium is the transducer structure, which includes the piezoelectric layer, bonding films, backing layers, and matching network. Recent advances in this area have made

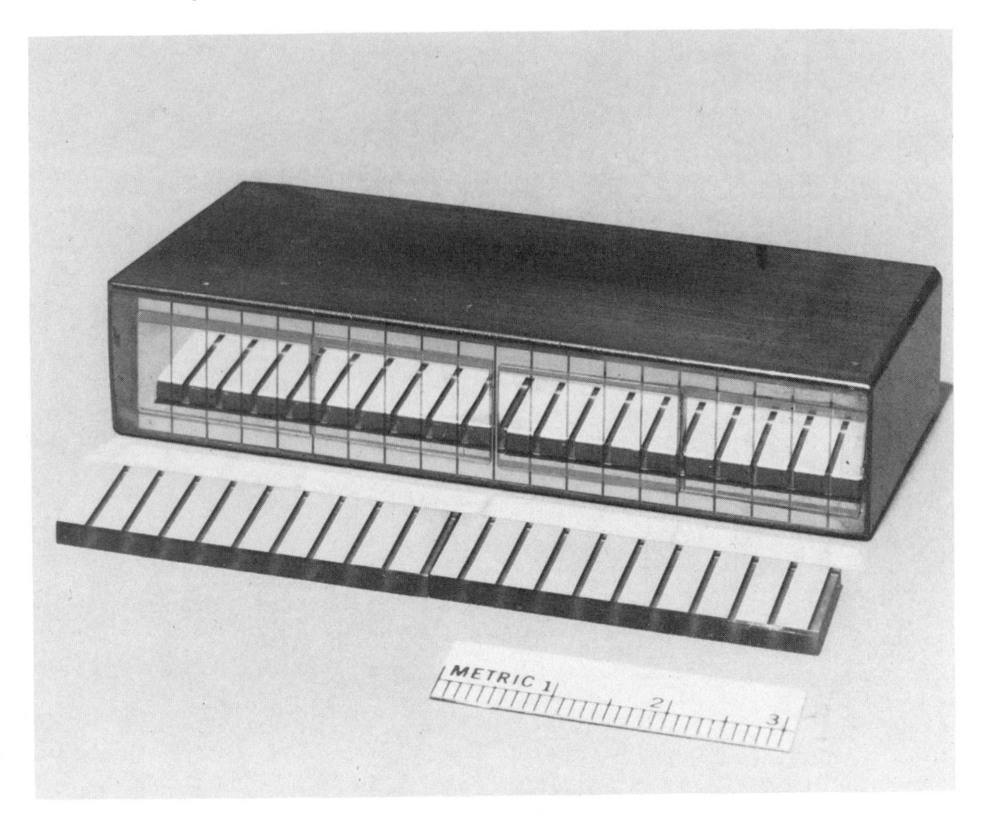

Figure 11.18. Transducer structure on TAS acoustooptic device, with matching heat sink for high RF power operation.

available a number of new piezoelectric materials of very high electromechanical conversion efficiency, and bonding techniques that permit this high conversion efficiency to be maintained over a large bandwidth. Furthermore, the design of high-performance transducer structures utilizing this new technology has been facilitated by new analytical tools [36, 37] which lend themselves to computer programs for optimizing this performance.

The most elementary configuration of a thickness-driven transducer structure is shown in Figure 11.20. It consists of the piezoelectric layer, thin film or plate, excited by metallic electrodes on both faces, and a bonding layer to acoustically couple the piezoelectric to the delay medium, or optical crystal. The backing is applied to mechanically load the transducer for bandwidth adjustment, but may simply be left as air. The thickness of the transducer is about half an acoustic wavelength at the resonant frequency, and the thickness of the bonding layer is chosen to allow high, broadband acoustic transmission. The most efficient op-

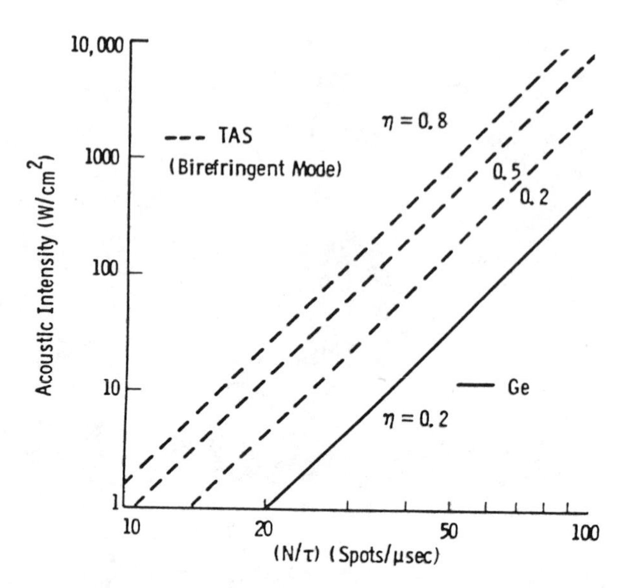

Figure 11.19. Acoustic intensity for various deflection efficiencies at the Bragg angle as a function of bandwidth using shear acoustic waves in TAS.

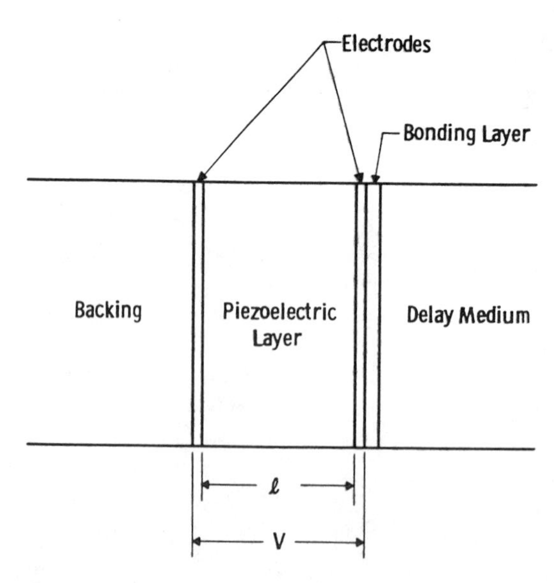

Figure 11.20. Transducer structure.

eration of the transducer is obtained when the mechanical impedances of all the layers are equal. The mechanical impedance is

$$Z = \rho v \tag{79}$$

and in general there is not sufficient choice of available materials to satisfy this condition. When the impedances are unequal, reflection occurs at the interfaces, reducing the efficiency of energy transfer. The reflection and transmission coefficients at the boundary between two media of impedances Z_1 and Z_2 are

$$R = \frac{(Z_1 - Z_2)^2}{(Z_1 + Z_2)^2} \tag{80}$$

$$T = \frac{4Z_1Z_2}{(Z_1 + Z_2)^2} \tag{81}$$

The electromechanical analysis is generally carried out in terms of an equivalent circuit model, first proposed by Mason [38]. Several variations of the equivalent circuit have since been developed, but the one due to Mason is shown in Figure 11.21. The fundamental constants of the transducer are permittivity s, acoustic velocity v, and electromechanical coupling factor k. The other parameters are transducer thickness l and area S. With these parameters, the circuit components shown in Figure 11.21 are

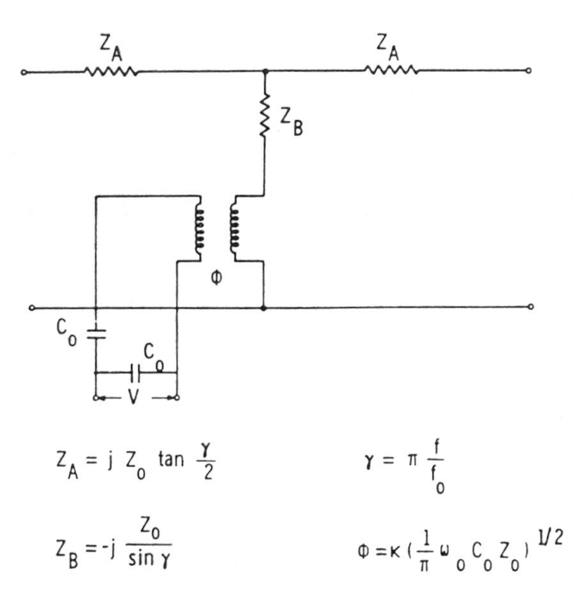

Figure 11.21. Equivalent circuit of Mason.

$$C_0 = \epsilon \frac{S}{l} \tag{82}$$

$$\phi = \kappa\left(\frac{1}{\pi}\,\omega_0 c_0 Z_0\right)^{1/2} \tag{83}$$

$$Z_A = jZ_0 \tan\frac{\gamma}{2} \tag{84}$$

$$Z_B = -j\frac{Z_0}{\sin\gamma} \tag{85}$$

where

$$\omega_0 = \frac{\pi v}{l} \tag{86}$$

$$\gamma = \pi\frac{\omega}{\omega_0} \tag{87}$$

$$Z_0 = S\rho v \tag{88}$$

This equivalent circuit was used by Sittig [36] and Meitzler and Sittig [37] to analyze the propagation characteristics of acoustic energy between a piezoelectric and a delay medium. This was done in terms of a two-port electromechanical network, described by the chain matrix

$$\begin{pmatrix} A & B \\ C & D \end{pmatrix} = \prod_m \begin{pmatrix} A_m & B_m \\ C_m & D_m \end{pmatrix} \tag{89}$$

If the equivalent circuit of Figure 11.21 is terminated at the input with a voltage source V_s and impedance Z_s, and at the output with a transmission medium of mechanical impedance Z_t, output voltage V_l, and load impedance Z_l, as shown in Figure 11.22, then the insertion loss is

$$L = 20\log\frac{V_s}{V_l} + 20\log\left|\frac{Z_s + Z_l}{Z_l}\right|\ dB \tag{90}$$

The impedances Z_s and Z_l are assumed to be purely resistive and

$$\frac{V_l}{V_s} = \frac{2Z_l Z_t}{\{AZ_t + B + Z_s(CZ_t + D)\}\{AZ_t + B + Z_l(CZ_t + D)\}} \tag{91}$$

The two-port transfer matrix was obtained by Sittig [39], with the result

$$A = \frac{1}{\phi H}\begin{vmatrix} A' & B' \\ C' & D' \end{vmatrix}\begin{vmatrix} \cos\gamma + jz_b\sin\gamma & Z_0(z_b\cos\gamma + z\sin\gamma) \\ \dfrac{j\sin\gamma}{Z_0} & 2(\cos\gamma - 1) + jz_b\sin\gamma) \end{vmatrix} \tag{92}$$

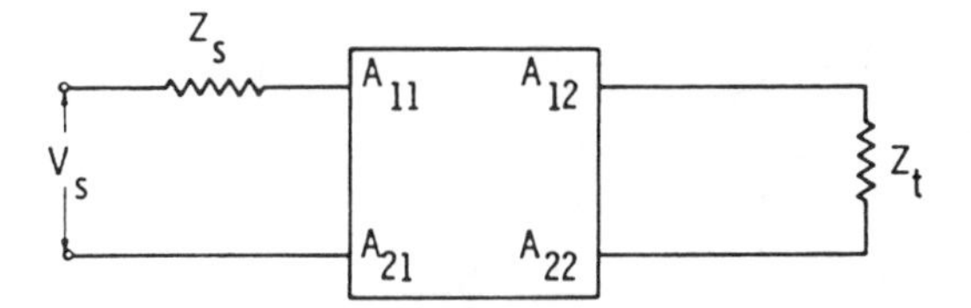

Figure 11.22. Terminated two-port transducer.

where

$$z_b = \frac{Z_b}{Z_0}, \qquad H = \cos\gamma - 1 + jZ_b \sin\gamma \tag{93}$$

and

$$A' = 1, \quad B' = j\frac{\phi^2}{\omega C_0}, \quad C' = j\omega C_0, \quad D' = 0 \tag{94}$$

The impedance Z_b represents the mechanical impedance of layers placed on the back surface of the transducer for loading, $Z_b = S\rho_b v_b$. In case the transducer is simply air-backed, $Z_b \simeq 0$. Electrical matching may be done at the input network by adding inductors either in parallel or in series in order to be electrically resonant with the transducer capacity C_0 at midband, $\omega = \omega_0$. If no inductances are added, the minimum loss condition is achieved for

$$R_s = \frac{1}{\omega_0 C_0} \tag{95}$$

where R_s is the source resistance. The inductance, if added, is chosen so that

$$L = \frac{1}{\omega_0^2 C_0} \tag{96}$$

A result of the matrix analysis shows that when piezoelectric materials with large values of the coupling constant κ are used, it is possible to achieve large fractional bandwidths without the necessity for electrical matching networks. As an example of the results obtained with this formalism, several plots of the frequency dependence of transducer loss for different values of the coupling constant are shown in Figure 11.23.

11.4.2 Transducer Bonding Methods

For transducers in the frequency range for which crystal plates are bonded to the delay medium, the bonding procedure is probably the most critical and most

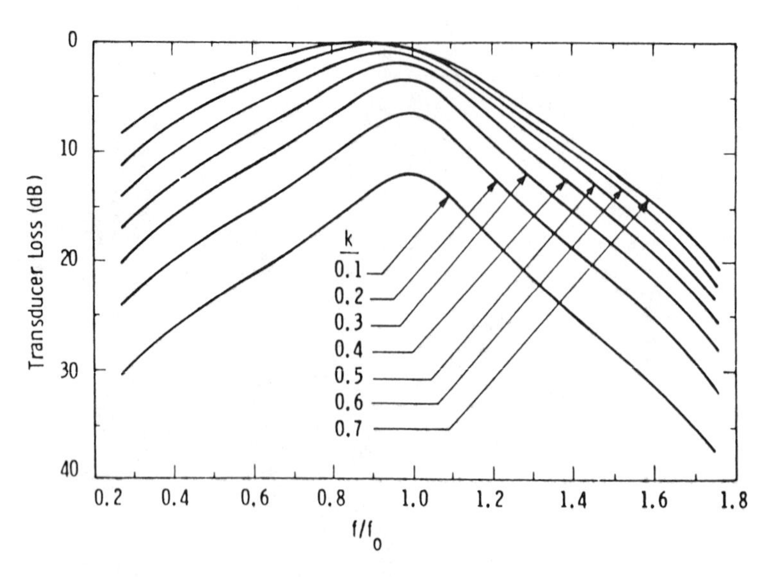

Figure 11.23. Transducer loss for various values of k; $z_{0t} = 0.4$ and $R_s = (w_0C_0)^{-1}$.

difficult step in fabricating the structure. The bonding layer can drastically modify the transmission of acoustic energy between the piezoelectric and the delay media; this is because the bond layer must provide molecular contact between the two surfaces, which will otherwise result in incomplete transfer, and because the mechanical impedance of the bond layer may produce a large acoustic mismatch with low transmission. In addition to these considerations, if the bond material is acoustically lossy, further decrease in transmission will result.

Because of the special properties required, there is only a very limited number of known bonding materials available. For temporary attachments, a commonly used agent is "salol," phenyl salicylate. It is easily applied as a liquid, which is crystallized by addition of a small seed. It is reliquified by gentle heating, and therefore is useful for various test measurements, but does not yield wide bandwidth or efficient coupling. A more satisfactory bond is made with epoxy resin, mixed to a very low viscosity which may be compressed to a layer less than one micrometer thick before setting. Such thin layers require a high degree of cleanliness to avoid inclusion of any dust particles. Because of the low impedance of epoxy compared with such transducer materials as lithium niobate, thicker bonding layers would cause serious impedance mismatch problems around 100 MHz, where this technique has been successfully used.

Good results can also be obtained with a low-viscosity, ultraviolet, light-cured cement. For frequencies higher than about 100 MHz, other techniques, capable of yielding still thinner bond layers, which must be kept to a small fraction of an acoustic wavelength, must be used. Vacuum-deposited metallic layers are well suited for this purpose, since their thickness can be very accurately controlled down to the smallest dimensions, and impedances much closer to those of commonly used piezoelectric materials are available.

Very good results were first obtained with indium bonds [40] which are deposited to a thickness of several thousand angstroms on both surfaces, and without removal from the vacuum systems are mated under a pressure of about 100 psi. This technique yields a cold-welded bond which has excellent mechanical properties with large acoustic bandwidth, if properly designed, and low insertion loss at frequencies of hundreds of megahertz.

The greatest fabrication difficulty is due to the necessity of maintaining the deposited films under vacuum to prevent oxidation. This requires a vacuum system with rather elaborate fixtures to bring the two surfaces together after film deposition and to apply the hydraulic pressure. The inside of a vacuum system in which this procedure is carried out is shown in Figure 11.24. The substrates are held on either side of the evaporation filament sources during film deposition and are quickly brought into contact before contamination can occur. Using this technique, compression bonds of indium, tin, aluminum, gold, and silver are routinely made. It is essential for these procedures to be carried out in a dust-free atmosphere in order to avoid contaminating the interface with particulates. Even the smallest particles will prevent good acoustic contact between the transducer and the cell, so that fabrication must be done in a clean-room facility. A typical acoustooptic device clean room is shown in Figure 11.25.

In a modification of the indium compression bond [41], which allows the freshly deposited indium surfaces to be removed from the vacuum system for handling, the work is then placed in an oven under a pressure of several hundred psi raised in temperature to slightly below the melting point of indium (156°C), and slowly cooled. This procedure forms a molecular bond in spite of the oxidation that may occur, and gives results similar to the vacuum bond. The principal drawback is that upon cooling, differential thermal expansion coefficients between the delay line material and the transducer material may set up unacceptable strains in the optical path. For some systems, this may not be a problem; for example, quartz or even lithium niobate transducers on fused or crystal quartz delay lines can be routinely made by this method. On the other hand, such crystals as tellurium dioxide require a great deal of care in handling, since they are extremely sensitive to thermal shock and strains. Differential contraction between the crystal and transducer for a bond made in this fashion may easily be severe enough to fracture the crystal. Therefore, its applicability will depend upon the materials and sizes involved and upon the degree of freedom from residual strain required.

For frequencies approaching 1 GHz, the attenuation of indium layers may become excessive, and better results can be achieved with metals with lower acoustic loss constants. Among such metals are gold, silver, and aluminum.

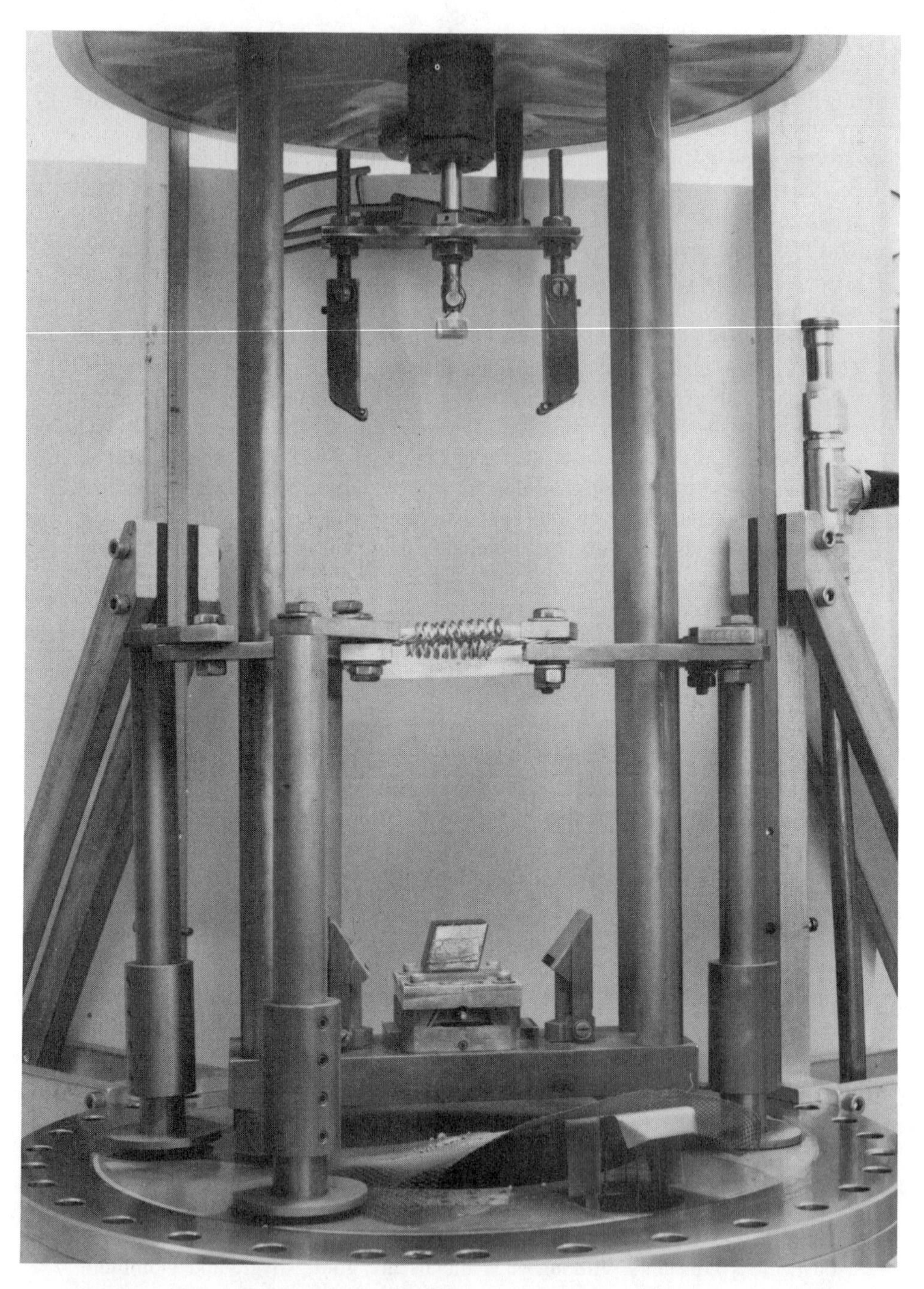

Figure 11.24. Vacuum compression bonding system. Metal films are deposited on transducer and delay line surfaces, which are then brought into contact.

Figure 11.25. A clean room for acoustooptic device fabrication.

Although these are made by the vacuum compression method, they generally require higher pressure. Still another method that has been used with these, as well as indium, is ultrasonic welding [42]. The chief advantage to be gained is that the procedure is carried out in normal atmosphere, since the ultrasonic energy breaks up the oxidation layer that forms on the surface. Some heating occurs as a result, but the temperature remains well below that required in the indium thermocompression method, with much lower residual strains. The technique requires the simultaneous application of pressures up to 3000 psi; this may be excessive for easily fractured or deformed materials or where odd-shaped samples are involved. A summary of the important properties of a few bonding materials, also used for electrodes and intermediate impedance-matching layers, is given in Table 11.5.

At lower frequencies, the effects of thin electrode and bonding layers on the performance of the transducer may be entirely negligible, but near 100 MHz, they become increasingly large, and even for layers less than 1 μm thick the

Table 11.5 Acoustic Properties of Bond Layer Materials

	Longitudinal waves			Shear waves		
Material	Velocity cm/s	Impedance g/s-cm^2	Attenuation dB/μm @ 1 GHz	Velocity cm/s	Impedance g/s-cm^2	Attenuation dB/μm @ 1 GHz
Epoxy	2.6×10^5	2.86×10^5	Very large	1.22×10^5	1.34×10^5	Very large
Indium	2.25×10^5	16.4×10^5	8	0.91×10^5	6.4×10^5	16
Gold	3.24×10^5	62.5×10^5	0.02	1.2×10^5	23.2	0.1
Silver	5.65×10^5	38×10^5	0.025	1.61×10^5	16.7×10^5	
Aluminum	6.42×10^5	17.3×10^5	0.02	3.04×10^5	8.2×10^5	
Copper	5.01×10^5	40.6×10^5		2.11×10^5	18.3×10^5	

effect may not be negligible if the impedance mismatch to the rest of the structure is large. The effects of the electrode layer can be determined by setting $Z_b = 0$ in equation (92), and the entire effect of the back layers will be due to the impedance of the electrode z_{b1} of thickness t_{b1}, so the normalized impedance

$$z_b = jz_{b1} \tan(t_{b1} \gamma) \equiv j \tan \delta \tag{97}$$

and the matrix of equation (92) becomes more complex.

The effect of the bond layer and front electrode is even more complex, but an interesting illustrative example of varying the bond layer thickness is shown in Figure 11.26. For this example, the normalized impedance of the bond layer is taken to be rather low, $z = 0.1$, and it can be seen that even for a fairly small thickness, the effect on the transducer loss is quite marked. Such a low value of impedance would correspond to the nonmetallic bond materials, but for the metallic bond materials the impedance mismatch would not be as severe, and the curve of transducer loss would be correspondingly less influenced. This influence of intermediate layers on the shape of the transducer loss curve can be used to determine the bandpass characteristics of the transducer structure. Such impedance transformers can be used, for example, to make the response symmetric about the band center f_0 by making the intermediate layer thickness one-quarter wavelength at f_0. By choosing other values for the thickness, the bandwidth can be enlarged, ripples smoothed, or various distortions introduced. In general,

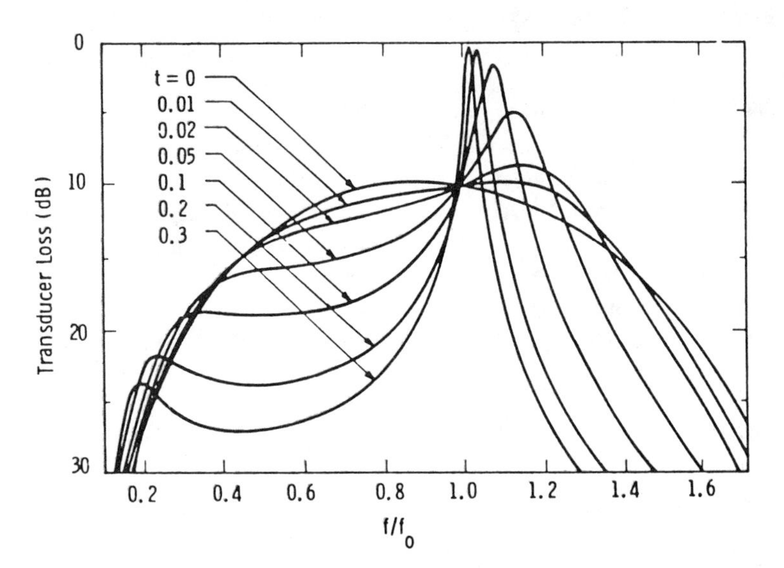

Figure 11.26. Transducer loss for various values of normalized transducer thickness t and intermediate layer normalized thickness 0.1. $R_s = (w_0 C_0)^{-1}$, $z_{0t} = 1$, and $k = 0.2$.

however, any such objectives are achieved at the expense of increased transducer loss.

11.4.3 Transducer Materials

The piezoelectric material itself is perhaps the single most important factor governing the efficiency with which electrical energy can be converted to acoustic energy, this through the electromechanical coupling factor κ. The coupling efficiency is equal to κ. Prior to the discovery of lithium niobate, quartz was the most commonly used high-frequency transducer material, although its coupling factor, even for the most efficient crystal orientations, is rather small. The very high efficiency transducers were introduced with the discovery of various new ferroelectrics, such as lithium niobate, lithium tantalate, and the ceramic PZT materials, lead-titanate-zirconate. While the PZT transducers have among the highest values of κ, up to 0.7, they are not suitable for high-frequency applications since they cannot be polished to very thin plates. The most suitable piezoelectric transducer materials for high-frequency applications and their important properties are listed in Table 11.6, which is based on a compilation of Meitzler [43].

In order to produce transducers in the high-frequency range, say larger than 100 MHz, the piezoelectric crystal must be very thin (<20–30 μm). There are three well-established techniques for fabricating such thin transducers. In the first method, the piezoelectric plate is lapped to the desired thickness by the usual optical shop methods and then bonded to the delay medium. This method becomes impossibly difficult for transducers of even small area as their frequency increases, because such thin plates cannot be manipulated. A much more convenient technique is to bond the piezoelectric plates with a convenient thickness, say several tenths of a millimeter, to the delay medium and then lap the plate to the final thickness. In both methods, one electrode is first deposited on the delay medium, and in the case of thinning the piezoelectric after bonding, the second electrode and back layers are deposited as the final step. Care is required in lapping the bonded transducer so that the base electrode is not damaged by the polishing compound. If a chemically active compound, such as Cyton is used, the delay medium as well as the electrode may be attacked and must be protected by some appropriate coating, such as photoresist. The final electrical connection to the top electrode must be made in some fashion which does not mass-load the transducer and distort its bandpass characteristics, or be so small as to cause hot spots from high current densities. The usual method is to bond thin gold wire or ribbon onto electrode tabs, as is done for electronic circuit chips.

The most successful method for fabricating very high frequency transducers for longitudinal wave generation is by deposition of thin films of piezoelectric materials by methods which yield a desired crystallographic orientation [44, 45]. The materials used are CdS and ZnO, whose properties are shown in Table 11.6.

Table 11.6 Properties of Transducer Materials

Material	Density	Mode	Orientation	K	ϵ_{rel}	V (cm/s)	Z (g/s-cm^2)
$LiNbO_3$	4.64	L	36° Y	0.49	38.6	7.4×10^5	34.3×10^5
		S	163° Y	0.62	42.9	4.56×10^5	21.2×10^5
		S	X	0.68	44.3	4.8×10^5	22.3×10^5
$LiTaO_3$	7.45	L	47° Y	0.29	42.7	7.4×10^5	55.2×10^5
		S	X	0.44	42.6	4.2×10^5	31.4×10^5
$LiIO_3$	4.5	L	Z	0.51	6	2.5×10^5	11.3×10^5
		S	Y	0.6	8	2.5×10^5	11.3×10^5
$Ba_2NaNb_5O_{15}$	5.41	L	Z	0.57	32	6.2×10^5	33.3×10^5
		S	Y	0.25	227	3.7×10^5	19.8×10^5
$LiGaO_2$	4.19	L	Z	0.30	8.5	6.3×10^5	26.2×10^5
$LiGeO_3$	3.50	L	Z	0.31	12.1	6.5×10^5	22.8×10^5
α-SiO_2	2.65	L	X	0.098	4.58	5.7×10^5	15.2×10^5
		S	Y	0.137	4.58	3.8×10^5	10.2×10^5
ZnO	5.68	L	Z	0.27	8.8	6.4×10^5	36.2×10^5
		S	39° Y	0.35	8.6	3.2×10^5	18.4×10^5
		S	Y	0.31	8.3	2.9×10^5	16.4×10^5
CdS	4.82	L	Z	0.15	9.5	4.5×10^5	21.7×10^5
		S	40° Y	0.21	9.3	2.1×10^5	10.1×10^5
$Bi_{12}GeO_{20}$	9.22	L	(111)	0.19	38.6	3.3×10^5	30.4×10^5
		S	(110)	0.32	38.6	1.8×10^5	16.2×10^5
AlN	3.26	L	Z	0.20	8.5	10.4×10^5	34.0×10^5

Such piezoelectric thin films generally cannot be grown with values of κ as high as that of the bulk material, but in the best circumstances κ may approach 90%. Thin-film transducers with band center frequencies up to 5 GHz can be prepared by these techniques.

A problem that arises with large-area transducers, or even with small-area transducers at very high frequencies, is that of matching the electrical impedance to the source impedance. It is especially true for the ferroelectric, piezoelectric materials of very high dielectric constant that the impedance of the transducer may be so low that it becomes difficult to efficiently couple electrical power from the source to the transducer. This problem can be largely overcome by dividing the transducer into a series connected mosaic, as reported by Weinert and deKlerk [46]. A schematic representation of such a mosaic transducer is shown in Figure 11.27. If a transducer of given area is divided into N elements, which are connected in series, the capacity of the transducer will be reduced by a factor of N^2. As an example, a 1-GHz lithium niobate transducer of 0.25-cm^2 (0.4-in^2) area would represent a capacitive impedance of only 0.038 Ω; if this area were divided into a 16-element mosaic, the impedance would be increased to 10 Ω. A 40-element thin-film transducer is shown in Figure 11.28. The same considerations will apply at lower frequencies for transducers with large areas, about 1 cm^2 or more. Because most ferroelectric transducer materials, such as the PZTs or lithium niobate, have high dielectric constants, the large areas will lead to very large capacitance values for frequencies far below 100 MHz. Thus, large-area transducers are usually divided into multiple elements, which are then wired in series to obtain the desired 50-Ω impedance to match to the RF driver. A large-area transducer which has been so wired is shown in Figure 11.29.

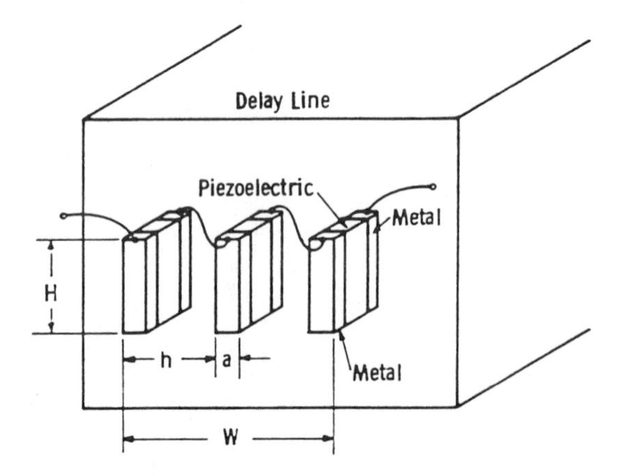

Figure 11.27. Schematic of mosaic transducer.

Figure 11.28. Forty-element thin-film mosaic transducer array.

Figure 11.29. Four-element, series-connected lithium niobate transducer metal-bonded to Bragg cell.

11.4.4 Acoustic Beam Steering Methods

One of the serious limitations of normal (i.e., isotropic) Bragg acoustooptic deflectors is that imposed by the bandwidth as limited by the Bragg interaction. The most straightforward method of enlarging the interaction bandwidth is simply to shorten the interaction length in order to increase the acoustic beam diffraction spread. This is generally not a very desirable method to increase bandwidth for systems in which the light to the Bragg cells is collimated because, it wastes acoustic power; only those momentum components of the acoustic beam which can be phase-matched to incident and diffracted light momentum components are useful. Furthermore, as the interaction length shrinks, the transducer becomes increasingly narrow, with a corresponding increase in power density. This increase in power density may produce heating at the transducer, which can cause thermal distortion in the deflector due to gradients in the acoustic velocity and refractive index.

An ideal solution to this difficulty would be one in which the acoustic beam changes in direction as the frequency is changed, so that for every frequency the Bragg angle is perfectly matched. The first approximation to such acoustic beam steering was carried out by Korpel [47] for a television display system. This transducer consisted of a stepped array, as shown in Figure 11.30. The height of each step is one-half an acoustic wavelength at the band center $\Lambda_0/2$, and the spacing s between elements is chosen so as to optimize the tracking of the Bragg angle. Each element is driven π rad out of phase with respect to the adjacent elements, and the net effect of such a transducer is to generate an acoustic wave with corrugated wavefronts, which are tilted at an angle with respect to the transducer surfaces when the frequency differs from the band center frequency f_0. For this transducer configuration, the acoustic beam steers with frequency but matches the Bragg angle only imperfectly.

To understand the steering properties of such an acoustic array, which was analyzed in detail by Coquin [48], consider the somewhat simpler arrangement shown in Figure 11.31, in which each transducer element is driven Ψ rad out of phase with respect to the next one, and Ψ may be electrically varied. This causes the effective wavefront to be tilted by an angle θ_e with respect to the

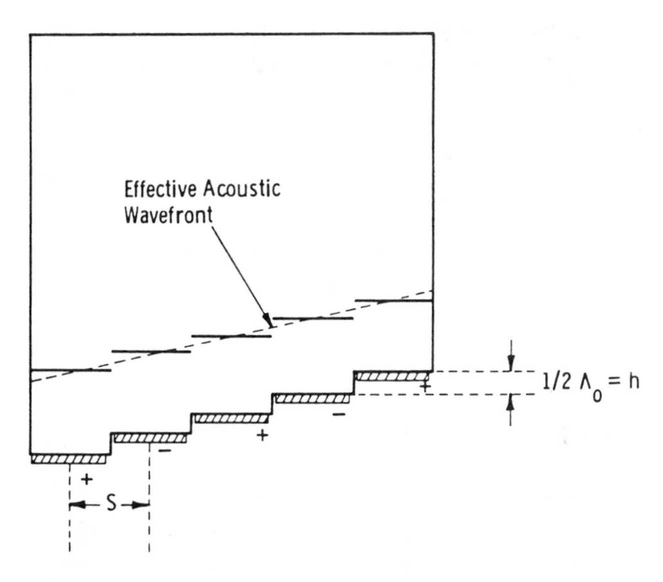

Figure 11.30. Stepped transducer array.

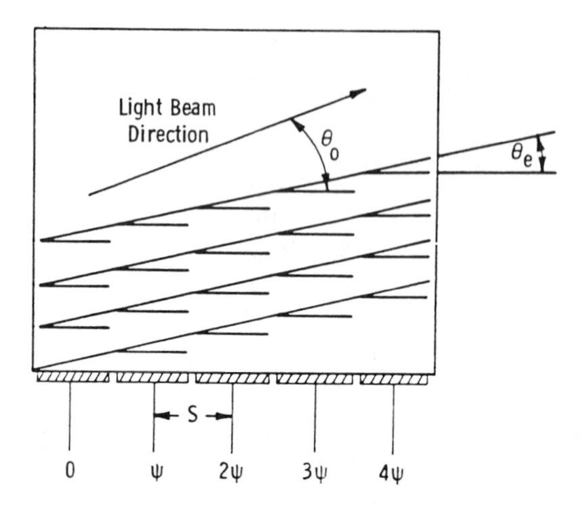

Figure 11.31. Steering of an acoustic beam by a phased array transducer.

piecewise wavefronts radiating from the individual elements. If θ_e is small, it can be approximated by

$$\theta_e \approx \tan \theta_e = \frac{\Psi}{2\pi}\frac{\Lambda}{s} = \frac{\Psi}{Ks} \tag{98}$$

If the incident light beam makes an angle θ_0 with the plane of the transducer and if the Bragg angle is $\theta_B = K/2k$, then the angular error from perfect matching is

$$\Delta\theta = (\theta_0 - \theta_e) - \theta_B = \left(\theta_0 - \frac{\Psi}{Ks}\right) - \frac{K}{2k} \tag{99}$$

The condition for perfect beam steering is that $\Delta\theta = 0$ for all values of K; setting $\Delta\theta = 0$, the required phase for perfect beam steering is

$$\Psi_p = \theta_0 Ks - \frac{K^2}{2k}s \tag{100}$$

from which it can be see that the phase must be a quadratic function of the acoustic frequency.

Most of the work done on acoustic beam steering has involved making various approximations to this condition. One such approximation is obtained by making Ψ a linear function of frequency, with $\Psi = 0$ at f_0, the midband frequency. This was accomplished in the step transducer method of Figure 11.30, as described in Ref. 47. For this case, the angle that the effective wavefronts make with respect to the transducer plane is

$$\theta_e \approx \frac{\pi}{Ks} - \frac{h}{s} \tag{101}$$

where h is the step height, and there is 180° phase shift between adjacent elements. The resulting beam steering error is

$$\Delta\theta = \left(\theta_0 - \frac{K}{2k}\right) + \left(\frac{h}{s} - \frac{\pi}{Ks}\right) \tag{102}$$

which can be made zero at the midband frequency f_0 by choosing

$$\begin{aligned} h &= \frac{1}{2}\Lambda_0 \\ s &= \frac{{\Lambda_0}^2}{\lambda} \end{aligned} \tag{103}$$

and

$$\theta_0 = \frac{1}{2}\frac{\lambda}{\Lambda_0}$$

A further improvement can be achieved by noting from equation (101) that θ_e varies as $1/f$, whereas perfect beam steering should lead to a linear variation of θ_e with f. Therefore, the constants h, s, and θ_e may be chosen to agree with the perfect beam steering case at two frequencies, rather than only one, as shown in Figure 11.32. This first-order beam steering can yield substantial improvements in performance for systems requiring less than one octave bandwidth [49], but bandwidths larger than this require a better approximation to the quadratic dependence of the phase on the acoustic frequency.

The next higher approximation to perfect beam steering was carried out by Coquin [48] for a 10-element array, as shown in Figure 11.33. If the phase applied to each transducer corresponds to that for perfect steering, $\Psi_l = l\Psi_p$, and the element spacing is $s = {\Lambda_0}^2/\lambda$, the bandwidth extends from 0 to about $1.6f_0$, the high-frequency dropoff being determined by the finite element spacing. Coquin pointed out that the deflector performance is very tolerant of errors in the individual phases; for example, if the phase applied to each transducer is within 45° of the perfect beam steering phase, there is a loss of only 0.8 dB in diffracted light intensity. If the phase error is increased to 90°, the loss increases to 3 dB. Thus, for deflectors, in which this degree of ripple is permissible, the transducer array may be driven by logic circuitry which sets digital phase shifters. This requires prior knowledge of the input frequency, or analog phase shifters, which accomplish the same function without the need for logic circuits.

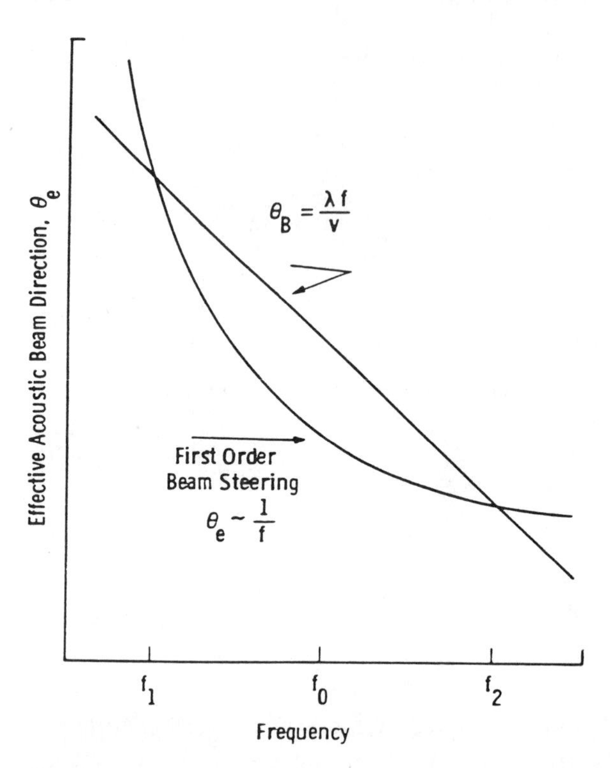

Figure 11.32. First-order beam steering with exact match at two frequencies.

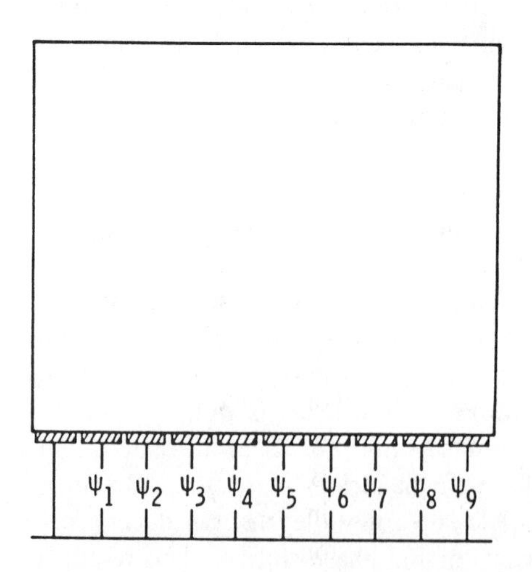

Figure 11.33. Ten-element phased array transducer, in which ψ = 0°, 90°, 180°, or 270°, leading to diffracted intensity less than 0.8 dB lower than for perfect beam steering.

An entirely different approach to broadband Bragg acoustooptic interaction matching is the use of the tilted transducer array, first reported by Eschler [50]. A tilted transducer array consists of two or more transducers electrically connected in parallel and tilted in angle with respect to each other, as illustrated in Figure 11.34. Each transducer element in the array is designed to cover some fraction of the entire bandwidth, and its angle with respect to the incident light direction is chosen to match the Bragg angle at the center of its subband. For frequencies near the midband of any of the transducer elements, the incident light will interact strongly only with the sound wave emanating from that element; interaction will be weak from the other elements both because the angle of incidence will be mismatched and the frequency will be far from the resonance frequency of those elements.

On the other hand, for frequencies which are midway between the resonance frequencies of adjacent elements, i.e., $(f_{01} + f_{02})/2$ or $(f_{02} + f_{03})/2$, the contributions to the acoustic fields from both elements are about equal and the effective wavefront direction lies midway between those of the components. Thus, the array behaves very much as if the acoustic wave were steering with frequency, although this is not true in a strict sense. The diffraction efficiency of the tilted

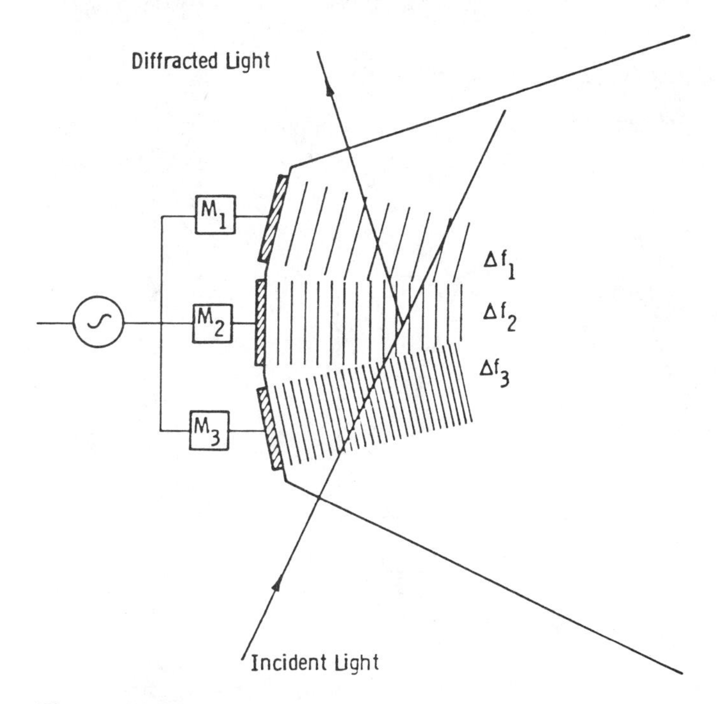

Figure 11.34. Tilted transducer array, in which each element is optimized for part of the entire frequency band.

transducer array is shown in Figure 11.35, in which the solid curves represent the efficiency of the individual elements and the dotted curve represents the overall efficiency. There will typically be about 1 dB of ripple across the full band, which is acceptable for most applications.

There are two additional advantages to the tilted array transducer. First, it is obviously relatively simpler to design a larger overall acoustic bandwidth, since each element of the tilted array need be only about one-third of the total bandwidth. Second, tilted-array transducers can generally be operated more deeply in the Bragg mode, as the combined acoustic wavefronts from adjacent elements are twice the length of that from a single element. If the second-order diffracted light is sufficiently low, then operation over a frequency range larger than one octave is possible. The elements of the array can be connected in parallel since they will tend to behave as bandpass filters, the power being directed to the element with the closest frequency range. In practice, it is generally necessary to provide impedance-matching networks, as indicated in Figure 11.34, because of the low reactance obtained with a parallel network.

11.5 APPLICATIONS OF ACOUSTOOPTIC SCANNERS

11.5.1 Scanning Systems Requirements

There are many applications of lasers requiring scanning devices with a very wide range of performance specifications regarding resolution and speed, and

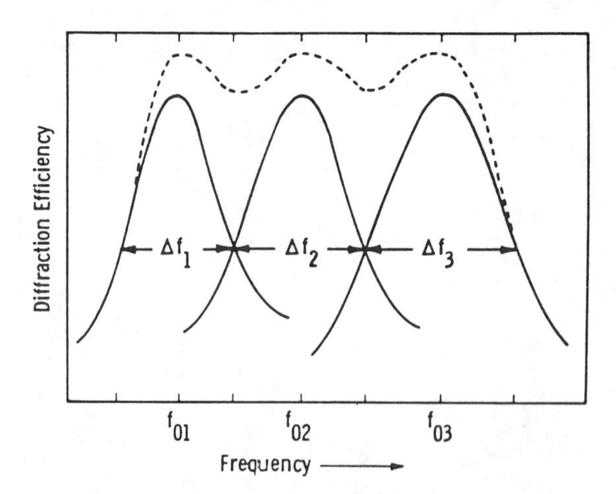

Figure 11.35. Diffraction efficiency of a three-element tilted transducer array. Solid curves represent efficiency of individual elements, and dotted curve represents efficiency of entire array.

with the need for either linear scanning or random access addressing. There are a number of manufacturers of acoustooptic Bragg cells in the United States from which devices with a wide range of characteristics may be purchased off-the-shelf basis and in large quantities. For the systems designer with nonstandard requirements, several of these manufacturers will provide consulting services and produce custom scanners. As may be deduced from the previous sections, acoustooptics offers a wide range of possibilities to satisfy diverse system needs, and these may apply to a large variety of applications. Each application will have its own set of optical requirements with respect to beam shaping and imaging, vertical deflection, intensity modulation, and blanking.

The acoustooptically scanned television display was probably the first such application, and incorporates most of the required elements of the other systems. Although acoustooptic television displays were described as early as 1939, the first practical system was described by Korpel et al. [51] in 1966. A more recent television rate scanner, incorporating higher-performance materials, was demonstrated by Gorog et al. [52, 53], in which the vertical deflection is done by a galvanometer or a rotating prism. The scan of a standard television operates at a rate of 525 lines per frame, with 30 frames per second, which corresponds to a scan time for one line of 6.35×10^{-5}s and 1.58×10^{4} lines/s. These requirements are easily within the capability of acoustooptic scanners. The most widely used acoustooptic scanner is tellurium dioxide; a typical one is shown in Figure 11.36, with an aperture of 25 mm (1 in.) and 1200 spot resolution. It will operate with very high diffraction efficiency with an inexpensive RF drive amplifier of 1 W.

11.5.2 Frequency Chirp Scanning

Most systems for which acoustooptic scanners may be considered will require linear scanning rather than the more demanding random access addressing. A linear scan is produced with an acoustooptic Bragg cell by applying a linearly swept frequency drive to the transducer. Such a linear frequency sweep is sometimes referred to as a chirp, for the obvious reason. The analysis of the chirp scanner follows easily from the basic principles already discussed. In the chirp scanner, a frequency gradient is produced across the optical aperture. This gradient will be, in effect, a cylindrical lens whose focal length is determined by the chirp rate. This is the focusing property of such diffraction, which is easily understood by reference to Figure 11.37.

The focal point of the scanner is located at a distance F given by

$$F \cong \frac{D}{\theta_1 - \theta_2} \tag{104}$$

where D is the aperture and θ_1 and θ_2 are the diffraction angles corresponding to the high and low frequencies, f_1 and f_2, within the aperture,

Figure 11.36. A tellurium dioxide acoustooptic scanner with RF drive.

$$\theta_1 = \frac{\lambda f_1}{v}$$

and (105)

$$\theta_2 = \frac{\lambda f_2}{v}$$

If the acoustic travel time across the aperture is $\Delta t = D/v$, then it follows easily that

$$F = \frac{v^2}{\lambda(df/dt)} \tag{106}$$

where df/dt is the frequency scan rate. This result was first obtained by Gerig and Montague [54], who also pointed out that the time duration required for the

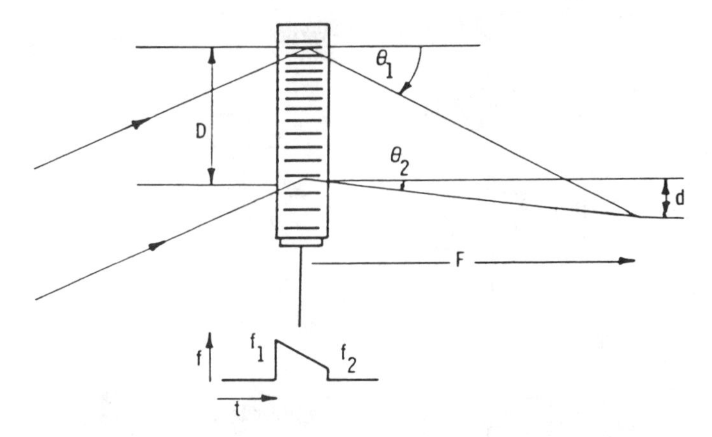

Figure 11.37. Convergence of light diffracted by linearly swept acoustic frequency.

focal spot of width a to traverse a narrow slit in the focal plane is a/v. This phenomenon was utilized in optical processing of radar signals for performing pulse compression [55, 56]. Obviously, since df/dt can be either positive or negative, the diffracted light can be either convergent or divergent. The most straightforward method of compensating for the cylindrical lensing due to this effect is the placement in the optical path of a cylinder lens with the negative of the focal length calculated from equation (106).

11.5.3 Representative Acoustooptic Scanning Systems

The principles of operation and design of acoustooptic Bragg cells discussed in the previous sections can be applied to a wide variety of systems requirements, in which the scanner will generally be incorporated into a more complex optical system. It will be apparent that the presence of the acoustooptic components will greatly influence the surrounding optics and may itself dictate the addition of certain optical components. In addition, various electronic functions must be performed in conjunction with the Bragg cell for most systems, such as display, recording, etc. In this section, we will describe two representative systems which embody many of these elements, and this may serve to illustrate some of the important features of acoustooptic scanning systems.

Acoustooptic Television Display System

The scanning system described in Ref. 52 for TV display exemplifies many of the considerations for various types of scanning systems, and will be described in some detail here. The optical system, incorporating a galvanometer-type vertical deflector, is shown in Figure 11.38, and the system incorporating the rotating

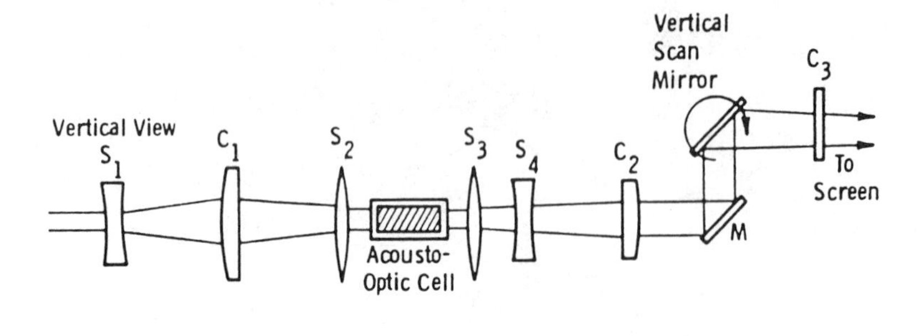

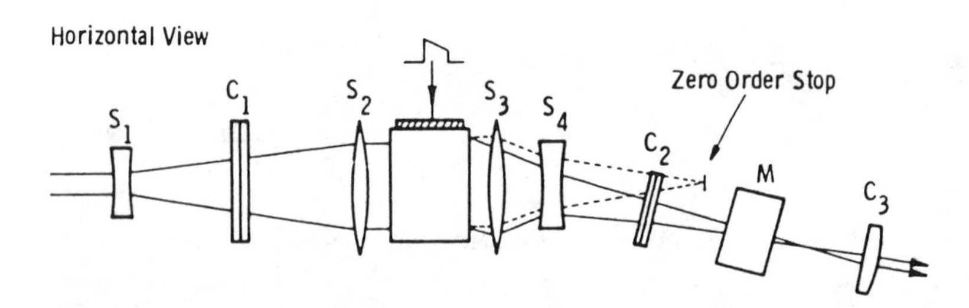

Figure 11.38. Vertical and horizontal views of scanning system for television display. S refers to spherical lenses, and C refers to cylinder lenses.

prism utilizes essentially the same optical train. The components s_1, c_1, and s_2 are used to reshape the small, circular-cross-section input laser beam into a large elliptical-cross-section beam which provides a good match to the rectangular aperture of the acoustooptical cell. A long, narrow rectangle is required to provide both high efficiency and many resolution elements. Thus, this combination of lenses is a beam expander, in which the cylindrical lens c_1 focuses the light to accommodate the narrow dimension of the acoustooptic cell. The combination of lenses s_3 and s_4 is a telephoto arrangement, whose long effective focal length produces a magnified image of the scanner at the viewing screen. The cylindrical lens c_2 is used to provide the proper convergence of the light beam to fill the vertical scan mirror aperture, and the last cylindrical lens c_3 is used to project the final image onto the screen. A zero-order stop following the post-scan optics prevents unwanted, unmodulated light from reaching the viewing screen. The cylindrical lensing effect described by equation (106) is compensated by adjustment of the positions of the cylindrical lenses backward or forward along the axis.

The electronic system used by Gorog to drive the acoustooptic scanner was designed for a standard black-and-white television receiver, as shown in the block diagram of Figure 11.39. The video chain consists of the processor, into which

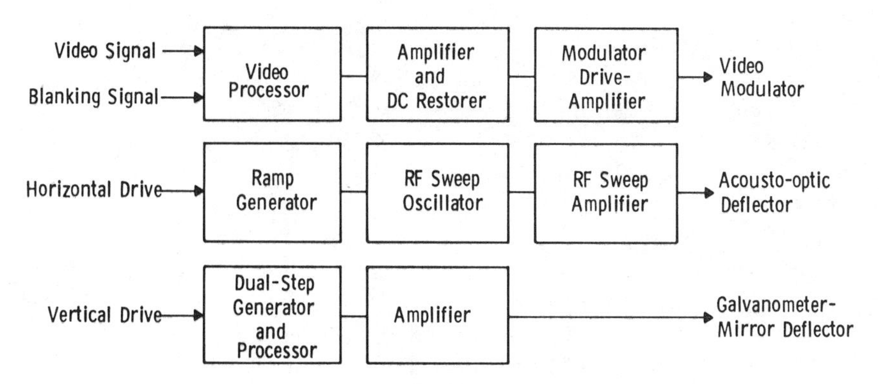

Figure 11.39. Electronic system for television display.

the video and blanking signals are fed, which extracts pure video signal, followed by amplification and dc restoration. The horizontal drive may consist of any number of commercially available swept frequency oscillators followed by a wideband power amplifier, capable of delivering output from 1 to 10 W, depending upon the requirements of the acoustooptic deflector material. The details of the vertical drive will depend upon the type of scanner used. For a galvanometer-type scanner, the high mechanical Q would normally result in overshoot and ringing at the end of the trace. The dual-step generator and processor provide current at two intervals and of such amplitude that the galvanometer comes to rest at the end of each sweep.

An important characteristic to consider in acoustooptic scanned displays is the ratio of the flyback time τ to the active line scan time T, since this will affect the resolution of the system. In order to avoid stray light or other disturbances at the end of each line, the laser beam is blanked at the start of the next line scan until the acoustic wave fills the optical aperture. The minimum flyback time for an acoustooptic scanner is the time required for the acoustic wave to cross the optical aperture, typically about 10 μs, much larger than the electronic flyback time. Because the laser is blanked for a time duration τ after initiation of the frequency sweep, a fraction of the total frequency sweep is lost, equal to $\tau/(T + \tau)$. Then it is easy to show that the number of resolution elements is reduced from $\tau\,\Delta f$ to

$$N = \left(1 - \frac{\tau}{T + \tau}\right)(\tau\,\Delta f) \tag{107}$$

As an example, if the transit time is one-tenth of the scan time, the resulting resolution is 91% of that computed from the time-bandwidth product.

Flying-Spot Microscope

Another type of system to which acoustooptic scanners have been applied with good success is the flying-spot laser microscope. Such systems have been reported by Lekavich et al. [57], using visible light, and Sherman and Black [58] at infrared wavelengths. The basic function of this device is to scan the focus of the laser beam over the field to be examined, detect the reflected or transmitted light, and display the magnified image on a television receiver. The advantage of a scanned-laser flying-spot microscope over other types of TV microscopes, such as a vidicon microscope, is that the high-intensity laser beam allows the imaging to be done with potentially a much greater signal-to-noise ratio. In the latter system, the entire field is illuminated, and the image is projected by conventional optics onto the face of a vidicon. A scanned electron beam converts this image into an electric signal for transmission to the television receiver.

A block diagram of a scanned-laser flying-spot microscope, as reported in Ref. 57, is shown in Figure 11.40. The beam from a 1-mW helium-neon laser is expanded by the telescope to 1.5 cm diameter, and the cylinder lens c_1 is used to shape the beam to fill the long, narrow aperture of the first acoustooptic scanner. This scanner produces the horizontal deflection, and the spherical lens following the first scanner rotates the beam to match the aperture of the vertical scanner. The cylinder lens c_2 restores the beam to its original circular cross section, after

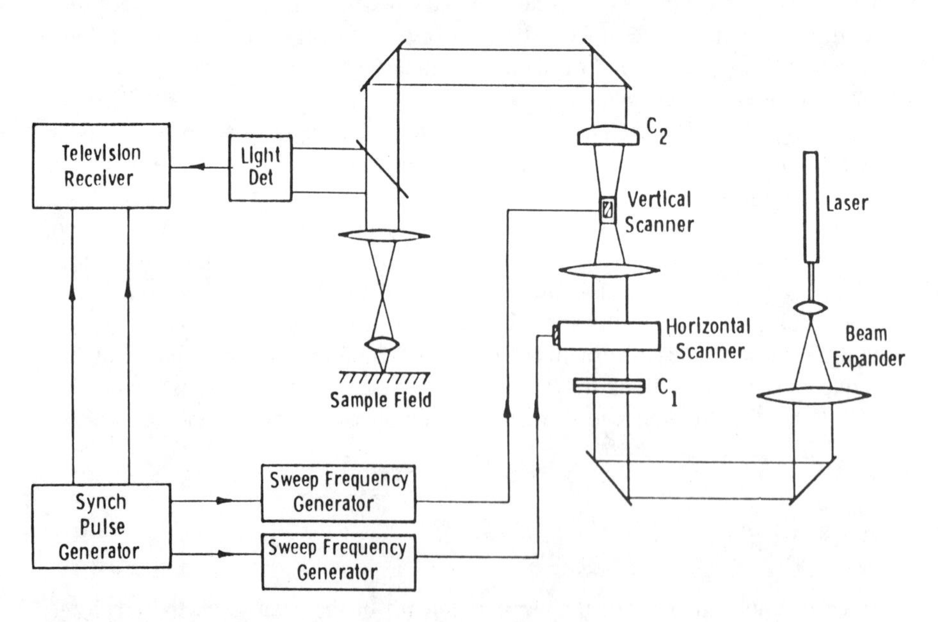

Figure 11.40. Flying-spot laser microscope system.

which the microscope optics bring it to a focus in the plane of the sample field, which is scanned by the focused spot.

The horizontal Bragg scanner operates at the line rate of the television receiver, 15.75 kHz, and the vertical Bragg scanner operates at the standard frame rate of 1/30 s. The light reflected by the sample field falls on the photodetector, where it is converted to an appropriate electric signal for the television receiver. The synch pulse generator provides the timing pulses for the television receiver and for the sweep frequency oscillators which drive the horizontal and vertical Bragg scanners. In the work reported in Ref. 57, the horizontal scan frequency ran from 18 to 38 MHz in 53.5 μs with a 10-μs blanking time. The vertical scan frequency ran from 20.5 to 35.5 MHz in 15.3 ms with 1.3 ms blanking time. Because relatively low bandwidths and frequencies were required, water was used as the acoustooptic medium.

The arrangement for viewing the object, as shown in Fig. 11.40, uses a beam splitter for on-axis reflection. Other viewing methods include off-axis reflection, in which no beam splitter is required, resulting in lower light losses. Improved depth perception is reported, but with lower signal-to-noise ratio, because of the small-angle light scattering. The magnification of the microscope is the ratio of the width of the TV screen to the width of the scanned field

$$M = \frac{W}{\alpha F} \tag{108}$$

where $\alpha = \lambda N/D$ is the maximum scan angle, and F is the focal length of the final lens. If the f-number of the lens is f, then

$$M = \frac{W}{f\lambda N} \tag{109}$$

To illustrate the limitations on the magnification, if $W = 40$ cm (16 in), the maximum aperture diffraction-limited lens might corrrespond to an f-number of 1.5, and the number of resolution elements for good image quality is no less than 200, so that if $\lambda = 0.6328$ μm, the resulting magnification is $M = 2100$. In practice, much larger f-numbers are used, with correspondingly lower magnification.

The x-y deflection system described here utilizes two Bragg cells conveniently displaced along the optic axis so that the oblong beam cross section could be rotated through 90°. There may, however, be applications where the required resolution and deflection efficiency can be obtained with a circular optical beam profile. For such a case, there is no reason for any separation between the x and y deflectors, and it may be more economical to combine both x and y deflectors in a single piece of material rather than two separate ones. It had been believed that the simultaneous x and y deflections of a beam light was necessarily less efficient than cascaded deflectors, in which no acoustic fields could overlap. An analysis by LaMachia and Coquin [59] showed that simultaneous and cascaded deflections are capable of identical efficiency, and that the intensity of diffracted light in overlapping acoustic beams is exactly that from two physically separated acoustic beams. A problem may result from different acoustic velocities and diffraction efficiencies in orthogonal directions in the same crystal, but generally, these will not be so drastically different as to make operation impossible. The choice of cascaded or simultaneous deflection can then be made entirely on the basis of effective design and economy.

Acoustooptic Array Scanners

There are many applications of scanned laser systems where the requirements on the number of resolution elements or bandwidth are greater than the state of the art capability of single acoustooptic devices. In such cases it may be possible to satisfy the system requirements by channelizing the device into a multiplicity of scanners. There are obviously many ways in which this may be done, depending upon the functions of the system. Two examples will be described here, one in which the bandwidth is multiplied, and the other in which the resolution is multiplied by array configurations.

A linear acoustooptic array was reported by Young and Yao [60] for application as a page composer for wideband recording. This device consists of a large number of individually driven transducers mounted on an acoustooptic medium, as illustrated in Figure 11.41. Such units were built with up to 138 channels, with total recorder bandwidths up to 750 Mbits/s. In order for a large number of channels to be placed on a crystal of convenient size, the transducers must be of small width and located very close to each other. This will result in the acoustic fields of neighboring channels overlapping each other by acoustic diffraction a short distance from the transducer. In order to avoid such crosstalk, the incident light must be focused onto the modulator array within a distance d from the transducers, given approximately by

$$d = \frac{1}{2}\frac{sHf}{v} \tag{110}$$

where H is the width of the transducers and s is the spacing between them. This requirement can be rather severe, as seen by the following example. Suppose the modulator crystal is 25 mm (1 in.) long, and we wish to place 100 channels on the crystal; if the center frequency is 100 MHz, then for an acoustic velocity of 3×10^5 cm/s (1.2 in/s) d is 2.6×10^{-2} cm (1.04×10^{-2} in). Careful design is required to achieve a high optical throughput; in particular it is important to match the acoustic and optical beam spreads. Results were reported in Ref. 60 for devices fabricated from SF-8, SF-59, and TeO_2 glasses and from $PbMoO_4$

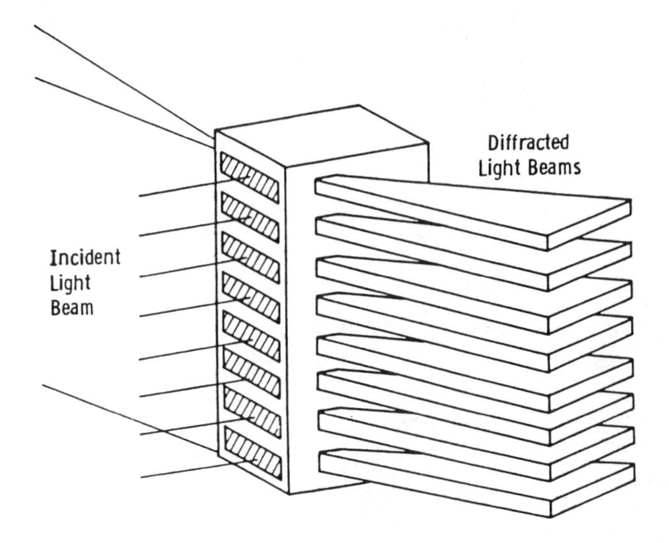

Figure 11.41. Linear acoustooptic scanner array for page composer.

and TeO_4 crystals. Electric power as low as 26 mW/channel was obtained using the fast shear mode in TeO_2.

In another type of acoustooptic array, the objective is to multiply the number of resolution elements available from a single channel, so that the elements of the array are arranged in series rather than in parallel, as was the case with the previous example. Many recording applications require resolution of 1500 points per line or better, and it is difficult to achieve this performance with a single scanner. An optical arrangement has been proposed [61] in which an array of scanners, each with a resolution capability of 500 points per line, yields 1500 points per line. The optical system for producing a linear scan from an array is rather complex, with the input light beam following a helical path so as to pass through the acoustic field of each of the transducers on successive passes through the device. The deflected beam from each scanner is extracted and aligned by a set of prisms so as to give a single continuous scan on the recording medium. Very high resolutions are, in principle, possible if the optical complexity can be dealt with.

REFERENCES

1. P. Debye and F. W. Sears, *Proc. National Acad. Sci.*, *18*:409 (1932).
2. R. Lucas and P. Biquard, *J. Phys. Rad.*, *3*(7):464 (1932).
3. C. F. Raman and N. S. N. Nath, *Proc. Indian Acad. Sci. I*, 2:406 (1935).

4. E. I. Gordon, *Proc. IEEE*, *54*:1391 (1966).
5. R. W. Dixon, *IEEE J. Quantum Electronics*, *QE-3*:85 (1967).
6. S. E. Harris, S. T. R. Nieh, and D. K. Winslow, *Appl. Phys. Lett.*, *15*:325 (1969).
7. R. Mertens, *Meded. K. Vlaam. Acad. Wet. Lett. Schone Kunsten Relg., Kl. Wet.*, *12*:1 (1950).
8. R. Exterman and G. Wannier, *Helv. Phys. Acta*, *9*:520 (1936).
9. W. R. Klein and E. A. Hiedemann, *Physica*, *29*:981 (1963).
10. O. Nomoto, *Jpn. J. Appl. Phys.*, *10*:611 (1971).
11. W. R. Klein and B. D. Cook, *IEEE Trans. Sonics Ultrason.*, *SU-14*:723 (1967).
12. A. Korpel, *J. Opt. Soc. Am.*, *69*:678 (1979).
13. A. Korpel and T. Poon, *J. Opt. Soc. Am.*, *70*:817 (1980).
14. I. C. Chang, *IEEE Trans. Sonics Ultrason.*, *SU-23*:2 (1976).
15. M. Cohen and E. I. Gordon, *Bell Syst. Tech. J.*, *44*:693 (April 1965).
16. R. W. Dixon, *J. Appl. Phys.*, *38*:5149 (1962).
17. E. H. Young and S. K. Yao, *Proc. IEEE*, *69*:54 (1981).
18. D. A. Pinnow, *IEEE J. Quantum Electronics*, *QE-6*:223 (1970).
19. J. H. Gladstone and T. P. Dale, *Phil. Trans. Roy. Soc. London*, *153*:37 (1964).
20. S. H. Wemple and M. DiDomenico, *J. Appl. Phys.*, *40*:735 (1969).
21. N. Uchida and N. Niizeki, *Proc. IEEE*, *61*:1073 (1973).
22. T. O. Woodruff and H. Ehrenreich, *Phys. Rev.*, *123*:1553 (1961).
23. H. Mueller, *Phys. Rev.*, *47*:947 (1935).
24. R. L. Abrams and D. A. Pinnow, *J. Appl. Phys.*, *41*:2765 (1970).
25. R. W. Dixon, *J. Appl. Phys.*, *38*:5149 (1967).
26. D. A. Pinnow, L. G. Van Uitert, A. W. Warner, and W. A. Bonner, *Appl. Phys. Lett.*, *15*:83 (1969).
27. G. A. Coquin, D. A. Pinnow, and A. W. Warner, *J. Appl. Phys.*, *42*:2162 (1971).
28. Y. Ohmachi and N. Uchida, *J. Appl. Phys.*, *40*:4692 (1969).
29. M. Gottlieb, T. J. Isaacs, J. D. Feichtner, and G. W. Roland, *J. Appl. Phys.*, *40*:4692 (1969).
30. G. W. Roland, M. Gottlieb, and J. D. Feichtner, *Appl. Phys. Lett.*, *21*:52 (1972).
31. T. J. Isaacs, M. Gottlieb, and J. D. Feichtner, *Appl. Phys. Lett.*, *24*:107 (1974).
32. J. D. Feichtner and G. W. Roland, *Appl. Optics*, *11*:993 (1972).
33. S. E. Harris and R. W. Wallace, *J. Opt. Soc. Am.*, *59*:744 (1969).
34. N. Uchida and Y. Ohmachi, *J. Appl. Phys.*, *40*:4692 (1969).
35. A. W. Warner, D. L. White, and W. A. Bonner, *J. Appl. Phys.*, *43*:4489 (1972).
36. E. K. Sittig, *IEEE Trans. Sonics and Ultrasonics*, *SU-16*:2 (1969).
37. A. H. Meitzler and E. K. Sittig, *J. Appl. Phys.*, *40*:4341 (1969).
38. W. P. Mason, *Electromechanical Transducers and Wave Filters*, Van Nostrand Reinhold, Princeton, NJ, 1948.
39. E. K. Sittig, *IEEE Trans. Sonics and Ultrasonics*, *16*:2 (1969).
40. E. K. Sittig and H. D. Cook, *Proc. IEEE*, *56*:1375 (1968).
41. W. F. Konog, L. B. Lambert, and D. L. Schilling, *IRE Int. Conv. Rec.*, *9*(6):285 (1961).
42. J. D. Larson and D. K. Winslow, *IEEE Trans. Sonics and Ultrasonics*, *SU-18*:142 (1971).

43. A. H. Meitzler, *Ultrasonic Transducer Materials* (O. E. Mattiat, ed.), Plenum, New York, 1971.
44. J. deKlerk, *Physical Acoustics*, W. P. Mason, ed., Vol. IV, Chap. 5, Academic Press, New York, 1970.
45. J. deKlerk, *IEEE Trans. on Sonics and Ultrasonics*, *SU-13*:100 (1966).
46. R. W. Weinert and J. deKlerk, *IEEE Trans. on Sonics and Ultrasonics*, *SU-19*:354 (1972).
47. A. Korpel et al., *Proc. IEEE*, *54*:1429 (1966).
48. G. Coquin, J. Griffin, and L. Anderson, *IEEE Trans. on Sonics and Ultrasonics*, *SU-7*:34 91971).
49. D. A. Pinnow, *IEEE Trans. on Sonics and Ultrasonics*, *SU-18*:209 (1971).
50. H. Eschler, *Optics Communications* *6*:230 (1972).
51. A. Korpel, R. Adler, P. Desmares, and W. Watson, *Appl. Optics*, *5*:1667 (1965).
52. I. Gorog, J. D. Knox, and P. V. Goedertier, *RCA Review*, *33*:623 (1972).
53. I. Gorog, J. D. Knox, P. V. Goerdertier, and I. Shidlovsky, *RCA Review*, *33*:667 (1972).
54. J. S. Gerig and H. Montague, *Proc. IEEE*, *52*:1753 (1964).
55. M. B. Schulz, M. G. Holland, and L. Davis, *Appl. Phys. Lett.*, *11*:237 (1967).
56. J. H. Collins, E. G. Lean, and H. J. Shaw, *Appl. Phys. Lett.*, *11*:240 (1967).
57. J. Lekavich, G. Hrbek, and W. Watson, *Proc. Electro-Optical Syst. Design Conf.*, p. 650, Sept. 1970.
58. B. Sherman and J. F. Black, *Appl. Opt.*, *9*:802 (1970).
59. J. T. LaMachia and G. A. Coquin, *Proc. IEEE*, *36*:304 (1971).
60. E. H. Young and S. K. Yao, *Proc. IEEE Ultrasonics Symp.*, Annapolis, p. 666, Sept. 1976.
61. J. T. McNaney, *Laser Focus*, June 1979, p. 84.

12

Electrooptical Scanners

Clive L. M. Ireland

Lumonics Ltd., Rugby, Warwickshire, England

John Martin Ley

Leysop, Basildon, Essex, England

12.1 INTRODUCTION

The main stimulus for the development of electrooptic (E-O) devices has been the invention of the laser. Although the observation of the electric field dependence of the optical properties of certain materials was reported by Kerr and Pockels around the turn of the century, it was the development of the laser, with its high-intensity and near diffraction-limited beam, that allowed practicable applications of these E-O effects in optical switching, modulating, and deflecting devices.

By 1960, when the first laser was operated, a considerable body of information had been collected on the E-O properties of many materials. In particular, Zwicker and Scherrer (1943) had reported on the materials potassium dihydrogen phosphate (KH_2PO_4) and deuterated potassium dihydrogen phosphate (KD_2PO_4), which are still among the most important for use in commercial E-O devices. As a result of the impetus received from the development of the laser, a wealth of information on the E-O properties of many other crystals and liquids has since been obtained. Unfortunately, as yet, few materials exhibit a strong enough E-O effect to make them of use in practicable E-O devices. The E-O industry still awaits the "ideal" material. Nevertheless, E-O deflectors have been fabricated and successfully employed in a wide range of applications, albeit often with somewhat complicated geometry or less than ideal performance.

Over the past decade a detailed theoretical understanding of the origin of the E-O effect in terms of the crystal atomic structure and electronic bonding has been developed and applied to the known simpler E-O crystal structures with reasonable success. Although it is not currently possible to predict with certainty the magnitude of the E-O effect in a particular crystal, this work gives cause for optimism that, before long, bulk E-O crystal engineering will be possible. In the future, new crystals showing a large E-O effect will be identified and grown to usable size. The development of a more efficient electrooptic material would greatly accelerate the application of bulk electrooptic deflectors. For surface wave deflectors and in particular switchable directional couplers existing electrooptic materials are capable of producing useful devices requiring only a few volts drive for operation. In this new area of work further improvement in technology is required to reduce crosstalk between spot positions and for the integration of these components into optical circuits.

The advantage of the electrooptic deflector is that in most applications it can be treated as a capacitor, with the speed of operation dependent on the output characteristics of the drive circuitry. In the most advantageous situations nanosecond sweep rates can be obtained with picosecond resolution and this has led to the construction of prototype streak cameras using bulk multiple-prism or gradient analog deflectors. A commercially available crystal streak camera having 40-ps resolution and a visual display produced by a photodiode array and video processor is shown in Figure 12.1.

Perhaps the most spectacular application for electrooptic deflectors has been in the area of digital laser displays. Digitally scanned continuous-wave (CW) laser systems are well suited for alphanumeric and graphic high-brightness displays. Real-time large-screen dislays have already been used in the presentation of information for military use, and it is suggested that they may eventually become more widely used as a visual aid in traffic control. Only bulk electrooptic digital deflectors have achieved the necessary resolution for the presentation of information in this way. An impressive display indicating the performance achieved by Philips' digital laser display system is shown in Figure 12.2.

An increasing interest is being directed to the development of electrooptic deflectors for integrated optics. These are planar devices where a guided beam travels just below a set of surface electrodes and is deflected by the change in refractive index produced by an applied electric field. Being quasi-two-dimensional, they can be made using manufacturing techniques developed for integrated optics and can be prepared as an integrated component on a crystal surface. Surface wave prism and Bragg deflectors were first investigated, but more recent work has centered on switchable optical directional couplers. A switchable optical directional coupler is formed by fabricating two parallel waveguides in proximity in order that light entering one guide can couple to the other. The coupled energy

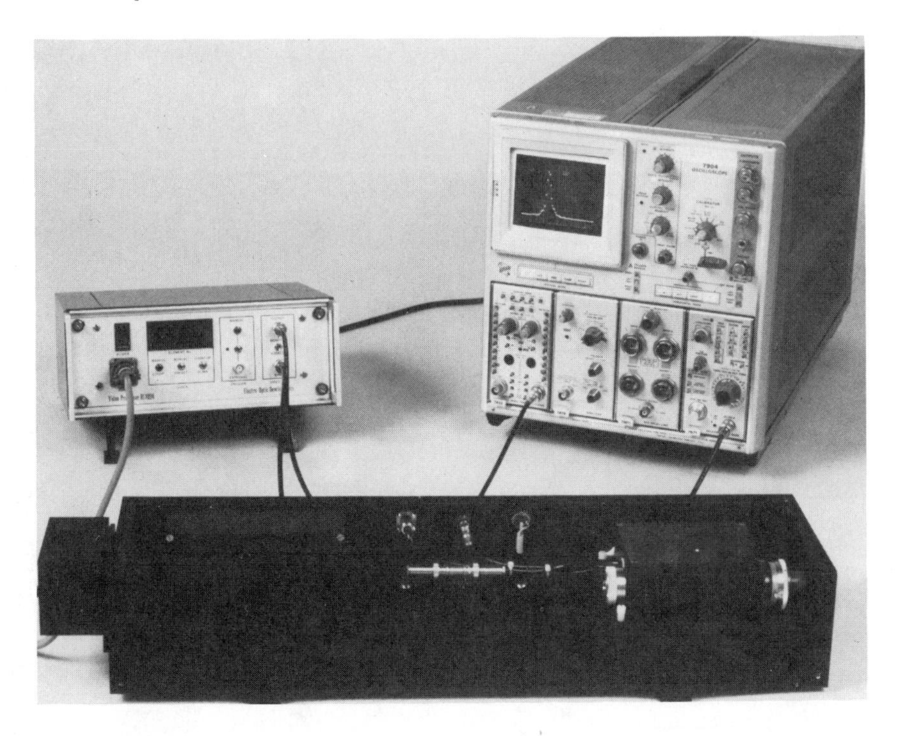

Figure 12.1 Crystal streak camera with photoiode array and video processor. (Published with permission of Electro-Optic Developments Ltd., Basildon, England.)

is then controlled via a refractive index change produced by a voltage applied to strip electrodes adjacent to the guides.

Devices with moderate resolutions have already been built with base bandwidths approaching 1 GHz; consequently they are attractive devices for many fiber-optic communication and integrated-optics applications. These include multiport optic deflection, ultrahigh-data-rate optic multiplexing, and demultiplexing and high-capacity optic connecting networks. Surface prism deflectors can also be applied directly with diode array detectors to form analog to digital convertors, and they have the potential there to produce a novel range of high-speed devices. In the area of computer development optic memories using holographic or localized storage techniques are also being investigated as potential solutions to the requirement for high-capacity stores of short access time. High-speed deflection of the reference read-in/read-out beam is a fundamental requirement for any optic memory system, and the electrooptic deflector is capable of achieving

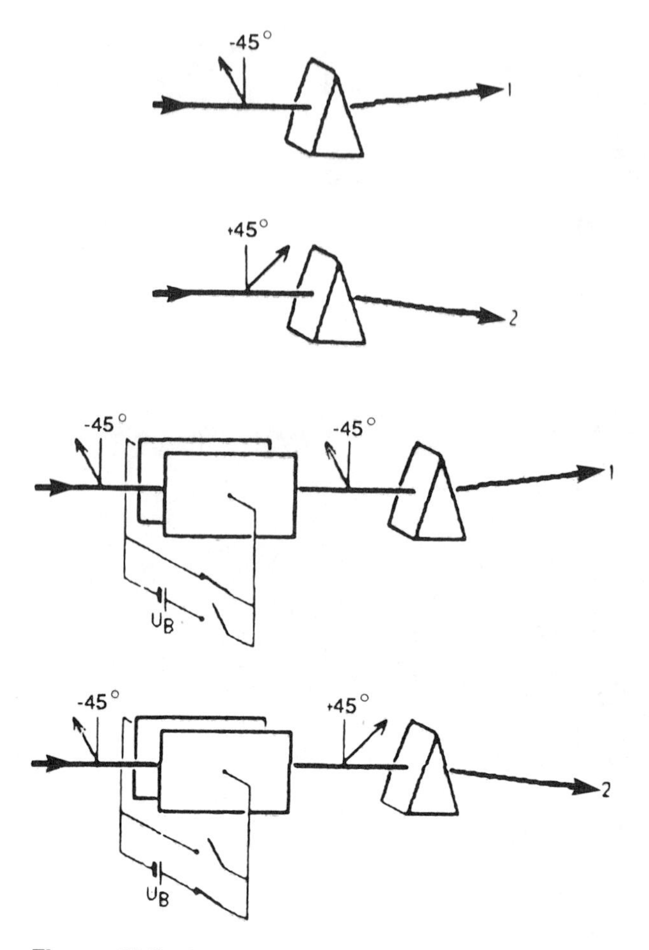

Figure 12.2 Example of the excellent performance achieved by Philips GmBH Hamburg for a real-time scanned laser display employing bulk nitrobenzene E-O digital deflectors. The image comprised 40,000 resolved points. The scanning rate was 12.5 Hz. (See Ref. 74 for a description of the system.)

a suitably short access time. It is considered by those involved in this work that optic memory systems could soon become commercially available.

Finally, in the area of laser control, E-O deflectors could be used as intracavity Q spoilers and mode lockers or externally as displacement modulators or pulse shapers. In the latter case the generation of shaped and short optic pulses in the picosecond regime is of fundamental interest in quantum electronics and of possible value in laser fusion work. For these applications back-to-back syn-

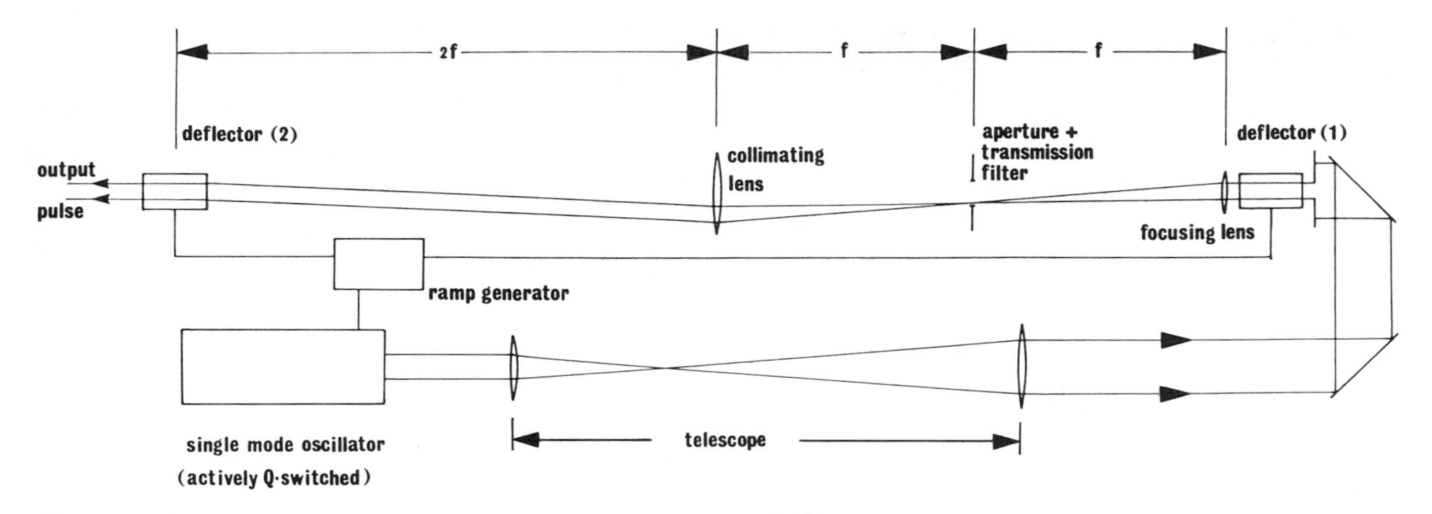

Figure 12.3 Picosecond pulse shaping with E-O deflectors. (From Ref. 88.)

chronized deflectors can be used to realign an optic beam after it has been deflected across an intermediate aperture. The aperture could be a pinhole of Airy disk size to give a very short optic pulse or a slit containing a suitable graded transmission filter to give a desired temporal profile. One such arrangement is shown in Figure 12.3.

Here a pair of back-to-back synchronized deflectors are used to realign an optical beam after it has been deflected across an intermediate far-field aperture. The aperture can be either a pinhole of approximately Airy disk size to give a very short optical pulse or a slit containing a suitably graded transmission filter to give a desired temporal profile. Using the $LiNbO_3$ deflector described in Section 12.5, Ireland demonstrated that this arrangement had the potential of yielding pulses shaped on a approx 100-ps time scale.

There is a common requirement in all the above E-O deflector applications for high-speed optic deflection. In all cases a random access time of less than 1 μs is necessary, and in certain cases less than 10 ns is required. It is in this area that the E-O deflector has unique advantages. Apart from the digital Kerr cell deflector, E-O devices have not yet achieved linear resolutions of above 1000 spot positions, and it would seem that at least for the immediate future their use will be confined to moderate-resolution high-speed or very high speed applications.

12.2 THEORY OF THE ELECTROOPTIC EFFECT

12.2.1 The Electrooptic Effect

An understanding of the nature of the E-O effect in materials is a prerequisite to understanding the principles of operation of E-O deflectors. In all materials an applied electric field results in a change in polarization. These changes are generally very small and except in a relatively few cases do not result in a measurable change in the optical properties of the material. Even if a measurable change results, the smallness of the effect restricts the number of materials from which practical E-O switching, modulating, or deflecting devices can be made.

For a description of the E-O properties of materials at low frequencies it is usual to express directly the change in the refractive index n as a function of the applied electric field **E** by

$$\frac{1}{n^2} = \frac{1}{n_o^{\,2}} + r \cdot \mathbf{E} + h{:}\mathbf{E}^2 + \cdots \tag{1}$$

Here n_o is the zero field value of the refractive index, and r and h are linear and quadratic E-O coefficients, respectively.

The reason for describing the field-induced changes in the optic properties of the material through the parameter $1/n^2$ is related to the fact that the most useful

E-O materials are crystalline. In the following section it will be shown how an equation with coefficients of $1/n^2$ can be used to characterize electromagnetic (EM) wave propagation in a general anisotropic crystalline material. With (1), use of the equation is extended to cover the optical properties of the material when it is subject to an external applied field.

12.2.2 EM Wave Propagation in a Crystal

For the general case of an anisotropic crystal the electric displacement **D**, resulting from an applied field **E**, is expressed by the second-rank tensor ϵ_{kl} through the relationship

$$D_k = \epsilon_{kl} E_l \tag{2}$$

Here the subscripts refer to a Cartesian coordinate ($k,l = x_1, x_2, x_3$), where x_1, x_2, x_3 are fixed with respect to the crystal axes and the Einstein convention of summation over repeated indices is observed. By equating the net power flow into the material (given by Poynting's vector) with the rate of change of electrical stored energy, it can be shown [3,4] that $\epsilon_{kl} = \epsilon_{kl}$. Consequently, ϵ_{kl} has, in general, only six independent elements.

Noting that the stored electrical energy density w in the crystal is given by

$$w = \frac{1}{8\pi} \mathbf{E} \cdot \mathbf{D} = \frac{1}{8\pi} E_k \epsilon_{kl} E_l \tag{3}$$

We therefore have

$$8\pi w = \epsilon_{x_1x_1} E_{x_1}^{\ 2} + \epsilon_{x_2x_2} E_{x_2}^{\ 2} + \epsilon_{x_3x_3} E_{x_3}^{\ 2} + 2\epsilon_{x_2x_3} E_{x_2} E_{x_3} + 2\epsilon_{x_1x_3} E_{x_1} E_{x_3} + 2\epsilon_{x_1x_2} E_{x_1} E_{x_2} \tag{4}$$

With a suitable choice of new coordinate axes, the last three terms in (4) can be eliminated. The new axes are called the *principal dielectric axes* of the crystal. In the new coordinate system (x,y,z), equation (4) becomes

$$8\pi w = \epsilon_x E_x^{\ 2} + \epsilon_y E_y^{\ 2} + \epsilon_z E_z^{\ 2} \tag{5}$$

In this coordinate system the tensor ϵ_{kl} is diagonal, so that $D_x = \epsilon_x E_x$, etc. Thus (5) can be written as

$$8\pi w = \frac{D_x^{\ 2}}{\epsilon_x} + \frac{D_y^{\ 2}}{\epsilon_y} + \frac{D_z^{\ 2}}{\epsilon_z} \tag{6}$$

Consequently, the constant energy density surfaces (w = constant) in the space D_x, D_y, D_z are ellipsoids. Here ϵ_x, ϵ_y, and ϵ_z are the principal dielectric constants of the crystal. Further, if we replace $D_k/\sqrt{8\pi w}$ by x_k and define the principal indices of refraction n_x, n_y, and n_z of the crystal by $n_k^2 = \epsilon_k$ ($k = x, y, z$), equation (6) becomes

$$\frac{x^2}{n_x^2} + \frac{y^2}{n_y^2} + \frac{z^2}{n_z^2} = 1 \tag{7}$$

This is the equation of a general ellipsoid with axes parallel to the x, y, and z directions whose respective lengths are $2n_x$, $2n_y$, and $2n_z$. The ellipsoid, which is known as the index ellipsoid, indicatrix, or sometimes the ellipsoid of wave normals, characterizes EM wave propagation in the crystal. If all three principal refractive indices are equal, the crystal is isotropic; if only two are equal, it is termed uniaxial; and if none are equal, it is called biaxial.

The major use of the index ellipsoid is to determine the value of the refractive index associated with the polarization (direction of **D**) of the two normal modes of a plane wave propagating along an arbitrary direction **s** in the crystal. A plane normal to the direction **s** and passing through the center of the ellipsoid cuts it in an ellipse whose major and minor axes define the directions of the two independent planes of polarization. The half-lengths of these axes are equal in value to the two indices of refraction.

As an example of the use of the index ellipsoid we shall consider beam propagation in a uniaxial crystal. In this case (7) becomes

$$\frac{x^2}{n_o^2} + \frac{y^2}{n_o^2} + \frac{z^2}{n_e^2} = 1 \tag{8}$$

where n_o and n_e designate the two principal (ordinary and extraordinary) refractive indices, respectively, and where the convention of taking the z crystal axis as the axis of symmetry has been adopted. For the particular case of a beam propagating at an angle θ of $\pi/2$ to the z crystal axis, the plane normal to **s** through the center of the ellipsoid intersects the ellipsoid to form an ellipse with semiaxes of length n_o and n_e. This is illustrated in Figure 12.4a. The more general case where $\theta \neq \pi/2$ is shown in Figure 12.4a. Here the length of the semiaxes of the ellipsoid are n_o and $n_e(\theta)$, where $n_e(\theta)$ is given by

$$\frac{1}{n_e^2(\theta)} = \frac{\cos^2\theta}{n_o^2} + \frac{\sin^2\theta}{n_e^2} \tag{9}$$

Consequently, the extraordinary refractive index varies from $n_e(\theta) = n_o$ for $\theta = 0°$ to $n_e(\theta) = n_e$ for $= \pi/2$. Crystals with $n_e > n_o$ are called positive uniaxial and with $n_e < n_o$ are called negative uniaxial. Useful examples of these two types of crystal are quartz and calcite, respectively. It is clear that any of these birefringent materials, cut in the form of a prism and orientated in a manner similar to that in Figure 12.4a, would act as a good polarizer, since it would introduce an angular separation between the two independent polarizations of

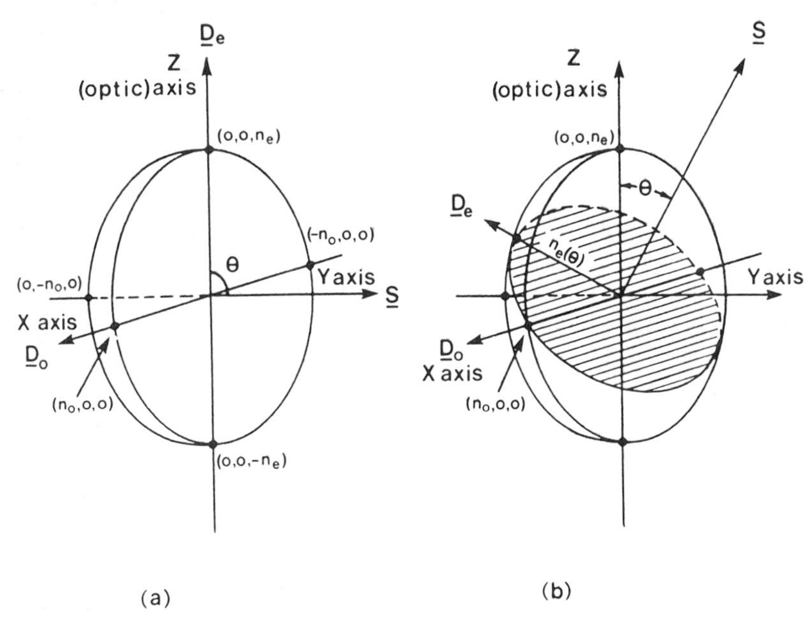

Figure 12.4 Use of the index ellipsoid for finding the refractive index for two normal mode planes of polarization associated with an *EM* wave propagating in a uniaxial crustal. (a) In this simple example, the direction of beam propagation **s** is at $\theta = \pi/2$ to the Z (optic) axis, and the two refractive indices are n_o and n_e. (b) As in (a) but the more general case where $\theta \neq \pi/2$. Here the refractive indices (the semiaxes of the hatched elliptic cross section) are n_o and n_e (θ), where $n_e(\theta)$ is given by (9).

the beam. Since calcite is available in large, high-optic-quality pieces and has a large birefringence, it is very commonly used as a polarizer material. In particular, it has found wide use in this role in digital-type deflectors.

The majority of crystals occurring in nature are biaxial. For the purpose of determining the refractive indices in these materials, the second (azimuthal) direction ϕ of the beam propagation, relative to the crystal axes, has to be defined since the index ellipsoid is no longer symmetric about z. The dielectric axes may or may not be determined by crystal symmetry and consequently can be wavelength-dependent.

12.2.3 The Linear Electrooptic Effect

The preceding discussion has shown how beam propagation in an anisotropic crystal can be characterized by the index ellipsoid. In a general (i.e., nonprincipal) coordinate system the equation of this ellipsoid can be obtained from (4), namely,

$$\frac{1}{n_1^2}x_1^2 + \frac{1}{n_2^2}x_2^2 + \frac{1}{n_3^2}x_3^2 + \frac{1}{n_4^2}x_2x_3 + \frac{1}{n_5^2}x_1x_3 + \frac{1}{n_6^2}x_1x_2 = 1 \tag{10}$$

Interchanging the indices in (4) leaves the equation unchanged; consequently, the contraction has been made here of writing $1 \leftrightarrow x_1x_1$, $2 \leftrightarrow x_2x_2$, $3 \leftrightarrow x_3x_3$, $4 \leftrightarrow x_2x_3$, $5 \leftrightarrow x_1x_3$, and $6 \leftrightarrow x_1x_2$. As (10) is a general equation characterizing beam propagation in a crystal, it is natural that changes in the optic properties of the crystal due to an applied field should be expressed as a change in its coefficients. The linear E-O effect (Pockels effect) [1] relates the linear variation of these coefficients to the field and is described by the E-O tensor r_{ij} through the relationship

$$\Delta\left(\frac{1}{n^2}\right)_i = r_{ij}E_j \tag{11}$$

where $\Delta(1/n^2)_i$ is the change in the coefficient $(1/n^2)_i$ and where, as usual, we sum over repeated indices.

In centrosymmetric crystals, that is, those possessing an inversion symmetry, the optical characteristics must remain the same when the sign of the applied field is reversed. It follows from (11) that $\Delta\,(1/n^2)_i = r_{ij}\,(-E_j)$, which can be satisfied only if $r_{ij} = 0$. Consequently, the linear E-O effect is found only in crystals lacking an inversion symmetry. The lack of inversion symmetry is also the prerequisite for the existence of a linear piezoelectric effect which is described by a 3×6 tensor d_{ji}. The form of both the r_{ij} and d_{ji} tensors is determined uniquely by the point group symmetry of the crystal. The form of the r_{ij} tensor is obtained from the d_{ji} tensor by conjugation, that is, by replacing terms d_{ji} by r_{ij}. This converts the 3×6 piezoelectric tensor into the 6×3 form describing the linear E-O effect. Although a large piezoelectric effect does not necessarily imply a large E-O effect, the relationship has been used widely to aid the selection of potentially useful E-O materials [5].

By combining (10) and (11), the equation of the index ellipsoid in the presence of the field E_j is obtained; that is,

$$\begin{aligned}&\left(\frac{1}{n_1^2} + r_{1j}E_j\right)x_1^2 + \left(\frac{1}{n_2^2} + r_{2j}E_j\right)x_2^2 + \left(\frac{1}{n_3^2} + r_{3j}E_j\right)x_3^2 + 2\\&\left(\frac{1}{n_4^2} + r_{4j}E_J\right)x_2x_3 + 2\left(\frac{1}{n_5^2} + r_{5j}E_j\right)x_1x_3 + 2\left(\frac{1}{n_6^2} + r_{6j}E_j\right)x_1x_2 = 1\end{aligned} \tag{12}$$

In general, it is not necessary to completely determine the surface represented by this equation. To find the two independent refractive indices for the beam, it is only necessary to determine the lengths of the semiaxes of the ellipse, normal to the direction of beam propagation **s**, through the center of the ellipsoid. For example, if **s** and the applied field $\mathbf{E}_j$ are directed along the x_2 and x_3 axes, respectively, (12) reduces to

$$\left(\frac{1}{n_1^2} + r_{13}E_3\right) x_1^2 + \left(\frac{1}{n_3^2} + r_{33}E_3\right) x_3^2 + 2\left(\frac{1}{n_5^2} + r_{53}E_3\right) x_1x_3 = 1 \quad (13)$$

In the special case where the crystal is oriented so that x_1, x_2, and x_3 correspond to the crystal axes x, y, and z, respectively, then we also have $1/n_4^2 = 1/n_5^2 = 1/n_6^2 = 0$. This is a geometry that has been used widely in deflector designs. Consequently,

$$\left(\frac{1}{n_x^2} + r_{13}E_z\right) x^2 + \left(\frac{1}{n_z^2} + r_{33}E_z\right) z^2 + 2r_{53}E_z xz = 1 \quad (14)$$

Depending on the symmetry of the particular crystal used, n_x and n_z may or may not be equal, and some r_{ij} terms may be equal or vanish.

Ignoring the term in r_{53} for the moment and making use of the fact that $\Delta\,(1/n^2) = -2\,\Delta\,n/n^3$ for small $\Delta\,n$, we see that (14) represents an ellipse with semiaxes along x and z equal to $n_x + \Delta\,n_x$ and $n_z + \Delta\,n_z$, respectively, where

$$\Delta\,n_x = \frac{-n_x^3}{2} r_{13}E_z \qquad \text{and} \qquad \Delta\,n_z = \frac{-n_z^3}{2} r_{33}E_z \quad (15)$$

These are the increments in refractive index, resulting from the application of the field $\mathbf{E}_z$, for the x and z polarization components of the beam, respectively.

The term in xz in (5) has the effect of rotating the axes of the elliptical cross section by an angle α, where

$$\tan 2\alpha = \frac{2r_{53}E_z}{1/n_1^2 - 1/n_2^2 + (r_{13} - r_{33})E_z} \quad (16)$$

In crystals where the natural birefringence is appreciable, that is, $|1/n_1^2 - 1/n_2^2| \gg (r_{13} - r_{33})E_z$, α is very small, and the principal axes lie close to the crystal axes. The lengths of the principal axes (and hence the refractive indices) are changed only by an amount of second order in $R_{53}E_z$ that is usually negligible. However, the radius vector of the ellipse at $\pi/4$ to x and z is changed linearly in E_z by an amount $n'^3/2r_{53}E_z$, where n' is an effective index intermediate between the principal indices [5,6].

If the particular crystal used in the example was, say, the important ferroelectric material $LiNbO_3$, then $r_{53} = 0$, and the principal axes would remain the crystal axes. For this material $n_x \neq n_z$, and so, cut in the form of a prism, it behaves as a polarizer like the calcite example considered earlier. But unlike the earlier example, the results expressed by (15) show that, through the linear E-O effect, the power of the prism can be changed linearly with applied field. This is the basic principle behind the operation of analog deflectors that use the linear E-O effect.

There is an additional effect that requires consideration when designing E-O devices. In E-O crystals, the application of an electric field leads, via piezoelecric and/or electrostrictive coupling (effects proportional to E and E^2, respectively), to strain which can distort the index ellipsoid. In the particular case of strain resulting from the inverse piezoelectric effect (usually the dominant effect for linear E-O crystals), it can be shown that these refractive index changes depend on crystal symmetry in the same way as the E-O effect itself. Hence,

$$\Delta \left(\frac{1}{n^2}\right)_i = p_{ij}{}^e S_j + r_{ik}{}^s E_k \tag{17}$$

Here k runs from $1 \rightarrow 3$, but i and j are the reduced indices discussed earlier and run from $1 \rightarrow 6$. S_j is the strain tensor, and $p_{ij}{}^e$ is the strain-optic (or elastooptic) tensor component measured under the conditions of constant electric field. $r_{ik}{}^s$ is the E-O coefficient measured at constant strain. The superscripts are used to indicate these two conditions. If the applied electric field is at high frequency, inertia prevents the material straining macroscopically. The crystal is therefore *clamped* and operates under the condition of constant strain. In this case, the first term on the right-hand side of (17) is zero. For low-frequency modulation the crystal is *free*, and the effect of the strain-optic tensor components, in general, cannot be ignored. The terms representing this effect are usually incorporated into the E-O coefficients, which at low frequency are generally written $r_{ik}{}^T$ to indicate that they are applicable to conditions of constant stress. For the case where the strain is wholly piezoelectric in origin, the high- and low-frequency E-O coefficients are related by

$$r_{ik}{}^T = r_{ik}{}^S + p_{ij}\, d_{kj} \tag{18}$$

in which p_{ij} and d_{kj} are the strain-optic and piezoelectric tensor coefficients, respectively.

The E-O coefficients of a large number of materials, under conditions of constant strain and constant stress, have been determined by many workers. Useful reviews of published data on E-O materials include Kaminow and Turner [5], Pressley [7], Milek and Neuberger [8], and Kuz'minour et al. [9]. Table 12.1 lists some of the more important materials and their main optic and E-O properties.

Table 12.1 Properties of 10 Important E-O Materials near Room Temperature[a]

Material	Transition temp. (K)	Principle r_{ij}'s ($\times 10^{-12} m/V$)		Refractive indices (~ 600 nm)	Dielectric const. (ϵ/ϵ_o)	Transmission range (μm)
KDP	123	$r_{41}{}^T \sim 8.6$	$r_{63}{}^T \sim 10.4$	$n_o \sim 1.51$	$\epsilon^T \parallel C \sim 20$	~ 0.2–1.5
(KH_2PO_4)				$n_e \sim 1.47$	$\epsilon^S \parallel C \sim 20$	
		$r_{41}{}^S \sim$	$r_{63}{}^S \sim 9.0$		$\epsilon^T \parallel C \sim 42$	
					$\epsilon^S \parallel C \sim 44$	
KD*P	222	$r_{41}{}^T \sim 9.0$	$r_{63}{}^T \sim 25$	$n_o \sim 1.51$	$\epsilon^T \parallel C \sim 50$	~0.2–2.1
(KD_2PO_4)					$\epsilon^S \parallel C \sim 48$	
		$r_{41}{}^S \sim$	$r_{63}{}^S \sim 23$	$n_e \sim 1.47$	$\epsilon^T \perp C \sim$	
					$\epsilon^S \perp C \sim 57$	
ADP	148	$r_{41}{}^T \sim 24$	$r_{63}{}^T \sim 8.5$	$n_o \sim 1.52$	$\epsilon^T \parallel C \sim 15$	~0.2–1.4
($NH_4H_2PO_4$)					$\epsilon^S \parallel C \sim 14$	
		$r_{41}{}^S \sim$	$r_{63}{}^T \sim 4.5$	$n_e \sim 1.48$	$\epsilon^T \perp C \sim 75$	
					$\epsilon^S \perp C \sim 74$	
KDA	97	$r_{41}{}^T \sim 12.5$	$r_{63}{}^T \sim 11$	$n_o \sim 1.57$	$\epsilon^T \parallel C \sim 21$	~0.3–1.6
(KH_2AsO_4)					$\epsilon^S \parallel C \sim 19$	
		$r_{41}{}^S \sim$	$r_{63}{}^S \sim$	$n_e \sim 1.52$	$\epsilon^T \perp C \sim 54$	
					$\epsilon^S \perp C \sim 53$	

Table 12.1 (Continued)

Material	Transition temp. (K)	Principle r_{ij}'s ($\times 10^{-12} m/V$)		Refractive indices ($\sim$ 600 nm)	Dielectric const. (ϵ/ϵ_o)	Transmission range (μm)
RDP (RbH_2PO_4)	147	$r_{41}^T \sim 9.1$	$r_{63}^T \sim 15.5$	$n_o \sim 1.50$	$\epsilon^T \parallel C \sim 25$ $\epsilon^S \parallel C \sim$	~0.2–1.4
		$r_{41}^S \sim$	$r_{63}^S \sim 14$	$n_e \sim 1.47$	$\epsilon^T \perp C \sim 40$ $\epsilon^S \perp C \sim$	
$LiNbO_3$	1470	$r_{33}^T \sim 32$	$r_{42}^T \sim 32$	$n_o \sim 2.30$	$\epsilon^T \parallel C \sim 32$ $\epsilon^S \parallel C \sim 28$	~0.4–5.0
		$r_{33}^S \sim 31$	$r_{42}^S \sim 28$	$n_e \sim 2.21$	$\epsilon^T \perp C \sim 80$ $\epsilon^S \perp C \sim 44$	
$LiTaO_3$	890	$r_{33}^T \sim 22$	$r_{42}^T \sim$	$n_o \sim 2.18$	$\epsilon^T \parallel C \sim 45$ $\epsilon^S \parallel C \sim 43$	~0.4–6.0
		$r_{33}^S \sim 30$	$r_{42}^S \sim 20$	$n_e \sim 2.19$	$\epsilon^T \perp C \sim 51$ $\epsilon^S \perp C \sim 41$	
$BaTiO_3$	390	$r_{33}^T \sim 108$	$r_{42}^T \sim 1640$	$n_o \sim 2.44$	$\epsilon^T \parallel C \sim 120$ $\epsilon^S \parallel C \sim 60$	
		$r_{33}^S \sim 23$	$r_{42}^S \sim 820$	$n_e \sim 2.37$	$\epsilon^T \perp C \sim 4000$ $\epsilon^S \perp C \sim 2300$	
CuCl		$r_{41}^T \sim 3.6$	$r_{41}^S \sim 2.4$	$n_o \sim 2.0$	$\epsilon^T \sim 10.0$ $\epsilon^S \sim 7.5$	~0.4–19.0
ZnTe		$r_{41}^T \sim 4.4$	$r_{41}^S \sim 4.3$	$n_o \sim 2.9$	$\epsilon^T \sim 10$ $\epsilon^S \sim 0$	

[a] Data are from references cited in the text. Agreement between authors for E-O coefficients is typically $\sim \pm 15\%$.

12.2.4 The Quadratic Electrooptic Effect in Crystals

In centrosymmetric crystals the linear E-O effect vanishes, but both noncentrosymmetric and centrosymmetric crystals can exhibit a quadratic effect. If the quadratic term dominates, equation (1) reduces to

$$\Delta\left(\frac{1}{n^2}\right)_{ij} = h_{ijkl}E_kE_l \tag{19}$$

where h is a fourth-rank tensor and i, j, k, and l are Cartesian coordinates. Again we sum over repeated indices. Usually this equation is recast with the refractive index expressed in terms of the induced polarization:

$$\Delta\left(\frac{1}{n^2}\right)_{ij} = g_{ijkl}P_kP_l \tag{20}$$

where $P_l = \psi_{lg}E_g$. Consequently,

$$h_{ijsg} = g_{ijkl}\psi_{ks}\psi_{lg} \tag{21}$$

The refractive index change expressed as a function of the polarization is generally preferred for expressing the quadratic E-O effect, since experiment has shown that the tensor components in (20) are nearly independent of temperature. In contrast, those in (19) embody the coefficients of the susceptibility tensor, which can be strongly temperature-dependent, particularly for E-O materials that are operated near a phase transition.

Since in (20) both ij and kl are interchangeable, the tensor elements $gijkl$ can be written in contracted notation as g_{lm}, where l represents ij and m represents kl, and both l and m run from $1 \rightarrow 6$, for example $g_{x_1x_2x_3x_1} \leftrightarrow g_{65}$. g is, in general, a 6×6 tensor identical in form to that of the strain-optic tensor discussed earlier. As with the linear E-O effect, two or more coefficients g_{lm} may be equal, or terms vanish, depending on the crystal symmetry. For example, in the important cubic crystals with point group symmetry $m3m$, g is given by [7]

$$g = \begin{bmatrix} g_{11} & g_{12} & g_{12} & 0 & 0 & 0 \\ g_{12} & g_{11} & g_{12} & 0 & 0 & 0 \\ g_{12} & g_{12} & g_{11} & 0 & 0 & 0 \\ 0 & 0 & 0 & g_{44} & 0 & 0 \\ 0 & 0 & 0 & 0 & g_{44} & 0 \\ 0 & 0 & 0 & 0 & 0 & g_{44} \end{bmatrix} \tag{22}$$

Consequently, the refractive index changes expressed by (20) are

$$\Delta\left(\frac{1}{n^2}\right)_1 = \begin{bmatrix} g_{11} & g_{12} & g_{12} & 0 & 0 & 0 \\ g_{12} & g_{11} & g_{12} & 0 & 0 & 0 \\ g_{12} & g_{12} & g_{11} & 0 & 0 & 0 \\ 0 & 0 & 0 & g_{44} & 0 & 0 \\ 0 & 0 & 0 & 0 & g_{44} & 0 \\ 0 & 0 & 0 & 0 & 0 & g_{44} \end{bmatrix} \begin{bmatrix} P_{x_1}^2 \\ P_{x_2}^2 \\ P_{x_3}^2 \\ P_{x_2}P_{x_3} \\ P_{x_1}P_{x_3} \\ P_{x_1}P_{x_2} \end{bmatrix} \tag{23}$$

In the absence of an applied field, cubic crystals are isotropic, and the index ellipsoid is a sphere of radius n_0. If we take for this example a field along the z crystal axis (i.e., $P_x = P_y = 0$), then the index ellipsoid becomes

$$\left(\frac{1}{n_0^2} + g_{12}P_z^2\right)(x^2 + y^2) + \left(\frac{1}{n_0^2} + g_{11}P_z^2\right) z^2 = 1 \tag{24}$$

In this case, the principal axes remain x, y, z, that is, aligned with the crystal axes. The semiaxes of the ellipsoid that give the new refractive indices become

$$n_x = n_y = n_0 - \frac{1}{2} n_0^3 g_{12} P_z^2 \qquad \text{and} \qquad n_z = n_0 - \frac{1}{2} n_0^3 g_{11} P_z^2 \tag{25}$$

For the particular case of a beam propagating in the x direction and polarized in the z direction, the refractive index change due to the quadratic E-O effect would therefore be

$$\Delta n_z = \frac{1}{2} n_0^3 g_{11} \left(\frac{\epsilon_z - 1}{4\pi}\right)^2 E_z^2 \tag{26}$$

where P_z has been written as $\psi E_z = |(\epsilon_z - 1)/4\pi| E_z$.

Some of the largest changes in refractive index occur in crystals of this $m3m$ symmetry group operated in the cubic phase above their transition temperatures. Although there is little difference in the g_{11} coefficients between these crystals, the dependence of ϵ on the nearness of the operating (room) temperature to the transition temperature causes some to exhibit a much larger refractive index change than others. These crystals are obvious candidates for possible use in analog deflectors.

12.2.5 The Quadratic Electrooptic Effect in Liquids

Refractive index changes proportional to the square of the applied field are permitted by symmetry in all materials. Besides the crystals discussed in the last section, liquids that are strongly polar are of particular E-O interest since they can exhibit a high anisotropic, optic polarizability. By applying a strong external field, the molecules of these substances partially align with the field, causing the bulk material to become birefringent. Due to the thermal motion of the molecules, the alignment, and hence the birefringence, is temperature-dependent.

The direction of main polarizability is usually nearly parallel to the dipole moment in these molecules. Consequently, the component of a beam polarized parallel to the main polarizability of the molecule (i.e., the applied field direction) sees an increased refractive index relative to that of the orthgonal polarization. This effect, which was observed by Kerr in glass and other materials, is generally described by the following simple equation:

$$n_p - n_s = B\lambda E^2 \tag{27}$$

Here λ is the vacuum wavelength of the beam, B is the Kerr constant for the material, and n_p and n_s are the parallel and orthogonal refractive index components, respectively.

A variety of Kerr substances have been investigated by several workers. See, for example, Lee and Hauser [10] and Kruger et al. [11]. Most of these substances are liquids, but some are solids. One of the earliest polar liquids investigated was nitrobenzene, and it is still the most popular for E-O applications. Switches using this liquid have been incorporated in digital beam deflectors.

12.3 PROPERTIES AND SELECTION OF ELECTROOPTIC MATERIALS

12.3.1 General

By its nature, the E-O effect in materials is very small. The change in optical susceptibility as a result of applied electric field is due to molecular, ionic, or electronic polarization. The polarization changes that can be achieved are small as the applied field is generally small in comparison with the field already existing within the material. The maximum refractive index change that can be achieved is generally only $\sim 10^{-3}$. Nevertheless, these small changes can be sufficient to allow practical E-O modulating or deflecting devices to be made providing the bulk material is of good optic quality.

Currently, the most useful E-O materials fall into three main groups. They are $NH_4H_2PO_4$ (ADP) and KH_2PO_4 (KDP) and its related isomorphs, ferroelectric materials related to perovskite ($CaTiO_3$), and a new ferroelectric family with $KTiOPO_4$ (KTP) as its most widely used member. Other materials with more limited E-O applications include AB-type binary compounds, Kerr effect liquids, and ceramics in the (Pb,La) $(Zr,Ti)O_3$ system. The E-O properties of these materials are briefly reviewed here.

12.3.2 ADP, KDP, and Related Isomorphs

Although it is now over 40 years since the E-O properties of ADP and KDP were first investigated [12–14], these materials remain the most widely used in bulk E-O devices. This is mainly due to their fine optic quality. Good single crystals in these materials have been grown in sizes up to >10 cm in diameter.

The growth process is slow, and high-quality large crystals can take up to six months to produce. Although the crystals are water soluble and fragile, they can be handled, cut, and polished without difficulty. At room temperature they are in an unpolarized phase and, in the absence of an applied electric field, are uniaxial. The atoms K, H, and P in KDP can be replaced by some of the atoms from corresponding columns in the periodic table without changing the crystal structure, for example, RbH_2PO_4 (RDP), KH_2AsO_4 (KDA), etc. The only non-vanishing E-O coefficients for this crystal class are $r_{41} = r_{52}$ and r_{63}.

ADP has an antiferroelectric transition at 148 K and KDP a ferroelectric transition at 123 K. In both materials the hydrogen can be replaced by its heavier isotope deuterium (D^*) to form AD*P and KD*P, respectively. This has the effect of raising the transition temperatures of the materials to 242 and 222 K, respectively. In the ferroelectric material, the E-O coefficients, based on dielectric polarization, are the same temperature-independent constants for both KDP and KD*P. As a consequence, the higher transition temperature in KD*P leads to an increase in the room temperature linear E-O coefficients by a factor of ~2.5. The dielectric constant is also higher in KD*P. In fact, the E-O coefficient and dielectric constant increase in such a way that the quantity $r_{63}/(\epsilon_r - 1)$ is roughly the same temperature-independent constant for all the KDP isomorphs despite the rapid increase in ϵ_r near T_c.

The refractive indices, and UV absorption which is associated with electronic transitions in the oxygen ions, are about the same for all materials in this group. The crystals are transparent down to ~0.18 μm [15]. At the long-wavelength end, the infrared absorption is the result of O—H or O—D vibrations, and the transmission cutoff for the deuterated salt occurs at roughly $\sqrt{2}$ times the wavelength of that for the undeuterated salts; for example, the cutoff wavelengths for ADP, KDP, and KD*P are 1.4, ~1.55, and 2.15 μm, respectively [16].

The resistivity of these materials is very high, typically $>10^{10}$ Ω cm. In the visible, the optical loss is small and has been measured as only ~0.5 dB m^{-1} in KDP at 632.8 nm, which is about as good as that found in the best fused quartz. The E-O coefficient r^{63} is practically independent of wavelength in the transparent region for both ADP and KDP. It is, of course, possible to obtain larger r^{63} by operating near the Curie temperature in KDP or its isomorphs, but both the loss tangent and dielectric constant increase toward infinity as T_c is approached. These two effects limit the high-frequency usefulness of the crystal near T_c except for low duty operation. The former effect results in excessive heating, and the latter leads to a large drive current requirement.

12.3.3 Ferroelectrics Related to Perovskite

A large class of ferroelectrics, having a structure of the form $A^{1+}B^{5+}O_3$ and $A^{2+}B^{4+}O_3$, are related to the mineral perovskite ($CaTiO_3$). These materials all have a centrosymmetric oxygen octahedron BO_6 as a central building block. The cubic (paraelectric) form, which is usually the high-temperature phase, belongs to the nonpiezoelectric, nonferroelectric point group $m3m$. In the low-temperature phase, the most useful crystals are either tetragonal (4 mm), with the crystal axis along one of the original cube edges, or rhombohedral (3 m), with the crystal axis along one of the cube body diagonals.

Below the transition temperature (T_c), the linear E-O effect in the perovskites can be regarded as fundamentally a quadratic effect biased by the spontaneous polarization of the crystal; that is, polarization (P_k) can be written as

$$P_k = P_k^s + \delta P_k \tag{28}$$

in which P_k^s denotes the spontaneous polarization and

$$\delta P_k = \epsilon_0(\epsilon_k - 1)E_k \tag{29}$$

is the field-induced polarization. Consequently, using (20), we have

$$\Delta\left(\frac{1}{n^2}\right)_{ij} = g_{ijkl}\,P_k^sP_l^s + 2g_{ijkl}P_l^s\,\delta P_k \tag{30}$$

The first term in (30) describes the field-free birefringence of the crystal, while the second term describes the linear E-O effect. This equation with (29) shows that the linear E-O coefficient and the quadratic E-O coefficient are related by

$$r_{ijl} = 2g_{ijkl}P_k^s\epsilon_0(\epsilon_k - 1) \tag{31}$$

Above the transition temperature a large dc field is often applied to the crystal to remove the center of symmetry and allow a linear E-O effect to be obtained. The field induces a large dc bias polarization P^{dc} so that a big refractive index change can be obtained by the use of a relatively modest modulating field. That is, above T_c, the induced polarization P^{dc} simply takes the role of P^s.

Generally g_{mn} is about the same for all the perovskite-related ferroelectrics. Most reported measurements of g_{mn} have been obtained under conditions of constant stress, that is, at low frequencies. Consequently, as discussed in Section 12.2 they include a secondary contribution arising from electrostrictive strain. This contribution can be comparable in magnitude to the primary E-O effect if undamped acoustic resonances are allowed to occur. For frequencies substantially higher than the fundamental longitudinal acoustic mode of the crystal being used, the effect decreases rapidly.

All the perovskite-related ferroelectrics are insoluble in water. They are also more rugged and have larger refractive indices and dielectric constants than the E-O crystals in the former group. As a rule, they are transparent between ~0.4 and ~6.0 μm. The infrared absorption is caused largely by vibrations of the BO_6 octahedra and the UV absorption by electronic transitions in the oxygen ions.

Currently, the most important materials are $KTa_{0.65}Nb_{0.35}O_3$ (KTN), $LiNbO_3$, $LiTaO_3$, and $BaTiO_3$.

KTN is a solid solution of two perovskites, $KTaO_3$ and $KNbO_3$, which have very nearly the same unit cell size (~4 Å) in their cubic phase but very different transition temperatures. These are ~4 and ~698 K, respectively. For the composition $KTa_{0.65}Nb_{0.35}O_3$, T_c is ~10°C; thus at room temperature KTN is just above the transition temperature and in its paraelectric phase. The properties of KTN and the other perovskites operated slightly above T_c are very temperature sensitive, since, as with the materials in the previous group, ϵ_r varies as $(T - T_c)^{-1}$. That is, T must be very carefully controlled. Even with samples of relatively high resistivity and low photoconductivity, the application of a large dc bias field leads to space charge effects which eventually reduce the internal biasing field.

If the temperature is reduced below T_c and the crystal poled into a single ferroelectric domain, the spontaneous polarization P^s removes the need for P^{dc}. Although P^s is usually several times greater than the induced P^{dc}, ϵ_r in the ferroelectric phase is much smaller than it is just above T_c. Well below T_c, r_{ij} and ϵ_r are insensitive to the temperature [17].

Barium titanate ($BaTiO_3$) was one of the earliest ferroelectric perovskites to be studied [18,19]. The crystals have a tetragonal (4 mm) phase between ~390 and 273 K and an orthorhombic phase below ~273 K. The only nonvanishing E-O coefficients are $r_{13} = r_{23}$, r_{33} and $r_{42} = r_{51}$. Both r_{13} and r_{33} are nearly temperature independent, but r_{42} increases rapidly as the tetragonal to orthorhombic phase transition is approached. At room temperature, r_{42}, at constant strain, is about 30 times bigger than r_{33}. Crystals with good optic and electrical properties with dimensions of $\gtrsim$1 cm can be grown routinely.

$LiNbO_3$ and $LiTaO_3$ are currently the most widely used of the perovskite group of ferroelectrics. Both are negative uniaxial crystals with transition temperatures of ~1470 and ~890 K, respectively. The E-O tensor has the nonvanishing components $r_{13} = r_{23}$, r_{33}, $r_{22} = -r_{12} = -r_{61}$, and $r_{42} = r_{51}$. Crystals [20] of up to ~5 cm in diameter of good optic quality and high resistivity can be grown and if necessary poled into a single domain while near T_c by application of a small field (~1 V cm^{-1}). Because of the large T_c, considerable mechanical energy would be required to depole these crystals at room temperature. Hence, unlike both KTN and $BaTiO_3$, these crystals may be cut, polished, and roughly handled without creating additional domains. This ease of handling makes both $LiNbO_3$ and $LiTaO_3$ more widely used in practical (commercial) E-O devices than either KTN or $BaTiO_3$.

12.3.4 Potassium Titanyl Phosphate (KTP)

In 1976 the Du Pont Company reported [21] on the growth and properties of a new crystalline material $K_xRb_{1-x}TiOPO_4$. This material exists in a solid solution for $0.0 \leq x \leq 1.0$ and all members are orthorhombic and belong to the acentric point group mm2. Although it is believed that the material is ferroelectric, the structure of the crystal family is characterized by chains of TiO_6 octohedra linked at two corners with alternating long and short bonds which makes the polarisation state extremely stable. The coercive force of the Ti-O chains is so large that polarisation reversal to demonstrate the ferroelectric nature of the material has not been reported.

Although the family of materials first excited interest for optical harmonic and parametric generation, they also have excellent E-O characteristics. Crystals are rugged and nonhygroscopic, can be grown with good optical quality to dimensions of approximately 1 cm, have an optical damage threshold $\geqslant 250\ MW$ cm^{-2}, resistively $\geqslant 5.5 \times 10^{11}$ ohm cm, and a transmission range of approximately 0.35–4.5 μm. Accurate refractive index measurements of $KTiOPO_4$ (KTP) have been made [22] at 16 wavelengths between 404.7 nm and 1.064 μm. Measured values of refractive index ranged between approximately 1.96 and 1.74, depending on crystal orientation and optical wavelength. The r_{23} and r_{33} E-O coefficients were measured by Massey et al. [23] and found to be $7.3 \times 10^{-12}\ mV^{-1}$ and $22.2 \times 10^{-12}\ mV^{-1}$, respectively. A measurement of optical phase retardation showed that r_{23} and r_{33} had opposite signs. A further experiment to measure the magnitude of any piezoelectric or electrooptic effects by these workers using applied electric pulses of 0.1 μs risetime and 40 μs duration was negative. The optical modulation was seen to follow the applied signal accurately with no ringing or other artifacts of mechanical response being observed.

12.3.5 AB-Type Binary Compounds

These compounds crystallize into either the cubic (43-m) zinc blende structure at room temperature or the hexagonal (6-mm) wurzite structure at higher temperatures. In the cubic phase there is only one E-O coefficient, that is, $r_{41} = r_{52} = r_{63}$. In the hexagonal phase there are three coefficients: $r_{13} = r_{23}$, r_{33}, and $r_{54} = r_{42}$.

The main interest in these materials has been for E-O devices in the infrared, particularly at 10.6 μm for use with the technologically important CO_2 laser. GaAs, ZnTe, ZnS, CdS, and CdTe are among the most widely used materials in this group that transmit out beyond 10 μm and are available as single crystals with dimensions $\gtrsim 1$ cm. Although the E-O coefficients of these materials are small, typically $\lesssim 10\%$ of the values for the crystals in the previous groups, their refractive indices are high (~2 to 4) so that the important parameter n^3r from (15) is comparable.

The fact that the 43m crystals are cubic and hence optically isotropic makes them potentially attractive for light beam deflection applications because they have a large acceptance angle. The hexagonal crystals are birefringent and consequently, from this point of view, are not so attractive.

The group covers a wide spectrum of materials with widely different optic and mechanical properties [24].

12.3.6 Kerr Effect Liquids

The advantages of using a fluid (potentially the ideal homogeneous medium) in E-O deflectors has led to renewed interest in Kerr effect liquids. There have been recent extensive studies of the behavior of nitrobenzene Kerr cells [25,26] and attempts to find Kerr liquids more suitable than nitrobenzene [27].

For the simple case of a polarization switch providing $\pi/2$ rotation at visible frequencies, a Kerr cell ~1 cm long using nitrobenzene requires a field of ~35 kV cm^{-1}. Such high fields can lead to heating and optic inhomogeneity in the media and, in extreme cases, to electrical breakdown. These effects can be prevented only if the solution is extremely pure. For high-field dc operation an ultralow charge carrier concentration (below $\sim 2.10^{11}$ cm^{-3}) is required. Larger residual currents lead to space charge, due to polarization effects at the electrodes, which lowers and makes inhomogeneous the field distribution in the bulk material.

Electrochemical processes at the electrodes and hydrodynamic current caused by injected charge carriers also contribute strongly to the residual current. Further, the accelerated charge carriers cause turbulence, which tends to disorientate the alignment of the molecules. Filippini [25] has shown that these effects can be mitigated to some extent by the use of electrodialytic electrodes. For pulsed high fields of duration less than $\sim 10^{-4}$ s, the conductivity of nominally pure nitrobenzene is not usually a problem, but it can prohibit the cell being dc biased to reduce the modulating (signal) power requirement.

Although there are theoretical models that allow Kerr constants to be estimated for gases, there is no satisfactory means of quantitatively predicting the values for liquids, because the internal field in a liquid deviates strongly from the external applied field.

Recent studies of a variety of polar substances by Blanchet [28] and Kruger et al. [27] have not revealed any with appreciably superior properties to nitrobenzene for use in Kerr cells. As a consequence of this and of the problems outlined above, and the difficulty of purifying highly polar molecules, Kerr effect liquids remain an unattractive option for many E-O applications.

12.3.7 Electrooptic Ceramics in the (Pb,La) (Zr,Ti)O_3 System

Lanthanum-modified lead zirconate titanate (PLZT) ceramic materials have been investigated since 1969 for their electrooptic properties, and by controlling the La/Zr/Ti ratio both the linear and quadratic [29,30] electrooptic effects have been obtained. PLZT is ferroelectric when exhibiting the linear and quadratic electrooptic effect and Δn follows a hysteresis loop with applied electric field, although this is less apparent in the quadratic cubic phase. It has so far proved difficult to produce material of good quality and experimenters have resorted to the use of thin plates in a longitudinal quadratic electrooptic modulation mode. Large apertures have been obtained [31] using interdigital electrodes for light modulation applications such as eye protective goggles, photographic shutters, and spatial light modulators [32]. Although the refractive index change can be large (e.g., $\sim 0.5 \times 10^{-8}$ m/V for the quadratic material), the response time is generally slow because of the ferroelectric domain orientation processes. The fastest switching time reported to date [31] for an experimental device has been 10 μs, and this in particular makes the material unattractive for most electrooptic beam deflection applications.

12.3.8 Material Selection

Currently, there are only a handful of materials that are suitable for use in E-O deflectors. The requirements are stringent, and many are common to all devices. Material requirements generally include high resistivity ($\gtrsim 10^8$ Ω cm), good homogeneity (refractive index variations of less than $\sim 10^{-6}$), and a large E-O effect ($\Delta n \gtrsim 10^{-4}$).

There are further material requirements for devices operated at high frequency. These often come down to minimizing the effects of heating. Crucial parameters include (1) the dielectric loss, which must be small to minimize heating; (2) the temperature dependence of birefringence, which must also be small for good stability; and (3) the heat conductivity, which must be large to reduce thermal gradients. Often there is a *figure of merit* appropriate to a material when used in a particular device. This figure of merit is used to weight the various material parameters to reflect their relative importance for a particular application and allow selection of the most suitable material available. Where such figures of merit exist, they will be discussed in subsequent sections along with the particular deflector designs and applications for which they are relevant.

12.4 PRINCIPLES OF ELECTROOPTIC DEFLECTORS

12.4.1 Digital Light Deflectors (DLDs)

General

DLDs have discrete deflection positions. The basic deflection cell consists of a polarization modulation element (E-O switch) in conjunction with a birefringent discriminator (polarizer). In operation, the switch can be activated to change the polarization state of an incoming light beam. This results in two possible beam directions from the polarizer corresponding to the ordinary (o) and extraordinary (e) ray directions. Depending on the type of polarizer used, the two beams can be either angularly or linearly displaced.

DLDs possess the following important features:

1. N stages can be cascaded to produce 2^N discrete beam positions.
2. They require only a two-state control voltage and as a consequence can convert binary coded digital voltage pulses directly into optic displacement.
3. The addressable beam directions are independent of fluctuations in control voltage because the directions are established by the polarizers; that is, steering and switching functions are separated.
4. Any beam position can be obtained at random as all switches can be addressed simultaneously. There is no *flyback* as on a CRT (cathode ray tube).
5. A two-dimensional displacement can be achieved with two sets of deflection stages with mutually perpendicular beam displacements.

Digital Deflector Unit

Polarizers. The basic digital deflector unit consists of a polarization discriminator and an E-O switch. Several types of polarization discriminator have been suggested and used in DLD systems. They include a split-angle birefringent plate, a total internal reflection polarizer, a Wollaston prism, and a simple birefringent wedge. These polarizers are shown in Figure 12.5.

Split-Angle Birefringent Plate. This polarizer is a birefringent plate orientated so that the o-ray passes through undeviated, whereas the e-ray is refracted and leaves the element displaced and parallel with the o-ray. With the notation used in Figure 12.5 it can be seen that the separation (b) of the two possible output beams is $l \tan \epsilon$. It can be shown [33,34] that the maximum value of ϵ is given by

$$\epsilon_{\max} = \arctan \frac{n_o^2 - n_e^2}{2n_o n_e} \tag{32}$$

where n_o and n_e are the refractive indices in the birefringent plate for the o- and e-rays, respectively. The values of $\epsilon_{\max}$ are fairly modest. For example $\epsilon_{\max}$ is ~5.9° in calcite and ~9.17° in sodium nitrite [35] at 632.8 nm. The corresponding value of ζ for the orientation of the optic axis is

$$\zeta_{\max} = \arctan \frac{n_e}{n_o} \tag{33}$$

For a noncollimated beam, the extraordinary output component suffers from aberrations in passing through the polarizer. Kulcke et al. [34] have shown that there is an OA orientation (other than $\zeta_{\max}$) that minimizes this aberration.

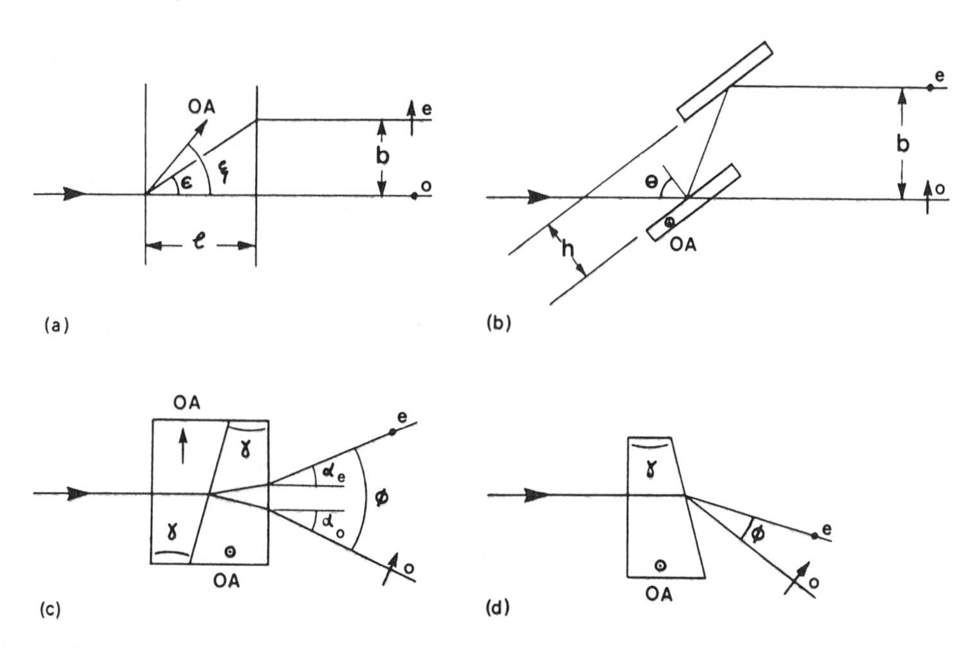

Figure 12.5 Four polarization elements proposed or used in DLD systems: (a) split-angle birefringent plate, (b) total internal reflection polarizer, (c) Wollaston prism, and (d) simple birefringent wedge. Polarization directions (E-vector) are indicated in the figure for the case of a negative uniaxial material, for example, calcite.

Total Internal Reflection (TIR) Polarizer. In this case the polarizer consists of a thin birefringent plate immersed in an optic-quality oil of refractive index (n_m) chosen to match that of the higher of the two indices of the plate. Total internal reflection occurs when the input beam angle (θ) is equal to or larger than the critical angle θ_c, where

$$\sin\theta_c = \frac{n_e}{n_m} \qquad \text{for } n_e < n_o = n_m \tag{34}$$

A second plate, which is either a mirror or an optic element with an index of refraction $n' < n_e$, reflects the TIR beam parallel to the transmitted one. The beam path displacement (b) is $2h \sin\theta$, and the path length difference between the polarization components is $n_m b \tan\theta$.

Wollaston Prism. A Wollaston prism consists of two birefringent wedges cemented together but with their respective OAs orthogonal. As a result an o-ray in the first prism becomes an e-ray in the second and vice versa. This leads to different angles of refraction at the interface for the two input polarizations.

For normal incidence the angle of divergnce ϕ between the two exiting components of the beam from the polarizer is given by [36]

$$\phi \simeq \alpha_o + \alpha_e \simeq 2(n_e - n_o)\tan\gamma \tag{35}$$

For angles of incidence $\alpha_i \simeq 0$, the angle ϕ is very nearly the same for small prism angles (γ), but the two angles α_o and α_e change to $\alpha_o \pm \alpha_i$ and $\alpha_e \pm \alpha_i$, respectively. In a cascaded DLD system these output angles are the angles of incidence for the following deflection stage.

Simple Birefringent Wedge. A birefringent prism that introduces a small angular separation between the o and e beam components can be used as a polarization discriminator. For a right-angle prism of apex angle γ, the beams are separated by ϕ, where

$$\phi \simeq (n_o - n_e)\sin\gamma \tag{36}$$

That is, in the small-angle limit the angular separation is half that of the Wollaston prism. The simple wedge polarizer has the advantage that it reduces the volume requirement of birefringent material. This is particularly important from a cost point of view in the construction of large-aperture deflection cells. A disadvantage is that the output beam components are not deviated symmetrically around the line of the input beam. If required, this can be corrected for by using a glass compensating wedge of refractive index intermediate between that of n_e and n_o of the polarizer [37].

All the polarization discriminators above introduce, to a lesser or greater extent, optic path length differences for the o and e polarization components. Consequently, if the beam in a DLD system is focused, there is a shift in the focal plane for each beam position, the magnitude of which depends on the beam path. Techniques of path length compensation exist so that the source point is projected as an equal-sized image at every output point [34].

If a polarized beam entering a birefringent crystal is not fully collimated, the beam is split into two components. The major component propagates through the crystal according to the polarization direction of the incident beam, whereas the minor part propagates as the complementary beam and appears as background light in the unwanted position, that is, as *crosstalk*. Figure 12.6 is an example of the dependence of this crosstalk on convergence angle (β') for an optimally oriented calcite split-angle polarizer used at 632.8 *nm*. The figure is taken from the paper by Kulcke and co-workers [34]. For most applications deflectors are designed around a maximum crosstalk level of ~1% per stage. From Figure 12.6 we see that this restricts a calcite split-angle polarizer to use with a beam of convergence angle less than ~2.7°.

The calculations used for crosstalk by Kulcke and co-workers are given in an earlier paper [38] for a light beam of oblique incidence on a *z*-cut calcite plate. In this analysis the coordinate system of the incident light beam is given by

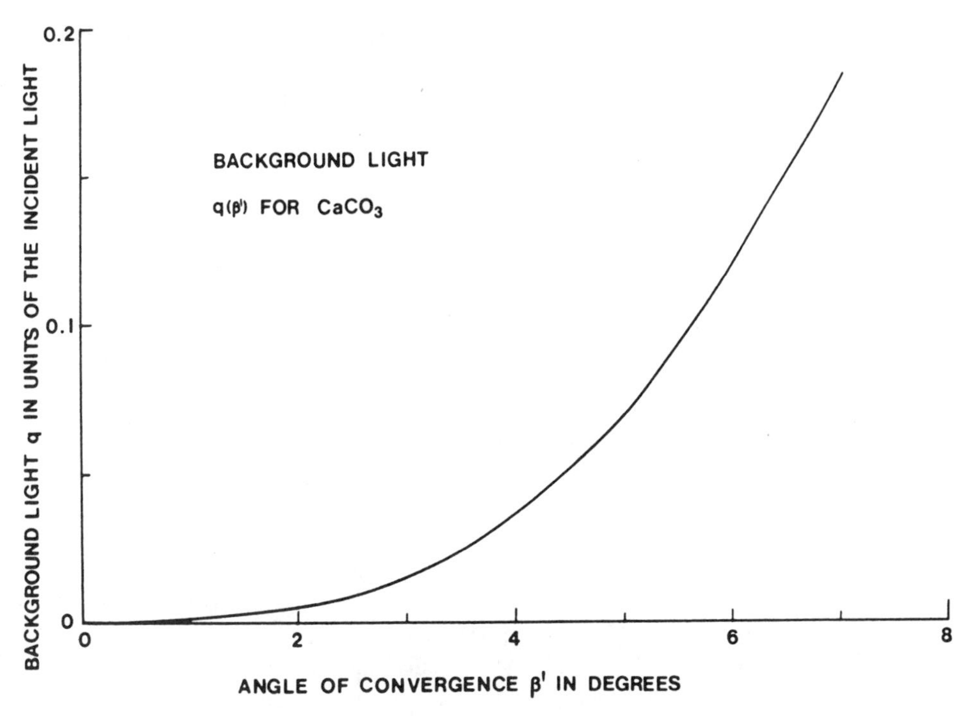

Figure 12.6 Background light q as a function of the maximum convergence angle β' (β' equals light cone semiangle) at 632.8 nm in a calcite split-angle polarizer at optimum orientation of the *OA*. (From Ref. 34).

$x'y'z'$, where x' is parallel to the x axis of the crystal surface and ν is the angle between the surface normal and the z' axis of the light beam. For light polarized along the y' axis the fractional intensity in the wrong channel is given by

$$Ix(\beta') = \frac{1}{2\pi\beta'} \int_0^{2\pi} \int_0^{\beta'} \sin^2 \zeta \, d\eta \, d\beta \tag{37}$$

The angular coordinates $\beta\eta$ form a principal plane which intersects the circular cut of the index ellipsoid of the birefringent crystal under an angle ζ relative to the x axis and

$$\tan \zeta = \frac{-\sin \beta \cos \eta}{\sin \beta \sin \eta \cos \nu + \cos \beta \sin \nu} \tag{38}$$

Electrooptic Switch. The number of E-O materials that can be used in practical DLD systems for the switch element is severely limited. Some Kerr liquids, for example, nitrobenzene, and some linear E-O crystals, such as KD*P, have been used. As with the polarizers discussed earlier, the natural birefringence of the latter class of materials can give rise to crosstalk when used with an angled or convergent beam. A second cause of crosstalk, common to both isotropic and anisotropic switch materials, is error in the control voltage.

For XDP Z-cut linear E-O crystals, Ley et al. [39] have shown that for small input angles (ζ) to the Z crystal axis the background light approximates to

$$\frac{\Delta I}{I} = \left[\frac{\pi l}{2\lambda} \frac{n_o^2 (n_o^2 - n_e^2)}{n_e^2} \zeta^2\right]^2 \tag{39}$$

Here l is the crystal thickness (the other symbols have their usual meaning).

Mechanical strength usually limits the aspect ratio of E-O crystals to less than ~10:1; for example, a 25-mm-aperture crystal will have a minimum thickness of ~2.5 mm. Using the data for KD*P in Table 12.1 for a crystal of this thickness, equation (39) shows that the beam angle ζ must be $\leqslant 0.8°$ for the crosstalk to be $\leqslant 1\%$.

Kulcke et al. [34] have made a similar calculation but for a convergent beam directed along the Z crystal axis. In this case the crosstalk contributions of all the rays at angles ζ, up to the maximum beam convergence angle β', need to be summed. The results of the calculations made by Kulcke and co-workers for KD*P are shown in Figure 12.7 for the particular case where $\beta' = 1.5°$. It can be seen that 1.5° is about the maximum convergence angle that can be tolerated if the crosstalk is to be <1% for a 2.5-mm-thick crystal. This is a more stringent collimation requirement than that for the polarizers.

The crosstalk as a function of control voltage error ($\Delta V / V_{\lambda/2}$) for a DLD with a linear E-O switch is given by

$$\frac{\Delta I}{I} = \sin^2 \left(\frac{\pi}{2} \frac{\Delta V}{V_{\lambda/2}}\right) \simeq \left(\frac{\pi}{2} \frac{\Delta V}{V_{\lambda/2}}\right)^2 \tag{40}$$

That is, the voltage error must be less than ~6% for the beam crosstalk not to exceed 1%. This is not a very stringent requirement. For the case of a material exhibiting the quadratic E-O effect the requirement is even more relaxed; that is,

$$\frac{\Delta I}{I} = \sin^2 \left[\frac{\pi}{2} \left(\frac{\Delta V}{V_{\lambda/2}}\right)^2\right] \simeq \left[\frac{\pi}{2} \left(\frac{\Delta V}{V_{\lambda/2}}\right)^2\right]^2 \tag{41}$$

For $\Delta I/I \simeq 1\%$, the tolerable voltage error is ~25%.

Quadratic E-O materials have the further advantage that they do not exhibit significant piezoelectric effects. These can be quite large in linear E-O materials and lead to cross-talk at high switching frequencies.

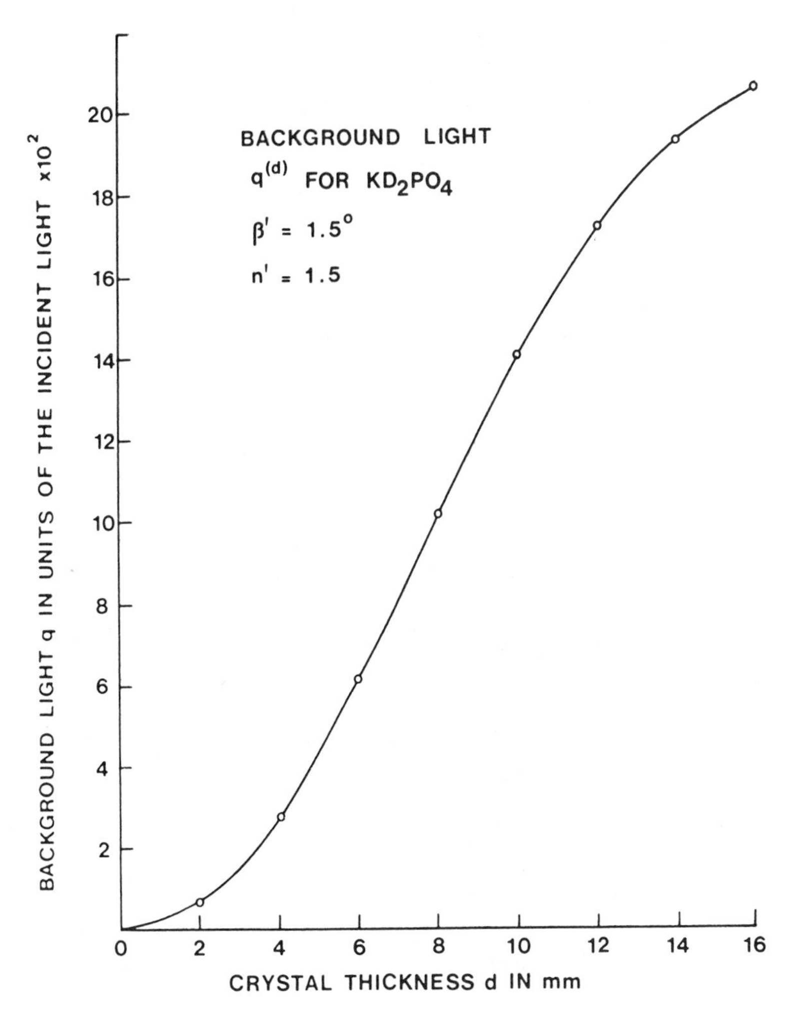

Figure 12.7 Background light q due to a beam with a maximum convergence angle passing through a KD*P crystal immersed in a liquid with a refractive index $n' = 1.5$. (From Ref. 34.)

DLD Arrays

Resolution Limit. In a large DLD system with Wollaston- or calcite-type wedge polarizers the angle ζ between the beam and the switch (e.g., Pockels cell) optic axis increases by a factor of 2 per stage in order to conserve angular resolution. Consequently, crosstalk generally sets an upper limit to the number of stages that can be used.

For an $n \times m$ DLD array (i.e., one with n elements giving horizontal deflection and m giving vertical deflection), $2^n \times 2^m$ accessible output positions are achieved. Assuming that they are equally separated, the nth displacement b_n is given by

$$b_n = 2^{n-1} b_1 \tag{42}$$

where b_1 is the displacement obtained in the stage producing the smallest deflection. From an availability and cost point of view it is important to employ the smallest possible crystals in the deflector system. Therefore, the output spot size and spacing are generally chosen to be of the order of the resolution limit obtainable with a convergent light beam. This requires the beam to be focused on the output face of the last deflector element. Optical relaying is then used for display purposes. Assuming that 1.5° is the maximum beam convergence angle that can be tolerated in the DLD system, the diffraction-limited focal spot diameter (which sets the resolution limit for a perfect DLD system) is ~ 30 μm at 632.8 nm. For a deflector system with an output aperture limited to 30 mm, equation (42) gives 11 as the maximum value of n; that is, a two-dimensional DLD array of this aperture can have a maximum of $2^{11} \times 2^{11} > 10^6$ resolved beam positions.

Optic Performance. To achieve uniform electric fields in DLDs employing large-aspect-ratio Z-cut crystals, resistive electrodes can be applied to the faces through which the optical beam passes. Typically, electrodes of resistivity ~500 Ω square are used. The optical losses that result can be reduced by applying antireflection coatings to the surfaces and by immersion of the crystals in an index-matching fluid. These techniques can result in practically the entire transmission loss of a DLD system being due to electrode absorption. An n-stage system having $2n$ electrodes will have an overall transmission limited to t^{2n}, where t is the transmission of each electrode. Clearly, a DLD system of this design with a large number of deflection stages will have modest transmission even if the electrode absorption is only a few percent.

At low switching frequencies the fraction of the transmitted beam that can be focused within a diffraction-limited spot size depends on the quality of the optical components that comprise the DLD system. Kruger et al. [40] have analyzed the case where an imput beam of Gaussian spatial profile is used with a deflector system that suffers from a Gaussian distribution of refractive index inhomogeneities which are small in size compared to the beam diameter and are evenly distributed across the aperture. They assumed that the index fluctuations were statistically independent of each other. For N inhomogeneities per square centimeter of average cross section σ cm^2, they found that $\sqrt{N}\,\sigma$ was the mean phase variation impressed on the beam and that the output intensity was reduced by a factor $\exp(-N\sigma^2)$. The "lost" radiation was found to be scattered rather broadly. As an example, they calculated that 10% of the input beam energy was scattered if the refractive index variation in a 20-stage DLD system distorted the beam wavefront by more than $\lambda/70$ per stage. For a 50% overall scattering loss, this figure reduces to $\lambda/30$ per stage [41]. These results imply that the very finest optic-quality components must be used in a large DLD system.

High-Frequency Operation. The performance of DLDs at high frequency can be limited by several factors. These include the geometry, heating (i.e., tan δ losses in the E-O material and resistive effects at the electrodes), and the onset of piezoelectric resonances.

Geometry. The aperture limitation of DLD systems requires the total length of the device to be restricted in order to prevent beam *walk-off* and the outermost beam positions losing intensity. Based on the assumption that a loss of 1% is tolerable for these beam positions, Kurtz [42] has derived an expression for the limitation of the capacity-speed product (Rf) of a DLD system using KD*P E-O switches. Here R is the number of light beam directions in one dimension (linear capacity of the DLD) and f is the fundamental frequency of the square-wave voltage by which the polarization switch is operated and which corresponds to a switching frequency of $2f$ light beam positions per second. Kurtz obtained

$$Rf \leqslant \frac{2 \times 10^{-2}}{\sqrt{3}} \frac{lP_r}{\gamma_f \beta l_m \lambda V_{\lambda/2}{}^2 \epsilon_0 \epsilon_z} \tag{43}$$

where β is the ratio of the basic angular unit of deflection to the diffraction angle (ideally $\beta \sim 1$), l_m is the length of a deflection stage, P_r is the reactive drive power for a polarization switch, and l is the thickness of the KD*P crystal plate of relative dielectric constant ϵ_z. (γ_f is a numerical factor, which for square-wave operation is ~2.4.) Application of this expression to a typical example, that is, $P_r \sim 200\ W$, $l \sim 1$ mm, $l_m \sim 7$ mm, $\lambda = 632.8$ nm, $\beta = 1$, and $\epsilon_z = 50$, gives

$$Rf < 2^8 \times 10^4\ s^{-1} \tag{44}$$

That is, an eight-stage DLD system using KD*P switches is limited by its geometry to a maximum operating frequency of ~10 kHz.

Heating. The capacity-speed product given by (43) can only be calculated assuming a value for the maximum permissible reactive drive power P_r to each stage of the DLD. The maximum of P_r is determined by the onset of unacceptable heating and consequent strain effects in the E-O switch material. This limit can be set either by the tan δ losses in the bulk material or by heat generated in resistive electrodes if they are used. For thin crystals in which the temperature gradient is only radially (x-y plane) significant, Pepperl [37] has calculated the maximum tolerable reactive drive power following a model due to Kaminow [43]. For the case of KD*P, a depolarization of ~1% resulting from shear strain occurs if the unit volume power dissipation exceeds ~0.27 W. Since tan $\delta \sim 10^{-3}$ for KD*P, this corresponds to a reactive power of 0.27/tan $\delta \sim 270\ W$.

The resistivity of the electrodes is a second source of heat in switches using linear Z-cut E-O crystals. The power dissipated in this case is given by

$$P = \frac{1}{2}\omega^2 C^2 V_{app}{}^2 R_e \tag{45}$$

where R_e is the square resistance of both electrodes, C is the cell capacitance, and V_{app} the applied voltage of angular frequency $\omega = 2\pi f$. Rearranging (45) yields

$$f = \frac{1}{2\pi C V_{app}} \sqrt{\frac{2P}{R_e}} \tag{46}$$

A reasonable criterion that can be used in setting the upper limit for the acceptable power dissipation in the electrodes is that it should not exceed that due to the tan δ losses in the bulk material. For the example above this gives $P < 4\ mW$. Assuming that $C \sim 100\ pF$, $V_{app} \sim 2\ kV$, and $R_e \sim 500\ \Omega$ are values typical for a switch, (46) gives $f_{\max} \simeq 4$ kHz, that is, a frequency limit similar to that imposed by the tan δ losses.

Piezoelectric Effects. A further constraint on the high-frequency operation of switches using linear E-O materials is their piezoelectric nature. DLD systems with linear E-O switching elements which exhibit significant piezoelectric effect must be operated at low frequencies under the condition of constant stress so that piezoelectric resonances are avoided. Ley et al. [39] have shown that the typical strain relaxation times for ADP and KD*P crystals of order ~1 cm in dimension are in the range 1 to 10 μs. Consequently resonances are avoided by switching at frequencies below ~10 kHz.

The preceding discussion has mainly concerned switches employing linear E-O, Z-cut crystals. Specifically, KD*P has been discussed, as this is a widely used material. Despite this, from a speed and heating point of view, Kerr cells are considerably more attractive; for example, nitrobenzene has a low ϵ and does not exhibit piezoelectric effects so that access times <0.5 μs are obtainable.

Optimizing DLD Design. The capacity-speed product (CSP) which is commonly used as a figure of merit for DLD devices has been defined by (43) for the maximum linear resolution (R) and maximum deflection rate (f) on the assumption that the deflection rate is equal for each stage of the DLD system. Schmidt et al. [41] have considered the case where the system geometry is optimized for minimum drive power to the last deflector stage (N). Their analysis is for switches driven in the transverse mode so that it is applicable to both linear and quadratic E-O materials. For minimum drive power the relations linking the height (d), width (w), and length (l) of the last deflector were found to be

$$d_N = 2.82\sqrt{R\beta l\gamma} \tag{47}$$

and

$$w_N = 3.53\sqrt{R\beta l\lambda} \tag{48}$$

The corresponding capacity-speed product for a DLD system with linear switches driven in the transverse mode configuration was found to be

$$Rf = \frac{P}{4.95\epsilon\epsilon_0 V_{\lambda/2}\beta\lambda} \tag{49}$$

In deriving this last expression it was assumed that the DLD comprised an equal number of x and y stages and that the beam-splitting angle increased in the direction of beam propagation.

For liquid E-O materials that always exhibit the quadratic effect, $V_{\lambda/2}$ in (49) should be replaced by $B/2l$, where B is Kerr's constant defined by (27). In this case a bias voltage can be applied to the polarization switch so that it operates between the mth and $(m + 1)$th $\lambda/2$ wave points. Then the CSP becomes [41]

$$Rf = P\left(\sqrt{m + 1} - \sqrt{m}\right)^{-2} \frac{2B}{\epsilon\epsilon_0} \frac{1}{4.95\beta\lambda l} \tag{50}$$

The advantage of applying a bias voltage is apparent if one considers the simplest case where $m = 1$. This reduces the power required to drive the switch by a factor of 6.

In the limit where $m \gg 1$, (50) becomes

$$Rf = P\frac{(4B)^2}{\epsilon\epsilon_0\beta\lambda}\frac{V_m^{\ 2}}{d^2}\frac{1}{4.95} \tag{51}$$

Here V_m denotes the applied voltage required for an $m\lambda/2$ waves phase difference.

For solid-state quadratic materials with a bias polarization P_b and E-O coefficient g_{ij}, (51) becomes

$$Rf = P\left(\frac{4a}{\lambda}P_b\right)^2 \frac{\epsilon\epsilon_0}{\beta\lambda}\frac{1}{4.95} \tag{52}$$

where, for example, $a = \frac{n^3}{2}(g_{11} - g_{12})$ if the polarization is in the x direction.

For display applications, the stages of small splitting angles in a DLD usually have to be switched at higher rates than those of large splitting angles. In this case Schmidt et al. [41] suggest that it might be of greater importance to design the deflector for higher resolution rather than minimum drive power in the last stage. An optimization of the geometry from this point of view for a two-dimensional deflector led Schmidt et al. to (1) a stage configuration where the y stages precede the x stages and (2) a capacity factor for E-O liquids of

$$R = \frac{V_{app}^{\ 2}}{2} \frac{B}{\beta\lambda} \tag{53}$$

A comparison of the relative CSPs of solid-state DLD systems with performance limited by tan δ heating has been carried out by Kruger et al. [40]. In a DLD system with $R = 2^{10}$ (i.e., 10 stages), they calculated that KTN or $LiNbO_3$ electrooptic switches could be operated at rates of $\sim 10^6$ addresses per second and $LiTaO_3$ at $\sim 10^5$ addresses per second. In contrast the high drive voltage required for KD*P electrooptic switches leads to high dissipated power and (as we have seen) maximum switching rates restricted to the range $< 10^4$/s.

Finally, since the CSP is proportional to $1/\lambda V_{\lambda/2}^{\ 2}$, it is worth noting that there is a considerable advantage in operating near the short-wavelength cutoff of the DLD optic components. In this case, the minimum laser wavelength and beam power are limited by absorption, causing bulk heating. For a DLD system using nitrobenzene Kerr cells and an He-Ne laser operating at a wavelength of 632.8 nm, this heating effect limits the beam power to ~70 to 100 mW [40].

12.4.2 Analog Light Deflectors

Prism and Multiple-Prism Deflectors

Basic Prism Unit. The angle (φ) through which a prism deflects an optic beam is a function of its refractive index (n). Consequently, a change in n, due to the E-O effect, leads to a change in φ. This is the basis of the E-O prism deflector.

The deflection produced by a prism can be calculated by applying Snell's law at the boundaries. The change Δφ in the deflection angle produced by a change Δn in refractive index is given to first order by carrying out the proper differentiation of the resultant equation and yields [44]

$$\Delta\phi = \frac{\Delta n(\tan\theta_1 \cos\theta_1' + \sin\theta_1)}{n_2 \cos\theta_2} \tag{54}$$

$$= \frac{\Delta n(L_a - L_b)}{n_2 W_2} \tag{55}$$

Here n_2 is the refractive index of the medium in which the prism is immersed, and the angles θ_1, θ_2, and θ_1' and the distances L_a, L_b, and W_2 are defined in Figure 12.8. It is usual to express the deflection Δφ in terms of *number of resolvable spots* (N_R). For the ideal case of a beam with diffraction-limited divergence (θ_R), the Rayleigh criterion gives

$$\theta_R = \frac{\epsilon\lambda}{n_2 W_2} \tag{56}$$

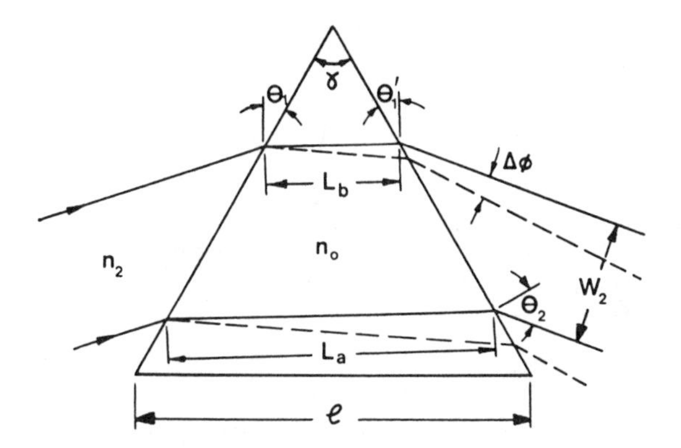

Figure 12.8 Beam deflection by an E-O prism element. (From Ref. 44.)

where λ is the free-space optic wavelength and ϵ is a factor ($\gtrsim 1$) that depends on the beam spatial intensity distribution. Consequently, N_R is given by

$$N_R = \frac{\Delta\phi}{\theta_R} = \frac{\Delta n(L_a - L_b)}{\epsilon\lambda} \tag{57}$$

We see that under the most favorable conditions ($\epsilon = 1$), N_R is numerically equal to the induced optic path difference for the marginal rays of the beam in the prism, expressed in number of waves.

It can be shown that N_R is maximized by using the prism at minimum deviation and at full aperture [45]. In this case, $L_a - L_b$ equals the prism base length (l) and N_R is independent of the prism apex angle (γ). To minimize the E-O material requirement, γ should be large but kept below $2 \sin^{-1}(n_2/n)$ to avoid TIR losses.

The deflection changes provided by a single prism of E-O material are generally rather modest. For example, in a typical case $\Delta n \simeq 10^{-4}$, $l \simeq 1$ cm, $\lambda \simeq 0.5$ μm, and (57) gives $N_R \simeq 2$. Consequently, for practical deflectors, multiple-prism devices are generally constructed.

Multiple-Prism Deflector. Equation (57) shows that the number of resolved beam positions from a prism deflector is proportional to the induced optic path difference for the marginal rays of the beam. This quantity can be increased by a series arrangement of E-O deflector prisms. By alternating the crystal orientation or field in consecutive prisms, the zero-field deflection of the device can be eliminated. A schematic of such an arrangement is shown in Figure 12.9. In this case, the total internal accumulated deflection can be regarded as the sum of the separate deflections resulting from the upright prisms of refractive index $n_0 + \Delta n$ that are located in a medium of index $n_0 - \Delta n$. The smallness of the E-O deflection usually allows the beam to be treated as near parallel to the base, and

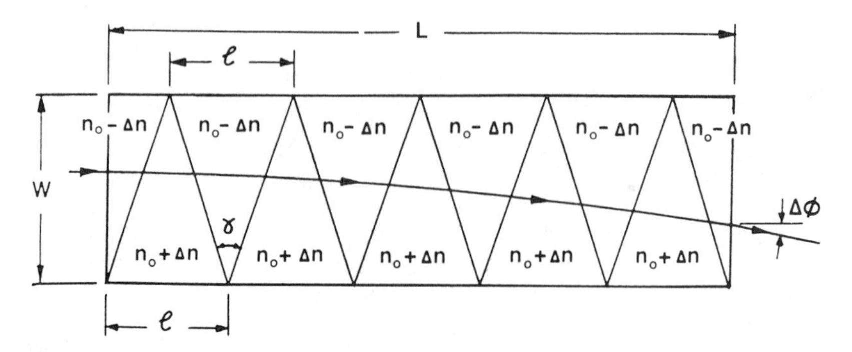

Figure 12.9 Multiple-prism deflector comprising a series arrangement of E-O prism elements.

hence the minimum deviation condition applies in each prism. As a result, $\Delta\phi$ is given by [46]

$$\Delta\phi = \frac{2L\Delta n}{Wn_2} \tag{58}$$

where L is the deflector length and W the aperture. Consequently, the maximum number of resolved beam positions N_R is

$$N_R = \frac{\Delta\phi}{\theta_R} = \frac{2L\,\Delta n}{\epsilon\lambda} \tag{59}$$

That is, in a cascaded deflector, N_R is increased by a factor $2L/l$ over that for a single-prism device.

Optimized Design. The limit to the number of prisms that can be cascaded in a series deflector is set by the criterion that the beam must not hit the edge of the exit aperture of the deflector. As with DLDs, this requirement naturally leads to the use of weakly focused optic beams. For the case of a deflector in which the total length of the light path is L and which has mirror symmetry about $L/2$, the rays of the deflected beam, when extended backwards, intersect with the undeflected rays at the $L/2$ plane. This is indicated schematically in Fig. 12.10, where, for convenience, the deflector has been considered to be immersed in an index-matching medium. With the beam focused at a distance L_1 beyond the exit aperture of the deflector, the deflection distance is $(L/2 + L_1)\,\Delta\phi$, and the resolved spot size is $\theta_R(L + L_1)$. Consequencly, for a deflector used with a focused beam, N_R is given by

$$N_R = \frac{\Delta\phi}{\theta_R}\left[1 - \frac{L}{2(L + L_1)}\right] \tag{60}$$

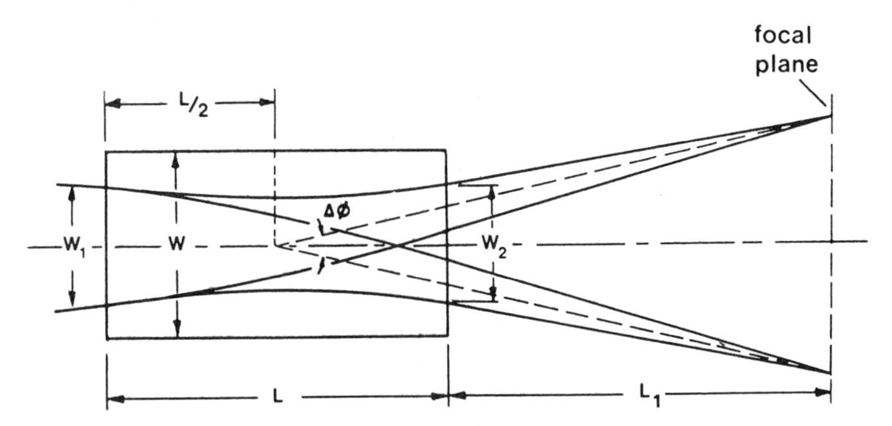

Figure 12.10 Prefocusing to avoid vignetting in an E-O deflector. (From Ref. 44.)

That is, as a result of focusing the beam to prevent vignetting, the number of resolvable spots is reduced by the factor in brackets.

The relationship between $\Delta\phi_m$ (the maximum value of $\Delta\phi$) and L_1, which through (60) sets the maximum value of N_R, has been investigated by Lee and Zook [44]. From Figure 12.10, ray optics gives

$$W_2 = \frac{\Delta\phi_m L}{2} + \frac{W_1 L_1}{L + L_1} \tag{61}$$

By eliminating L_1 from (60) and (61) and using $\theta_R = \lambda\epsilon/nW_1$, they obtained

$$N_R = \frac{n\Delta\phi_m}{2\epsilon\lambda}\left(W_1 + W_2 - \frac{\Delta\phi_m L}{2}\right) \tag{62}$$

This expression shows that N_R is a maximum when W_1 and W_2 are a maximum, that is, when $W_1 = W_2 = W$. Therefore,

$$N_R(\max) = \frac{\Delta\phi_m}{\theta_R}\left(1 - \frac{\Delta\phi_m L}{4W}\right) \tag{63}$$

Additionally, (61) can now be rearranged to give the beam focusing required to achieve $N_R(\max)$; that is,

$$\frac{L + L_1}{W} = \frac{2}{\Delta\phi_m} \tag{64}$$

Generally, only very weak focusing is required. As an example, if the deflector is capable of achieving $\Delta\phi_m \simeq 50$ mrad, equation(64) gives the focusing required as $\sim F/40$. For a typical deflector with an aspect ratio (L/W) of say 10, equation (63) shows that this focusing of the beam reduces N_R by only $\sim 12.5\%$ from that which could be obtained by using a collimated beam of width W in a deflector not limited in aperture.

Equations (63) and (64) optimize the deflector performance for a given maximum deflection $\Delta\phi_m$. If this angle is not limited by the maximum field that can be applied to the deflector, it is limited by the deflector aperture under the condition $L_1 = 0$–that is, when the beam is focused on the deflector exit aperture. Although under this geometrical limit the beam can still be considered quasi-parallel in each prism element, the beam diameter reduces such that in the last prism, where the beam is focused, the width is near zero. Hence, the induced optical path difference for the marginal rays tends toward zero in the prism elements near the deflector exit face. In terms of this limit, Lee and Zook [44] showed that N_R was given by

$$N_R = \frac{\Delta\phi_m}{\theta_R}\left(1 - \frac{1}{2}\beta^2\right) \tag{65}$$

Here $\beta = (L/W)/(L/W)_G$, where $(L/W)_G$ is the geometrically limited apect ratio for the deflection $\Delta\phi_m$.

To maximize N_R in (65) it was found necessary to fix some dimensional parameter for the deflector, for example, the aperture (W), area (LW), or volume. For these three specific cases, Lee and Zook [44] showed that N_R was maximized when β equaled $\sqrt{2/3}$, $\sqrt{2/5}$, and $\sqrt{1/3}$, respectively. In the last case, where N_R is maximized for a given volume, the deflector cost for a given performance is minimized, and the energy required to deflect the beam in a given time (which is proportional to volume but independent of shape) is also minimized.

In a deflector where $\Delta\phi_m$ is field- rather than geometry-limited, multiple passing can be used to increase N_R by increasing the deflector angle up to the optimum (β) limit. In a practical device, other factors, such as optical aberrations, reflection losses, or extractability of the deflected beam, will generally limit the number of passes that can be made.

A further parameter that affects N_R, and over which some control is possible, is the beam spatial intensity distribution. Equation (56) gives the diffraction-limited resolution angle θ_R in terms of the factor ϵ. For the case of a beam of uniformly intense spatial distribution used with a square aperture, $\epsilon = 1.0$. If the aperture is circular, $\epsilon \simeq 1.22$. Uniform illumination maximizes N_R, but in the far field diffracted energy produces crosstalk between adjacent beam positions. For the above apertures, this crosstalk is ~4.7 and ~1.8%, respectively [47]. The effect is less severe if a truncated gaussian beam is used. In particular, for a beam apertured at the $1/e^2$ intensity points, the crosstalk figures reduce to ~0.8 and ~0.5% for the two cases, respectively [44,48]. The penalty that has to be paid is an increase in ϵ to ~1.25 and ~1.27 for the respective apertures and consequent proportional decreases in N_R.

The above discussion has considered a perfect optic system used in conjunction with a focused beam of diffraction-limited divergence. In the real case, aberrations result from the use of a convergent beam with a deflector, and Beiser [49] has examined the effect of these on the resolution. Using ray optics, Beiser derived an expression for the focal spot size (d_a) due to the aberrations alone in terms of the *F/No.* and deflector refractive index. With the deflector immersed in an index-matching medium, this reduces to

$$d_a = \frac{n\Delta\phi(L/2 + L_1)}{16F^2} \tag{66}$$

Where the device is not in an index-matching medium, additional aberration occurs at the deflector boundaries, and for this case Beiser found that d_a was twice as large.

The beam aberration will have little effect on the resolution so long as d_a is considerably smaller than the diffraction-limited spot size. If we let this ratio be δ, then under conditions where (63) applies, we have

$$\delta = \frac{n^2\,\Delta\phi_m W}{16F^2\epsilon\lambda} \tag{67}$$

For the particular example given earlier, where $\Delta\phi_m = 50$ mrad, $F/No. = 40$, and using $n \simeq 1.5$, $\epsilon \simeq 1$, and $\lambda = 632.8\ nm$, we obtain

$$\delta = 0.07\ W\ \text{cm}^{-1} \tag{68}$$

We see from this result that use of a weakly focused beam with a deflector of 1 or 2 cm in aperture does not significantly degrade the resolution. Additional care is needed when working with smaller *F/No.*'s, as besides causing loss of resolution, all aberrations adversely affect the crosstalk level of the deflector.

Figures of merit have been derived for both spatial resolution (M_R) and temporal response (M_τ) for linear E-O prism deflectors by Thomas [50]. For devices of square cross section he obtained

$$M_R = \frac{\epsilon_r\lambda N_R z}{wV_{\text{app}}} \quad \text{and} \quad M_\tau = \frac{\epsilon_r\lambda\epsilon_0 z}{I\tau} \tag{69}$$

where V_{app} is the potential applied across the deflector (in the z direction) to produce N_R resolved deflection positions, w is the usable aperture (i.e., $w < z$), and I is the current required by the device to allow adjacent spots to be resolved in the time τ. Thomas reviewed available linear E-O materials for use in a deflector, and his results are presented in Table 12.2.

For quadratic E-O materials, orientated so that the effective E-O coefficient was $g_{11} - g_{12}$, he derived comparative figures of merit:

$$M_R = \epsilon_0^2\,(\epsilon_r - 1)^2 n_0^3\,(g_{11} - g_{12})E_z$$

and (70)

$$M_\tau = \epsilon_0^2\epsilon_r n_0^3\,(g_{11} - g_{12})E_z$$

Table 12.2 Summary of Some Possible Materials for Linear Deflector Use

Material	E-O coefficient r (10^{-12} m/V)	Relative dielectric constant ϵ_r or ϵ_1, ϵ_3	n_0	Figure of merit: M_R, 10^{-9} m/V >1.08	Figure of merit: M_τ, 10^{-12} m/V >9.5	Comment
ADP	24.5 (r_{41})	56, 15	1.5	0.083	1.5	
KDP	10.3 (r_{63})	44, 21	1.5	0.035	0.79	
KD*P	26.4 (r_{63})	58, 50	1.5	0.089	1.5	
$LiNbO_3$	30.8 (r_{33})	80, 30	2.2	0.328	4.1	
$LiTaO_3$	30.3 (r_{33})	42.8	2.14	0.297	6.9	
KTP	35 (r_{33}	13	1.86	0.225	17.3	
BBO	2.5 (r_{22})	?	1.67	0.012	?	
$BaTiO_3$ const. strain (clamped)	840 (r_{42})	1970, 11	2.4	11.6	5.9	Cannot now be reliably made ($BaTiO_3$, AMO, KTN)
AMO	327 (r_{52})	17	1.5	1.1	65	
KTN	14,000 (r_{42})	Very high, ~5500	2.3	170	31	
HIO_3	?	20, 11	2	?	?	Very little known

Source: Refs. 50, 51.

Here ϵ_r is the relative dielectric constant for the crystal material, and E_z is a field of the same strength as used in the linear E-O material case. Generally, quadratic (including Kerr effect) materials were shown to have lower figures of merit than the best linear E-O materials in Table 12.2.

Finally it should be noted that, as with digital devices, two-dimensional deflection can be obtained with analog deflectors [44,45,52]. In this case, the second-stage deflection element, producing orthogonal deflection to the first, needs to be of comparable aperture in both deflection directions. As a consequence, a considerably larger deflection voltage and drive power are needed than in the first stage. Although Lee and Zook suggest a technique involving the use of thin crystal elements with interspersed electrodes to overcome this problem in the second stage, digital deflectors are more attractive for many two-dimensional applications, particularly those requiring a large number of resolved beam positions. As a consequence, little work has been published in the literature on comparable two-dimensional analog devices.

Refractive Index Gradient Deflectors

Operating Principle. Whereas homogeneous refractive index changes only result in optical deflection at prism interfaces, gradiential changes also produce deflection in the bulk material. Two common examples of the latter effect are the bending of radio waves in the ionosphere and the refraction of the sun's rays in the earth's atmosphere.

Deflection by an E-O crystal, due to a refractive index gradient, was first reported by Fowler et al. [53]. They showed that a linear refractive index gradient, transverse to the direction of optical beam propagation, results in a deflection that is proportional to both the index gradient and interaction length. This can be seen with the help of Figure 12.11.

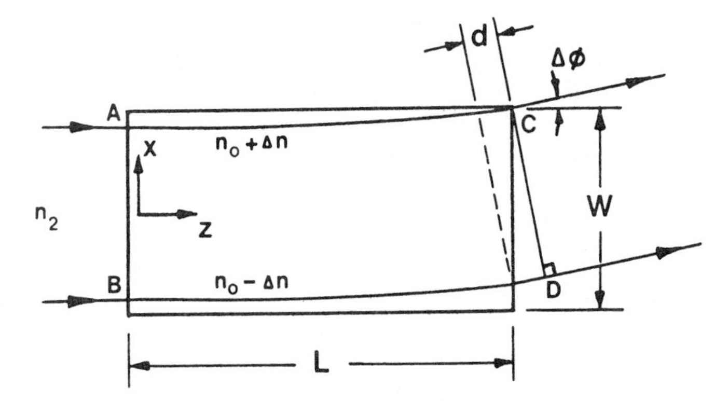

Figure 12.11 Passage of an optical beam through a gradient deflector. (From Ref. 53.)

Fowler et al. assumed that the refractive index changes induced in the medium were so small that the light rays traveled in nearly straight lines. For a beam incident normally on the end face of the deflector, rays *A* and *B* arrive in phase but travel through the medium with different velocities. If the refractive indices for the two marginal rays are $n_0 + \Delta n$ and $n_0 - \Delta n$, respectively, a phase difference $\Delta\gamma$ between them of $4\pi\ \Delta nL/\lambda$ will have accumulated at the output face. As a result, the output beam wavefront becomes *CD*, where *d* in the figure is a distance equivalent to the phase lag $\Delta\gamma$; that is,

$$\Delta\gamma = \frac{2\pi n_2 d}{\lambda_0} = \frac{4\pi\ \Delta nL}{\lambda_0} \tag{71}$$

Therefore,

$$d = \frac{2L\ \Delta n}{n_2} \tag{72}$$

The above analysis considers the *small-angle* limit, and consequently the deflection $\Delta\phi$ is given by

$$\Delta\phi \simeq \sin \Delta\phi \simeq \frac{2L\ \Delta n}{n_2 W} \tag{73}$$

By using (56), the number of resolved beam positions (N_R) is therefore

$$N_R = \frac{\Delta\phi}{\theta_R} = \frac{2L\ \Delta n}{\epsilon\lambda} \tag{74}$$

It can be seen that the last two results are identical to those given by (58) and (59) for the deflection and number of resolved beam positions, respectively, of an E-O prism-type deflector.

A more rigorous analysis of beam propagation in a gradient deflector has been carried out by Grieb et al. [54] using the calculus of variations (see, for example, Born and Wolf [47] and Stavroudis [55]). This approach is a generalization of ordinary geometrical optics and is the mathematical basis for Fermat's principle. That is, in the coordinates used here, it essentially reduces the problem to finding the ray path for which the integral I given by

$$I = \int_0^L n \left(1 + \left|\frac{dx}{dz}\right|^2\right)^{1/2} dz \tag{75}$$

is an extremum (i.e., maximum or minimum).

Grieb et al. considered the two distinct cases where the *E* vector of the incident beam was either perpendicular to or in the plane of deflection (in the former case, the bending of the beam does not affect the polarization orientation, but in the latter case it does). Grieb and co-workers showed that, in the small-angle limit, both beam polarizations resulted in near parabolic ray trajectories, confirming the accuracy of the linear result given by (73). They deduced the ratio (δ) of the output to input beam divergence and obtained

$$\delta = \left[1 - \left(\frac{\Delta nL}{n_0 W}\right)^2\right] \tag{76}$$

For achievable refractive index changes (i.e., $\Delta n \lesssim 10^{-3}$) this ratio is within ~0.01% of unity. Consequently, a gradient deflector has negligible effect on the beam divergence. This result is implicit in (73), where it can be seen that beam deflection is only a function of L. That is, in the small-angle limit, deflection does not depend on ray position in the wavefront.

Grieb et al. generalized their analysis of the gradient deflector and considered the case where the refractive index was some arbitrary function of the transverse coordinate x. They found that the resultant deflection was determined only by the relative change in refractive index over the beam width. Specifically, for the case where the refractive index varied according to x^q ($q > 1$), they calculated that the deflector dimensions required to obtain a given deflection became smaller than in the linear gradient ($q = 1$) case but that the beam aberrations were worse.

Electrode Profile. Electrodes in a quadrupole arrangement are generally employed with a gradiential deflector. To provide a linear refractive index gradient in a linear E-O material, they need to be of hyperbolic profile [46, 56]. Electrodes of this section are difficult to fabricate with the accuracy required. Consequently, calculations have been carried out to determine the effect on field linearity of replacing them by simple cylindrical electrodes. Figure 12.12 shows some computed curves of the field error ($\Delta E/E$) resulting from the use of cylindrical electrodes (radius R) to match a hyperbolic profile [56]. Although the results presented in the figure relate specifically to the hyperbola $xz = 1$ (where x and z are in millimeters), the dimensions can be scaled linearly to fit a deflector of arbitrary aperture. The curves show that use of cylindrical electrodes with a gradient deflector can result in a substantial (>10%) field error. As a result of this error the wavefront of the optical beam is distorted as it propagates through the deflector. For a beam deflected through N_R spots, the accumulated wavefront error ($\Delta\lambda$) after transit through the crystal is $(2N_R \epsilon x/w)(\Delta E/E)$ wavelengths. Figure 12.13 is a plot of this error for the specific case discussed above with $R = 2.0$ mm, $w = 3.0$ mm, and N_R taken as 40 [56]. Although the field distortion leads to a wavefront error of several wavelengths, the figure shows that it can be represented by a linear component (that simply adds extra deflection to the beam) and a much smaller component, which in this example is only of order $\lambda/2$.

Similar conclusions concerning the relative merits of cylindrical and hyperbolic electrodes for a gradient deflector have been drawn by Sevruk and Gusak [57] but for a different reason. They examined in detail the effect of finite deflector dimensions on the field linearity. The calculations, based on the numerical solution

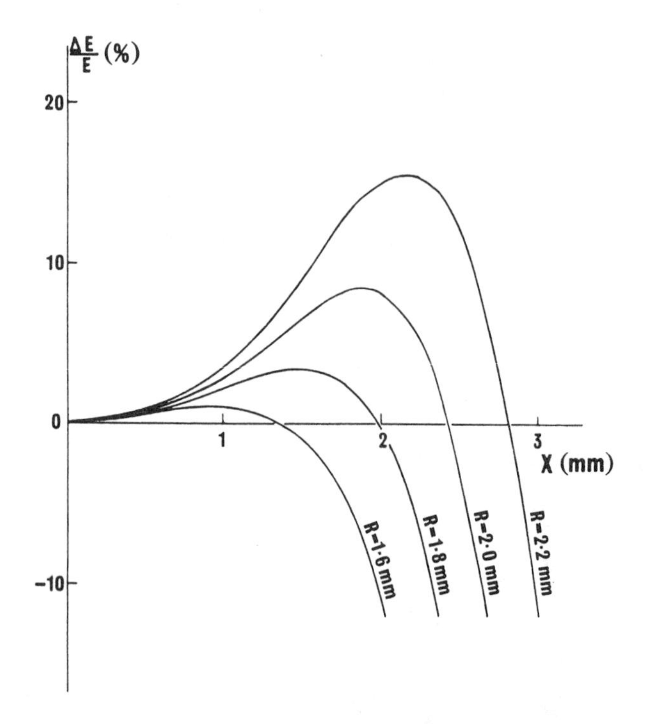

Figure 12.12 Field error ($\Delta E/E$) as a function of X arising from the use of electrodes of cylindrical section (radius R) rather than of hyperbolic profile to form a quadrupole field for a crystal deflector. (From Ref. 56.)

of the Laplace equation, showed that replacement of hyperbolic electrodes by ones of circular section did not necessarily lead to appreciable loss of deflector ones of optimum circular section. For the particular case of a deflector performance. This conclusion is a result of edge effects in a deflector of limited transverse dimension distorting the (ideal) field of hyperbolic electrodes to such an extent as to render them of no advantage over ones of optimum circular reaction. For the particular case of a deflector in air and with transverse dimensions limited to 1.5 times the clear aperture, they computed the radius of circular section electrodes that minimized the average field error over the circular beam aperture. For three commonly used crystal orientations, Sevruk and Gusak showed that an appropriate choice of electrode radius could reduce the average field error to only a few percent.

Frequency Response. In a deflector used for scanning or streak applications the refractive index is a function of both transverse coordinate and time; that is, $n = n(xt)$. The calculus of variations can be used to show that for streak applications the linear approximation for the deflection is still valid. This is not

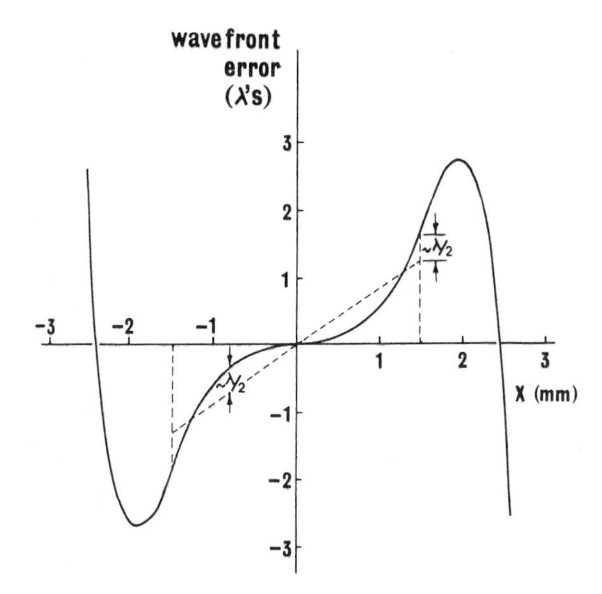

Figure 12.13 Wavefront error resulting from single transit of a beam through a gradient deflector with cylindrical electrodes. In this case, $R = 2.0$ mm, $w = 3.0$ mm, and N_R has been taken as 40. The error results in a small additional beam deflection plus a $\sim\lambda/2$ wavefront aberration. (From Ref. 56.)

intuitively the case, for at high sweep speeds the expected time resolution (time to sweep through one resolvable spot) can be less than the transit time of the beam through the deflector, and this could possibly be significant.

For a deflector with a linear refractive index gradient which is also a linear function of time, the refractive index can be written as

$$n(xt) = n_0(1 + \alpha xt) \tag{77}$$

where n_0 and α are constants. If $t = t_0 + t_z$, where t_0 is the time at which a ray reaches the input face of the deflector after the start of the linear field ramp and t_z is the time taken by the ray to reach the point xz in the crystal, then in the small-angle limit $t_z = zn_0/c$. Consequently, (77) becomes

$$n(xz) = n_0 \left[1 + \alpha x \left(t_0 + \frac{n_0 z}{c} \right) \right] \tag{78}$$

Here the refractive index is only a function of spatial coordinates. The integral I given by (75) can be formed and the integrand used to set up the Euler equation from which the external space curve, representing the ray trajectory in the deflector, can be found. The gradient at the output face is equal to $(n_2/n_0)\ \Delta\phi$.

Consequently, differentiation with respect to t_0 at $z = L$ gives the linearity of the deflector response. This procedure yields

$$\frac{d(\Delta\phi)}{dt_0} = \frac{n_0\alpha L}{n_2}\left(1 - \frac{1}{2}\alpha^2L^2t_0^{\,2}\right) \tag{79}$$

To assess its significance, the last term in (79) can be determined for a typical deflector. For a ~5-cm-long device of a few millimeters in aperture achieving a refractive index change of 10^{-3} in say 5 ns, the term is <0.01. This result shows that a gradient deflector can be operated at high streak speeds corresponding to a few picoseconds of resolution without the optical transit time leading to nonlinear performance.

In the case of an oscillatory applied field, as in a device used for scanning, the deflector cannot respond in times which are comparable to, or less than, the optical transit time through the crystal. This limit arises since the final deflection is the result of an *integrating* effect as the optic beam passes through the crystal.

For the example of the ~5-cm-long deflector discussed above and made (say) from $LiNbO_3$, the maximum scanning frequency would be restricted to <500 MHz. Correspondingly, in a streak application, the linear field ramp must be applied for a period long compared with that of the optic transit time to achieve the linear performance limit of (79).

In the case of a prism deflector a similar constraint on scanning frequency applies. The equivalent streak case has been examined in detail by Elliot and Shaw [58]. The fact that the deflection is localized at the interface of the two prisms means that different parts of the wavefront of the propagating beam are refracted at different times after they enter the deflector. As a result, application of a voltage ramp produces the greatest deflection in the parts of the wavefront reaching the interface latest. This leads to wavefront curvature. In scanning applications, the same frequency limit as that given above for the gradient deflector applies, since otherwise the wavefront curvature becomes excessively aberrated.

As a result of the finite transit time, attempts to operate a gradient deflector well above the maximum frequency limit would result in a static, undisplaced spot. For a simple prism deflector the result would be an extended line image over the total scan range.

The field transit time is a further important consideration with implications for the high-frequency performance of an analog deflector. Table 12.2 shows that relative dielectric constants of ~50 are typical of linear E-O crystals. Consequently, the field propagation velocity is usually considerably slower than that of the optic beam. Techniques, such as matching the deflector into a broadband transmission line, can help distribute the current more uniformly at high frequencies. For some E-O materials it is possible to achieve collinear and synchronous propagation of the optic and electric fields [58–60]. This makes a device which is potentially very broad band.

Optimized Design. The comments made earlier concerning optimization of the design of a prism-type analog deflector are generally applicable to a gradient device. One difference is the figure of merit for temporal response given in (69). In a gradient deflector, the capacitance depends on the particular electrode geometry used and is usually greater than in the equivalent iterated prism device. In a quadrupole arrangement, the electrodes generate a field gradient in the direction orthogonal to the deflection field, and consequently the relative dielectric constant of the E-O material in this direction is equally important when assessing the temporal response of the device and drive power requirement.

Analog-Digital Array Deflectors

In the subsection "Prism and Multiple-Prism Deflectors" it was shown how a series arrangement of analog deflector elements can be used to increase N_R by increasing $\Delta\phi$. From the definition of N_R in that section it can be seen that an alternative means of increasing N_R is to reduce the beam diffraction angle θ_R. As the minimum value of θ_R is determined by the beamwidth, this approach implies an increase in deflector aperture.

Figure 12.14 is an example showing how a parallel array of prism-type deflector elements can be used to make a large-aperture device [61]. In this case, each deflector element produces a deflection of $\Delta\phi_1$ given by equation (58):

$$\Delta\phi_1 = \frac{2\,\Delta n l}{n_2 W} \tag{80}$$

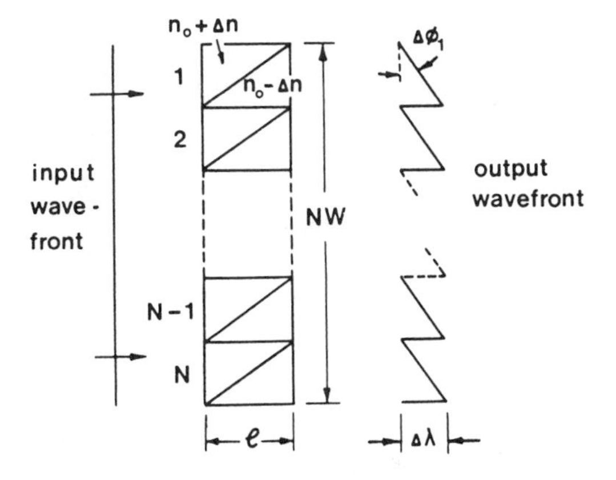

Figure 12.14 Analog-digital array deflector using prism-type deflector elements. (From Ref. 61.)

Since there are N parallel elements, the minimum diffraction angle θ_R is given by

$$\theta_R = \frac{\epsilon\lambda}{NWn_2} \tag{81}$$

Consequently, the maximum number of resolved beam positions that it is possible to achieve is

$$N_R = \frac{\Delta\phi_1}{\theta_R} = \frac{2\,\Delta nl}{\epsilon\lambda} N \tag{82}$$

that is, N times that of a single-prism element.

It can be seen from the figure that the parallel array deflector behaves as a phase grating, and in general the phase of the transmitted beam is discontinuous. Ninomiya [61] calculated the Fraunhofer (farfield) diffraction pattern expected from this deflector and came to the following conclusions:

1. Without bias voltages on the deflector elements, the deflection $\Delta\phi$ is discontinuous.
2. When $\Delta\lambda = M\lambda$ (M equal to a positive integer), the beam is deflected perfectly in the direction $\Delta\phi$, given by

$$\Delta\phi = \frac{M\lambda}{Wn_2} \tag{83}$$

3. Continuous deflection requires a bias voltage, increasing in increments, to each element in the array such that for the mth element it produces a retardation $\Delta\lambda_m$ given by

$$\Delta\lambda_m = (m - 1)\,\Delta\lambda - q\lambda \tag{84}$$

where q is zero or an integer.

4. Discrete deflection angles $\Delta\phi_t$, given by

$$\Delta\phi_t = \frac{t\lambda}{NWn_2} \tag{85}$$

where t is an integer, result if the bias voltage produces a retardation $\Delta\lambda_m$, where

$$\Delta\lambda_m = \frac{t}{N}(m - 1)\,\lambda - q\lambda \tag{86}$$

In the digital case (4), the angle between adjacent deflection positions is λ/NWn_2 and can be considered a *digital unit*. However, since the smallest resolved angle is θ_R, given by (81) with $\epsilon > 1$, an angle of two-digit units is resolvable, but generally one is not.

Parallel array deflectors are potentially attractive in applications requiring either high deflection sensitivity or a large number of deflection positions, either analog or digital. In particular, the planar geometry makes these devices suitable for use in integrated optic circuits. In this application, diffusion techniques can be used to make a shallow waveguide layer of depth ~ 100 μm in E-O materials and surface electrodes used to provide the deflection field [62, 63] (see Section 12.5).

In a recent review, Bulmer et al. [64] have examined some possible geometries for waveguide array deflectors. They found that the far-field intensity distributions are given by the product of two terms. One is an array function (*AF*), which is determined by the period of the array and the number of elements it contains and is independent of the phase distribution within any one element. The other is the element function (*EF*), which is the square of the magnitude of the Fourier transform of the transmission function at the output aperture of one element. Whether the device produces a deflection that is continuous or not depends on whether the *EF* or *AF*, or both, is displaced as a function of voltage. If a phase slope is created across the wave propagation through one element, as for the deflector in Figure 12.14, the resulting element pattern is deflected as a function of the voltage inducing the phase slope. If there is a phase difference between consecutive elements in an array, the array pattern is deflected as a function of that phase difference. This is the case for the prism array when bias voltages are applied to the individual deflector elements. Bulmer and co-workers tabulated the type of deflection expected from the different geometries that they investigated.

For a prism array, Revelli [65] has shown that the geometric limit to the number of resolvable spots $N_R(\max)$ is set by total internal reflection at the ouput face. He obtained

$$N_R(\max) = \frac{2}{\theta_R}\left(\frac{F-1}{F+1}\right)^{1/2} \tag{87}$$

where $F = n(n^2 - 1)^{1/2}$ and n is the refractive index of the crystal. When N_R is large, the small-angle minimum deviation approximation of the section on prism and multiple-prism deflectors is no longer valid and (56) must be used for θ_R. This makes the effective aperture of the device a function of the deflection angle. As an example of the limit represented by (87), Revelli considered an $LiNbO_3$ deflector with aperture $NW = 1$ cm. For $\Delta n/n = 10^{-3}$, he found $N_R(\max) \simeq 10^3$.

12.4.3 Other E-O Deflection Techniques

Although the digital and analog light deflectors discussed in the previous sections have received the most attention for high-frequency deflection and scanning applications, several other E-O deflector schemes have been demonstrated or proposed in recent years. The following are some of the more interesting.

Analog Deflector Using Frequency Shifting

A novel deflector involving frequency shifting has been proposed and demonstrated by Wilkerson and Casperson [66]. In this device an E-O crystal is used to impress a frequency shift on an optical beam which is subsequently deflected by a dispersive element, that is, a prism, diffraction grating, or Fabry-Perot etalon. Although similar to a *DLD* in that it is a two-element device with the switching and deflection functions separated, the deflection in this case can be continuous since arbitrary shifts are possible.

In a linear E-O material a frequency shift is obtained by applying a linear voltage ramp. The resultant refractive index sweep causes the phase of an optical pulse propagating through the material to vary linearly with time. Since a linear phase variation is equivalent to a frequency shift, the pulse leaving the crystal is centered at a different frequency from that when it entered. The dispersive element following the frequency shifter deflects the beam by an amount which depends on the frequency shift achieved.

For a frequency shifter with a weak time-dependent index of refraction of the form

$$n = n_0 - \Delta n \sin \omega t \tag{88}$$

where w is the angular frequency of the modulator, it can be easily shown that for a portion of the light pulse crossing the center of the crystal at time t the instantaneous frequency shift is

$$\Delta\nu(t) = \frac{\omega\nu d\,\Delta n}{c}\,\frac{\sin(2\omega T)}{2\omega T}\cos \omega t \tag{89}$$

where ν is the central frequency of the optical beam, d is the length of the crystal, and T is the transit time. If the light pulse is short compared to the *RF* modulation period and is synchronized to cross the crystal when ωt is an integral multiple of 2π, then $\cos \omega t$ may be approximated by unity, and the entire pulse is upshifted by the maximum amount. The factor $\sin(2\omega T)/2\omega T$ is also equal to unity if the optic transit time is short compared to the modulation period or if appropriately matched traveling-wave modulation fields are employed. Consequently, the maximum frequency shift is

$$\Delta\nu \simeq \frac{\omega\nu d\,\Delta n}{c} \tag{90}$$

If the dispersive element is chosen to just resolve one full bandwidth of the optical pulse spectrum, then the number of resolvable spots for the beam scan is just the number of spectral bandwidths that can be fitted into the total frequency shift given by (90). For an optic pulse of duration t_p that is transform-limited, that is, $\nu_p t_p \simeq 1$, the number of resolvable spots (N_R) becomes

$$N_R = \frac{\Delta\nu}{\nu_p} \simeq \Delta\nu tp \simeq \frac{\omega\nu dt_p\,\Delta n}{c} \tag{91}$$

For the example of 1-*ns* duration bandwidth-limited pulses from an He-Ne laser operating at 633 *nm* being frequency-shifted in a ~5-cm-long crystal modulated at 50 MHz and achieving a maximum refractive index change of 10^{-3}, (91) gives $N_R \simeq 25$. In fact, since both $\pm\Delta n$ refractive index changes are possible, the direction of frequency shift can be reversed, and the maximum number of resolved beam positions becomes $2N_R$. It can be seen from (91) that further increase in the number of resolved positions is possible by maximizing the crystal length, modulating at a higher frequency, or employing multipass techniques, provided, of course, that the approximations resulting in (90) remain valid. The maximum theoretical access time of the deflector is the crystal transit time and would be achieved with a *sawtooth* applied field of this period.

Combined E-O and Acoustic Digital Deflector

In Section 12.4.2 the operating principles of prism deflectors were discussed. It was shown that in bulk single-element devices the achievable refractive index change produced only limited deflection capability. Nevertheless, in the form of miniature surface-wave devices these simple deflectors have important potential applications, because the scaling down of their size for inclusion in integrated optic circuits results in the required deflection field being achieved by only a modest applied voltage (see the later subsection "Surface-Wave and E-O Deflectors"). For these integrated circuit applications there is a continuing search for techniques to minimize the required drive voltage to allow operation of the deflector by simple transistor logic. The localized nature of the power dissipation in the drive transistors and deflector elements can also present problems for high-frequency operation if the voltage is not minimized.

A simple scheme to reduce the deflection drive voltage has been proposed by Kotani et al. [67]. These workers proposed the combination of an E-O prism deflector with an acoustooptic grating deflector and showed that potentially an order-of-magnitude reduction in drive voltage for switching between two beam directions was achievable. A schematic of the proposed deflector arrangement is shown in Figure 12.15.

A surface corrugation grating diffracting in the Bragg regime couples the maximum energy into the diffracted beam when the *Bragg condition* is satisfied, that is, when

$$\Lambda \sin\theta = \frac{\lambda}{2n} \tag{92}$$

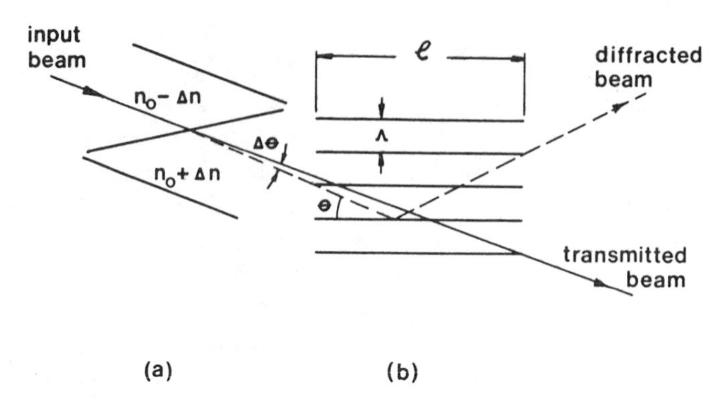

Figure 12.15 Combined E-O and acoustic digital deflector: (a) E-O prism deflector, (b) acoustic deflector operating in the Bragg regime.

Here Λ is the grating period, θ the incident ray angle, λ the free-space wavelength, and n the effective index of the guided optical mode.* From (92) it can be seen that the coupling of energy into the diffracted beam can be controlled by either a change in n or θ. Kotani et al. [67] showed that the diffracted energy was much more sensitive to changes in θ than in n. For example, they calculated that for a grating with spacing Λ of 1 μm and length l of 5 mm the required angular change $\Delta\theta$ for complete switching was 170 μrad. In principle, this deflection can be achieved in a single-element surface-wave $LiNbO_3$ prism deflector with an applied voltage of only ~1.5 V. In comparison, the same deflector would require 14 V if it operated alone to produce two well-resolved beam positions.

To achieve the maximum performance from the combined prism and grating deflector, Kotani and co-workers calculated that the grating spacing would need to be very tightly controlled and the grating refractive index very homogeneous. For the above example, they calculated that the grating spacing error $\Delta\Lambda$ should be $\ll$ 12 Å and that the refractive index inhomogeneity Δn should be $\ll 2.6 \times 10^{-3}$.

To avoid crosstalk, the diffraction beam spread must be less than the change in angle. Since ~170 μm corresponds to the divergence from a diffraction-limited gaussian beam of ~100-μm diameter at 633 nm, this sets the limit to the miniaturization of a device based on this principle.

Analog Interference Deflectors

In principle, the deflectors discussed in Sections 12.4.1 and 12.4.2 could be used with any approximately monochromatic light source so long as it was spatially

---

* See Chapter 11 for a more complete discussion of these devices and the derivation of the equations that characterize their operation.

filtered to produce a beam of divergence close to the diffraction limit. In practice, beams from lasers are generally used since only they can provide the intensity required in most deflection and scanning applications. In common with the analog deflector using frequency shifting, the interference deflector makes use of the other principal laser property, high monochromaticity. In principle, the use of multiple-beam interference effects allows scanning devices to be built which require a change in optical path of only half a wavelength per transit to scan through all the resolved positions. This contrasts with other types of E-O deflectors, which for a similar change in refractive index in the active material produce only a deflection of one resolved position.

The interference deflector is based on the Fabry-Perot interferometer and is shown schematically in Figure 12.16. The far-field beam pattern from the interferometer is a series of narrow fringes whose intensity distribution I(x) is proportional to the following function:

$$I(x) \propto \left(1 + F \sin \frac{\psi}{2}\right)^{-1} \tag{93}$$

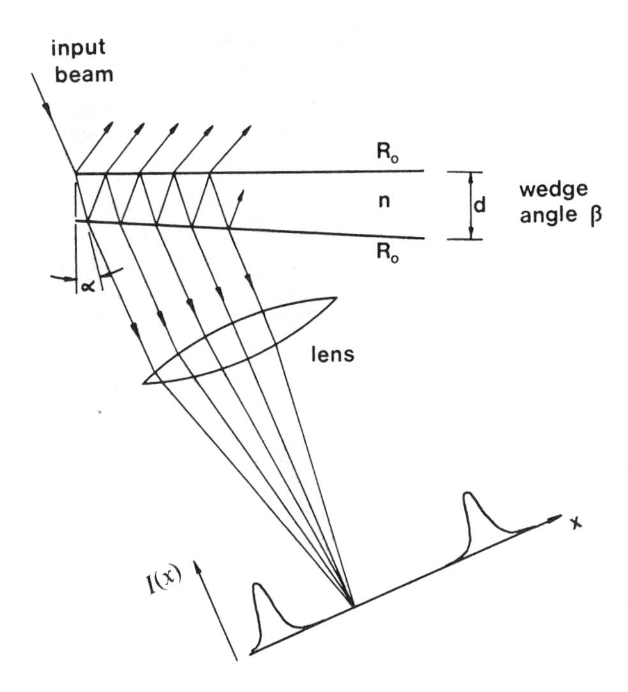

Figure 12.16 Fringes formed by beams multiply reflected and transmitted by an interferometer comprising two high-reflectivity mirrors at a small wedge angle. The etalon is the basis of an interference deflector.

where the phase difference between consecutive transmitted beams $\psi = 4\pi/\lambda(d + \beta x)n \cos \alpha$ and F is the finesse of the interferometer. It can be seen from (93) that the spacing between adjacent fringes in the output distribution depends on the wedge angle β. This angle can be chosen so that only one fringe appears across the desired linear field at the lens focal plane. A change in phase corresponding to one-half wavelength in the cavity (i.e., ψ changes by 2π) results in the fringe being scanned completely across the linear field range. This phase change can be produced either by varying the etalon spacing d or by changing the refractive index n. The number of resolvable elements N_R in the scan is determined by the ratio of the fringe width to the fringe spacing. This ratio can be either aperture- or loss-limited [68]. For a sufficiently monochromatic beam Korpel [68] has shown that the maximum value of N_R is equal to the etalon finesse:

$$N_R = \frac{\pi\sqrt{R_0}}{1 - R_0} \tag{94}$$

where R_0 is the reflectivity of each of the mirrors. For mirrors with $R_0 \sim 99.7\%$ we see that N_R can exceed 10^3.

Buck and Holland [69] have noted that the scanning rates achievable by these devices are ultimately limited by the time needed to reestablish the field in the etalon after deflection to a new position. This time increases with the etalon finesse and consequently with the resolution of the deflector. In principle, the etalon spacing (d) can be made arbitrarily small to minimize this time.

Analog Deflector Using Frequency Tuning

While the scanning technique in the section "Analog Deflector Using Frequency Shifting" relies on a narrow bandwidth (ideally transform limited) laser for the attainment of high resolution, a similar deflection scheme, but appropriate for use with wide-band width lasers (e.g., dye or Alexandrite lasers) has recently been proposed by Filinski and Skettrup [70]. In this case it has been proposed that the frequency shifting be achieved by the use of an intracavity E-O tuning element [71] rather than a frequency modulator. The result is a narrow bandwidth optical output that is tunable over the broad laser bandwidth. As in Wilkerson and Casperson's scheme, scanning is achieved by following the laser with a high-dispersion element.

For the specific case of an Alexandrite laser with an assumed useful tuning range of 120 nm, Filinski and Skettrup calculated that a deflection of >3000 resolvable spots could be achieved by use of a suitable dispersive grating and that a deflection rate of up to $\sim 10^{10}$ spots per second was possible. The rate attainable being ultimately limited by the time taken for the laser to reestablish a steady-state E-M field distribution and suitably narrow bandwidth at the new frequency.

12.5 ELECTROOPTIC DEFLECTOR DESIGNS

12.5.1 Digital Light Deflectors (DLDs)

General

The DLD uses a birefringent element to give spatial separation between switched orthogonal polarization states of an incident light beam. The birefringent elements used in DLDs have been calcite rhombs or prisms and the polarization switches electrooptic cells of the liquid Kerr or the solid-state Pockels type. For this important deflector there is only a limited choice of components. Proposed or used birefringent elements are shown in Figure 12.5 and polarization switches in Figure 12.17.

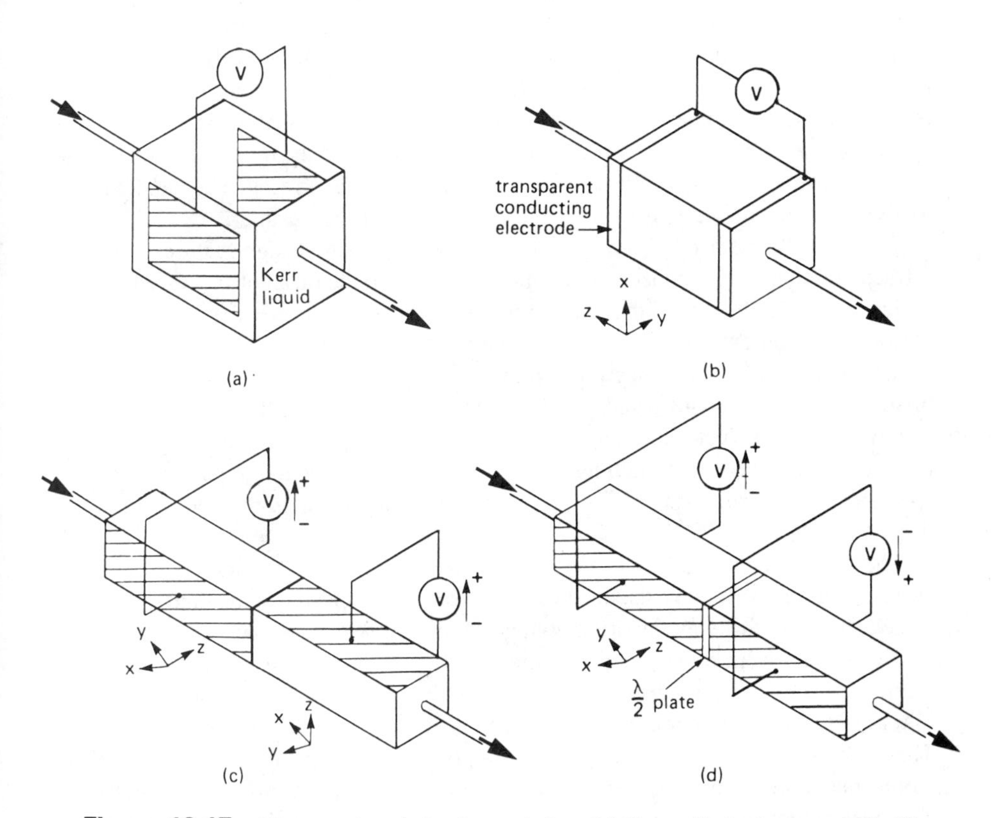

Figure 12.17 Electrooptic polarization switches: (a) Kerr cell nitrobenzene [72]; (b) longitudinal Pockels cell, KD*P Z-cut with conducting transparent electrodes; (c) transverse composite without half-wave plate, KD*P 45° Z-cut [73]; (d) transverse composite with half-wave plate, KD*P 45° Z-cut.

A purely theoretical approach to the selection of optic components for the DLD is unlikely to achieve a practical device, because the optic components may not be readily available and the problems associated with their use may not be well documented or necessarily well understood. DLDs for which published information exists have all required the practical development of the chosen optic components. Consequently, work in this field has required access to many practical skills and in particular those necessary for the production of crystal and liquid optics.

Kerr Cell Deflectors

Probably the most advanced and successful work on DLDs has been carried out at Philips GmbH Forschungslaboratorium Hamburg by Schmidt [74] and others as part of a program to develop a large-screen laser display facility. This work started in 1964, and by 1972 a 20-stage light beam deflector giving a two-dimensional raster of 1024 × 1024 positions had been successfully operated [75]. This system has not been outdated in performance by any later design and is therefore still of considerable interest. Schmidt's 20-stage deflector is also of special interest in that it is an integrated design using only nitrobenzene and calcite as the active optic components. There are no air interfaces. The deflector is divided into two sections of 10 stages and is shown in Figure 12.18.

The deflector built by Schmidt embodied the following design features. The refractive index of nitrobenzene, which is very close to the mean of the refractive indices of calcite, removed the requirement for compensating wedges to balance the ordinary and extraordinary ray paths about the systems axis.

The first 10 stages of deflection achieved a 32 × 32 block scan using calcite prisms of 6′ 12″, 24′ 48″, and 1°36′. (Refer to Table 12.3.) This design was achieved with an electrode separation of only 1.4 to 1.5 mm and consequently required a relatively low switching voltage of 2.1 kV. The deflector gave a high-speed localized writing facility of particular value in alphanumeric and graphic displays.

The second deflection stage of 10 units was coupled to the first by a lens system designed to achieve the most favorable electrode spacing at the final stages of deflection. In this way the drive voltage of the last two stages was kept below 10 kV for an electrode spacing of 6 mm. A degree of compensation for beam wander, due to the variation of refractive indices of calcite and particularly nitrobenzene with temperature, was achieved by inverting the last two prisms. The deflection system involved very little optics but required a cell design that would maintain the nitrobenzene in an ultrapure contamination-free condition. This was achieved by an all-glass structure with glass-to-metal seals as shown in Figure 12.19. The optic transmission for the complete system was 40% at 520.8 nm. Two-thirds of the loss was attributed to uncoated connecting optics with only one-third of the loss in the two deflector stages. The ratio of the total

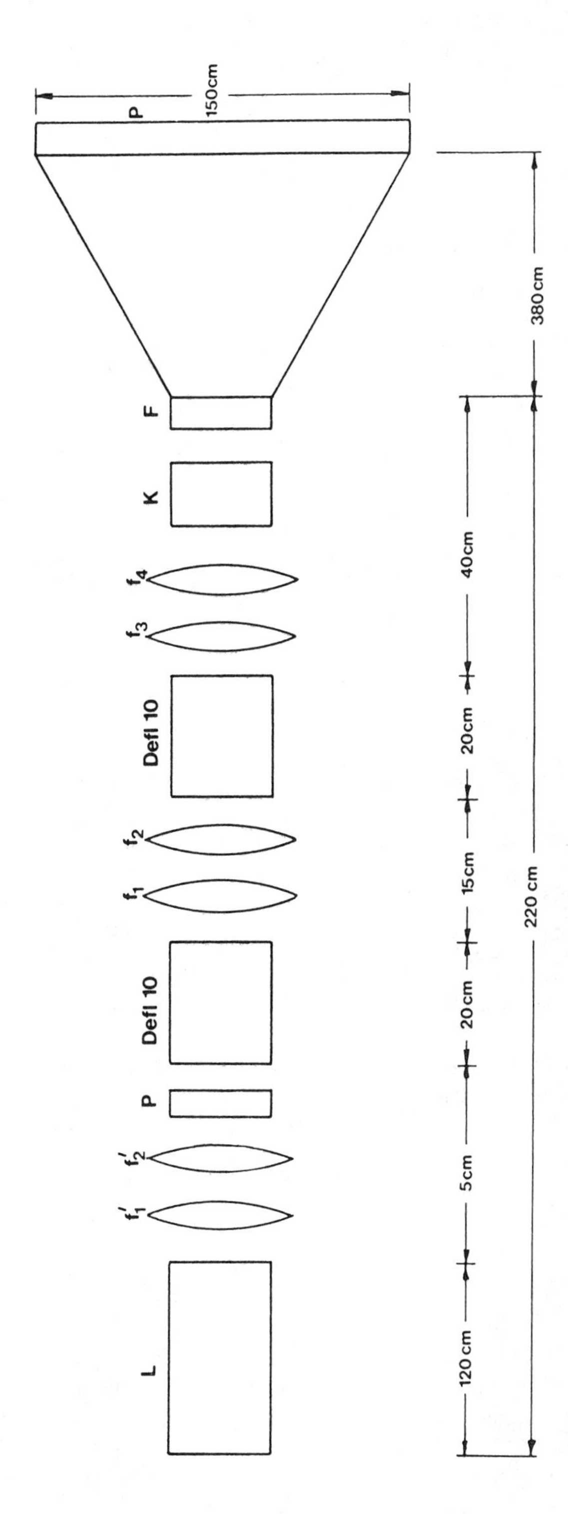

Figure 12.18 Optic section of the 20-stage deflector used in an experimental large-screen projector for alphanumeric data. L, laser; P, polarizer; Defl 10, deflector with 10 stages; f'_1 (40 mm) and f'_2 (80 mm), telescope pair; f_3 (90 mm) and f_4 (35 mm), projection optics; K, Kerr cell; F, polarization filter; f_1 and f_2, matching lenses; P, projection screen. (From Ref. 74.)

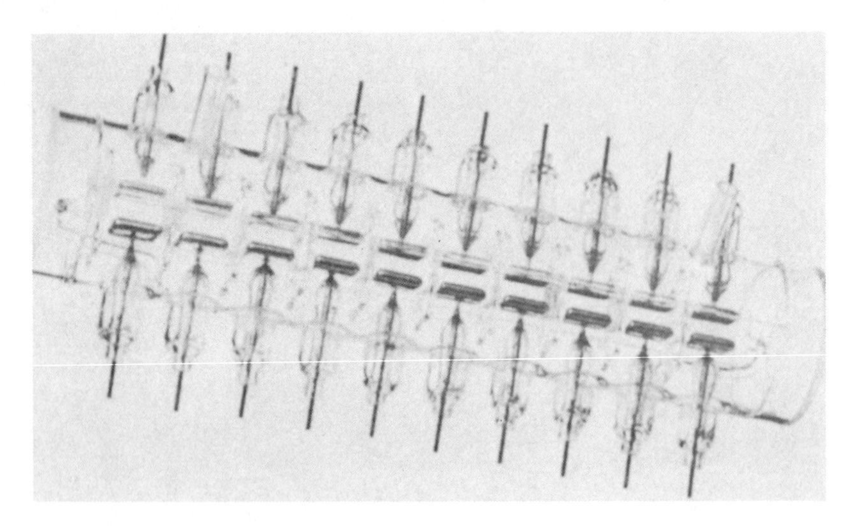

Figure 12.19 Nitrobenzene glass trough holding 10 sets of electrodes and 10 calcite prisms. (From Ref. 74.)

background light to the signal light was 0.25, which was reduced by a final polarization filter stage to 0.033. The brightest of the individual unwanted spots was 0.3% of that of the picture element.

The beam switching times obtained with the transistorized circuits employed ranged from 250 ns at voltages of 2.5 kV to 900 nsec at 8.5 kV. To take advantage of the square-law nature of the Kerr effect, each stage was dc biased to a prerotation through 90° (see page 719). To prevent the occurrence of Schlieren effects, the electrode current was kept to below 0.4 μA per stage.

There are two further important advantages that this deflector has that must be mentioned. First, the polarization switches are optically isotropic and have large angular apertures; consequently they will operate at more than one wavelength, although the drive voltages must be altered to suit. Second, they do not suffer from piezoelectric resonances.

Pockels Cell Deflectors

The problems associated with the use of Kerr liquids, in particular with their ability to collect contamination and the high voltages required in the final stages of deflection, have encouraged other investigators to develop deflectors using solid-state electrooptic polarization switches. Unfortunately, the crystal electrooptic polarization switch has its own set of material problems which have to date prevented these deflectors from achieving the performance of the nitrobenzene Kerr deflector.

The theory of the linear electrooptic effect and its application to light mod-

Table 12.3 Design Data for a 20-Stage Deflector

Stage no.	Prism angle γ		Aperture		Electrode spacing (mm)	Voltage (for λ = 520.8 nm)	
	d stage	h stage	d (mm)	h (mm)		Bias (kV)	Control (kV)
1		6′	1.4	1.4	1.6	5.3	2.1
2		12′	1.4	1.4	1.6	5.3	2.1
3		24′	1.4	1.4	1.6	5.3	2.1
4		48′	1.4	1.4	1.6	5.3	2.1
5		1°36′	1.4	1.5	1.6	5.3	2.1
6	6′		1.4	1.6	1.6	5.3	2.1
7	12′		1.4	1.7	1.6	5.3	2.1
8	24′		1.4	1.8	1.6	5.3	2.1
9	48′		1.4	1.9	1.6	5.3	2.1
10	1°36′		1.5	2.0	1.6	5.3	2.1
11		1°36′	3.0	4.0	4.5	14	6.5
12		3°11′50″	3.0	4.1	4.5	14	6.5
13		6°22′20″	3.0	4.2	4.5	14	6.5
14	1°36′		3.1	4.6	4.5	14	6.5
15	3°11′50″		3.1	4.9	4.5	14	6.5
16	6°22′20″		3.3	5.3	4.5	14	6.5
17		12°35′15″	3.6	5.6	4.5	14	6.5
18	12°35′15″		4.0	6.4	4.5	14	6.5
19[a]		−25°11′30″	4.7	7.0	6.0	20	8.8
20[a]	−25°11′30″		5.5	8.6	6.0	20	8.8

[a] These two prisms are upside down to eliminate temperature effects.

Source: Ref. 74.

ulation is discussed in Section 12.2. In choosing an electrooptic modulator for a deflection system the following design details must be considered:

1. Polarization ratio required for each stage
2. Angular aperture required for each stage
3. Linear aperture required for each stage
4. Capacitance of stage and dielectric loss
5. Voltage required after consideration of details 1 to 4
6. Drive power required for chosen address rate
7. Effect of piezoelectric resonances on picture definition
8. Operation for single or multiple wavelengths
9. Availability of chosen electrooptic material

Longitudinal XDP modulators have been fabricated using transparent conducting electrodes evaporated on thin Z-cut plates. This is illustrated in Figure 12.17b. They have the advantage that large diameters and angular apertures are possible without increasing the voltage required to drive subsequent stages.

For KD*P, a 1-mm-thick Z-cut longitudinal modulator using transparent electrodes would have a half-wave voltage of approximately 3.5 kV at 632.8 nm. The capacitance would be 40 pF/cm^2 aperture. The half-wave voltage would be independent of the aperture of the cell, and a deflector using these modulators would have the important advantage that the switching voltage would be identical for all stages.

Z-cut KD*P modulators using transparent electrodes were used by Pepperl [76] at Philips GmbH Forschungslaboratorium Hamburg in a prototype three-stage DLD. Each calcite prism was combined with a glass prism to form a deflection element with parallel input-output faces. The refractive index of the glass was chosen to give symmetrical deflection of light with respect to the system axis, and all components were antireflection-coated to reduce reflection losses. The device is shown in Figure 12.20.

It seems likely that some difficulty was encountered in the development of a suitable transparent conducting electrode. Both tan δ and electrode ohmic resistance limited the maximum drive frequency to 35 kHz (see Section 12.4). This limitation occurs at the onset of thermally induced optical strain and results in a deterioration of the polarization ratio. Furthermore, the switching frequency was limited by piezoelectric resonances. These resonances were severe as no attempt was made to achieve a degree of mechanical damping. For the 1-mm crystals fundamental shear and longitudinal modes occurred at 80 and 274 kHz. The exceptionally slow access time was attributed to the driver output impedance. This gave a 30- to 50-μs charging time for a cell capacitance of 20 pF. The optical loss of each polarization cell was given as 2%, and it would be reasonable to assume that the optic loss of each composite birefringent element was about the same value. For a 20-stage deflector the transmission losses would therefore

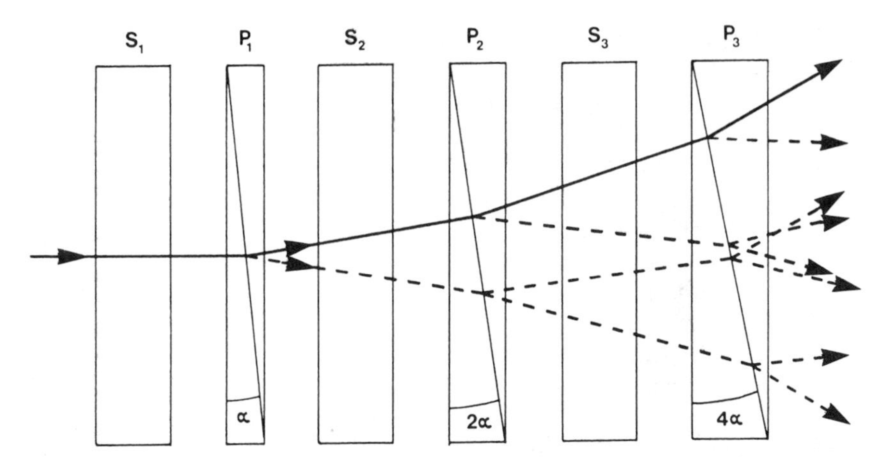

Figure 12.20 Optical arrangement for a three-stage longitudinal KD*P line deflector. (From Ref. 76.)

have been considerably higher than those of Schmidt's nitrobenzene deflector.

Transverse modulators using XDP materials require crystal pairs of precise orientation and of identical optic length. Exceptional skills are required in their preparation, and their delicate nature makes this optic work even more difficult. Special mounting techniques are necessary to maintain their relative orientation and to minimize strain and maximize piezoelectric damping. It has been suggested that the only problem not present is the purification requirement pertaining to the liquid Kerr cell, and yet if devices are required to have lifetimes acceptable for commercial applications, then the crystal grower must achieve similar purity levels.

At the present time there are a number of transverse modulators that could be used for DLDs.

1. Z-cut KD*P [73]
   a. Two-crystal composite without half-wave plate
   b. Two-crystal composite with half-wave plate
2. 45° Y-cut ADP [77]
   a. Two-crystal composite with half-wave plate
   b. Four-crystal composite without half-wave plate
3. Z-cut transverse [78]
   Single-crystal lithium niobate

For transverse electrooptic modulators we have seen that the half-wave switching voltage $V_{\tau/2}$ is proportional to d/L, where d is the distance between the electrodes and L is the composite crystal length. For a line deflector, the beam can be deflected parallel to the electrodes and the aperture increased in that direction without increasing the electrode separation. This can be seen by inspection of Figure 12.17d. This argument also applies to the liquid Kerr cell. Increasing the composite crystal length L reduces the drive voltage, but a practical limit is reached when either the physical aspect ratio is too small for the deflected beams or the angular aperture defined by an acceptable polarization is exceeded.

In general, transverse modulators employing half-wave plates have greater angular acceptance angles than those without. Also, transverse modulators with half-wave plates show interference figures between crossed polarizers of twofold symmetry rather than four, and this can be used with advantage to achieve large acceptance angles for line deflectors.

A line deflector using composite 45° Z-cut KD*P modulators with half-wave plates was described by Hepner [79]. This deflector, which is illustrated in Figure 12.21, gave a single line scan of 64 positions, having six stages, with an input modulator and an output polarization filter stage. The system was immersed in an index-matching oil of refractive index 1.49, and the total optic losses for the 34 optical components used was 50% at the operating wavelength of 632.8 nm. An excellent cross-path performance was achieved for this design with a relative intensity of false positions being about 10^{-4}. The input beam diameter chosen

Figure 12.21 View of six-stage transverse KD*P line deflector. (From Ref. 79.)

set the electrode spacing at 2.5 mm. The calcite single prisms were cut to give 10 resolution angles per deflection position, and for the 64 deflection positions the last deflector gave an acceptable polarization ratio for an angular field approaching 10°. The KD*P crystals each of 11-mm length required 880 V to half-wave-switch at 632.8 nm. The maximum switching frequency achieved was 100 kHz with a random access time of 10 μs. No mention of piezoelectric resonances was made, although they were undoubtedly present. However, the resonances may have been effectively damped by the index-matching oil.

The problems associated with the fabrication of transverse modulators prevented the further development of these deflectors. More recently, improvements have been made in the design and fabrication of these devices, and it is now considered that it would be possible to build useful solid-state electrooptic deflector systems. Certainly, transverse modulators of the 45° Z-cut KD*P type can now be made to a higher optic standard simply because of 10 years further manufacturing experience, and improvements have obviously occurred in the performance of high-voltage drive circuits. Consequently, switching rates of 1 to 10 MHz are now possible with access times of less than 0.1 μs at polarization ratios of $1:10^4$ or more [72]. It would also be possible to build a deflector using 45° Y-cut transverse modulators, and these would be free from piezoelectric resonances.

The development of new electrooptic crystals with improved characteristics could change this situation. The availability of a cubic optically isotropic crystal, capable of producing a modulator requiring a drive voltage of a few hundred volts or lower, would most certainly revive interest in these potentially useful devices.

12.5.2 Analog Beam Deflectors

Device Designs

For both prism and refractive index gradient deflectors the number of resolvable beam positions obtained is proportional to the induced refractive index change Δn, which is related to the electrooptic coefficient r_{ij} and the third power of refractive index by

$$\Delta n \propto r_{ij} n^3 = M_R \qquad (95)$$

M_R was used by Thomas [50] as the figure or merit for spatial resolution. Also, as the speed of deflection is for many applications as important as the number of resolvable beam positions and since the speed of operation depends on the device capacitance, a material figure of merit for temporal resolution is given by

$$M_\tau = \frac{r_{ij} n^3}{\epsilon_{ij}} \qquad (96)$$

Figures of merit are listed in Table 12.2 and can be used as a basis for material selection. The performance of a particular deflector can be optimized by choice of crystal orientation and dimensions and by careful design of the electronic drive circuits.

Although ADP is the least attractive electro-optic material, judged by the figures of merit M_R and M_τ given in Table 12.2, it has some advantages and has been used by some workers for constructing analog deflectors.

It has been shown by Magdich [80] that the 45° Y-cut ADP crystal is non-piezoelectric, and it is for this reason that it has become the most widely used material for low-voltage light modulators. It has found many industrial applications where lasers require amplitude control, and an example of a six-channel 45° Y-cut ADP light modulator is shown in Figure 12.22. This modulator requires 300-V drive at 488 nm to give full amplitude control, and the modulated light is resonance free. It is for these reasons that ADP was chosen by Beasley [81] for his deflector in an experimental large-screen TV projection display. The electrooptic multiple-prism scanner he constructed is shown in Figure 12.23. The scanner had an aperture of 40 × 2.5 mm and therefore required only 10 prisms of 80-mm base length to give a light path of 400 mm. The price paid for the relatively few number of prisms involved was a high capacitance (3000 pF). The line scan at 60 Hz needed 14-kV drive to give 160 resolvable spots for single-

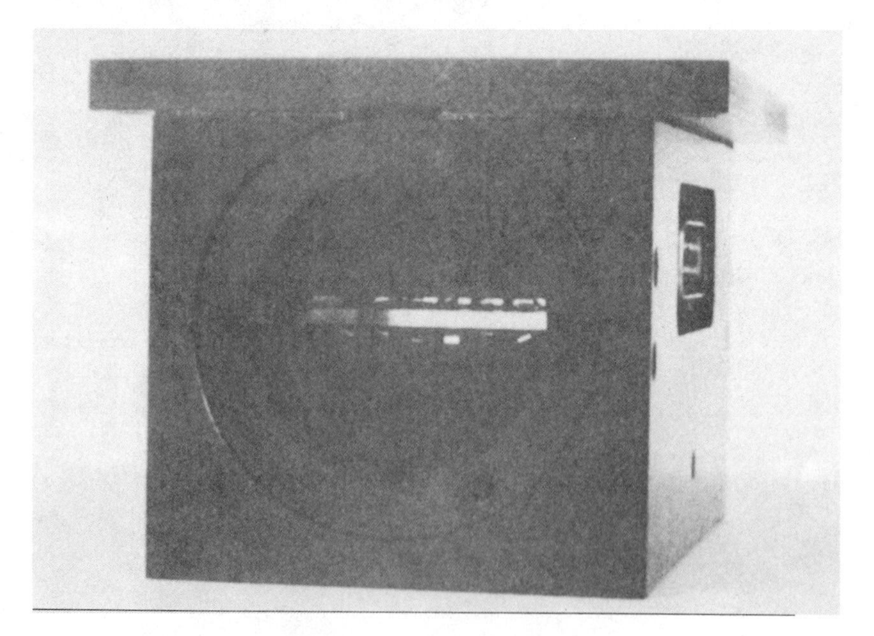

Figure 12.22 Six-spot *ADP* modulator.

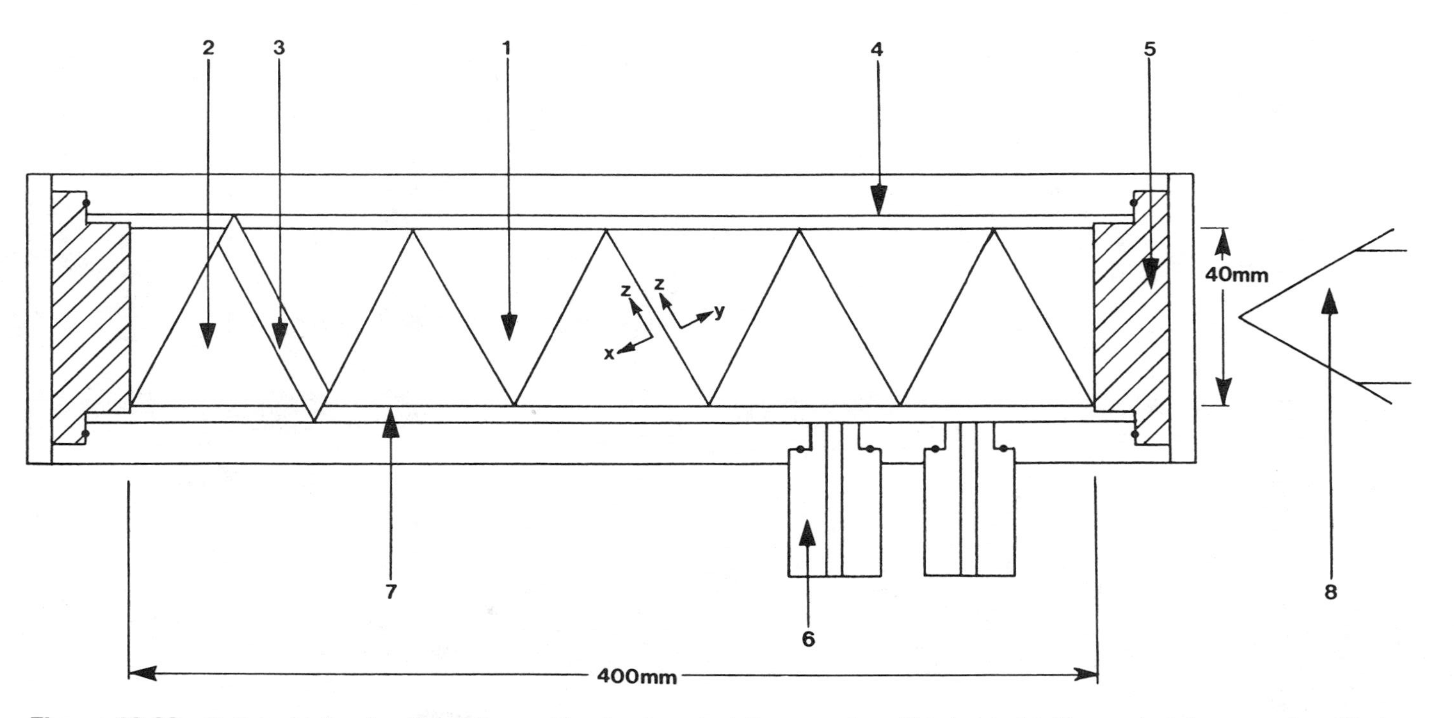

Figure 12.23 Bulk multiple-prism line deflector: (1) prism in main deflector section; (2) prism in third harmonic deflector section, (3) glass spacer, (4) outer case, (5) end windows with O-ring seals, (6) high-voltage connector, (7) Teflon spacer, (8) polarized light beam. (From Ref. 81.)

pass operation. The field strength of 2.8 MV/m could not be sustained without electrical breakdown, and it is probable for this range of ionic crystals that this field strength is too high. An electrooptic modulator of ADP would today be given a maximum permitted operational field strength of 0.5 MV/m. An alternative ADP multiple-prism design used by Thomas [82] was built by Coherent Associates (USA) and had an aperture of 2 × 2 mm and an overall crystal length of 200 mm. This deflector was evaluated for use in a crystal streak camera. The maximum recommended voltage of ±2 kV reduced the field strength to 1 MV/m and no doubt improved the lifetime. This streak camera required low-repetition-rate impulses of between 1- and 100-ns rise time, which would not have overstressed the material, and it is probable in this application that this deflector would have had an acceptable lifetime. The capacitance of 100 pF prevented very fast switching, and the circuits employed achieved a 20-ns sweep, giving a temporal resolution of approximately 2.5 ns per resolvable spot at 1.06 μm. A krytron sweep circuit [83] would have improved this resolution to possibly 500 ps, but this is still two orders of magnitude slower than currently available with image convertor streak cameras [84]. It therefore seems unlikely that ADP would produce anything other than an experimental deflector for this application.

We have seen that the M_R and M_τ values for $LiNbO_3$ indicate that it is more suited to analog deflection than ADP. For analog deflectors using $LiNbO_3$ it is useful to apply a field parallel to the z axis to obtain a refractive index change $\Delta n = (1/2)n_e^3 r_{33} E_z$ for light polarized in the direction of the applied field. The refractive index change is four times that obtained from ADP when the field is applied in the y direction and for which

$$\Delta n = \frac{\sqrt{2}\, n_o^3 n_e^3 r_{41} E_y}{(n_o^2 + n_e^2)^{3/2}} \tag{97}$$

Figure 12.24 shows an $LiNbO_3$ quadrupole deflector of dimensions 3.3 × 3.3 × 18 mm built by Gisin et al. [85]. Electrodes of width equal to half the face width were deposited on each of the side faces to produce a gradient field within the crystal. As the deflector was used for on-off light modulation by beam displacement from a detector, high resolution and a linear deflection were not necessary, and the simple electrode geometry sufficed. Nevertheless, it was stated that the deflection increased linearly with the voltage throughout the investigated range of −2.8 to +2.8 kV. The electrode geometry gave the device a low capacitance, and for a 3-mm beam at 632.8 nm the drive voltage required would have been approximately 200 V. This is in itself a reasonable specification for a modulator. The on-off modulation that the deflector provided at approximately 1 MHz did not, apparently, suffer piezoelectric resonance effects, but the techniques employed to achieve the necessary damping were not described. The $LiNbO_3$ gradient deflector was further developed by Ireland [86–88] using a

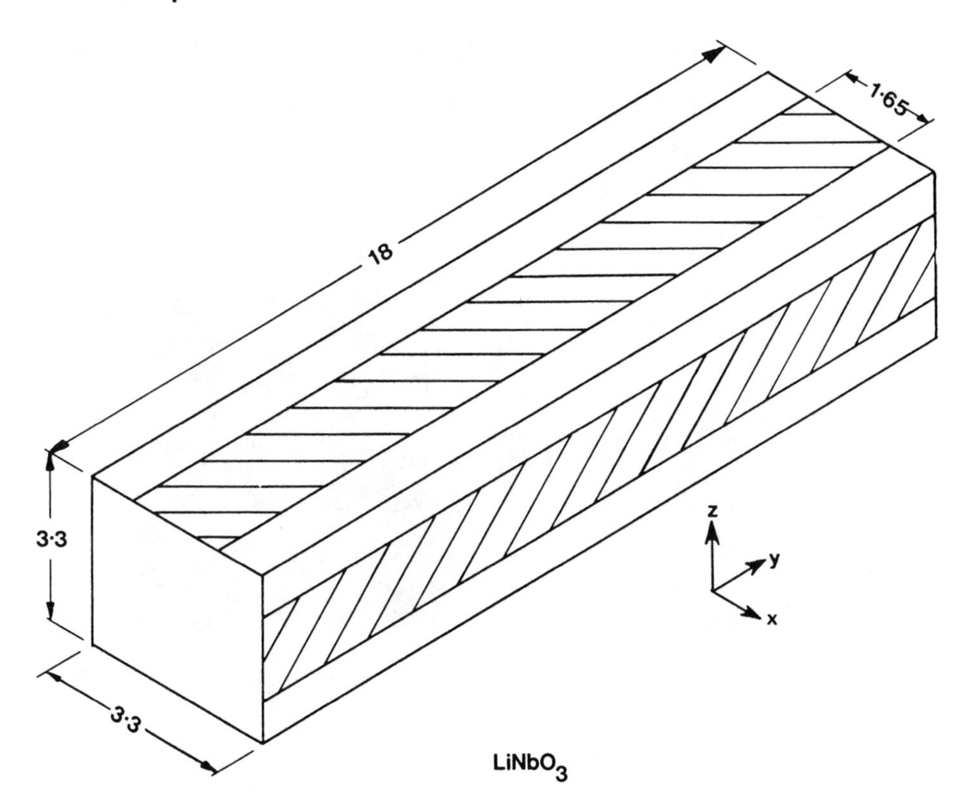

Figure 12.24 Quadrupole deflector using planar electrodes. (From Ref. 85.)

cylindrical electrode geometry to produce a linear field gradient. The crystal used was a 5 × 5 × 32 mm Y-cut rotated 45° to the z axis. This configuration produces, to a close approximation, a linear field gradient in the x and z directions. In the second publication an aperture defined by the hyperbola $xy = \pm 1$, where x and y are in millimeters was chosen and examined theoretically to see how best-fit cylindrical electrodes would degrade its operation. Ireland showed that in using cylindrical electrodes of radius 2 mm the field error at $N_R = 40$ degraded the diffraction-limited condition by only a few percent, this degradation being less than that produced from a ± 12-μm positional tolerance on the electrodes. Each of the 2-mm-radius electrodes were stopped 2 mm away from the polished end faces of the crystal to avoid surface breakdown problems. Additionally, the tracking distance around the ''arrowheads'' (see Figure 12.25) was 3.4 mm so that the device would withstand a voltage of ± 8 kV when immersed in a high dielectric fluorocarbon fluid which was used to reduce the surface reflection losses.

Static optic tests were performed with the deflector using a collimated beam

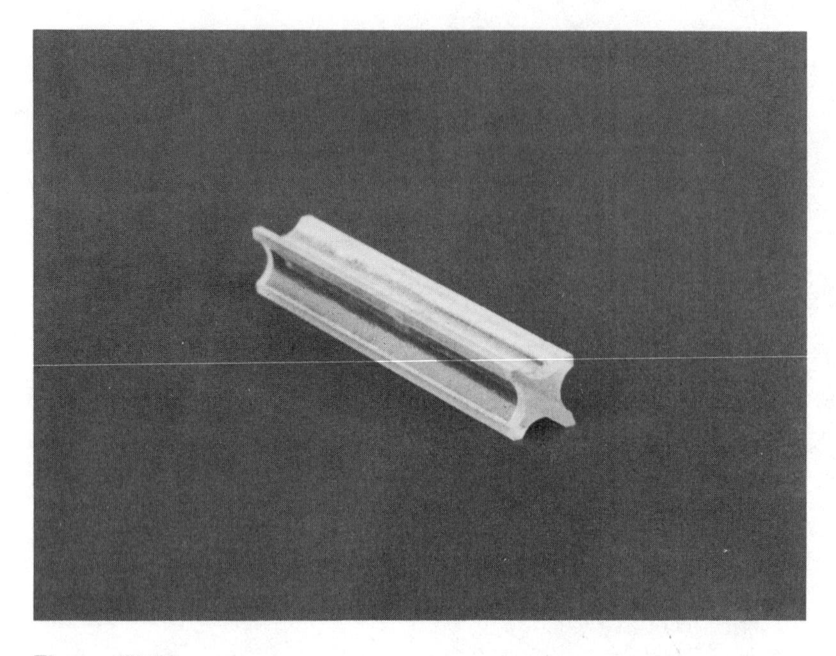

Figure 12.25 Lithium niobate crystal fabricated and used by Ireland and Ley in crystal streak camera experiments. Crystal dimensions, 5 × 5 × 32 mm.

at 632.8 nm from an He-Ne laser. This beam was magnified and spatially filtered so that the deflector aperture was uniformly illuminated. A 50-μm slit was positioned parallel to the x axis and immediately after the deflector. By using a 50-cm-focal-length lens, the far-field spatial distribution was examined with voltages in the range 0 to 5 kV applied to the device. The distributions were very close to being diffraction-limited over the whole voltage range. Photographic records showed that the main difference between the distributions and the $((\sin x)/x)^2$ function expected for perfect diffraction at a slit was an increase in the intensity of the subsidiary diffraction maxima. For example, at 5 kV the first such maximum was 8% of the peak intensity rather than 4.7%. This "spreading" of energy is consistent with a wavefront aberration of $\lambda/2$ [89]. The deflector produced a static deflection that was a linear function of voltage to within better than 2%. It required 100 V to produce one resolvable spot at 632.8 nm. By using (74) with $r_{33} = 30.8 \times 10^{-12}\ mV^{-1}$ and $n_e = 2.20$, this corresponds to an E_z(max) of 4.5 MV/m at 5 kV. Elctrical tests on the deflector showed that it had a capacitance of 45 pF, so that driven by two 50-Ω lines it had a potential rise time (10% to 90% amplitude) of ~2.5 ns. TDR measurements with a 30-ps resolution Tektronix type 7S12 pulse-sampling unit confirmed this result.

For operation in the streak mode the deflector was driven by a pair of 50-Ω lines from an EG&G KN22B Krytron switch. With the transmission lines and the Krytron envelope carefully screened, an 8-kV voltage ramp could be applied to the deflector in 4.0 ns. By backing off a pair of the deflector electrodes to −4 kV, the interelectrode pd was swept from +4 to −4 kV when the Krytron conducted. This allowed the beam to be swept symmetrically through the zero-field (low-aberration) position. To verify the deflector could achieve the temporal resolution and dynamic range implied by the optic and electrical tests outlined above, it was used in conjunction with frequency-doubled pulses from a mode-locked Nd-YAG laser. These pulses had previously been measured with an IMACON 675 image convertor streak camera (resolution ~5 ps) and found to be of 27-ps duration. Before entering the deflector, these 0.53-μm pulses were passed through a spatial filter and 2.0-cm BK-7 etalon with 70% reflective coatings. The etalon was used to calibrate the streak records. It produced multiple pulses, temporally separated by 200 ps, and of 2:1 intensity ratio. Figure 12.26a shows a microdensitometer scan of a far-field streak record. It can be seen that the recorded FWHM intensity width for the pulses is 50 ps. Assuming a quadratic temporal folding, this implies that the camera resolution was 42 ps. By comparing this with earlier records, Ireland showed that the use of the slit immediately after the deflector led to a considerable improvement in the dynamic range of the camera. These results indicated that satellite pulses of 5% of the main pulse intensity could be detected.

To achieve better temporal resolution, the camera was set up with the deflector double-passed. In this case the 50-μm slit was positioned immediately before the deflector and a 100% reflectivity mirror after it. A partially reflecting mirror in the beam was used to deflect the double-passed pulses for focusing and recording. Figure 12.26b gives the microdensitometer record obtained with this experimental arrangement. From this and other records it was found that double-passing the deflector improved the resolution to 20 ps as expected while adversely affecting the dynamic range.

Comparisons Between Multiple-Prism and Gradient Deflectors

From an examination of the equations for both multiple-prism and gradient deflectors it can be concluded that for devices of identical aperture and crystal length the deflection angle for a given value of Δn will be the same; that is,

$$\Delta\phi = 2L\frac{\Delta n}{Wn_2} \qquad (58)/(73)$$

In calculating Δn for a further comparison, we shall assume identical-length deflectors of the same aperture and of diameter d, as shown in Figure 12.27.

Obviously the field strength for case a is V_{app}/d, and for case b, using $zx =$

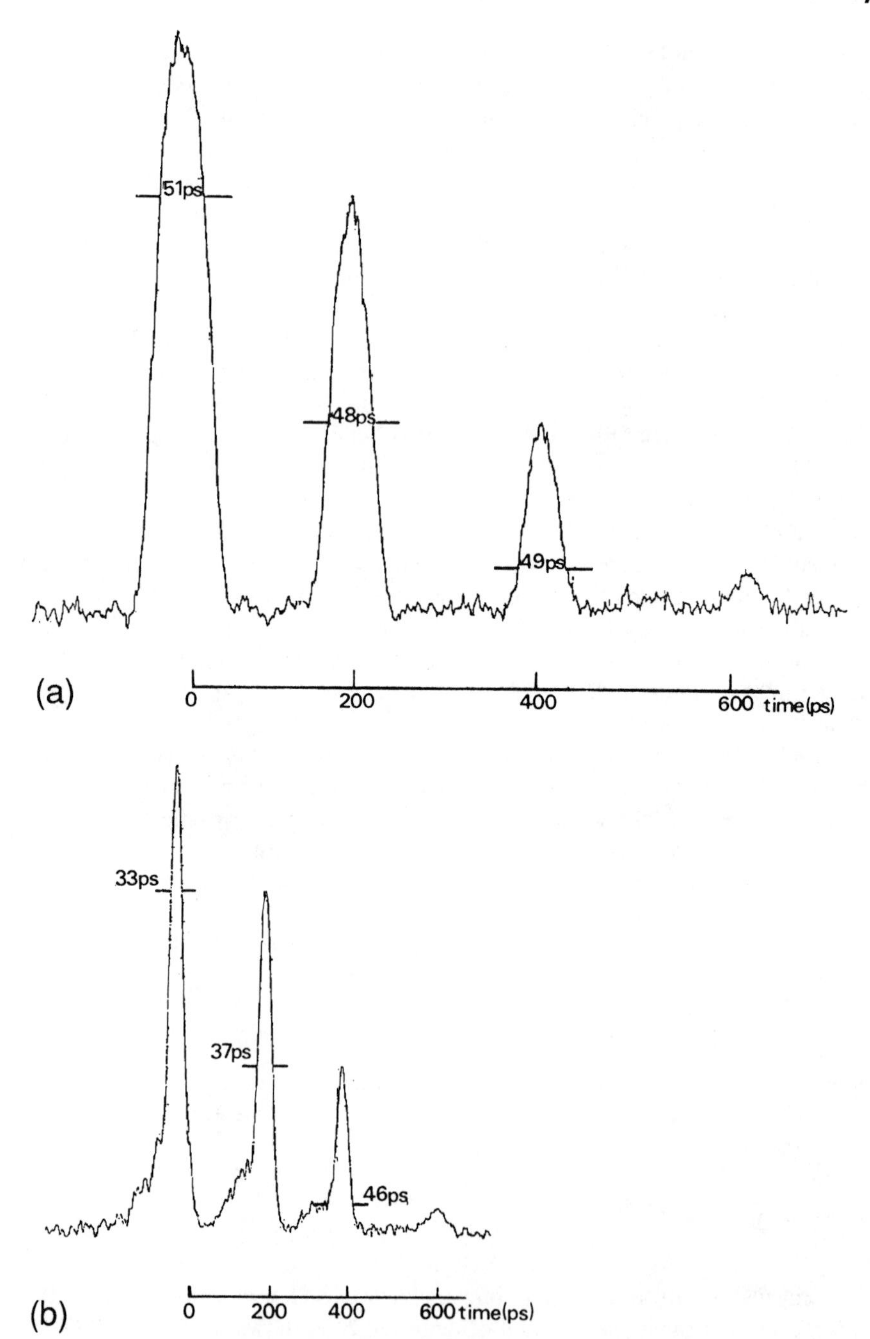

Figure 12.26 Multiple image streak records of 27-ps frequency doubled Nd-YAG laser pulse: (a) quadrupole deflector, single pass; (b) quadrupole deflector, double pass.

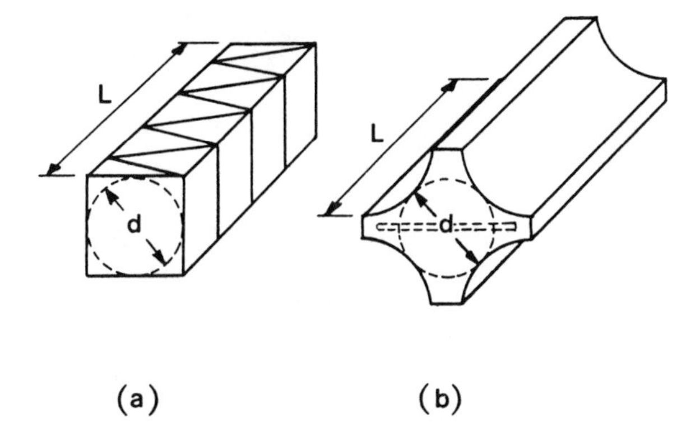

Figure 12.27 Analog prism deflectors: (a) multiple prism, (b) gradient quadrupole.

$d^2/8$, the field strength at the aperture edge is $2V_{app}/d$. The gradient deflector will therefore give twice the deflection for the same aperture since $\Delta\ n \alpha V_{app}/d$.

It is interesting to make a direct comparison for one possible E-O material, and for this purpose $LiNbO_3$ has been chosen, since it can be readily obtained with high optical quality.

By taking information from Table 12.2,

$$r_{33} = 30.8 \times 10^{-12}\ \text{m/V}, \quad n \simeq 2.2$$

and by choosing

$$\begin{aligned} a = d = W &= 2 \times 10^{-3}\ m \\ h &= 60\ mm \\ V_{app} &= 10 \times 10^{3} \\ \lambda &= 632.8\ \text{nm} \end{aligned}$$

and by using equations (58) and (59) we find for the multiple-prism deflector $\Delta\phi = 48$ mrad, and $N_R = 151$ and for the gradient deflector, $\Delta\phi = 96$ mrad, $N_R = 302$.

The situation can be further improved in favor of the gradient deflector by extending the beam width using a slit aperture, as indicated in Figure 12.27b. This will lead to an improvement in effective aperture by utilizing the gradient field well into the arrowheads of the crystal. (This technique was used by Ireland [87] for his crystal streak camera, as maximum resolution was required only in the deflection plane.) Such large deflection angles would require beam focusing to prevent vignetting, but to date deflectors have not been constructed where vignetting has required this preventative measure.

It is also necessary to examine the performance of both deflectors as crystal dimensions are altered. In Section 12.4 it has been shown that

$$N_R = \frac{2L\ \Delta n}{\epsilon\ \lambda} \tag{59)/(74}$$

For a fully filled aperture the spatial resolution is proportional to the crystal length and to the induced refractive index change but is independent of aperture. The value of Δn, however, is inversely proportional to the aperture for both deflectors, and, surprisingly, for a given drive voltage, small apparently means good. There are unfortunately several physical limitations which prevent this conclusion from being realized in practice. First, it is extremely difficult to manufacture from fragile electrooptic crystals optic components when dimensions are reduced to below 2 mm. Second, vignetting occurs for quite moderate deflection angles in this situation. For example with a 2-mm aperture, $L = 60$ mm and $\phi = 96$ mrad, an impractical situation results for which prefocusing is no longer a solution. Third, and again taking the example of a 2-mm aperture, the electrode spacing gives for the multiple-prism and gradient deflectors field strengths of 3 and 6 MV/m, respectively. This is close to the known breakdown strength of the XDP range of materials [90] and could well lead to refractive index damage in the oxygen octahedra ferroelectrics. Refractive index damage believed to be of this type has recently been observed in $LiNbO_3$ by the authors.

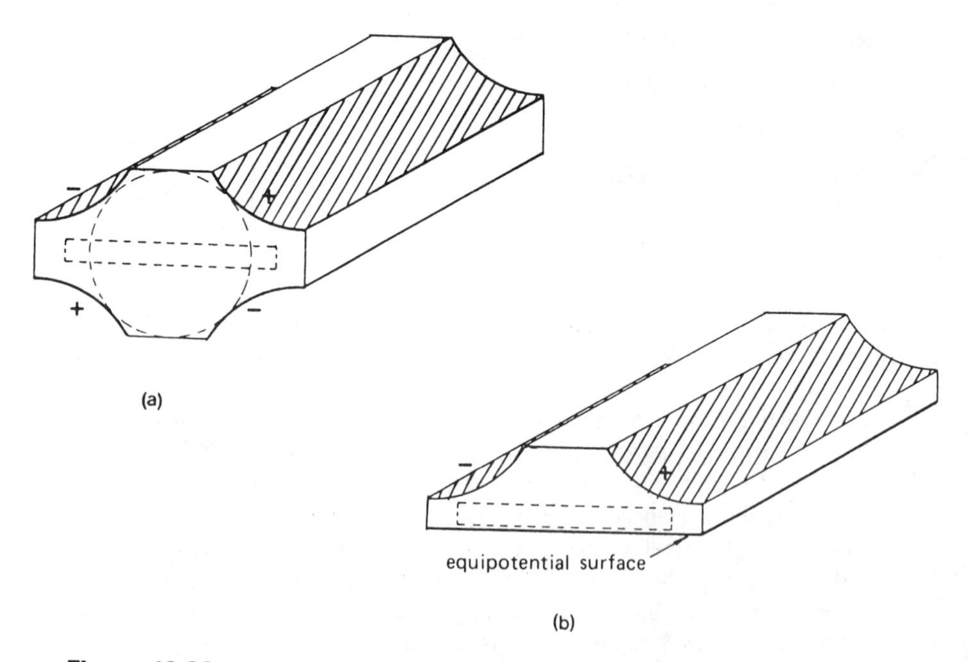

Figure 12.28 Variation to the quadrupole design: (a) truncated quadrupole deflector, (b) optimized gradient deflector.

Finally, although N_R increases linearly as the aperture is reduced, the device capacitance also increases with the same relationship; therefore, for a given voltage source the temporal resolution remains constant and is not affected by device aperture. For a display system this may be an acceptable price to pay for an increased deflection angle, but for a crystal streak camera no advantage would be gained.

Within these constraints there are some possible variations. The aperture of the multiple-prism deflector can be widened in the plane of the electrodes [81], and this reduces the prism elements required and eases the manufacturing difficulties. The spatial resolution is unaffected, and the capacitance is increased, but for display systems this could be tolerated. If a slit aperture is used, the gradient deflector can be trimmed to a minimum capacitance value by removing material no longer required, as shown in Figure 12.28a. This truncated design can reduce the capacitance by a factor of 2. A further reduction of 2 can be obtained if the device is reshaped as shown in Figure 12.28b. The aperture is further restricted in height but not in width, and for a streak camera application this could be acceptable. By these two methods the capacitance of the gradient deflector could be reduced to that of the equivalent aperture prism deflector while retaining its resolution advantage. (See Table 12.4)

12.5.3 Analog-Digital Array Deflectors

Bulk Crystal Deflectors

The concept of an array of prism deflectors was introduced in a paper by Ninomiya in 1973 [91] and further expanded by the same author in 1974 [92]. Both of these papers dealt with bulk prism array deflectors, and it is significant that they were developed for the possible application to optic data processing and optic memory systems. For these applications work in integrated optics has now led to the development of a range of deflectors using the array concept of Ninomiya but involving surface-induced electrooptic refractive index effects rather than the bulk effect discussed so far.

The basic electrooptic prism array deflector is shown in Figure 12.29a, where the type B prisms are used for index-matching purposes only. This basic design has the advantage that it can be built on a single-crystal slab without interprism spacing, provided that the deflector is used in a digital mode for main diffracted orders only. Nevertheless, this simple arrangement does give the closest approach to a sawtooth phase grating and will, at the main diffracted-order positions, give a high signal-to-noise ratio.

For ease of manufacture, Ninomiya [91] described a deflector manufactured from a single plate of Z-cut lithium niobate, as shown in Figure 12.29b.

To achieve continuous deflection, adjacent prisms were electrically isolated

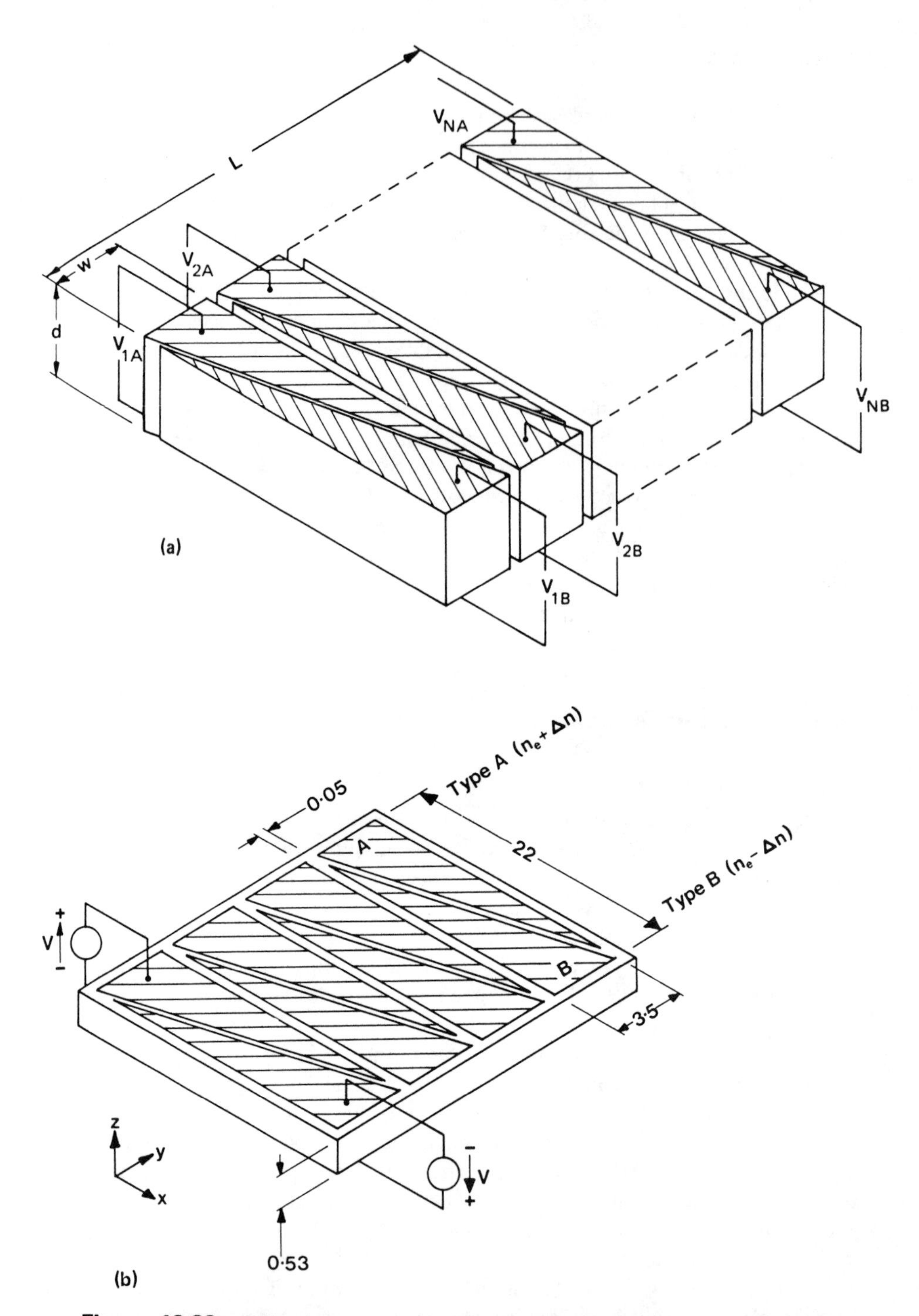

Figure 12.29 Bulk prism array deflectors: (a) generalized prism array deflector, (b) lithium niobate slab deflector [91], (c) nine-crystal $LiNbO_3$ prism array deflector [92], (d) nonlinear field prism deflector.

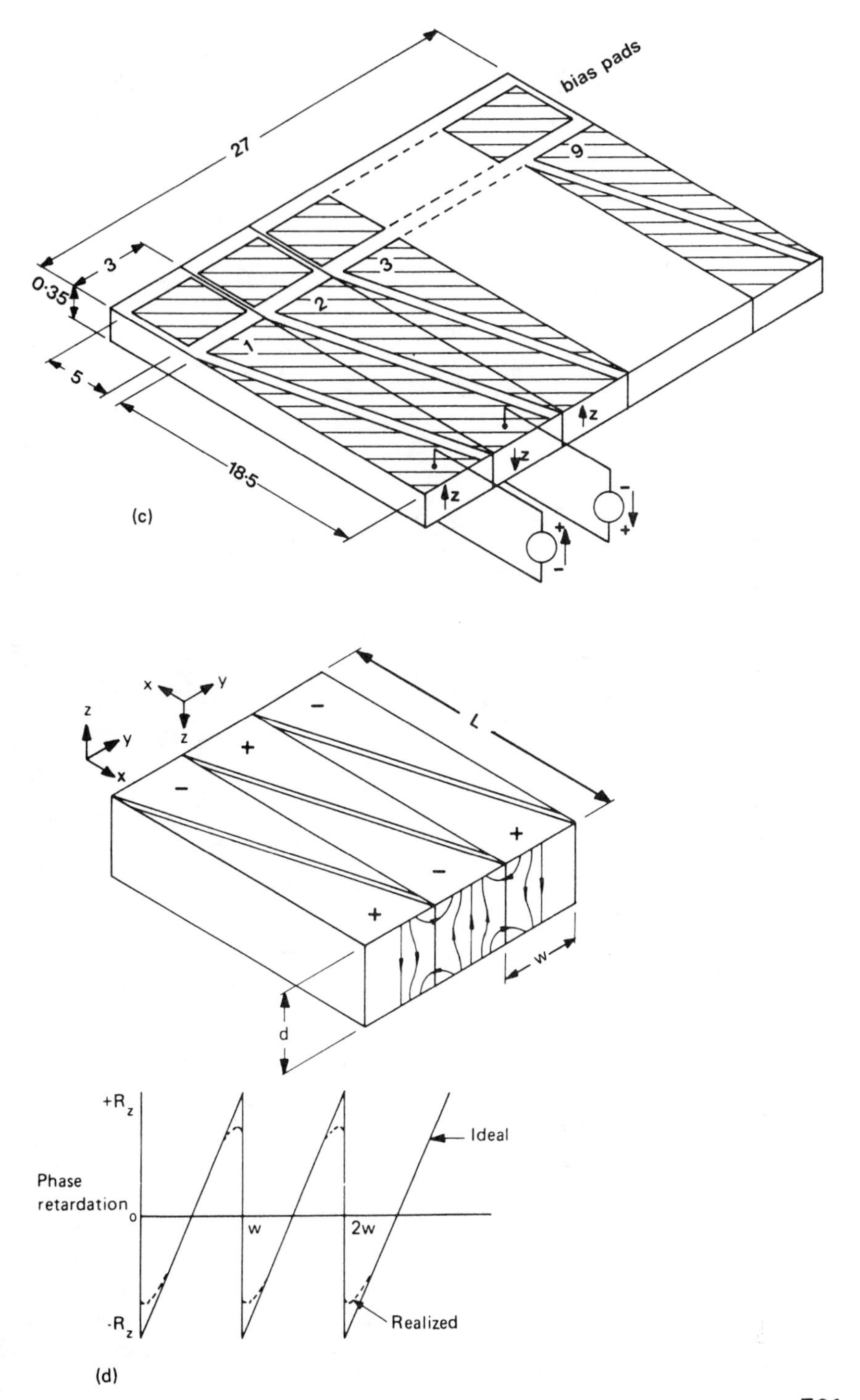

27
bias pads
9
3
3
2
1
0·35
5
18·5
z
z
z
(c)
x
y
z
z
y
x
L
w
d
+R
z
Ideal
Phase
retardation
o
w
2w
-R
z
Realized
(d)

Table 12.4 Analog Deflector Characteristics Taken from Reviewed Publications

Deflector type	Reference	Application	Material dimensions (mm)	Number of resolvable spots (N_R) wavelength (nm)	Deflection per kilovolt (mrad/kV)	Voltage drive (kV), rise time (ns)	Capacitance (pF)
9 multiple prisms	J. Beasley	TV line scan	ADP: 2.5 × 40 × 400	160 568	0.33	±7	3000
Multiple prisms	S. W. Thomas	Streak camera	ADP: 2.0 × 2.0 × 200	18 532	1.3	±2 20	100
Quadrupole cylindrical electrodes	C. L. M. Ireland	Streak camera	$LiNbO_3$: 5 × 5 × 32	100 632.8	1.8	±5 4.0	45
20 Multiple prisms (proposed)	S. W. Thomas	Streak camera	AMO: 2 × 2 × 60	153 1060	16	±3 0.5	10
Quadrupole plane electrodes	B. V. Gisin	Light modulator	$LiNbO_3$: 3.3 × 3.3 × 18	42 632.8	1.85	±2.8	16
Quadrupole cylindrical electrodes (proposed)	Text example	Streak camera	$LiNbO_3$: 3 × 3 × 60	300 632.8	10	±5	100 unshaped 25 shaped

Table 12.5 Bulk Array Deflector Characteristics

Deflector	Resolvable positions (analogue drive) N_R	*Dimensions of single prisms (mm)*			$V_{\lambda/2}$ (volts)	V applied (volts)	C (pF)	V_{app}/N_R	Material
		w	d	l					
4-section array slab design	50	3.5	0.53	23.5	48	±597	140	22	$LiNbO_3$
20-section array individual prisms (proposed)	630	2	0.5	25	40	±800	500	2.5	$LiNbO_3$
9-section array individual prisms	180	3	0.35	18.5	38.5	±580	358	6.5	$LiNbO_3$
54-section array individual prisms (proposed)	1080	0.8	0.2	21	16.4	±164	1750	0.34	$LiTaO_3$
3-section array slab design	50	0.5	0.02	9	44	±150	150	6	$LiNbO_3$

Note: All data given for 632.8 nm.
Source: From Refs. 91, 92, 93.

so that bias voltages could be separately applied. However, the discontinuities produced by these isolation spaces resulted in a poor signal-to-noise ratio, and because of this, Ninomiya considered this slab design to be an early prototype. He then proposed an alternative design of a 20-prism array of oppositely oriented prisms not requiring interprism spacing and capable of giving 630 resolvable positions for a drive voltage of ± 800 V. (See Table 12.5.)

Ninomiya's paper of 1974 gave details of a nine-prism array of the above design, shown in Figure 12.29c. This design allowed adjacent prisms to be joined electrically with both type *A* and *B* prisms operating electrooptically. It did, however, preclude the application of individual bias voltages to the prisms, and to overcome this problem, bias pads were added.

Although this design did give an improved signal-to-noise ratio, sidelobes still occurred at the main diffraction order positions owing to a fringing field effect at the regions of inverted polarity. This is illustrated in Figure 12.29d.

Ninomiya calculated the field across the aperture (w) for a line midway between the electrodes to take into account the fringing field effect. In these calculations he defined the effective thicknesses as

$$p = \frac{\epsilon_x}{\epsilon_z} \frac{d}{w} \tag{98}$$

For the nine-prism array constructed, p was equal to 0.28, and the calculations indicated a respectable sawtooth wavefront. By using a definition for signal-to-noise ratio of

$$S/N = \max_v \left(\frac{I_0}{\max\{I_+, I_-\}} \right) \tag{99}$$

where $\max_v$ means the maximum of the function V and I_0 is the intensity of the main lobe and I_+ and I_- are the intensities of the plus and minus first-order sidelobes, respectively, these calculations gave the signal-to-noise ratio as 10:1 for the tenth order of diffraction. In his experimental work the deflector achieved a slightly better signal-to-noise ratio than his calculations indicated. Ninomiya also proposed a 54-crystal 1080-spot deflector using $LiTaO_3$, but it is not known whether this device was built.

In a novel design, Dikaeu et al. [93] exploited the advantage obtained in reducing d to the smallest practical dimension by fusing a z-plate of lithium niobate on to a glass substrate using a gold bond. The lithium niobate plate was reduced to 20 μm in thickness, polished, and three sets of sawtooth electrodes deposited, with light coupled in and out of the deflector by prism elements. The deflector gave 50 resolvable spot positions for ± 150-V drive. The construction of the deflector is shown in Figure 12.30. The dimensions of the upper sawtooth electrodes were $L = 9$ mm and $a = 0.5$ mm.

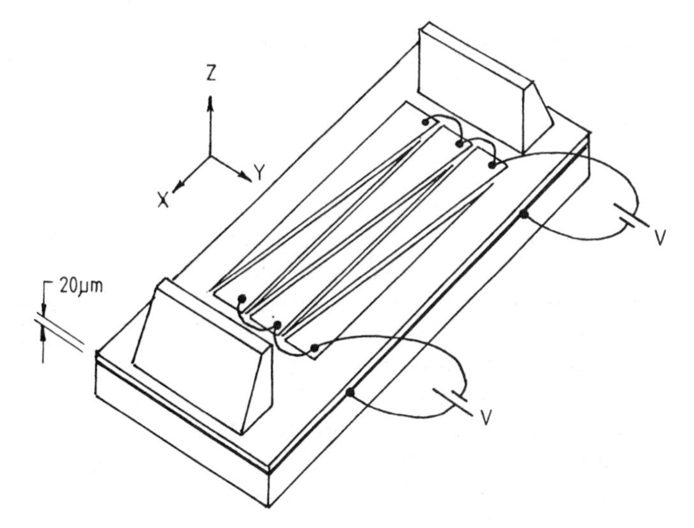

Figure 12.30 Construction of Dikae deflector.

Surface-Wave E-O Deflectors

To take maximum advantage of the electrooptic effect, by using the principles discussed for single-crystal prism deflectors, it is necessary (for given beam width) either to increase the interaction length L or to increase Δn by increasing the electric field strength (see (59)).

In practice it is more difficult to double a crystal length than to halve its thickness. With bulk deflectors the tendency has been to increase Δn by reducing aperture height, with a geometric limit set on this from a maximum acceptable diffraction condition and, also, with a practical limit set by fabrication and polishing constraints. Consequently bulk prism deflectors of moderate resolution have been built with slit apertures of 0.5 mm in height, and any further attempt to reduce this to 0.1 mm has been frustrated.

There is therefore a good argument for building a surface waveguide deflector if, as a result, higher field strengths can be achieved. We have also seen that reducing the aperture width W for a single-prism deflector leaves the resolution unaltered but that again fabrication difficulties prevent very narrow crystals from being produced for array deflectors. In surface waveguide deflectors using two-dimensional electrode systems the possibility exists that novel designs would allow a large number of narrow-aperture prisms to be arrayed as an integrated component. Such a device would apparently achieve the ideal conditions for E-O beam deflection, as high field strength, reasonable interactional length, and simple array configurations are possible.

The hope of designing a deflector with an improved resolution and the attractive commercial possibility of developing a device that could form a critical part of many integrated waveguide systems has resulted in the recent publication of a number of novel electrooptic surface-wave prism deflector (*ESP*) devices.

A promising planar configuration device based on a *Y*-cut diffused $LiNbO_3$ waveguide was devised independently by Kaminow and Stulz [94] and Tsai and Saunier [95]. The basic deflector is formed by narrow electrodes deposited on the surface of an optic waveguide. Two electrodes are parallel to the direction of beam propagation, and the third is tilted to give a surface field gradient. This effectively simulates a thin-film electrooptic prism deflector. The electrode arrangement is shown in Figure 12.31a.

An example of an uncompensated $LiNbO_3$ waveguide array structure built by Tsai and Saunier with $A = 150$ μm, $L = 10$ mm, and $N = 4$ required 8 *V* per beam position. These were completely resolved at 632.8 nm. With a capacitance of 4 *pF* the device had a subnanosecond switching capability or as a modulator, driven from a 50-Ω source, a base bandwidth of 1.66 GHz.*

An alternative approach by Sasaki [96] and Saunier et al. [97] shown in Figure 12.32 uses an array of interdigital finger electrodes to give a linear variation of phase retardation across an array of wave-guide channels.

The electrooptically induced phase for light entering the device at aperture coordinate z is given by

$$n(Z) = \frac{-\pi n_e^3 r_{33}}{\lambda_0} \int_0^L E_z zy \, dy \tag{100}$$

Kaminow and Stulz have shown that the linear phase shift produced across the aperture causes the wavefront to rotate through a scan angle ϕ given by

$$\tan \phi = \frac{2\lambda_0 n_o}{\pi n_e A} \tag{101}$$

where

$$n_o = \frac{2n_e^3 r_{33} V_0 L}{\lambda_0 A} \tag{102}$$

When a number of basic prism deflectors are placed side by side along a line perpendicular to the direction of propagation of the light beam, an ESP array is produced, as shown in Figure 12.31b. Recently, Revelli [98] has shown that the periodicity involved in arraying prisms in parallel increases the number of resolvable spots by only a factor of 2 on that of a single element alone. To overcome

---

* The lithium niobate surface waveguide was formed by diffusing a layer of titanium approximately 10-Å thick into the surface of *Y*-cut $LiNbO_3$. This required heating to approximately 980°C in an inert atmosphere for a period of 4 h [96].

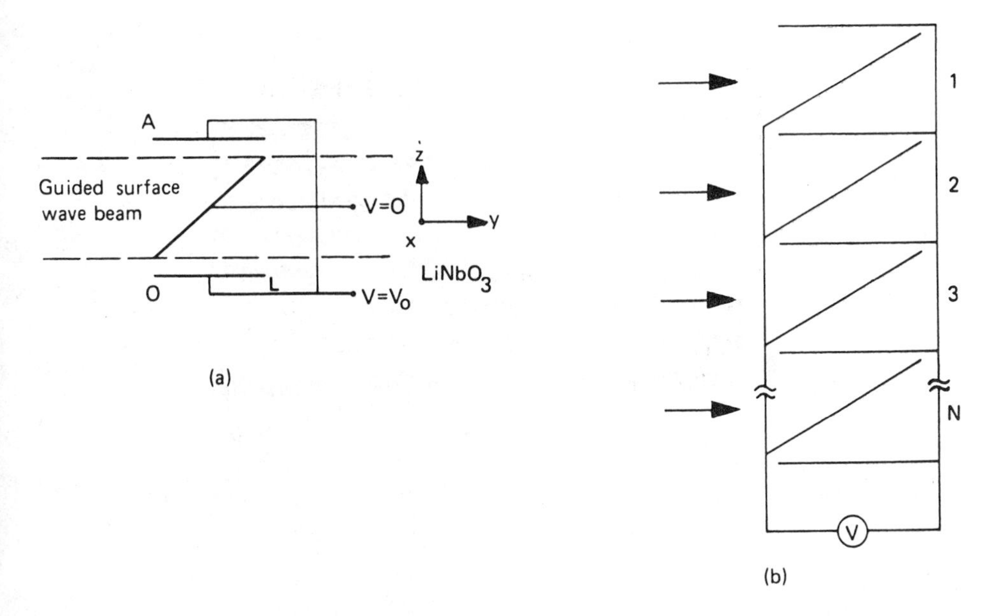

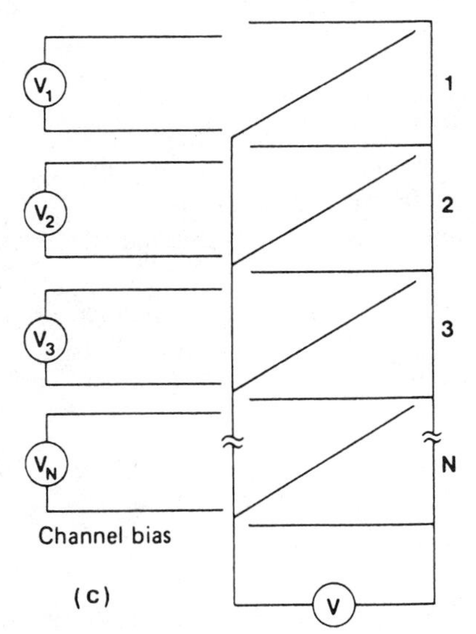

Figure 12.31 Schematic patterns for ESP devices: (a) single-prism element, (b) array of prism elements, (c) surface array with phase linearization.

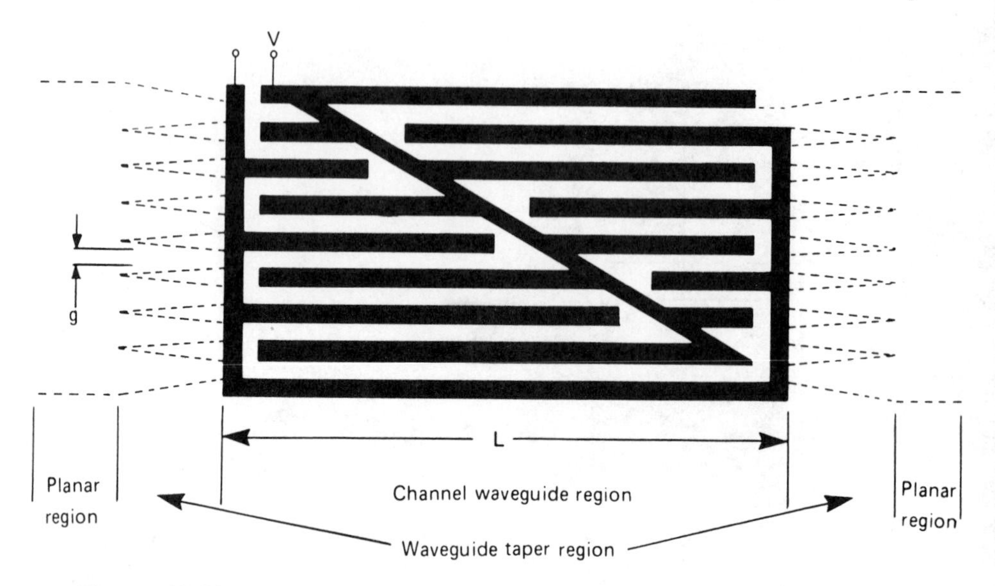

Figure 12.32 Schematic multichannel waveguide deflector.

this difficulty, a surface-wave analog to Ninomiya's [92] bulk device might be constructed in waveguide form, as indicated in Figure 12.31c. By this bias technique the number of resolvable spots would approach the theoretical value of N times that of the single-element resolution. In practice the realization of such a solution is somewhat complicated, and no experimental details are available for phase-corrected *ESP* arrays.

If ϕ is the incremental phase change produced between two adjacent waveguide channels, then it has been shown by Bulmer et al. [99] that the number of spots is

$$N_R = \frac{N\,\Delta\gamma}{\pi} \tag{103}$$

and that by taking the electric field midway between the electrodes as

$$E_z = \frac{2V_{app}}{\pi g} \tag{104}$$

where g is the spacing between electrodes, then

$$\Delta\phi = 2N_0{}^3\frac{r_{33}V_{app}L}{\lambda_0 gN} \tag{105}$$

and the number of resolvable spots becomes

$$N_R = 2N_0^3 \frac{r_{33} V_{app} L}{\pi \lambda_0 g} \tag{106}$$

or the voltage required to deflect the spot to its first position is given by

$$V_1 = \frac{\pi \lambda_0 g}{n_e^3 r_{33} L} \tag{107}$$

The deflection is continuous, but the maximum useful deflection is through N_R resolvable spots, which is limited to the number of channels N.

For the device built by Sasaki [96] where $N = 20$, $L = 18$ mm, and $g \approx 10$ μm, (107) gives a voltage of 1.7 V for one resolvable spot position. The experimental device achieved about 16 resolvable positions for a voltage swing of ± 16 V, although the signal-to-noise ratio was poor at voltage levels above 5 resolvable spot positions. At 148 pF the capacitance of this device was high for the resolution obtained.

A similar five-channel device constructed by Saunier et al. [97] of dimensions $L = 3.6$ mm and $g \approx 7.5$ μm required 3.1 V for one resolvable spot position. Saunier demonstrated the recycling property of this type of deflector and operated his device up to 45.5 V, finishing the fourth cycle of five resolvable positions. The capacitance was only 4.3 pF, which demonstrates the critical depndence of the type of deflector on the design of the electrode array. This device is being studied for application to gigahertz-rate *A/D* convertors.

A Bragg deflector can be made using the E-O effect to produce a refractive index grating in an optical waveguide [100]. As with acoustic Bragg deflectors, light enters the grating region at an angle

$$\sin \theta_B = \frac{\lambda}{2\Lambda} \tag{108}$$

and when $Q = 4l\lambda/\Lambda \gg 1$, that is, the Raman-Nath parameter is large, light can be reflected with high efficiency, and periodic exchange takes place between the undeflected and deflected beams with complete power transfer to the deflected beam occurring when

$$\frac{l\,\Delta n}{\cos \theta_B} = \frac{n\lambda_0}{2} \tag{109}$$

Auracher et al. [100] used *Y*-cut $LiNbO_3$ as the E-O substrate but produced a waveguide by out-diffusion of LiO_2 for 0.5 h at 980°C in a flowing O_2 atmosphere. The interdigital electrode structure used was designed for a beam width of 0.5 mm, and the three electrode configurations tested are shown as deflector designs *i*, *ii*, and *iii* in Table 12.6. Clearly these fixed-grating Bragg deflectors are digital deflectors as the deflection angle cannot be altered without changing

Table 12.6 Electrooptic Bragg Surface-Wave Deflector Using $LiNbO_3$

	i	*ii*	*iii*	*iv*
Finger gap width, a (μm)	3	4	6	13.33
Length of fingers, *l* (mm)	1.6	3	6.6	
Number of pairs of fingers, *N*	42	31	21	15
Drive voltage for maximum intensity in deflected beam, V_m	8.3	41	2.6	9.5
Deflection angle, $2\theta_B$	3.03	2.26	1.51	1.24
Capacitance, *C* (*pF*)	42	58	88	
Cutoff frequency, *fc* (*MHz*)	151	110	72	
Deflection efficiency, η (%)	96	93	94	95

the electrode configuration. They are, however, potentially very interesting devices, particularly as the active component for modulation and deflection in integrated optics. Already a 32-element integrated-optic spatial modulator IOSM has been built by Veber and Kenan [101] and computer logic elements proposed using Bragg E-O surface-wave deflectors. These devices are shown in Figure 12.33. There is a variety of possible parallel and serial combinations for EOBDs (electrooptic Bragg deflectors) that would achieve multispot deflection, and many

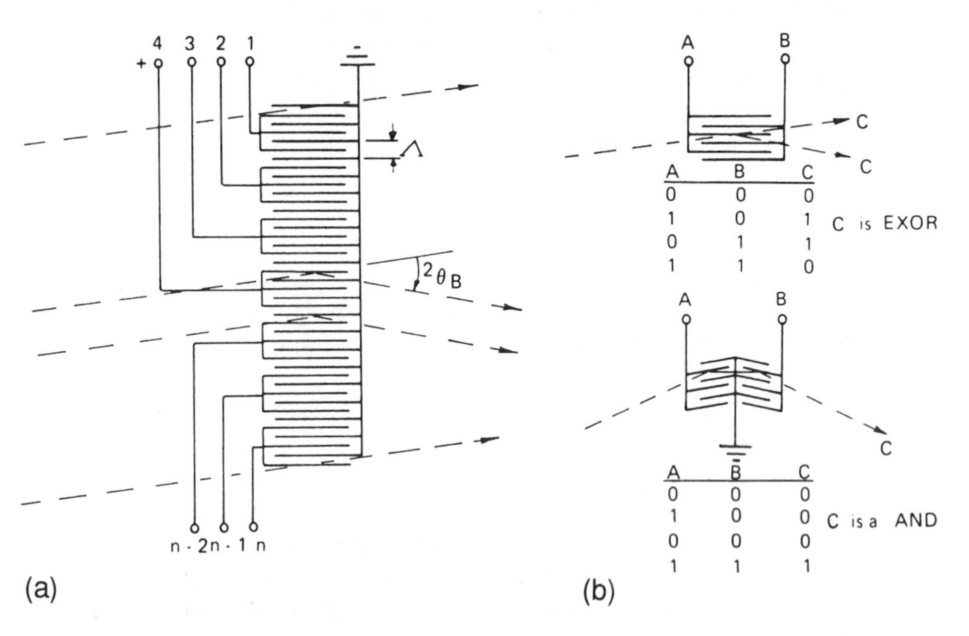

Figure 12.33 (a) E-O Bragg surface-wave spatial modulator. (b) E-O Bragg surface-wave computer logic elements.

integrated devices will no doubt be built and evaluated. Nevertheless, it does seem from the devices already evaluated that this simple and efficient deflector may form a building block for future integrated-optic systems.

It is also of interest to those working in this area that a single-spot EOBD could be used as an efficient high-speed modulator, replacing many of the bulk E-O and A-O modulators now in use. For discrete component applications tapered-gap prism couplers have recently reduced insertion losses to below 0.5 *dB* (Sarid et al. [102]) and piezoelectric induced acoustic resonances familiar to those who have experimented with E-O $LiNbO_3$ have been successfully damped by simple mechanical means (Ramachandran [103]).

Electrooptically Switched Directional Couplers in Integrated Optics

An integrated optics directional coupler is formed by fabricating two parallel waveguides in close proximity so that light in one waveguide can couple to the

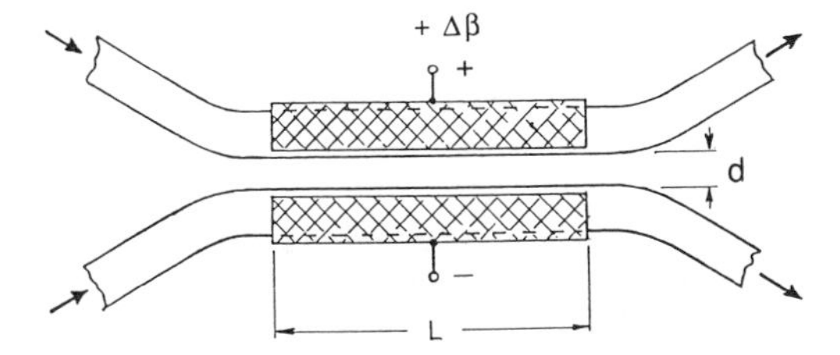

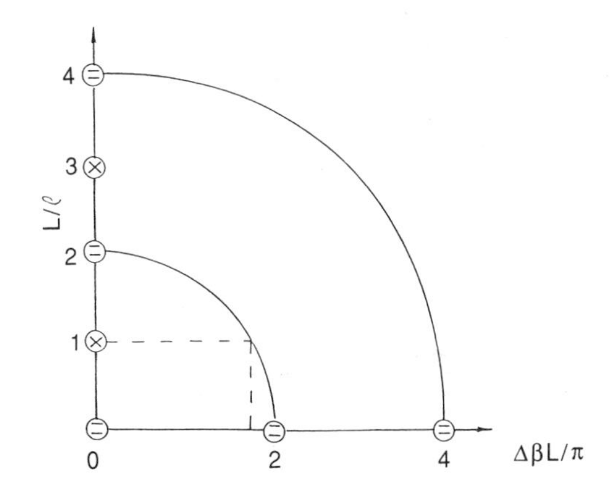

Figure 12.34 Crossbar switching diagram for a switch with uniform electrodes. The isolated points marking the conditions required for complete energy crossover are marked by ⓧ and arcs indicating the conditions required for the straight-through state are marked by ⊜.

other via the evanescent fields. Such a waveguide directional coupler is shown in Figure 12.34.

If the waveguides have the same propogation constants β_r and β_s and energy is incident in only one guide, then light will transfer completely to the other guide in a distance $l = \pi/2K$, where K is the coupling coefficient given by $K = K_0 \exp(-d/\lambda)$. (K_0 and λ depend on waveguide parameters, and d is the separation between the guides.)

Much of the early work on static waveguide coupling was carried out at Bell Telephone Labs by Miller [104] and Marcatili [105]. The first published work where it was suggested that the coupled energy could be controlled by fabricating the coupler in an electrooptic material occurred in 1972, and the first practical implementation of such switches using diffused strip guides in $LiNbO_3$ [106] and metal-gap strip guides in GaAs [107] was reported in 1975.

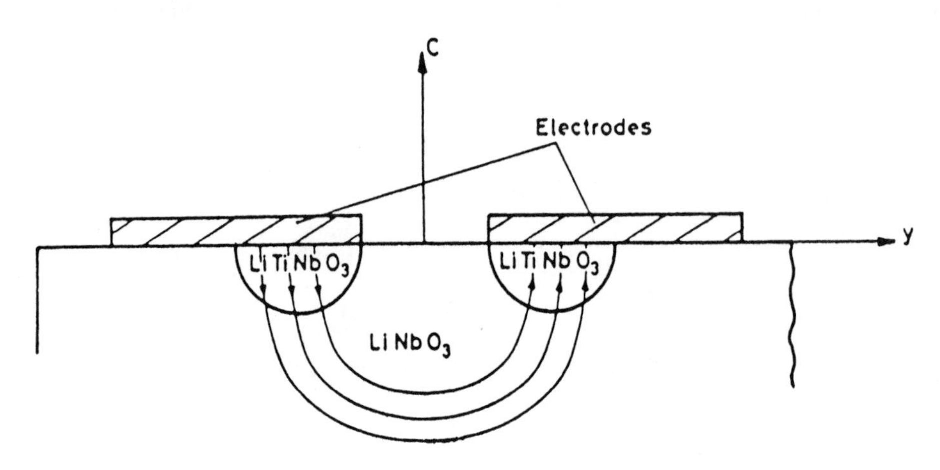

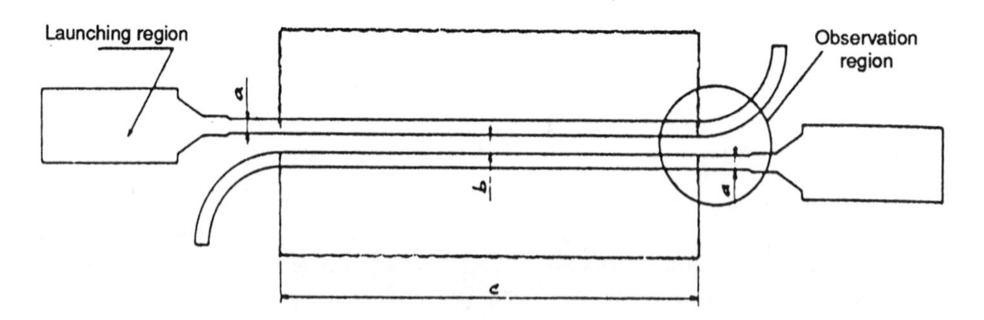

Figure 12.35 Crossbar diagram of a stepped $\Delta\beta$ switch.

The most successful of these optical devices used lithium niobate as the substrate with titanium-diffused strip guides. The first type of directional coupler switch that was experimentally demonstrated is illustrated in Figure 12.35. The switch is designed so that it is in the crossover state with no voltage applied to the electrodes. In this condition $L = l$ or $L = (2n + 1)l$. The applied electric fields, which are normal to the surface, are of opposite sense and produce by the electrooptic effect an optical phase mismatch $\Delta\beta = \beta_r - \beta_s$. The required phase mismatch to switch completely to the bar state is $\Delta\beta \cdot L = \sqrt{3}\pi$. To obtain the cross state, however, exact odd-integer values of L/l must be attained during device fabrication, and voltage adjustment cannot compensate for fabrication errors. Many switches of this basic type have been constructed, but it has proved difficult to achieve a good crossover state because of this practical requirement.

To obtain a cross state with better than $-20dB$ crosstalk requires manufacturing techniques holding L to better than $\pm 6\%$. Nevertheless, recent work on a

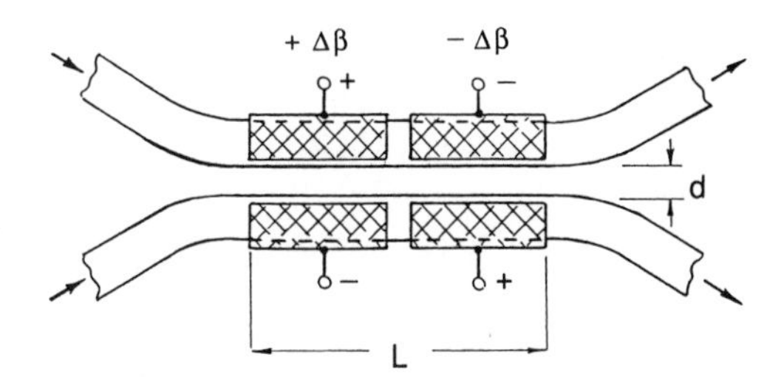

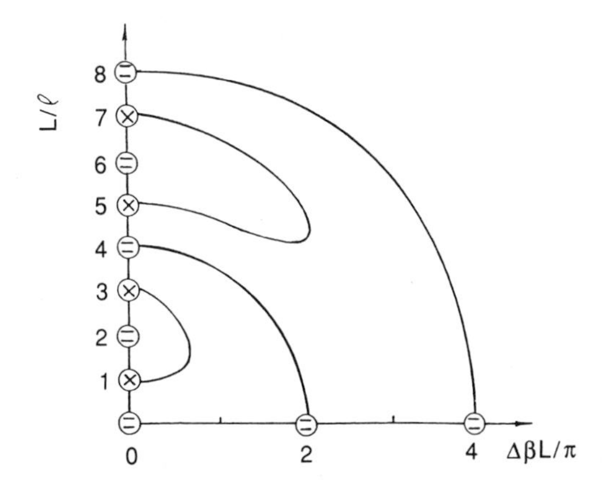

Figure 12.36 The crossbar diagram of a stepped $\Delta\beta$ switch.

4×4 crossbar switch array has [108] demonstrated extincion ratios of between -12 to -35 *dB* for the 16 possible states. For this uniform $\Delta\beta$ design bar state voltages were 10.5 ± 1.5 *V*.

A directional coupler switch that allows electrical adjustment to achieve both switch states thus eliminating stringent fabrication requirements is illustrated in Figure 12.36. In this configuration the interaction length L is divided into two equal sections, and the electrodes split to achieve opposite senses for $\Delta\beta$ in the two sections when applying voltages of opposite polarity.

The phase-reversal coupler was first proposed by Kogelnik et al. [109], and the first successful operation of a device was reported by Schmidt et al. [110].

The behavior of the $\Delta\beta$ reversal switch is described by the switching diagram of Figure 12.36. The diagram indicates that the stepped $\Delta\beta$ configuration makes available design ranges for L/l values (e.g., the range $L/l = 1$ to 3) in which complete crossover can be obtained by applying a suitable voltage to the electrodes, while further increase in the voltage produces a bar state. Many of these switches have been built with optical crosstalk below -20 *dB*, and recent work by McCaughan et al. [111] would suggest that -30 *dB* should now be obtained.

LIST OF SYMBOLS

B	Kerr constant for quadratic E-O liquid
$\mathbf{D}$	Electric displacement vector
d_{ij}	Linear piezoelectric coefficient
$\mathbf{E}$	Electric field vector
g_{ijkl}	Induced polarization coefficient
h_{inkl}	Quadratic E-O coefficient
M_R	Figure of merit for spatial resolution of deflector
M_τ	Figure of merit for temporal resolution of deflector
N_R	Number of resolvable optic spots
$\mathbf{P}$	Polarization vector
P_r	Reactive drive power
p_{cj}	Strain-optic coefficient
R	Linear capacity of digital light deflector (DLD)
R_e	Square resistance of electrodes
r_{ij}	Linear E-O coefficient
S/N	Signal-to-noise ratio
T_c	Transition temperature
$\tan\delta$	Loss tangent
V_{app}	Applied voltage
ϵ_0	Permittivity of free-space susceptibility
θ_R	Diffraction-limited optic beam divergence
γ	Plane angle of harmonic wave

REFERENCES

1. F. Pöckels, *Lehrbuck Der Kristalloptic,* B. Teubner, Leipzig, 1906.
2. G. D. Boyd and D. A. Kleinman, *J. Appl. Phys., 39*:3597 (1968).
3. A. Yariv, *Quantum Electronics*, Wiley, New York, 1967.
4. M. Born and E. Wolf, *Principles of Optics*, 5th ed., Pergamon, London, 1975.
5. I. P. Kaminow and E. H. Turner, *Proc. IEEE, 54*:1374 (1966).
6. M. J. Dore, *SRDE Report 68022*, HMSO, Oct. 1968, p. 31.
7. R. J. Pressley, ed., *Handbook of Lasers*, Chemical Rubber Co., Cleveland, 1971.
8. J. T. Milek and M. Neuberger, Linear E-O modulator materials. In *Handbook of Electronic Materials*, IFI/Plenum, New York, 1972, Vol. 8.
9. Yu S. Kuz'minour, V. V. Osiko, and A. M. Prokhorov, *Sov. J. Quantum Electron., 10*:941 (1981).
10. S. M. Lee and S. M. Hauser, *Rev. Sci. Instrum., 35*:1679 (1964).
11. U. Kruger, R. Pepperl, and U. Schmidt, *Proc. IEEE, 61*:992 (1973).
12. B. H. Billings, *J. Opt. Soc. Am., 39*:797 (1949).
13. B. Zwicker and P. Scherrer, *Helv. Phys. Acta, 16*:214 (1943).
14. R. Carpenter, *J. Opt. Soc. Am., 40*:225 (1950).
15. W. L. Smith, *Appl. Opt., 16*:1798 (1977).
16. I. P. Kaminow and E. H. Turner, *Proc. IEEE, 54*:1374 (1966).
17. W. Haas and R. Johannes, *Appl. Opt., 6*:2007 (1967).
18. A. R. Johnson and J. M. Weingart, *J. Opt. Soc. Am., 55*:828 (1965).
19. A. R. Johnson, *Appl. Phys. Lett., 7*:195 (1965).
20. K. Nassau, H. J. Levinstein, and G. M. Loiacono, *J. Phys. Chem. Solids, 27*:989 (1966).
21. F. C. Zumsteg, J. D. Bielein, and T. E. Gier, *J. Appl. Phys., 47*:4980 (1976).
22. Tso Yee Fan, C. E. Huang, B. Q. Hu, R. C. Eckardt, Y. X. Fan, R. L. Byer, and R. S. Feigelson, *Appl. Opt., 26*:2390 (1987).
23. G. A. Massey, T. M. Loehr, L. J. Willis, and J. C. Johnson, *Appl. Opt., 19*:4136 (1980).
24. R. J. Pressley, ed., *Handbook of Lasers*, Chemical Rubber Co., Cleveland, 1971.
25. J. C. Filippini, *J. Phys. D., 8*:201 (1975).
26. H. Krause and K. Barner, *J. Phys. D., 10*:2429 (1977).
27. U. Kruger, R. Perrerl, and U. Schmidt, *Proc. IEEE, 61*:992 (1973).
28. M. Blanchet, in *Proc. 8th Int. Symp. on High Speed Photography, Stockholm, Sweden*, 1968, p. 64.
29. G. H. Haertling, *J. Am. Ceram. Soc., 3*:269 (1972).
30. G. H. Haertling and C. E. Land, *IEEE Trans. Ultrason., 3*:269 (1972).
31. T. J. Cutchen, J. O'Harris, and G. R.Laguna, *Appl. Opt., 14*:1866 (1975).
32. K. Ueno and T. Saku, *Appl. Opt., 19*:164 (1980).
33. T. J. Nelson, *Bell Syst. Tech. J., 43*:821 (1964).
34. W. Kulche, K. Kosanke, E. Max, M. A. Haerger, T. J. Harris, and H. Fleischer, *Appl. Opt., 5*:1657 (1966).
35. J. C. Bass, *Radio Electron. Eng.*, 345 (Dec. 1967).
36. S. Flugge, *Handb. Phys., 24*:431 (1956).
37. R. Pepperl, *Opt. Acta, 24*:413 (1977).

38. W. Kulcke, K. Kosanke, E. Max, H. Fleisher, and J. J. Harris, *Optical and Electro-Optical Information Processing*. M.I.T. Press, Cambridge, Mass., 1965, Chap. 33.
39. J. M. Ley, T. M. Christmas, and C. G. Wildey, *Proc. IEEE, 117*:1057 (1970).
40. U. Kruger, R. Pepperl, and U. Schmidt, *Proc IEEE, 61*:992 (1973).
41. U. J. Schmidt, E. Schroder, and W. Thust, *Appl. Opt., 12*:460 (1973).
42. S. K. Kurtz, *Bell Syst. Tech. J., 45*:1209 (1966).
43. I. P. Kaminow, *Appl. Opt., 3*:511 (1964).
44. T. C. Lee and J. D. Zook, *IEEE J. Quantum Electron., QE-4*:442 (1968).
45. F. S. Chen, J. E. Geusic, S. K. Kurtz, J. G. Skinner, and S. H. Wemple, *J. Appl. Phys., 37*:388 (1966).
46. J. F. Lotspeich, *IEEE Spectrum*, 45 (Feb. 1968).
47. M. Born and E. Wolf, *Principles of Optics*, 5th ed., Pergamon, London, 1975.
48. A. L. Buck, *Proc. IEEE* (*Lett.*), 448 (March 1967).
49. L. Beiser, *J. Opt. Soc. Am., 57*:923 (1967).
50. S. W. Thomas, *Proc. 13th Int. Congr. on High Speed Photogr. and Photonics, Tokyo, 1978*, Vol. 189, SPIE, 1978, p. 499.
51. C. A. Ebbers, *Appl. Phys. Lett. 52*:1948 (1988).
52. J. D. Beasley, *Appl. Opt., 10*:1934 (1971).
53. V. J. Fowler, C. F. Buhrer, and L. R. Bloom, *Proc. IEEE* (*Corres.*), 1964 (Feb. 1964).
54. B. N. Grieb, P. A. Korotkov, and V. N. Mal'nev, *Sov. Phys. J. USA, 12*:1207 (1976).
55. O. N. Stravroudis, *The Optics of Rays, Wavefronts and Caustics*, Academic, New York, 1972.
56. C. L. M. Ireland, *Opt. Commun., 30*:99 (1979).
57. B. B. Sevruk and N. A. Gusak, *Opt. Spektrosk, 45*:910 (1978).
58. R. A. Elliot and J. B. Shaw, *Appl. Opt., 18*:1025 (1979).
59. I. P. Kaminow and E. H. Turner, *Proc. IEEE, 54*:1374 (1966).
60. A. A. Basov, A. A. Vorob'yev, and I. G. Katayev, *Radio Eng. Electron. Phys.,* 22:77 (1977).
61. Y. Ninomiya, *IEEE J. Quantum Electron., QE-9*:791 (1973).
62. I. P. Kaminow and L. W. Stulz, *IEEE J. Quantum. Electron., QE-11*:633 (1975).
63. C. S. Tsai and P. Saunier, *Appl. Phys. Lett., 27*:248 (1975).
64. C. H. Bulmer, W. K. Burns, and T. G. Giallorenzi, *Appl. Opt., 18*:3282 (1979).
65. J. F. Revelli, *Appl. Opt., 19*:389 (1980).
66. J. L. Wilkerson, and L. N. Casperson, *Opt. Commun., 13*:117 (1975).
67. H. Kotani, S. Nambo, and M. Kawabe, *IEEE J. Quantum Electron., QE-15*:270 (1979).
68. A. Korpel, *Proc. IEEE* (*Corres.*), *53*:1666 (1965).
69. W. E. Buck and T. E. Holland, *Appl. Phys. Lett., 8*:198 (1966).
70. I. Filinski, and T. Skettrup, *IEEE J. Quantum Electron., QE-18*:1059 (1982).
71. J. M. Telle and C. L. Tang, *Phys. Lett., 24*:85 (1974).

72. S. Sullivan, *Proc. E-O '80 Int. (Brighton).*
73. I. P. Kaminow and E. H. Turner, *Proc. IEEE, 54*:1374 (1966).
74. U. J. Schmidt, *Philips Tech. Rev., 36*:117 (1976).
75. H. Meyer, D. Riekmann, K. P. Schmidt, U. J. Schmidt, M. Rahlff, E. Schroder, and W. Thust, *Appl. Opt., 11*:1732 (1972).
76. R. Pepperl, *Opt. Acta, 24*:413 (1977).
77. J. M. Ley, *Electron. Lett.*, 2:138 (1966).
78. E. H. Turner, *Appl. Phys. Lett, 8*:303 (1966).
79. G. Hepner, *IEEE J. Quantum Electron., QE-8*:169 (1972).
80. L. N. Magdich, *Opt. Spectrosc.*, 248 (1969).
81. J. D. Beasley, *Appl. Opt., 10*:1934 (1971).
82. S. W. Thomas, *Proc. Conf. on High Speed Photogr., Toronto, 1976*, SPIE, Vol. 97, 1976, p. 73.
83. J. M. Ley, T. M. Christmas, and C. G. Wildey, *Proc. IEEE, 117*:1057 (1970).
84. B. Cunin, J. A. Miehe, B. Sipp, M. Ya. Schelev, J. N. Serduchenko, and J. Thebault, *Rev. Sci. Instrum., 51*:103 (1980).
85. B. V. Gisin, O. K. Sklyarov, and O. A. Herdochnikov, *Sov. J. Quantum Electron., 5*:248 (1975).
86. C. L. M. Ireland, *Opt. Commun., 27*:459 (1978).
87. C. L. M. Ireland, *Opt. Commun., 30*:99 (1979).
88. C. L. M. Ireland, *Proc. IVth Nat. Quantum Electron. Conf., Edinburgh UK, Sept. 1979*, Wiley, New York, 1980, p. 87.
89. M. Born and E. Wolf, *Principles of Optics*, 5th ed., Pergamon, London, 1975.
90. H. Koester, *Electron. Lett., 3*:54 (Feb. 1967).
91. Y. Ninomiya, *IEEE J. Quantum Electron., QE-9*:791 (1973).
92. Y. Ninomiya, *IEEE J. Quantum Electron., QE-10*:358 (1974).
93. Yu. M. Dikaev, L. Kopylev, I. M. Kotelyaneki, and V. B. Kravchenko, *Sov. J. Quant. Phys., 11*:1235 (1981).
94. I. P. Kaminow and L. W. Stulz, *IEEE J. Quantum Electron., QE-11*:633 (1975).
95. C. S. Tsai and P. Saunier, *Appl. Phys. Lett., 27*:248 (1975).
96. H. Sasaki, *Electron. Lett., 18*:295 (1977).
97. P. Saunier, C. S. Tsai, I. W. Yao, and Le T. Nguyen, *Tech. Digest of Opt. Soc. Am. Meeting on Integrated and Guided Wave Optics, Washington D.C.*, paper TuC2, 1978.
98. J. F. Revelli, *Appl. Opt., 19*:389 (1980).
99. C. H. Bulmer, W. K. Burns, and T. G. Giallorenze, *Appl. Opt., 18*:3282 (1979).
100. F. Auracher, R. Keil, and K. H. Zeitler, *Sixth Europ. Conf. on Opt. Commun., IEEE* (Sept. 1980), pp. 272–275.
101. C. M. Veber and R. P. Kenan, *Sixth Europ. Conf. on Opt. Commun., IEEE*, pp. 124–125.
102. D. Sarid, P. J. Cressmann, and R. L. Holman, *Appl. Phys. Lett. 33*:514 (1978).
103. V. Ramachandran, *J. Phys. D., 12*:2223 (1979).
104. S. E. Miller, *Bell Syst. Tech. J., 48*:2059 (1969).
105. E. A. J. Marcatili, *Bell Syst. Tech. J., 48*:2071 (1969).

106. M. Papachon et al., *Appl. Phys. Lett.*, *27*:289 (1975).
107. J. C. Cambell et al., *Appl. Phys. Lett.*, *27*:202 (1975).
108. L. McCaughan et al., *Appl. Phys. Lett.*, *47*:348 (1985).
109. H. Kogelnik et al., *IEEI J. of Qu. Elec.*, *12*:396 (1976).
110. L. V. Schmidt et al., *Appl. Phys. Lett.*, *28*:503 (1976).
111. L. McCaughan et al., *J. of Lightwave Tech.* *4*:1342 (1986).

13

Optical Disk Scanning Technology

Tetsuo Saimi

Matsushita Electric Industrial Co., Ltd.,
Kodoma, Osaka, Japan

13.1 INTRODUCTION

13.1.1 Progress in Optical Disk Technology

The fundamental concept of an optical disk dates back to 1961 when Stanford Research Laboratories developed a video disk using a photographic technique. However, the low luminance of available light sources yielded reproduced images of low quality. Columbia Broadcasting System announced the EVR (Electronic Video Recorder) system in 1967, but enormous costs ultimately forced them to discontinue development. The invention of the laser by Maiman et al. in 1960 provided the light source considered to be most suitable for optical disks.

Lasers have good temporal and spatial coherence, which enable one to obtain the small, diffraction-limited beam spot necessary for high-quality information retrieval from optical disks. After many approaches were considered, the basic design of optical disks, the ''bit-by-bit'' recording method, was developed in the 1970s. The first optical video disk system for commercial use, the VLP (video long play), was released in 1973 by Philips of Holland and MCA (Music Corporation of America) of the USA. In early systems, the He-Ne laser was the preferred light source. The introduction of many new optical disk systems soon followed. The 12-cm-diameter digital audio disk (DAD), later called the CD (compact disk), was announced in 1978. Standardized CD products from several manufacturers became available in December 1982. CD players use semiconductor lasers to allow the design of small and lightweight players. These playback-only systems marked the inception of optical disk products.

Write-once optical disk systems were first introduced by Philips in 1978 [1]. Meanwhile, development of erasable optical disk systems began as the performance of reversable media progressed. Magnetooptical disks that utilize a magnetic field reversal for recording and the Kerr effect for playback have been reported at technical meetings and are now ready for commercial production, as are phase-change disks that utilize an amorphous-to-crystalline phase change for recording and a reflectivity change for playback.

13.1.2 Characteristics of Optical Disks

Optical disks are now used in various applications, including computer memory devices, picture files, and document files. The advantages of optical disks over other known memory devices are:

1. Large capacity/high information density
   The information capacity of the optical disk is more than 10 times as large as that of rigid magnetic disks.
   The recording density of optical disks is about 0.5 to 1 bit/μm^2 (645 bit/mil^2), and its cost per bit is much lower than magnetic media.
2. Fast random-access mechanism that allows access to large mass memories.
3. Reliability
   The information surface of the disk is covered with a protective layer which ensures a long archival life.
   Information retrieval is achieved without physical contact between the head and disk, which increases the reliability of stored information.
4. Replication
   Mass production by injection molding or other high-volume techniques is possible, providing replica memories at low cost.
5. Removability
   A large quantity of data can be handled by exchanging disks. Using an autochanging system, gigabytes of memory can be accessed within a very short time.

Applications for computer use are being developed by utilizing these advantages of optical disks.

13.1.3 Principles of Optical Read/Write [2–5]

In many optical disks, as in the normal audio disk, information is recorded in a spiral groove, referred to as the "track." The information cells shown in Figure 13.1 are called "pits," which are discontinuous small bumps or differential reflectivity patterns or phase-shifting patterns with differential reflectivity. Information signals are derived from changes in luminance caused by diffraction of the laser beam by the pits (which are about 1 μm^2 diffraction cells.).

The laser beam emerging from the objective lens is focused to a spot on the disk. The spot size is proportional to the wavelength λ of the laser beam and inversely proportional to the numerical aperture NA of the objective lens. The

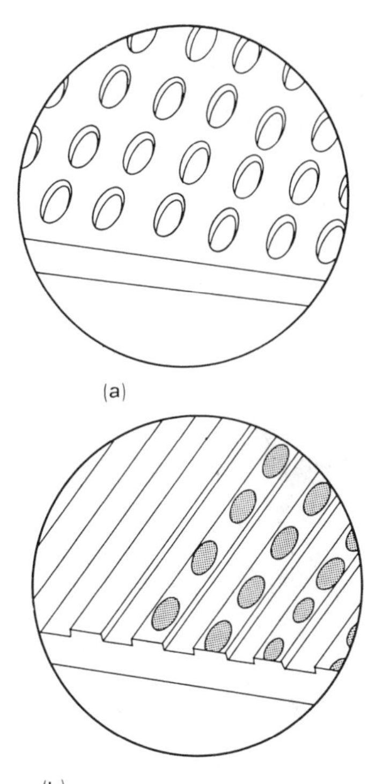

Figure 13.1 Pits pattern: (a) phase pit, (b) amplitude change pit.

numerical aperture is given by the sine of the angle θ between the optical axis and the marginal rays [3].

$$NA = n \sin(\theta) \tag{1}$$

where n represents the refractive index of the medium in object space. The full width at half-maximum (FWHM) intensity diameter (Ds) of the beam spot on the disk is expressed as

$$Ds = k \frac{\lambda}{NA} \tag{2}$$

where k represents a constant dependent upon the light amplitude distribution at the objective lens pupil. If a plane wave is incident on the objective lens, $k = 0.53$. When the incident beam is gaussian or contains some aberrations, k becomes large. Consequently, the information density of the disk is inversely proportional to the square of Ds. This means that k must not be too large. Supposing $k = 0.53$, $\lambda = 0.83$ μm (32.7 μin), and NA = 0.53, we obtain the beam spot diameter Ds = 0.83 μm (32.7 μin). Using this beam spot, more than 1.3 ×

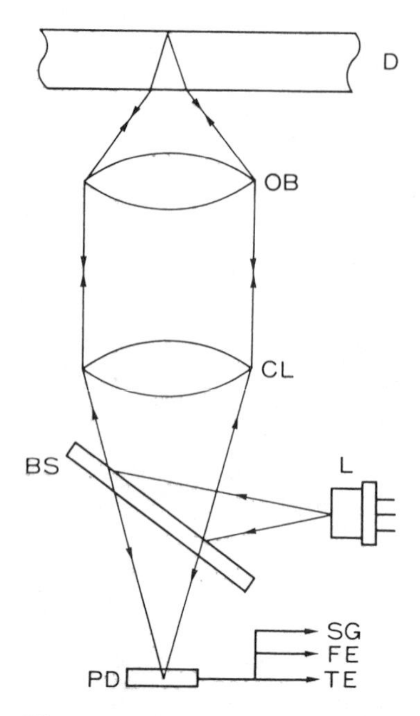

Figure 13.2 Playback optics.

10^{10} information bits can be stored on one side of a 5.25-in optical disk. Therefore the total information capacity is about 10^9 to 10^{11} bits on one side of an optical disk.

Figure 13.2 shows the playback optics for a reflective optical disk. The laser emission from the semiconductor laser (L) is reflected by a beam splitter (BS) and is incident on the objective lens (OB) through a collimating lens (CL). The wavelength (λ) generally used for optical disks is 0.78 to 0.83 μm. The numerical aperture of the lens is generally from 0.45 to 0.55 when used for read-only memories and 0.5 to 0.6 when used for read/write memories. The beam spot formed on the disk (D) has a FWHM diameter of about 1 μm. The objective lens aperture limits the spatial frequency response of the optical system. The laser beam reflected from the disk is intensity-modulated by the pits prior to a second pass through the objective lens. Part of the beam is transmitted through the beam splitter and is incident on the photodetector (PD). The information signal (SG), focusing error signal (FE), and tracking error signal (TE) are generated from the photodetectors. Reflective or transmissive mode systems can be constructed, but the reflective mode is used in most optical disk systems. In the transmissive mode, a second optical head with the photodetectors must be positioned on the other side of the disk, complicating the design of the drive. Another problem is that the pits must be so deep that degradation of signals in the replication stage is enhanced. A third problem with the transmission mode is difficulty in obtaining a good focus error signal with a satisfactory S/N ratio. Simple focus error detection methods are possible in reflective mode.

13.2 APPLICATIONS OF OPTICAL DISK SYSTEMS

13.2.1 Read-Only Optical Disk Systems

Three types of players are available for read-only optical disks: video disk (laser vision), digital audiodisk (compact disk), and data file disk (CD-ROM). Among the advantages of read-only optical disks are (1) mass replication, (2) a relatively simple optical layout as compared with the write-once head, and (3) ease of commercialization due to its use as a stand-alone unit. In read-only optical disk systems, signals are generally recorded with pulsewidth modulation (PWM), resulting in exceptionally high recording density.

Video Disk

Optical video disk systems are commerically available. They have the international standard name of LV (laser vision). Two types of disks with diameters of 30 cm and 20 cm are utilized. Rotational speed is constant at 1800 rpm for the constant-angular-velocity (CAV) mode, and a variable rotational speed of 600 to 1800 rpm is used for the constant-linear-velocity (CLV) mode. The disk substrate is 1.25 mm thick polymethyl methacrylate (PMMA). The playback time is 10 to 30 min for the CAV mode and 60 min for the CLV mode. With the CAV mode, features such as variable-speed playback, still and slow playback, etc., are feasible. This extends applicability to encyclopedias, picture files, how-to software, educational applications, and game uses. The CLV mode is used mostly for movies and live recordings.

CD

The digital audio disk system has been standardized, using the name of compact disk. The diameter of the disk is 12 cm (4.7 in), and the thickness of the polycarbonate protective layer is 1.2 mm. The linear velocity can vary from 1.2 to 1.4 m/s (3.9 to 4.6 ft/s) in the CLV mode. The maximum playback time is about 75 min, long enough to accommodate a fairly long classical music selection on a single disk. Audio signals are quantitized using 16 bits, allowing a dynamic range of 96 dB in playback. The production of CDs already exceeds that of conventional audio records, and market dominance is expected in the near future.

CDV

The CDV is an integration of CD and LV. The outer diameter section of the 12-cm disk is used for 5 min of video and CD, with a variable rotational speed of

1800 to 2700 rpm. The inner diameter section of the disk is used for 20 min of ordinary digital audio. Possible applications of CDV include video clips, music videos, etc.

CD-ROM Systems

Using an ordinary CD system, the coded data signal can be used for a read-only memory of a personal computer. More than 550 Mbytes of information can be stored on one side of a disk. This is enough capacity to store the entire text of *Encyclopedia Britanica* on one side of a CD-ROM disk.

13.2.2 Write-Once Disk Systems [6]

Write-once disks have already been commercially implemented in various applications such as (1) memory devices for the personal computers, (2) document filing systems, and (3) analogue video disk filing systems. Both PMMA and polycarbonate (PC) are being used as disk substrate materials. The thickness of the injection molded substrate is 1.2 mm. The recording mechanism of the disk may be (1) phase-changing, (2) hole-burning, or (3) bubble-forming.

In Figure 13.3, the pits formed by these different recording methods are shown. Signal pits are recorded by irradiation with a semiconductor laser focused to a spot about 1 μm in diameter. This irradiation increases the surface temperature of the recording medium to about 400 to 600°C (848 to 1272°F), and the recording takes place as the result of the consequent physical or chemical change of the medium.

Data File

Erasable optical disk systems hold great promise for application as peripheral computer memories. However, by taking advantage of their large capacity, write-once disk systems can be used in substantially the same manner as ordinary hard disks. The removability of optical disks promises a broad range of application, but standardization of products is an important ingredient. Typical specifications of write-once disk systems for personal computer application are shown in Table 13.1.

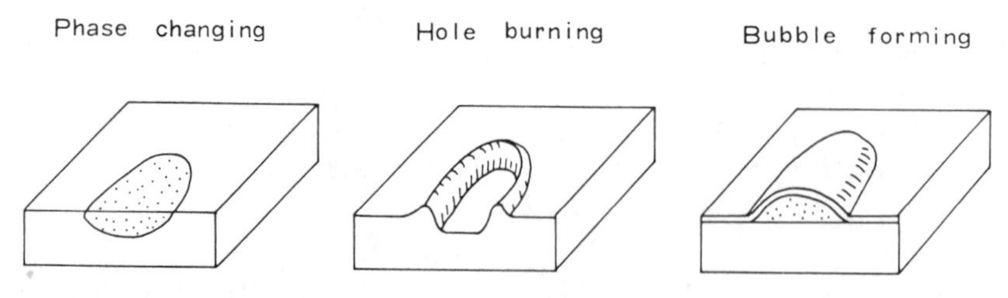

Figure 13.3 Recorded pit.

Table 13.1 Specifications of the IBM 3363 Write-Once Optical Disk Drive

Recording capacity	200 Mbytes
Sector size	512 bytes
Disk diameter	5.25 inches
Number of sectors/track	23
Number of tracks	17100
Track pitch	1.60 μm
Data transfer rate	2.5 Mbps
Bit length of the signal	$>$ 1.2 μm
Rotational speed	875 rpm
Coding method	2–7 coding
Average access time	$<$ 230 ms
Latency	34 ms
Disk substrate	Polycarbonate

Document File Systems

Document files have slightly greater redundancy than coded memory. With a resolution of 8 dots/mm, about 20,000 A4-size documents can be stored on one side of a 200-mm disk. Although the information capacity is smaller than that of the coded disk by an order of magnitude, the error rate is lower by two orders of magnitude. A document file only needs a corrected bit error rate of 10^{-9} for sufficient reliability.

Ease of recording complicated figures such as Japanese characters (Kanji) has ensured the successful commercial use of document files in facsimile systems in Japan.

Analog Video Filing Systems [7, 8]

More than 13 min of video information can be recorded in NTSC format on one side of a 200-mm disk. With 300-mm disks, 30 min of video recording is possible. The two-channel audio and video retrieval available with this system make it ideally suited for expositions in museums and libraries, shopping guides, and product information programs for dealers. The recording density is much higher for analog signals than for digital signals, but the analog signal is easily affected by dropouts caused by the medium. These dropouts cannot be corrected in the analog system as compared with digital systems. The current NTSC format has relatively low resolution, but high-definition video signal systems are under development. These will be 1.6 times higher in horizontal resolution and 2 times higher in vertical resolution. The quality of these images will be comparable to ordinary photographic film, allowing encyclopedic and other printed pictures to be successfully stored therein. See Figure 13.4.

Figure 13.4 Analog video file system.

13.2.3 Erasable Optical Disk Systems

Erasable optical disk systems will embody all of the attributes of write-once systems and have the capability of direct file replacement of current magnetic systems. Full realization of the potential of optical disk systems awaits the availability of erasable systems, particularly when used as computer peripherals. Two major families of erasable media are available: (1) phase-change erasable (PCE) system and (2) magnetooptical (MO) erasable system. Data recording on phase-change erasable media relies on the transition from a crystalline phase to an amorphous phase. Magnetooptical recording utilizes the polarization change of the laser beam magnetically modulated in media which display the Kerr effect. The principal characteristics of erasable disks are shown in Table 13.2.

With a PCE disk, the overwrite mechanism is easy to design; however, reversibility is only fair compared to the MO disk. The MO disk drive requires a complicated system for applying the writing magnetic field with a polarity opposite to the erasing magnetic field.

PCE Disk

Figure 13.5 shows the principle of the direct overwrite PCE disk. The laser intensity at the disk is modulated in correspondence with the pit pattern to be

Table 13.2 Characteristics of Erasable Disks

	PCE	MO
Recording and erasing mode	Phase change	Change of magnetization
Read	Change of amplitude	Change of polarization
Material of medium	Te-Ge-Sb	Tb-Fe-Ni-Co
Overwrite mechanism	Simple	Complicated
Magnetic field	Not required	Required
Reversibility	Fair	Good
Required laser power	High	Medium

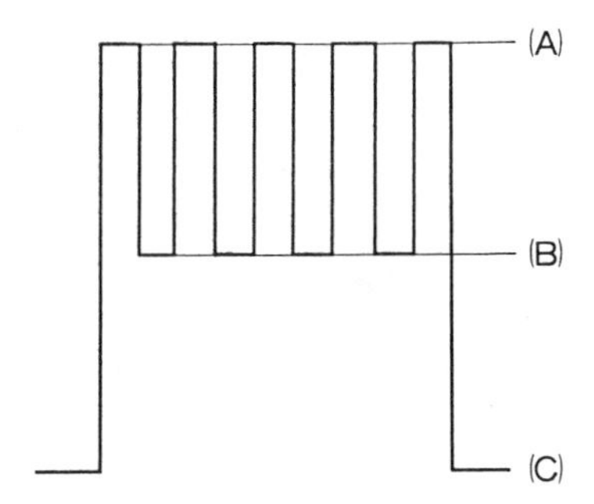

(a)

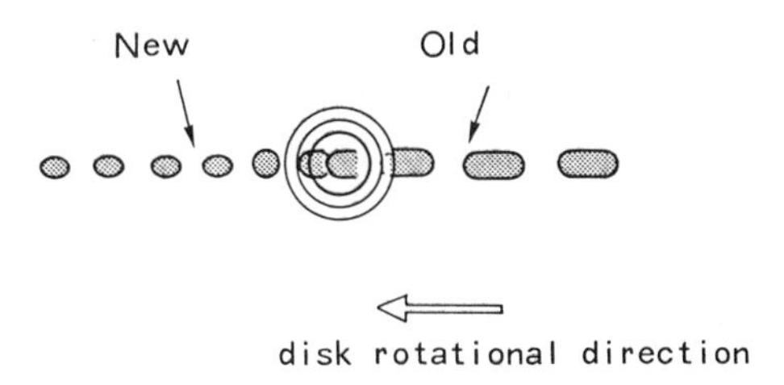

(b)

Figure 13.5 PCE overwrite method: (a) optical beam modulation for overwriting, (b) overwritten marks.

recorded between the maximum level (A) and the intermediate level (B) as illustrated in view (a). At exposure level A, the material reaches a melting temperature of over 600°C. Rapid quenching forces the material to remain in the amorphous phase, giving low surface reflectivity. Exposure level B heats the material to about 400°C, allowing rapid crystallization to proceed, giving an increased reflectivity. View (b) is a schematic illustration of the overwriting operation.

MO Disk [9, 10]

In magnetooptical disks, a light beam is directed at a material showing the Kerr effect, to record or erase information. The underlying principle is the utilization of temperature-dependent change of magnetic properties. There are several methods, including Curie point recording and compensation point recording, which can be used for recording.

Figure 13.6 is an elementary illustration showing the principle of Curie point recording. First, the magnetization of the recording layer is uniformly oriented in a given direction, as shown in view (a). When a limited area of the recording layer is irradiated with light sufficient in intensity to heat it to a temperature above the Curie point T_c, the magnetization of the local area is lost, as shown in view (b). When the exposure is discontinued, the temperature of the recording layer falls below T_c. The exposed area is remagnetized, but the direction of this magnetization coincides with the direction of the applied external magnetic field. Therefore, if the external magnetic field is applied in a direction opposite that of the magnetization of the recording layer, as shown in view (b), a magnetic domain different from the surrounding area remains, as shown in view (c), enabling the recording of binary information. For reading the signal, the recording layer is irradiated with a laser light of low power. The polarization rotations of the reflected beam from the signal surface and the land surface are in opposite directions, as shown in view (c). These beams are detected with a polarization analyzer to obtain a signal. For erasing the information, the selected area is heated to a temperature above the Curie point, as shown in view (d). In this case, the direction of the external magnetic field is reversed from that for recording.

In signal readout, the linear polarization angle of the incident beam is set at θ. The Kerr rotation angle is given by $\pm\Phi_k$. Referring to Figure 13.7, the differential output ΔI of the analyzer between the x and y directions is given by

$$\begin{aligned}\Delta I &= I_0 R[\cos^2(\theta - \Phi_k) - \cos^2(\theta + \Phi_k)] \\ &= \frac{I_0}{2} R \sin(2\theta)\sin(2\Phi_k)\end{aligned} \tag{3}$$

where R is the disk reflectivity.

Since the liner polarization angle of the incident beam is $\pi/4$ and $\Phi_k << 1$,

$$\Delta I \cong 2 I_0 R \Phi_k \tag{4}$$

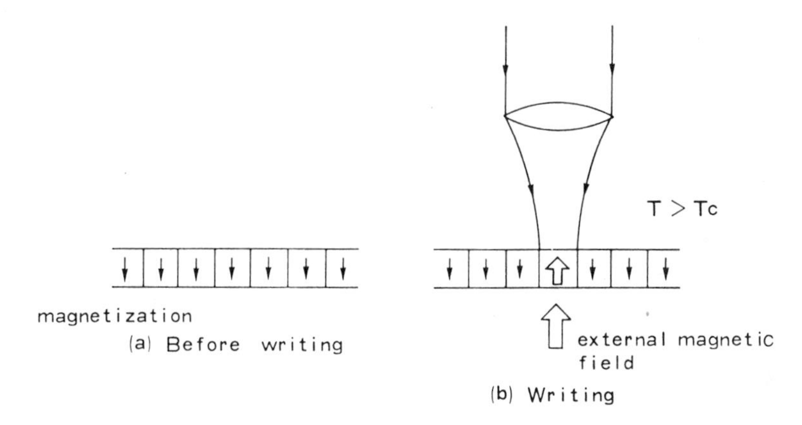

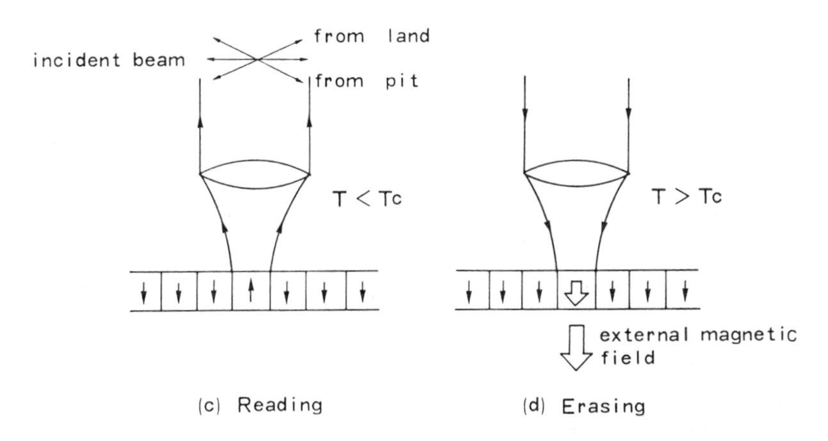

Figure 13.6 Magnetooptical disk method: (a) before writing, (b) writing, (c) reading, (d) erasing.

Thus, the playback signal level is proportional to the incident light intensity, the disk reflectivity, and the Kerr rotation angle.

13.3 BASIC DESIGN OF OPTICAL DISK SYSTEMS

13.3.1 Head Optics

As described in the preceding chapters, there are many types of optical disks for various applications, and the corresponding optical heads logically have their own optimal optics. In this section, the methods of design for the optics and mechanics of a write-once head will be described.

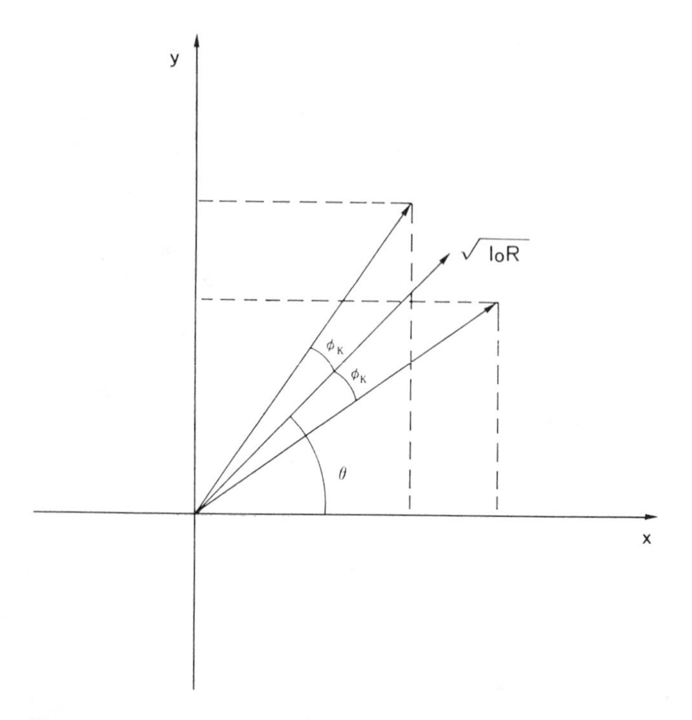

Figure 13.7 Readout of MO signal.

The following factors determine the quality of read/write signals.

1. Frequency characteristics of signals
2. Crosstalk from the adjacent tracks, which degrades read/write signals
3. Carrier-to-noise ratio (CNR) of read/write signals
4. Errors in read/write signals

Factors 1 and 2 are mainly dependent on the wave aberrations of the optics, factor 3 is as much associated with the characteristics of elements such as the semiconductor laser, detector, etc., as with wave aberrations, and factor 4 is mainly dependent on defects in the disk.

Optical Layout

The schematic construction of the optics for a write-once head is shown in Figure 13.8. In this example, the astigmatic method is used for detecting the focusing signal, and the push-pull method is used for detecting the tracking signal. The laser beam emitted from the semiconductor laser (LR) has a near-field pattern elongated in the direction of the active layer of the laser and is polarized in the

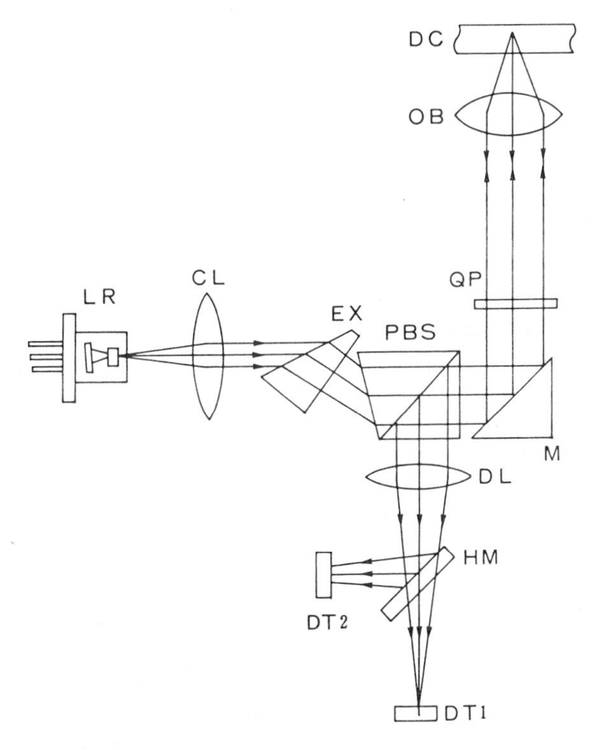

Figure 13.8 Optical layout for write-once disk.

same direction. The beam waist in this direction lies within the laser, and the beam waist in a direction perpendicular to the above direction is situated at the end facet of the laser active layer. The beam emergent from the laser is therefore anamorphic, and its far-field distribution is elliptical in cross section with an ellipticity of 2 to 3. To correct this elliptical distribution, it is necessary to use an afocal system after the collimator consisting of two cylindrical lenses or a wedge prism. With a single wedge prism, the designed incident angle of the laser beam must be approximately 69 to 72°. Since the wedge prism has chromatic dispersion, a change of the wavelength of the laser results in an angular deviation of the beam. Taking this angular deviation as $\Delta\theta$ and the focal length of the objective lens (OB) as f_0, the beam spot moves approximately by $\Delta\theta \cdot f_0$ on the disk (DC). Using a single BK7 wedge prism with the incident angle 72°, an objective lens with focal length 4.5 mm and wavelength 0.78 μm, the beam spot displacement on the disk is approximately 0.073 μm for a change of 1 nm in wavelength. Therefore, the optics should be designed such that the direction of this movement will not cause a track offset. For this purpose it is good practice to use two wedge prisms as illustrated in Figure 13.8. In the case of the playback-only optical head, the influence of the elliptical and astigmatic beam can be diminished at the cost of beam utilization efficiency. In Figure 13.8, the laser beam transmitted through a polarizing beam splitter (PBS) as a P-polarized beam passes through the $\lambda/4$ plate (QP) to become a circularly polarized beam which is incident on the objective lens (OB). The beam emerging from the objective lens is incident on the disk (DC) to form a beam spot for recording and reproducing the signals. The beam reflected at the disk enters the objective lens and again passes through the $\lambda/4$ plate to become an S-polarized beam for the polarizing beam splitter and is reflected to the detection lens (DL). The beam emergent from the detection lens is partially reflected by a half-mirror (HM) and incident on the detector (DT2) for push-pull tracking signal detection. Since the convergent beam passing the half-mirror is astigmatic, it is received by a quadrant detector (DT1) to give a focusing signal. The data signal is retrieved by sum of the output from both detectors (DT1 and DT2). The astigmatic focusing and push-pull tracking methods will be described in detail later.

Influence of Intensity Distribution

The intensity distribution of the beam incident on the objective lens is dependent on the beam divergence angle distribution of the semiconductor laser. With the objective lens aperture radius being standardized as unity and the intensity distribution of the incident beam assumed to be $\exp(-\alpha r^2)$, the amplitude distribution $g(w)$ of the beam spot on the disk is expressed by the Fourier-Bessel transform:

$$g(W) = \int_0^1 \exp(-\alpha r^2) J_0(Wr)\, d(r^2) \tag{5}$$

with $W = 2\pi nR/\lambda f_0$, where f_0 is the focal length of the objective lens and R is the polar coordinate in the focal plane. Integration gives (Appendix A equation (A4))

$$g(w) = \sum_{n=0}^{\infty} 2^n \alpha^n e^{-\alpha} \left(\frac{2J_{n+1}(W)}{W^{n+1}} \right) \tag{6}$$

Since $\alpha = 0$ for plane-wave incidence,

$$g(w)|_{\alpha=0} = \frac{2J_1(W)}{W} \tag{7}$$

This is the well-known Airy distribution. When $\alpha = 1$, the beam intensity distribution around the objective lens aperture is $1/e^2$.

Figure 13.9 shows the intensity distribution of the $|g(w)|^2$ beam spot with various values of α. It is apparent from Figure 13.9 that when $\alpha = 1$, the FWHM of the beam spot is increased by about 10% (relative to $\alpha = 0$), and the peak of the sidelobe diffraction ring is made sufficiently small.

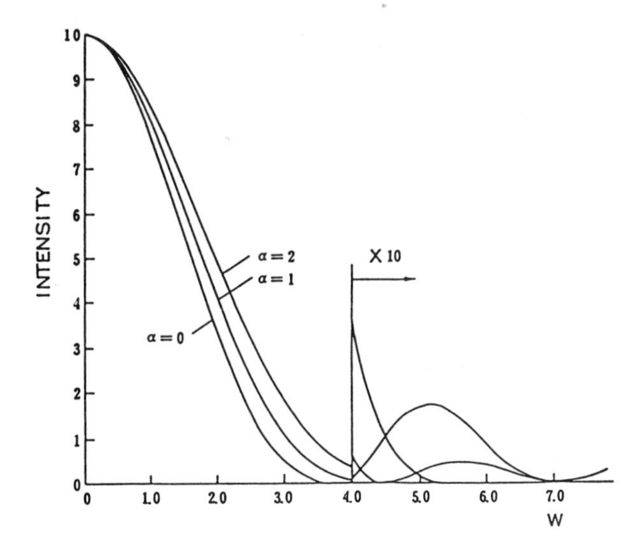

Figure 13.9 Intensity distributions of laser beam spot.

In order for the reproduced signal to have satisfactory frequency characteristics, the value of α in the signal direction must be in the range of $0 < \alpha < 1$.

On the other hand, in the direction perpendicular to the signal direction, the crosstalk from the adjacent track must be minimized. This crosstalk can be diminished by using α close to 1. Therefore, the spot on the disk need not be truly round, but an improved frequency characteristic is sometimes obtained when the beam spot is elliptical with an ellipticity of about 10%.

13.3.2 Wave Aberrations [11]

When root-mean-square wave aberration W exists, the on-axis energy density Strehl definition (SD) of the beam spot is expressed by

$$SD = 1 - k^2W^2 \qquad \text{with } k = \frac{2\pi}{\lambda} \tag{8}$$

where λ is the wavelength.

Figure 13.10 shows the relation between rms wave aberration and on-axis energy density. The on-axis energy density SD is a factor directly associated with reproduced signal SNR or record/reproduced signal SNR. The allowable rms wave aberration for the whole optical disk system is subject to Maréchal's criterion that the rms wave aberration is 0.070λ when SD has decreased about 20% from

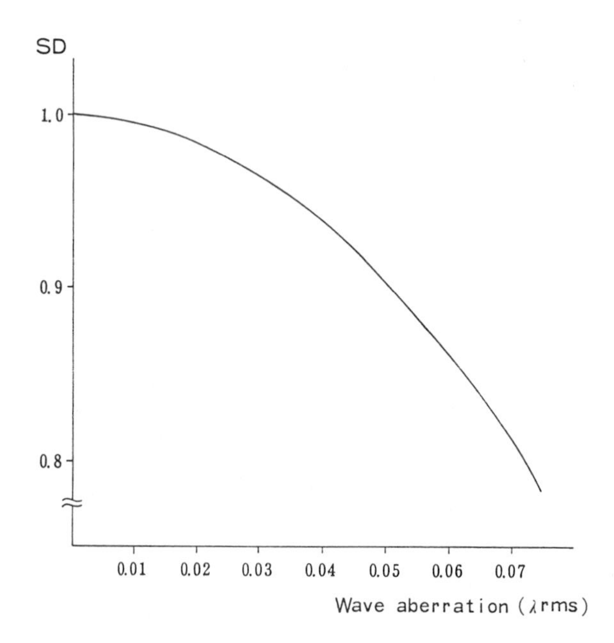

Figure 13.10 Wave aberrations versus energy density on axis.

the level at no aberration. The validity of the criterion has been endorsed by read/write experiments. This allowable wave aberration for the whole system must be allocated to disk thickness error and tilt error, initial optical aberration, and allowable defocus value.

Aberration Derived from Disk Substrate

The aberration originating from the disk substrate is composed of the aberration W_{ST} due to the error Δt of substrate thickness t and the aberration W_{TL} due to the inclination θ of the substrate. These aberrations, when small, are expressed by the following equations (equations A11) and (A14)):

$$W_{ST} = \frac{\Delta t(n^2 - 1)(NA)^4}{8\sqrt{180}\, n^3} \tag{9}$$

$$W_{TL} = \frac{t(n^2 - 1)\theta(NA)^3}{2\sqrt{72}\, n^3} \tag{10}$$

where NA is the numerical aperture of the objective lens and n is the refractive index of the disk substrate.

Figure 13.11 shows the relation between the disk thickness error and wave aberration W_{ST}. Figure 13.12 shows the relation between disk tilt angle θ and

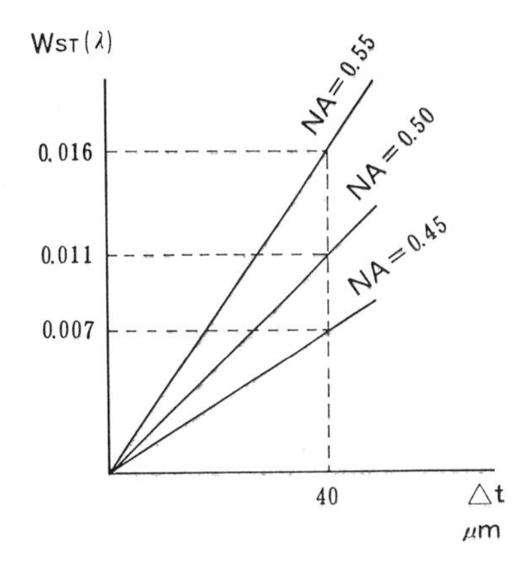

Figure 13.11 Wave aberrations versus disk thickness error.

wave aberration W_{TL} with NA as the parameter. In the usual optical disk, the practical values are $NA = 0.5$, $t = 1.2$ mm, $n = 1.51$, $\lambda = 780$ nm, $\Delta t = 40$ μm, and $\theta = 4$ mrad. Substituting these values, we obtain $W_{ST} = 0.011\lambda$ and $W_{TL} = 0.017\lambda$.

Wave Aberrations of Optical Components

Because mass-produced items are used for the disk optical components, the influence of variations in wave aberrations cannot be disregarded. Of all the components of the optical head, the objective lens and the collimating lens have the largest wave aberrations. Both the objective lens and the collimating lens

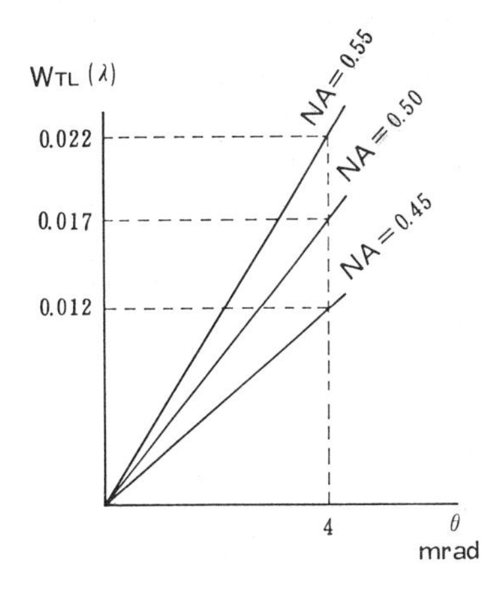

Figure 13.12 Wave aberrations versus disk tilt angle.

usually comprise three to four single lenses. Recently, however, aspherical pressed glass (APG) objective lenses have been developed and are now available on a mass production basis. Figure 13.13 shows an example of a mass-produced aspherical objective lens. Figure 13.14 shows the wave aberrations of a typical APG objective lens as measured with a Fizeau interferometer. The wave aberrations of prism systems are generally small, but as the number of prisms increases the allowance for the entire head is consumed.

Aberration Due to the Semiconductor Laser

Semiconductor lasers are generally astigmatic, and as this astigmatism is propagated and focused on the disk, variations in frequency characteristics may occur according to relative directions on the disk, or there may be reduced focusing latitude. In the record-playback optics, the electromagnetic emission from a semiconductor laser is passed through a collimating lens to yield a beam of substantially parallel rays, which is then converted by an anamorphic beam expander to a beam having substantially isotropic distribution. The correction for astigmatism is carried out concurrently in this stage. If a stationary prism is used for correction, the astigmatism generated at an angle of 45° with the prism cannot be corrected. Therefore, when the semiconductor laser is mounted at an angle of θ with the horizontal direction of the anamorphic expander prism, there

Figure 13.13 Aspherical objective lens for optical disk.

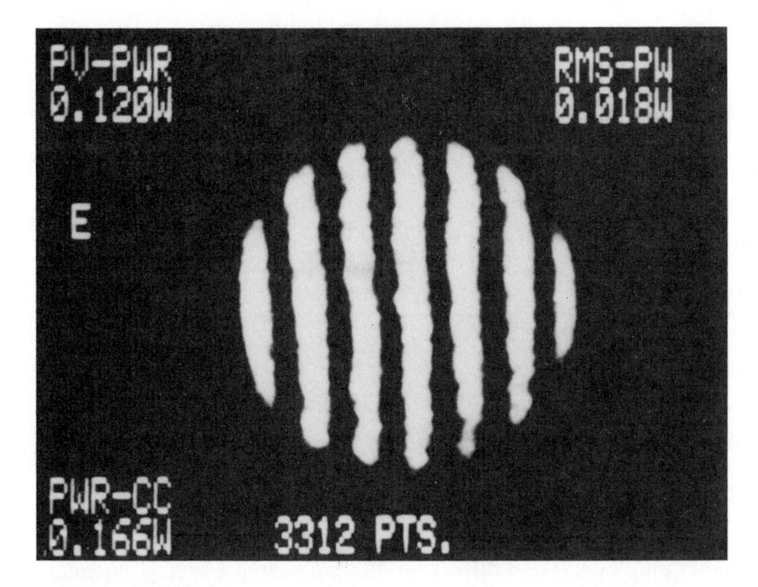

Figure 13.14 Wave aberrations of aspherical objective lens.

occurs a residual astigmatism. The wave aberration W_{LA} due to this astigmatism is expressed as (equation (A20))

$$W_{LA} = \frac{\tan\theta}{\sqrt{6}\cos^2\theta}\,\Delta L(NA_c)^2 \tag{11}$$

where NA_c is the NA of the collimating lens and Δ_L is the astigmatism of the semiconductor laser. Assuming that the allowable wave aberration dependent on the semiconductor laser is 0.010λ, the NA of the collimating lens is 0.25, and the astigmatism of the semiconductor laser is 8 μm (0.32 mil), the allowable tilt angle θ of the semiconductor laser is found from equation (11) to be

$$\theta \leqq \pm 4^\circ \tag{12}$$

Defocus

The factors responsible for defocus in an optical system may be classified as in Table 13.3. The relationship between the amount of defocus ϵ and the maximum optical path difference Δ_{DF} of the wavefront can be found from Figure 13.15:

$$\Delta_{DF} = \frac{\epsilon}{2}(NA)^2 \tag{13}$$

Table 13.3 Defocusing Factors

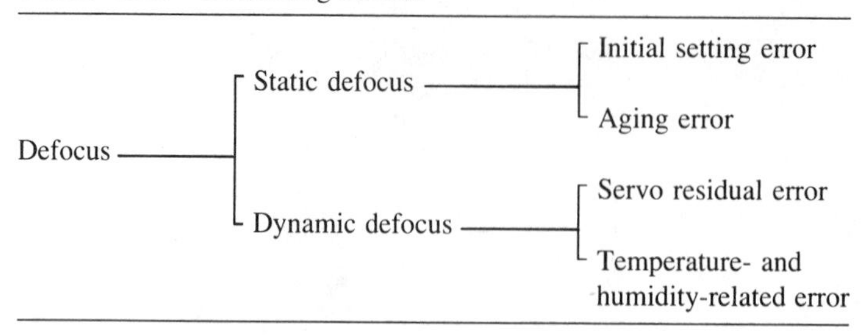

Defocus	Static defocus	Initial setting error
		Aging error
	Dynamic defocus	Servo residual error
		Temperature- and humidity-related error

Using the relationship between maximum optical path difference Δ_{DF} and wave aberration W_{DF}, wave aberration can be expressed as

$$W_{DF} = \frac{\epsilon}{4\sqrt{3}}(NA)^2 \tag{14}$$

For initial focus setting, the use of a diffraction grating having a space frequency near one-half of the cutoff frequency NA/λ of the disk optics is advantageous, for the influence of defocus is then the most pronounced. Since the track pitch of the optical disk is usually the space frequency in the vicinity, the position of best focus is where the modulation by the track is maximal. By this adjustment,

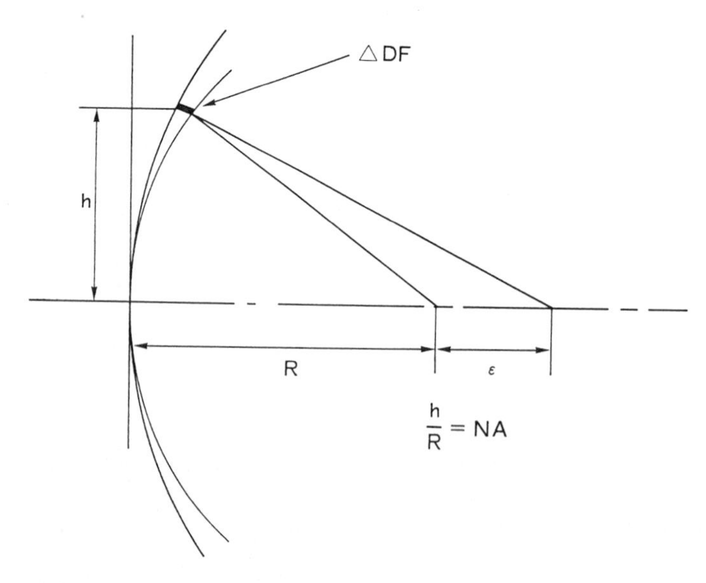

Figure 13.15 Optical path difference versus defocus.

the setting error can be reduced to less than ± 0.2 μm. Thus, with an optical disk of NA = 0.5 and λ = 0.78 μm and a defocus of ± 0.24 μm, the wave aberration is $W_{DF} = 0.011\lambda$.

Allowable Wave Aberration

Table 13.4 shows the wave aberration classified by causative factors. These wave aberrations can be integrated into the system allowance limit of 0.070λ rms as a totality. Since most of the factors responsible for wave aberrations are independent by nature, it is possible, in the actual design of an optical head, that the allowable aberration value of each optical component be fairly liberal, as shown in the table.

13.3.3 Head Mechanism

Head Construction

The optical head generally consists of an optical base forming the optics assembly and an actuator for allowing the objective lens to follow the disk plane and tracking groove.

A typical optical head construction is shown in Figure 13.16. The environmental resistance and reliability of the system are much enhanced when the number of reflective surfaces is minimized throughout the optical path from the laser to the objective lens.

The optical base has a three-point support structure which enables a two-

Table 13.4 Factors Responsible for Wave Aberrations and the Amounts of Aberrations

System allowance limit 0.070λ			
Disk 0.028λ	Thickness error	$< \pm 40$ μm	$\leq 0.011\lambda$
	Tilt	$< \pm 4$ mrad	$\leq 0.017\lambda$
Head 0.054λ	Semiconductor laser		$\leq 0.010\lambda$
	Objective lens		$\leq 0.035\lambda$
	Collimating lens		$\leq 0.025\lambda$
	Wedge prism		$\leq 0.014\lambda$
	PBS		$\leq 0.020\lambda$
	$\lambda/4$ plate		$\leq 0.020\lambda$
Defocus 0.036λ	Perpetual change		$\leq 0.011\lambda$
	Initial setting error		$\leq 0.011\lambda$
	Servo residual error		$\leq 0.023\lambda$
	Temperature-dependent error		$\leq 0.023\lambda$

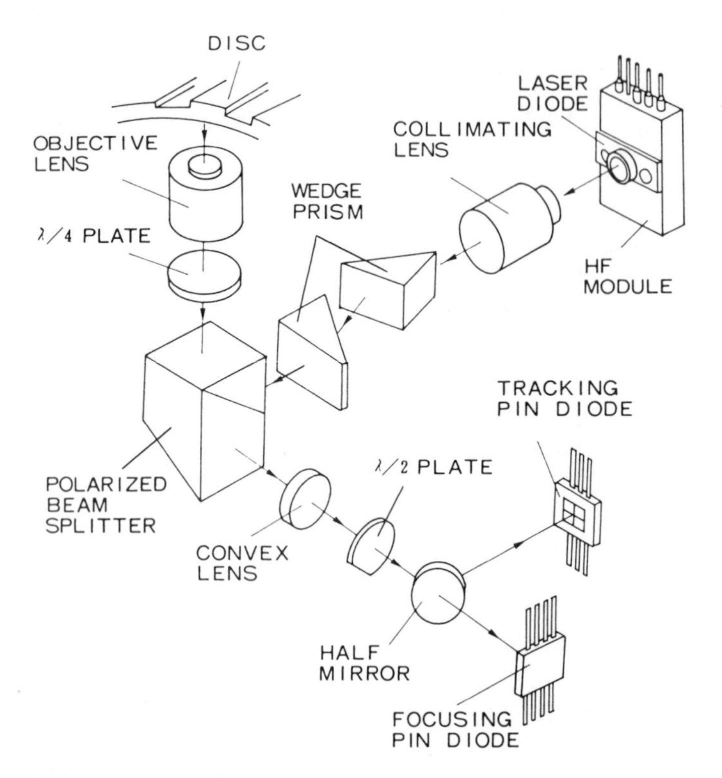

Figure 13.16 Diagram of optical head.

dimensional adjustment of the tilt angle of the head. Actually, the axis of the objective lens cannot be aligned perpendicular to the disk plane by mere mechanical means, but the tilt mechanism of the optical base or the angle of inclination of the actuator can be adequately controlled. Figure 13.17 is a schematic view showing a method of actuator tilt correction.

As shown, the actuator and the optical base are provided with a convex and a concave spherical surface, respectively, whereby tilt correction can be made two-dimensionally by means of a couple of screws and springs. As the center of the sphere is aligned with the focal point of the objective lens, there is no transverse shift of the beam passing through the objective lens.

Actuator

The actuator not only encompasses a focusing drive mechanism but has, in combination therewith, a tracking drive mechanism for following the track on the disk. Thus, the actuator is a balanced combination of a focusing mechanism and a tracking mechanism, which must be designed in such a manner that there

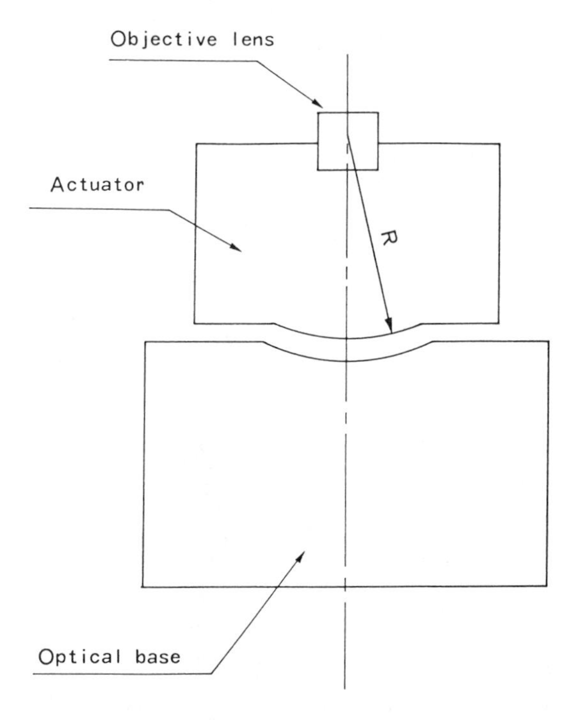

Figure 13.17 Tilt mechanism of actuator.

will be a minimum of mutual interference. The essential conditions that must be satisfied in the design of an actuator are

1. Satisfactory frequency characteristics
2. High acceleration characteristics
3. High current sensitivity
4. Broad focusing and broad tracking dynamic ranges

Typical actuator constructions satisfying these criteria are shown in Figure 13.18.

The wire-suspended actuator [12] (a) is quite simple in construction and can be moved in the focusing and tracking directions by driving the center of gravity of its movable segment. Moreover, reliability is high because the four wires can be utilized as leads to the coils.

The rotational actuator (b) is characterized in a small tilting angle of the optical axis and a large focusing dynamic range.

Figure 13.19 contains the Bode diagrams of a rotational actuator. The first-order resonant frequency f_0 of the actuator is

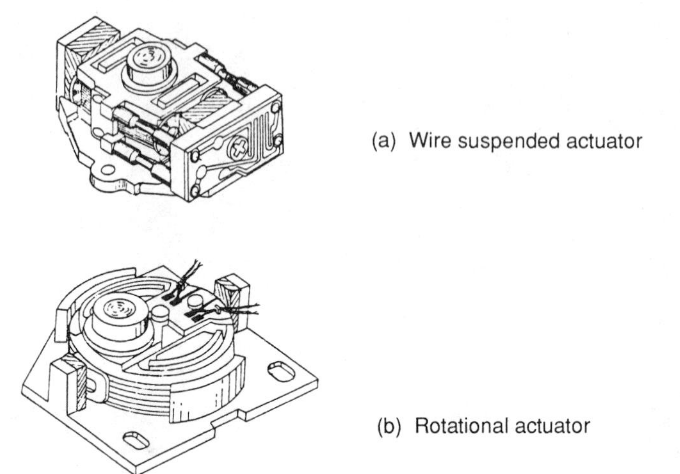

Figure 13.18 Two different types of actuator: (a) wire-suspended, (b) rotational.

$$f_0 = 2\pi \sqrt{\frac{K}{m}} \tag{15}$$

where K is a spring constant and m is a movable mass. As a rule of thumb, the dynamic frequency range of an actuator is approximately the range from the level of the basic disk rotation frequency to the peak level of high-order resonant frequency. The larger this dynamic range value is, the larger is the servo gain that can be obtained.

13.4 SEMICONDUCTOR LASER

13.4.1 Laser Structure

Al-Ga-As Double Heterojunction Laser [13]

The operating principle of a double heterojunction semiconductor laser is shown in Figure 13.20. This laser consists of three layers having dissimilar energy gaps E_g, with increased energy gaps for the n-type and p-type cladding layers on both sides of the active layer.

As a photon $h\nu g_2$ corresponding to the active layer energy gap E_{g2} (E_{g2} = $h\nu_{g2}$) passes through the active layer, the electrons in the conduction band drop into the positive holes in the valence band to trigger a stimulated emission in phase with the incident photon. As a current I_p in the normal direction is passed through this diode, the probability of the presence of electrons in the active layer 2 is increased by the energy barrier ΔE_c in the conduction band. On the other

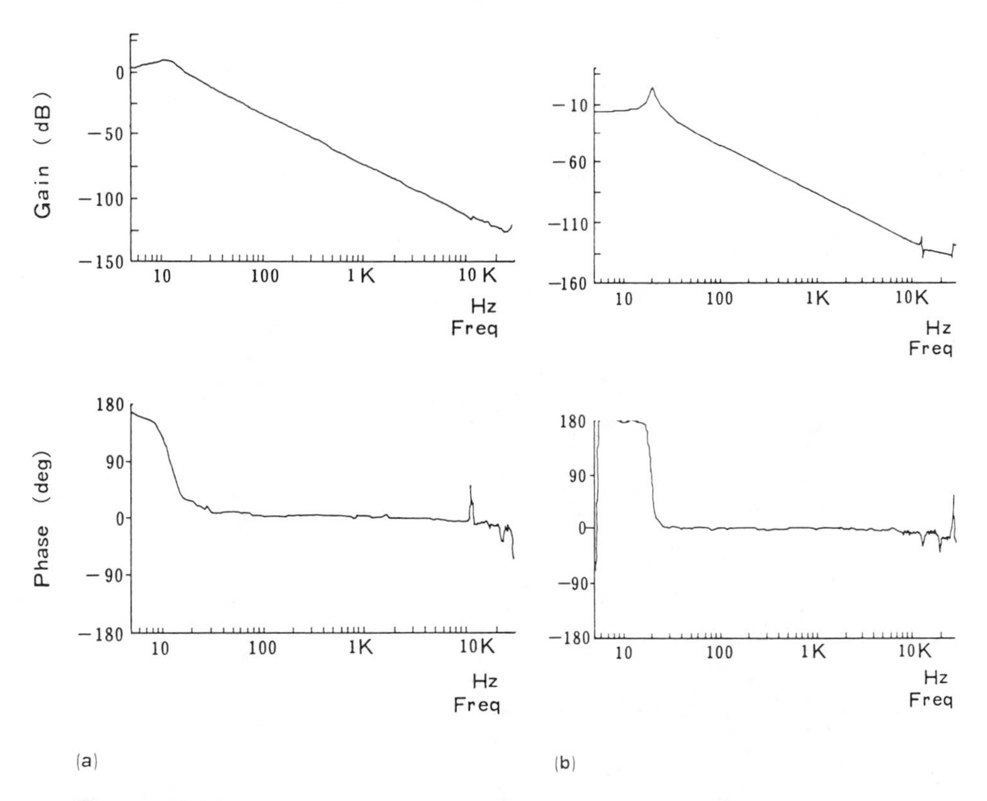

Figure 13.19 Bode diagrams of actuator: (a) focusing characteristics, (b) tracking characteristics.

hand, in the valence band, the probability of the presence of positive holes in the active layer 2 is increased by the energy barrier ΔE_v to cause a population inversion in the active layer 2. In the active layer, therefore, the conduction band becomes full of electrons normally absent at thermal equilibrium, and the probability of recombination of electron-hole pairs with stimulated emission is increased. An incoming photon into the active layer is thereby amplified. The feedback mirrors at both ends of the active layer constitute a resonant cavity, and as the amplification surpasses the losses within the resonance cavity laser emission takes place.

Fabricating Technology [14]

The method of fabricating a buried twin-ridge substrate (BTRS) is illustrated in Figure 13.21. The BTRS laser is produced by two steps of liquid phase epitaxial (LPE) growth. Prior to the initial LPE growth, the rectangular mesa structure

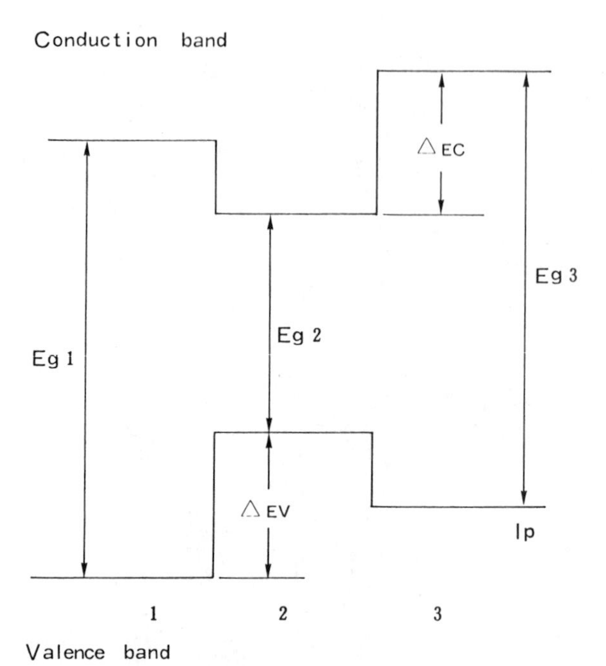

Figure 13.20 Energy band diagram.

extending in the <01$\bar{1}$> direction of the crystal is formed on the (100) surface of a p-type GaAs substrate. In this initial LPE growth, an n-GaAs current blocking layer is formed to present a planar surface. Then a couple of rectangular ridges are formed, and the second LPE growth is carried out. The second LPE growth occurs in four layers. These are a p-type $Ga_{0.6}Al_{0.4}As$ cladding layer, a 0.04-μm-thick $Ga_{0.9}Al_{0.1}As$ active layer, an n-type $Ga_{0.6}Al_{0.4}As$ cladding layer, and an n-type GaAs contact layer.

13.4.2 Astigmatism of the Laser

Semiconductor lasers may be classified into two categories: gain-guided and index-guided. In a gain-guided laser, the direction of beam propagation is not perpendicular to the wavefront. Some lasers classified as index-guided also have weak evanescent waves. As a result, the beam waist in the horizontal direction is situated inwardly by Δ_L from the plane of beam emergence to produce astigmatism. Whereas this astigmatism Δ_L is as large as 10–50 μm in the gain-guided laser, it is about 5–10 μm (0.2–0.4 mil) in the index-guided laser. Figure 13.22

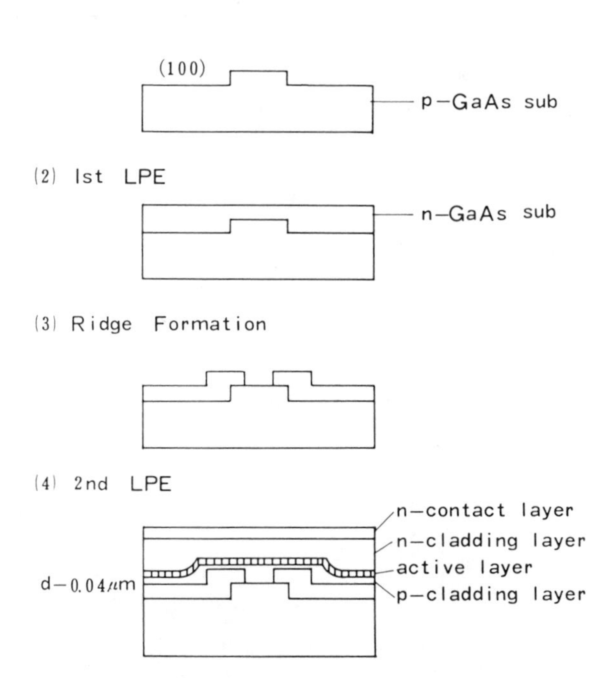

Figure 13.21 The fabrication process of the BTRS laser.

shows a typical distribution of astigmatism in index-guided lasers. Generally speaking, the astigmatism of a laser tends to decrease as the laser output increases.

13.4.3 Laser Noise [15]

When the temperature of the injected current is varied during the operation of a semiconductor laser, there occurs a mode-hopping noise associated with a wavelength change. Moreover, when a slight return beam is directed into the semiconductor laser, the laser emission becomes unstable, yielding mode hopping and excessive noise. Figure 13.23 shows the temperature dependency of a semiconductor laser wavelength.

As the temperature of the semiconductor laser increases, the longitudinal mode of the laser is shifted toward longer wavelengths. Substantial noise appears in this type of mode hopping. Figure 13.24 shows the relative intensity of noise versus the laser heat sink temperature.

Figure 13.24a shows the characteristic of the element itself, with the broken line representing the noise level allowable for an optical head. If the level of

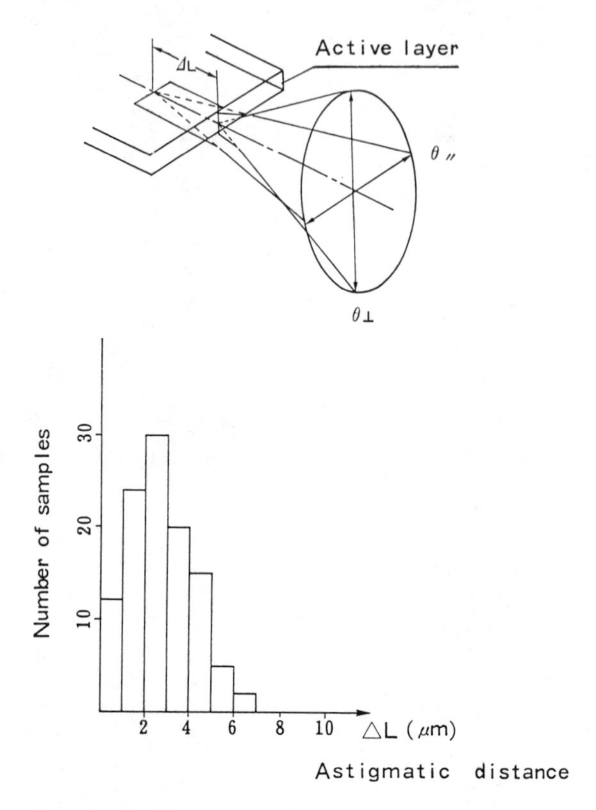

Figure 13.22 Distributions of astigmatic distance.

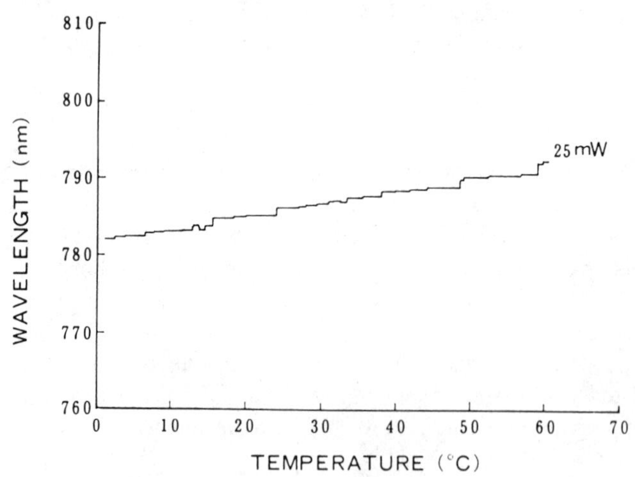

Figure 13.23 Temperature dependency of wavelength.

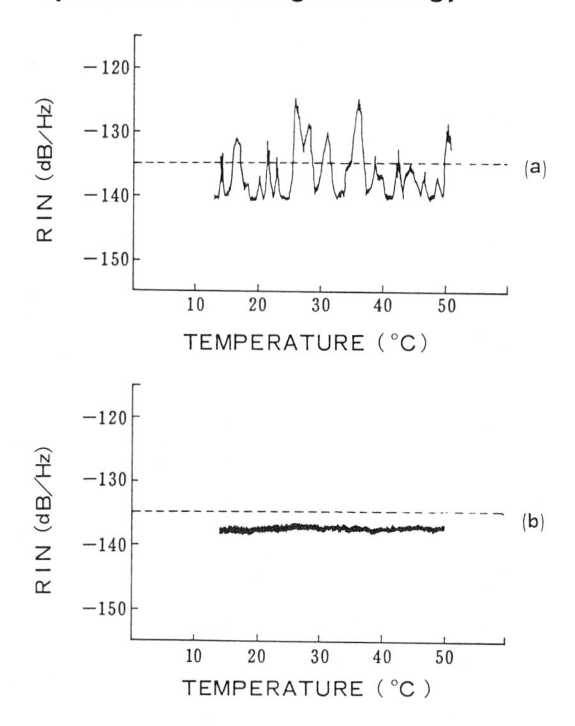

Figure 13.24 Noise characteristics of semiconductor laser.

return light is about 0.5%, the relative intensity noise (RIN) is invariably above the noise level allowable for an optical head. Noise of this magnitude not only leads to a decrease in disk recording/reproduction SNR but may render unstable focus and tracking servos.

To preclude this unwanted level of return light, it is common practice to incorporate an optical isolator consisting of a $\lambda/4$ plate and a polarizing beam splitter. However, the return light cannot be sufficiently controlled due to the birefringence of the disk and the variation in isolator performance. As a consequence, the generation of noise due to return light is inevitable in semiconductor lasers of longitudinal single mode. Within the constraints of the basic aspects of the optical head design, return light noise can be best controlled by broadening the emission spectral linewidth of the semiconductor laser to reduce the coherence of light. The emission spectral linewidth can be broadened by making the longitudinal mode a multiple mode. In the index-guided laser, the transverse mode becomes a single mode at the emission output of 1 mW or more, due to the confining effect on the transverse mode. On the other hand, in the gain-guided laser, the transverse mode is confined by the gain corresponding to the carrier density so that it is generally a multiple mode. In such a multiple-mode laser, the influence of return light is small, and hence laser emission is unaffected, but the inherent noise level is higher than in the single-mode laser. As shown in Figure 13.24b, when the index-guided laser operating single mode is modulated with a high-frequency carrier, the longitudinal mode becomes a multiple mode with the result that the noise level is lowered.

Figure 13.25 shows the return light noise levels of various lasers [16]. Noise level RIN is calculated as

$$\mathrm{RIN} = \frac{\langle \Delta P^2 \rangle}{P^2} \frac{1}{\Delta f} \tag{16}$$

where $\langle \Delta P^2 \rangle$ is the mean square of noise power, P is the output power, and Δf is the noise bandwidth. Figure 13.26 shows a laser equipped with an HF module.

When the HF oscillation frequency is set at 600–700 MHz and the modulation level is set below the laser emission threshold, the light output becomes a pulse emission, providing a multiple mode. Figure 13.27 shows an example of (a) the read signal obtained without HF modulation and (b) the read signal with HF

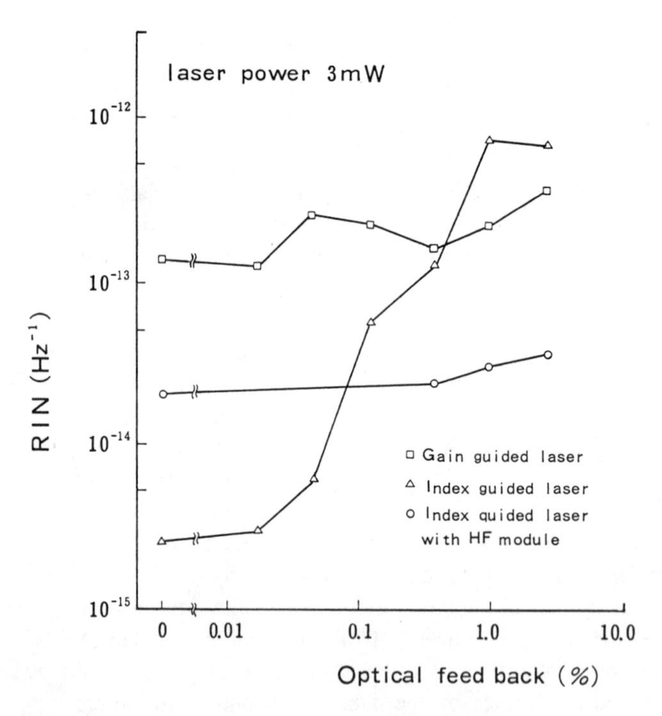

Figure 13.25 Laser noise versus optical feedback.

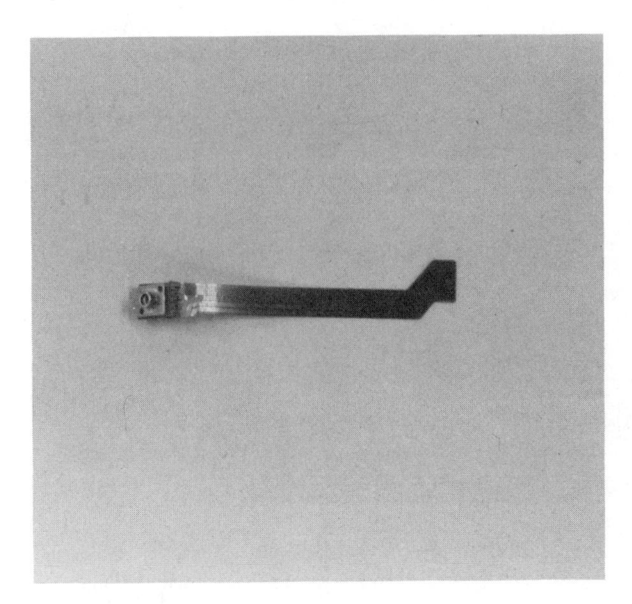

Figure 13.26 Laser with an HF module.

modulation. With the addition of the HF modulation, the CNR of the carrier signal is improved by about 8 dB.

13.5 FOCUSING AND TRACKING TECHNIQUES

13.5.1 Focusing Servo System and Method of Error Signal Detection

In an optical disk system, the laser beam is focused on the disk surface while the disk revolves at a high speed. An optical disk spinning at a high speed typically has an undulation of tens to hundreds of micrometers, and it is necessary that the objective lens follows this undulation so as to focus the beam on the signal plane of the disk within the allowable limits of defocus of the optics. The focusing mechanism generally used for this purpose is a moving-coil actuator employing a magnet and a coil. The required frequency response of the system is from several kilohertz to more than 10 kHz. Figure 13.28 shows a block diagram of the optical disk focusing system.

The focus servo loop comprises a focus error signal detection unit, a circuit for amplification, a phase correction of the detected error signal, and an actuator for driving the objective lens. This actuator is designed to follow the disk undulation in response to a servo signal, the external noise associated with the

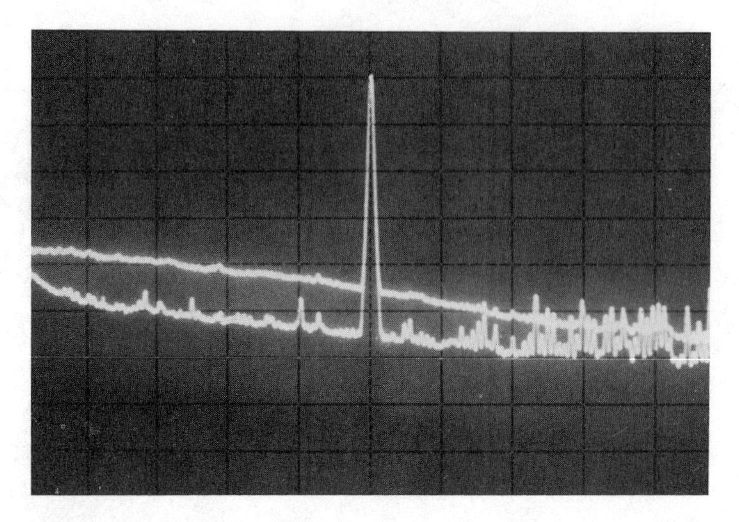

Figure 13.27 Recording/reproduction carrier signals: (a) Upper baseline shows the noise level without HF module. (b) Lower baseline shows the noise level with HF module.

movement of the actuator, and the track beam interference noise. In designing an autofocusing mechanism for the optical disk, external light noise must be reduced as much as possible. A balanced design must be employed and take into consideration, (1) tracking interference which occurs when the beam traverses the track, (2) mutual interference between the focusing drive and tracking actuator, and (3) incidence of false focus error signals associated with movement of the beam on the detector in the course of tracking. The focus errors arise from the undulation of the disk, vibrations of the device, and other causes.

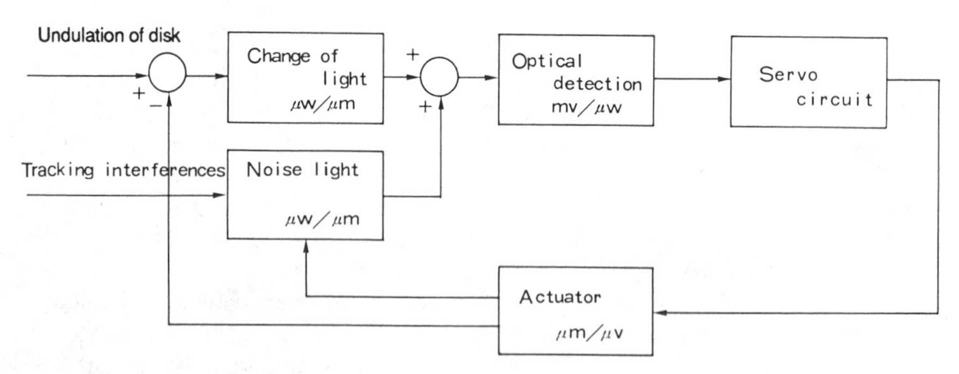

Figure 13.28 Block diagram of focusing servo.

The focus error information contained in the laser beam reflected from the disk can be transformed into intensity or phase differences for detection. The following beam characteristics can be utilized for focus error detection.

1. Change of beam shape
2. Transverse movement of the beam
3. Phase of the modulated waveform of the beam

Beam Shape Detection Method

Two separate techniques can be used to detect the beam shape to obtain a focus error signal. These are the astigmatic focus detection method and the spot size detection method.

Figure 13.29 shows a basic optical system for the astigmatic focus detection method using a tilted parallel plate. Compared with the conventional method of using a cylindrical lens, this method is advantageous in the simplicity of the optics. Assuming that the magnification of the objective lens is constant, the sensitivity of focus error signal detection is dependent on the thickness and refractive index of the parallel plate. The greater the thickness of the plate and/or the larger the refractive index, the larger the astigmatism and, hence, detection sensitivity. Figure 13.30 shows a typical focusing error signal in the optimum design. When the detection sensitivity is relatively low, defocus due to false signals caused by dropouts in the disk or movement of the objective lens during tracking is relatively high. When the detection sensitivity is too high, the dynamic range of the focusing servo mechanism is diminished and the stability of the servo is decreased.

Figure 13.31 shows the operating principle of the spot size detection method.

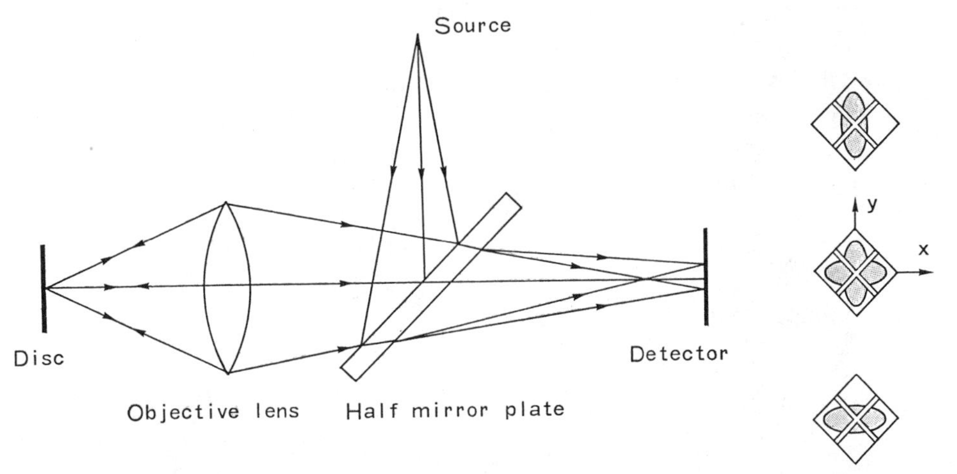

Figure 13.29 Astigmatic focusing method with plate.

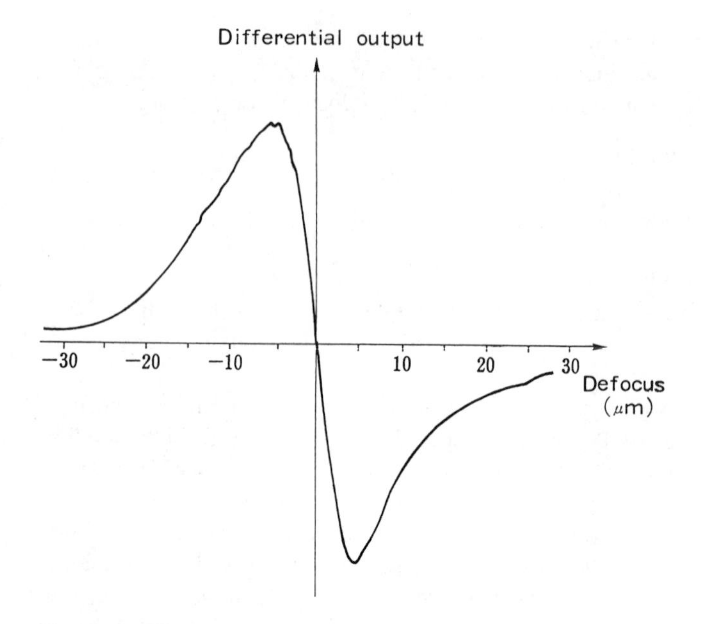

Figure 13.30 Astigmatic focus detection signal.

The central part of the beam is received in front of and beyond the focal point of the beam by two three-segment detectors. A focus error signal is derived from the intensity difference on the central and outer segments. The beam shape detection method generally has a large allowance for detector offset and has good temperature characteristics and aging stability.

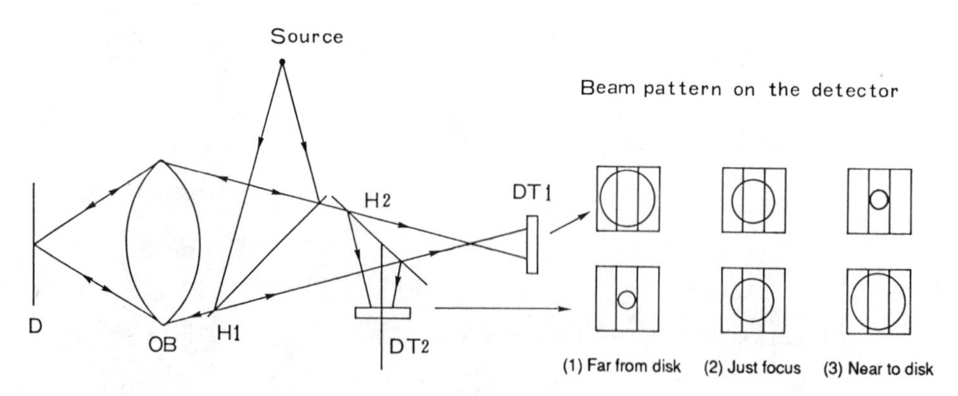

Figure 13.31 Spot size detection method.

Beam Position Detection Method

The beam position detection method converts the movement parallel to the optical axis—such as the undulation of the optical disk—to a beam movement in a plane perpendicular to the optical axis to obtain a focus error signal. This detection method uses relatively simple hardware construction for detection and gives a broad-focusing dynamic range.

Figure 13.32 shows a focus error signal detection system using a biprism. This is an example of the Foucault focus detection method. As the distance between the disk and the objective lens decreases, the intensity on the inner side of the respective split detectors is increased, while an increasing distance between the disk and the objective lens results in increased intensity on the outer sides of the split detectors.

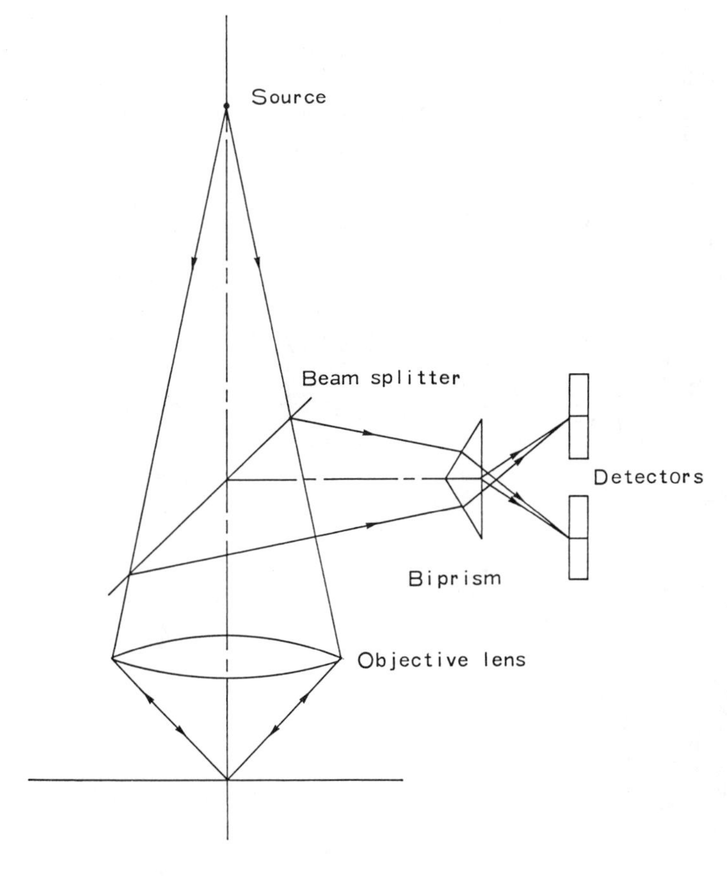

Figure 13.32 Foucault focusing method.

Figure 13.33 shows a focus detection system using the critical angle of a prism. The beam of rays reflected from the disk enters the prism as a divergent beam when the distance between the disk and the objective is small, or as a convergent beam when the distance is large. When the prism is set at the critical angle, and the rays are not parallel, the beam will be partially transmitted through the prism to establish differential intensities on the detectors. In the case of a divergent beam, the near detector receives less light; in the case of a convergent beam the far detector receives less light. There are other methods for focus detection, such as a skew beam focus detection system, a system wherein the incident beam is eccentric with respect to the objective lens axis, a system using a single knife-edge, a beam rotation focus detection system, and others.

Beam Phase Difference Detection

There are two methods to detect beam phase difference: the spatial phase difference detection method and the temporal phase difference detection method. In the spatial phase difference detection method, illustrated in Figure 13.34, the phase of the beam located in the far-field pattern of the reflected beam diffracted by a

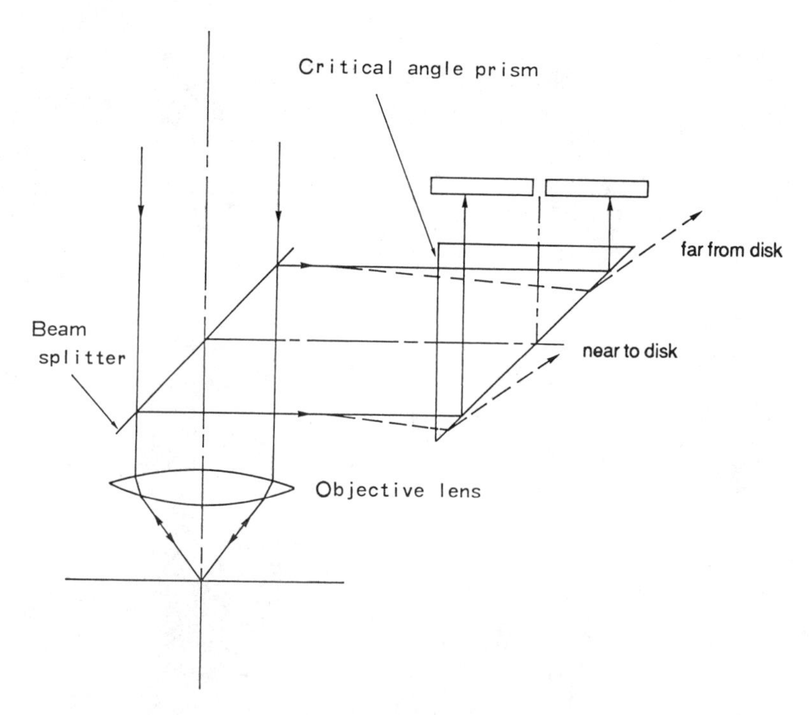

Figure 13.33 Critical angle focusing method.

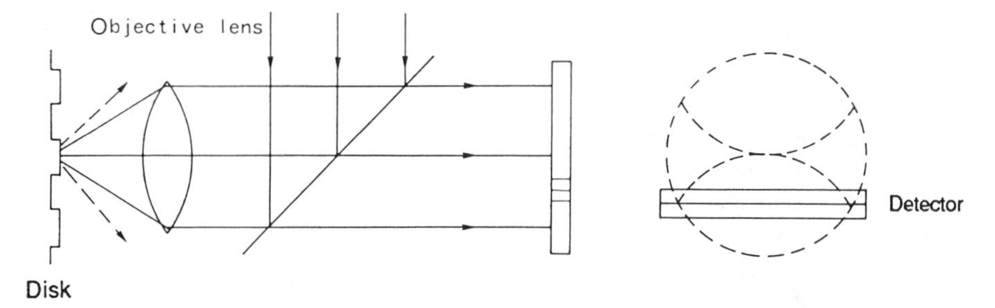

Figure 13.34 Spatial phase detection method.

given pattern in the optical disk (for example, the pregrooved track pattern) is detected. This method is dependent on beam wavelength, and the dynamic range of focus error signal is narrow. The temporal phase difference detection method is also known as the wobbling method. In this method, the focal point of the beam irradiating the optical disk is modulated along the optical axis with a wobbler. The phase of the modulated signal obtained with a detector is compared with the phase of the modulated drive signal of the wobbler to obtain a focus error signal proportional to the phase difference.

13.5.2 Track Error Signal Detection Method

Various Detection Methods

The signal track on an optical disk has a pitch of 1.6 μm, and the beam spot must follow this signal track with an accuracy of $<\pm 0.1$ μm. This tracking movement is affected by driving the objective lens in the actuator described in the subsection "Actuator." The following methods are available for optical detection of the track error signal.

1. Detection of the far-field distribution of the read/write beam reflected from the disk
2. Detection from the difference of two signal levels obtained with sample pits disposed at an offset of $\pm 1/4$ pitch from the track
3. Detection using two auxiliary beams
4. Detection of the phase difference between the playback signal obtained by a slight induced displacement of the beam in the direction perpendicular to the track and the phase of the corresponding drive signal

In the method using auxiliary beams (3), two first-order beams obtained with a diffraction grating in the laser beam are focused to positions on the disk about plus and minus one-quarter of the track pitch apart from the track center and received by two detectors to obtain a track error signal. While this method is well suited to the read-only optical head, it is unsuited to the read/write optics. In this case the beam intensity is increased during the writing mode, introducing the risk of harmful recording by the auxiliary beams.

In the wobbling method (4), the track error signal corresponds to the phase difference between the transducer signal inducing slight displacement of the beam in the direction perpendicular to the track and the beam intensity modulated by track edge diffraction. This method has not been implemented except in certain applications. This is partly due to the poor stability of the wobbling frequency and partly due to the 0.1-μm wobble displacement to obtain a good tracking error signal with a satisfactory S/N ratio. Wobble displacement of 0.1 μm is close to the allowable tracking error value.

Push-Pull Track Error Signal Detection Method

The simplest method for obtaining a track error signal in read/write applications is called the "push-pull method." A split detector is inserted in the far field of the beam in such a manner that the line of division of the detector is lined up with the track.

Figure 13.35 shows a basic optical system for signal detection using the push-pull method. The overlapped areas of the two first-order beams with the zero-order beam, due to diffraction caused by the track on the disk, are incident on the split detector. The difference between the intensities on the two photodetectors generates the track error signal. Figure 13.36 shows the far-field beam distributions according to the beam spot position on the track. The asymmetry of the far-field beam intensity distribution and the track error signal level are maximum when the track groove depth is $\lambda/8$. (When the depth is $\lambda/4$ multiplied by an integer, asymmetry disappears and no tracking signal can be obtained.)

Slit Detection Method [17]

When the beam spot is located in the center of the track, the phase difference ψ between the zero-order beam and the two first-order beams in the push-pull system is dependent on the diffraction by the track and on the defocus ΔZ and can be expressed as [18]

$$\psi = \frac{\pi}{2} + \frac{2\pi}{\lambda}\left[\sqrt{1 - \left(\frac{\lambda}{p} - \sin\alpha\right)^2} - \cos\alpha\right]\Delta Z \tag{17}$$

where α is an angle between the optical axis and an arbitrary point in a far-field image.

From equation (17) for $\alpha = \sin^{-1}(\lambda/2p)$ in the far-field image, the phase difference is constant and independent of defocus but exclusively dependent on

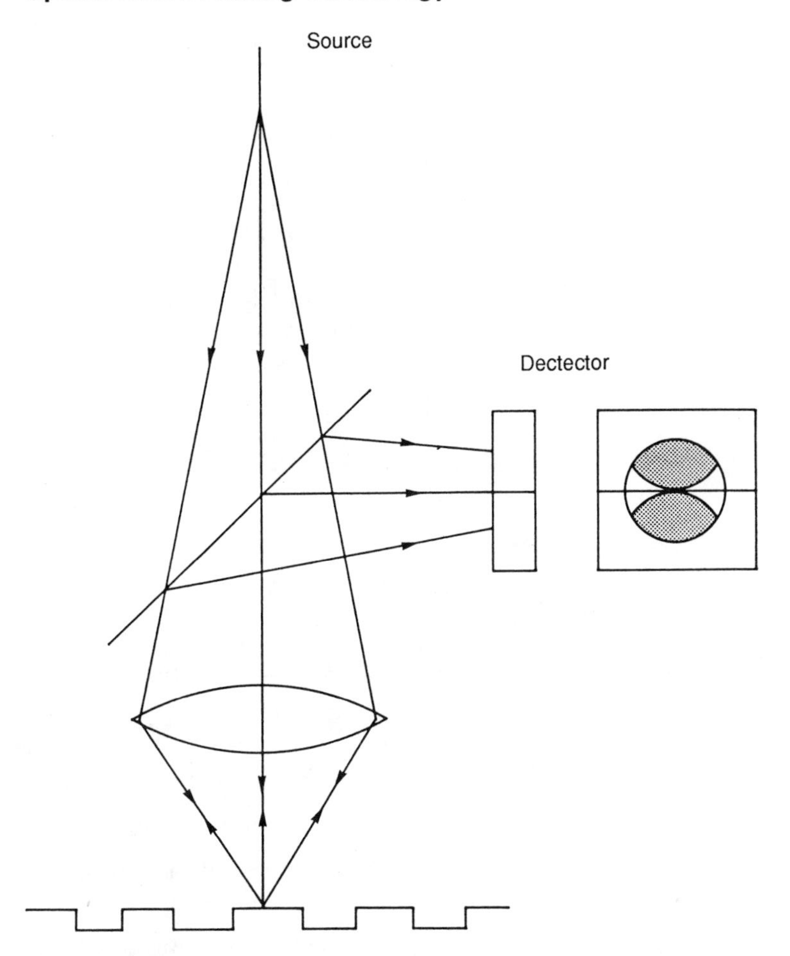

Figure 13.35 Push-pull method.

diffraction by the track. Figure 13.37 shows a typical far-field beam distribution in the presence of defocus. By utilizing this property in the far field, the control range of tracking with respect to defocus can be expanded. Thus, the defocus characteristic of the track error signal can be improved by providing slits symmetrically in the centers of overlaps between the zero-order beam and the two first-order beams, as illustrated in Figure 13.38.

Figure 13.39 shows the change in track error signal level according to defocus at various slit widths. It is seen that if the slit width is about 20% of the far field, the proportion of change due to defocus is improved by about a factor of 2. If the slit width is too narrow, the *S/N* of the track error signal will decrease.

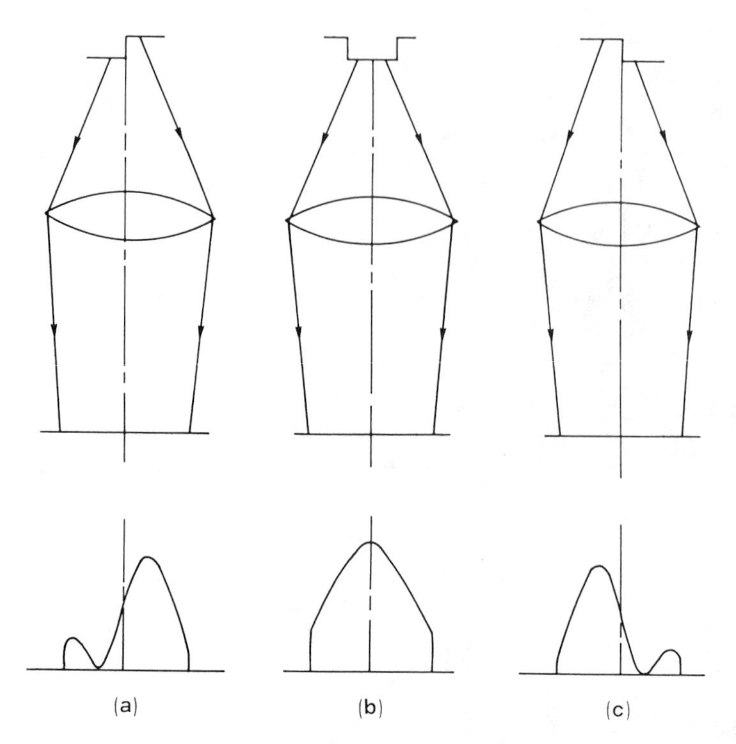

Figure 13.36 Far-field beam distribution.

The optical parameters in these experiments and theoretical calculations are as follows:

Objective lens: $NA = 0.5$
Laser wavelength: $\lambda = 830$ nm
Track pitch: $t = 1.6$ μm
Slit width: W mm

Sampled Tracking Method [19]

In the sampled tracking system, track error signal detection pits are provided in lieu of the continuous groove in the usual disk. The sample pit actually consists of two pits displaced from track center by about $\pm\frac{1}{4}$ of the track pitch and one pit centered on the track. Figure 13.40 shows the principle of sampled track error signal detection. When the beam spot is (a) off-track upwardly from track center, the first pit output is large and the second pit output is small. In the on-track condition, the first pit output level and the second pit output level are equal (b). If the beam spot is downwardly off-track, the first pit output is small and the

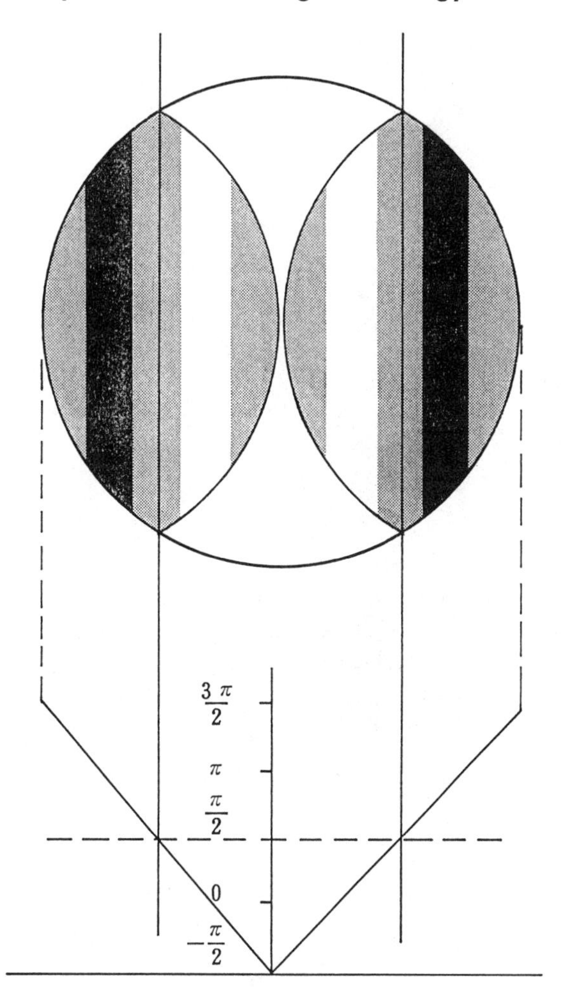

Figure 13.37 Far-field pattern when defocus occured.

second pit output is large (c). The off-track condition is diagnosed by comparing these pit output levels. The third pit is used for making a sampling clock, and the track error signal detection is made by using these outputs. Figure 13.41 shows a block diagram of the detection circuit.

In the sampled tracking system, each set of pits provided for track error signal detection uses a track length equivalent to that used to store 1 byte of information. The sampling frequency must be higher than 10 times the cutoff frequency of the tracking servo. This reduces the size of the usable data area, but the overall system performance improves because there is less degradation of data signals

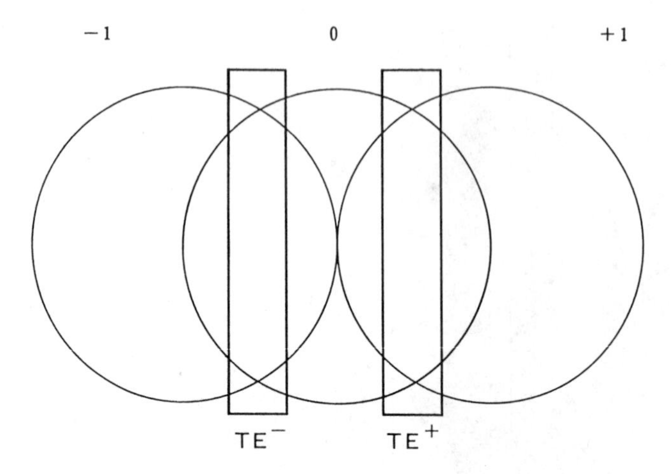

Figure 13.38 Slit detection method.

and interference effects on the focusing servo caused by the track groove. Moreover, in the push-pull system, an inclination of the disk induces a track offset. For example, a disk inclination of 0.7° causes a 0.1-μm lateral shift in the quiescent operating position of the tracking servo. This type of systematic track error is decreased by about a factor of 5 when the sampled tracking method is used.

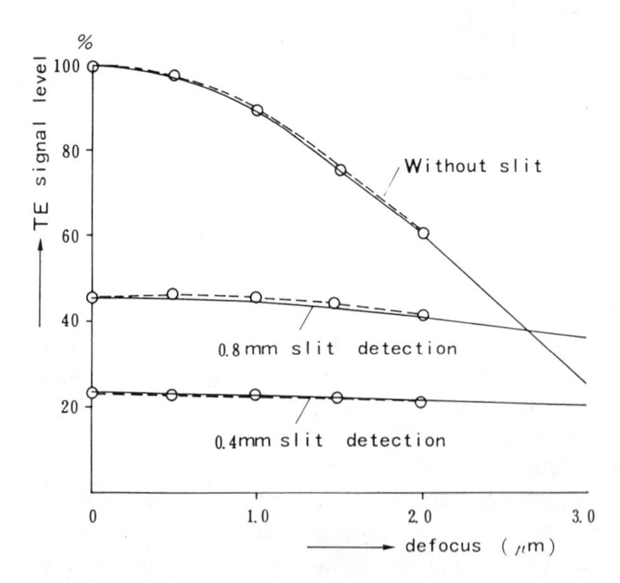

Figure 13.39 Defocus versus track error signal level.

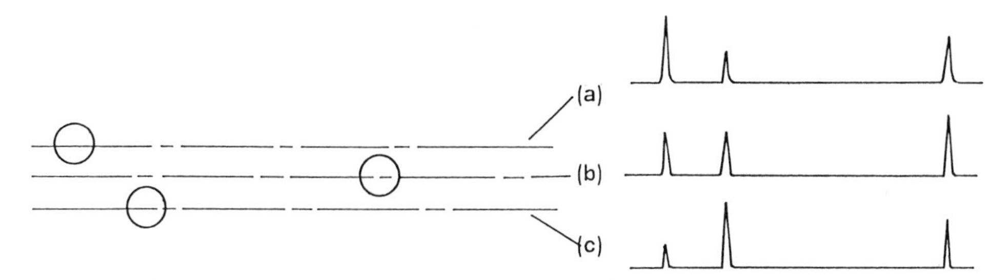

Figure 13.40 Track error signal detection from sampled pits.

13.6 RADIAL ACCESS AND DRIVING TECHNIQUE

13.6.1 Fast Random Access

A critical aspect of an optical disk memory system is fast random access to the stored information. This random access is accomplished via two mechanisms: optical head motion for rough positioning and the tracking actuator for precise positioning. Figure 13.42 is a perspective view showing a linear actuator used as the coarse positioning means. To minimize access time, it is necessary to (1) develop a small, lightweight optical head, (2) increase the resonant frequency of the linear actuator, and (3) develop a transfer mechanism with a minimum of friction. In the actuator shown in Figure 13.42, the transfer segment weighs only 78 g (0.17 Ib) and has a thrust of 3.0 *N/A*. This linear actuator gives an average access time of 75 ms or less. The optical head base is provided with roller bearings so that it freely moves on the guide rods. The low-frequency component of the track error signal is fed to the linear actuator so that the center of drive of the objective lens will lie at the center of the tracking mechanism motion range.

Figure 13.43 shows the modes of access by the linear actuator/tracking actuator combination [20]. First, the tracking actuator is disabled. Then the linear actuator is accelerated and decelerated at the maximum speed to position the optical head in the vicinity of the correct track (*A* to *B*). The tracking actuator is reactivated and the track address is read (*B* to *C*). The number of tracks between the actual and desired address is calculated, and access is completed by executing a multitrack jump.

Figure 13.44 is a block diagram of the tracking servo circuit and linear actuator circuitry. The track error signal obtained by the procedure described in Section 13.5.2 is fed to an amplification circuit, a switching circuit, and a drive circuit in succession to drive the tracking actuator. The tracking drive signal is also used to drive the linear actuator [8]. During random access, the drive signal is removed from the tracking actuator and a voltage corresponding to the access signal is generated and supplied to the linear actuator drive circuit. Two methods are

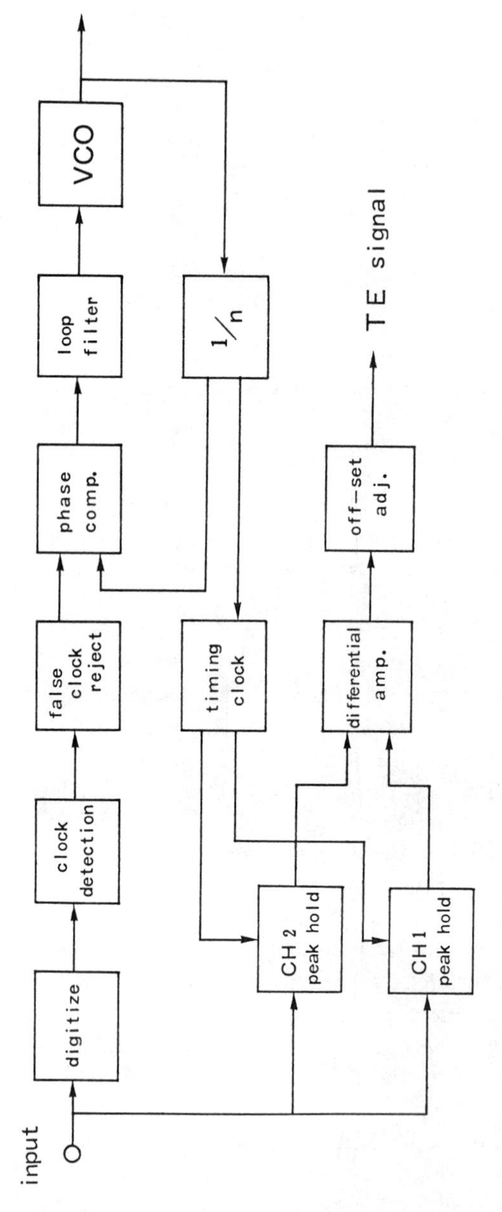

Figure 13.41 Sampled servo tracking method.

Figure 13.42 Perspective view of linear actuator.

available for ascertaining the actual position with respect to the target track. The first method involves calculating the number of tracks between the current and desired positions, then detecting and counting optically each pregroove as it is passed over in the radial scan. The scan is stopped when the correct number of tracks have been crossed.

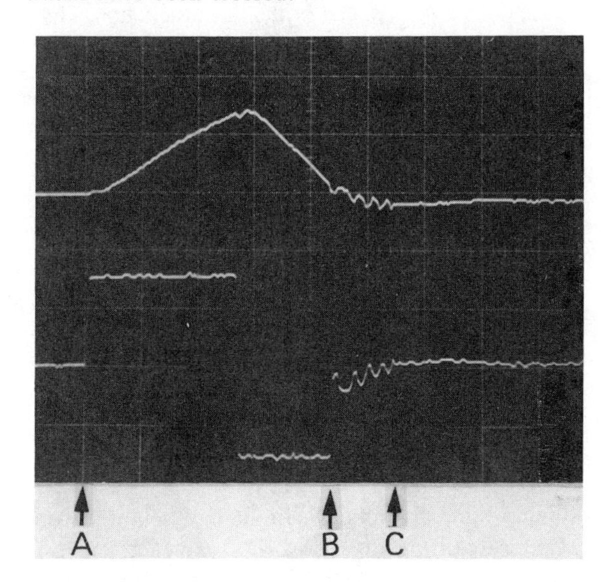

Figure 13.43 Access speed and drive current.

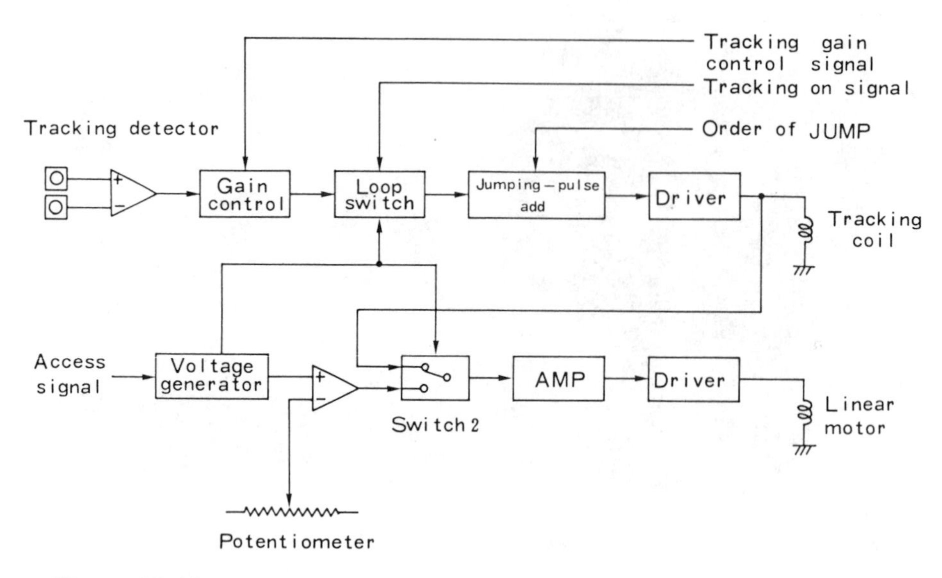

Figure 13.44 Block diagram of tracking and accessing servo.

Since this method counts the tracks themselves, the distance to the target track can be accurately computed. The track detection bandwidth must be broad enough to prevent miscounts of the tracks during peak speed of the linear actuator. This method is less applicable when using sampled format disks; even track addresses can complicate the track-counting process in continuous format disks. A second method provides the optical head with a position sensor for detecting the current position. This provides a stable position signal, and the access servo can be damped using the output signal from this sensor. Examples of position sensors include linear scale sensors, optical position sensors, and slide resistance sensors. Figure 13.45 shows a typical optical position sensor.

13.6.2 Optical Drive System

The optical disk system consists of hardware comprising the disk, optical head, accessing circuit, signal processing circuit, error correction circuit, microcomputer, etc., and software for processing the various signals. The height of the 5.25-in optical disk drive is either full-height (82 mm) or half-height (41 mm), corresponding to standardized magnetic disk products. In the half-height drive, design goals include the use of a disk cartridge that is inserted and clamped and low profiles of the component parts for the access mechanism. The height of the optical head must be less than 15–16 mm, and an ingenious design is required.

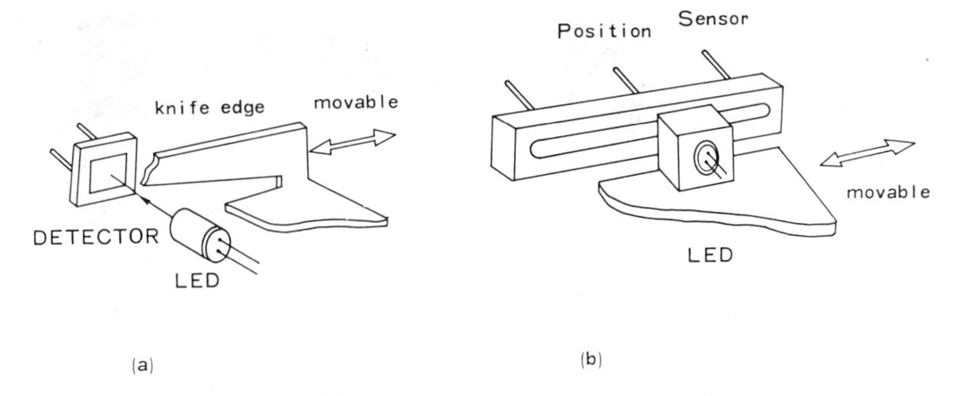

Figure 13.45 Optical position sensor: (a) knive-edge type, (b) PSD type.

For example, the linear actuator must be at the same level as the optical unit base.

The removability of the optical disk affects differences in the amounts of eccentricity and undulation each time a disk is mounted. Moreover, since the disk substrate is made of plastic, the amount of undulation increases with the age of the disk. The dynamic balance of the disk is affected by these factors, and vibrations are induced. It is necessary, therefore, to design the various actuators such that these vibrations do not degrade performance.

Figure 13.46 is a perspective view showing an optical drive. The disk clamping system is generally mechanical or magnetic.

APPENDIX A

When the amplitude distribution in the pupil is given by $f(r^2)$ and the radius of the pupil is one unit, the integral of the Fourier-Bessel transform is written as

$$g(W) = \int_0^1 f(r^2) J_0(Wr)\, d(r^2) \tag{A1}$$

A plurality of solutions exist for this integral. However, the solution given by A. Boivin is easily understood. Thus, the amplitude of the Fourier spectrum $g(w)$ is written in the form of a Bessel series:

$$g(W) = \sum_{n=0}^{\infty} (-1)^n 2^{n+1} f_{(1)}{}^n \frac{J_{n+1}(W)}{W^{n+1}} \tag{A2}$$

Figure 13.46 Optical disk drive. (From Ref. 21)

where $f^n(r^2)$ denotes the nth differential of the function $f(r^2)$.

For calculating the Fourier spectrum of a truncated gaussian, $f(r^2)$ is expressed by $\exp(-\alpha r^2)$. Then the integral of Fourier-Bessel transform is written as

$$g(W) = \int_0^1 \exp(-\alpha r^2) J_0(Wr) r \, dr \tag{A3}$$

and the result becomes [22]

$$g(W) = \sum_{n=0}^{\infty} 2^n \alpha^n e^{-\alpha} \left[\frac{2J_{n+1}(W)}{W^{n+1}} \right] \tag{A4}$$

APPENDIX B

Thickness variations, index changes, and tilts of the disk substrate all cause wavefront aberrations. Here, we calculate the optical path differences Δ_0 between two rays: the first is the on-axis ray, and the second is the outermost ray which determines the numerical aperture of the objective lens. From Figure 13.47, the following relations can be easily calculated:

$$\sin(\psi - \theta) = \mathrm{n} \sin r_1 \tag{A5}$$

$$\sin \theta = n \sin r_0 \tag{A6}$$

$$\Delta_0 = nt \left\{ \frac{1}{\cos r_1} - \frac{1}{\cos r_0} \right\} + \frac{t}{n} \left\{ \frac{\cos(\psi - \theta)}{\cos r_1 \cdot \cos \theta} - \frac{1}{\cos r_1} \right\} \tag{A7}$$

Developing the power series of ψ and θ, the next quadratic terms are obtained:

$$\Delta = \frac{t(1 - n^2)}{8n^3} [\psi^4 - 4\psi^3\theta + 8\psi^2\theta^2 + 8\psi\theta^3] \tag{A8}$$

where ψ is the *NA* of the objective lens and t is the thickness of the disk substrate.

Here, each term denotes the Siedel aberration. When only a thickness error Δt exists, the spherical aberration S_1 is generated:

$$S_1 = \frac{(1 - n^2)\psi^4}{8n^3} \Delta t \tag{A9}$$

The relation between the wave aberration W_{ST} and the spherical aberration S_1 can be calculated from Maréchal's equation [23]:

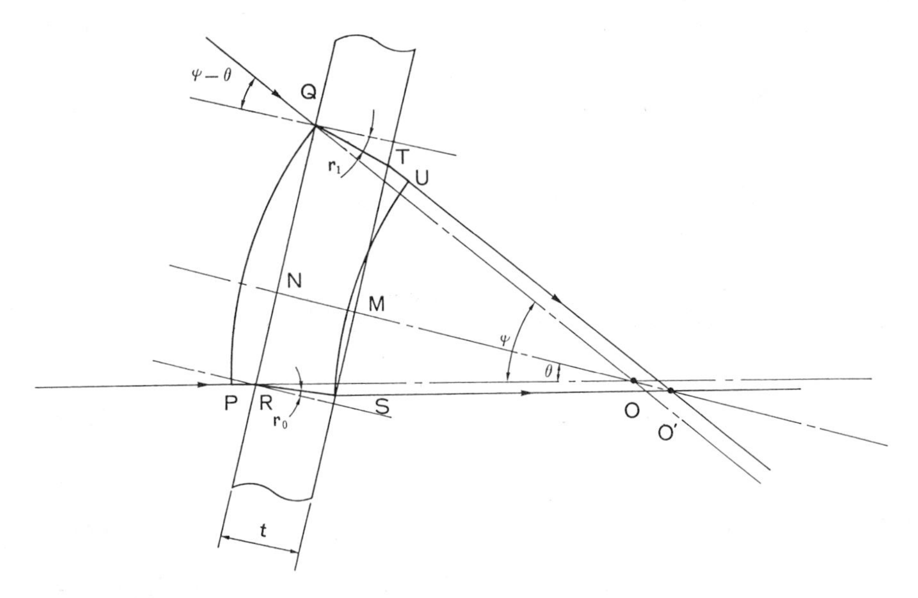

Figure 13.47 Optical pass differences between two rays.

$$W_{ST}{}^2 = \frac{d^2}{12} + \frac{dS_1}{6} + \frac{4S_1{}^2}{45} = \frac{1}{12}(d + S_1)^2 + \frac{S_1{}^2}{180} \tag{A10}$$

where the wave aberration W_{ST} is minimal when the defocus d is equal to the third spherical aberration S_1. Thus, the wave aberration due to the spherical aberration becomes

$$W_{ST} = \frac{S_1}{\sqrt{180}} = \frac{(1 - n^2)\psi^4 \, \Delta t}{8\sqrt{180}\, n^3} \tag{A11}$$

For a small tilt θ of the disk substrate, high orders can be disregarded and the only important aberration is the coma C_1:

$$C_1 = \frac{t(1 - n^2)\psi^3}{2n^3}\theta \tag{A12}$$

The relationship between the wave aberration W_{TL} and the coma C_1 can be calculated, again from Maréchal's equation [23]

$$W_{TL}{}^2 = \frac{K^2}{12} - \frac{KC_1}{6} + \frac{C_1{}^2}{18} = \frac{1}{12}(K - C_1)^2 + \frac{C_1{}^2}{72} \tag{A13}$$

where the wave aberration W_{TL} is minimal when the tilt of wavefront k is equal to coma C_1. Then the wave aberration W_{TL} due to a tilt of the disk substrate becomes

$$W_{TL} = \frac{C_1}{\sqrt{72}} = \frac{t(n^2 - 1)\theta\psi^3}{2\sqrt{72}\, n^3} \tag{A14}$$

APPENDIX C

The allowable limit of laser-mounting angle is calculated. When the wavefront of the beam emerging from a laser is astigmatic and its axis is not in agreement with the x-y axis of the optics, there exists a residual astigmatism. With the y-axis as a reference, the phase difference $\psi(x)$ along the x-axis within the pupil plane of the collimating lens with an astigmatic distance of Δ_L and a focal length of f_c is expressed by

$$\psi(x) = \frac{\Delta_L}{2f_c{}^2}x^2 \tag{A15}$$

Assuming that the wavefront is inclined through an angle θ with respect to the y-axis, the phase difference $\psi_1(x, y)$ may be expressed as

$$\psi_1(x, y) = \frac{1}{\cos^2\theta}\frac{\Delta_L}{2f_c^{\,2}}(x\text{-}y\tan\theta)^2$$
$$= \frac{\Delta_L}{2f_c^{\,2}\cos^2\theta}(x^2 + y^2\tan^2\theta - 2\tan\theta x, y) \tag{A16}$$

By focusing the optics, the term $x^2 + y^2 \tan^2\theta$ can be zero. Therefore, the wavefront aberration assumes a maximum value in the direction of $x = y = h$:

$$\psi_0 = \frac{\tan\theta}{f_c^{\,2}\cos^2\theta}\Delta_L h^2 \tag{A17}$$

Since h/f_c is the *NA* of the collimating lens, the above equation may be rewritten as follows:

$$\psi_0 = \frac{\tan\theta}{\cos^2\theta}\Delta_L(NAc)^2 \tag{A18}$$

Since Maréchal's equation gives the relationship between maximum astigmatism ψ_0 and wavefront aberration W_{LA} as

$$W_{LA}^{\,2} = \frac{\psi_0^{\,2}}{6} \tag{A19}$$

we obtain

$$W_{LA} = \frac{\tan\theta}{\sqrt{6}\cos^2\theta}\Delta_L(NAc)^2 \tag{A20}$$

REFERENCES

1. Philips Press Information, Nov. 7, 1978.
2. H. H. Hopkins, Diffraction theory of laser readout systems for optical video discs, *J. Opt. Soc. Am.*, *69*:4–24 (1979).
3. J. W. Goodman, *Introduction to Fourier Optics*, Chap. 6.3, McGraw Hill, New York, 1968.
4. J. Braat et al., *Principles of Optical Disc System*, Adam Hilger Ltd., New York, pp. 7–85, 1985.
5. A. H. Firester et al., Optical read out of RCA video disc, *RCA Review*, *39*(3):392–407 (1978).
6. T. Yoshida et al., Tellurium sub-oxide thin film disc, *Optical Disks Systems and Applications*, pp. 79–84, SPIE, Vol. 421, 1983.
7. T. Saimi, Compact optical pick-up for three dimensional recording and playing system, CLEO'82 Phoenix, April 1982.
8. R. Imanaka and T. Saimi et al., Recording and playing system having a compatibility with mass produced replica disc, *IEEE Consumer Electronics CE-29*(3):135–140 (1983).

9. M. Hartmann, B. A. J. Jacobs, and J. J. M. Braat, Erasable magneto-optical recording, *Philips Technical Review*, *42*(2):37–47 (1985).
10. T. Deguchi et al., Digital magneto-optical disk drive, *Appl. Opt.*, *23*:(22):3972–3978 (1984).
11. M. Born and E. Wolf, *Principles of Optics*, Pergamon Press, Oxford, 1970.
12. H. Nakamura et al., *Fine Focus 1-Beam Optical Pick-up System*, National Technical Report, pp. 72–80, 1986.
13. J. C. J. Finck et al., A semiconductor laser for information read-out, *Philips Technical Review*, *139*(2):37–47 (1980).
14. K. Hamada et al., A 0.2W CW laser with buried twin-ridge substrate structure, *IEEE Journal of Quantum Electronics*, Vol. QE-21, No. 6:625–628 (1985).
15. H. Shimizu et al., Noise and longitudinal characteristics of GaAlAs TS-laser with reduced facet reflectivities, *IEEE Journal of Quantum Electronics*, Vol. QE-19, No. 3:470–474 (1983).
16. N. Chinone et al., A semiconductor laser below allowance of noise due to the optical feedback by adding the high frequency generating circuit, Nikkei Electronics, 10.10, pp. 173–194, 1983.
17. T. Saimi, Amelioration of tracking signals by using slit-detection method, Conference of Japan Society of Applied Physics, Tokyo, March 1987.
18. Ad Oudenhuysen and Wai-Hon Lee, Optical component inspection for data storage applications, *Optical Mass Data Storage II*, pp. 206–214, SPIE, Vol. 695, 1986.
19. Y. Tsunoda et al., On-land composite pregroove method for high track density recording, *Optical Mass Data Storage*, pp. 224–229, SPIE, Vol. 695, 1986.
20. A. Saito et al., Fast accessible optical head, *O plus E*, No. 76:84–87 (1986).
21. Panasonic New Product Information, Preliminary LF-5000S 5.25″ WORM Optical Disk Drive.
22. A. Boivin, *Théorie et Calcul des Figures de Diffractions*, Press de l'Université Laval, Québec, pp. 118–122, 1964.
23. A. Maréchal and M. Françon, *Diffraction Structure des Images*, Masson & Cie, Paris, pp. 105–112, 1970.

Afterword

The Foreword, as a prologue to this volume, relates scanning to information handling and outlines the diverse contributions provided by these chapters. This Afterword, an epilogue to this volume, summarizes the options in view of the design disciplines, and anticipates some trends toward continued progress.

How is this 13-chapter resource to be utilized? How does one create a system which optimizes performance and cost-effectiveness? Viewing its requirements, (Chapter 3), the challenge of each system relates primarily to its unique stature within that field. Equally astute ingenuity and skill must be exercised in selecting and refining the options for each system.

WHAT ARE THE OPTIONS?

Surely, the most critical is the selection and design of the scanner—which influences the objective lens—which combination must blend with the other components within the package. Respective of this integrated responsibility, some fundamental criteria can be expressed.

A leading determinant is the system resolution. Because the scanner is to convey a total number of distinguishable image points, we are concerned first with that total number N elements per scan. Appearing in several alternative notations, the basic expression for angularly scanned resolution is $N = \Theta/\Delta\Theta$, which leads directly to

$$N = \frac{\Theta D}{a\lambda} \text{ elements per scan line} \tag{1}$$

in which Θ is the full useful scan angle and $\Delta\Theta$ is the diffraction spread of the scanning beam at wavelength λ derived from an aperture width D having an aperture shape factor $a(1 \approx a < 2)$. This simple relationship for the diffraction-limited resolution applies to all angular deflectors.* The numerator ΘD may be recognized as a form of the Lagrange invariant, in which the complementary factors Θ and D retain the resolution N with optical magnification and demagnification of the image. Resolution N is determined at the scanner. Thus, a flat-field lens accepts the input beam size D over the scanned angle Θ and focuses it over an arbitrarily selected format with the same number of N diffraction-limited spots. As in any good system design, N must be derated by some factor, to allow for reasonable aberration and imperfection. This can be done by effectively increasing the denominator $a\lambda$.

Taking $a\lambda \approx 1$ μm in the mid-visible region, several mileposts can be expressed for the value of N. Above $N = 1000$, the nonmechanical techniques, having known acoustooptic (Chapter 11) and electrooptic (Chapter 12) materials limitations, often become impractical. Multiplexing may be used to extend the total resolution. But matching (''stitching'') subrasters accurately can be an heroic task! Up to about $N = 10{,}000$, the galvanometer (Chapter 10) is very effective, though limited in speed for a large ΘD product. At $N = 20{,}000$, we are exercising very fancy footwork at any speed. Here, the resonant scanner (Chapter 10) limits at under 200 Hz, again to sustain a large Θ with a large D. The polygon (Chapters 6 and 7) and holographic scanners (Chapters 4 and 5) continue at higher resolutions and speeds. For flat-field scan, they are limited by the flat-field lens to approximately $N = 25{,}000$. Internal drum scanners, which have curved image fields, are capable of resolutions to $N = 100{,}000$, while sustaining positional integrity at high data rates.

The flat-field lens (Chapter 2) serves typical resolutions ranging from $N = 2000$ to 20,000. Below 2000—with Θ small enough and the focusing cone shallow enough (high *f*-number)—the curved image field (absent a lens) may match a flat surface adequately. Above $N = 20{,}000$, with large ΘD product, the lens is challenged to maintain a flat focal surface with high linearity and low aberration. The designer then usually adds lens elements—providing more degrees of freedom, with the concomitant complexity and cost; this explains why some exotic designs fill the entire optical path with glass!

Extra elements often appear in systems that require telecentric imaging (Chapter

---

* The digital deflector (Chapter 12) of n binary stages develops resolution $N(n) = 2^n$. It executes translational, not angular scan. In general, translational scan, when combined with angular scan, forms augmented resolution. This develops from certain polygon and hologon scanners, discussed by L. Beiser in *Holographic Scanning*, John Wiley & Sons, NY (1988).

2). While the conventional flat-field lens launches the focused beam to land on the image plane over a range of incident angles, the telecentric lens directs the beam essentially normal to a flat image surface throughout the entire scan. This may require not only more lens elements, but the output element must be at least as wide as the final image!

Another optical option is anamorphic beam handling, which reduces the effect of cross-scan errors (perpendicular to the scan line), as denoted in several chapters and references. This error derives from the scanner. Mirror wobble, for example, will misdirect the beam through a minute angle $\Theta_\perp$ in the cross-scan direction, generating (from equation (1)) a cross-scan elemental error

$$N_\perp = \frac{\Theta_\perp D_\perp}{a\lambda} \text{ elements} \tag{2}$$

where $D_\perp$ is the beam height on the scanner in the cross-scan direction. If $D_\perp$ is compressed, compared to the original beam height D, the elemental error will be reduced by the ratio $D_\perp/D$. A cylinder may be used to compress the input beam on the scanner. Another cylinder can restore the beam diameter after deflection so that it focuses to a nominally round spot, with significantly reduced cross-scan error. The control of aberration which may result from the interaction of these components with the lens is a key design responsibility, especially at high resolutions. Some holographic deflectors utilize the option of operating in the transmission Bragg regime, where the effect of wobble tends to be nulled, thereby reducing the cross-scan errors.

Holographic deflectors (Chapters 4 and 5) represent another resource. They rotate like polygons, with holograms replacing mirror facets, and are governed by the same resolution equations. They differ profoundly, however, in detail. While they can be designed to perform similar tasks, they require exercise of a specialized and maturing technology. Some of the attractive features are reduced cross-scan error per above; reduced aerodynamic loading and windage via a smooth rotating substrate; increased facet position accuracy via indexed non-contacting holographic exposure; and potential cost-effectiveness via some simplified components and possible grating replication. However, some limitations are special demands on fabrication and quality control; diffractive shifts due to wavelength shifts and, in some forms, performance shifts during scan; interrelated system components to reduce scan bow and nonlinearity; and variation of light efficiency with incident beam polarization. They can accommodate many system requirements.

There are many more options. For example, double-pass operation (Chapter 2), pyramidal versus prismatic polygon (Chapters 6 and 7), ball versus gas bearings versus other suspensions (Chapters 9 and 10), captured versus cantilevered load (Chapters 6 and 7), in normal air versus controlled atmosphere (Chapter 8), preobjective versus postobjective scan (Chapters 2, 6, and 10), over versus underillumination (over versus underfill), aperture shape factor versus resolution (Chapter 2), traveling-wave acoustic lens, vector versus raster scan (Chapter 10), piezoelectric vibrational scanning, and more. Then appear the options within each discipline, such as optical disk focusing and tracking (Chapter 13) optimal monogon/polygon facet number and critical dimensions. It is impractical to render perspective on these and other alternatives and techniques in this brief overview. Suffice the message that option selection is our most important task, aided with the resources of this volume. The components and techniques must then be integrated into a system which, as motivated at the outset, optimizes performance and cost-effectiveness.

WHAT TRENDS CAN BE ANTICIPATED?

The experience of two decades has compressed the alternatives into the options available now. Thus, one can regard the current technology as a surviving set to serve the next two decades, each technique in its optimal niche. Where there is overlap, usually special features, or even predispositions, determine direction. Interestingly, there is enough variety such that distinctly different approaches can satisfy similar operations and can coexist.

We can expect advances in production techniques and quality control. The field of precision optical replication seeks to produce polygons, hologons, mirrors, lenses, prisms, cylinders, and toroids. Some are already replicated and used. Integrated assemblies of aspherics and anamorphs can render, according to our more advanced computational capability, corresponding advances in manufacture. Diamond and numerical control machining has made, and will continue making, cost reductions in precision fabrication while maintaining operational integrity.

Look toward increased adaptation of the semiconductor laser. Having served the optical disk scanner for many years (Chapter 13), the trend toward shorter-wavelength (visible) diodes has already infused some bar code scanners and laser printers. Attractive as they are, care must be exercised to shape the beam properly and to adapt to some deficiency in beam modal integrity and stability. Their currently long (red and near-*i r*) wavelengths are especially burdensome where gas lasers now serve at $\lambda < 600$ nm, benefiting from the smaller diffraction-limited spot for a given *f*-number. To offset this longer wavelength, per equation (1), one can compensate for it with a larger ΘD product. This burdens everything else—more reason to optimize the design.

Leo Beiser

Index

$1/e^2$ diameter, 6, 10, 11, 33–34, 39, 71, 176, 177, 218, 224
$1/e^2$ irradiance level, 1, 4, 26, 33–34, 224, 225, 414
AB-type compounds, 703, 707
ABEC (Annular Bearing Engineers' Committee), 501–502
Aberrations, 66–68, 127–128, 186–187, 192, 194, 198, 202, 710, 725, 794
  first order chromatic, 66
  low and high order, 37
  spherical and coma, 67
  third order, 67
  wave, 616, 793
Access time, 615, 718, 746, 749, 821
Acoustic beam deflector, 263, 343
Acoustic microscopes, 526
Acoustic waves, 617
  longitudinal, 617
  shear (transverse), 617, 627, 646
Across scan error, or Cross scan error, 46, 203, 214, 218, 253, 311, 315, 320, 324
Actuator, 800
Addressability, 36, 116, 217
ADP (ammonium dihydrogen phosphate), 703, 726, 750
Advantages of polygonal mirrors, 331, 351
AIAG (Automotive industry action group), 184, 208, 209
Air drag, rotor, 275, 352
Air dynamics, 560, 578
Air motion, 560

Air turbulence, 218, 220, 578, 609
Air disc, 35, 557
Airy distribution, 36, 792
Alexandrite laser, 740
Aliasing, 96–99, 102–104, 114, 132, 137, 153
Alignment, 317, 321, 331–334, 345, 580
AMO (ammonium oxalate), 726
Analog deflectors, designs (*see also* Deflectors), 720–733, 749–760
Analog to digital converter, 88, 91, 123–124, 127, 136, 139
Analog video filing system, 785
Anamorphic, 48–49, 215, 248, 291, 796
Angle doubling effect, 411
Angle of incidence, 67, 191, 194, 201, 241, 243, 287, 291, 265, 558
Angle of view, 13, 20
Angle
  apex, 287, 288, 289, 416, 417, 421
  attitude, 404
  Bragg, 191, 623, 628, 647
  critical, 814
  facet, 56, 411, 415–416
  hologon incident, 241, 243
  plane, 13–17
  scan (*see also* Scan angle), 39, 174, 179, 186–187, 227, 229–230, 243–244, 308
  solid, 13, 14–16
Angles, datum, 377
Angular tolerances, 191, 377
Angular, constant velocity (CAV), 783
Animation, 554
Annealing, 547
Apertures, 7, 85–86, 96, 101–103, 112–113, 127–128, 138, 224–225, 273, 274
  clear, 224–225
  effective clear, 383
  stop, 7–8, 10–11, 32, 33, 41, 224, 273
Apex angle, 287, 288, 289, 416, 417, 421
Applications for regular polygons, 354
Archard, J.F., 509
Armature suspension, 585
Array
  CCD, 85, 213
  linear detector, 9
  scanners, 682
    analog/digital, 733–735, 758–771
    digital, 715–720
    parallel, 683
    series, 683
  two-dimensional detector, 10
Artificial horizon, 526
Assembly
  construction, 371
  motor-polygon, 215
Astigmatism, 67, 331
Asymmetry, 404
Attenuation, 121, 632
  acoustic, 632, 637, 640, 646
Audio coupling, 572
Automative Industry Action Group (AIAG), 184, 208–209
Axes, 242–243
  of symmetry, 694
  principal dielectric, 693
Axial irradiance, 3, 6, 22, 24
Axial wobble (*see also* Wobble), 52, 253, 515

Background (*see also* Signal-to-noise ratio), 118–120
Baffles, effect on windage, 470, 472
Balance
  dynamic, 195
  perfect, 402
Balancing, 220, 580
Ball bearings, 220, 294, 394, 395, 479, 498–511, 513, 517, 519, 521, 586
  advances in technology, 508–511
  advantages of, 519
  angular contact, 501
  cages, 503, 508
  cause of failure, 396
  compared with self-acting gas bearings, 521–524
  disadvantages of, 519
  elastohydrodynamic lubrication, 499
  failure of, 508
  fatigue life of, 507
  geometrical errors in, 499, 502
  Ion implantation of, 509
  life of, 220, 507, 521
  lubricants for, 395, 504–506
    hydrolysis of, 504–506
    properties of, 505
    types of, 506
  optical axis errors in, 515
  performance of, 509, 518–519, 521–524
  preload, 502
  radial deep groove, 500
  seals, 507
  shields, 507
  spin axis error in, 515, 521
  standards of, 502
  terminology for, 499
[Ball bearings]
  titanium carbide coated, advantages, 219, 509–510
Band limited signal, 98–99
Bandwidth, 90, 96, 99, 570, 597, 631, 649
  interaction, 633, 658
  modulation, 633
Bar codes, 159, 160, 163–165, 170, 173–178, 180, 182–192, 195, 197, 204–209, 526
  3 of 9, 163
  UPC, 160, 163, 197, 205
BaTiO3 (Barium Titanate), 700, 706, 726
BBO (B-Barium Borate), 726
Beam, 159, 162, 165–166, 168–170, 172–178, 180–181, 183, 185–187, 194–195, 197–204, 224
  cross section, 2, 10, 21, 24, 224, 229, 248, 273
  diameter, 134–136, 198, 224–225, 352
    $1/e^2$ excluded power, 5, 6, 11, 225
    50% power, 5, 218, 225
    full power, 3, 6
    FWHM, 6, 10, 781
    half-power, 5, 10
  irradiance, 5, 231, 270, 271
    axial, 3, 270
  path distortion, 561
  pointing, 541
  power, 3–9, 18, 21–24, 221, 231, 270, 352
    encircled, 23, 24
    excluded, 23
    total, 3, 5, 22–26
  quality, 12
  steering, 668

[Beam]
acoustic, 668
perfect, 671
translation, 363
waist, 2, 6, 11–13, 20–23, 176, 633
wavelength, 224, 241, 352
spatial profile, 721, 724
truncation ratio, 224, 226, 229
walk-off, 717
Bearing, 220, 294
errors
optomechanical, 514–517
reduction of, 516
gap, convergence, 479–480, 484
life, 220, 396
temperature dependence, 396
performance comparison, 517–524
preload, 571–572, 587
systems, 393
technology, advances in, 508–511
wobble, 515, 586
Bearings for instruments, 479
Bearings (*see also* Ball, Externally pressurized, Gas, Hydrodynamic, Hydrostatic, Journal, Plain journal, Self-acting, and Spiral groove)
air, 393
angular contact
performance of, 501, 518, 521
preload, 502
separable and nonseparable, 502
conical gas, self-acting, 487–489
fixed geometry, 485
[Bearings]
fluid film, 478, 479
foil, 484
hybrid, 480, 493
hydrodynamic gas, 415, 421, 446
linear, 478
magnetic, 512, 519
pivoted pad, 484
procurement of, 512–514
radial deep groove, 500–501
resilient conforming, 484
rolling element, 478–479
rotating, 478
silicon nitride ball, 510
squeeze film, 519
thrust, 479, 484, 487, 498
load on, 395
Beryllium, 375, 415, 418, 420, 421
Bessel functions, 619, 625
Bias voltage, 719, 734, 744–745
Binary image, 89–91, 93–94, 100, 104–105, 109, 115, 155
Birefringence, 197, 626, 697
plate, 314, 322, 711, 710
wedge, 711
Blur, 89–91, 93, 95, 99, 103–104, 111–112, 114, 120–121, 126–128
Bode diagrams, 801
Boundary lubrication, 478, 489
Bow, 221, 233, 240, 248, 283, 285, 287–288, 296, 299, 310, 317
correction, 46, 54, 56
tie error, 49
Bragg angle, 191, 623, 628, 647
Bragg deflector (*see also* Deflectors), 737, 770
Bragg diffraction, 623, 737

Bragg planes, 194–195, 623
Brinelling, 396
Brushless DC motor (*see also* Motors), 388–390, 432–433, 435, 439, 443
Bulk modulus, 624
Burgdorfer, A., 491

Cages (*see also* Bearings)
for ball bearings, 503, 508
centrifuging of, 504
materials for, 504
porous, 504
vacuum impregnation of, 504
Calcite, 694–695, 711, 742
Calculus of variations, 728
Calibration, input scanners, 123, 124, 136
Cantilevered design, 397–398
Captured design, 397–398
CaTiO3 (Calcium titanate), 703–704
CAV (Constant angular velocity), 783
CD (Compact disc), 779, 783
CD-ROM, 783
CdS (Cadnium sulphide), 664, 707
CdTe (Cadnium telluride), 707
CDV (Compact disc, vision), 783
Centrifugal stress, 276, 378
Chalcogenide crystals, 641
Character distortion, 119–121
Characteristic curve, 143
Charles Stark Draper Laboratory, 499
Chemical reaction, 547, 550
Chief ray, 41, 67
CIE (Commission International de l'Eclairage), 3, 17
Clip levels, 11
Closed loop (*see also* Loop), 432, 435
CLV (Constant linear velocity), 783
Coating
enhancement, 385
vacuum-deposited, 386
Codabar, 163–164
Code V, optical design program, 63
Collective surfaces, 68
Coma (*see also* Aberration), 67
Commission International de l'Eclairage (CIE), 3, 17
Communication, 526
Compact disc (CD), 779, 783
Compressibility, 616
number, 490
of gas, 480, 484
Computer output microfilm, 213, 336, 526–527, 532
Computerized tomography, 532
Conductivity, in Kerr cells, 708
Conical scans, 362, 364, 487–489
Conjugate change, 49
Contrast, 103–104, 106, 109, 112, 118, 120, 132, 146, 163–165, 171, 196, 204–205
Control voltage, 714, 742, 745, 760
Convergence in bearing gap, 479–480, 484
Copy disk, 192, 194–195
Counter, 543
Coupling, 257
forces, 402
electrostrictive, 698
piezoelectric, 698
Cowking, E.W., 499
Critical angle, 814
Critical frequency, 148–149
Cross flexures, 587

Cross scan error, 46, 203, 214, 218, 253, 311, 315, 320, 324
Cross talk, 682, 712
Cross-axis, 534, 573
  resonant, 578, 606, 608
Crystal streak camera, 689, 754, 757
CuCl (Copper Chloride), 700
Current, hydrodynamic, 708

Daily, J.W., 475
Damping, 578
  piezoelectric, 746, 771
Darchuk, J.M., 26
Data file, 784
DCG (Dichromated gelatin), 191–195
De Ferranti, S.Z., 480
Decentering, 403
Deep groove grating, 269
Deflection (*see also* Linearity of deflection), 166, 171–175, 201, 202
  geometric limit of, 717
  linear, 227, 230
  two-dimensional, 339, 716, 727, 742
  XY, 681
Deflector, 166, 169, 171, 173–174, 184–185, 187–188, 191–192, 197
Deflector geometry
  bulk crystal, 758
  multiple prism, 721, 757
  parallel array, 733–734, 762–763
  prism, 758
  series array, 722
Deflector system, hologon, 213, 216, 223, 232, 239, 246, 282, 306, 331, 336
Deflector wobble, hologon, 253, 256, 257, 259, 311
Deflector
  acoustic beam, 263, 343
  analog, 720–733, 749–760
  analog/digital, 720, 733, 758
    application of, 213, 216, 687–692
  Bragg, 737, 770
  capacitance of, 733, 746, 750, 752, 763
  capacity speed product of, 718
  designs and considerations, 214, 741, 746–771
  digital light (DLD), 709–720, 741–749, 769
  frequency shift, 736, 740
  gradient, 752–758, 727–732
  interference, 738–740
  line, 722
  rotation limitation, 275
  rupture strength, 275
  series, 722
  two dimensional, 339, 719
Deflectors
  analog, characteristics of, 755–758, 760
  bulk array, 763
  Kerr cell, 692, 742, 744
  surface wave, 764–771
Defocus, 67, 226, 797
Demand function, 144, 146–148
Density wedge (grey wedge), 124
Depth of field, 163, 166, 169, 170–171, 175–185, 189–190, 203, 205–206
  large, 171, 179, 183, 205, 206, 208
Depth of focus, 38, 203, 226, 369

Design, 556, 579
  aperture, 34
  differences, scan motors, 391
  trade-offs, 336, 339, 369
Designators, 526
Detective quantum efficiency (DQE), 142–143
Detectors, 219, 257, 313, 345
Diameter, $1/e^2$, 6, 10–11, 33–34, 39, 71, 176–177, 218, 224
Diameters (*see also* Beam diameter and Inscribed diameter), 1–10, 14
Diamond machining, 371, 585
Dichromated gelatin (DCG), 188, 191–195, 209
Dielectric
  loss, 717
  principal axes, 693
  relative constants, 693, 699–700
Diffraction (*see also* Resolution limit), 232, 240, 243, 269, 272, 296, 618
  angle, hologon, 234, 240, 243
  anisotropic, 625, 642
  Bragg, 623, 737
  collinear, 630
  Debye-Sears, 620
  degeneracy in, 649, 651
  efficiency, 188–192, 194, 203, 267, 269, 623, 625
  limited, 33, 779, 832
  limited resolution, 218, 239
  of light, 267, 269
  orders, 241, 272
  Raman-Nath, 249, 621
  spread
    of light, 621, 631, 634
    of sound, 634
Digital images, 88, 90, 100, 107, 112, 151, 153–157
Digital light deflector (DLD), 709–720, 741–749, 769
Dimensional analysis, 461
Dimensionless groups, 461
Dimensions, range of, 354
Directional cosines, hologon, 243
Dispersive surfaces, 69
Distortion, 67, 227, 283, 285
  local, 31
  scan lens, 283, 285
  third order, 68
Distribution
  profile, 2–3, 5–6, 21
    normalized irradiance, 25–26
Dithering
  spatial, 111
    clustered dot, 111
    dispersed dot, 111
    with blue noise, 111
Divergence, 1, 3, 7, 11–14
DLD (Digital light deflector) (*see also* Deflector), 709–720, 741–749, 769
Dorfman, L.A., 476
Dots per inch (dpi), 217
Double pass, 29, 54, 62
Drag, 452–453
Drift, 574, 608–609
Drilling, 547
Dust particles, 582
Duty cycle, 227, 229, 244, 248, 252, 292
  angular, 229
Dynamic, 380, 570–573, 574–575, 590
  balance, 358, 399, 570–573
  deformations, 579, 581
  imbalance, 400, 570–573, 586
    radial, 570–573, 586, 608

E-O effect (*see also* Electrooptic effect), 692

Edge profile, 128
Elastic stiffness, 624
Elastohydrodynamic lubrication, 499
Elastooptic (*see also* Photoelastic), 616
  effect, 616, 698
  coefficient, 616
Electrical breakdown, conductivity, 708
Electrodes, 654, 716
  interdigital, 709, 766
  quadrupole, 729, 752
  transparent conducting, 716, 746
Electroless
  nickel, 418, 421
Electromagnetic radiant power, 16
Electromechanical coupling factor, 655
Electronic controls, 404, 603
Electrooptic effect (E-O effect), 692
  coefficients, 692, 698–700
  materials
    properties of, 699–700, 703
    selection of, 703–704
  theory
    linear, 695
    quadratic, 701, 702
  switch, 713, 771
  transverse modulators, 747
Electrostriction
  coupling by, 698
  strain due to, 705
Ellipticity, 170, 248, 273, 274, 277
Elrod, H., 498
EM propagation in crystals, 693
Encircled beam power, fraction of, 4–5, 9–10, 23, 25
Encircled energy, 68
Energy gap, 639
Enhancement, 137, 152
  coating, 383
Entrance pupil, 60
Entrance scanning (*see also* Scanning), 360–361
Epoxy resin, 372, 374
Erosion, 582
Excluded beam power, fraction of, 2–6, 10–11, 23, 25
Exit scanner, scanning, 362, 369
Externally pressurized bearings (*see also* Journal bearings), 413, 423, 443, 445, 480, 492–498, 513–514, 518, 520
  advantages of, 492, 520
  air hammer in, 498
  bearing stiffness of, 494, 497
  conical, 498
  design theory for, 480
  disadvantages of, 492, 520
  gas consumption of, 492, 494, 497
  half speed whirl in, 498
  instability in, 498
  performance of, 518, 520
  resonance of, 498
  restrictors for, 494–496
  thrust surfaces for, 498
Eye frequency response, 123, 152–153

F-number, F/No, F#, 32, 224
F-theta (F-θ) condition, 29
F-theta lens (*see also* Lens), 29, 360, 413, 443, 563
F-theta optical system, 29, 415
F-theta scan lens, 29, 227, 230, 283, 413, 415, 443

Facet, 46–47, 53, 214, 219, 225, 228–229, 232, 247, 277, 306, 324
  angle, 53, 411, 415–416
  error, 219, 261
  flatness, 378, 381
    distortion, 331, 378
    dynamic and static, 378
  grating lines, 247–248
  identification, 174, 179, 183, 209
  size, 201, 229, 274, 370
    effect on windage, 468
  wobble, 50
Facsimile, 94, 526
Faddy, D., 486
Feed beam, 33, 46
Feedback correctors, 365
Feedback loop, 392
Fermat's principle, 728
Ferranti Ltd., 452
Ferro fluids (*see also* Bearings), 511, 519
Ferroelectric crystals, 666, 704
Field flattening lenses (*see also* Lens), 558
Figure of merit, 139, 583, 625
  acoustooptic, M2, 636, 647
  bandwidth-efficiency, M1, 636
  electrooptic, 709, 718, 725, 733, 749
  modulation bandwidth, M3, 636
Filter
  polarization, 744
  reconstruction, 99
Finesse of etalon, 740
Finite impulse response filter, 136–137
First order design (layout), 40
First resonance, 608
Flare light, 125, 131
Flat field (*see also* Lens, field flattening), 28, 223, 558
Flat image plane, 223, 226, 578
Flat-field scanning system, 223, 226, 246, 315, 528
Flatbed scanners (*see also* Scanners and Scanning), 55, 527, 529
Flatness (*see* Facet)
Fleischer, J.M., 26
FLIR (Forward-looking infrared), 539
  sensors, 539–541
Fluid boundary layer, 453
Flurorescence emission, 542
Flux, 11, 17, 18
Flyback time, 679
Focal length, 30, 224, 227, 681
  calibrated, 31, 227, 230
Focal points, 40–41
Focal surface, plane, 40, 361
  multiple, 205, 207–208
Focal-point, multiple, 337
Focus
  depth of, 226
  servo system, 809
  sagittal, 68
  tangential, 68, 722
Focusing equipment, 722
Foil bearings (*see also* Bearings), 484
Form Drag, 453
Fourier transform, 97, 98, 128, 129, 133, 156–158, 634
  of acoustic profile, 635
Fourier-Bessel transform, 792
Frequency
  critical, 148–149
  fundamental, 95, 541

[Frequency]
response, 100, 127, 128, 132, 135, 136, 152, 730–733, 746, 749
phase-lock loop, 413, 433
tuning, 611
Fringe, optical, 739
Fringing field, electric, 764
Frown, 569
Fujitsu, 171, 202, 211
FWHM (Full width at half maximum), 6, 781, 792

GaAs (Gallium Arsenide), 204, 707
Gain guided laser, 804
Galvonometer scanner, 44
Gamma, 146
GaP (Gallium Phosfide), 641
Gas bearings (*see also* Externally pressurized and Self-acting), 479–498
hemispherical
externally pressurized, 448–449, 498
self-acting, 487, 489
theoretical analysis of, 480
unique advantages of, 482
Gas compressibility, 462
Gas turbines, 352
Gases
advantages of, 482
as lubricants, 482
compressibility effects on, 480, 484
effect of temperature on, 480
mean free paths, effect in, 491
viscosity of, 482
Gaussian beam, 1–26, 32, 176, 217, 224, 556, 557, 633
coefficient, 35
[Gaussian beam]
focusing of, 224, 633
truncation of, 224–225
Gaussian characteristics, 2, 10–12
Gel swell (Gelatin swell), 194–195
Germanium, 641, 651
Gladstone-Dale equation, 560, 639
Glass, 66, 173, 188, 192–193, 195
Graininess, 83
Granularity, 87, 101–102, 112, 124, 146, 148–149, 154
Grassam, N.S., 480
Grating, 173, 196, 232, 239, 240, 264, 268, 283, 291, 296, 372, 527, 601
equation, 240, 272
periodicity error, 261
refractive index, 272, 734, 738
Gray image, 91, 94, 121
Gray levels, distinguishable, 99–102
Gray scale, 86, 123, 125, 140, 146, 532
Gray tone, 531
Gross, W.A., 480
Ground paths, 576
Grunneisen constant, 640
Gyroscopes, spin axis bearing for, 499, 503

Half-power, 5, 10
Halftone, 100, 101, 104–105, 107–111, 114, 116–117, 121, 155, 214, 218, 529
digital, 107, 109, 111
Hanson, R.A., 510
Harmonic, 96, 129, 134, 218, 541
motion, 525
Harrison, W.J., 480

Hayosh, T.D., 452, 459
Heating, 651
 in bulk material, 708, 717, 720
 in resistive electrodes, 717–718
Helium, use of, 472
Helium-neon (HeNe), (*see also* Laser, HeNe), 1, 177, 193, 198, 202
Heterodyne interferometers, 602–603
High order chromatic aberrations, 67
HOE (Holographic optical element), 171, 173
Holmes, J., 503
Hologon deflector system, (*see also* Deflectors), 213, 216, 223, 232, 239, 246, 282, 306, 331, 336
 centered facet disk, 277
 Cindrich-type, 232, 239
 concentric configuration, 236
 dipole disk deflector, 279, 319
 non-flat deflector, 234, 306
 plane grating
  disk deflector, 239, 282
  nondisk deflector, 306
  pyramidal deflector, 324
 quadrapole disk deflector, 279
Holographic deflector disk, 174, 191, 213, 232, 239
Holographic disk, 173, 175, 178–180, 182–183, 185–186, 188–190, 192–193, 195, 197–198, 201–203, 208–209, 232, 239
Holographic optical element (HOE), 171, 173, 205, 213
Holographic scanning, 54
Housing, effect on windage, 468, 472
Human eye, 99, 135, 152
Hunting (*see also* Phase jitter), 219, 430, 433, 435
Hydrodynamic (*see also* Self-acting gas), 423, 484–492
 air bearing, 393, 423
  design parameters, 394
 currents, 708
 fluid film bearings, 480
 gas bearings, 415, 421, 446
Hydrostatic air bearings, 393, 419, 423

IBM (International Business Machines), 81, 159, 160, 168, 171, 197, 208–210, 785
Image
 banding, 218
 digitizers, 526
 distortions, 559–569
 information, 126
 input terminal (IIT), 86–87, 89
 output terminal (IOT), 87, 89
 processing, 86–89, 91, 155
 projection, 534
 spot ellipticity, 273
 spot size, 218, 224, 239
 velocity, 231, 558
Imaging science, 83–86, 99, 122, 149, 155
Imbalance, produced by asymmetries, 404, 570–572
Impedance, 655
 acoustic, 658
 electrical, 658, 666
 matching, 666, 674
Index ellipsoid (indicatrix), 617, 694–696
Index of refraction
 of air, 559, 601
 of dielectric materials, 386
Index pulse, 413, 417

Indicatrix (index ellipsoid), 617, 694–696
Indices of refraction, principal, 694, 699
Indium bonding, 659
Induced moving coil, 593
Inductive encoders, 598
Inertia, 698
Information
  capacity, 114, 149, 150, 154
  content, 114, 149–151, 154
Infrared (IR) viewing, 354
Inhomogeneity, 404, 708
Injection molding, 371
Input scanner, 84, 90, 94, 102–105, 112, 117, 121–126, 128–129, 132–133, 136, 138–140, 145–146, 150
Inscribed diameter, 415
Inspection systems, 213, 216, 354
Instability
  damping to limit, 498
  externally pressurized bearings, 492, 498
Instability, self-acting gas bearings, 491
Integrated optics, 769, 771
Integrated scatter, 382
Interference deflector (*see also* Deflector), 738–740
Interferometers, 526, 601–603
  dual-path, 603
Interferometric measurements, 601
Interferometric zone plate (IZP), 202
Interleaved 2 of 5 code, 164
International Business Machines (IBM), 81, 159, 160, 168, 171, 197, 208–210, 785
Inverted telephoto, 42
Irradiance, 2–6, 10–13, 17, 20–23, 25–26, 32, 85, 105
[Irradiance]
  level, $1/e^2$, 1, 4, 26, 33–34, 224–225, 414
ISO (International Standards Organisation), 502

Jagged appearance, 120, 218
Japanese Article Number (JAN), 202
Jenkins, J. A., 26
Jitter (*see also* Wobble), 112, 119, 122, 219, 256, 261, 262, 267, 313, 325, 367, 417–418, 444, 446, 561, 574, 578, 579, 608
Journal bearings
  description of, 484
  externally pressurized, 493–497
  lifting efficiency of, 494
  load coefficient for, 494
  restrictors for, 494–496

KDA (potassium dihydrogen arsenate), 699, 704
KDP (pottasium dihydrogen phosphate), 687, 699, 703, 713–714, 717–718, 726, 746, 748
Kerr cell, 708, 741
  constant, materials for, 703
Kerr effect, liquids, 708, 744
Kingsbury, A., 480
Klein-Cook parameter, 622
Knife-edge, 7–11, 225
Komstar, 532
Kryton switch, 754
KTN (potassium tantalate niobate), 706, 720, 726
KTP (potassium titanyl phosphate), 706, 726

Lambertian emitter, 19
Lambertian radiator, 18–20
Lapping, 374
Laser, 779, 802
  Alexandrite, 740
  beam, 1–3, 5–8, 11–12, 14, 20–22, 25
  carbon dioxide, 650, 707
  display, 553, 688
  helium-neon (HeNe), 177, 193, 198, 680, 720, 753, 779
  modes, 1
  Nd-YAG, 755–756
  reflectometers, 383
  scan lenses, 70
Laurin, T.C., 26
Lead bonders, 526
Lead molybdate, $PbMoO_4$, 641
Lens, 20, 99, 104, 125, 127–128
  cylindrical, 48, 366, 678
  F-theta, 29, 227, 230, 360, 563
  field flattening, 28, 223, 558
  relay, 45, 339, 341, 566
  thin, 40
  toroidal, 49
  translator, 532
Licht, L., 498
Life, polygon, 387
Lift off, 421–422
Light, 88, 103–105, 109, 120, 124–125, 132, 136, 142, 155
  for entertainment, 526, 553
Light valve, 536
$LiNbO_3$ (lithium niobate), 641, 644, 664, 692, 697, 706, 726, 757, 772
Lindeman relation, 640
Line deflector (*see also* Series deflector), 722
Line spread function (*see also* Spread function), 35, 126–128, 137
Line straightness, 527
Linear E-O effect, 695
Linearity, 126, 591–594
  of deflection, 227, 230
Linotype, 527
Liquid crystal, 526, 536
Liquid E-O materials, 702
Lithium niobate, 641, 644, 664, 692, 697, 706, 726, 757, 772
Lithium tantelate ($LiTaO_3$), 726
Lithography, 526, 550
Loop, 430, 432–433, 435, 441
Lorentz-Lorenz relation, 616, 626
Lubricant leakage, evaporation, 396

Magnetic resonance, 532
Manufacturing methods, 79
Markers, 526, 549–551
Marking, 104, 111–112, 121, 547, 549–550
Marshall, G.F., 26
Massey, B.S., 476
Master disk, 192–194
  holographic, 195
Material selection, 375, 583
Materials, 546, 637, 641
  acoustooptic, 637, 643–645
  amorphous, 643
  birefringent, 694
  bond, 662
  E-O (electrooptic), 696, 699–700
  infrared, 645
  piezoelectric, 664–665
  polarizer, 710–711

Mechanical constraints, 575, 578
Mechanical duty cycle, 227
Mechanical resonances, 572
Mechanical Technology Inc., 480
Mechanical transducer, 590
Memory repair, 549
Meridional plane, 68
Metal mirrors (*see also* Mirror), 583
Metresites, 598
Microphonic, 560
Microscope, 526, 541–542, 680
  flying spot, 680
  vidicon, 680
Microsyns, 598
Minimum deviation condition, 254, 259, 291, 293, 299, 308
Mirror
  mounting of, 579–580, 609
  off axis, 561
  surface of, 583
Mirror bonding, 580
Mirror deformation, 331, 580–581, 609
Mirror dynamic rigidity, 608
Mirror laws, first and second, 362
Mirrors, 45, 578, 583
  metal, 583
Model testing
  advantages of, 462, 463
  compared to real scanners, 464
  for windage prediction, 461–468
  in liquid, 462
  results of, 466–468
  rig for, 463
Modulation, 112, 120, 128–129, 131–135, 140, 144, 146–148
  amplitude, 104
[Modulation]
  minimum visually perceivable, 120
  phase, 619, 624
Modulation transfer function (*see also* MTF), 35, 103, 128–129, 132–133, 152, 218
Modulator, 342, 532, 746
  acoustooptic, 342
Moire, 104, 114, 117
Moment of inertia, 570
Momentum, 618, 622, 629
Motion nonuniformity, 117
Motor bearing life, deflector, 220
Motor drives, 388
Motor hunting, deflector, 219
Motor, synchronous, 406
Motor-polygon assembly, 215, 353
Motors
  brushless, 388–390, 432–433, 435, 439, 443
  direct-current (DC), 389
  hysteresis synchronous, 294, 388, 424, 426–428, 432–433, 435, 445, 446
  induction
    slipped, 388, 390
    synchronous, 388, 390
  synchronous, 392
Moving coil, and iron, 591–594
Moving magnet, 596
MTF (*see also* Modulation transfer function), 35, 87, 99, 102, 104, 126, 128, 132–137, 140, 144, 146–151, 154
Muijderman, E.A., 511
Multi-configuration, 32, 56, 61
Multiple focal-plane, 205, 207–208
Multiple focal-point, 337
Munsell value, 105

Nd-YAG laser, 755–756
NEC (Nippon Electric Company), 171, 202–203, 211
Nece, R.C., 475
Nicodemus, F.E., 26
Nippon Electric Company (NEC), 171, 202–203, 211
Nitinol, 612
Nitrobenzene, 742, 744
  Kerr cell, 713, 741
  purity of, 742
Noise, 98, 102, 103, 111–113, 118, 121, 124, 136–141, 145–147, 149–152, 155, 805
  additive, 136, 138, 141–142, 151
  electronic, 112
  fixed, 136
  multiplicative, 136, 141–142
  paper, 138, 180, 207
  quantization, 116, 138, 153
  relative intensity (RIN), 807
  removal filters, 112
  sampling, 111
  sources, 136, 138, 146
  white, 136, 138
Noise equivalent quanta (NEQ), 142–144
Nonlinearity, 87, 102, 104, 113, 127, 129, 131
Numerical aperture (NA), 32
Nyquist frequency, 96, 98, 132, 154

Obstructions, 468
Omni-directional, 166–167, 203, 208
Open loop (*see also* Loop), 430, 432–433
Ophthalmoscopes, 526, 545–546
Optic axes, 647, 710
Optical
  constraints, 556
  encoder, 413, 417, 600–603
  homogeneity and inhomogeneity, 708, 716
  modulators, 746
Optical transfer function (OTF), (*see also* MTF), 127–128
Optomechanical errors, 514–517
Oslo, optical design program, 36
OTF (Optical transfer function), (*see also* MTF), 127–128
Overlapping focal-zones, 174, 179, 181, 208

Pan, C.T.H., 486
Passband, 148, 149, 153
Pell box, 530
Performance
  of deflector, 606
    capacity-speed product, 717
    comparisons, 755–758
    resolution, 715–716
  of galvonometric scanners, 606–609
  of resonant scanners, 609
Periodic scanning artifacts, 526
Peripheral velocity, 375
Perovskites, 706
Petzval curvature or radius, 38, 41–42, 69, 75
Petzval role, 44
Phase lock loop (PLL), (*see also* Loop), 413, 433, 441
Phase modulation, 619, 624
Phonons, 618, 622
Photocomposition, 213, 216, 526
Photocompositor, 530–531
Photoelastic constant, 616
  tensor components of, 617, 640

Photoelastic effect (*see also* Elastooptic), 616
Photoplotter, 527, 532–536
Photopolymer, 190–192
Photoreceptor, curved surface, 362
Photoresist, 188, 189, 195–196, 203, 268
Phototransmissions, 527, 530–531
Phototypesetters, 213, 216, 526, 527
Pictorial quality, 147–148
PID (Proportional-integral-derivative), 433, 435
  controller, 433, 435
Piezoelectric
  coupling, 653, 698
  effect, 714, 718
  resonances, 717, 744, 746
Pincushion errors, 565–568
Pinhole, 7–8, 10–11
Pixel (Pel), 39, 86, 90–91, 93–95, 99–116, 121, 123–124, 126–127, 137–138, 141, 149, 150, 217, 231, 415, 532
  density, 217, 231
  diameter, 218, 415–416
  elements, 9–10
  frequency, 415
  grating clock, 220
  placement error, 112, 118, 119, 121, 218, 219, 532
  registration, 413, 417, 433
Pixels, number of, 413, 417, 433
Plain journal bearings
  bearing efficiency of, 478, 482
  drag power of, 482
  load capacity of, 482
  wedge effect in, 484
Plating and polishing, 583
PLZT ceramic, 708
Pneumatic drives, 388
Pockels cell, 741
  deflectors, 744
Pockels effect
  coefficients, 698
  definition, 696
Point spread function (*see also* Spread function), 37
Point-of-sale (POS), 197, 202, 213
Pointing, and errors, 579–580
Poisson's ratio, 276, 418, 420
Polar molecules, 702, 708
Polarization, 269, 270, 314, 318, 626
  circular, 270, 314, 319, 646
  electric field on, 692
  extraordinary, 626
  field induced, 703
  filter, 744
  planes of, 694
  ratio, 749
  spontaneous, 705
Polarizers, 710–712
Polishing, 371
  compound, 375
Polygon, 121, 128, 410, 528
  durability of, 387
  efficiency of, 52
  modifications, effect on windage, 472
  pyramidal and random errors in, 46, 52, 515–516
Polygon-motor interface, 397
Polygonal mirrors, advantages of, 351
Polygonal scanner system, 45, 214, 369, 418
Polygonal scanners, 410, 417–418, 442
Polygons
  cleaning of, 386
  commercial grade of, 352

[Polygons]
fabrication methods of, 370
inverted, 356, 359
irregular, 354, 358
pyramidal, 54, 352, 359, 410, 443
regular, 352–353, 410
stacking of, 354
Position transducers, 597–603
Postobjective scanning, 532, 563–565
Powell, J.W., 480
Power density (*see also* Beam irradiance), 2–6, 10–13, 20–22, 24–26, 624
Power, (*see also* Beam power), 624
acoustic, 624
dissipated, 718
reactive, 717
Preobjective deflector system, 223
Preobjective scanning (*see also* Scanning), 28, 223, 550
Principal points, 41, 311
Print contrast signal (PCS), 164–165
Print plate, 527
Printers, 88, 526, 528, 529
Printing, 526
Prism, 54, 283, 287, 288, 291
birefringent, 709
deflector, 720
Porro, 27, 368
retro reflecting, 367
Wollaston, 711
Profile, 2
Proprietary methods, 371
Psychophysical evaluation, 154
Pulse compression, 677
Pulsewidth modulation (PWM), 104, 108, 439–441, 783
Quadratic E-O effect (*see also* E-O effect), 701, 714
Quadrupole electrodes, 752, 757
Quantization, 91, 98–99, 101–103, 111, 124, 126, 132, 137–139, 150, 154
Quartz, 633, 694
Quiet zone, 180

Radial force, 570, 571
Radiance, 17–19, 112
Radiant flux, 8, 17–18, 20–25
Radiant intensity, 17–20
Radiometric scan uniformity, hologon, 270, 282, 294, 314, 315, 318
Radiometric terms, 3, 13–22, 270
Raggedness, 119–120, 122
Raman-Nath parameter, 622
Rare earth magnets, 591
Raster
distortion, 112, 118, 227
input scanner (RIS), 88–89
output scanner (ROS), 88–89, 112
scan, 171, 184, 185, 208, 209, 214, 217
scanning, 121, 179, 183, 208, 214, 217, 526
separation errors, 116, 217–218
Ray, 40, 710
extraordinary, 710
Rayleigh, 557–558
limit, 720
range, 2, 11–13
RDP (rubidium dihydrogren phosphate), 704
Reaction torque, testing, 454–458
Reactive drive power, 717
Reconstruction filter, 99
Reference system, 242, 412, 418

Reflectance, 88, 105, 107–109, 122–126, 129, 131, 136, 139, 142, 148, 151, 382
comparative, 384
measurement, 383
specific, 383
test, 384
Reflection losses, 746
Refraction, 287
index of, 694
principal indices, 694, 699–700
Repeatability, 529, 574, 578, 608
Replication, 371, 585
Resolution, 86, 93, 99, 102, 104, 114, 127, 149, 157, 218, 526, 528, 529, 631, 679, 715
of a system, 831
spatial, 715, 721, 735, 749
temporal, 749, 755, 758
Resolved beam positions, 721, 728
Resolved spots, 631, 637, 720, 736
geometric limit to, 724
number of, 637, 720–721, 735, 754, 764
Resonances, 572–573, 575, 580, 608
magnetic, 532
mechanical, 572
piezoelectric, 717, 750, 771
torsional, 573
Resonant scanners, 525, 532, 560, 577–578
Responsivity, 124, 126
Retina, 545
Retrodiffracted beam, 319, 334
Retroreflective, 48, 50, 332–333
Reynolds, 479
Reynolds number (Re), 459, 462, 467, 475, 560
RMB, 479
RMS roughness, 382
Rotary variable differential transformer (RVDT), 598
Rotational speeds, 352
Rotor
definition of, 475
resonant frequency, 402

Safety, 277, 375
Sagittal focus, 68
Sampled tracking method, 818
Sampling, 85–86, 88, 94, 96–99, 103, 104, 117, 127, 132, 150
frequency, 93, 95, 97–99, 127, 131–133
phase, 93–95, 127, 133
theorem, 95, 98, 102
Satellite communication, 541
Scan, 83–86, 104, 114, 116, 122, 125, 127, 131, 136, 140, 153, 217
addressability, 217
angle, 174, 179, 186–187, 227, 229, 243, 308, 352, 411–412, 651, 832
efficiency, 227, 229, 577
beam angle, hologon, 243, 308
beam tracking error, 214, 218, 253, 261, 263, 311
bow compensation, 244, 248, 283, 285, 287, 296
grating element, 283, 296, 299, 301
prism element, 283, 287–288, 291, 293, 301
efficiency, 227, 229, 352, 577
frequency, 113–116, 231
jitter, 219, 256, 261, 267, 313, 325

[Scan]
length, 38, 39, 167, 187, 227, 230–231, 249, 369
lens distortion, 227, 283, 285
linearity, 227, 230, 352
radius, 369
raster, 171, 184–185, 208–209, 214, 217
rate, 231, 352, 369, 468, 675–676
resolution, 218, 231, 352, 674
velocity, 167, 231, 369
Scan-angle magnification, 174, 179, 186, 187, 230, 339
Scan-line bow, 221, 233, 240, 248, 283, 310, 317
Scan-line spacing, 217, 231
Scanner
definition of, 475
exit, 369
raster input (RIS), 88
raster output (ROS), 88, 112
tunable resonant, 532, 536, 541, 578, 609, 611, 612
XY, 43, 337, 339, 681
Scanners, 83–87, 99, 101–105, 112, 118, 121–128, 131, 136–140, 145–147, 150–151, 525, 650
flatbed, 55, 527, 529
frequency chirp, 675
procurement of rotary, 512–514
Scanning, 650
dual axis, 61
dual beam, hologon, 321
entrance or preobjective, 28, 223, 360–361, 550
exit, 362
laser acoustic microscope, 526, 543, 544
microscope, 541, 542
periodic artifacts, 526
[Scanning]
raster, 121, 179, 183, 208, 214, 217, 526
straight line, 221, 362
system
flat-field, 223, 226, 246, 315, 528
internal drum, 29, 55, 223, 306, 308
vector, 579
Scatter losses, 369
Scattering of light in paper, 111
Scratch and dig, 381
Seals
materials used, 507
purpose of, 507
use of ferro fluids, 507
Second focal point, 40
Second nodal point, 41
Second principal point, 41
Self-acting bearings, 421–423
Self-acting gas bearings (*see also* Spiral groove gas bearings), 479, 484–492, 521
attitude angle, 484
boundary lubrication of, 489
built-in wedge effect, 484, 485, 487
centrifugal effects in, 491
characteristics of, 489
compared with ball bearings, 521–524
compressibility effects in, 484, 489
contamination, 523
cost of, 524
description of, 484
eccentricity, 484
gas mean free path effects in, 491
half speed whirl in, 491
hard surface for, 489

[Self-acting gas bearings]
instability, 491, 492
life of, 521
load capacity of, 489
spin axis error in, 521
thrust of, 484, 487, 490
vibration and shock capacity of, 523
Self-acting liquid bearings (*see also* Spiral groove liquid bearings), 515, 519
advantages of, 519
disadvantages of, 520
performance of, 511, 512, 519
use of ferro fluids with, 511, 519
Semiconductor laser, 177, 204, 263, 295, 312
Sensitivity, 10
Servo control, 573, 574
Shaft critical speed, 482
Sharpness, 83, 95, 104
Shepherd, J., 503
Signal-to-noise ratio (S/N), 114, 140–141, 369, 712
Silver halide, 188, 190, 192, 194, 202
Sioshansi, P., 509
Sixsmith, H., 498
Skin friction, 453
Slit, 7–11, 35, 154
Smile, X., 569
Snell, J.F., 26, 287
Snell's law, 287, 720
Spatial resolution, 715, 721, 735, 749
Spectrometer, 283, 537
Spectrophotometers, 383
Spherical aberration, 67
Spiral groove gas bearings
advantages of, 520
configurations of, 486, 488
[Spiral groove gas bearings]
disadvantages of, 520
for gyroscopes, 486
performance of, 521–524
theoretical analysis of, 486
Spiral groove liquid bearings
advantages of, 519
barrier film coating for, 511
disadvantages of, 520
performance of, 511, 512, 519
use of ferro fluids with, 511, 519
Split detection method, 820
Spot diameter, 35
Spot overlap, 218
Spot size, 35, 104, 217, 224, 226, 369, 556
detection, 333, 811
Spread function, 35, 37, 114, 126–128, 131, 150, 155
Stability, angular velocity, 361, 519
Standard deviation, 5–6, 21, 24–25, 101–103, 112–113, 141, 143, 150–151
Standing pressure wave, 394
Stationarity, principle of, 132
Step response, 105, 570, 607
Stepping time, 570, 607
Stitching, 547
Stops (*see also* Aperture stop), 7–8, 10–11
Strain, 617
acoustic, 617
cartesian components of, 617
electroductive, 705
electrostrictive, 705
piezoelectric, 696
tensor, 696
Strainoptic (electrooptic) tensor, 696
Streak camera, 689, 754, 757

Strobing, 402
Structured background, 118–120
Subjective evaluation, 87, 144
Subjective quality factor, 148–154
Submaster, 196–197
Substrate wedge, hologon, 261, 311, 324
Surface
  quality, 380
  relief, 188, 195, 196, 202, 203, 241, 268
    grating, 241, 268
Surface-relief phase media, 188
Switch
  E-O (electrooptic), 713, 771
  kryton, 754
Symmetry
  axis of, 233, 237, 239, 308, 694
  inversion, 696
Sync generator, 413
System dynamics, 569
System resolution, 831
System, polygonal scanner, 214
Systems, 83–84, 86–89, 89, 99–100, 104–105, 112–114, 116, 122, 124–128, 136–140, 142, 144, 149, 154–155, 157

Tangent error, 578
Telecentric, 41
Telephoto, 42
Television
  display system, 86, 534, 539, 675, 677
  scanning, 677
Tellurium dioxide, 641
  anisotropic diffraction in, 642, 648
TEMoo mode, 1, 12, 25
Temperature, 578
  Curie, 704
  gradient, 582, 609
  transition, 699, 700, 702
Temporal resolution, 749, 755, 758
Testing for performance, 79, 406
Thallium arsenic selenide, 641, 651
Thermal conductivity, 579, 582, 640
Thermal deformations, 581
Thermal stabilization, 583
Thermally stable mounting, 399
Thin lens, 40
Threshold, 89–95, 104, 107–110, 116, 125, 145–146
  detectability curve, 144, 146
  quality factor, 147
Time-bandwidth product, 637
Timing errors, 219, 364
Tolerance stack-ups, 408
Tone reproduction, 106, 107, 122, 126, 148
Tone reproduction curve (TRC), 109, 111, 124, 147–149
Torque, 570, 590–597
Torque coefficient, 462
Torsion bar, 589
Tower, B., 479
Tracking errors, 214, 218, 253, 261, 263, 311, 364, 816
  acoustic compensation, 263, 343, 365
  correction, 215, 259, 263–264, 292, 324, 343, 364
    active, 263, 343, 364
    passive, 215, 259, 264, 292, 324, 364, 366
Transducer, 590, 597, 651–652
  bandwidth, 597–600, 674
  bonding of, 657

[Transducer]
characteristics, 664
equivalent circuit, 655–657
insertion loss of, 658
materials, 664–666
mechanical, 590
multiple element mosaic, 666
phased array, 670–672
stepped array, 669
structure, 654
thinning of, 664
tilted array, 673
thin film, 666
Transit time, 732
Transition, ferroelectric and antiferroelectric, 704
Translation, beam, 363
Transmission efficiency, 231
Transmission range, optical, 699–700
Transverse modulators, 747
Trimmers, 526, 547–548
Trimming, 547
Truncated gaussian, 224
Truncation, 33–35, 224, 226, 229
Tunable filter, acoustooptic, 642
Turbines, gas, 352
Turbulence in molecules, 708
Two-dimensional deflection, 339, 710, 716, 727, 742

Unbalanced armature, 578
Uniformity, 270, 382
Universal Product Code (UPC), 160–162, 165–170, 197, 205, 354, 409

Variable light-collection aperture, 174, 178–179, 181, 183, 208
Vector scanning, 579
Velocity ratio, 578
Vignetting, 722, 757
Viscosity
of gas compared to liquids, 482–483
of lubricating oil, 483
Visual darkness for lines, 120
Visual system, 100, 102, 104, 106, 134–135, 145, 150–152
Vohr, J.H., 486
Voltage
bias, 719, 734, 744–745
control, 714, 742, 745, 760
error in, 714
ramp, 732
Volume-phase material, 190
Volume-phase media, 188, 190

Waist (*see also* Beam waist), 176–177
Waveguide, optic, deflector geometry, 765–771
Wavelength, 1, 12, 16, 21, 224, 241, 352
Wavelength shift induced scan error, hologon, 263, 295, 304, 312
Wavy lines, 120
Wedge effect, built-in, 484, 486
Welding, 547
Whipple, R. T. R., 486
Wiener spectrum, 139, 142, 150–153
Wilcock, D.F., 480
Windage (*see also* Model testing), 220, 275, 451–476
benefits of proper analysis, 451–452
definition of, 452

[Windage]
design optimisation, 468, 472–474
effect of atmosphere on, 472
equations
derivation of, 459
use of, 458
form drag, 453
measurement of, 453, 464
method of predicting, 458
of rotating cylinders, 452
of rotating disks, 220, 452
parameters affecting, 468–472
skin friction, 453
sources of, 453
torque, 452, 455
Window holes, 470, 473
effect on windage, 470, 473
Witness sample, 383
Wobble (*see also* Axial wobble, and Jitter), 253, 256–257, 259, 311, 324, 421, 528–530, 534, 560, 573, 574, 579–581, 586, 606, 608, 816
Wollaston prism, 711, 715
Woodward, W., 26
Wraparound, 169, 208
Write black, 110, 118, 120
Write white, 118, 120

XY deflection, 681
XY scanner, 43, 337, 339, 681

Young, T., 573, 603, 612
Young, M., 26

ZnS (zinc sulphide), 707
ZnTe (zinc tellunide), 707

Biographies

Brian J. Thompson *Series Editor*

Brian J. Thompson received his education at Manchester University in England, obtaining his B.Sc. and Ph.D. degrees. He served for 4 years as a faculty member at the University of Manchester and then 4 years at the University of Leeds. He and his wife, Joyce, came to the United States in 1962.

Thompson is a leading researcher in the fields of coherent optics, holography, phase microscopy, and image processing; and he held research and management positions at Technical Operations, Inc. for several years. He joined the University of Rochester, Rochester, New York, in 1968 as Director of The Institute of Optics and Professor of Optics. He has served as Dean of the College of Engineering and Applied Science since 1975. In 1982 he was named the University's first William F. May Professor of Engineering and was appointed Provost of the University in 1984.

He is also the editor of the *Journal of Optical Engineering* and the general editor of the International Society for Optical Engineering* (SPIE) Milestone Series of Selected Papers. A recent book, which he coauthored, is entitled *The New Physical Optics Notebook: Tutorials in Fourier Optics* (1990). The book is a mixture of the fundamentals of optics and the applications of those fundamentals.

Thompson's experimental studies on partially coherent light and its effects are the standard works in the literature of this field; his illustrations of these and other optical phenomena are widely used in optical texts and monographs dealing with coherent optics.

Thompson's awards include four from the SPIE: the Gold Medal, the President's Award, the Pezzuto Award, and the Rudolf Kingslake Medal.

---

* Formerly the Society of Photo-Optical Instrumentation Engineers.

Gerald F. Marshall *Volume Editor*

Gerald F. Marshall is a consultant, specializing in optical design and engineering, optical scanning, and display systems. This is the second book he has edited for Marcel Dekker, Inc.—the first being *Laser Beam Scanning* (1985) and to which he has contributed chapters.

Marshall was born in England and received his education in England and Scotland. Before and after being conscripted for national service in the Royal Air Force, during which time he was schooled in ground and airborne electronic systems technology, he studied physics and mathematics, receiving his B.Sc. degree from London University. He served for 5 years at Morganite International Ltd. in England as an optical Physicist and as a Technical Adviser in marketing and sales. Marshall joined Ferranti Ltd. of Edinburgh, Scotland (1959), where he specialized in the design and development of airborne navigational display systems.

In 1967, with his wife and children, he emigrated from Scotland to the United States. He served 2 years with Diffraction Limited Inc.; since then he has been involved in many aspects of optics, including scanning devices and systems, X-ray optics, and helmet-mounted displays (HMDs). Marshall has held senior positions with Kaiser Electronics, Energy Conversion Devices, Inc., and Speedring Systems, Inc. He writes regularly, is the author of many papers, and holds a number of patents.

He is a Fellow of both the International Society for Optical Engineering* (SPIE) and the Institute of Physics (IOP); a board member of SPIE (1991–1993) and the Optical Society of Northern California (OSNC, 1990–1992); and a member of the Optical Society of America (OSA). Marshall and his wife, Irene, live in Morgan Hill, which is near San Jose, California. He has had a lifelong keen interest in flying and aviation.

---

* Formerly the Society of Photo-Optical Instrumentation Engineers.

Leo Beiser

Before forming his consulting and research company 15 years ago, Leo Beiser served as Director of the Dennis Gabor Laboratory for Advanced Image Technology and as Staff Scientist at CBS Laboratories, Stamford, Connecticut. In 1973, he received the IR-100 Award for his Holofacet Optical Scanner milestone invention, which is now in the permanent collection of the Smithsonian Institution. In 1978 Leo received the Society for Information Display (SID) Special Recognition Award for Outstanding Contributions to Laser Scanning and Recording.

Recognized internationally as a pioneer and specialist in laser beam scanning systems, Beiser is invited often to lead seminars and to edit technical publications. He holds B.S. and M.S. degrees in physics from Hofstra University, Hempstead, Long Island, New York, graduated with honors the EE course of RCA Institutes, graduated the Business Administration Course of Alexander Hamilton Institute, and completed special laser courses at University of California, Los Angeles (UCLA). Beiser is Adjunct Professor at the Institute of Imaging Sciences at Polytechnic University, New York, New York.

LeRoy D. Dickson

LeRoy Dickson holds B.E.S. (1960), M.S.E. (1962), and Ph.D. (1968) degrees, all in electrical engineering, from The John Hopkins University. He has 27 years of experience with lasers and laser optical systems and has published many papers and holds 12 patents in the field of laser optics.

Dickson joined IBM in 1968 and has worked on the applications of lasers, including optical character recognition (OCR), document and holographic scanning, and bar code scanners; holographic nondestructive testing and precision measurement; liquid crystal displays; and optical storage. He was responsible for the design of the laser scanner used in the first IBM supermarket scanner. From 1972 to 1988 he was directly involved in the development of bar code scanning systems for IBM. He was also responsible for the design and development of the holographic deflector disk used in the IBM supermarket scanner and led the development effort for the IBM holographic industrial bar code scanner. Dickson is presently involved in the development of optical storage technology.

Milton Gottlieb

Milton Gottlieb is a consultant scientist in the Applied Physics Department at the Westinghouse Science and Technology Center. He received his B.S. degree from the City College of New York, New York, and his M.S. and Ph.D. in physics from the University of Pennsylvania, where he completed his thesis on infrared absorption in the alkali halides.

Gottlieb's work in industry began as a research associate at General Atomic in San Diego, where he was involved with a project on radiation damage in semiconductors. He joined Westinghouse in 1959 and did research in thermoionic energy conversion, superconductivity, and other areas of cryogenics until 1970. Since then he has been involved primarily in acoustooptic phenomena and their application to signal processing and spectroscopic devices. He has worked extensively in the development of new materials for acoustooptic devices and their application to acousto-optic tunable filters and Bragg cells.

Other research interests have included the development of fiber optic sensors, particularly for temperature and pressure measurements. Gottlieb has published extensively in these fields, including coauthoring four books and 40 patents.

Robert E. Hopkins

Robert E. Hopkins was born in Cambridge, Massachusetts, in 1915. He entered Massachusetts Institute of Technology in 1933. A. C. Hardy and David MacAdam initiated his interest in the science and engineering of optics.

After graduation from Massachusetts Institute of Technology (MIT) in 1937, Robert Hopkins undertook graduate studies at the Institute of Optics at the University of Rochester, Rochester, New York. During World War II Hopkins learned the art of optical design while, under the leadership of Brian O'Brien, the Institute contributed to the war effort with optical research and the development of instrumentation.

Hopkins joined the Institute of Optics' faculty upon receiving his Ph.D. in 1945; he served as Director of the Institute from 1953 to 1964. Since optical design involves extensive calculations, he and his students became pioneers in computer-aided design. In 1968 he left the University to become President of Tropel, an optical instrument company. He returned in 1974 to the University of Rochester Laboratory of Laser Energetics and participated in the design and construction of the 24-beam system for laser fusion.

After retirement from the University of Rochester, in 1980, he formed a small consulting service and has concentrated on optical lens design as a profession and a hobby.

Clive L. M. Ireland

Clive L.M. Ireland has worked since 1980 at Lumonics Ltd., Rugby, Warwickshire, England, where he has held a number of posts in development and technical management related to the company's laser and laser applications business. Currently, he is Director—New Technology, responsible for advanced product and concept development.

Ireland was born in London in 1946. Before joining Lumonics he had been involved in the development of lasers for inertial confinement fusion studies and research on the ionization mechanisms responsible for optical absorption and breakdown of gases irradiated by intense laser beams. Ireland is a Fellow of the Institute of Physics (IOP) and a Fellow of the International Society for Optical Engineering* (SPIE).

Charles J. Kramer

Charles J. Kramer is President and founder (1982) of Holotek Ltd., a Rochester, New York–based company that manufactures hologon laser beam deflector components and laser scanning systems for the graphic arts and machine vision industries. Before founding Holotek he worked 8 years at Xerox Corporation, where he performed research and development work in the area of electrooptical reading and printing devices. Kramer has worked in holography since 1968 and is recognized for major theoretical and practical contributions to the field of holographic laser beam scanning.

Born and raised in New Jersey, Kramer holds B.S. and M.S. degrees in physics from Fairleigh Dickenson University and M.S. and Ph.D. degrees in optics from the University of Rochester's Institute of Optics, Rochester, New York. He holds 24 U.S. patents and has presented and published many technical and scientific papers. Prior to his doctoral studies, he served as an engineer at Bendix Corporation, where he worked on holographic head-up displays, holographic nondestructive testing, laser machining, ring laser gyrosystems, and optical detectors.

---

* Formerly the Society of Photo-Optical Instrumentation Engineers.

Donald R. Lehmbeck

Donald R. Lehmbeck is Manager of Imaging Science Engineering and Technology of the Systems Reprographic Development Unit in the Xerox Corporation, where he is also a Principal Engineer. Lehmbeck has been with Xerox since the late 1960s and for the last 15 years has worked on various aspects of imaging science and image quality for electronic imaging systems.

He holds a B.S. degree in photographic science from the Rochester Institute of Technology (RIT), Rochester, New York, and an M.S. in optics from the University of Rochester, Rochester, New York. He has published articles and book chapters, has been awarded patents, and has given many talks on imaging science. Lehmbeck has also taught at, and worked for, Cornell Aeronautical Labs, the National Institute of Standards and Technology* (NIST), and the National Microfilm Association. He is a former Vice President of the Society for Imaging Science and Technology† (IS&T) and a President of the Rochester Chapter.

Born in Buffalo, New York, Lehmbeck received his college education in Rochester. He and his wife, Jane, live in the Rochester suburb of Penfield, where they raised a daughter, Kim, and a son, Jason. Lehmbeck is an avid sailor and photographer.

John Martin Ley

John Martin Ley was born in London in 1935. He lived and was educated in the Greater London area until 1954, when he was conscripted for national service into the Royal Air Force for 2 years during which time he was trained in ground radar techniques. After national service he joined the Atomic Energy Authority and worked principally on nuclear instrumentation, thereby gaining early experience in laser technology including optical modulation techniques.

In 1961 Ley was awarded a first class honors degree in electrical engineering from the City University, London, and in 1966 he earned a Ph.D. in optoelectronics at Imperial College. In 1966 he became a lecturer in optoelectronics at the City University, London, but left in 1971 to start Electro-Optic Developments Limited, where he remained as Owner–Director until 1986.

For the past 5 years he has continued to work in the area of optoelectronics and now directs two new companies manufacturing electrooptic and acousto-optic modulators, deflectors, and associated products.

---

* Formerly the National Bureau of Standards (NBS).

† Formerly the Society of Photographic Scientists and Engineers (SPSE).

Jean I. Montagu

Jean I. Montagu was born in France. He is Chairman of the Board and Co-Founder of General Scanning Inc., a diversified manufacturer of proprietary optical scanning components, subsystems, and full systems. The major application for its products is found in vector scanning, as well as raster scanning systems for the graphics markets and the laser micro machining markets. He is also the founder of Mechanics for Electronics (MFE) Ltd., which manufactures biomedical instruments.

Montagu was Assistant Professor in the Department of Physics at the Catholic University in Rio de Janeiro, Brazil, where he taught laser theory. He has also taught a course on design of electromagnetic devices at North Eastern University, Boston, and he is an occasional lecturer on oscillatory scanning methods.

A graduate of Massachusetts Institute of Technology (MIT), Cambridge, Massachusetts, with a B.S. and M.S. in mechanical engineering, Montagu holds 20 patents on heat transfer techniques, magnetic devices, and mirror construction. He lives and works in Boston.

Gerald A. Rynkowski

Gerald A. Rynkowski has been associated with Speedring Systems since 1968 and is presently serving on staff as Chief Engineer. His duties and responsibilities include servo systems design for state-of-the-art rotating optical scanners, gyro-stabilized and slaved positioning platforms, phase-lock control systems, and proprietary instrumentation techniques used in optical metrology. He also provides engineering support to the marketing and sales departments for proposal evaluation and technical presentations.

Rynkowski is a graduate of Electronics Institute of Technology, Detroit, where he majored in communications and design engineering.

Previous optical design and development experiences include the electronic circuit design for an electro-optical autocollimator (optical-synchro) and an infrared image target lock-on system. While at Speedring he has contributed to the design and development of many polygonal scanning systems, including multispectral passive scanners, cryogenic vacuum, and high-speed (120,000 rpm) imaging systems.

Rynkowski was born in 1932 in Detroit, and married in 1968. He and his wife, Virginia, have two children attending Wayne State University.

Tetsuo Saimi

Tetsuo Saimi is the manager of the optical team for the research and development of optical disk drives at the Disc Systems Division, Matsushita Electric Industrial Co., Ltd., in Osaka, Japan.

Saimi was born in 1947 and educated in Hiroshima, Japan. He received the B.S. degree (1970) and M.S. degree (1974) from Hiroshima University, Hiroshima. From 1971 to 1973 he studied at the Institute d'Optique in Paris, where he was engaged in research of incoherent holograms. In 1974 Saimi joined the First Development Division, Matsushita Electric Industrial Co., Ltd. He is a member of the Society of Applied Physics (Japan) and the Optical Society of America (OSA).

Joseph Shepherd

Born in Blairgowrie in the Scottish Highlands, Joseph Shepherd trained as a shipwright and served his national service in Kenya, Africa, before entering Strathclyde University, Scotland, to study naval architecture. Owing to the uncertainty in the shipbuilding industry, Shepherd switched his major to mechanical engineering and received an honors degree in 1961.

Joining Ferranti Ltd. in Edinburgh, Scotland, as a Research Engineer to work on advances to inertial gyroscopes, he has since been involved in many phases of precision instrument technology, including design, development, and production. Specializing in inertial gyro spinners and the performance of ball bearings, he later also led a team developing a new gyro based on self-acting (self-pumping) gas bearings. This resulted in a high-performance gas bearing scanner in 1971 and the development of a group to exploit and advance Ferranti ultraprecision technology. He was recently Design Manager for Rotating Devices, which included control of gyro spinner design and all aspects of rotary scanner business, where he enjoyed the challenge of new technical disciplines and business for commercial markets.

In 1990 he took early retirement from GEC Ferranti Defense Systems Ltd. to concentrate on innovative ideas for future products.

Randy J. Sherman

Randy J. Sherman is the President and General Manager of Lincoln Laser Company in Phoenix, Arizona. The company, of which he was a founder in 1975, develops and manufactures scanners and scanning systems, chiefly for the laser printer, bar code, and inspection markets.

Sherman was born and raised in Chicago. Following his high-school interest and training in machining, he attended the University of Colorado in Boulder and subsequently joined the University's High Altitude Observatory as an instrument maker.

Sherman's career experiences include machining, drafting, designing, engineering, and managing, while working primarily for Technical Operations, Inc., and Hughes Aircraft Co. He contributed a chapter to *Laser Beam Scanning* and has published several articles; he has authored and coauthored eight patents in scanner and scanning technology.

Sherman and his wife, Doris, live in Phoenix and enjoy the recreational opportunities that abound in Arizona.

Glenn T. Sincerbox

Glenn T. Sincerbox received the B.S. degree in physics from Rensselaer Polytechnic Institute in 1959 and the M.S. degree in physics from the University of Illinois in 1960. He continued graduate studies at the University of Illinois until 1962 when he joined IBM. Since 1980, Sincerbox has held several management positions and is currently manager of the Optical Storage Heads department at the IBM Almaden Research Center in San Jose, California.

Sincerbox's primary research contributions have been in the areas of holography, novel recording processes, and optical devices, with emphasis on their application to information storage, display, scanning, and inspection. His current research interests include the use of integrated optics and holographic elements in optical storage heads. He has published and presented over 70 papers and two book chapters, holds 34 patents, has two patents pending, and over 65 patent publications.

Sincerbox is a Fellow of the Optical Society of America (OSA) and has been an active participant in the optics community since 1978. He has held many positions in society and conference management. He is currently a member of the Executive Committee of the OSA Technical Council, Chair of the 1991 OSA Annual Meeting, and Vice President of the International Commission for Optics (ICO).

David Stephenson

David Stephenson is Engineering Manager at Melles Griot, Optics Division, in Rochester, New York, where he has been employed since 1981. His educational experiences—B.S. in medical photography (1982) and M.S. in imaging science (1990), at Rochester Institute of Technology (RIT), Rochester, New York—and previous work in the medical field combine to give him a background in both the theoretical design considerations and practical end-user concerns that are necessary for the production of successful products. Stephenson's technical leadership has contributed to Melles Griot becoming a recognized leader in the production of optical components for scanning systems and system designs intended for quantity production.

John C. Urbach

Since 1987 John C. Urbach, Vice President of Technology, Strata Systems Inc., has been working on input image scanning, digital image processing, and image printing. From 1985 to 1987 he was Manager of Color Systems Technology in the Xerox Corporate Research Group, pursuing research in color printing.

In 1963 he joined the Xerox Corporation's Research Center in Webster, New York, where he investigated thermoplastic electrophotography for recording images and holograms. Urbach also managed research in acoustic imaging, image processing and evaluation, and displays. In 1970 Urbach helped to form the Xerox Palo Alto Research Center (PARC) in California, where he initiated research in laser printing, optical data storage, and optical communication. Subsequently, as Manager of the PARC Optical Science Laboratory, he directed programs in these fields as well as in high-quality laser xerographic color copying and printing.

Urbach received a Ph.D. in Optics from the University of Rochester, Rochester, New York, and studied photographic image quality while on a postdoctoral fellowship at the Royal Institute of Technology, Stockholm, Sweden. He is the author of many scientific papers and reports and holds patents in optics and image recording. He lives with his wife, Mary, in Portola Valley, California; they have three children, Thomas, Michael, and Katherine.